...NTROL ENGINEERING

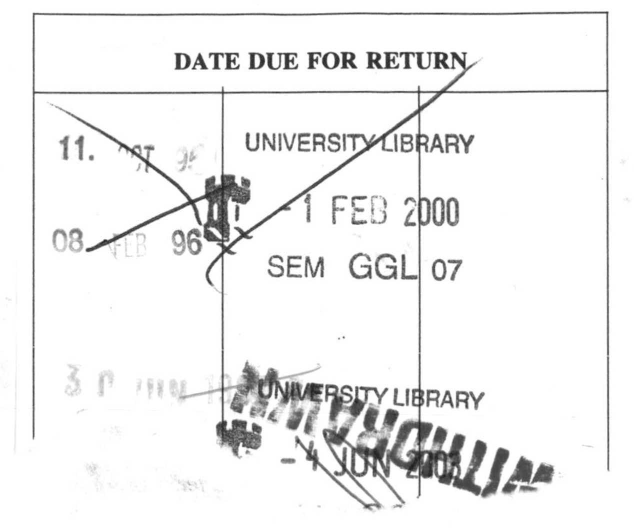

McGraw-Hill Series in Mechanical Engineering

CONSULTING EDITORS

Jack P. Holman, *Southern Methodist University*
John R. Lloyd, *Michigan State University*

Anderson: *Modern Compressible Flow: With Historical Perspective*
Arora: *Introduction to Optimum Design*
Bray and Stanley: *Nondestructive Evaluation: A Tool for Design, Manufacturing, and Service*
Burton: *Introduction to Dynamic Systems Analysis*
Culp: *Principles of Energy Conversion*
Dally: *Packaging of Electronic Systems: A Mechanical Engineering Approach*
Dieter: *Engineering Design: A Materials and Processing Approach*
Eckert and Drake: *Analysis of Heat and Mass Transfer*
Edwards and McKee: *Fundamentals of Mechanical Component Design*
Gebhart: *Heat Conduction and Mass Diffusion*
Gibson: *Principles of Composite Material Mechanics*
Hamrock: *Fundamentals of Fluid Film Lubrication*
Heywood: *Internal Combustion Engine Fundamentals*
Hinze: *Turbulence*
Holman: *Experimental Methods for Engineers*
Howell and Buckius: *Fundamentals of Engineering Thermodynamics*
Hutton: *Applied Mechanical Vibrations*
Juvinall: *Engineering Considerations of Stress, Strain, and Strength*
Kane and Levinson: *Dynamics: Theory and Applications*
Kays and Crawford: *Convective Heat and Mass Transfer*
Kelly: *Fundamentals of Mechanical Vibrations*
Kimbrell: *Kinematics Analysis and Synthesis*
Krieder and Rabl: *Heating and Cooling of Buildings*
Martin: *Kinematics and Dynamics of Machines*
Modest: *Radiative Heat Transfer*
Norton: *Design of Machinery*
Phelan: *Fundamentals of Mechanical Design*
Raven: *Automatic Control Engineering*
Reddy: *An Introduction to the Finite Element Method*
Rosenberg and Karnopp: *Introduction to Physical Systems Dynamics*
Schlichting: *Boundary-Layer Theory*
Shames: *Mechanics of Fluids*
Sherman: *Viscous Flow*
Shigley: *Kinematic Analysis of Mechanisms*
Shigley and Mischke: *Mechanical Engineering Design*
Shigley and Uicker: *Theory of Machines and Mechanisms*
Stiffler: *Design with Microprocessors for Mechanical Engineers*
Stoecker and Jones: *Refrigeration and Air Conditioning*
Ullman: *The Mechanical Design Process*
Vanderplaats: *Numerical Optimization: Techniques for Engineering Design, with Applications*
White: *Viscous Fluid Flow*
Zeid: *CAD/CAM Theory and Practice*

AUTOMATIC CONTROL ENGINEERING

Fifth Edition

Francis H. Raven

Professor of Mechanical Engineering
University of Notre Dame

McGraw-Hill, Inc.

New York St. Louis San Francisco Auckland Bogotá
Caracas Lisbon London Madrid Mexico City Milan
Montreal New Delhi San Juan Singapore Sydney Tokyo Toronto

AUTOMATIC CONTROL ENGINEERING
International Editions 1995

1 2 3 4 5 6 7 8 9 0 KKP KKP 9 8 7 6 5 4

This book was set in Times Roman.
The editors were John J. Corrigan and Margery Luhrs;
the production supervisor was Paula Keller.
The cover was designed by Joseph Gillians.

Library of Congress Cataloging-in-Publication Data

Raven, Francis H. (Francis Harvey), (date).
Automatic control engineering / Francis H. Raven. – 5th ed.
p. cm. – (McGraw-Hill series in mechanical engineering)
Includes bibliographical references and index.
ISBN 0-07-051341-4
1. Automatic control. I. Title. II. Series.
TJ213.R38 1995 94-10701
629.8'3–dc20

When ordering this title, use ISBN 0-07-113639-8

Printed in Singapore

ABOUT THE AUTHOR

Francis H. Raven is Professor of Aerospace and Mechanical Engineering at the University of Notre Dame. He is the author of *Mathematics of Engineering Systems* and *Engineering Mechanics,* both published by McGraw-Hill, Inc. Dr. Raven originated the method of independent position equations, a generalized analytical method for kinematic analysis. Presently he is involved in research in the Robotics and Programmed Automation Laboratory at the University of Notre Dame.

Professor Raven received his Ph.D. in Mechanical Engineering from Cornell University. As a control engineer at the Hamilton Standard Division, United Technologies, he obtained extensive experience in the design of hydraulic and pneumatic control systems. He now does consulting for a variety of industries, some of which includes the design of computer controlled systems.

Dr. Raven is listed in the international *Who's Who in Engineering*, *Engineers of Distinction, and Who's Who in the Midwest.* He is the recipient of the AT&T award for excellence in teaching, presented by the American Society for Engineering Education.

To Therese

CONTENTS

Preface xiii

1 Introduction to Automatic Controls 1
- 1.1 Historical Development 1
- 1.2 Feedback Control Systems 3
- 1.3 System Representation 5
- 1.4 Modern Control Systems 7

2 Representation of Control Components 12
- 2.1 Operational Notation 13
- 2.2 Mechanical Components 16
- 2.3 Electrical Components 22
- 2.4 Series and Parallel Laws 23
- 2.5 Analogies 31
- 2.6 Scale Factors 34
- 2.7 Thermal Systems 38
- 2.8 Fluid Systems 40

3 Representation of Control Systems 55
- 3.1 Linearization of Nonlinear Functions 55
- 3.2 Linearization of Operating Curves 66
- 3.3 Hydraulic Systems 70
- 3.4 Pneumatic Systems 79
- 3.5 DC Motors 85
- 3.6 AC Motors 91
- 3.7 Block-Diagram Algebra 92
- 3.8 Speed Control Systems 95
- 3.9 Generalized Feedback Control System 103

4 Steady-State Operation 123
- 4.1 Steady-State Analysis 123
- 4.2 Equilibrium 131

4.3 Proportional Control Systems 134
4.4 Integral Control Systems 139
4.5 Proportional Plus Integral Control Systems 141
4.6 Modes of Control 144
4.7 Summary 147

5 Laplace Transforms 162
5.1 Classical Methods 163
5.2 Laplace Transform Method 167
5.3 Transform Properties 173
5.4 Initial Conditions 185
5.5 General Procedures 191
5.6 Piecewise Continuous Functions 198
5.7 Convolution Integral 200
5.8 Error Coefficients 204

6 Transient Reponse 216
6.1 Inverse Transformations 218
6.2 Complex Conjugate Zeros 221
6.3 Damping Ratio and Natural Frequency 227
6.4 Computer Solution 233
6.5 Transient Response Specifications 243
6.6 General Form of Transient Response 250
6.7 Response to an External Disturbance 252
6.8 Routh's Stability Criterion 257
6.9 Summary 262

7 The Root-Locus Method 279
7.1 Significance of Root Loci 279
7.2 Construction of Loci 285
7.3 General Procedure 295
7.4 Newton's Method and the Remainder Theorem 307
7.5 Loci Equations 310
7.6 Variation of Parameters 312
7.7 Computer Solution 318
7.8 Sensitivity 319

8 Analog Computers 337
8.1 Computer Operations 338
8.2 Direct Programming 341
8.3 Time Scale 353
8.4 Simulation 358

9 State-Space Methods 377
9.1 System Representation 377
9.2 Signal Flow Graphs 391
9.3 Solution of State-Space Equations 398
9.4 Methods of Computing $\Phi(t)$ 401
9.5 Forced Response 411

9.6 Transfer Functions 416
9.7 Multivariable Systems 421

10 Digital Control Systems 439
10.1 Sampled-Data Systems 440
10.2 The z Transform 441
10.3 Inverse z Transforms 450
10.4 Block-Diagram Algebra 458
10.5 Transient Response 461
10.6 Filters 470
10.7 Discrete Data Systems 476
10.8 Sampled-Data Control Systems 479
10.9 Computer-Controlled Systems 486

11 Frequency-Response Methods 502
11.1 Frequency Response 502
11.2 Logarithmic Representation 508
11.3 Evaluating the Gain K 518
11.4 Equivalent Unity-Feedback Systems 520
11.5 Polar Plots 521
11.6 M and α Circles 524
11.7 Correlation between Transient and Frequency Response 528
11.8 Determining the Gain K to Yield a Desired M_p 537

12 System Compensation 553
12.1 Nyquist Stability Criterion 553
12.2 Gain Margin and Phase Margin 562
12.3 Lead Compensation 568
12.4 Lag Compensation 575
12.5 Lag-Lead Compensation 580
12.6 Internal Feedback 581
12.7 Inverse Polar Plots 583
12.8 Stability Criteria in the Inverse Plane 590

Appendix A Equilibrium Flow 601

Appendix B Fourier Series, Fourier Integral, and the Laplace Transform 606

B.1 Fourier Series 606
B.2 Fourier Integral 609
B.3 Laplace Transform 610

Index 613

PREFACE

In recent years, automatic control systems have been rapidly increasing in importance in all fields of engineering. The applications of control systems cover a very wide range, from the design of precision control devices such as delicate electronic equipment to the design of massive equipment such as that used for the manufacture of steel or other industrial processes. Microprocessors have added a new dimension to the capability of control systems. New applications for automatic controls are continually being discovered.

This text is the outgrowth of notes developed by the author to teach control engineering at the University of Notre Dame. The author has endeavored to give the principles a thorough presentation and yet make them clear and easy to understand. It is assumed that the reader has the general maturity and background of a third- or fourth-year engineering student, but no previous training in control engineering.

Although the principles of feedback control systems are presented in a manner which is appropriate to the interests of mechanical engineers, this text has also been successfully used to teach students in other fields of engineering. In addition, the author has taught night courses for practicing engineers. In the light of their enthusiastic comments, it is felt that this book will be of much value to the engineer in industry who did not have the opportunity to take such a course while in college.

The study of control systems is begun by showing how to obtain the equation of operation for each component in a system. It is then shown how each of these equations may be represented as a block diagram. Each of these block diagrams is connected to form the overall block-diagram representation for the entire system, just as the actual components are connected to form the complete control system. This overall block diagram is a very helpful representation of the differential equation that describes the operation of the system. Because actual control systems frequently contain nonlinear components, considerable emphasis is given to such components. This material is presented in the first three chapters. In Chap. 4, it is shown that much important information concerning the basic or inherent operating characteristics of a system may be obtained from knowledge of the steady-state behavior.

This introduction to control theory differs from the usual blackbox approach, in which the block diagram for a system is given outright. The blackbox approach permits the introduction of Laplace transforms and other methods of system analysis at an earlier stage. However, it has been the author's experience that students achieve deeper understanding of the various techniques used in system analysis if students are first familiarized with the physical significance of feedback controls by knowing how to obtain the differential equation that describes the operation of a system.

In Chap. 5, it is shown how the linear differential equations which describe the operation of control systems may be solved algebraically by the use of Laplace transforms. It is seen that this method reveals directly much interesting information about systems. General characteristics of transient behavior are described in Chap. 6. It is pointed out that the transient response is governed by the location of the roots of the characteristic equation. It is also shown how to program a digital computer for investigating the performance of control systems. The application of the root-locus method to the design of control systems is the topic of Chap. 7. The use of the analog computer to simulate control systems is explained in Chap. 8.

State-space methods are presented in Chap. 9. It is shown how the concepts of classical control theory are combined with state-space concepts to yield the "modern" control theory. By seeing how modern control theory builds on classical methods, the reader attains both a better understanding of modern methods and a deeper appreciation of the classical formulation. Digital computers and microprocessors are becoming increasingly important as elements of control systems. As explained in Chap. 10, most of the methods (both classical and modern) used in the design of continuous-data systems may be extended to digital control systems. The term "digital control system" comprises a broad class of systems including computer-controlled systems, discrete data systems, sampled-data systems, and timesharing systems. The use of frequency-response techniques to evaluate dynamic performance is explained in Chaps. 11 and 12.

The author wishes to express his appreciation for the many fine suggestions made by teachers who have used the previous editions. These suggestions have been of great value in the preparation of the fifth edition. Special gratitude is owed to Anuradha Annaswany, Michigan Institute of Technology; Allen Arthur, University of Cincinnati; Frank D'Souza, Illinois Institute of Technology; David Hullender, University of Texas at Arlington; Suhada Jayasuriya, Texas A&M University; Scott Kimbrough, University of Utah; Richard Klafter, Temple University; Lee J. LaFrance, New Mexico State University; Ching Li, Columbia University; Ronald A. Perez, University of Wisconsin, Milwaukee; Nader Sadegh, Georgia Tech; F. H. Speckhart, University of Tennessee; Gerald Whitehouse, Louisiana State University; and William Wainwright, University of Colorado, who reviewed the entire manuscript for this edition. Students have also been very helpful in their comments. Particular recognition is due to Linda F. Raven.

The author also wishes to express his gratitude for the continued encouragement of his colleagues at the University of Notre Dame, especially to Dr. Thomas J. Mueller, chairman of the Department of Aerospace and Mechanical Engineering.

Thanks are also due to Kelly Marie Keller, Martha Van Overberghe, and Lisa Tranberg for typing the numerous revisions of the notes from which this textbook has been developed.

The author's wife, Therese, has faithfully worked with him throughout the development of this text. She has made innumerable suggestions and has been a constant source of encouragement.

Francis H. Raven

AUTOMATIC CONTROL ENGINEERING

CHAPTER 1

INTRODUCTION TO AUTOMATIC CONTROLS

1.1 HISTORICAL DEVELOPMENT

Early people had to rely upon their own brute strength or that of animals to supply energy for doing work. By the use of simple mechanical devices such as wheels and levers, people accomplished such feats as the building of high pyramids and Roman highways and aqueducts. They first supplemented their energy and that of beasts by utilizing power from natural sources, such as the wind for powering sailing vessels and windmills and waterfalls for turning waterwheels. The invention of the steam engine was a milestone in human progress because it provided people with useful power that could be harnessed at will. Since then, people have devised many different means of obtaining abundant and convenient sources of energy. Engineering effort is primarily concerned with the practical applications of using power to serve human purposes. That is, the engineer designs and develops machines and equipment by which people can utilize power.

A manually controlled system is shown in Fig. 1.1*a*. It is desired to maintain the fluid in the tank at a fixed height H so that the pressure at the outlet will remain constant. The quantity of flow out from the tank Q_{out} varies with the demand. The

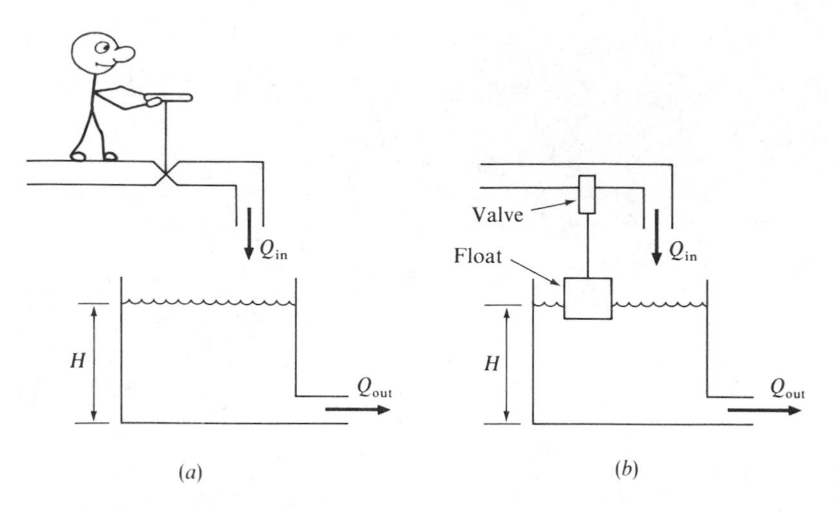

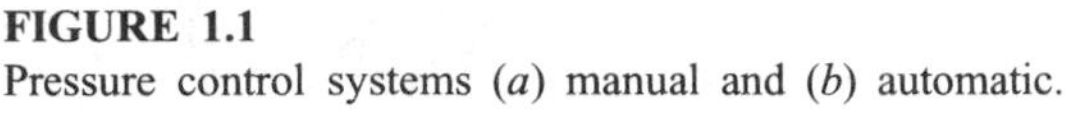

FIGURE 1.1
Pressure control systems (*a*) manual and (*b*) automatic.

operator opens or closes a valve to regulate the rate of flow into the tank Q_{in} and thus maintain the desired height of fluid in the tank.

The same system automatically controlled is shown in Fig. 1.1*b*. The height of fluid in the tank is measured by the float. When the level rises, so does the float, which in turn raises the valve to decrease the amount of flow going into the tank. Similarly, when the level drops, then the float and the valve are lowered, which increases the amount of flow going into the tank. Thus, the rate of flow into the tank is automatically regulated so as to maintain the desired height of fluid in the tank. By placing a turnbuckle on the rod connecting the float and the valve, the length of the rod can be changed. Thus, the desired height of fluid in the tank can be adjusted.

Early machines and equipment had controls which were predominantly of a manual nature, and the adjustments had to be reset frequently to maintain the desired output or performance. The design of newer equipment with greater usefulness and capabilities is bringing about an ever-increasing growth in the development of control equipment. There are two reasons. First, automatic controls relieve people of many monotonous activities so that they can devote their abilities to other endeavors. Second, modern complex controls can perform functions which are beyond the physical abilities of people to duplicate. For example, an elaborate automatic control system operates the engine of a modern jet airplane with only a minimum amount of the pilot's attention so that the pilot is free to maneuver and fly the airplane.

It is interesting to note that as the applications and uses for controls have increased, so also have the demands upon the performance of these systems increased. There is no doubt that a major concern of the engineer today, and even more so in the future, is the design and development of automatic control systems.

1.2 FEEDBACK CONTROL SYSTEMS

Various forms of transportation are illustrated in Fig. 1.2. All the basic concepts of feedback control systems are contained in each of these means of transportation. For the basic form of transportation, walking, the desired speed at which the walker wishes to go is the reference input. When the walker takes a leisurely stroll through a park, the desired speed is slow. When the person is in a hurry, the desired speed is fast. The actual speed is the controlled variable (i.e., the quantity being controlled). The part of a system which compares the reference input (desired value) with the controlled variable (actual value) in order to measure the error is called the *comparator*. The brain serves as the comparator for the walker. Typically, the error signal goes to a power-amplifying device (muscles in the legs) which actuates the system to be controlled so as to reduce the error to zero. Thus, the actual speed is the same as the desired speed.

The driver of a car compares the actual speed with the desired speed. Again the brain serves as the comparator. In response to the error signal, the driver changes the position of the accelerator pedal. The engine serves as the power-amplifying device. That is, a little motion of the accelerator pedal causes a substantial change in the power produced by the engine.

For cruise control, a magnet is attached to the driveshaft which rotates at some constant times the speed of the wheels. A magnetic pickup fixed to the body of the car generates a pulse every time the driveshaft makes one complete rotation. The rate at which pulses are being generated is a measure of the actual speed of the car. At the instant the cruise control is set, the rate at which pulses are being generated is stored so as to serve as the reference input (i.e., desired speed). A vacuum-operated device is attached to the accelerator linkage. When the actual speed is greater than the reference speed, the rate at which pulses are being generated is greater than the reference value. This error signal goes to the vacuum device which serves as a power amplifier to change the position of the accelerator linkage so as to decrease the actual speed.

FIGURE 1.2
Various forms of transportation.

The most difficult problem which the Wright brothers encountered in the process of making the airplane into a useful machine was the control problem. After much experimentation, they developed an ingenious pulley system to coordinate the motion of the aileron and the rudder so as to produce a smooth turning effort. Note that the pilot (Orville or Wilbur) would compare the actual path of the plane with the desired path, and then in response to the error signal he would actuate the pulley system. Aerodynamic forces acting on the wings and rudder would serve as the power-amplifying device to change the trajectory of the plane. Although modern airplanes have much more sophisticated means for controlling the position of the ailerons and rudder, the principles are still the same.

The controlling of temperature is a typical example of a feedback control system. The position of the temperature dial sets the desired temperature (i.e., the reference input). The actual temperature of the system is the controlled variable (i.e., the quantity which is being controlled). The thermostat, or comparator, compares the actual temperature with the desired temperature in order to measure the error. This error signal is the actuating signal, which is then sent to the heating units in order to correct the temperature. For example, if the actual temperature is less than the desired temperature, the actuating signal causes the heating elements to supply more heat. If there is no error, the actuating signal does not change the amount of heat which is being supplied. When the actual temperature is greater than the desired value, then the actuating signal calls for a decrease in the amount of heat.

For a system to be classified as a feedback control system, it is necessary that the controlled variable be fed back and compared with the reference input. In addition, the resulting error signal must actuate the control elements (power-amplifying device) to change the output so as to minimize the error. A feedback control system is also called a *closed-loop system*. Any system which incorporates a thermostat to control temperature is a feedback, or closed-loop, system. Well-known examples are electric frying pans, irons, refrigerators, and household furnaces with thermostatic control.

For speed control systems, the device which subtracts the feedback signal from the reference input (i.e., the comparator) is often a centrifugal governor. The compression of a spring sets a force which is a measure of the desired speed. The centrifugal force of the flyweights is a measure of the actual speed. The difference in these forces is a measure of the error. The governor serves the same purpose that the thermostat does for temperature controls. That is, the governor compares the actual speed with the desired value and measures the error. This error signal then goes to a power amplifier such as a hydraulic servomotor that controls the position of a flow valve which in turn determines the rate of fuel flow to the engine. The same basic concepts apply to all types of feedback control systems, whether the controlled variable is temperature, speed, pressure, flow, position, force, torque, or any other physical quantity.

In an open-loop system, there is no comparison of the controlled variable with the desired input. Each setting of the input determines a fixed operating position for the control elements. For example, for a given input temperature setting, the heating units are positioned to supply heat at a fixed rate. Note that there is no comparator, or thermostat, which measures the error and resets the heating units. The disadvantage of such a system is illustrated by the fact that if a fixed rate of heat is supplied to a house, the inside temperature will vary appreciably with changes in the outside temperature. Thus, for a given set input to an open-loop system, there may be a big variation of the controlled variable depending on the ambient temperature.

In this example, the ambient temperature is an external disturbance. By an external disturbance is meant something external to the system which acts to change or disturb the controlled variable. A major advantage of employing feedback control is that, because of the comparator, the actuating signal continually changes so that the controlled variable tends to become equal to the reference input regardless of the external disturbance. Another consideration is that with feedback one can generally use relatively inexpensive components and yet obtain better control than is possible with very expensive components in an open-loop system. The primary focus of this text is feedback control systems.

1.3 SYSTEM REPRESENTATION

The mathematical relationships of control systems are usually represented by block diagrams. These diagrams have the advantage of indicating more realistically the actual processes which are taking place, as opposed to a purely abstract mathematical representation. In addition, it is easy to form the overall block diagram for an entire system by merely combining the block diagrams for each component or part of the system.

A comparator subtracts the feedback signal from the reference input r. For the case in which the controlled variable c is fed back directly (i.e., for unity-feedback systems), the signal coming from the comparator is $r - c$, which is equal to the actuating signal e. The mathematical relationship for this operation is

$$e = r - c \tag{1.1}$$

As illustrated in Fig. 1.3, the circle is the symbol used to indicate a summing operation. The arrowheads pointing toward the circle indicate input quantities, while the arrowhead leading away signifies the output. The sign at each input arrowhead indicates whether the quantity is to be added or subtracted.

The portion of a system between the actuating signal e and the controlled variable c is called the *control elements*. The relationship between the actuating signal e, which enters the control elements, and the controlled variable c, which is the output of the control, is expressed by the equation

$$c = G(D)e \tag{1.2}$$

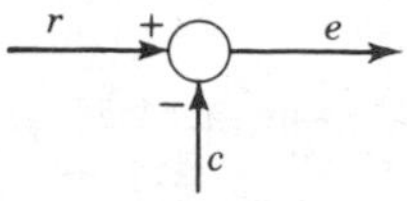

r + e − c

FIGURE 1.3
Block diagram of a comparator.

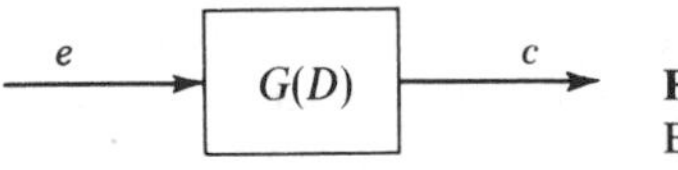

FIGURE 1.4
Block diagram of the control elements.

where $G(D)$ represents the operation of the control elements. In Chaps. 2 and 3, it is shown how the actual values of $G(D)$ for specific control systems are obtained. The block-diagram representation for the preceding equation is shown in Fig. 1.4. A box is the symbol for multiplication. In this case, the input quantity e is multiplied by the function in the box $G(D)$ to obtain the output c. With circles indicating summing points and with boxes, or blocks, indicating multiplication, any linear mathematical expression may be represented by block-diagram notation.

The complete block diagram for an elementary unity-feedback control system is obtained by combining Figs. 1.3 and 1.4 to yield Fig. 1.5. This diagram shows the controlled variable c being fed back to the summing point, where it is compared with the reference input r. This diagram pictorially shows why a feedback control system is also called a closed-loop system.

When the controlled variable is fed back to the comparator, it is usually necessary to convert the form of the controlled variable to a form that is suitable for the comparator. For example, in a temperature control system, the controlled temperature is generally converted to a proportional force or position for use in the comparator. This conversion is accomplished by feedback elements $H(D)$. The block-diagram representation for this more general case of a feedback control system is shown in Fig. 1.6. The signal which is fed back is

$$b = H(D)c \tag{1.3}$$

The elements represented by $H(D)$ are called the *feedforward elements* because they are located in the feedback portion of the control. The control elements represented by $G(D)$ are the *feedforward elements* because of their location in the feedforward portion of the loop. The actuating signal e is now $r - b$. The actuating signal e is a measure or indication of the error.

The term *feedback control system* is a general term which applies to any system in which the controlled variable is measured and fed back to be compared with the reference input. The terms *servomechanism* and *regulator* are distinguished as follows. A servomechanism is a particular type of feedback control system in which the controlled variable is a mechanical position (e.g., the angular position of a shaft). A regulator is distinguished as a feedback control system in which the reference input, although adjustable, is held fixed, or constant, for long periods (e.g., most temperature controllers).

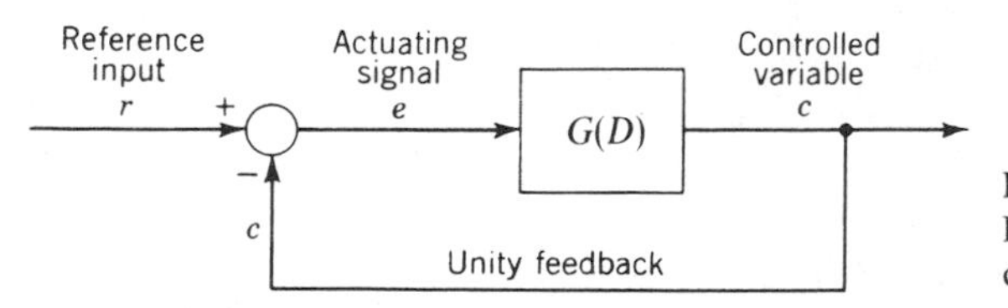

FIGURE 1.5
Block diagram of a unity-feedback control system.

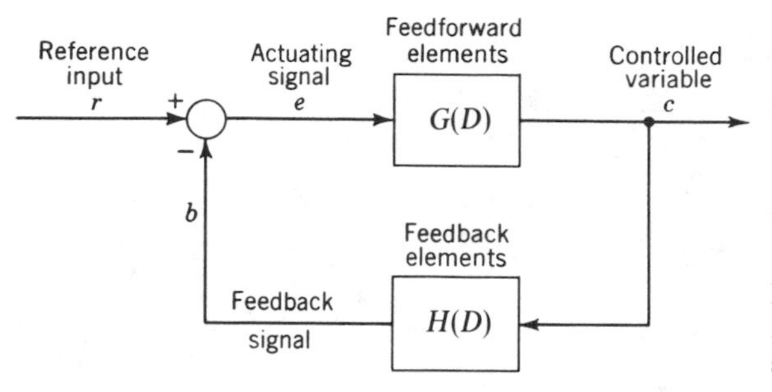

FIGURE 1.6
Block diagram of a feedback control system.

1.4 MODERN CONTROL SYSTEMS

Commercial aviation as we know it today would not be possible without modern control systems. In a blind landing (i.e., an instrument landing), the pilot gets an error signal which indicates the actual position of the plane relative to the ideal landing path. The pilot then maneuvers the plane so as to keep it on the ideal landing path. Some modern airplanes such as the Boeing 747 shown in Fig. 1.7 have a completely automated landing system (instrument landing system) which

FIGURE 1.7
The Boeing 747 airplane. (Courtesy the Boeing Company.)

FIGURE 1.8
The cockpit of the Boeing 747. (Courtesy the Boeing Company.)

automatically flies the vehicle along the ideal path of the landing beam. A picture of the cockpit is shown in Fig. 1.8.

The inertial guidance system for the space shuttle contains a stable platform which provides the reference orientation. Three gyroscopes are mounted on the stable platform so that their axes are mutually perpendicular. One axis is in an east-west direction, and another axis is in a north-south direction. The third axis is always directed along a line from the vehicle to the center of the earth. An accelerometer is mounted along each of the axes. A computer automatically integrates each of these accelerations to yield the corresponding components of velocity. Integration of these velocities yields the position of the vehicle relative to the earth. This actual position is then compared to the ideal position. When the error is sufficiently great, the astronaut fires the rocket so as to get back "on course."

The engine control system for a modern high-performance jet engine incorporates many subsystems which perform automatically to ensure the safe, efficient operation of the engine. For example, under certain conditions a jet engine can stall. The control system senses these conditions and automatically prevents the engine from running at a condition that would cause stall. At higher altitudes the air is considerably thinner. For the same amount of fuel flow, the air-fuel ratio would decrease considerably with altitude. The engine control system automatically compensates for changes in altitude. The control system also controls the nozzle jet area (this affects thrust), the position of the inlet guide vanes (this directs the air into the compressor for the best angle of attack), etc.

When the comparator is a piece of hardware such as a thermostat or a governor, the equations that describe the operation of the hardware are part of the control loop. Such controllers are referred to as *analog-type controllers*. The use of microprocessors is becoming increasingly popular in control systems. The microprocessor is in effect a digital computer and is programmed to yield the relationship that the designer desires. The microprocessor thus provides a flexibility that is not possible with analog-type controllers. With a microprocessor, a design change is accomplished by a simple change in the software, whereas with an analog controller a design change necessitates a hardware change. Because the microprocessor requires digital signals and the system to be controlled requires analog signals, the use of a microprocessor necessitates both an analog-to-digital (A/D) converter and a digital-to-analog (D/A) converter.

The microprocessor of a robot is programmed such that the robot can perform a variety of operations. Encoders are used to determine the actual position of a robot. The microprocessor compares the desired position with the actual position and sends out a signal to the motors controlling the robot so as to make the actual position the same as the desired position. The repeatability and accuracy of robots are extremely high. Robots come in a wide variety of sizes ranging from large industrial robots to small robots used for the assembly of delicate electronic equipment. They have found extensive use in hazardous environments such as painting and welding in the automotive industry. Robots were sent into high-radiation areas after the accident at the Three Mile Island nuclear power plant to help with the decontamination. Recent advances in robot technology have primarily been made possible by recent advances in control methods.

Manufacturing is being revolutionized by computer-controlled machines. The desired path of the cutter is preprogrammed into the computer (microprocessor). The actual position of the cutter is determined by an encoder. The desired position is compared to the actual position to determine the error. The actuating signal causes the cutting tool to move so as to reduce the error. Computer-controlled machines can manufacture complex shapes much more accurately than was heretofore possible. A program can be written to manufacture an entire piece. With a library of programs, it is simply a matter of selecting the program to produce the desired piece.

In Fig. 1.9 is shown a picture of South Bend Lathe's Magna Turn 1218 lathe, which is a computer-controlled lathe. The personal computer (PC) shown in the upper left corner continuously reads axis position using feedback from magnetic linear scales and commands appropriate axis motion through simple electronic drives. This complete closed-loop system results in *computer numerical control* (CNC) accuracy and consistency, combined with hydraulic power which provides the fastest possible machining time.

At the National Institute of Standards and Technology (NIST), the Humvee "high-mobility multipurpose wheeled vehicle" shown in Fig. 1.10 travels along the

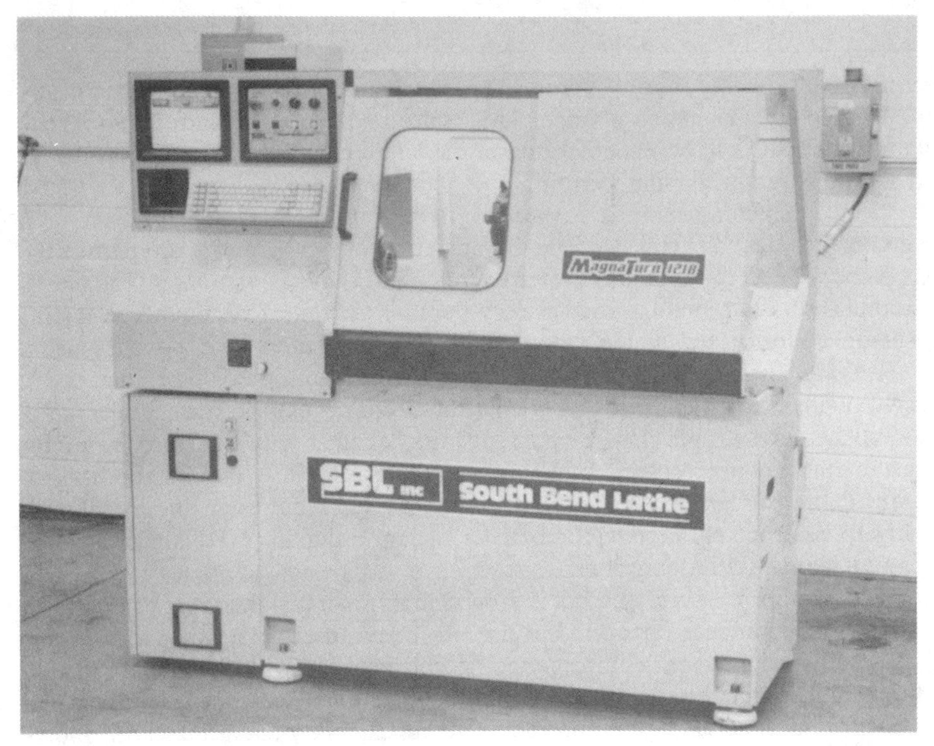

FIGURE 1.9
The Magna Turn 1218 lathe. (Courtesy South Bend Lathe.)

FIGURE 1.10
The Humvee. (Courtesy the National Institute of Standards and Technology.)

institute's back roads, being driven by a computer. The Humvee is a camera-equipped vehicle which takes the video image of the standard traffic lanes painted along the side and the center of the road. It latches onto the converging lines and automatically keeps the vehicle between the lane markings. An automatic control system controls the steering, brakes, throttle, and transmission. It is expected that this research will eventually lead to a national automated highway system. The automated vehicles would have the capability of driving to a preprogrammed destination, and they would also have a crash avoidance system that would sense an impending collision and would automatically brake or swerve to avoid it.

The U.S. Navy's new Advanced Unmanned Search System (AUSS) is shown in Fig. 1.11. The AUSS is a 17 foot long, 2800 pound Autonomous Underwater Vehicle (AUV) which can explore the ocean to depths of 20,000 feet. The main vehicle control system performs onboard navigation, and controls all motion functions (heading, pitch, depth, etc.) so that the position of the vehicle is precisely controlled at all times. It also controls maneuvers such as search patterns and hovering over a target. The sensor control system handles all sensor data and provides the information needed for image interpretation. Operators evaluate the images and supervise the operation. If they wish to investigate an image, they can order the vehicle to swim over for a closer look to obtain a more detailed image. Such vehicles cost far less than manned submersibles, are safer, and can stay down far longer amid the crushing pressures and inky darkness miles beneath the ocean surface to explore the planet's last, largest, and most mysterious frontier.

FIGURE 1.11
The U.S. Navy's Advanced Unmanned Search System. (Courtesy the Naval Command, Control, and Ocean Surveillance Center.)

CHAPTER 2

REPRESENTATION OF CONTROL COMPONENTS

To investigate the performance of control systems, it is necessary to obtain the mathematical relationship $G(D)$ relating the controlled variable c and the actuating signal e of the feedforward elements. This is accomplished by first obtaining the mathematical representation for each component between the actuating signal and the controlled variable and then expressing each of these equations as a block diagram. The combination of the block diagrams for each component yields the desired representation for $G(D)$. The value of $H(D)$ is obtained by applying the same technique to the components in the feedback portion of the control.

The quantity $G(D)$ could be obtained by writing the mathematical equation describing the operation of each component between e and c and then combining these individual equations algebraically to obtain the overall relationship between e and c. However, for all but the simplest systems, this procedure proves cumbersome because of the interaction between the various components in a typical control system. In addition, the block-diagram method gives one a better understanding of the system because of its visual representation.

The obtaining of block diagrams for typical elements used in control devices is illustrated in this chapter. In the next chapter, it is shown how these individual diagrams are combined to form entire control systems.

2.1 OPERATIONAL NOTATION

In writing equations for control systems, it is convenient to use the operational notation

$$D^n = \frac{d^n}{dt^n} \qquad n = 1, 2, 3, \ldots \tag{2.1}$$

The operator D is a symbol which indicates differentiation with respect to time. For example, if x and y are functions of time, then

$$D(x + y) = \frac{d}{dt}(x + y) = \frac{dx}{dt} + \frac{dy}{dt} = Dx + Dy$$

This shows that the operator D obeys the distributive law, i.e.,

$$D(x + y) = Dx + Dy$$

Similarly, the commutative law

$$Dx + Dy = Dy + Dx$$

and the associative law

$$Dx + (Dy + Dz) = (Dx + Dy) + Dz$$

hold.

Any operator such as D^n which has the properties

$$D^n(x + y) = D^n x + D^n y \qquad D^n(ax) = aD^n x$$

is called a *linear operator.*

Consider the differential equation

$$\frac{d}{dt}x(t) + ax(t) = f(t)$$

In operational notation, this differential equation is

$$(D + a)x(t) = f(t) \tag{2.2}$$

To solve this differential equation, first multiply both sides by the integrating factor e^{at}. That is,

$$e^{at}\left[\frac{d}{dt}x(t) + ax(t)\right] = \frac{d}{dt}[e^{at}x(t)] = e^{at}f(t)$$

Note that the left side may be written as a perfect differential, as is shown by the representation in the middle. Integration yields

$$e^{at}x(t) = \int f(t)e^{at}\,dt + C$$

where C is the constant of integration. Multiplying through by e^{-at} yields the function $x(t)$ which satisfies the differential equation

$$x(t) = e^{-at}\left[\int f(t)e^{at}\,dt + C\right] \tag{2.3}$$

To check that this function $x(t)$ does satisfy the differential equation, note that

$$Dx(t) = \frac{d}{dt}x(t) = -ae^{-at}\left[\int f(t)e^{at}\,dt + C\right] + e^{-at}[f(t)e^{at}]$$

and that

$$ax(t) = ae^{-at}\left[\int f(t)e^{at}\,dt + C\right]$$

Adding verifies that $Dx(t) + ax(t) = f(t)$. Solving Eq. (2.2) for $x(t)$ yields

$$x(t) = \frac{1}{D+a}f(t) \tag{2.4}$$

Equations (2.2) and (2.4) are equivalent forms of the same differential equation. Equating the right sides of Eqs. (2.3) and (2.4) shows that

$$\frac{1}{D+a}f(t) = e^{-at}\left[\int f(t)e^{at}\,dt + C\right] \tag{2.5}$$

Note that the operator $1/(D + a)$ is a weighted type of integration. It has previously been shown that the derivative d/dt of the right side plus a times the right side yields $f(t)$. Thus, operating on both sides of Eq. (2.5) by $D + a$ yields

$$(D+a)\frac{1}{D+a}f(t) = f(t)$$

Since the right side is $f(t)$, it follows that the operators on the left side may be canceled. In general, it may be shown that

$$(D+a)^n\frac{1}{(D+a)^m}f(t) = (D+a)^{n-m}f(t) \tag{2.6}$$

The meaning of the reciprocal of D is obtained by noting that the integral of the derivative of a function is

$$y(t) = \int f'(t)\,dt = \int [Df(t)]\,dt = f(t) + C \tag{2.7}$$

where C is the constant of integration. The derivative of both sides with respect to time is

$$Dy(t) = Df(t)$$

Solving for $y(t)$ gives

$$y(t) = \frac{1}{D}[Df(t)] \tag{2.8}$$

Comparison of Eqs. (2.7) and (2.8) shows that

$$\frac{1}{D}[Df(t)] = \int [Df(t)]\,dt$$

Therefore, the symbol $1/D$ indicates integration with respect to time.

In general, it may be shown that cancellation of operators is valid. That is,

$$\frac{1}{D^m} D^n f(t) = D^{n-m} f(t) \tag{2.9}$$

and

$$\frac{1}{(D+a)^m} (D+a)^n f(t) = (D+a)^{n-m} f(t) \tag{2.10}$$

In deriving the differential equation of operation for a system, the operator D may be treated as an algebraic quantity.

Illustrative example 2.1. For $f(t) = e^t$ and $y(0) = y'(0) = 2$, evaluate

$$y(t) = \frac{1}{D^3} Df(t)$$

Solution. Cancellation of operators gives

$$y(t) = \frac{1}{D^2} f(t) = \iint e^t \, dt \, dt = \int (e^t + c_1) \, dt = e^t + c_1 t + c_2$$

Differentiating gives

$$y'(t) = e^t + c_1$$

Evaluation of $y(t)$ and $y'(t)$ at time $t = 0$ yields

$$y(0) = 1 + c_2 = 2$$
$$y'(0) = 1 + c_1 = 2$$

Hence, $c_1 = c_2 = 1$. The desired result is

$$y(t) = e^t + t + 1$$

Illustrative example 2.2. For $f(t) = t$ and $y(0) = 1$, evaluate

$$y(t) = \frac{1}{(D+1)^2} (D+1) f(t)$$

Solution. Cancellation of operators gives

$$y(t) = \frac{1}{D+1} f(t)$$

Application of Eq. (2.5) in which $a = 1$ yields

$$y(t) = e^{-t} \left[\int te^t \, dt + C \right] = e^{-t}[e^t(t-1) + C]$$
$$= t - 1 + Ce^{-t}$$

Evaluation at time $t = 0$ yields

$$y(0) = -1 + C = 1$$

Hence $C = 2$. The desired result is

$$y(t) = 2e^{-t} + t - 1$$

2.2 MECHANICAL COMPONENTS

The load-deflection characteristics for a mechanical spring are shown in Fig. 2.1*a*. The spring force F_s required to deflect a spring a distance X from its free length is given by the equation

$$F_s = KX \tag{2.11}$$

where K, the spring rate, is a constant which is equal to the slope of the curve of the load F_s versus deflection X. The input to a spring is usually the force F_s, and the output is the deflection X, so that the block-diagram representation for Eq. (2.11) is as shown in Fig. 2.1*b*.

For a viscous damper as illustrated in Fig. 2.2*a*, the force F_d required to move one end of the dashpot at a velocity V relative to the other end is equal to the product of the damping coefficient B and the velocity, i.e.,

$$F_d = BV = B\frac{dX}{dt}$$

Substitution of the operator symbol $D = d/dt$ into the preceding expression yields

$$F_d = BDX \tag{2.12}$$

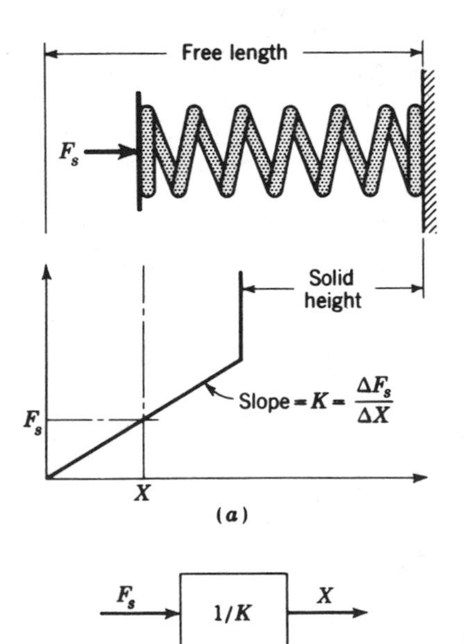

FIGURE 2.1
Spring characteristics.

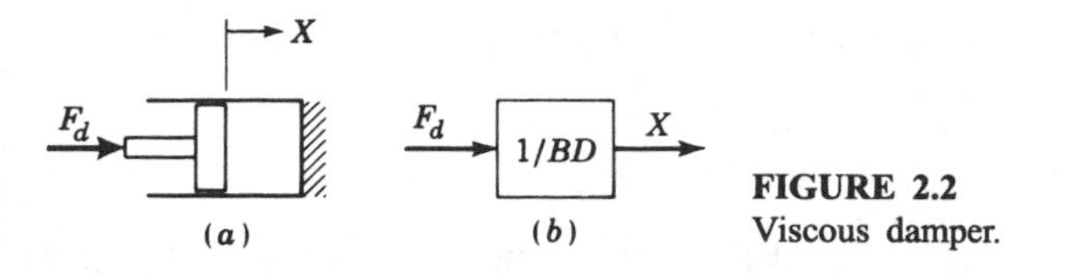

FIGURE 2.2
Viscous damper.

With the force F_d as the input and the displacement X as the output, the block-diagram representation for Eq. (2.12) is as shown in Fig. 2.2*b*.

By Newton's second law of motion, it follows that the summation of the external forces $\sum F_e$ acting on a mass is equal to the product of the mass and acceleration.

$$\sum F_e = MA = M\frac{d^2X}{dt^2} = MD^2X$$

The displacement X is given by the equation

$$X = \frac{1}{MD^2}\sum F_e \tag{2.13}$$

This is represented diagrammatically in Fig. 2.3.

The left side of Fig. 2.4*a* shows a spring at its free length. Next a mass M is placed on top of the spring. The deflection of the spring due to the gravitational force Mg is δ_{ST} (read "delta static"), where δ_{ST} is the static deflection. A constant force F_i is then applied to the top of the mass. The additional deflection of the spring due to this constant force is X_i. Finally, as shown at the far right, a variable force f

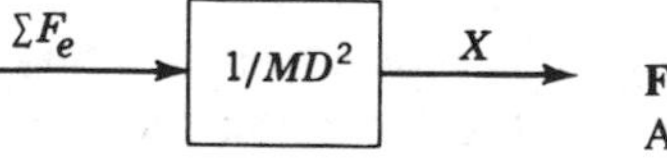

FIGURE 2.3
Acceleration of a mass.

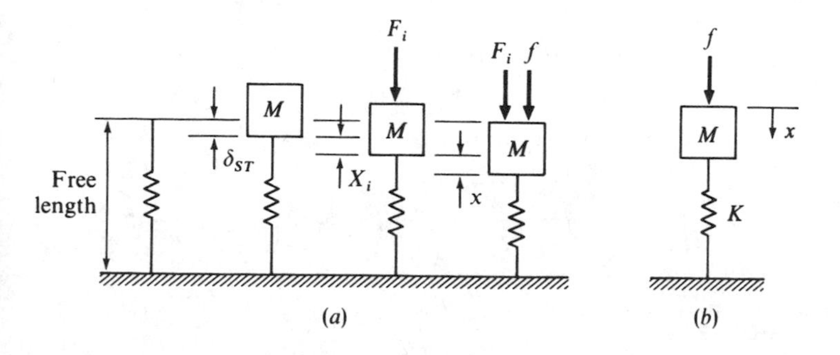

FIGURE 2.4
Spring with successive loads.

is applied. The deflection due to the variable force f is x. The summation of the total forces acting on the mass is

$$Mg + F_i + f - K(\delta_{ST} + X_i + x) = MD^2(\delta_{ST} + X_i + x) = MD^2x$$

where $\delta_{ST} + X_i + x$ is the total deflection of the spring. The reference position is the equilibrium position in which the deflection due to Mg and F_i is $\delta_{ST} + X_i$. The force equation at this reference position is

$$Mg + F_i - K(\delta_{ST} + X_i) = 0$$

Subtracting the equation for the forces at the reference position from the equation for the total forces yields

$$f - Kx = MD^2x$$

or

$$f = (MD^2 + K)x \tag{2.14}$$

In writing the equations for control systems, it is customary to consider only changes from some reference or equilibrium point of operation. In Fig. 2.4*b* the change in force from the reference value is f, and the change in position from the reference value is x. The net change in force acting on the mass is $f - Kx$, which is equal to the mass times the acceleration (MD^2x). Hence Eq. (2.14) is obtained directly from Fig. 2.4*b*. Note that forces which remain constant, such as the gravitational force and F_i, have no change and thus do not appear.

In obtaining the equation of operation for control components, it is customary to obtain the equation of operation with respect to some convenient reference point rather than to use total values. When total values are desired, it is an easy matter to add the reference value to the variation. Lowercase letters are used to represent changes from reference values, and capital letters are used to represent total values.

For the mass-spring-damper combination shown in Fig. 2.5*a*, the change in position from the reference position is x, and the change in force is f. The spring force Kx and the damper force BDx oppose, or resist, the motion caused by the applied force f. The summation of forces acting on the mass is

$$f - Kx - BDx = MD^2x$$

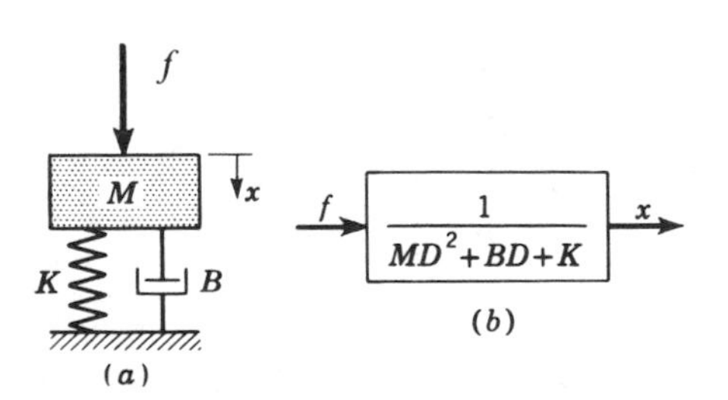

FIGURE 2.5
Series mass-spring-damper combination.

The preceding equation of operation with respect to the reference position thus becomes

$$f = (MD^2 + BD + K)x \tag{2.15}$$

Although x and f are measured from the reference point of operation, Eq. (2.15) is a general equation describing the dynamic behavior of the system. It is not necessary that the system be initially at this reference operating point or that the system be initially at rest. The block-diagram representation for Eq. (2.15) is shown in Fig. 2.5*b*.

Degrees of Freedom

By degrees of freedom is meant the number of coordinates required to specify the position of all the elements in a mechanical system. The system shown in Fig. 2.6 has 2 degrees of freedom (x and y), and the system shown in Fig. 2.7 has 3 degrees of freedom. The equation of operation for a mechanical system may be obtained by writing the force balance at each coordinate.

A mechanical system for obtaining phase lead compensation is shown in Fig. 2.6. The arrow on each coordinate indicates the positive sense of motion for that coordinate. The positive sense for velocity, acceleration, and force must be the same as that indicated by the position arrow at that coordinate. As indicated by the arrow at the x coordinate for the system of Fig. 2.6, the positive sense of motion is in the down direction. Thus, the positive sense for velocity $\dot{x}$, acceleration $\ddot{x}$, and force f is also down. When the force f is increased, the spring K_1 is compressed. As x increases, so does the spring force; but as y increases, the spring force decreases. The compression of spring K_1 is $x - y$. The damper force $B(\dot{x} - \dot{y}) = BD(x - y)$ is proportional to the relative velocity $\dot{x} - \dot{y}$ of the damper. Both the spring and viscous

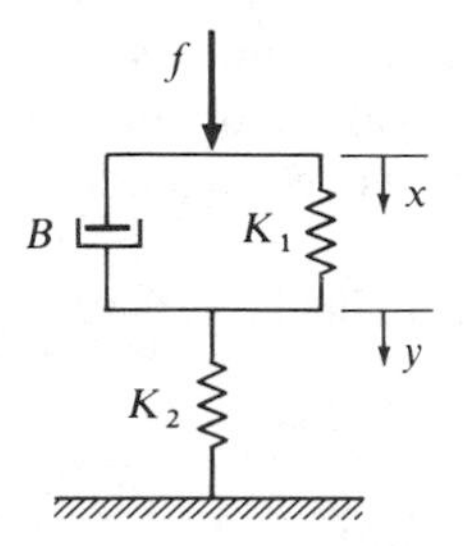

FIGURE 2.6
Mechanical system.

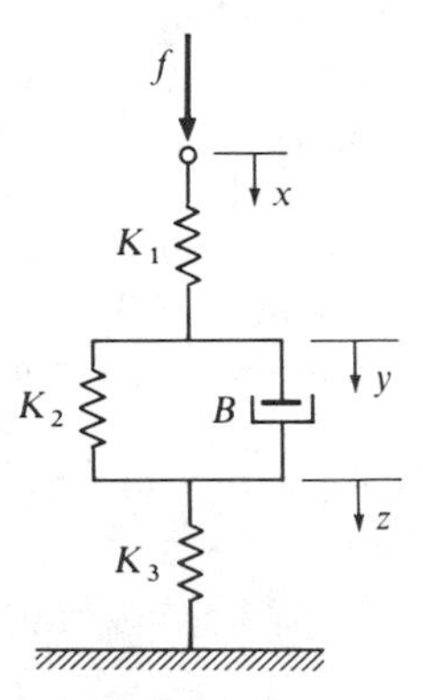

FIGURE 2.7
Mechanical system.

damper resist the downward motion of the x coordinate (i.e., they exert an upward force). The force equation at the x coordinate is

$$f - K_1(x - y) - BD(x - y) = 0$$

Because of action and reaction, the spring K_1 and viscous damper B exert a downward (positive) force at the y coordinate. The spring force K_2y resists the downward motion of the y coordinate (i.e., it exerts an upward force). The force equation at the y coordinate is

$$K_1(x - y) + BD(x - y) - K_2y = 0$$

The two preceding equations may be rewritten in the form

$$\begin{aligned} (K_1 + BD)x - (K_1 + BD)y &= f \\ (K_1 + BD)x - (K_1 + K_2 + BD)y &= 0 \end{aligned} \tag{2.16}$$

Solving these equations for x as a function of f, for y as a function of f, and for y as a function of x gives

$$\begin{aligned} x &= \frac{K_1 + K_2 + BD}{K_2(K_1 + BD)} f \\ y &= \frac{1}{K_2} f \\ y &= \frac{K_1 + BD}{K_1 + K_2 + BD} x = \frac{K_1(1 + BD/K_1)}{(K_1 + K_2)[1 + BD/(K_1 + K_2)]} x \\ &= \frac{\tau_2(1 + \tau_1 D)}{\tau_1(1 + \tau_2 D)} x \end{aligned} \tag{2.17}$$

where $\tau_1 = B/K_1$ and $\tau_2 = B/(K_1 + K_2) = [K_1/(K_1 + K_2)]\tau_1$. As is explained in Chap. 12, phase lead compensation is a common method for improving system performance.

Suppose it is desired to consider y positive for upward motion rather than downward (downward motion is shown in Fig. 2.6). This reversal would change the sign of y in all the force equations which would change the sign of y in the final answers, Eqs. (2.17). In effect, reversing the positive sense of a coordinate merely changes its sign. If the positive senses for all coordinates are reversed, the answers are unaffected. For example, if the positive senses for f, x, and y in Fig. 2.6 were all taken as up rather than down, the resulting force equations and answers would be the same.

For the system shown in Fig. 2.7, the force equations at the x, y, and z coordinates are

$$\begin{aligned} &f - K_1(x - y) = 0 \\ &K_1(x - y) - (K_2 + BD)(y - z) = 0 \\ &(K_2 + BD)(y - z) - K_3z = 0 \end{aligned}$$

These equations may be rewritten in the form

$$K_1x - K_1y + 0 = f$$
$$K_1x - (K_1 + K_2 + BD)y + (K_2 + BD)z = 0$$
$$0 + (K_2 + BD)y - (K_2 + K_3 + BD)z = 0$$

These equations may now be solved for any of the relationships between x, y, z, and f. For example, the equations relating y to f and z to y are

$$y = \frac{K_2 + K_3 + BD}{K_3(K_2 + BD)} f$$
$$z = \frac{K_2 + BD}{K_2 + K_3 + BD} y \tag{2.18}$$

Rotational Mechanical Components

A torsional spring is characterized by the equation

$$T_s = K_s\theta \tag{2.19}$$

where T_s = torque tending to twist spring
K_s = torsional spring rate
θ = angular displacement of spring

A well-known example of a torsional spring is a shaft, as shown in Fig. 2.8. The right end of the shaft is displaced an angle θ with respect to the left end because of the twisting torque T_s. For a straight shaft, the torsional spring rate is

$$K_s = \frac{\pi d^4 G}{32L}$$

where G = modulus of elasticity in shear
d = diameter of shaft
L = length of shaft

The torque T_d required to overcome the viscous friction of a rotating member is

$$T_d = B_v\omega = B_v\frac{d\theta}{dt} = B_vD\theta \tag{2.20}$$

where B_v = coefficient of viscous friction and ω = angular velocity.

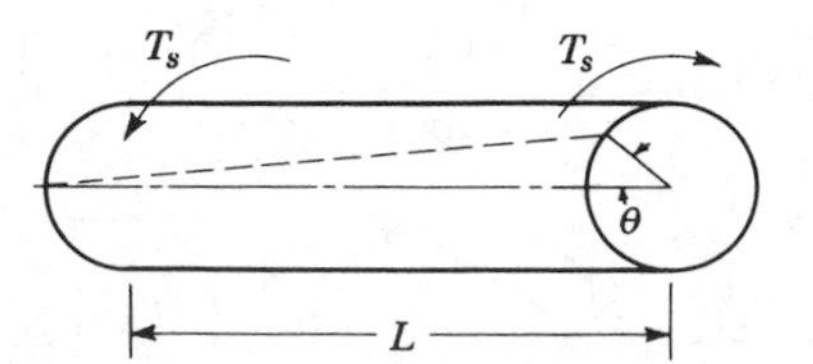

FIGURE 2.8
Shaft acting as a torsional spring.

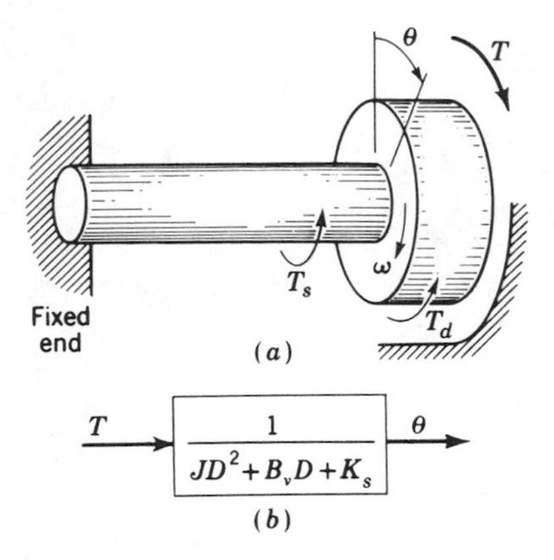

FIGURE 2.9
Torsional inertia-spring-damper combination.

A disk rotating in a viscous medium and supported by a shaft is shown in Fig. 2.9*a*. The applied torque tending to rotate the disk is T. The shaft torque and viscous friction oppose the motion, so that

$$\sum T_e = T - T_s - T_d = J\alpha = JD^2\theta \tag{2.21}$$

where $\sum T_e$ is the summation of external torques acting on the disk, J is the mass moment of inertia of the disk, and α is the angular acceleration. The substitution of T_s from Eq. (2.19) and T_d from Eq. (2.20) into Eq. (2.21) yields

$$T = (JD^2 + B_vD + K_s)\theta \tag{2.22}$$

The block-diagram representation for this system is shown in Fig. 2.9*b*.

2.3 ELECTRICAL COMPONENTS

The resistor, inductor, and capacitor are the three basic components of electric circuits. The equation for the voltage drop E_R across a resistor is

$$E_R = RI \tag{2.23}$$

where R is the resistance in ohms and I is the current flowing through the resistor in amperes.

For an inductor, the voltage drop E_L is given by the equation

$$E_L = L\frac{dI}{dt} = LDI \tag{2.24}$$

where L is the inductance in henrys.

Similarly, the voltage drop E_C across a capacitor is

$$E_C = \frac{1}{C}\int I\,dt = \frac{1}{CD}I \tag{2.25}$$

where C is the capacitance in farads.

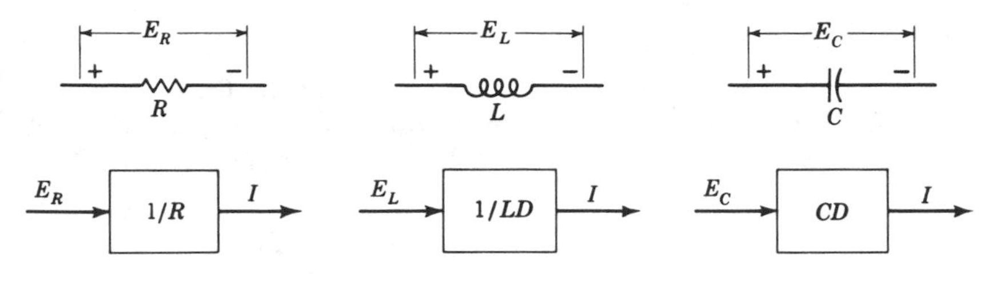

FIGURE 2.10
Representation of a resistor, inductor, and capacitor.

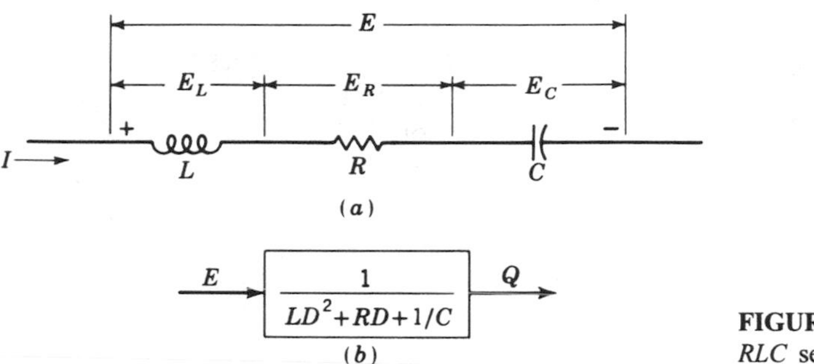

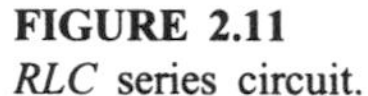

FIGURE 2.11
RLC series circuit.

The diagrammatic representations of Eqs. (2.23) to (2.25) are shown in Fig. 2.10.

For the series *RLC* circuit shown in Fig. 2.11*a*, the total voltage drop E is the sum of that across the inductor E_L, that across the resistor E_R, and that across the capacitor E_C.

$$E = E_L + E_R + E_C = \left(LD + R + \frac{1}{CD}\right)I \tag{2.26}$$

The charge Q is the time integral of the current; that is, $Q = (1/D)I$ or $I = DQ$. By noting that $LDI = LD^2Q$, $RI = RDQ$, and $(1/CD)I = (1/C)Q$, Eq. (2.26) becomes

$$E = \left(LD^2 + RD + \frac{1}{C}\right)Q \tag{2.27}$$

The overall block-diagram representation for this *RLC* circuit is shown in Fig. 2.11*b*.

2.4 SERIES AND PARALLEL LAWS

Elements are usually connected in either a series or a parallel arrangement. Arriving at the equation for such systems can be greatly simplified by the use of the laws for series and for parallel combinations.

Series Electric Circuits

A general series circuit is shown in Fig. 2.12*a*. In a series circuit, the total voltage drop E is the sum of the individual voltage drops across each element, and the same current I flows through each element. The equation for the summation of the voltage drops is

$$E = \left(L_1D + L_2D + R_1 + R_2 + \frac{1}{C_1D} + \frac{1}{C_2D}\right)I \tag{2.28}$$

The equation relating the voltage E and current I for any electric circuit can be expressed in the form

$$E = ZI$$

where Z is the impedance for the circuit, E is the voltage drop across the circuit, and I is the total current flowing through the circuit. For the case of Eq. (2.28), the impedance Z is

$$Z = L_1D + L_2D + R_1 + R_2 + \frac{1}{C_1D} + \frac{1}{C_2D} \tag{2.29}$$

For elements in series, the total impedance is the sum of the individual impedances of each element. The block-diagram representation is shown in Fig. 2.12*b*.

Parallel Electric Circuits

A general combination of electrical elements in parallel is shown in Fig. 2.13*a*. The distinguishing features of a parallel arrangement are that the voltage drop E across each element is the same and that the total current I flowing into the system is the sum of the currents flowing through each element. Thus

$$I = \frac{E}{L_1D} + \frac{E}{L_2D} + \frac{E}{R_1} + \frac{E}{R_2} + \frac{E}{1/(C_1D)} + \frac{E}{1/(C_2D)} \tag{2.30}$$

or

$$E = \frac{1}{1/(L_1D) + 1/(L_2D) + 1/R_1 + 1/R_2 + C_1D + C_2D}I = ZI$$

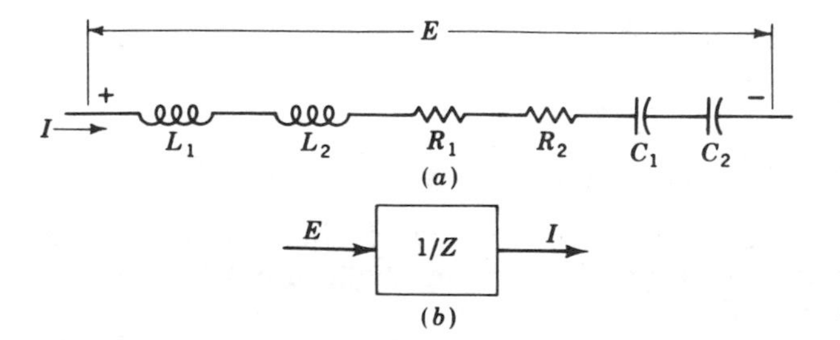

FIGURE 2.12
General series circuit.

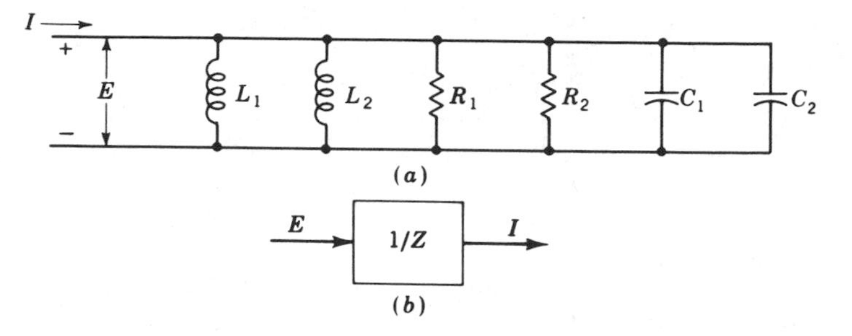

FIGURE 2.13
General parallel circuit.

The impedance Z is

$$Z = \frac{1}{1/(L_1D) + 1/(L_2D) + 1/R_1 + 1/R_2 + C_1D + C_2D} \tag{2.31}$$

For elements in parallel, the total impedance is equal to 1 divided by the sum of the reciprocal of the impedance of each element. The block-diagram representation is shown in Fig. 2.13*b*.

Illustrative example 2.3. For the circuit shown in Fig. 2.14, determine the equation relating the output voltage E_2 to the input voltage E_1.

Solution. The parallel combination of R_1 and C_1 is in series with R_2, so that the total impedance Z is

$$Z = Z_1 + R_2 = \frac{1}{1/R_1 + C_1D} + R_2 = \frac{R_1}{1 + R_1C_1D} + R_2 \tag{2.32}$$

The voltage E_1 is given by

$$E_1 = ZI = \frac{R_1 + R_2 + R_1R_2C_1D}{1 + R_1C_1D}I \tag{2.33}$$

Similarly E_2 is

$$E_2 = R_2I \tag{2.34}$$

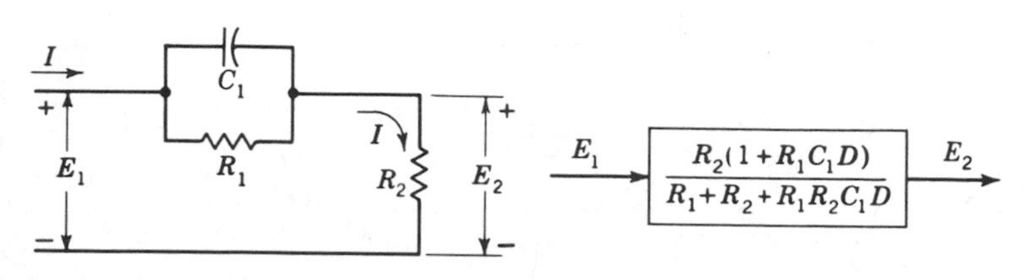

FIGURE 2.14
Electric circuit.

The substitution of I from Eq. (2.33) into Eq. (2.34) yields the desired answer:

$$E_2 = \frac{R_2(1 + R_1C_1D)}{R_1 + R_2 + R_1R_2C_1D}E_1 \tag{2.35}$$

Series Mechanical Elements

A series arrangement of mechanical elements is shown in Fig. 2.15*a*. In general, it is better to use the equivalent "grounded-chair" representation for a mass, as shown in Fig. 2.15*b*, rather than the more common representation of Fig. 2.15*a*. The fact that the mass is in series with the other elements is more readily seen from Fig. 2.15*b* than from Fig. 2.15*a*. In determining the inertial force, the acceleration of a mass is always taken with respect to ground. Thus, providing the grounded chair to indicate motion relative to ground is a more justifiable representation than Fig. 2.15*a*, which shows better the actual physical arrangement of the elements in the system. For series mechanical elements, the force f is equal to the summation of the forces acting on each individual component, and each element undergoes the same displacement. Thus,

$$f = (K_1 + K_2 + B_1D + B_2D + MD^2)x = Zx \tag{2.36}$$

where x and f are measured from a convenient reference operating point. The equivalent impedance is

$$Z = K_1 + K_2 + B_1D + B_2D + MD^2 \tag{2.37}$$

For mechanical elements in series, the total impedance is the sum of individual impedances of each element. The block-diagram representation for this system is shown in Fig. 2.15*c*.

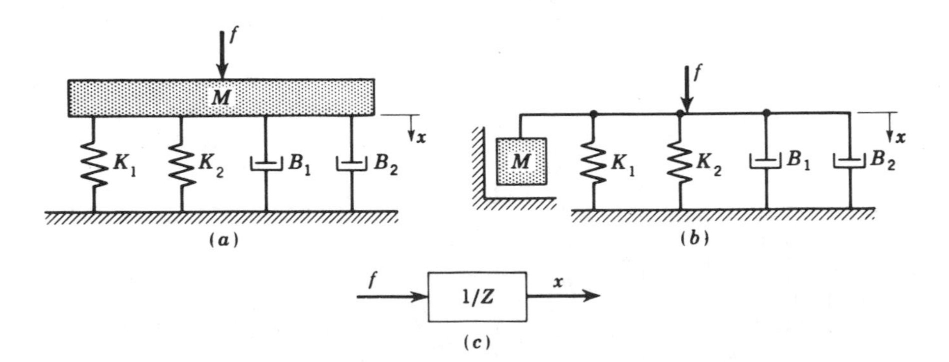

FIGURE 2.15
Mechanical elements in series.

Parallel Mechanical Elements

A parallel combination of mechanical elements is shown in Fig. 2.16*a*. For parallel elements, the same force f is transmitted through each element. In addition, the total deflection x is seen to be the sum of the individual deflections of each element. Thus

$$x = \frac{f}{K_1} + \frac{f}{K_2} + \frac{f}{B_1 D} + \frac{f}{B_2 D} \tag{2.38}$$

or

$$f = \frac{1}{1/K_1 + 1/K_2 + 1/(B_1 D) + 1/(B_2 D)} x = Zx$$

The impedance is

$$Z = \frac{1}{1/K_1 + 1/K_2 + 1/(B_1 D) + 1/(B_2 D)} \tag{2.39}$$

For mechanical elements in parallel, the total impedance is equal to 1 divided by the sum of the reciprocals of the individual impedances of each element. The block-diagram representation is shown in Fig. 2.16*b*.

A necessary condition for parallel elements is that the same force be transmitted through each element. Springs and dampers satisfy this condition because the force is the same on both sides. However, this is not the case for a mass such as that shown in Fig. 2.17*a*, because the difference in forces acting on both sides of a mass is utilized in acceleration. Thus, a mass located between other elements cannot be in parallel with them. A mass can be in parallel only if it is the last element, as shown in Fig. 2.18. For this system, the displacement x is

$$x = (x - y) + (y - z) + z = \left(\frac{1}{K} + \frac{1}{BD} + \frac{1}{MD^2}\right) f$$

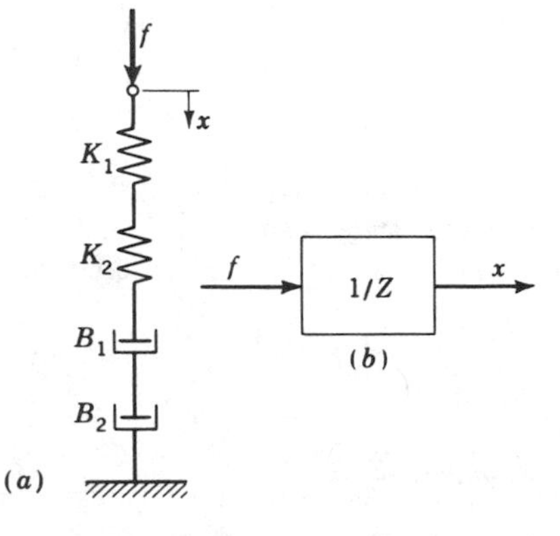

FIGURE 2.16
Mechanical elements in parallel.

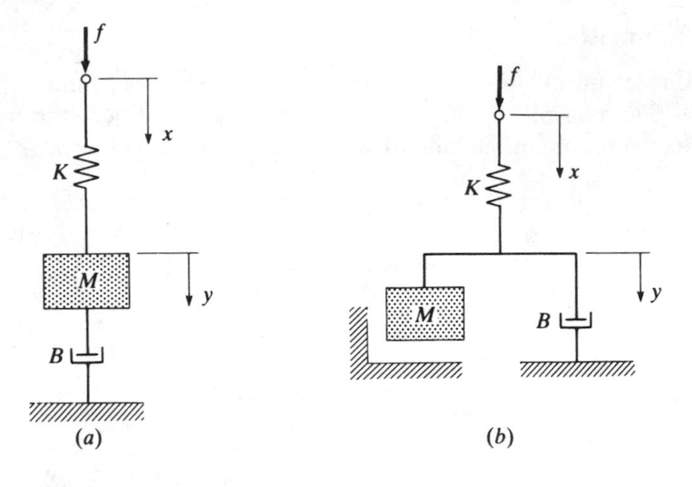

FIGURE 2.17
Mechanical circuit: (*a*) schematic representation, (*b*) grounded-chair representation.

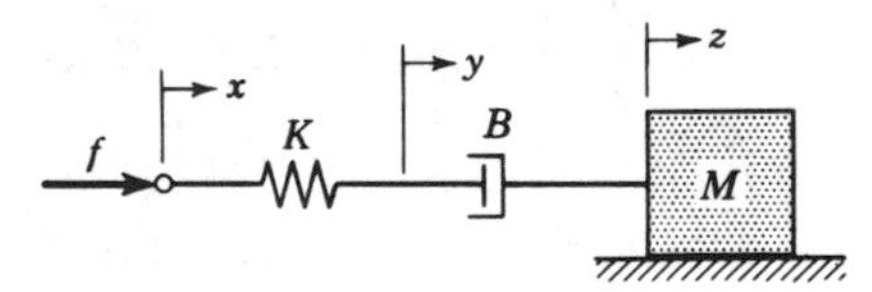

FIGURE 2.18
Parallel mass-spring-damper combination.

Thus,

$$f = \frac{1}{1/K + 1/(BD) + 1/(MD^2)} x \tag{2.40}$$

Parallel and series laws for rotational mechanical components may also be developed by extending the preceding techniques.

Illustrative example 2.4. For the mass-spring-damper combination shown in Fig. 2.17*a*, determine the equation relating f and x, the equation relating f and y, and the equation relating x and y.

Solution. The first step is to draw the equivalent grounded-chair system, in which the motion of the mass with respect to ground is clearly indicated, as shown in Fig. 2.17*b*. The spring K is in parallel with the series combination of M and B. Application of the laws for parallel and series elements gives

$$f = \frac{1}{1/K + 1/Z} x \tag{2.41}$$

The impedance Z of the series combination of M and B is

$$Z = MD^2 + BD$$

Thus,

$$f = \frac{1}{1/K + 1/(MD^2 + BD)} x = \frac{K(MD^2 + BD)}{MD^2 + BD + K} x \tag{2.42}$$

The force f is transmitted through the spring K and acts upon the series combination of M and B. Thus, the equation of motion for this part of the system which relates f and y is

$$f = (MD^2 + BD)y \tag{2.43}$$

The desired relationship between x and y is obtained by eliminating f from Eqs. (2.42) and (2.43). That is,

$$y = \frac{K}{MD^2 + BD + K} x \tag{2.44}$$

Grounded-Chair Representation

The general procedure for constructing the grounded-chair representation is as follows:

1. Draw coordinates such that the coordinate at which the force acts is at the top and ground is at the bottom.
2. Insert each element in its correct orientation with respect to these coordinates.

For the coupled mass-spring-damper system shown in Fig. 2.19*a*, the coordinates are x, y, and ground. Step 1 is carried out by drawing the coordinates as shown in Fig. 2.19*b*. To do step 2, we must note that for the spring K_1 and mass M_1 the coordinates are x and ground; for the spring K and damper B the coordinates are x and y; and finally for mass M_2 and spring K_2 the coordinates are y and ground. Inserting these elements between the proper coordinates as shown in Fig. 2.19*c* completes the grounded-chair representation. The series and parallel combinations are now recognized directly from this grounded-chair representation. The equation for the force f is

$$f = \left[M_1D^2 + K_1 + \frac{1}{1/(K + BD) + 1/(K_2 + M_2D^2)}\right] x$$

As another example, consider the dynamic vibration absorber shown in Fig. 2.20*a*. The grounded-chair representation is shown in Fig. 2.20*b*. The coordinate x at which the force f acts is drawn at the top, and ground is at the bottom. The coordinates for M_1, K_1, and B are x and ground; for K_2 the coordinates are x and y; and finally for M_2 the coordinates are y and ground. The equation of operation is

$$f = \left[M_1D^2 + BD + K_1 + \frac{1}{1/K_2 + 1/(M_2D^2)}\right] x$$

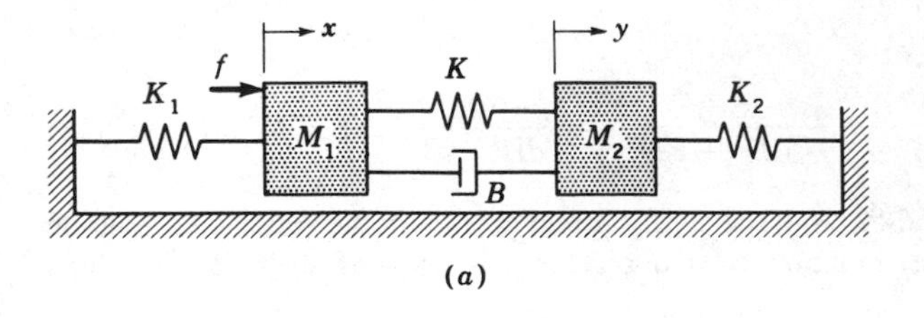

(a)

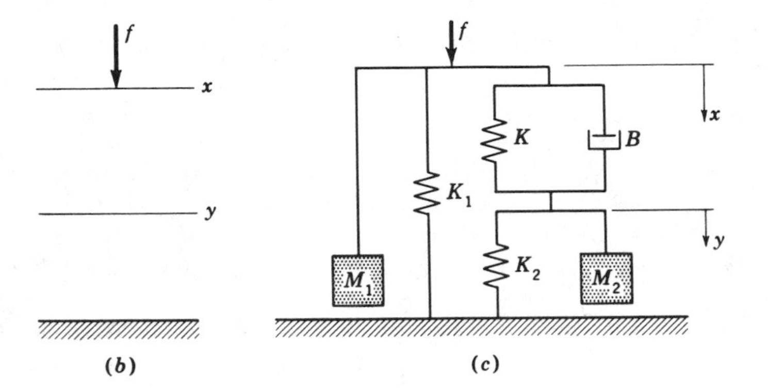

(b) (c)

FIGURE 2.19
General mechanical system.

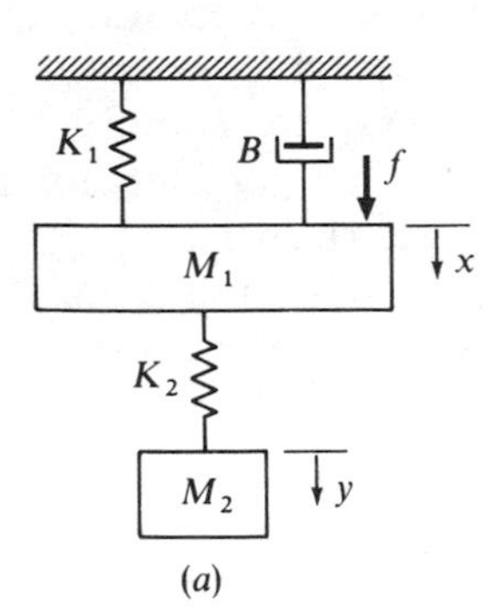

(a)

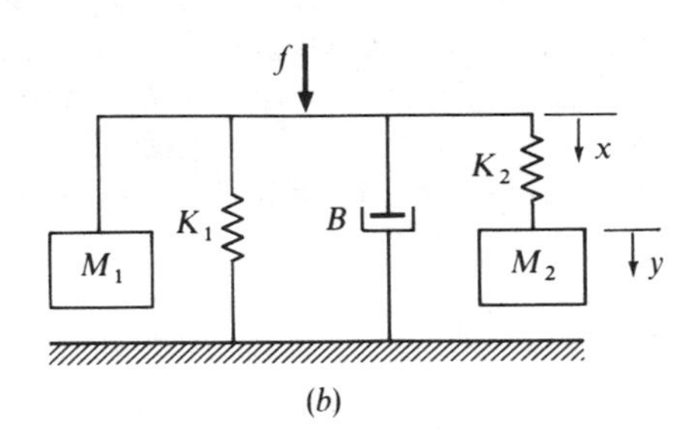

(b)

FIGURE 2.20
Dynamic vibration absorber.

2.5 ANALOGIES

The equation of operation for the series mechanical system of Fig. 2.5*a* is given by Eq. (2.15), and the equation for the series electric circuit of Fig. 2.11*a* is given by Eq. (2.27). That is,

$$f = (MD^2 + BD + K)x \tag{2.15}$$

$$E = \left(LD^2 + RD + \frac{1}{C}\right)Q \tag{2.27}$$

Comparison of corresponding terms in Eqs. (2.15) and (2.27) shows that the differential equation of operation for each system has the same form. The terms which occupy corresponding positions are called *analogous quantities*. This particular analog is referred to as the *direct analog*. The analogous quantities for a direct analog are shown in Table 2.1.

The total force acting on a group of mechanical elements in series is equal to the sum of the forces exerted on each element. Similarly, the total voltage drop across a group of electrical elements in series is equal to the sum of the voltage drops across each element. Thus, in constructing a direct analog, series mechanical elements are replaced by analogous series electrical elements.

For parallel mechanical elements, the force acting on each element is the same, and for parallel electrical elements the voltage drop across each element is the same. Thus, in a direct analog, parallel mechanical elements should be replaced by equivalent electrical elements in parallel. A direct analog is also called a *force-voltage analog* in that force and voltage are analogous quantities.

The other type of analog is the *inverse analog*. To construct an inverse analog, first note that the total current flowing through a group of electrical elements in parallel is the sum of the currents in each element. This is analogous to the fact that the total force acting on a group of mechanical elements in series is the sum of the forces acting on each element. Thus, to construct an inverse analog, series mechanical elements must be replaced by parallel electrical elements. Similarly, in an inverse analog, it can be shown that parallel mechanical elements

TABLE 2.1
Analogous quantities in a direct (force-voltage) analog

Translational mechanical system	Force	Velocity	Displacement	Mass	Viscous damping coefficient	Spring constant
	f	$\dot{x} = Dx$	x	M	B	K
Electrical system	Voltage	Current	Charge	Inductance	Resistance	Reciprocal of capacitance
	E	$I = DQ$	Q	L	R	$\frac{1}{C}$

should be replaced by series electrical elements. Thus, the arrangement of series and parallel elements is inverted in constructing an inverse analog (i.e., series elements are replaced by parallel elements, and vice versa). An inverse analog is also called a *force-current analog* in that force and current are analogous quantities.

Analogous quantities for an inverse analog may be determined by comparing the equation of operation for the parallel mechanical system of Fig. 2.18 with that for the series electrical system of Fig. 2.11*a*. The equation of operation for the parallel mechanical system of Fig. 2.18 is given by Eq. (2.40). Solving for x and then multiplying both sides by D gives

$$\dot{x} = \left(\frac{D}{K} + \frac{1}{B} + \frac{1}{MD}\right)f \tag{2.45}$$

The operation of the series electric circuit of Fig. 2.11*a* is described by Eq. (2.26), which has the same form as Eq. (2.45). That is,

$$E = \left(LD + R + \frac{1}{CD}\right)I \tag{2.26}$$

Comparison of corresponding terms in Eqs. (2.26) and (2.45) yields the analogous quantities for an inverse analog that are shown in Table 2.2.

Illustrative example 2.5. Suppose we want to determine the electrical analog for the mechanical system of Fig. 2.17 by using (*a*) the direct analog and (*b*) the inverse analog.

Solution

(*a*) The grounded-chair representation for the mechanical system of Fig. 2.17 is shown in Fig. 2.21*a*. The direct analog is shown in Fig. 2.21*b*. Note that the capacitor C is in parallel with the series combination of L and R, just as the spring

TABLE 2.2
Analogous quantities in an inverse (force-current) analog

Translational mechanical system	Force	Velocity	Spring constant	Damping coefficient	Mass
	f	$\dot{x}$	K	B	M
Electrical system	Current	Voltage	Reciprocal of inductance	Reciprocal of resistance	Capacitance
	I	E	$\frac{1}{L}$	$\frac{1}{R}$	C

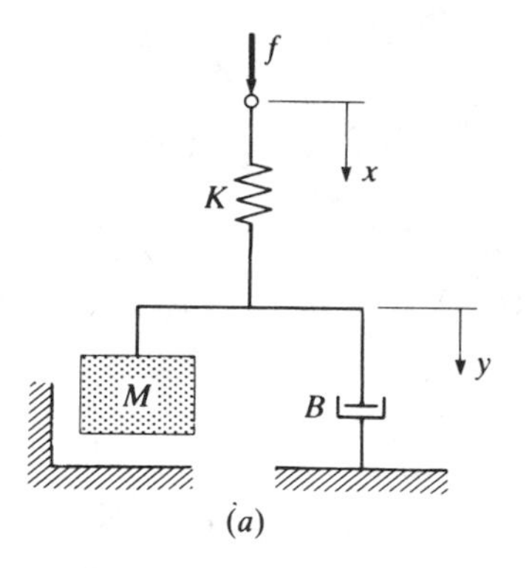

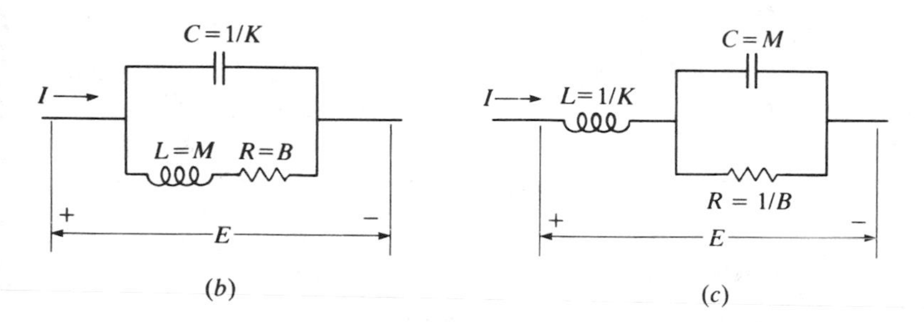

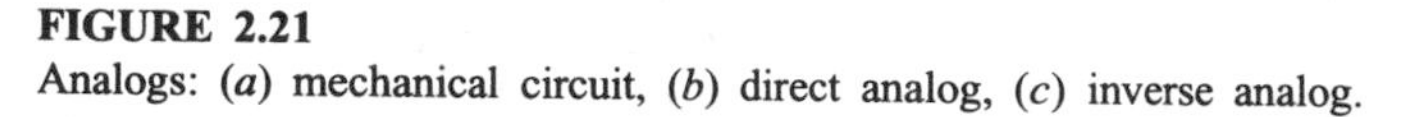

FIGURE 2.21
Analogs: (*a*) mechanical circuit, (*b*) direct analog, (*c*) inverse analog.

K of Fig. 2.21*a* is in parallel with the series combination of *M* and *B*. The equation of operation for the electric circuit of Fig. 2.21*b* is

$$E = \frac{1}{CD + 1/(LD + R)} I = \frac{1}{C + 1/(LD^2 + RD)} \frac{I}{D}$$

$$= \frac{1}{C + 1/(LD^2 + RD)} Q = \frac{(1/C)(LD^2 + RD)}{LD^2 + RD + 1/C} Q \tag{2.42a}$$

The equation of operation for the mechanical system of Fig. 2.21*a* is given by Eq. (2.42).

$$f = \frac{1}{1/K + 1/(MD^2 + BD)} x = \frac{K(MD^2 + BD)}{MD^2 + BD + K} x \tag{2.42}$$

Comparison of corresponding terms in Eqs. (2.42) and (2.42*a*) verifies the analogous quantities given in Table 2.1.

(*b*) The inverse analog is shown in Fig. 2.21*c*. Note that the inductor *L* is in series with the parallel combination of *C* and *R*, whereas in Fig. 2.21*a* the spring *K* is in parallel with the series combination of *M* and *B*. The equation of operation for the electric circuit of Fig. 2.21*c* is

$$E = \left(LD + \frac{1}{CD + 1/R} \right) I$$

or

$$I = \frac{1}{LD + 1/(CD + 1/R)} E = \frac{1}{L + 1/[CD^2 + (1/R)D]} \frac{E}{D}$$

$$= \frac{(1/L)[CD^2 + (1/R)D]}{CD^2 + (1/R)D + 1/L} \frac{E}{D} \tag{2.42b}$$

Comparison of corresponding terms in Eqs. (2.42) and (2.42*b*) verifies the analogous quantities given in Table 2.2. In the inverse analog, velocity is analogous to voltage, and thus displacement is analogous to the integral of voltage ($x = E/D$).

Comparison of corresponding terms in Eqs. (2.15) and (2.22) shows that a direct (torque-force) analog may be developed in which series translational mechanical elements are replaced by series rotational elements. Similarly, parallel translational elements should be replaced by parallel rotational elements.

2.6 SCALE FACTORS

Because of the convenience of working with electrical equipment, ordinarily an electrical analog is constructed for some other system. Scale factors ensure that the values of the voltages and currents in the electrical analog will be reasonable. For example, in a direct analog, analogous quantities are force-voltage and velocity-current; hence

$$f = k_E E \tag{2.46}$$

$$\frac{dx}{dt} = k_I I \tag{2.47}$$

Thus, if the maximum value of f is 1000 lb and the maximum desired voltage is 100 V, then

$$k_E = \frac{f}{E} = 10 \text{ lb/V}$$

In this case, 1 V corresponds to 10 lb of force.

To determine the other analogous relationships, first write the force equations for a mass, a damper, and a spring, i.e.,

$$f = M \frac{d}{dt}\left(\frac{dx}{dt}\right) \tag{2.48}$$

$$f = B \frac{dx}{dt} \tag{2.49}$$

$$f = Kx \qquad or \qquad \frac{df}{dt} = K \frac{dx}{dt} \tag{2.50}$$

The voltage equations for an inductor, resistor, and capacitor are

$$E = L\frac{d}{dt}I \qquad E = RI \qquad \frac{dE}{dt} = \frac{1}{C}I$$

The preceding expressions may be written in the forms

$$k_E E = \frac{k_E}{k_I} L \frac{d}{dt} k_I I \tag{2.51}$$

$$k_E E = \frac{k_E}{k_I} R k_I I \tag{2.52}$$

$$\frac{d}{dt} k_E E = \frac{k_E}{k_I}\frac{1}{C} k_I I \tag{2.53}$$

From a comparison of Eqs. (2.48) and (2.51), it follows that to have $f = k_E E$ and $dx/dt = k_I I$, then $M = (k_E/k_I)L$ or

$$L = kM \tag{2.54}$$

where $k = k_I/k_E$.

Comparison of Eqs. (2.49) and (2.52) shows that

$$R = kB \tag{2.55}$$

Comparison of Eqs. (2.50) and (2.53) yields

$$C = \frac{1}{kK} \tag{2.56}$$

For the case in which $k = 1$, the preceding analogous relationships are the same as those given in Table 2.1.

Time Scale

For systems which are extremely fast-acting, it may be desired to slow down the solution on the electrical analog. Similarly, for extremely slow systems, it may be desired to speed up the solution. By letting t represent actual time for a phenomenon to occur in the actual system and τ represent the time for the corresponding phenomenon to take place in the analog,

$$\tau = at \tag{2.57}$$

where t = actual time and τ = time in analog.

If a phenomenon takes 1 s to complete in actual time t and if $a = 10$, then $\tau = at = 10$, or the solution has been slowed down on the analog by a factor of 10. Hence, for $a > 1$ the solution is slowed down, and for $a < 1$ the solution is sped up.

When a time-scale change is effected, events in the electrical analog occur in analog time τ; hence, writing Eqs. (2.51), (2.52), and (2.53) in terms of τ gives

$$k_E E = \frac{L}{k}\frac{d}{d\tau}k_I I$$

$$k_E E = \frac{R}{k}k_I I$$

$$\frac{d}{d\tau}k_E E = \frac{1}{kC}k_I I$$

Differentiation of Eq. (2.57) shows that $d\tau/dt = a$; hence

$$\frac{dI}{d\tau} = \frac{dt}{d\tau}\frac{dI}{dt} = \frac{1}{a}\frac{dI}{dt} \qquad \text{and} \qquad \frac{dE}{d\tau} = \frac{1}{a}\frac{dE}{dt}$$

Expressing the equations for the electrical elements in terms of time t gives

$$k_E E = \frac{L}{ka}\frac{d}{dt}k_I I \tag{2.58}$$

$$k_E E = \frac{R}{k}k_I I \tag{2.59}$$

$$\frac{d}{dt}k_E E = \frac{a}{kC}k_I I \tag{2.60}$$

From a comparison of Eqs. (2.48) and (2.58), it follows that to have $f = k_E E$ and $dx/dt = k_I I$, then $M = L/(ka)$, or

$$L = akM \tag{2.61}$$

Comparison of Eqs. (2.49) and (2.59) shows that

$$R = kB \tag{2.62}$$

Comparison of Eqs. (2.50) and (2.60) shows that

$$C = \frac{a}{k}\frac{1}{K} \tag{2.63}$$

For the case in which $a = 1$ and $k = 1$, the preceding results are identical to those given in Table 2.1.

The corresponding relationships for an inverse analog are developed in a similar manner. In this analog, force and current are analogous quantities, as are velocity and voltage; hence, the desired scale factors are

$$f = k_I I \tag{2.64}$$

$$\frac{dx}{dt} = k_E E \tag{2.65}$$

Equations (2.58), (2.59), and (2.60) are modified forms of the basic relationships of an inductor, resistor, and capacitor, respectively. Solving Eqs.

(2.58), (2.59), and (2.60) for $k_I I$ gives

$$\frac{d}{dt}k_I I = \frac{ka}{L}k_E E \tag{2.66}$$

$$k_I I = \frac{k}{R}k_E E \tag{2.67}$$

$$k_I I = \frac{kC}{a}\frac{d}{dt}k_E E \tag{2.68}$$

Comparison of Eqs. (2.50) and (2.66) reveals that to have $f = k_I I$ and $dx/dt = k_E E$, then $K = ka/L$, or

$$L = ak\frac{1}{K} \tag{2.69}$$

Similarly, comparing Eq. (2.49) with Eq. (2.67) and comparing Eq. (2.48) with Eq. (2.68) show that

$$R = k\frac{1}{B} \tag{2.70}$$

$$C = \frac{a}{k}M \tag{2.71}$$

For the case in which $a = 1$ and $k = 1$, these relationships become identical to those given in Table 2.2.

Units

In the British gravitational system, force is measured in pounds (lb_f), length in feet (ft), time in seconds (s), and mass in slugs. From Newton's equation of motion $f = ma$, it follows that the unit relationship is

$$\text{lb}_\text{f} = \text{slug} \cdot \text{ft/s}^2$$

For a direct analog, the units of k_E are lb_f/V, and the units of k_I are (ft/s)/A = ft/(s · A). Thus, $k = k_I/k_E$ has units of ft · V/(lb_f · A · s). From Eq. (2.61), the resulting units for $L = akM$ are

$$L = \left(\frac{\text{ft} \cdot \text{V}}{\text{lb}_\text{f} \cdot \text{A} \cdot \text{s}}\right)\left(\frac{\text{lb}_\text{f} \cdot \text{s}^2}{\text{ft}}\right) = \frac{\text{V} \cdot \text{s}}{\text{A}}$$

Substitution of units into the relationship $E = L\,dI/dt$ shows that inductance L has units of V · s/A, which are called henrys (H). Similarly, from Eq. (2.62) it is found that resistance R has units of V/A which are ohms (Ω), and from Eq. (2.63) it is found that capacitance C has units of A · s/V, which are farads (F). For the inverse analog, k_I has units of lb_f/A, and k_E has units of (ft/s)/V = ft/(V · s). The resulting units for $k = k_I/k_E$ are lb_f · V · s/(ft · A). From Eq. (2.69), the resulting units for $L = ak/K$ are

$$L = \left(\frac{\text{lb}_\text{f} \cdot \text{V} \cdot \text{s}}{\text{ft} \cdot \text{A}}\right) \Big/ \left(\frac{\text{lb}_\text{f}}{\text{ft}}\right) = \frac{\text{V} \cdot \text{s}}{\text{A}}$$

These are the units for henrys. Similarly, Eqs. (2.70) and (2.71) yield the proper units for R and C, respectively.

In SI, force is measured in newtons (N), mass in kilograms (kg), length in meters (m), and time in seconds (s). The relationship among the units in the SI system is

$$\mathrm{N} = \frac{\mathrm{kg} \cdot \mathrm{m}}{\mathrm{s}^2}$$

The use of SI units in the preceding analysis yields the same results as those obtained with the British gravitational system. The only difference is that ft is replaced by m, $\mathrm{lb_f}$ is replaced by N, and slug = $\mathrm{lb_f} \cdot \mathrm{s}^2/\mathrm{ft}$ is replaced by kg = $\mathrm{N} \cdot \mathrm{s}^2/\mathrm{m}$.

2.7 THERMAL SYSTEMS

For small temperature differences, the rate of heat transferred into a body is proportional to the temperature difference across the body

$$Q = hA(T_1 - T) = \frac{T_1 - T}{R_T} \tag{2.72}$$

where Q = rate of heat flow
h = coefficient of heat transfer of surface of body
A = surface area
T = temperature of body
T_1 = temperature of surrounding medium

and the symbol $R_T = 1/(hA)$ means equivalent thermal resistance and will soon be shown to be analogous to the electrical resistance R.

The rate of change of temperature of the body dT/dt is related to the rate of heat transfer into the body by

$$Q = Mc\frac{dT}{dt} = C_T DT \tag{2.73}$$

where c = average specific heat of body
M = mass
$C_T = Mc$ = equivalent thermal capacitance

The equations for a resistor and a capacitor may be written in the form

$$I = \frac{1}{R}E \tag{2.74}$$

and

$$I = CDE \tag{2.75}$$

Comparison of Eq. (2.74) with Eq. (2.72) and of Eq. (2.75) with Eq. (2.73) shows the following quantities to be analogous.

$$T \sim E \qquad Q \sim I$$
$$R_T \sim R \qquad C_T \sim C \tag{2.76}$$

This is the direct, or temperature-voltage, analog.

In Fig. 2.22*a* is shown an insulated container. The liquid is stirred so that the temperature T of the liquid is constant throughout. The equation for the heat transfer into the liquid is given by Eq. (2.72). Because this equation has the same form as the resistor equation, it is represented as a thermal resistor R_T in the equivalent thermal circuit of Fig. 2.22*b*.

Similarly, the equation for the heat stored in the liquid is Eq. (2.73). Because this equation has the same form as the capacitor equation, it is represented as a thermal reservoir in Fig. 2.22*b*. The equation of operation for this thermal circuit is

$$Q = \frac{T_1 - T}{R_T} = C_T DT$$

Solving for T gives

$$T = \frac{T_1}{1 + (R_T C_T)D} \tag{2.77}$$

The electric circuit which is the direct analog of the thermal circuit is shown in Fig. 2.22*c*. The equation of operation for the electric circuit is

$$I = \frac{E_1 - E}{R} = CDE$$

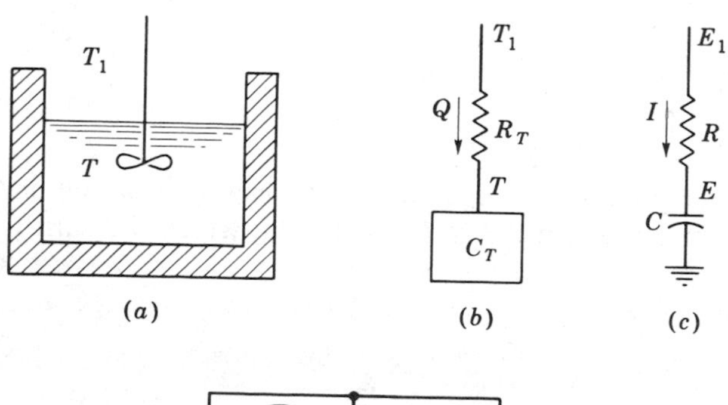

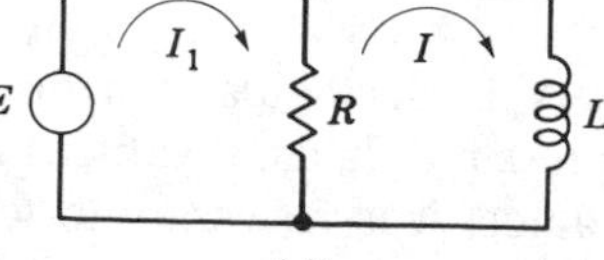

FIGURE 2.22
Thermal system.

Solving for E gives

$$E = \frac{E_1}{1 + (RC)D} \tag{2.78}$$

In the direct analog, series thermal elements are replaced by analogous electrical elements in series. Similarly, parallel thermal elements are replaced by analogous electrical elements in parallel. For thermal systems, there are only two elements.

The inverse, or temperature-current, analog is developed by first writing the equations for a resistor and an inductor in the forms

$$E = RI \tag{2.79}$$

$$E = LDI \tag{2.80}$$

Comparison with Eqs. (2.72) and (2.73) reveals the following analogous relationships:

$$\begin{aligned} T &\sim I & Q &\sim E \\ R_T &\sim \frac{1}{R} & C_T &\sim L \end{aligned} \tag{2.81}$$

The inverse analog of Fig. 2.22*b* is constructed by replacing the thermal resistor R_T and thermal capacitor C_T, which are in series, by an electric resistor R and inductor L in parallel, as shown in Fig. 2.22*d*. Because the thermal elements R_T and C_T have the same rate of heat flow Q, each electrical element R and L must have the same voltage E. The temperature of the thermal reservoir is T, and thus the inductor current is I. Similarly, because the temperature difference across the thermal resistor is $T_1 - T$, the electric resistor current is $I_1 - I$. The equation of operation for the inverse analog, Fig. 2.22*d*, is

$$E = R(I_1 - I) = LDI$$

Solving for I gives

$$I = \frac{I_1}{1 + (L/R)D} \tag{2.82}$$

For any circuit, there are but two possible analogs: the inverse analog in which the series-parallel arrangement is inverted and the direct analog in which the series-parallel arrangement is retained.

2.8 FLUID SYSTEMS

In working with fluid systems, it is necessary to distinguish whether the fluid is incompressible or compressible. For incompressible fluids, it suffices to work with the volume rate of flow. For compressible fluids, it is necessary to work with the mass rate of flow.

Incompressible Fluids

When the pressure difference across a flow restriction is small, the volume rate of flow Q is proportional to the pressure drop $P_1 - P$ across the restriction

$$Q = \frac{P_1 - P}{R_F} \tag{2.83}$$

where R_F is the equivalent fluid resistance.

The rate of flow into a tank, such as that shown in Fig. 2.23*a*, is equal to the cross-sectional area A of the tank times the rate of change of height. Thus,

$$Q = ADH = \frac{A}{\rho} DP = C_F DP \tag{2.84}$$

where $P = \rho H$ and ρ is density of fluid
$H =$ head
$C_F = A/\rho =$ equivalent fluid capacitance

The equation of operation for the fluid system of Fig. 2.23*a* is

$$Q = \frac{P_1 - P}{R_F} = C_F DP$$

Solving for P gives

$$P = \frac{P_1}{1 + (R_F C_F)D} \tag{2.85}$$

Replacing P by ρH yields the equation for the head H. Figure 2.23*b* shows the fluid circuit representation for this system.

Comparison of Eqs. (2.83) and (2.84) with Eqs. (2.74) and (2.75) shows the following quantities to be analogous:

$$P \sim E \qquad Q \sim I$$
$$R_F \sim R \qquad C_F \sim C \tag{2.86}$$

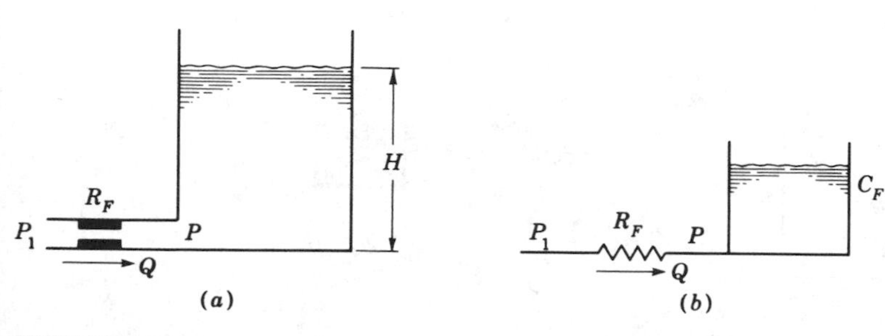

FIGURE 2.23
Fluid system.

These are analogous quantities for the direct, or pressure-voltage, analog. The electric circuit which is the direct analog for the fluid system of Fig. 2.23*b* is the same as that shown in Fig. 2.22*c*.

Comparison of Eqs. (2.83) and (2.84) with Eqs. (2.79) and (2.80) yields the analogous relationships for the inverse, or pressure-current, analog. Thus,

$$
\begin{aligned}
P &\sim I & Q &\sim E \\
R_F &\sim \frac{1}{R} & C_F &\sim L
\end{aligned}
\tag{2.87}
$$

The inverse analog for the fluid system of Fig. 2.23*b* is the same as that shown in Fig. 2.22*d*.

Compressible Fluids

For small pressure differences, the mass rate of flow M through a restriction is proportional to the pressure difference $P_1 - P$

$$
M = \frac{P_1 - P}{R_F} \tag{2.88}
$$

where R_F is the equivalent fluid resistance. Figure 2.24*a* shows a tank of constant volume V. The equation of state for the fluid in the tank is

$$
PV = WRT
$$

The flow into such a tank is usually isothermal. Thus, differentiating both sides of the equation of state with respect to time and solving for $M = dW/dt$ give

$$
M = \frac{dW}{dt} = \frac{V}{RT}\frac{d}{dt}P = \frac{V}{RT}DP = C_F DP \tag{2.89}
$$

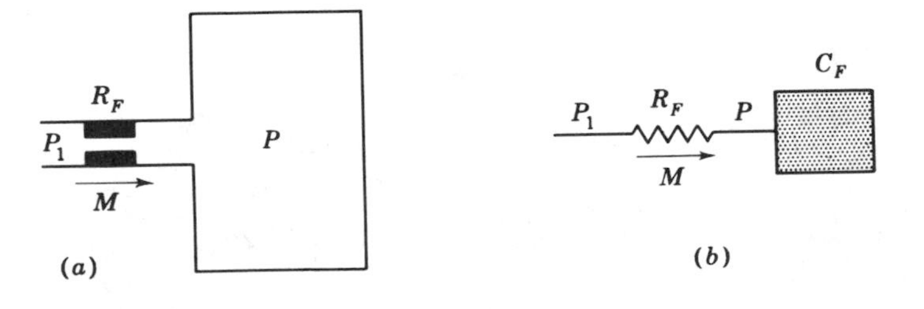

FIGURE 2.24
Fluid system.

where $C_F = V/(RT)$ is the equivalent fluid capacitance. Figure 2.24*b* shows the fluid circuit representation for this system. The equation of operation is

$$P = \frac{P_1}{1 + (R_F C_F)D} \tag{2.90}$$

The analogous relationships for the direct, or pressure-voltage, analog are the same as those given by Eq. (2.86) except that Q is replaced by M. The electric circuit which is the direct analog is the same as Fig. 2.22*c*.

Similarly, the analogous relationships for the inverse, or pressure-current, analog are the same as those given by Eq. (2.87) with Q replaced by M, and the electric circuit which is the inverse analog is the same as Fig. 2.22*d*.

A major use of analogs is that often it is easier to study experimentally one type of system rather than another. For example, it may be easier to change a resistor rather than a coefficient of viscous friction. Whenever possible, it is best to work with the system directly rather than to consider the operation of an analogous system. This eliminates the chance of error in construction of the analogy. Also, when they are carried far enough, analogies usually break down because things which are physically possible for one component may be impossible for the analogous component.

PROBLEMS

2.1. For $f(t) = 1$ and $y(0) = 0$, perform the indicated operations to evaluate $y(t)$ for each of the following cases.

(*a*) $y(t) = \frac{10}{D} f(t)$ (*b*) $y(t) = \frac{10}{D+1} f(t)$

2.2. For $f(t) = 4t$ and $y(0) = 0$, perform the indicated operations to evaluate $y(t)$ for each of the following cases.

(*a*) $y(t) = \frac{1}{D} f(t)$ (*b*) $y(t) = \frac{1}{D+2} f(t)$

2.3. Repeat Prob. 2.2 except $f(t) = 4 \sin 2t$.

2.4. For $f(t) = e^{-t}$ and $y(0) = 0$, perform the indicated operations to obtain $y(t)$ when

(*a*) $y(t) = \frac{1}{D} f(t)$ (*b*) $y(t) = \frac{1}{D-1} f(t)$

2.5. For $f(t) = \sin t$ and $y(0) = 0$, perform the indicated operations to obtain $y(t)$ when

(*a*) $y(t) = \frac{1}{D} f(t)$ (*b*) $y(t) = \frac{1}{D+1} f(t)$

2.6. For $f(t) = 1$ and $y(0) = \dot{y}(0) = 0$, perform the indicated operations to determine

(*a*) $y(t) = \frac{1}{D^2} f(t)$ (*b*) $y(t) = \frac{1}{(D+1)^2} f(t)$ (*c*) $y(t) = \frac{1}{D(D+1)} f(t)$

2.7. For each of the mechanical systems shown in Fig. P2.7, write the force equation at each coordinate and then determine the equation which relates

(*a*) x to f

(*b*) y to f

(*c*) y to x

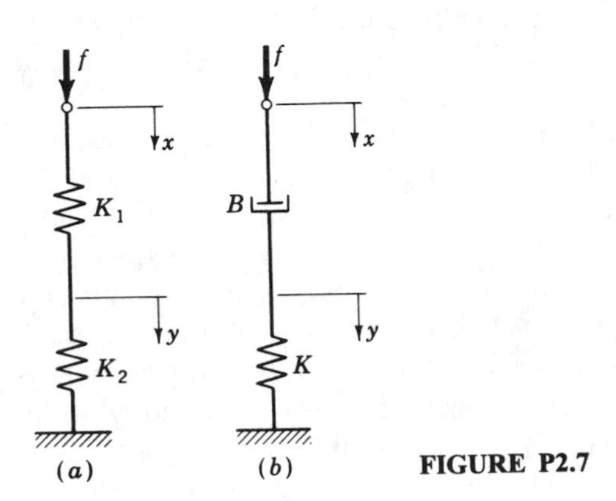

FIGURE P2.7

2.8. For the mechanical system shown in Fig. P2.8, write the force equation at each coordinate and then determine the equation which relates

(*a*) x to f

(*b*) y to f

(*c*) y to x

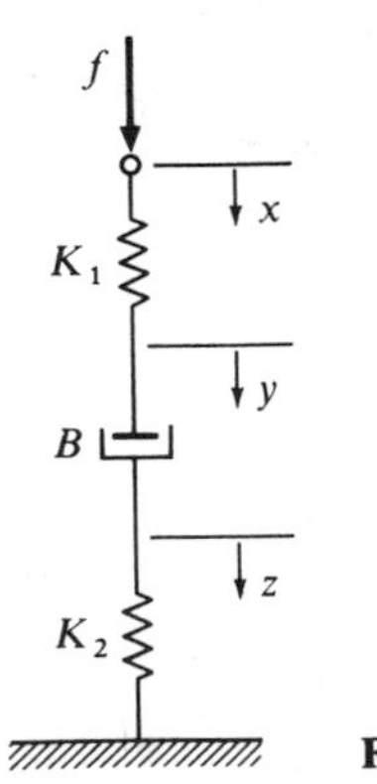

FIGURE P2.8

2.9. For each of the mechanical systems shown in Fig. P2.9, write the force equation at each coordinate and then determine the equation which relates

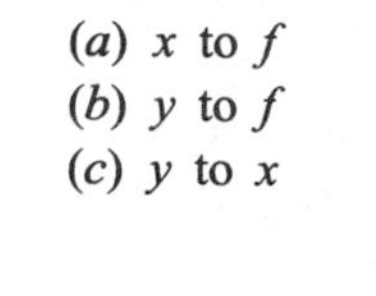

(*a*) x to f
(*b*) y to f
(*c*) y to x

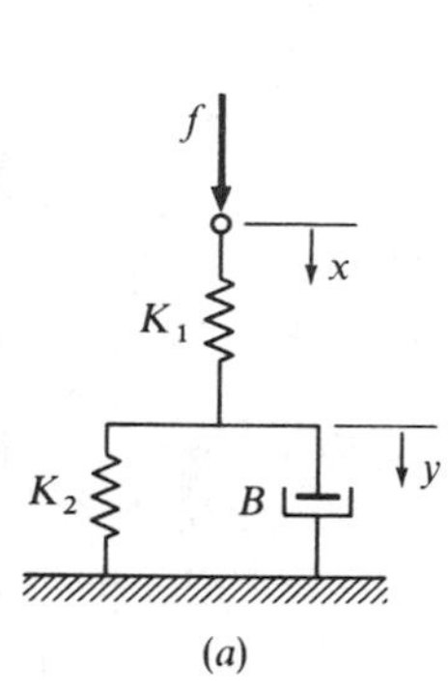

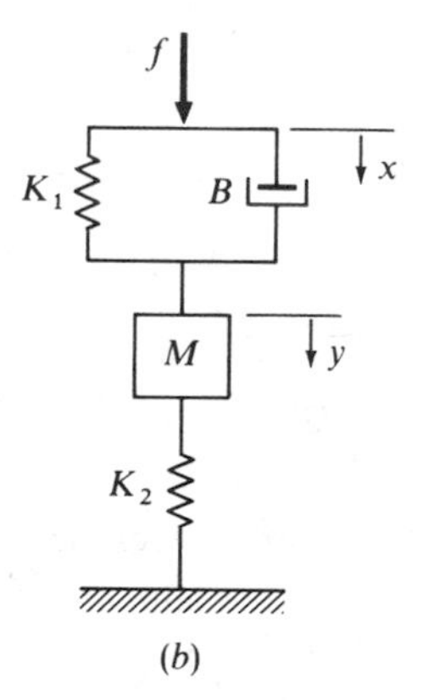

FIGURE P2.9

2.10. For the mechanical system shown in Fig. P2.10*a*, write the force equation at the x coordinate, and then determine the equation relating x to f. For the mechanical system shown in Fig. P2.10*b*, determine the equation which relates x to f, the equation that relates y to f, and the equation that relates y to x.

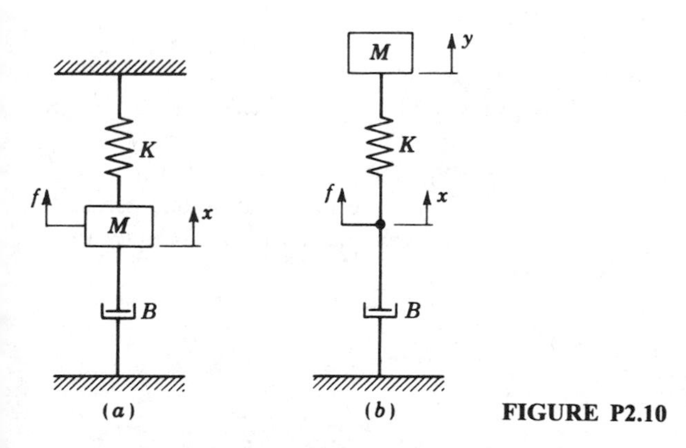

FIGURE P2.10

2.11. For each of the mechanical systems shown in Fig. P2.11, write the force equation at each coordinate and then determine the equation that relates (*a*) x to f, (*b*) y to f, and (*c*) y to x.

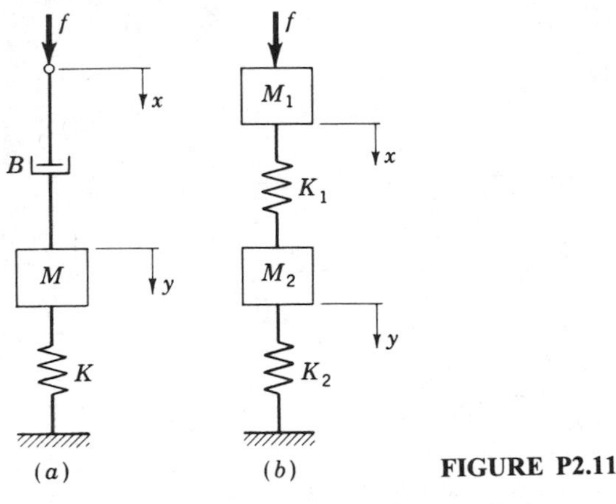

FIGURE P2.11

2.12. A schematic diagram of an accelerometer for measuring the linear acceleration d^2x/dt^2 is shown in Fig. P2.12. Determine the operational form for the differential equation which relates y (the change in the position of the mass relative to the frame) to the acceleration D^2x of the frame.

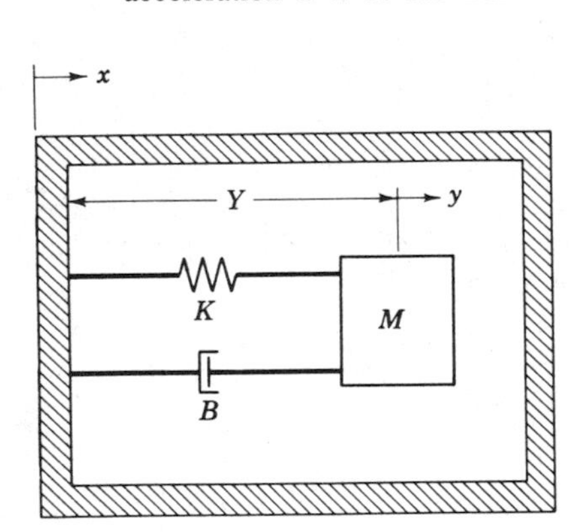

FIGURE P2.12

2.13. A viscous damper (shock absorber) is shown in Fig. P2.13. The rate of flow of fluid q is proportional to the pressure drop $q = C(p_1 - p_2)$. The force transmitted across the damper is $f = A(p_1 - p_2)$, where A is the cross-sectional area. The velocity of the

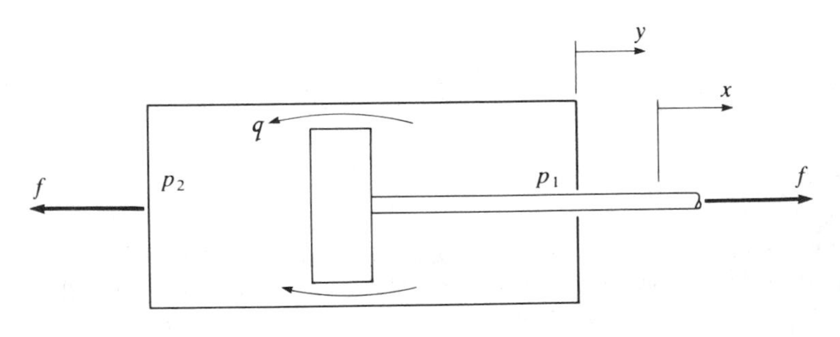

FIGURE P2.13

piston relative to the cylinder is $\dot{x} - \dot{y}$. The rate of change of volume is $q = A(\dot{x} - \dot{y})$. Show that the equation of operation of this viscous damper may be expressed in the form $f = B(\dot{x} - \dot{y})$. Determine the equation for the coefficient of viscous damping B.

2.14. For each of the mechanical systems shown in Fig. P2.7, apply the rules for mechanical elements in series and parallel to determine the equation relating x to f.

2.15. For the mechanical system shown in Fig. P2.8, apply the rules for mechanical elements in series and parallel to determine the equation relating x to f.

2.16. For each of the mechanical systems shown in Fig. P2.9, construct the grounded-chair representation and then determine the equation relating x to f.

2.17. For each of the mechanical systems shown in Fig. P2.10, construct the grounded-chair representation and then determine the equation relating x to f.

2.18. For each of the mechanical systems shown in Fig. P2.11, construct the equivalent grounded-chair representation and then determine the equation relating x to f.

2.19. An electric circuit for obtaining phase lead compensation is shown in Fig. P2.19.
(*a*) Determine the equation which relates E_1 and I.
(*b*) Determine the equation which relates E_2 and E_1.

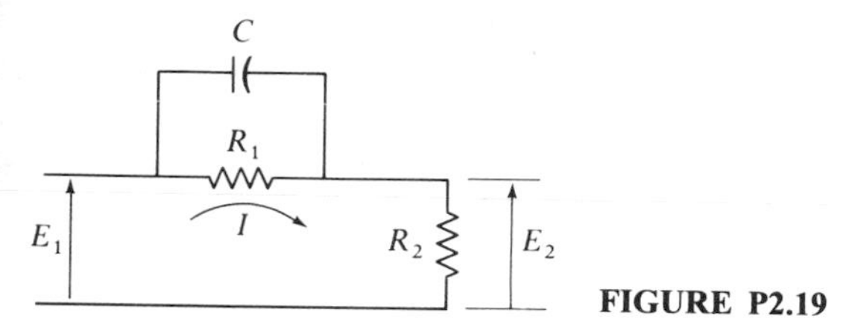

FIGURE P2.19

2.20. For each of the electrical networks shown in Fig. P2.20:
(*a*) Determine the equation which relates E_1 and I.
(*b*) Determine the equation which relates E_2 and E_1.

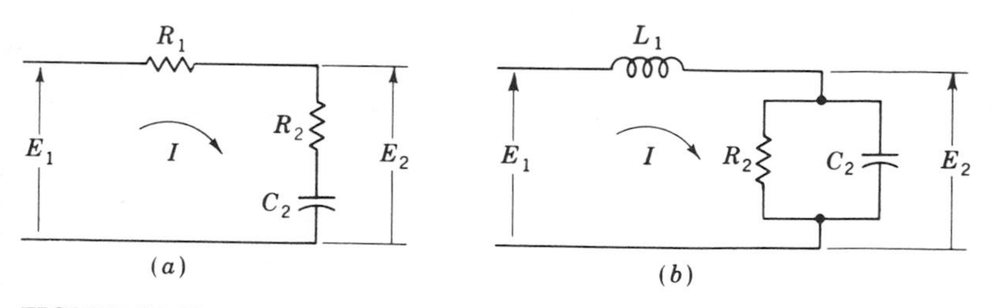

FIGURE P2.20

2.21. For the mechanical system shown in Fig. P2.9*a*, construct the electric circuit which is
(*a*) The direct analog. Determine the equation relating E and I for this electrical system, and compare this equation with that relating f and x of Fig. P2.9*a*.
(*b*) The inverse analog. Determine the equation relating E and I for this electrical system, and compare this equation with that relating x and f of Fig. P2.9*a*.

2.22. For the electrical system shown in Fig. P2.22, construct the mechanical circuit which is

(*a*) The direct analog. Determine the equation relating f and x for this mechanical system, and compare this equation with that relating E and I of Fig. P2.22.

(*b*) The inverse analog. Determine the equation relating f and x for this mechanical system, and compare this equation with that relating E and I of Fig. P2.22.

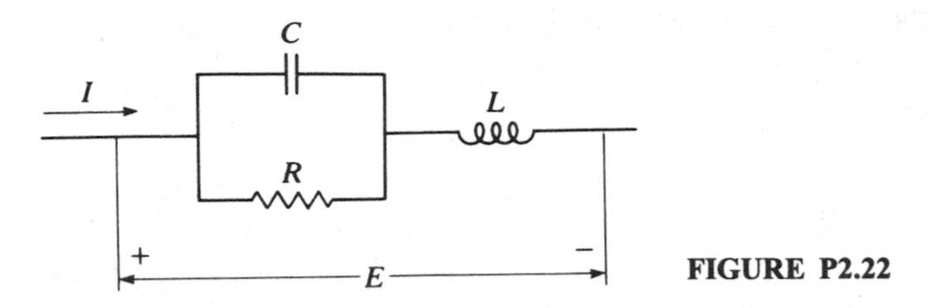

FIGURE P2.22

2.23. For the mechanical system shown in Fig. P2.23

(*a*) Determine the differential equation relating f and x.

(*b*) Construct the direct (force-voltage) analog.

(*c*) Construct the inverse (force-current) analog.

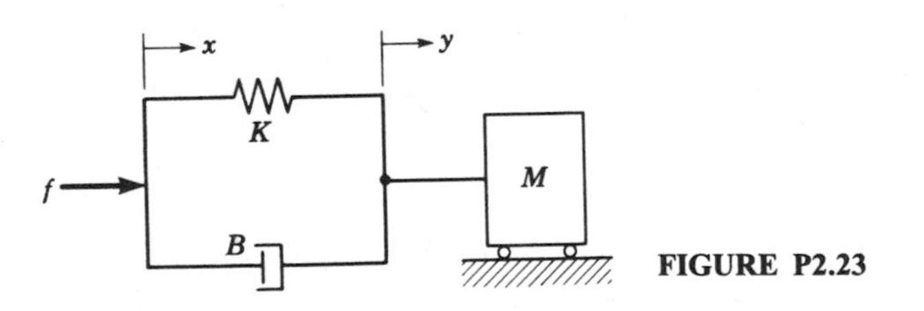

FIGURE P2.23

2.24. For the mechanical system shown in Fig. P2.24

(*a*) Construct the grounded-chair representation.

(*b*) Obtain the overall equation relating f and x.

(*c*) Construct the electric circuit which is the direct analog.

(*d*) Indicate what current in the analog corresponds to the velocity of the mass M.

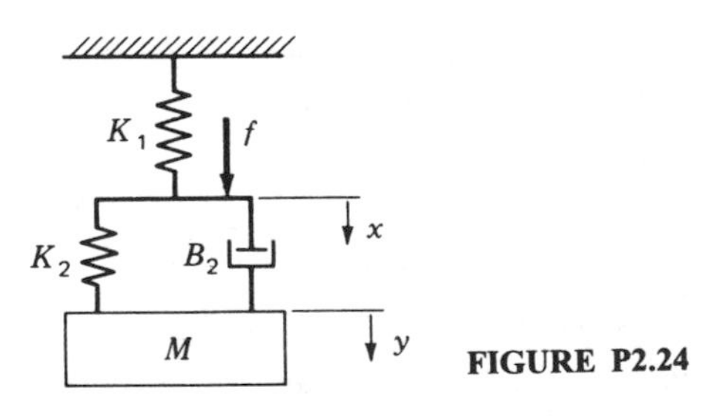

FIGURE P2.24

2.25. Repeat Prob. 2.24, except construct the inverse analog, and then indicate what voltage in this analog corresponds to the velocity of the mass M.

2.26. For the mechanical system shown in Fig. P2.26

(*a*) Draw the grounded-chair representation.

(*b*) Determine the equation relating f and x.

(*c*) Draw the electric circuit which is the direct analog.

(*d*) On the analog show where you would place a voltmeter to obtain a voltage proportional to the force across damper B_2.

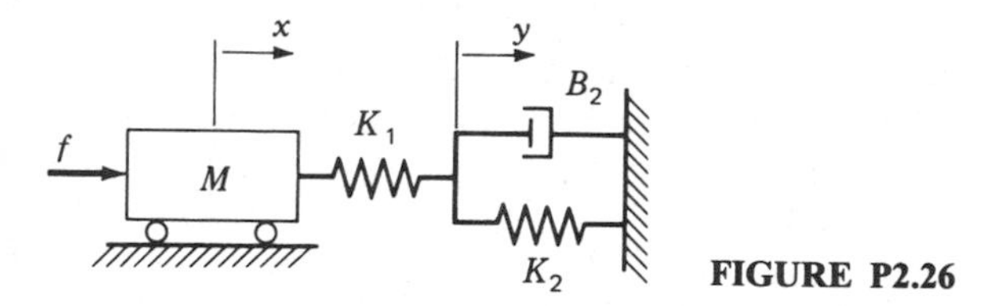

FIGURE P2.26

2.27. Repeat Prob. 2.26 except construct the inverse analog, and then indicate where you would place an ammeter to obtain a current proportional to the force across damper B_2.

2.28. The mass-spring-damper system shown in Fig. P2.28 represents the suspension for an automobile in which B is the shock absorber, K is the spring, and M is the mass of the car.

(*a*) Construct the grounded-chair representation, and then determine the equation relating f and x and that relating y and x.

(*b*) Construct the electric circuit which is the direct analog, and then show where you would place an ammeter to obtain a current proportional to the velocity $\dot{y}$ of the mass.

(*c*) Construct the electric circuit which is the inverse analog, and then show where you would place a voltmeter to determine a voltage proportional to the velocity $\dot{y}$ of the mass.

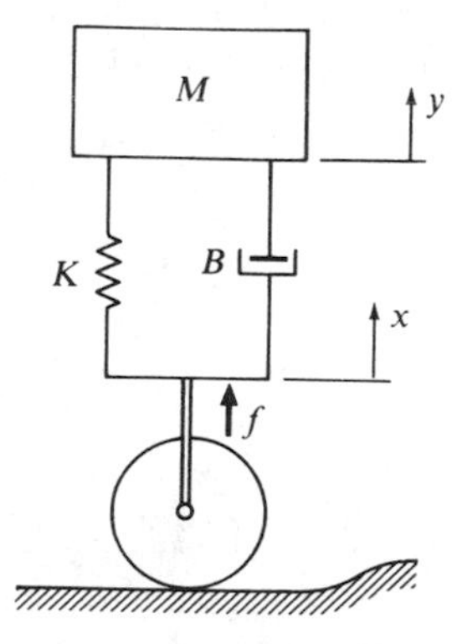

FIGURE P2.28

2.29. For the mechanical system shown in Fig. P2.23, determine the numerical values for the electrical elements of the direct analog when

(*a*) $M = 0.5$ slug, $B = 10$ lb$_f \cdot$ s/ft, $K = 50$ lb$_f$/ft, $k_E = 2$ lb$_f$/V, and $k_I = 200$ ft/(s $\cdot$ A)

(*b*) $M = 0.5$ kg, $B = 10$ N $\cdot$ s/m, $K = 50$ N/m, $k_E = 2$ N/V, and $k_I = 200$ m/(s $\cdot$ A)

2.30. For the mechanical system shown in Fig. P2.23, determine numerical values for the electrical elements of the inverse analog when

(a) $M = 0.5$ slug, $B = 10$ lb$_f$ · s/ft, $K = 50$ lb$_f$/ft, $k_E = 2$ ft/(s · V), and $k_I = 200$ lb$_f$/A

(b) $M = 0.5$ kg, $B = 10$ N · s/m, $K = 50$ N/m, $k_E = 2$ m/(s · V), and $k_I = 200$ N/A

2.31. Repeat Prob. 2.29 except slow down the solution by a factor of 2.

2.32. Repeat Prob. 2.30 except speed up the solution by a factor of 2.

2.33. The parameters for a mechanical system are $M = 10$ slugs, $B = 5000$ lb$_f$ · s/ft, and $K = 2500$ lb$_f$/ft. It is estimated that $f_{max} = 500$ lb$_f$ and $\dot{x}_{max} = 5$ ft/s.

(a) Determine what size resistor, inductor, and capacitor to use in the direct analog such that $E_{max} = 100$ V and $I_{max} = 10$ A.

(b) After the analog has been made, it is found that $E_{max} = 50$ V and $I_{max} = 4$ A. What were the actual values of f_{max} and $\dot{x}_{max}$?

(c) What size resistor, inductor, and capacitor should be used to speed up the solution of this problem by a factor of 10?

2.34. Repeat Prob. 2.33 except use an inverse rather than direct analog.

2.35. The differential equation for a series RLC circuit is

$$E = \left(LD + R + \frac{1}{CD}\right)I$$

(a) Construct the mechanical circuit which is the direct (force-voltage) analog for the series RLC circuit. Determine the equation for the force f as a function of velocity $\dot{x}$. (Label mass as M_a, spring K_a, and damper B_a.)

(b) Construct the mechanical circuit which is the inverse analog for the series RLC circuit. Determine the equation for velocity $\dot{x}$ as a function of force f. (Label mass as M_b, spring K_b, and damper B_b.)

(c) Compare the results of (a) and (b), and then suggest analogous quantities in constructing the dual of a mechanical circuit. (Duals are two different mechanical circuits whose differential equations have the same form.)

2.36. For the lever shown in Fig. P2.36, the variation in the applied force is f, and the variation in spring position is x. The horizontal line represents the reference position of the lever.

(a) Determine the equation relating f and x.

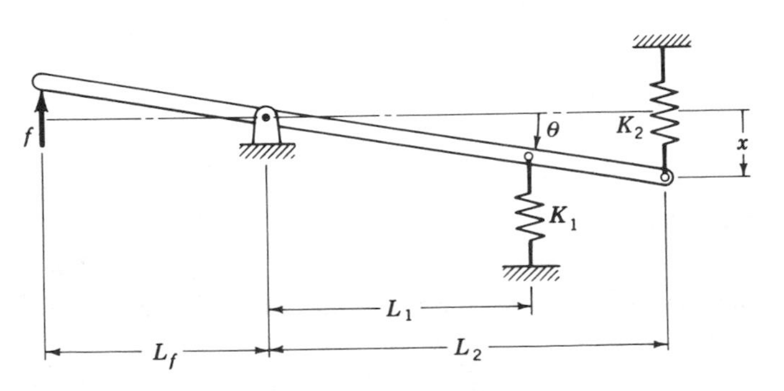

FIGURE P2.36

(*b*) Determine the relationship between t and θ (where $t = f L_f$ is the variation in applied torque and $x = L_2\theta$).

2.37. The lever system shown in Fig. P2.37 is drawn in its reference position. The variation in spring position is designated by x. The variation in applied force is designated by f (f and x are zero at the reference position).

(*a*) Determine the equation relating f and x.

(*b*) Determine the relationship between t and θ where $t = f L_f$ and $x = L\theta$.

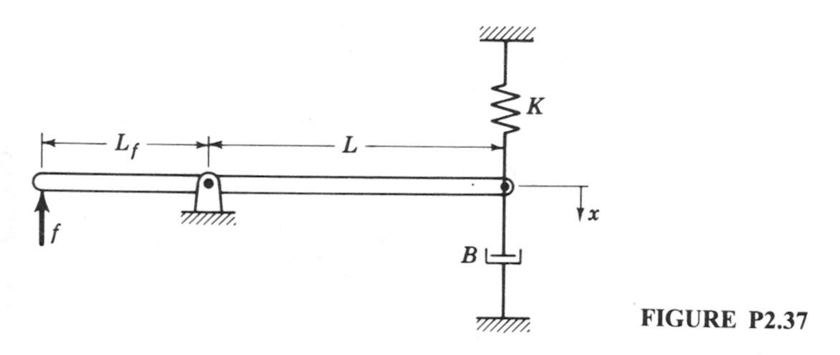

FIGURE P2.37

2.38. Often mechanical systems have gearing. Such systems may be represented by equivalent systems without gearing. Figure P2.38*a* shows a geared system, and Fig. P2.38*b* shows an equivalent system without gearing. The gear ratio $n = \omega_1/\omega_2$ is the ratio of the speed of the driving shaft ω_1 to that of the driven shaft ω_2. For the systems to be equivalent, the kinetic energy of the system without gearing $J_e\omega_1^2/2$ must be the same as that of the system with gearing $J\omega_2^2/2$. Thus, show that $J_e = J/n^2$

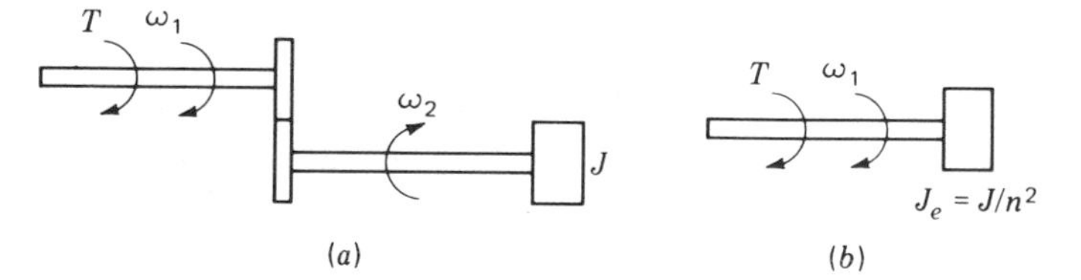

FIGURE P2.38

2.39. As shown in Fig. P2.39*a*, the torque T is transmitted through gearing to the load, which is viscous friction. The resisting torque due to viscous friction is $B\omega_2$, where

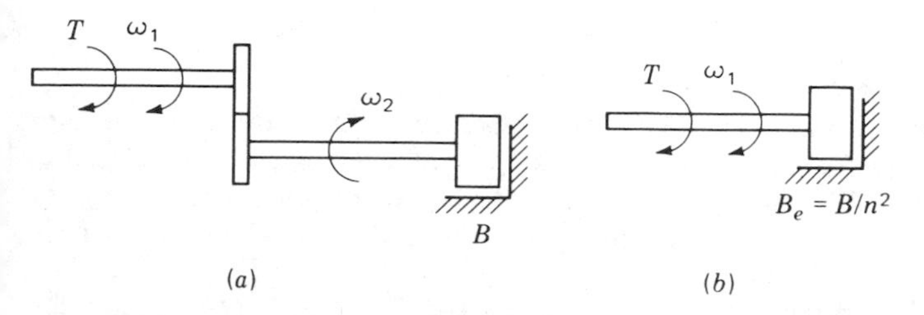

FIGURE P2.39

B is the coefficient of viscous friction. The rate at which power is being dissipated is equal to the product of the resisting torque and the angular velocity $P = (B\omega_2)\omega_2 = B\omega_2^2$.

The system without gearing is shown in Fig. P2.39*b*. For the systems to be equivalent, the rate of power dissipation of the system without gearing $B_e\omega_1^2$ must be the same as that with gearing. Thus, show that $B_e = B/n^2$.

2.40. As shown in Fig. P2.40*a*, the torque T is transmitted to the load, which is a torsional spring. The torque applied to the torsional spring is $K\theta_2$, where K is the torsional spring rate and θ_2 is the angle of twist. The potential energy stored in the spring is $K\theta_2^2/2$. The corresponding system without gearing is shown in Fig. P2.40*b*. For the systems to be equivalent, the potential energy stored in the system without gearing $K_e\theta_1^2/2$ must be the same as that for the system with gearing. Thus, show that $K_e = K/n^2$.

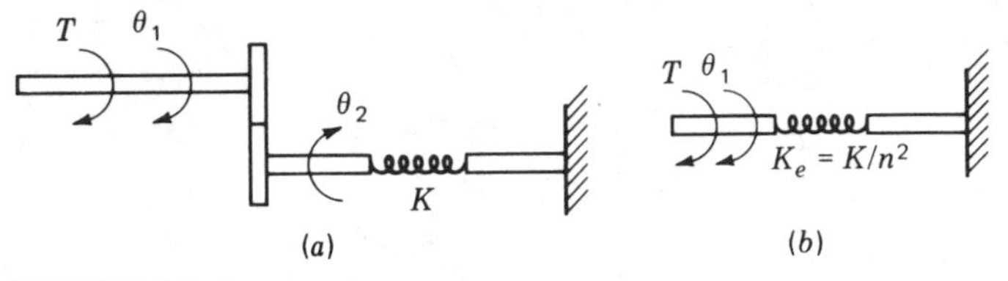

FIGURE P2.40

2.41. For the manometer shown in Fig. P2.41, determine the equation of motion relating the pressure P at the open end to the position x. The length of the measuring column is L, the cross-sectional area is A, and the density is ρ.

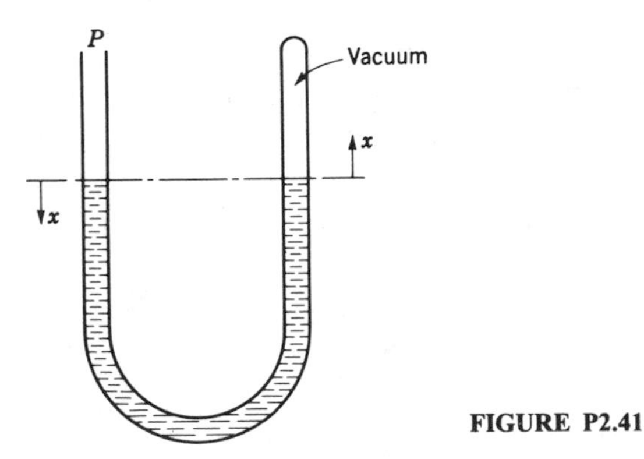

FIGURE P2.41

2.42. For the thermometer shown in Fig. P2.42, the temperature of the surrounding medium is T_1. The temperature of the glass enclosure is T_2, and the temperature of the fluid in the thermometer is T. The rate of heat flow from the surrounding medium to the glass is $Q_1 = (T_1 - T_2)/R_{T_1}$. The rate of heat flow from the glass to the fluid is $Q_2 = (T_2 - T)/R_{T_2}$.

The rate of temperature change of the glass is $DT_2 = (Q_1 - Q_2)/C_{T_1}$, and the rate of change of temperature of the fluid is $DT = Q_2/C_{T_2}$.

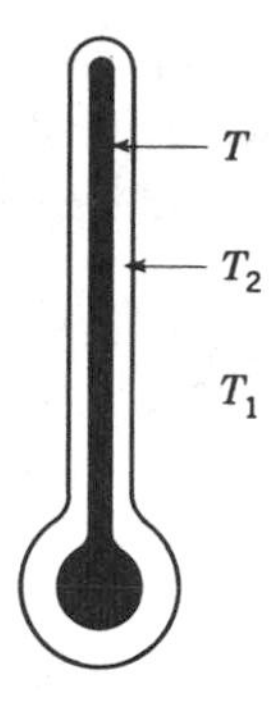

FIGURE P2.42

Construct the thermal circuit representation for this system, and then determine the equation for the temperature T of the fluid as a function of the surrounding temperature T_1.

2.43. Figure P2.43 shows an oven which is supplied with heat from an electric source. The rate of heat supplied is proportional to the voltage: $Q_s = KE_s$. The rate at which heat is lost through the walls is $Q = (T - T_1)/R_T$. The rate of change of temperature of the oven is

$$DT = \frac{Q_s - Q}{C_T}$$

Construct the thermal circuit representation for this system, and then determine the equation for the temperature T of the oven as a function of the applied voltage E_s and the surrounding temperature T_1.

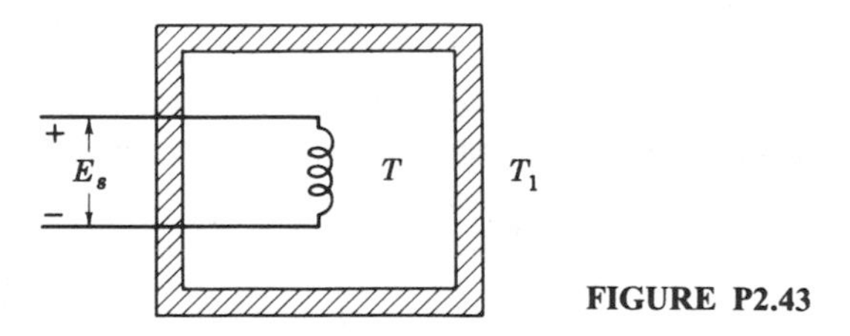

FIGURE P2.43

2.44. Figure P2.44 shows a fluid system of two tanks in series. Construct the fluid circuit representation for this system. Determine the equation for the pressure P (head $H = P/\rho$) as a function of the inlet pressure P_1 (P_2 should not appear in this equation).

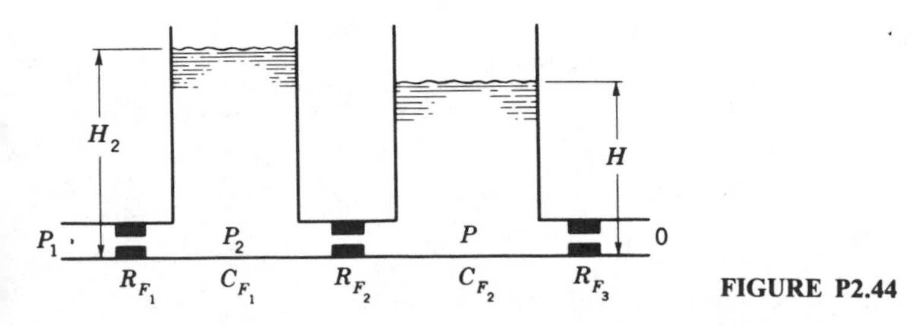

FIGURE P2.44

2.45. In Fig. P2.45 is shown a tank in which flow is supplied at a rate Q_s. Construct the fluid circuit representation for this system. Determine the equation for the pressure P (head $H = P/\rho$) as a function of P_1 and Q_s.

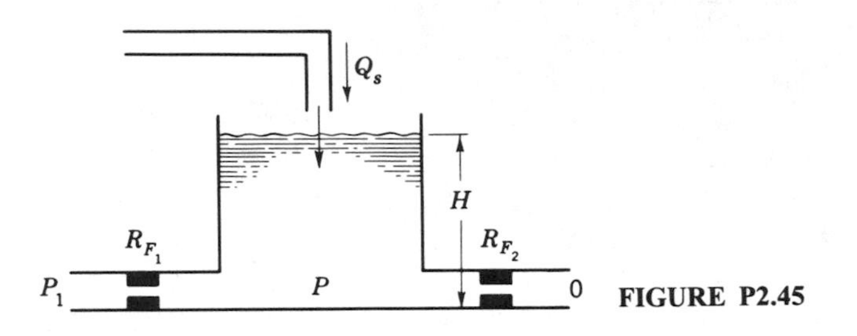

FIGURE P2.45

2.46. Figure P2.46 shows a tank with two inlets. The fluid is compressible. Determine the equation for the pressure P as a function of P_1 and P_2.

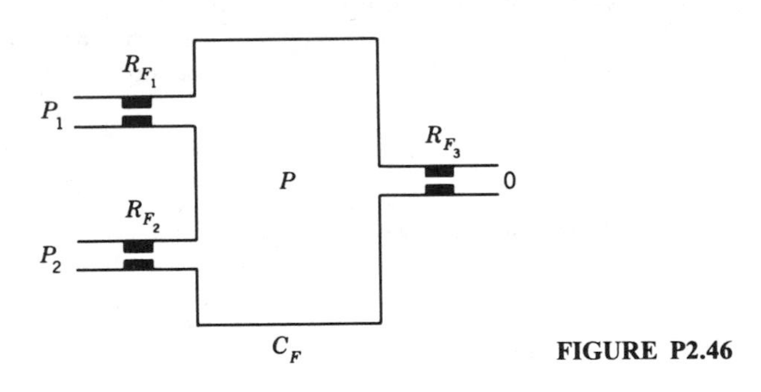

FIGURE P2.46

2.47. Figure P2.47 shows two tanks in series. For a compressible fluid, determine the equation for the pressure P as a function of the inlet pressure P_1 (P_2 should not appear in this equation).

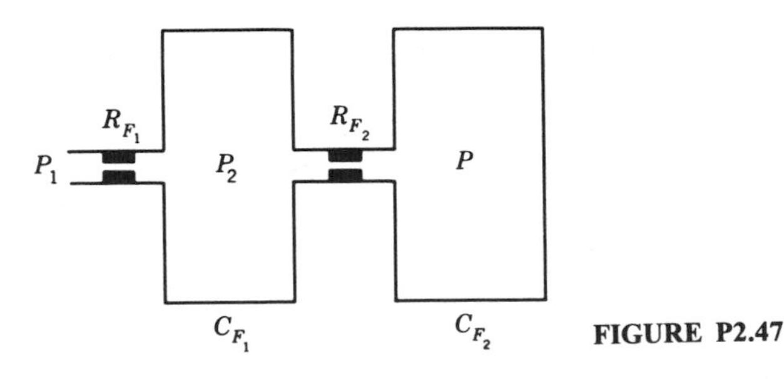

FIGURE P2.47

CHAPTER 3

REPRESENTATION OF CONTROL SYSTEMS

In this chapter we show how to obtain the overall block-diagram representation for some typical control systems. In brief, the method employed is to obtain the block diagram for each component or process and then "hook up," or connect, the corresponding inputs and outputs for each diagram to obtain the one overall representation for the system. The techniques which are presented in later chapters for determining the operating characteristics of control systems are based on a knowledge of the overall block-diagram representation for the system.

3.1 LINEARIZATION OF NONLINEAR FUNCTIONS

The most powerful methods of system analysis have been developed for linear control systems. For a linear control system, all the relationships between the variables are linear differential equations, usually with constant coefficients. The reason that differential equations rather than algebraic equations are obtained is that in feedback control systems, the variables are functions of time. For example, in controlling temperature, the actuating signal causes a change in heat flow, but time is required for this added heat to bring the temperature to its desired value. In speed control systems, the actuating signal causes a change in power of the prime mover, but time is required for the engine to accelerate or decelerate to its desired speed. Similarly, in pressure control systems, it takes time to bring the pressure in a chamber to some desired value.

Actual control systems usually contain some nonlinear elements. Such elements would in turn yield nonlinear differential equations for the system. In the following we show how the equations for nonlinear elements may be linearized. Thus, the resulting differential equation of operation for the system becomes linear. A linear equation in the n variables $x_1, x_2, \ldots, x_n$ is one which has the form

$$Y = Y_i + c_1x_1 + c_2x_2 + \cdots + c_nx_n$$

where $c_1, c_2, \ldots, c_n$ and Y_i are constants. Note that each of the n variables is to the first power and that no products of these variables appear. A nonlinear equation results if one or more of the variables is to some power other than the first power (for example, x_n^3) or if the equation contains products of the variables (for example, x_2x_5).

A plot of the nonlinear relationship

$$Y = X^2 \tag{3.1}$$

is shown in Fig. 3.1. In the vicinity of the point (X_i, Y_i), the function is closely approximated by the tangent. For example, consider the point (X, Y) on the curve of the nonlinear function. The abscissa X is displaced a distance x from X_i. This abscissa X intersects the nonlinear function a vertical distance $y + \varepsilon$ from Y_i, and it intersects the tangent a distance y from Y_i. The equation for Y is

$$Y = Y_i + y + \varepsilon \approx Y_i + y \tag{3.2}$$

Lowercase letters indicate the variation of the capital-letter parameters from the reference point. From Fig. 3.1, the slope of the tangent line is

$$\frac{y}{x} = \left.\frac{dY}{dX}\right|_i = \text{slope at point } (X_i,\ Y_i)$$

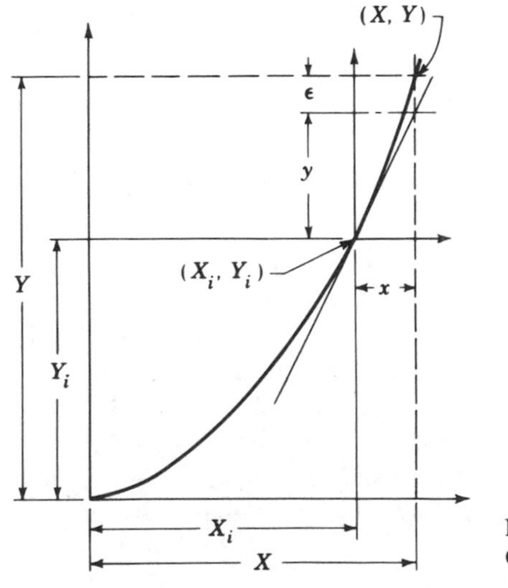

FIGURE 3.1
Graph of function $Y = X^2$.

The symbol $|_i$ means that the derivative is to be evaluated at the reference point. Thus

$$y = \left.\frac{dY}{dX}\right|_i x = \frac{d}{dX}(X^2)\,|_i\, x = 2X_i x \tag{3.3}$$

Substitution of y from Eq. (3.3) into Eq. (3.2) yields the following linear approximation for Y:

$$Y \approx Y_i + 2X_i x \tag{3.4}$$

Illustrative example 3.1. Effect a linear approximation for the equation $Y = X^2$ for values of X in the neighborhood of 10, and find the error when you use this approximation for $X = 11$.

Solution. The reference values are $X_i = 10$ and $Y_i = X_i^2 = 100$. The variation from the reference value is $x = X - X_i = 11 - 10 = 1$. Substitution of these values into Eq. (3.4) yields

$$Y \approx Y_i + 2X_i x = 100 + (2)(10)(1) = 120$$

The exact value is $Y = X^2 = 121$; thus the error is 1 part in 121, or less than 1 percent.

A general procedure for obtaining a linear approximation is to use the expression derived in calculus for approximating the variation ΔY for a function $Y = Y(X_1, X_2, \ldots, X_n)$ of n independent variables. That is,

$$\Delta Y = \left.\frac{\partial Y}{\partial X_1}\right|_i \Delta X_1 + \left.\frac{\partial Y}{\partial X_2}\right|_i \Delta X_2 + \cdots + \left.\frac{\partial Y}{\partial X_n}\right|_i \Delta X_n$$

Since lowercase letters represent variations from the reference values,

$$\begin{aligned} y &= \Delta Y = Y - Y_i \\ x_1 &= \Delta X_1 = X_1 - X_{1i} \\ x_2 &= \Delta X_2 = X_2 - X_{2i} \\ &\vdots \\ x_n &= \Delta X_n = X_n - X_{ni} \end{aligned}$$

Thus, the general expression for obtaining a linear approximation for a nonlinear function is

$$y = C_1 x_1 + C_2 x_2 + \cdots + C_n x_n \tag{3.5}$$

where

$$C_1 = \left.\frac{\partial Y}{\partial X_1}\right|_i \qquad C_2 = \left.\frac{\partial Y}{\partial X_2}\right|_i \qquad \text{etc.}$$

Evaluation of these partial derivatives at the reference condition yields constants. As is illustrated by Eq. (3.5), in a linear equation each variable term $x_1, x_2, \ldots, x_n$ is of the first power, and the contribution of each term is added independently.

The application of Eq. (3.5) to the nonlinear equation $Y = X^2$ is effected as follows: The independent variable is X, which corresponds to X_1 in the general equation. Thus,

$$C_1 = \left.\frac{\partial Y}{\partial X_1}\right|_i = \left.\frac{\partial Y}{\partial X}\right|_i = \left.\frac{\partial}{\partial X} X^2\right|_i = 2X_i$$

The variation is

$$y = C_1 x = 2X_i x$$

This is the result obtained by the preceding geometric interpretation and given by Eq. (3.3).

Illustrative example 3.2. For the triangle shown in Fig. 3.2, the length of the base is X, the length of the adjacent side is Y, and the included angle is θ. The altitude h is $Y \sin \theta$. The equation for the area of a triangle (one-half the altitude times the base) may thus be written in the form

$$A = \tfrac{1}{2} XY \sin \theta$$

Determine the linear approximation for the area A. The reference values are $X_i = 10$, $Y_i = 16$, and $\theta_i = 30°$. Determine the approximate area when $X = 9$, $Y = 18$, and $\theta = 33°$.

Solution. The reference value of the area is

$$A_i = \tfrac{1}{2} X_i Y_i \sin \theta_i = \tfrac{1}{2}(10)(16)(0.5) = 40$$

The change a in the area is

$$a = \left.\frac{\partial A}{\partial X}\right|_i x + \left.\frac{\partial A}{\partial Y}\right|_i y + \left.\frac{\partial A}{\partial \theta}\right|_i \Delta\theta$$
$$= (\tfrac{1}{2} Y \sin \theta)_i x + (\tfrac{1}{2} X \sin \theta)_i y + (\tfrac{1}{2} XY \cos \theta)_i \Delta\theta$$

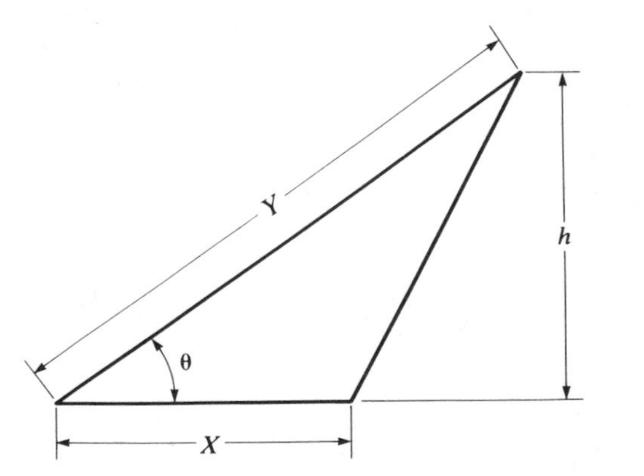

FIGURE 3.2
Triangle.

It is convenient to express linear approximations in nondimensional form. This is accomplished by expressing the partial derivatives in terms of the reference value of the function being approximated. For the preceding, the reference value is

$$A_i = (\tfrac{1}{2} XY \sin\theta)_i$$

Thus,

$$\begin{aligned} a &= (\tfrac{1}{2} XY \sin\theta)_i \frac{x}{X_i} + (\tfrac{1}{2} XY \sin\theta)_i \frac{y}{Y_i} + (\tfrac{1}{2} XY \sin\theta)_i \frac{\Delta\theta}{\tan\theta_i} \\ &= \left(\frac{x}{X_i} + \frac{y}{Y_i} + \frac{\Delta\theta}{\tan\theta_i}\right) A_i \end{aligned}$$

The variations are $x = X - X_i = 9 - 10 = -1$, $y = Y - Y_i = 18 - 16 = 2$, $\Delta\theta = \theta - \theta_i = (33 - 30)\pi/180 = \pi/60$ rad. Note that each of the ratios x/X_i, y/Y_i, and $\Delta\theta/\tan\theta_i$ are nondimensional. Thus, the change in the angle $\Delta\theta$ must be measured in radians.

$$a = \left(-\frac{1}{10} + \frac{2}{16} + \frac{\pi\sqrt{3}}{60}\right) 40 = 4.63$$

The approximation for the total area A is

$$A = A_i + a = 40.0 + 4.63 = 44.63$$

Illustrative example 3.3. For the triangle shown in Fig. 3.3, it follows from the law of sines that

$$\frac{\sin\alpha}{A} = \frac{\sin\beta}{B}$$

or

$$\sin\beta = \frac{B}{A} \sin\alpha$$

Effect the linear approximation for the change $\Delta\beta$ in angle β due to a change a in A, a change b in B, and a change $\Delta\alpha$ in α. For $\alpha_i = 30°$, $\beta_i = 120°$, and $A_i = 10$, determine the variation $\Delta\beta$ when $\Delta\alpha = 2°$, $a = 1$, and $b = 1$.

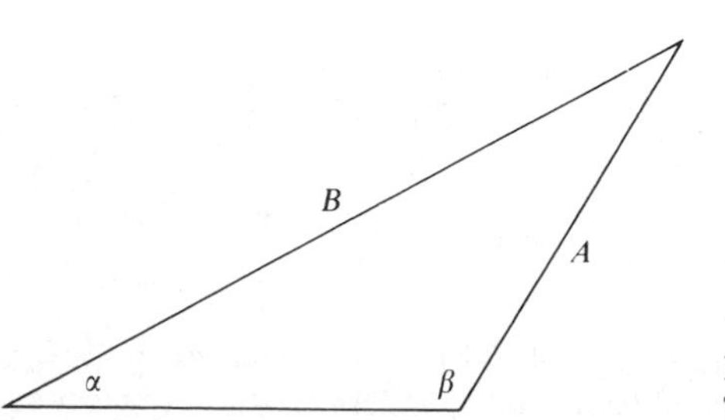

FIGURE 3.3
Triangle.

Solution. The left side is a function of β, and the right side is a function of A, B, and α. Thus, the general form of the linearized equation is

$$\frac{\partial}{\partial \beta}(\sin \beta)\Delta\beta = \frac{\partial}{\partial A}\left(\frac{B}{A}\sin\alpha\right)a + \frac{\partial}{\partial B}\left(\frac{B}{A}\sin\alpha\right)b + \frac{\partial}{\partial \alpha}\left(\frac{B}{A}\sin\alpha\right)\Delta\alpha$$

The resulting linearized equation is

$$(\cos\beta)_i\Delta\beta = -\left(\frac{B}{A^2}\sin\alpha\right)_i a + \left(\frac{\sin\alpha}{A}\right)_i b + \left(\frac{B}{A}\cos\alpha\right)_i \Delta\alpha$$

$$= \left(\frac{B}{A}\sin\alpha\right)_i \left(-\frac{a}{A_i} + \frac{b}{B_i} + \frac{\Delta\alpha}{\tan\alpha_i}\right)$$

Solving for $\Delta\beta$ yields

$$\Delta\beta = \frac{\sin\beta_i}{\cos\beta_i}\left(-\frac{a}{A_i} + \frac{b}{B_i} + \frac{\Delta\alpha}{\tan\alpha_i}\right)$$

Note that

$$B_i = \frac{A_i \sin\beta_i}{\sin\alpha_i} = \frac{10\sin 120^\circ}{\sin 30^\circ} = 10\sqrt{3}$$

and

$$\alpha_i = \frac{2\pi}{180} = 0.0349 \text{ rad}$$

Thus,

$$\Delta\beta = \tan 120^\circ\left(\frac{-1}{10.0} + \frac{1}{17.32} + \frac{0.0349}{\tan 30^\circ}\right)$$

$$= -0.314 \text{ rad} = -1.8^\circ$$

The need for linearizing nonlinear relationships is frequently encountered in control engineering. For example, most mechanical speed control systems incorporate a flyball governor for sensing the speed error. This is a centrifugal device, so that a force is obtained which is proportional to the square of the speed. In the design of hydraulic equipment in which the working medium is an incompressible fluid, one encounters the nonlinear equations which govern such fluid flow. Similarly, the working medium for pneumatic equipment is air, whose flow is described by nonlinear relationships.

Illustrative example 3.4. Effect the linear approximation for P in the equation of state $PV = WRT$. The reference conditions are $P_i = 100$ lb$_f$/ft^2, $V_i = 100$ ft^3, $W_i = 10/53.3$ lb$_m$, and $T_i = 1000$°R. Determine the percentage of error in using this approximation for P when $V = 110$ ft^3, $T = 1200$°R, and W remains the same. The constant R is 53.3 ft · lb$_f$/(lb$_m$ · °R).

Solution. From the equation of state and the fact that W remains constant, it is seen that P is a function of the independent variables T and V, or $P = P(T, V)$. Application of Eq. (3.5) to obtain the variation p of the pressure from its reference value yields

$$p = \left.\frac{\partial P}{\partial T}\right|_i t + \left.\frac{\partial P}{\partial V}\right|_i v \tag{3.6}$$

The partial derivatives are evaluated from the equation of state as follows:

$$\left.\frac{\partial P}{\partial T}\right|_i = \frac{\partial}{\partial T}\left(\frac{WRT}{V}\right)\bigg|_i = \left.\frac{WR}{V}\right|_i = \frac{(10)(53.3)}{(53.3)(100)} = 0.10$$

$$\left.\frac{\partial P}{\partial V}\right|_i = \frac{\partial}{\partial V}\left(\frac{WRT}{V}\right)\bigg|_i = \left.\frac{-WRT}{V^2}\right|_i = -1.0$$

The linearized approximation for P is

$$P \approx P_i + p = P_i + 0.1t - v$$

From the given information, it follows that

$$v = V - V_i = 110 - 100 = 10$$
$$t = T - T_i = 1200 - 1000 = 200$$

Thus

$$P \approx 100 + (0.1)(200) - 10 = 110 \text{ lb}_f/\text{ft}^2$$

To express the preceding approximation in nondimensional form, note that the reference value of the function being approximated is $P_i = (WRT/V)|_i$. Thus,

$$\left.\frac{\partial P}{\partial T}\right|_i = \left.\frac{WR}{V}\right|_i = \left.\frac{WRT}{V}\right|_i \frac{1}{T_i} = \frac{P_i}{T_i}$$

$$\left.\frac{\partial P}{\partial V}\right|_i = -\left.\frac{WRT}{V^2}\right|_i = -\left.\frac{WRT}{V}\right|_i \frac{1}{V_i} = \frac{-P_i}{V_i}$$

Substitution of these results into Eq. (3.6) gives

$$p = \left(\frac{t}{T_i} - \frac{v}{V_i}\right) P_i \tag{3.7}$$

Note that t/T_i and v/V_i are nondimensional ratios. Evaluation of the resulting linear approximation gives

$$P = P_i + p = 100 + 100\left(\frac{200}{1000} - \frac{10}{100}\right) = 100 + 20 - 10 = 110 \text{ lb}_f/\text{ft}^2$$

This checks with the previously obtained result. The exact value of P is

$$P = \frac{WRT}{V} = \frac{(10)(53.3)(1200)}{(53.3)(110)} = 109.1$$

Therefore, the percentage of error is

$$\frac{(110 - 109.1)(100)}{109.1} = 0.82\%$$

In SI, pressure has units of newtons per square meter (N/m^2), volume has units of cubic meters (m^3), mass has units of kilograms (kg), and temperature has units of kelvins (K). In SI units, the value of R in the equation of state ($PV = WRT$) is $R = 287\ \mathrm{N \cdot m/(kg \cdot K)}$.

Illustrative example 3.5. The equation for the flow of an incompressible fluid through a restriction is

$$Q = C_c A \sqrt{\frac{2g_c}{\rho}(P_1 - P_2)} \tag{3.8}$$

where Q = rate of flow, $\mathrm{ft^3/s}$
C_c = coefficient of discharge (dimensionless)
A = area of restriction, $\mathrm{ft^2}$
$P_1 - P_2$ = pressure drop across restriction, $\mathrm{lb_f/ft^2}$
ρ = density of fluid, $\mathrm{lb_m/ft^3}$
g_c = gravitational conversion constant = 32.2 $\mathrm{lb_m/slug}$

Determine the linear approximation for the variation q. Letting $g_c = (32.2)(12) = 386$ converts the length units in Eq. (3.8) from feet to inches. Thus, the flow units are cubic inches per second ($\mathrm{in^3/s}$), the area units are square inches ($\mathrm{in^2}$), the pressure units are pounds force per square inch ($\mathrm{lb_f/in^2}$), and the density units are pounds mass per cubic inch ($\mathrm{lb_m/in^3}$).

Solution. Because Q is a function of the area A and pressure drop $P_1 - P_2$,

$$Q = Q(A,\ P_1 - P_2)$$

The variation q is

$$q = \left.\frac{\partial Q}{\partial A}\right|_i a + \left.\frac{\partial Q}{\partial(P_1 - P_2)}\right|_i (p_1 - p_2)$$

The partial derivatives are

$$\left.\frac{\partial Q}{\partial A}\right|_i = C_c \left.\sqrt{\frac{2g_c}{\rho}(P_1 - P_2)}\right|_i = \frac{Q_i}{A_i}$$

and

$$\left.\frac{\partial Q}{\partial(P_1 - P_2)}\right|_i = \frac{C_c A}{2}\left.\sqrt{\frac{2g_c}{\rho}\frac{1}{P_1 - P_2}}\right|_i = \frac{1}{2}\frac{Q_i}{(P_1 - P_2)_i}$$

Thus,

$$q = \left[\frac{a}{A_i} + \frac{1}{2}\frac{p_1 - p_2}{(P_1 - P_2)_i}\right] Q_i \tag{3.9}$$

When SI units are used, the g_c term does not appear in Eq. (3.8); thus

$$Q = C_c A \sqrt{(P_1 - P_2)/\rho}$$

Substitution of units into this expression shows that the units of Q are

$$m^2\sqrt{(N/m^2)/(kg/m^3)} = m^2\sqrt{N \cdot m/kg} = m^2\sqrt{N \cdot m/(N \cdot s^2/m)}$$
$$= m^2\sqrt{m^2/s^2} = m^3/s$$

For fluid flow, it is customary to express mass in units of pounds mass (lb_m) rather than slugs. The mass of a body in slugs is equal to its mass in pounds mass divided by 32.2. The units relationship is slug $= \text{lb}_\text{m}/g_c = \text{lb}_\text{m}/32.2$, where $g_c = 32.2\ \text{lb}_\text{m}/\text{slug}$ is the mass conversion constant. The constant $g_c = 32.2$ is not to be confused with the local value of g, which is the acceleration of a freely falling body.

In the flow equation, the kinetic energy term $mv^2/2$ has units of (slug)(ft/s)^2 = ($\text{lb}_\text{f} \cdot \text{s}^2$/ft)(ft/s)^2 = ft $\cdot$ lb_f. When mass is expressed in units of pounds mass, the kinetic energy term is $mv^2/(2g_c)$. Note that dividing m by g_c converts mass from pounds mass to slugs.

Illustrative example 3.6. For sonic flow of air through a restriction, the mass rate of flow is

$$M = \frac{0.53}{\sqrt{T}} AP \tag{3.10}$$

where M = mass rate of flow, $\text{lb}_\text{m}/\text{s}$
T = inlet temperature, °R
A = area of restriction, ft^2
P = inlet pressure, $\text{lb}_\text{f}/\text{ft}^2$

Determine the linearized approximation for the variation m when the inlet temperature T is constant. It is interesting to note that because the length unit cancels in Eq. (3.10), any unit of length may be used. When inches are used, the area is in square inches and the pressure is in pounds force per square inch ($\text{lb}_\text{f}/\text{in}^2$).

Solution. Because the temperature is constant, the mass rate of flow is a function of A and P. Thus

$$m = \left.\frac{\partial M}{\partial A}\right|_i a + \left.\frac{\partial M}{\partial P}\right|_i p$$

The partial derivatives are

$$\left.\frac{\partial M}{\partial A}\right|_i = \left.\frac{0.53}{\sqrt{T}} P\right|_i = \frac{M_i}{A_i}$$

and

$$\left.\frac{\partial M}{\partial P}\right|_i = \left.\frac{0.53}{\sqrt{T}} A\right|_i = \frac{M_i}{P_i}$$

Thus,

$$m = \left(\frac{a}{A_i} + \frac{p}{P_i}\right) M_i \tag{3.11}$$

When SI units are used, Eq. (3.10) becomes $M = 0.04AP/\sqrt{T}$. The units for the mass rate of flow M are kilograms per second (kg/s), the units for area A are square meters (m^2), the units of pressure P are newtons per square meter (N/m^2), and the units for temperature T are kelvins (K).

Illustrative example 3.7. For subsonic flow of air through a restriction, the mass rate of flow is

$$M = \frac{1.05}{\sqrt{T}} A\sqrt{(P_1 - P_2)P_2} \tag{3.12}$$

where M = mass rate of flow, $\text{lb}_\text{m}/\text{s}$
T = inlet temperature, °R
A = area of restriction, ft^2
P_1 = inlet pressure, $\text{lb}_\text{f}/\text{ft}^2$
P_2 = outlet pressure, $\text{lb}_\text{f}/\text{ft}^2$

Determine the linearized approximation for the variation m when the inlet temperature is constant. As was the case for Eq. (3.10), any unit of length may be used in Eq. (3.12).

Solution. The mass rate of flow is a function of A, $P_1 - P_2$, and P_2. Thus

$$m = \left.\frac{\partial M}{\partial A}\right|_i a + \left.\frac{\partial M}{\partial(P_1 - P_2)}\right|_i (p_1 - p_2) + \left.\frac{\partial M}{\partial P_2}\right|_i p_2$$

Evaluating the partial derivatives yields

$$m = \left[\frac{a}{A_i} + \frac{p_1 - p_2}{2(P_1 - P_2)_i} + \frac{p_2}{2(P_2)_i}\right] M_i \tag{3.13}$$

The same result is obtained if P_1 and P_2 are regarded as the variables rather than $P_1 - P_2$ and P_2.

When SI units are used, Eq. (3.12) becomes $M = 0.08A\sqrt{(P_1 - P_2)P_2/T}$. The units are kilograms, meters, newtons, and kelvins.

An oscillating pendulum is shown in Fig. 3.4. The positive sense for θ is counterclockwise as shown. The torque about the point O is $-MgL \sin\theta$ which is equal to $I\alpha$, where $I = ML^2$ is the mass moment of inertia about point O and $\alpha = D^2\theta$ is the angular acceleration. Thus

$$-MgL \sin\theta = ML^2 D^2\theta$$

or

$$D^2\theta + \frac{g}{L}\sin\theta = 0$$

This is a nonlinear differential equation. The linear approximation about the equilibrium position $\theta_i = 0°$ is

$$\frac{\partial}{\partial\theta}\left(D^2\theta + \frac{g}{L}\sin\theta\right)_i \Delta\theta = \left(D^2 + \frac{g}{L}\cos\theta\right)_i \Delta\theta = \left(D^2 + \frac{g}{L}\right)\Delta\theta = 0$$

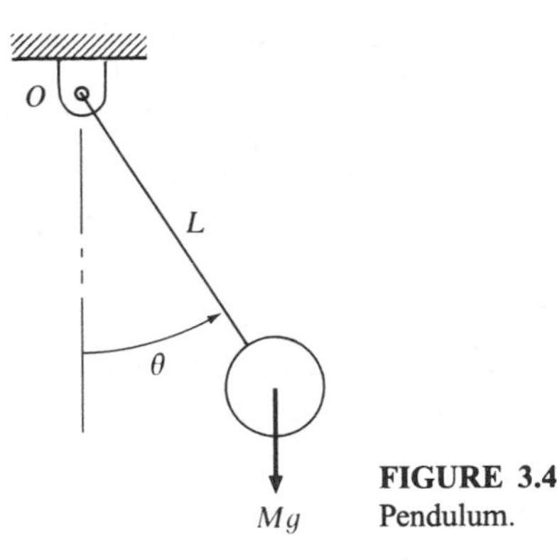

FIGURE 3.4
Pendulum.

where $\cos \theta_i = \cos 0° = 1$. Because $D = d/dt$ is a function of time t, it is regarded as a constant when we are taking the partial derivative with respect to θ. This approximation is quite accurate for $-\pi/4 \leq \Delta\theta \leq \pi/4$. Thus, a good approximation for the motion of a pendulum is

$$\left(D^2 + \frac{g}{L}\right)\theta = 0 \qquad -\pi/4 \leq \theta \leq \pi/4$$

Geometric Interpretation of Error Introduced by a Linear Approximation

From the linear approximation for the area of a rectangle, it is posssible to represent geometrically the error which is introduced. Figure 3.5 shows a rectangle in which the reference length is L_i and the width is W_i. The area of a rectangle is a function of the length L and the width W, so the variation in the area from its reference size is obtained as follows:

$$A = A(L,\ W)$$

$$a = \left.\frac{\partial A}{\partial L}\right|_i (l) + \left.\frac{\partial A}{\partial W}\right|_i w$$

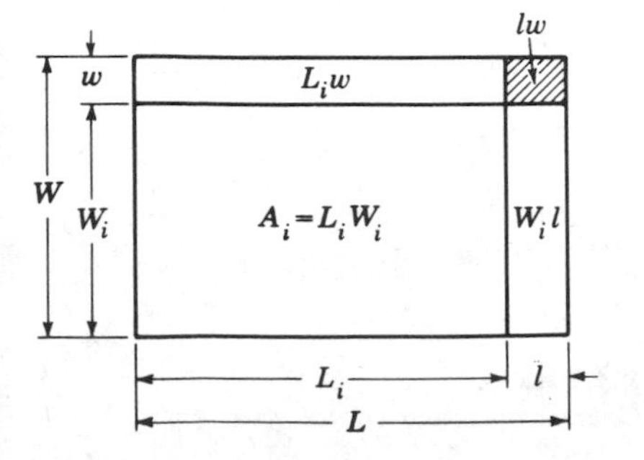

FIGURE 3.5
Geometric representation of error.

The preceding partial derivatives are evaluated from the equation $A = LW$:

$$\left.\frac{\partial A}{\partial L}\right|_i = W_i \qquad \text{and} \qquad \left.\frac{\partial A}{\partial W}\right|_i = L_i$$

Thus the linearized expression for the area A is

$$A \approx A_i + a = A_i + W_i l + L_i w \tag{3.14}$$

Each term in the preceding expression is represented by an area in Fig. 3.5. The difference between this approximation and the actual area LW is the small shaded portion lw.

3.2 LINEARIZATION OF OPERATING CURVES

In the preceding section, we showed how equations which are nonlinear could be linearized. For many components encountered in control systems, the operating characteristics are given in the form of general operating curves rather than equations. For example, Fig. 3.6 shows a family of curves of constant values of Z (that is, $Z = 15, 20$, and 25). For $X = 1000$ and $Z = 20$, the corresponding value of Y is 60. This point is indicated by A.

The general functional relationship is

$$Y = Y(X, Z)$$

Linearization about a reference point of operation gives

$$y = \left.\frac{\partial Y}{\partial X}\right|_i x + \left.\frac{\partial Y}{\partial Z}\right|_i z = \left.\frac{\partial Y}{\partial X}\right|_Z x + \left.\frac{\partial Y}{\partial Z}\right|_X z \tag{3.15}$$

In obtaining the partial derivative of Y with respect to X, all variables in Y are maintained constant except X. Because $Y = Y(X, Z)$, the notation $\partial Y/\partial X$ means

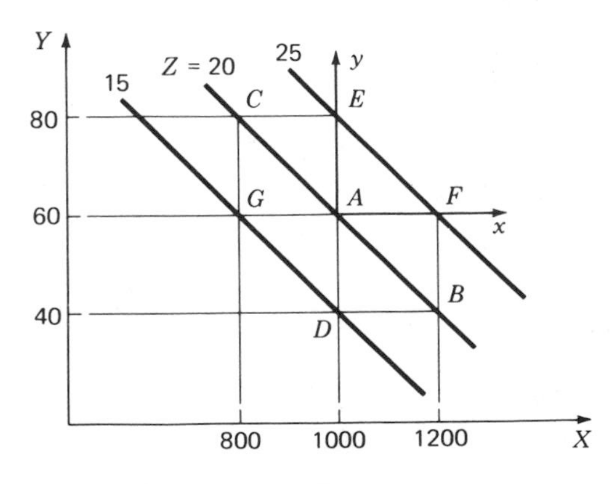

FIGURE 3.6
General operating curves.

that in differentiating Y with respect to X, variable Z is regarded as constant. The term

$$\left.\frac{\partial Y}{\partial X}\right|_i = \left.\frac{\partial Y}{\partial X}\right|_Z$$

is the change of Y per change in X with Z held constant at its reference value. Similarly,

$$\left.\frac{\partial Y}{\partial Z}\right|_i = \left.\frac{\partial Y}{\partial Z}\right|_X$$

is the change in Y per change in Z with X fixed at its reference value.

Thus, the term $\partial Y/\partial X|_Z$ is evaluated by dividing changes ΔY by corresponding changes ΔX in which Z is maintained constant at its reference value. For $Z_i = 20$, the value of Y at point B is 40 and the value at point C is 80. Similarly, the value of X at point B is 1200, and the value at point C is 800. Thus

$$\left.\frac{\partial Y}{\partial X}\right|_Z = \left.\frac{\Delta Y}{\Delta X}\right|_Z = \frac{Y_B - Y_C}{X_B - X_C} = \frac{40 - 80}{1200 - 800} = -\frac{40}{400} = -0.10$$

This partial derivative is the slope of the curve $Z_i = 20$ at the reference operating point.

The term $\partial Y/\partial Z|_X$ is obtained by dividing changes ΔY by corresponding changes ΔZ in which X is maintained constant at its reference value. For $X_i = 1000$, the value of Y at point D is 40, and the value of Z is 15. Similarly, at point E, the value of Y is 80, and the value of Z is 25. Thus

$$\left.\frac{\partial Y}{\partial Z}\right|_X = \left.\frac{\Delta Y}{\Delta Z}\right|_X = \frac{Y_D - Y_E}{Z_D - Z_E} = \frac{40 - 80}{15 - 25} = \frac{-40}{-10} = 4$$

This partial derivative, which is obtained from a vertical (X_i = constant) interpolation of the curves, determines the vertical spacing. That is,

$$\Delta Y = 4\Delta Z$$

As Z changes by 5 units, the vertical distance ΔY between lines of constant Z is $\Delta Y = 4(5) = 20$ units. Substitution of the preceding results into Eq. (3.15) yields

$$y = -0.1x + 4z \tag{3.16}$$

The origin of the (x, y) system is at point A, which is the reference point. Note that

$$X = X_i + x = 1000 + x \qquad \text{and} \qquad Y = Y_i + y = 60 + y$$

At point B,

$$x = X - X_i = 1200 - 1000 = 200$$

and

$$z = Z - Z_i = 20 - 20 = 0$$

Thus,

$$y = -0.1(200) + 0 = -20$$

The value of Y at point B is $Y = Y_i + y = 60 - 20 = 40$. At point E,

$$x = 1000 - 1000 = 0$$

and

$$z = 25 - 20 = 5$$

Thus,

$$y = 0 + 4(5) = 20$$

and

$$Y = Y_i + y = 60 + 20 = 80$$

Equation (3.16) may be solved for z. That is,

$$z = \frac{0.1x + y}{4} = 0.025x + 0.25y \tag{3.17}$$

The curves shown in Fig. 3.6 may be interpreted as expressing any one of the three variables (X, Y, or Z) as a function of the other two. Consider now that they represent Z as a function of X and Y. That is,

$$Z = Z(X, Y)$$

Linearization gives

$$z = \left.\frac{\partial Z}{\partial X}\right|_Y x + \left.\frac{\partial Z}{\partial Y}\right|_X y \tag{3.18}$$

The term $\partial Z/\partial X|_Y$ is obtained by dividing changes in Z by corresponding changes in X in which Y is maintained constant. For $Y_i = 60$, the value of Z at point F is 25, and the value of X is 1200. Similarly, at point G, the value of Z is 15, and the value of X is 800. Thus

$$\left.\frac{\partial Z}{\partial X}\right|_Y = \left.\frac{\Delta Z}{\Delta X}\right|_Y = \frac{Z_F - Z_G}{X_F - X_G} = \frac{25 - 15}{1200 - 800} = 0.025$$

This partial derivative, which is obtained from a horizontal (Y_i = constant) interpolation of the curves, determines the horizontal spacing. That is,

$$\Delta X = \frac{1}{0.025}\Delta Z = 40\ \Delta Z$$

As Z changes by 5 units, the horizontal spacing ΔX between lines of constant Z is $\Delta X = 40(5) = 200$ units. The term $\partial Z/\partial Y|_X$ is obtained by dividing changes in Z by corresponding changes in Y with X fixed. Thus

$$\left.\frac{\partial Z}{\partial Y}\right|_X = \left.\frac{\Delta Z}{\Delta Y}\right|_X = \frac{15 - 25}{40 - 80} = \frac{-10}{-40} = 0.25$$

Note that $\partial Z/\partial Y|_X = \frac{1}{4}$ is the reciprocal of $\partial Y/\partial Z|_X = 4$, which was previously evaluated. Substitution of the results for $\partial Z/\partial X|_Y$ and $\partial Z/\partial Y|_X$ into Eq. (3.18) verifies the result given by Eq. (3.17).

The functional relationship between X, Y, and Z represented in Fig. 3.6 may be expressed in the implicit form

$$G(X,\ Y,\ Z) = 0$$

From calculus, it is known that for an implicit function of n variables, the product of the n partial derivatives is $(-1)^n$. For this function of the three variables X, Y, and Z, this relationship is

$$\left.\frac{\partial X}{\partial Y}\right|_Z \left.\frac{\partial Y}{\partial Z}\right|_X \left.\frac{\partial Z}{\partial X}\right|_Y = (-1)^3 = -1 \tag{3.19}$$

Substitution of the values

$$\left.\frac{\partial X}{\partial Y}\right|_Z = \frac{1}{-0.1} = -10 \qquad \left.\frac{\partial Y}{\partial Z}\right|_X = 4$$

and

$$\left.\frac{\partial Z}{\partial X}\right|_Y = 0.025$$

verifies that $(-10)(4)(0.025) = -1$. In effect, if any two of the three quantities horizontal spacing $\partial X/\partial Z|_Y$, vertical spacing $\partial Y/\partial Z|_X$, or slope $\partial Y/\partial X|_Z$ are known, then the third may be determined in accordance with Eq. (3.19).

A typical family of operating curves for an engine is shown in Fig. 3.7. Usually such curves are determined experimentally, and it would be quite tedious and difficult to express them as equations. The linearized equation for the operation

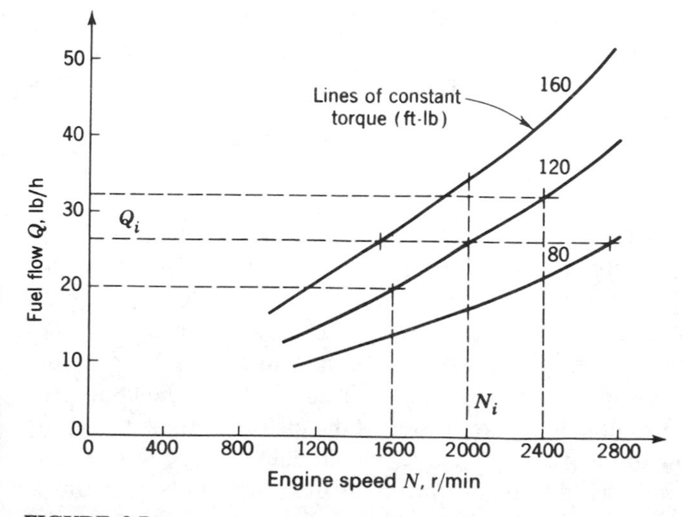

FIGURE 3.7
Characteristic curves for an engine.

of the engine about some reference operating point is obtained as follows: From Fig. 3.7, it is seen that the speed N is a function of the rate of fuel flow Q and the engine torque T; thus

$$N = N(Q, T)$$

Linearization gives

$$n = \frac{\partial N}{\partial Q}\bigg|_i q + \frac{\partial N}{\partial T}\bigg|_i t$$

The term $\partial N/\partial Q|_i$ is the change in speed per change in fuel flow with all other parameters held constant (in this case with T constant). This partial derivative is equal to the reciprocal of the slope of the line of constant torque evaluated at the reference point. That is,

$$\frac{\partial N}{\partial Q}\bigg|_i = \frac{2400 - 1600}{32 - 20} = 66.7$$

The partial derivative $\partial N/\partial T|_i$ is the change in speed per change in torque with Q held constant. This is evaluated from a horizontal interpolation of the characteristic operating curves as

$$\frac{\partial N}{\partial T}\bigg|_i = \frac{2730 - 1530}{80 - 160} = -15$$

The minus sign indicates that for a constant Q the speed decreases as the torque increases. Thus, for operation in the vicinity of the point $N_i = 2000$, $Q_i = 26$, and $T_i = 120$, the linearized approximation for N is

$$N \approx N_i + n = 2000 + 66.7q - 15t$$

The main difference in working with characteristic operating curves for a component rather than equations is that the partial derivatives are evaluated from a physical interpretation of the curves rather than mathematically from the equations.

3.3 HYDRAULIC SYSTEMS

A schematic diagram of a hydraulic amplifier is shown in Fig. 3.8*a*. Such amplifiers are very rapid-acting and are capable of producing very large forces. The position of the valve is designated by x, and the position of the large piston which moves the load is y. This type of valve is called a *balanced valve* because the pressure forces acting on it are all balanced so that little force is required to change its position. When the valve is moved upward, the supply pressure is connected to the upper side of the piston to admit oil. The fluid in the lower side of the piston is connected to the drain to permit return flow to the pump, where it is recirculated in the system. For the reverse process, the valve is moved downward so that the supply pressure is connected to the bottom side of the big piston. The upper side of this piston is connected to the upper drain to permit return flow to the pump.

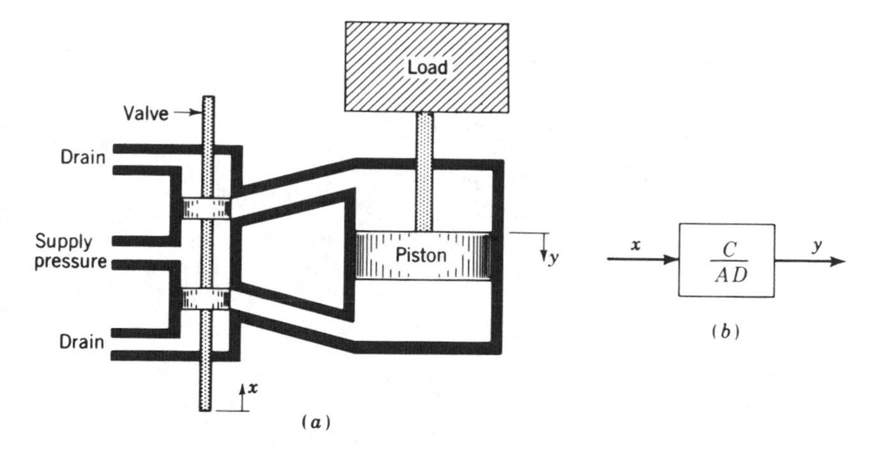

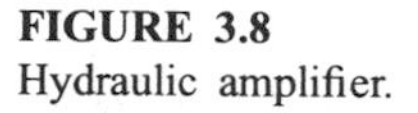
FIGURE 3.8
Hydraulic amplifier.

When the mass of the load is negligible, the pressure drop across the valve remains constant. For this case, the rate of flow to the piston is proportional to the area uncovered by the valve, which is seen to be proportional to the distance x. Thus,

$$q = Cx$$

where q is the rate of flow through the valve to the piston chamber. This rate of flow q into the piston chamber is equal to the rate of change of volume of the chamber which is equal to the piston velocity Dy times the cross-sectional area A of the piston.

$$q = ADy$$

Equating the preceding expressions for q and solving for y give

$$y = \frac{C}{AD}x = \frac{C}{A}\frac{1}{D}x \tag{3.20}$$

The $1/D$ term indicates that this hydraulic valve and piston combination in effect integrates hydraulically. That is, the position y is proportional to the integral of the position x. The block-diagram representation is shown in Fig. 3.8b.

The effect of load M upon the operation of this hydraulic amplifier may be ascertained from the operating curves shown in Fig. 3.9. The rate of flow Q to the cylinder is a function of the valve position X and the pressure drop P across the power piston. That is,

$$Q = Q(X, P)$$

Linearization gives

$$q = \left.\frac{\partial Q}{\partial X}\right|_i x + \left.\frac{\partial Q}{\partial P}\right|_i p = C_1 x - C_2 p \tag{3.21}$$

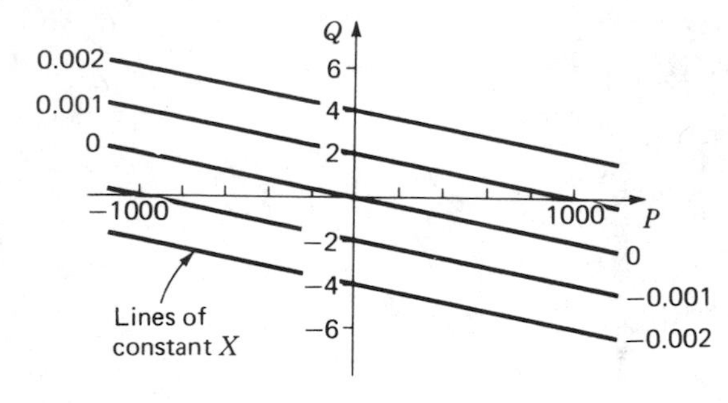

FIGURE 3.9
Operating curves for a hydraulic amplifier.

where $C_1 = \partial Q/\partial X|_P$ and $C_2 = -\partial Q/\partial P|_X$. The term $\partial Q/\partial X|_P$ is obtained from a vertical interpolation of the curves in which P is maintained constant at the reference value. Because Q increases as X increases, $\partial Q/\partial X|_P$ is a positive constant. The term $\partial Q/\partial P|_X$ is the slope of the curve at the reference value of X. Because Q decreases as P increases (negative slope), the term $\partial Q/\partial P|_X$ is negative. The constant $C_2 = -\partial Q/\partial P|_X$ is positive. In analyzing control systems, it is desirable that all constants be positive. This enables the control engineer to determine directly from the equations of operation and the resulting block diagrams whether the output increases or decreases when the input undergoes a certain change.

The force transmitted to the load by the power piston is equal to the product of the pressure drop across the piston and the cross-sectional area A of the piston. Thus,

$$pA = M\frac{d^2y}{dt^2} = MD^2y \tag{3.22}$$

where M is the mass of the load and d^2y/dt^2 is the acceleration. Substitution of p from Eq. (3.22) into Eq. (3.21) gives

$$q = C_1x - \frac{C_2M}{A}D^2y$$

The rate of flow q into the cylinder is equal to the rate of change of volume, which is the cross-sectional area A times the velocity dy/dt. Thus

$$q = A\frac{dy}{dt} = ADy$$

Equating the preceding expressions for q and solving for the output position y as a function of the input position x gives

$$y = \frac{(C_1/A)x}{D(1 + \tau D)} \tag{3.23}$$

where $\tau = C_2M/A^2$. Note that when the load is neglected ($M = 0$), then $\tau = C_2M/A^2 = 0$ and Eq. (3.23) reduces to Eq. (3.20). Usually C_2 is quite small, so that $\tau \approx 0$. In general, the approximation given by Eq. (3.20) is quite good.

Hydraulic Servomotor

A hydraulic servomotor is shown in Fig. 3.10. A linkage called a *walking beam* connects the input position x, the valve position e, and the piston position y. The centerline of the lever when the servomotor is in its reference position is indicated in Fig. 3.10. The variations in x, e, and y from their reference positions are also indicated. When e is zero, the valve is "line on line" and no flow can go to or from the big piston.

The operation of this servomotor may be visualized as follows. When the input x is changed from the reference position, the walking beam first pivots about the connection at y because the large forces acting on the piston hold it in place temporarily. This position of the walking beam is shown by the dashed line in Fig. 3.10. Because of the corresponding movement of e, the valve now admits fluid to the big piston to move it in the direction which makes e zero. The final position of the walking beam, in which e is again zero and the piston has moved a distance y, is indicated in Fig. 3.10. For steady-state operation ($e = 0$), the relationship between the input x and the ouput y is

$$\frac{y}{b} = \frac{x}{a} \qquad \text{or} \qquad y = \frac{b}{a}x$$

The overall block diagram which describes the dynamics as well as the steady-state operation of this servomotor is obtained as follows. Figure 3.11*a* shows the walking-beam linkage. For small variations about the reference position,

$$e = \left.\frac{\partial E}{\partial X}\right|_i x + \left.\frac{\partial E}{\partial Y}\right|_i y \tag{3.24}$$

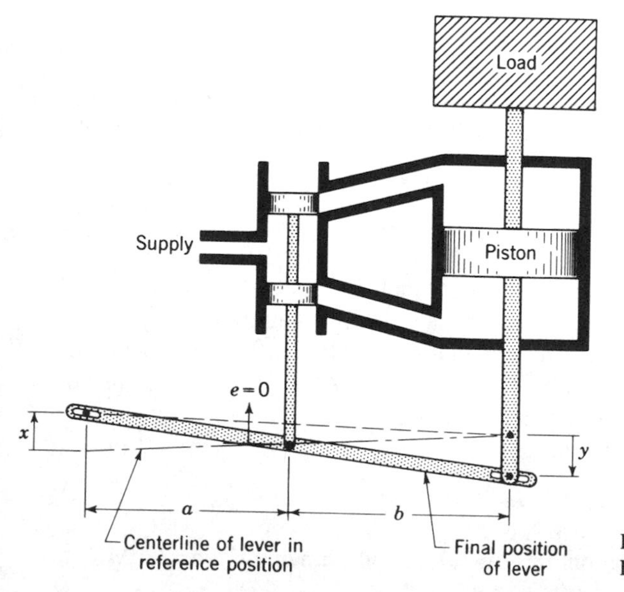

FIGURE 3.10
Hydraulic servomotor.

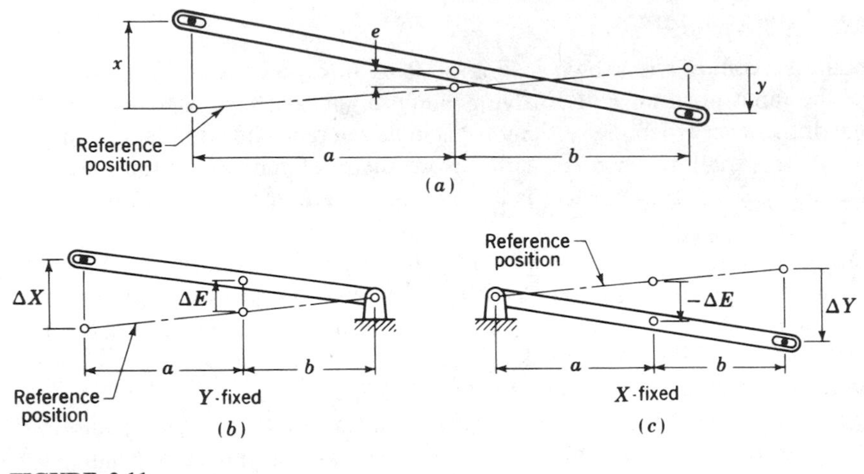

FIGURE 3.11
Walking-beam linkage.

The value of $\partial E/\partial X|_i$ is obtained by finding the ratio of the change in E to a change in X with all other parameters held constant at the reference position. Figure 3.11b illustrates the linkage with Y fixed in the reference position. From similar triangles,

$$\left.\frac{\partial E}{\partial X}\right|_i = \lim_{\substack{\Delta E \to 0 \\ \Delta X \to 0}} \frac{\Delta E}{\Delta X} = \frac{b}{a+b}$$

Similarly, from Fig. 3.11c in which X is fixed in the reference position,

$$\left.\frac{\partial E}{\partial Y}\right|_i = \lim_{\substack{\Delta E \to 0 \\ \Delta Y \to 0}} \frac{\Delta E}{\Delta Y} = \frac{-a}{a+b}$$

The minus sign occurs because e decreases as y increases. Substitution of the preceding results into Eq. (3.24) yields the following expression for the walking-beam linkage:

$$e = \frac{b}{a+b}x - \frac{a}{a+b}y \tag{3.25}$$

This equation shows that the walking-beam linkage is actually a comparator or summing point. The preceding result could have been obtained directly by closer examination of Fig. 3.11a. It is apparent that the motion of e is the sum of the contribution due to changing x with y fixed, that is, $[b/(a+b)]x$, and that due to changing y with x fixed, that is, $[-a/(a+b)]y$.

For the case in which $a = b$,

$$e = \frac{x-y}{2} \tag{3.26}$$

The block-diagram representation for Eq. (3.26) is shown in Fig. 3.12a.

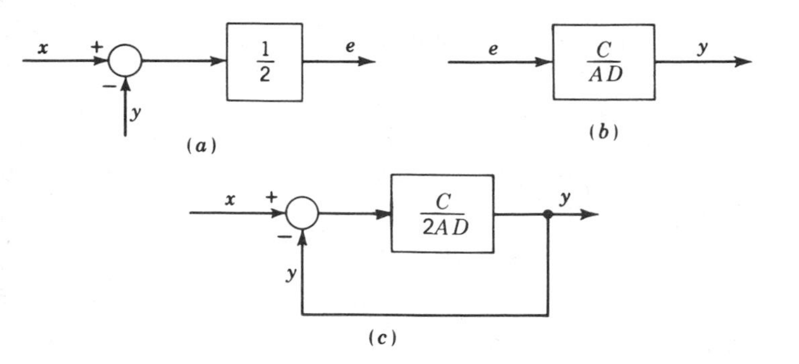

FIGURE 3.12
Block diagram for (*a*) walking-beam linkage, (*b*) valve and piston, (*c*) servomotor.

The equation for the valve-and-piston combination is given by Eq. (3.20) but with x replaced by e. Thus

$$y = \frac{C}{AD}e \tag{3.27}$$

The block-diagram representation for Eq. (3.27) is shown in Fig. 3.12*b*. Combining Fig. 3.12*a* and *b* yields the overall block diagram for the servomotor, as shown in Fig. 3.12*c*.

The overall relationship between the input x and the output y is obtained as follows from the block diagram of Fig. 3.12*c*:

$$(x - y)\frac{C}{2AD} = y$$

or

$$(1 + \tau D)y = x \tag{3.28}$$

where

$$\tau = \frac{2A}{C}$$

Equation (3.28) is the differential equation relating x and y. For steady-state operation, both x and y are constant. The quantity $Dy = dy/dt$ is zero when y is constant, and thus for steady-state operation Eq. (3.28) becomes

$$y = x$$

First-Order Systems

To determine the transient response of y for a given change in x, it is necessary to solve Eq. (3.28), which is a first-order linear differential equation with constant coefficients. Many elements encountered in control systems are described by a first-order differential equation of the form

$$(1 + \tau D)y = Kx \tag{3.29}$$

where K is a constant. For $K = 1$, this equation reduces to Eq. (3.28). To solve Eq. (3.29), first rewrite it in the form

$$\frac{dy}{dt} + \frac{1}{\tau}y = \frac{K}{\tau}x$$

Multiplying through by the integrating factor $e^{t/\tau}$ gives

$$e^{t/\tau}\frac{dy}{dt} + \frac{e^{t/\tau}}{\tau}y = \frac{d}{dt}\left(e^{t/\tau}y\right) = \frac{Ke^{t/\tau}}{\tau}x$$

Integration shows that

$$e^{t/\tau}y = \frac{K}{\tau}\int xe^{t/\tau}\,dt$$

Thus

$$y = \frac{Ke^{-t/\tau}}{\tau}\int xe^{t/\tau}\,dt$$

For a given input x, the preceding may be solved for the resulting response y. Consider the case in which the input x changes instantaneously from its initial value $x = 0$ at time $t = 0$ to a new or final value $x = 1$ for $t > 0$. A unit step change, as is illustrated graphically in Fig. 3.13a, has occurred.

Solving for y when $x = 1$ gives

$$y = \frac{Ke^{-t/\tau}}{\tau}\int e^{t/\tau}\,dt = \frac{Ke^{-t/\tau}}{\tau}\left(\tau e^{t/\tau} + C\right)$$

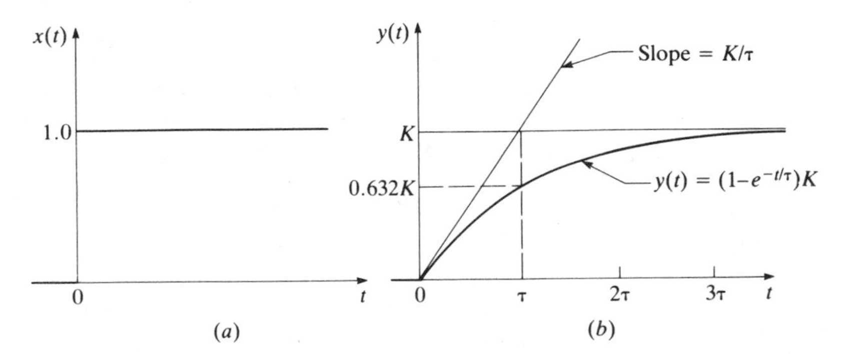

FIGURE 3.13
Response of first-order system to a unit step function.

The constant of integration C is evaluated from the initial condition $y(0) = 0$ when $t = 0$. Thus

$$0 = \frac{K}{\tau}(\tau + C)$$

or

$$C = -\tau$$

The resulting solution is

$$y = K(1 - e^{-t/\tau}) \tag{3.30}$$

A graph of the response $y(t)$ is shown in Fig. 3.13*b*. A characteristic feature of such an exponential response curve is that when $t = \tau$, then $y(t)$ has undergone 63.2 percent of its total change. This is proved by letting $t = \tau$ in Eq. (3.30); thus

$$y(\tau) = K(1 - e^{-1}) = K(1 - 0.368) = 0.632K$$

When $t = 2\tau$, $y(t)$ has undergone 86.5 percent of its total change, and when $t = 3\tau$, $y(t)$ has undergone 95 percent of its total change. The final steady-state value of $y(t)$ is K. Because y is constant for the steady state, $Dy = 0$ in Eq. (3.29). The resulting equation for steady-state operation is

$$y = Kx$$

Thus a 1-unit change in x results in a K-unit change in y. The parameter K is referred to as the *steady-state gain*. Another feature of an exponential response, as shown in Fig. 3.13*b*, is that the tangent to the curve at $t = 0$ intersects the final value at time $t = \tau$. This is proved by differentiating Eq. (3.30) to obtain the slope of the response curve and then evaluating this result at time $t = 0$. Thus

$$\left.\frac{dy}{dt}\right|_{t=0} = \left.\frac{K}{\tau}e^{-t/\tau}\right|_{t=0} = \frac{K}{\tau}$$

As shown in Fig. 3.13*b*, the slope of the tangent line at $t = 0$ is K/τ. The term τ is called the *time constant* and is a measure of the speed of response. When τ is small, the system approaches its new operating condition very rapidly; and when τ is large, more time is required for the change to occur.

Jet Pipe Amplifier

A schematic representation of a jet pipe amplifier is shown in Fig. 3.14*a*. The jet pipe pivots about a fixed point. The position of the jet pipe at the centerline of the springs is indicated by e, and the position at the nozzle end is indicated by w. The pipe is supplied with a high-pressure hydraulic fluid such that a steady stream is continually flowing out the nozzle end. The rate of flow q to the power piston is $q = Cw$, where C is a constant depending on the particular jet pipe and the supply pressure.

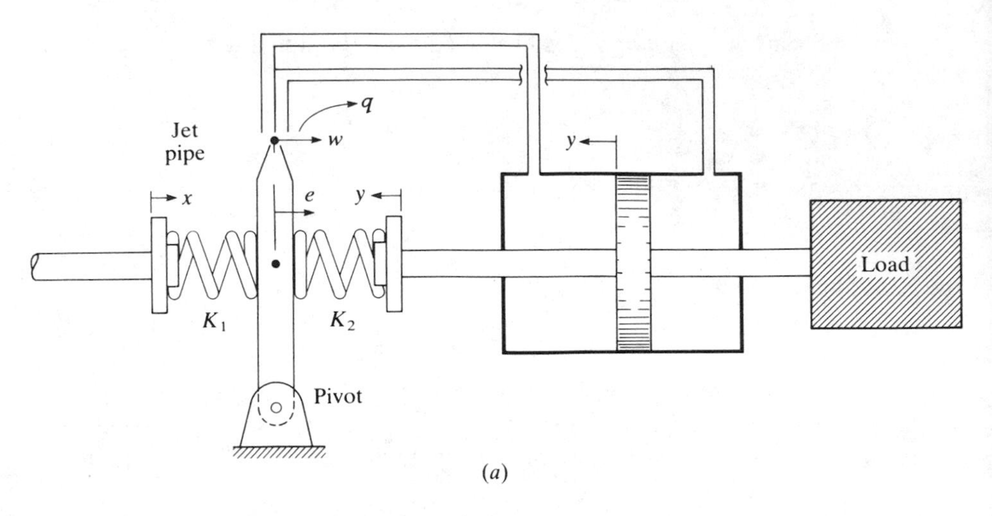

(a)

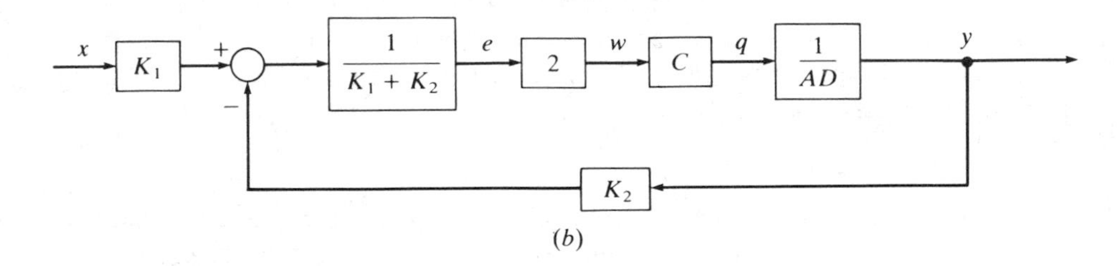

(b)

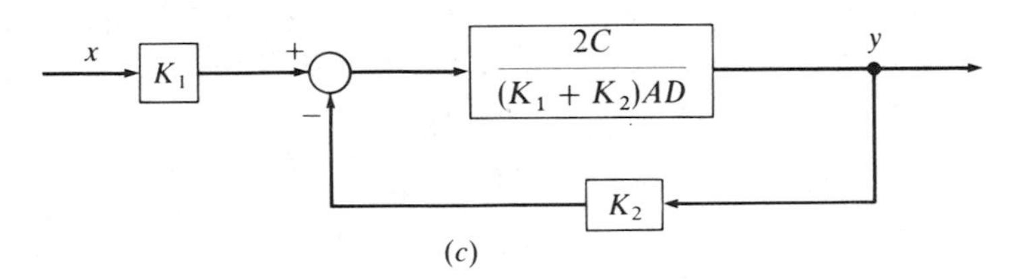

(c)

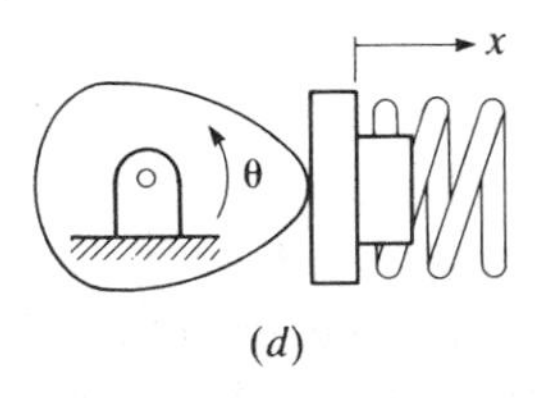

(d)

FIGURE 3.14
Jet pipe amplifier: (*a*) schematic diagram, (*b*) initial block diagram, (*c*) final block diagram, (*d*) position x determined by cam.

The position x is the input position, and y which is the position of the power piston is the output position. The compression of spring K_1 is $x - e$, and the compression of spring K_2 is $e + y$; hence $K_1(x - e) = K_2(e + y)$, or $K_1x - K_2y = (K_1 + K_2)e$. For a typical jet pipe, the distance from the fixed pivot to the end where w is measured is twice the distance from the fixed pivot to the center of the springs where e is measured; thus $w = 2e$. The rate of change of volume in the power piston is $q = A\, dy/dt = ADy$. Combining the preceding equations yields the block-diagram representation shown in Fig. 3.14*b*. Multiplying the feedforward elements together yields the block diagram shown in Fig. 3.14*c*. The differential equation of operation for the jet pipe amplifier is obtained from Fig. 3.14*c* as follows:

$$y = \frac{K_1/K_2}{1 + \dfrac{A}{2C}\dfrac{K_1 + K_2}{K_2}D}x$$

This has the form of Eq. (3.29), i.e.,

$$y = \frac{K}{1 + \tau D}x \tag{3.29}$$

Thus, the steady-state gain is K_1/K_2, and the time constant is $A(K_1 + K_2)/(2CK_2)$. Usually C is very large, in which case the time constant is very small. In many applications the position x is determined by a cam pressing against the bottom of the spring seat, as shown in Fig. 3.14*d*. For each angular position θ of the cam there is one position x. For steady-state operation, $y = (K_1/K_2)x$. A typical application is that in which the load is a metering valve. Each position y of the metering valve results in a different amount of fluid flow through the valve, hence a very accurate flow metering device results.

The jet pipe shown in Fig. 3.14*a* has force feedback. That is, spring K_2 provides a force proportional to the output position y. Spring K_1 provides a force proportional to the desired input. The difference causes a movement of the jet pipe. If the desired input x is connected to one end of a walking-beam linkage, the output y to the other end, and the jet pipe position e to the center, then a jet pipe amplifier with position feedback results.

3.4 PNEUMATIC SYSTEMS

The pneumatic valve shown in Fig. 3.15 is used in critical-care ventilator systems to regulate both the flow and the ratio of air to oxygen in a patient's breathing circuit. The valve also follows the command signal to determine the waveform and frequency of each breath and is sufficiently responsive to follow a patient's sigh, cough, or hiccough. A picture of a ventilator system in use is shown in Fig. 3.16.

A flapper valve, as shown in Fig. 3.17*a*, is one in which small changes in the position X of the flapper cause large variations in the controlled pressure P_2 in the

FIGURE 3.15
A pneumatic mixing and metering valve. (Courtesy South Bend Controls, Inc.)

chamber. When the flapper is closed off, the pressure P_2 in the chamber is equal to the supply pressure P_1. If the flapper is opened wide, the chamber pressure approaches the ambient pressure P_a. If a spring-loaded bellows which is free to expand as the chamber pressure increases is attached, this becomes a position control device as well as a pressure controller. An equivalent system which uses a piston rather than a bellows is shown in Fig. 3.17*b*.

Although flapper valves have numerous applications in hydraulic systems, they are more extensively used in pneumatic controls. A major reason is that spool valves do not lend themselves to pneumatic applications because of the excessive leakage of air that is the result of the very low viscosity of air. When the flapper valves of Fig. 3.17 are supplied with air, pneumatic amplifiers result.

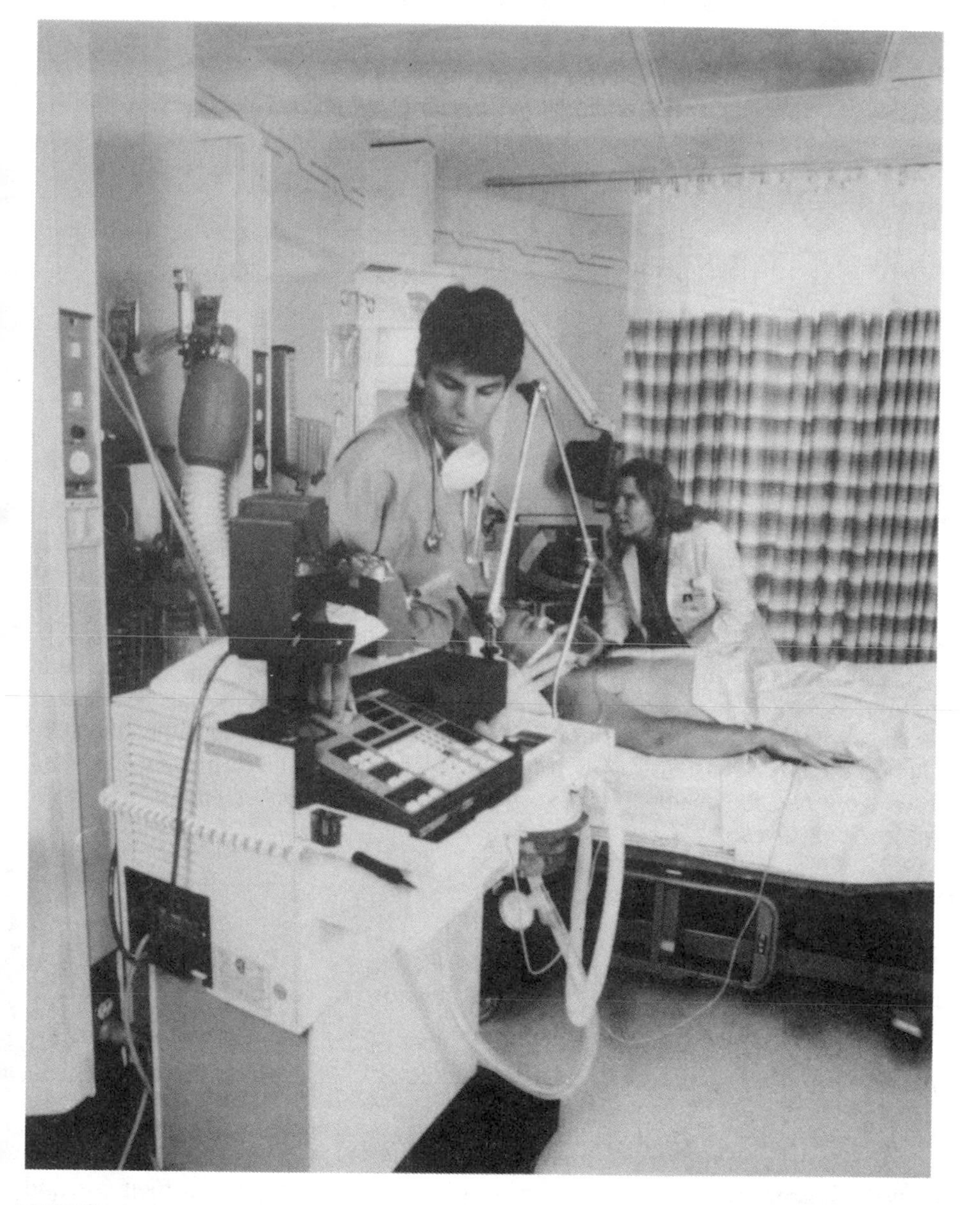

FIGURE 3.16
A ventilator system. (Courtesy Puritan-Bennett.)

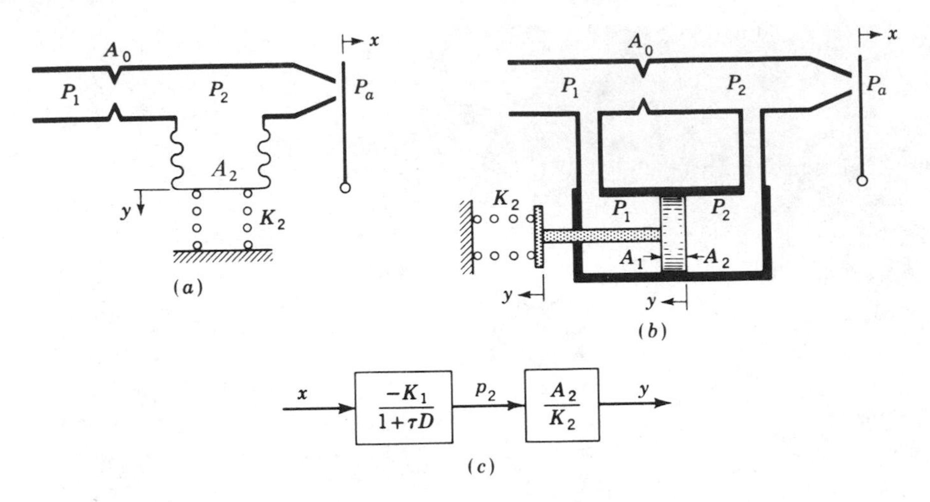

FIGURE 3.17
Pneumatic amplifier.

The procedure used to obtain the equation of operation for a pneumatic amplifier is similar to that used for a hydraulic amplifier, except that, because of compressibility effects, the mass rate of flow must be considered rather than the volume rate. For a constant supply pressure P_1 and fixed inlet orifice A_0, the mass rate of flow into the chamber M_{in} is a function of the chamber pressure P_2 only. Linearization gives

$$m_{\text{in}} = \left.\frac{\partial M_{\text{in}}}{\partial P_2}\right|_i p_2 = -C_1 p_2$$

The minus sign indicates that as p_2 increases, m_{in} decreases. For a constant ambient pressure P_a, the mass rate of flow out of the chamber M_o is a function of X and P_2. Thus,

$$m_o = \left.\frac{\partial M_o}{\partial X}\right|_i x + \left.\frac{\partial M_o}{\partial P_2}\right|_i p_2 = C_2 x + C_3 p_2$$

When the pressure ratio across an orifice is less than 0.528, the flow is sonic, and when it is greater than 0.528, the flow is subsonic. For sonic flow the mass rate of flow is given by Eq. (3.10), and for subsonic flow it is given by Eq. (3.12).

The change in mass w of air in the chamber is the integral of $m_{\text{in}} - m_o$. That is,

$$w = \frac{m_{\text{in}} - m_o}{D} = \frac{-C_1 p_2 - (C_2 x + C_3 p_2)}{D}$$

Multiplying through by D shows that

$$-C_1 p_2 - C_2 x - C_3 p_2 = Dw \tag{3.31}$$

From the equation of state, the total mass W of air in the chamber is

$$W = \frac{P_2 V_2}{RT_2}$$

where V_2 is the volume of the chamber and T_2 is the stagnation temperature of air in the chamber. For the usual case of adiabatic flow, T_2 is equal to the stagnation temperature of the supply, which is constant. Linearization yields, for the change in mass w of air in the chamber,

$$w = \left.\frac{\partial W}{\partial V_2}\right|_i v_2 + \left.\frac{\partial W}{\partial P_2}\right|_i p_2 = C_4 v_2 + C_5 p_2 \tag{3.32}$$

where

$$\left.\frac{\partial W}{\partial V_2}\right|_i = \left.\frac{P_2}{RT_2}\right|_i = \left.\frac{W}{V_2}\right|_i = C_4$$

and

$$\left.\frac{\partial W}{\partial P_2}\right|_i = \left.\frac{V_2}{RT_2}\right|_i = \left.\frac{W}{P_2}\right|_i = C_5$$

The term $C_4 v_2$ represents the change in mass required to account for the change in volume, and the term $C_5 p_2$ represents the change in mass required to account for the change in pressure. This is the compressibility effect. The change in volume of the chamber is equal to the cross-sectional area of the bellows or piston A_2 times the change in position:

$$v_2 = A_2 y$$

From Fig. 3.17*a*, the summation of forces acting on the bellows is

$$P_2 A_2 = K_2 Y$$

Similarly, from Fig. 3.17*b*, the summation of forces acting on the piston is

$$P_2 A_2 - P_1 A_1 = K_2 Y$$

Because $P_1 A_1$ is constant, linearization of either of the preceding gives the same result. That is,

$$y = \frac{A_2}{K_2} p_2$$

Substituting w from Eq. (3.32) into Eq. (3.31) and then using the preceding expressions to eliminate v_2 and y yield

$$p_2 = -\frac{K_1}{1 + \tau D} x \tag{3.33}$$

where $K_1 = C_2/(C_1 + C_3)$ and $\tau = (C_5 + A_2^2 C_4/K_2)/(C_1 + C_3)$. The minus sign indicates that as x increases, p_2 decreases. Because $y = (A_2/K_2)p_2$, it follows that

$$y = \frac{A_2}{K_2} \frac{-K_1}{1 + \tau D} x$$

The block-diagram representation is shown in Fig. 3.17*c*. For most flapper valves the time constant τ may generally be regarded as negligible.

A nondimensional family of curves (Fig. A.1) for determining the equilibrium chamber pressure P_2 is developed in App. A. Lines of constant area ratio A_2/A_1 are plotted, where A_1 is the area of the first orifice times its coefficient of discharge and A_2 is the area of the second orifice times its coefficient of discharge. The abscissa is the overall pressure ratio P_1/P_3, where P_1 is the supply pressure and P_3 is the discharge pressure, which is usually the ambient pressure P_a. The ordinate P_2/P_1 is the ratio of the chamber pressure to the supply pressure. If the area ratio A_2/A_1 and the overall pressure ratio P_1/P_3 are known, then the ratio P_2/P_1 may be determined. The desired chamber pressure is equal to this ratio (P_2/P_1) times the supply pressure P_1. A typical plot of P_2 versus X is shown in Fig. 3.18. Note that absolute pressure is used for pneumatic amplifiers, whereas gauge pressure is used for hydraulic amplifiers.

For steady-state operation, p_2 is constant. Therefore, $\tau Dp_2 = 0$ in Eq. (3.33). Thus, the equation for steady-state operation is

$$p_2 = -K_1 x \tag{3.34}$$

The constant $-K_1 = p_2/x$ is the slope of Fig. 3.18 at the reference point of operation. As shown in Fig. 3.18, this slope remains quite constant over a considerable portion of the operating range. Note that it is not necessary to evaluate C_1, C_2, and C_3 individually in order to obtain $K_1 = C_2/(C_1 + C_3)$.

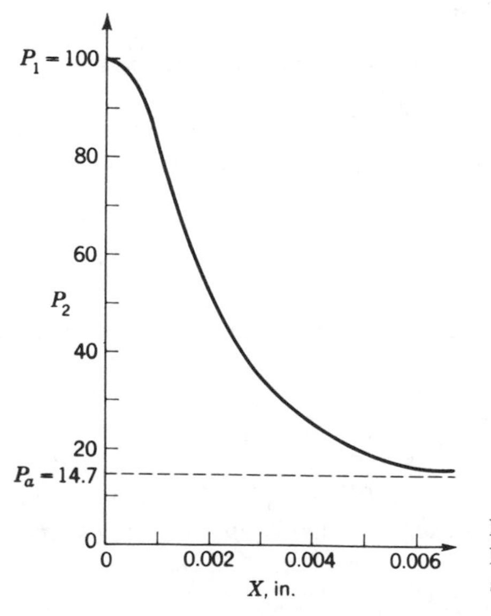

FIGURE 3.18
Equilibrium curve of P_2 versus X for a pneumatic amplifier.

3.5 DC MOTORS

A major reason for the use of dc machines in electromechanical control systems is the ease with which speed can be controlled. The polarity of the applied voltage determines the direction of rotation. Also, dc machines are capable of providing large power amplifications. An exploded view of a brushless dc servomotor is shown in Fig. 3.19.

The field and armature windings of dc motors may be shunt-connected, series-connected, compounded, or separately excited. The motors used in control systems are generally separately excited. There are two types of separate excitation: field control with fixed armature current and armature control with fixed field current.

Field Control

A separately excited motor in which the armature current I_a is maintained constant is shown in Fig. 3.20*a*. The constant current I_a may be supplied by a dc generator or from an ac line. The latter method requires the use of transformers and rectifiers to obtain the proper rectification. The voltage E_f applied to the field is obtained from the output of an amplifier in low-power applications or from a dc generator when

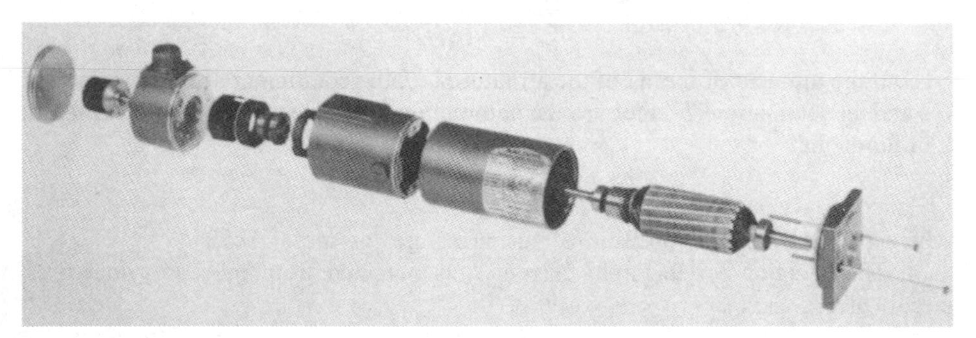

FIGURE 3.19
A brushless dc servomotor. (Courtesy Baldor.)

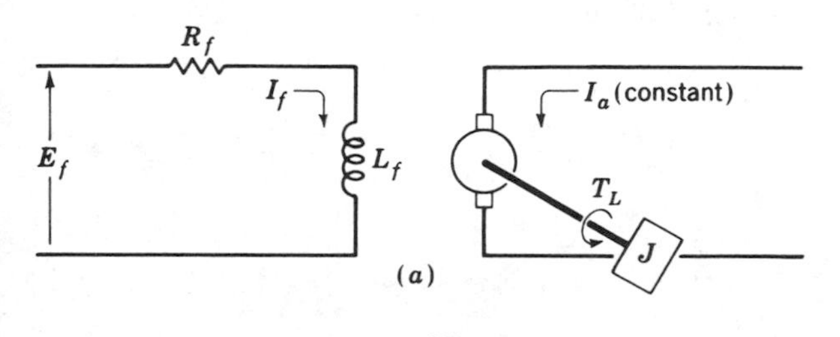

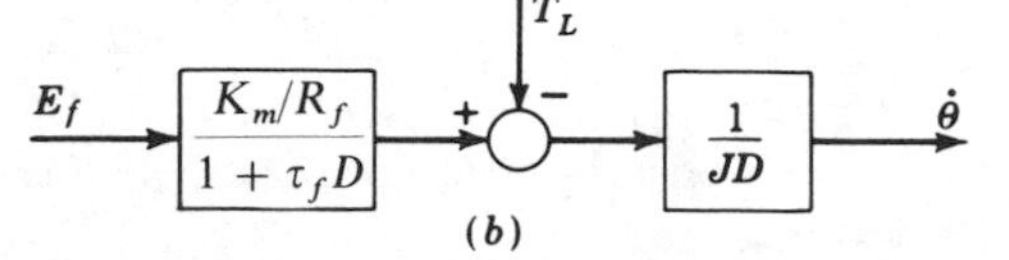

FIGURE 3.20
Field-controlled dc motor.

greater power is needed. In the field circuit, the resistance of the windings is R_f, and the inductance is designated by L_f. The torque T developed by a motor is proportional to the product of the armature current I_a and the magnetic flux ϕ of the field

$$T = K_1 \phi I_a \tag{3.35}$$

where K_1 is a constant for any motor and depends upon the total number of armature conductors, the number of field poles, etc.

A typical curve of flux ϕ versus field current I_f is shown in Fig. 3.21. When the field current I_f becomes great enough to cause the iron to saturate, the flux ϕ no longer increases linearly with the current. Motors used in control systems usually operate over the linear portion of this curve, in which case

$$\phi = K_2 I_f \tag{3.36}$$

where K_2 is the slope of the linear portion of the curve, as shown in Fig. 3.21. Substitution of the preceding result into Eq. (3.35) yields

$$T = K_1 K_2 I_a I_f = K_m I_f \tag{3.37}$$

where

$$K_m = K_1 K_2 I_a$$

If the moment of inertia of the armature is J, the coefficient of viscous friction B_v, and the load torque T_L, then from a summation of torques acting on the armature it follows that

$$T = (B_v D + JD^2)\theta + T_L \tag{3.38}$$

where θ is the angular position of the armature, or motor shaft.

The equation for the field current I_f is obtained from the equivalent field circuit of Fig. 3.20*a*

$$I_f = \frac{E_f}{R_f + L_f D} = \frac{E_f}{R_f(1 + \tau_f D)} \tag{3.39}$$

where $\tau_f = L_f/R_f$ is the time constant of the field circuit.

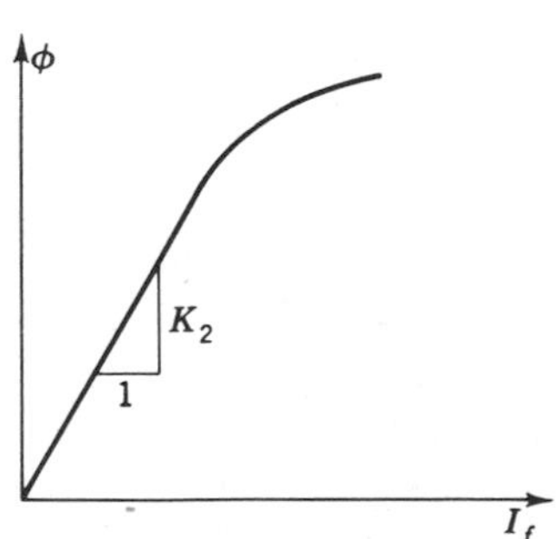

FIGURE 3.21
Plot of flux versus field current.

Substituting T from Eq. (3.38) and I_f from Eq. (3.39) into Eq. (3.37) and solving for θ give

$$\theta = \frac{1}{D(B_v + JD)}\left[\frac{(K_m/R_f)E_f}{1+\tau_f D} - T_L\right] \tag{3.40}$$

Multiplication of the preceding equation by D gives the angular velocity $\dot{\theta} = D\theta$:

$$\dot{\theta} = \frac{1}{B_v + JD}\left[\frac{(K_m/R_f)E_f}{1+\tau_f D} - T_L\right] \tag{3.41}$$

Generally, the damping B_v is negligible, so that the block-diagram representation for the speed of this field-controlled dc motor is as shown in Fig. 3.20*b*.

Armature Control

A dc motor with armature control is one in which the speed is controlled by the armature voltage E_a. An armature-controlled motor in which the field current I_f is kept constant is shown in Fig. 3.22*a*. The armature voltage E_a is usually supplied by a generator, which in turn may be supplied by an amplifier. The voltage E_b is the back emf induced by the rotation of the armature windings in the magnetic field. The back emf is proportional to the product of the armature speed $\dot{\theta}$ and the field strength ϕ. That is,

$$E_b = K_3\phi\dot{\theta} \tag{3.42}$$

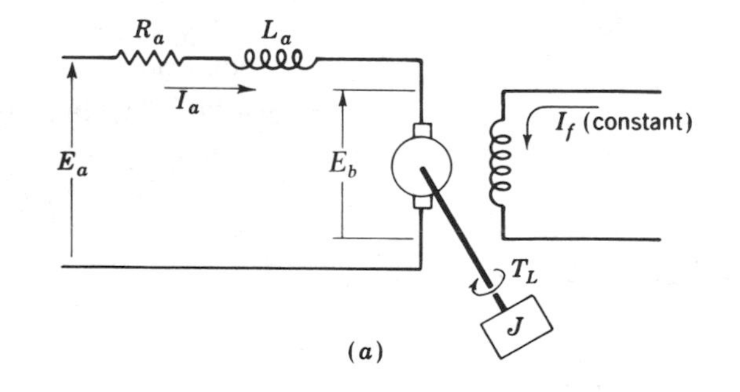

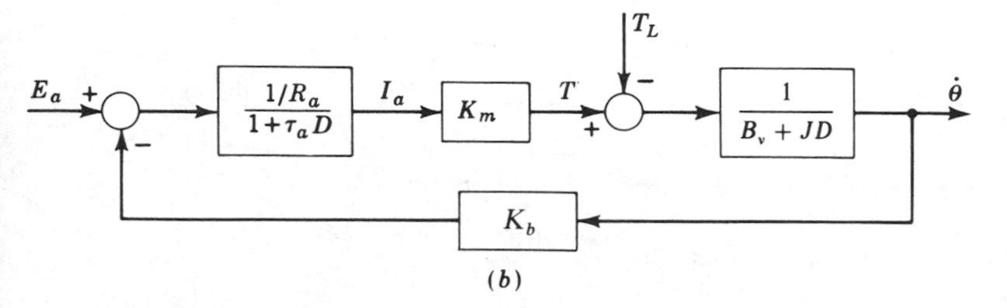

FIGURE 3.22
Armature-controlled dc motor.

where K_3 is a constant for any particular motor. Substitution of ϕ from Eq. (3.36) into the preceding equation gives

$$E_b = K_2 K_3 I_f \dot{\theta} = K_b \dot{\theta} \tag{3.43}$$

where $K_b = K_2 K_3 I_f$. The circuit equation for the armature portion of Fig. 3.22*a* is

$$E_a - K_b \dot{\theta} = R_a I_a + L_a D I_a = R_a (1 + \tau_a D) I_a \tag{3.44}$$

where $\tau_a = L_a/R_a$. The torque developed by the motor is given by

$$T = K_m I_a \tag{3.45}$$

and the torque balance for the output shaft is

$$T = (B_v D + J D^2)\theta + T_L = (B_v + JD)\dot{\theta} + T_L \tag{3.46}$$

The block-diagram representation for this armature-controlled motor is obtained by combining the block-diagram representations for Eqs. (3.44), (3.45), and (3.46) as shown in Fig. 3.22*b*. The *back emf*, which is also called the *counter emf*, is responsible for the feedback.

Usually the inductance L_a of an armature-controlled motor is negligible. Similarly, the coefficient of viscous friction B_v is ordinarily negligible. Thus, solving Eqs. (3.44), (3.45), and (3.46) for the output $\dot{\theta}$ as a function of the input E_a yields

$$\dot{\theta} = \frac{K}{1 + \tau D} E_a$$

where $K = 1/K_b$ and $\tau = R_a J/(K_b K_m)$. For any given motor R_a, K_b, and K_m are constants. Actually the quantity $R_a/(K_b K_m)$ does not vary appreciably with the size of the motor. The time constant τ depends primarily on J. For large inertial loads, values of τ can be 10 s. Choosing a larger motor is little help.

Figure 3.23*a* shows the block-diagram representation for this armature-controlled motor. Note that it is running as an open-loop system. The time

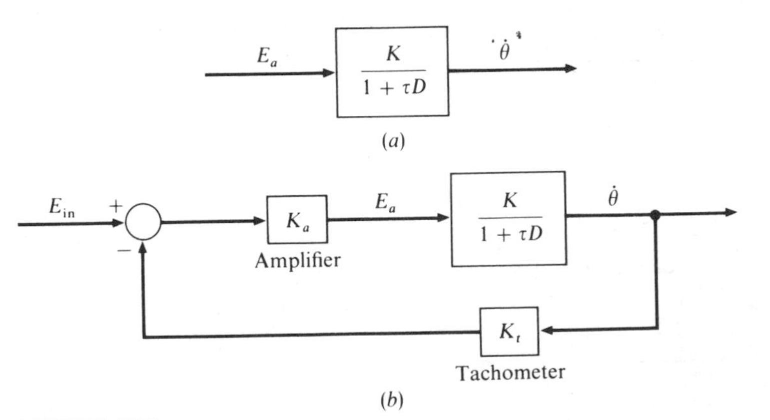

FIGURE 3.23
Block diagram representation for armature-controlled motor (*a*) without tachometer feedback, (*b*) with tachometer feedback.

constant can be significantly reduced by using a tachometer as shown in Fig. 3.23*b* to change this to a closed-loop system. The tachometer provides a feedback voltage $K_t\dot{\theta}$. A comparator subtracts this voltage from the reference input E_{in}. The resulting error signal is then amplified by K_a to yield the armature voltage E_a. The input-output relationship is

$$\dot{\theta} = \frac{KK_a/(1 + KK_aK_t)}{1 + [\tau/(1 + KK_aK_t)]D} E_{in}$$

The time constant for this closed-loop system is $\tau/(1 + KK_aK_t)$. Values of $1 + KK_aK_t$ of 100 are practical to obtain. Thus, with the addition of feedback, it is possible to obtain a rapid-responding system even with a very high inertial load.

A complete generator and armature-controlled motor combination is shown in Fig. 3.24. This is, in effect, a Ward-Leonard system, or motor-generator set. The voltage E supplied to the generator may be quite small, as in the case of that coming from an amplifier. The resistance of the field of the generator is R_{fg}, and the inductance is L_{fg}. The armature of the generator is driven at a constant speed by a prime mover. The output voltage of the generator E_g goes directly to the armature of the motor, so that $E_g = E_a$.

The circuit equation for the generator field is

$$E = (R_{fg} + L_{fg}D)I_{fg} \tag{3.47}$$

The voltage induced in the armature is the generated voltage $E_g = E_a$, which is

$$E_a = K_{cg}\dot{\theta}_g I_{fg} = K_c I_{fg} \tag{3.48}$$

where $\dot{\theta}_g$ is the angular velocity of the prime mover, which is constant, so that $K_{cg}\dot{\theta}_g = K_c$. The substitution of I_{fg} from Eq. (3.47) into Eq. (3.48) yields

$$E_a = \frac{K_c E}{R_{fg} + L_{fg}D} = \frac{(K_c/R_{fg})E}{1 + \tau_{fg}D} \tag{3.49}$$

where $\tau_{fg} = L_{fg}/R_{fg}$ is the time constant of the generator field.

The overall block-diagram relating the input voltage E and the velocity $\dot{\theta}$ of this armature-controlled motor-generator system is obtained by connecting the block-diagram representation for the output E_a from Eq. (3.49) to that for E_a in Fig. 3.22*b*.

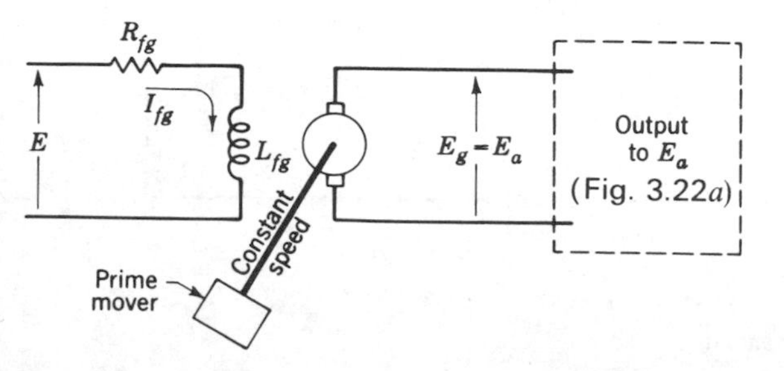

FIGURE 3.24
Complete generator and armature-controlled motor combination.

DC Tachometer

A dc tachometer is a generator in which the magnetic flux is usually supplied by a permanent magnet. Because the flux is kept constant, the equation of operation for the tachometer is obtained from Eq. (3.42) as follows:

$$E_c = E_g = K_3\phi\dot{\theta} = K_g\dot{\theta} \tag{3.50}$$

where $K_g = K_3\phi$ is a constant and E_g is the generated voltage.

Thus a tachometer supplies a voltage E_g which is proportional to the speed at which it is driven.

Remote-Control Positional Servomechanism

A remote-control positional servomechanism is shown in Fig. 3.25*a*. The wiper arm of the input potentiometer is positioned by the desired input position θ_r, so that the voltage E_r is proportional to θ_r (that is, $E_r = K_r\theta_r$). Similarly, the controlled shaft position θ_c determines the position of the wiper arm for the other potentiometer, so that $E_c = K_c\theta_c$. The error signal $E_e = E_r - E_c$ is amplified by the amplifier, and the resultant voltage is applied to the field of a field-controlled motor, so that $E_f = K_1E_e$. The operational representation of the differential equation for the motor is given by Eq. (3.40). The overall block diagram for this system is shown in Fig. 3.25*b*.

The motor must be located at the output shaft, while the input potentiometer is usually situated in any convenient location. A major advantage in using such electrical equipment for position control systems is the ease of connecting the input and output by wires.

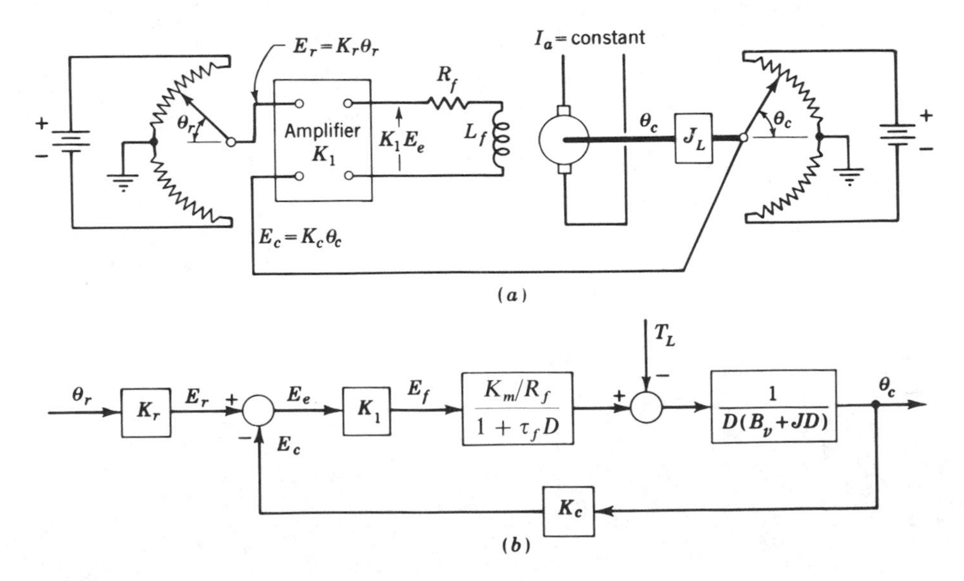

FIGURE 3.25
Remote-control positional servomechanism.

The preceding position controller may be converted to a speed control system by connecting the output shaft to a tachometer rather than to a potentiometer. In this case, the voltage signal E_c coming from the tachometer is proportional to the speed $\dot{\theta}_c$ (that is, $E_c = K_c\dot{\theta}_c$). Similarly, each wiper position of the input potentiometer corresponds to a desired speed setting $\dot{\theta}_r$, rather than position θ_r, so that the reference voltage is $E_r = K_r\dot{\theta}_r$.

3.6 AC MOTORS

An ac two-phase motor is used for simple low-power applications. One of the phases is supplied with a fixed ac voltage which acts as the reference voltage. The other phase is connected to the controlled voltage. A schematic representation of a two-phase motor is shown in Fig. 3.26*a*. Because the reference voltage E_R is constant, the speed depends upon the control voltage E. The direction of rotation is reversed by changing the polarity of the control voltage.

As shown in Fig. 3.26*a*, the reference and control windings are displaced by 90° in the stator of the motor. Thus, although the voltage applied to each winding has the same frequency, there is a 90° phase shift of one with respect to the other.

Typical performance curves relating the developed motor torque T and the angular velocity $\dot{\theta}$ for constant values of control current I are shown in Fig. 3.26*b*. The equation describing the operation of a two-phase motor about some equilibrium point of operation is derived as follows: From Fig. 3.26*b* note that the speed is a function of T and I:

$$\dot{\theta} = F(T, I)$$

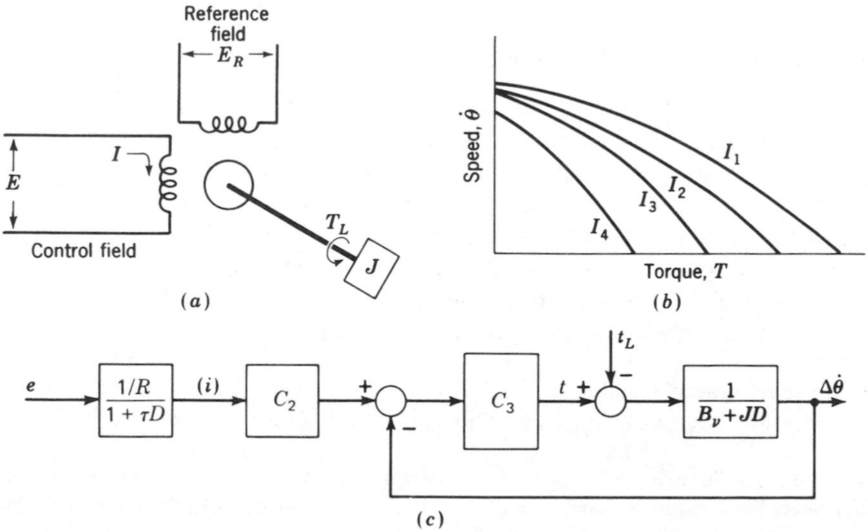

FIGURE 3.26
(*a*) Two-phase ac motor, (*b*) typical performance curves, (*c*) block-diagram representation.

Linearization about the reference point of operation gives

$$\Delta\dot{\theta} = \left.\frac{\partial\dot{\theta}}{\partial T}\right|_I t + \left.\frac{\partial\dot{\theta}}{\partial I}\right|_T i = -C_1 t + C_2 i \tag{3.51}$$

where $\Delta\dot{\theta}$ is the change in speed from the reference value, $C_1 = -\partial\dot{\theta}/\partial T|_I$, and $C_2 = \partial\dot{\theta}/\partial I|_T$. The term $\partial\dot{\theta}/\partial T|_I$ is the slope of the operating curves at the reference operating condition, which is negative. The minus sign converts C_1 to a positive constant. For a given torque, the speed increases as the current increases, so that $C_2 = \partial\dot{\theta}/\partial I|_T$ is positive. Solving Eq. (3.51) for t gives

$$t = \frac{1}{C_1}(C_2 i - \Delta\dot{\theta}) = C_3(C_2 i - \Delta\dot{\theta}) \tag{3.52}$$

where $C_3 = 1/C_1 = -\partial T/\partial\dot{\theta}|_I$. The torque balance for the armature is

$$T - T_L = (B_v + JD)\dot{\theta}$$

For small departures,

$$t - t_L = (B_v + JD)\Delta\dot{\theta} \tag{3.53}$$

The equation of operation for the control windings is

$$E = (R + LD)I = R(1 + \tau D)I$$

where $\tau = L/R$. For small departures, the preceding becomes

$$e = R(1 + \tau D)i \tag{3.54}$$

The block-diagram representation of this two-phase ac motor is obtained from Eqs. (3.52), (3.53), and (3.54), as shown in Fig. 3.26*c*.

3.7 BLOCK-DIAGRAM ALGEBRA

It is often desirable to rearrange the form of a block diagram. Figure 3.27 shows a number of rearrangements which are commonly employed. Note that in all cases the rearrangement does not affect the overall relationship between the input elements (i.e., elements with arrowheads pointing into the diagram) and the output elements (i.e., elements with arrowheads pointing away from the diagram). There are many possible rearrangements for systems. However, it is usually desirable to make the ultimate form of the block diagram the same as that shown in Fig. 3.28, which is the general form of the block-diagram representation for a control system.

Illustrative example 3.8. Use block-diagram algebra to put the block-diagram representation shown in Fig. 3.29*a* into the general form shown in Fig. 3.28.

Solution. The first step is to move the point where the upper feedback path (element H_1) enters the feedforward loop from the right side of element G_1 to the left side. This is accomplished by the application of the technique shown in Fig. 3.27*b*. As shown in Fig. 3.29*b*, the resulting upper feedback element is H_1/G_1. The next step is to move the point where the lower feedback path (element H_2) leaves the feedforward path from

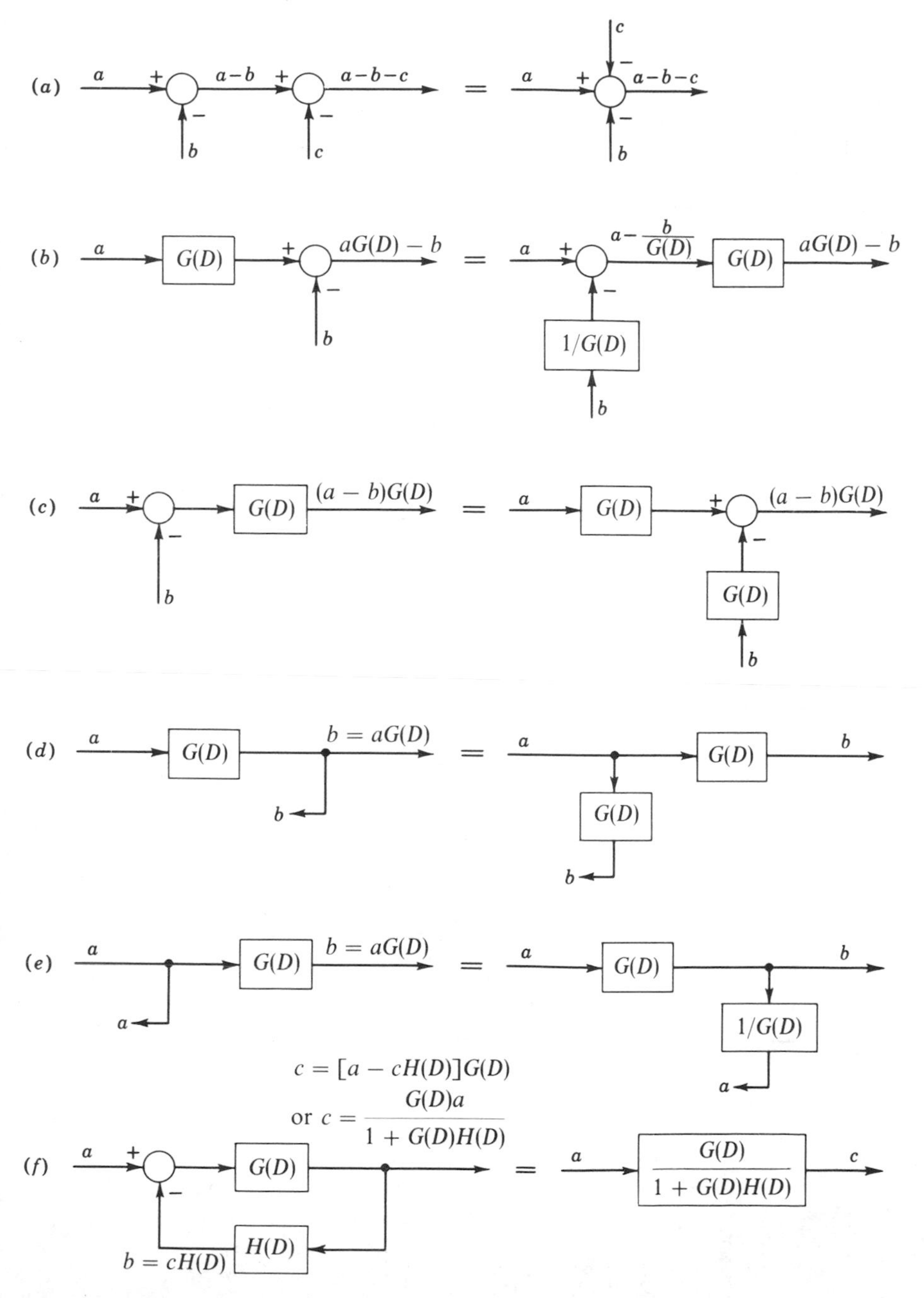

FIGURE 3.27
Equivalent block diagrams: (*a*) combining interconnected summing points, (*b*) moving a summing point behind an element, (*c*) moving a summing point ahead of an element, (*d*) moving a takeoff point behind an element, (*e*) moving a takeoff point ahead of an element, (*f*) eliminating a minor feedback loop.

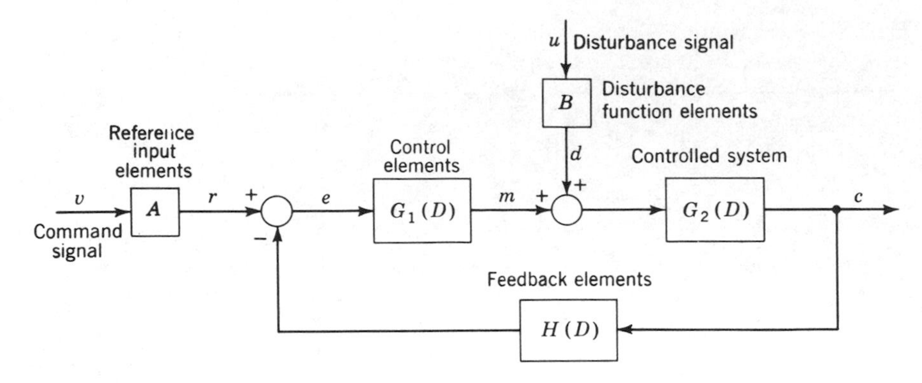

FIGURE 3.28
General block-diagram representation for a control system.

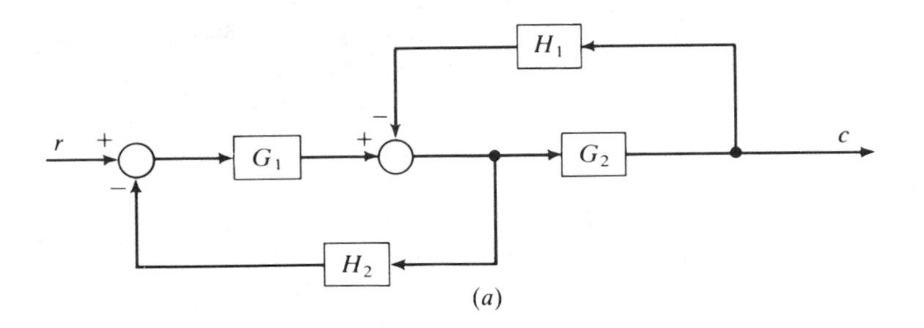

(*a*)

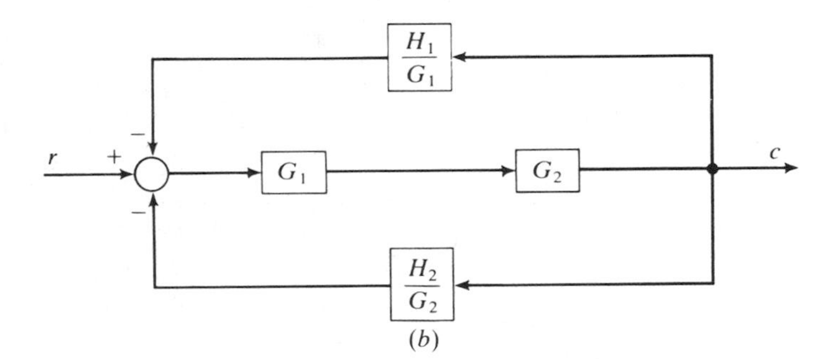

(*b*)

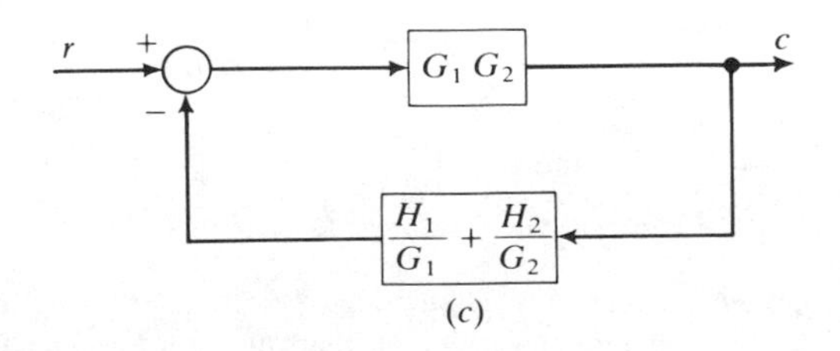

(*c*)

FIGURE 3.29
Application of block-diagram algebra.

the left side of element G_2 to the right side. This is accomplished by application of the technique shown in Fig. 3.27*e*. The resulting lower feedback element is H_2/G_2. The upper and lower feedback paths of Fig. 3.29*b* may now be combined into a single feedback path, as is illustrated by the block diagram shown in Fig. 3.29*c* which has the same form as Fig. 3.28.

3.8 SPEED CONTROL SYSTEMS

A picture of the engine control system for the Boeing 747 airplane (see Fig. 1.7) is shown in Fig. 3.30. All engines derive their power from some source of energy, i.e., fuel. The power developed by an engine is controlled by controlling the rate of flow

FIGURE 3.30
Engine control system. (Courtesy Allied-Signal Inc., Controls and Accessories.)

of fuel to the engine. When the power developed by an engine exceeds the power required (load on the engine), the excess power accelerates the engine, thus resulting in an increased speed. Conversely, when the load is greater than the power developed, a decrease in speed results. A speed governor automatically controls the speed of an engine.

Figure 3.31*a* shows a speed governor. The flyweights are pin-connected to a rotating ballhead. This ballhead is connected by gears to the engine so that the speed of rotation of the flyweights is equal to a constant times the engine speed. Because of the bearing located between the toes of the flyweights and the spring, the spring does not rotate. Figure 3.31*b* shows a schematic representation of the governor shown in Fig. 3.31*a*.

Figure 3.32 shows a typical speed control system for gas turbines, steam turbines, or diesel engines. The position of the throttle lever sets the desired speed of the engine. The speed control is drawn in some reference operating position so that

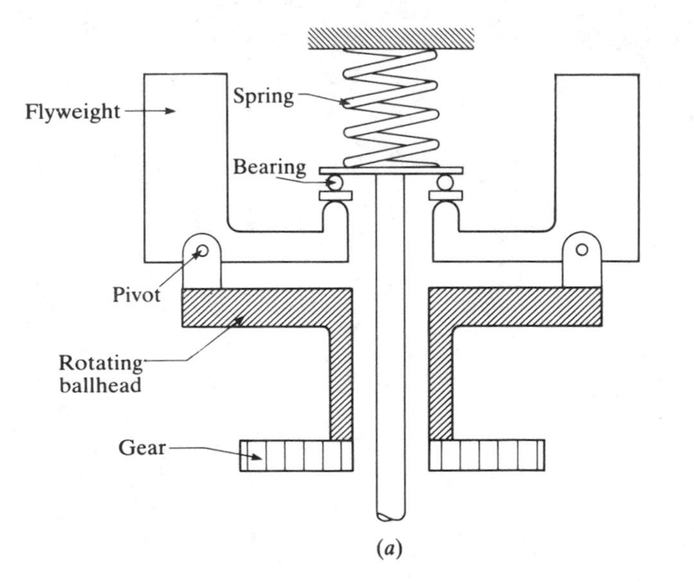

(*a*)

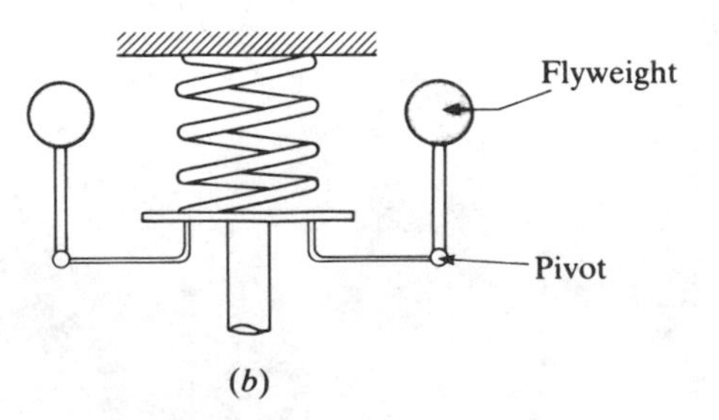

(*b*)

FIGURE 3.31
(*a*) Flyball governor, (*b*) schematic representation.

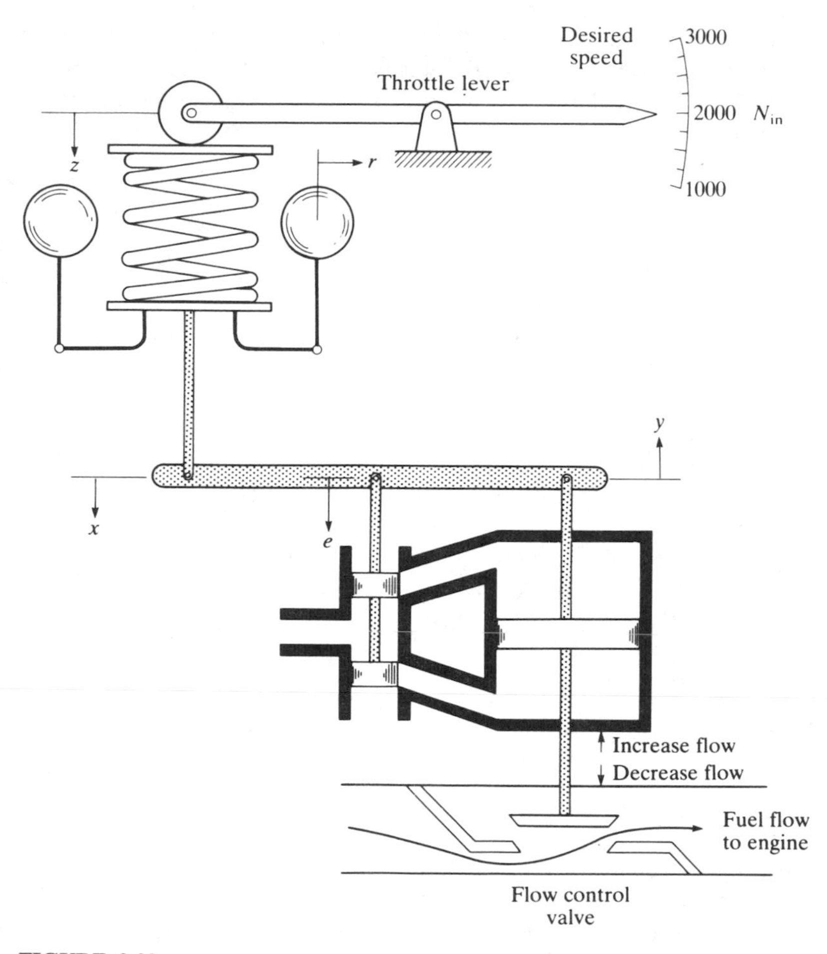

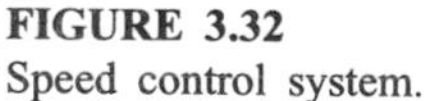
FIGURE 3.32
Speed control system.

the values of all the lowercase parameters are zero. The positive direction of motion of these parameters is indicated by the arrowhead on each.

If the speed of the engine were to drop below its reference value, then the centrifugal force of the flyweights would decrease, thus decreasing the force exerted on the bottom of the spring. This causes x to move downward, which in turn moves e downward. Fluid then flows to the bottom of the big piston to increase y and thus opens the flow control valve wider. When more fuel is supplied, the speed of the engine increases until equilibrium is again reached. For steam turbines, the flow control valve controls the flow of steam rather than fuel, as is the case with gas turbines and diesels.

Suppose that the throttle lever is moved to a higher speed setting, which in turn causes z to move downward. This in turn causes x to move downward. As just discussed, moving x downward opens the fuel flow valve, which increases the speed.

The overall block-diagram representation for this system is obtained as follows. The position Z of the top of the spring is a function of the desired speed only [that is, $Z = Z(N_{in})$]. Thus, the variation of the top of the spring z from its reference position is

$$z = \left.\frac{\partial Z}{\partial N_{in}}\right|_i n_{in} = C_2 n_{in} \tag{3.55}$$

where $n_{in} = N_{in} - N_i$ is the change in desired speed and $C_2 = \partial Z/\partial N_{in}|_i$ is the change in position Z per change in desired speed N_{in}.

Figure 3.33 shows the free-body diagram of the right flyweight of Fig. 3.31*b*. The centrifugal force F_c acting on the flyweight is

$$F_c = MR\omega^2$$

where M = mass of flyweight
$R = R_i + r$ = distance from center of rotation to center of mass of flyweight
ω = angular velocity of flyweight

Usually, the governor is geared directly to the output shaft such that ω is equal to the gear ratio times the output speed, i.e.,

$$\omega = C_g N_o$$

where C_g is the gear ratio. Substitution of this value for ω into the preceding expression for F_c gives

$$F_c = C_g^2 MRN_o^2 \tag{3.56}$$

Because there are two flyweights, half of the spring force $F_s/2$ acts on the toe of each flyweight. The perpendicular distance from the pivot to the line of action of the force $F_s/2$ acting on the toe is a. The centrifugal force F_c acts radially outward through the center of mass of the flyweight. The perpendicular distance from the

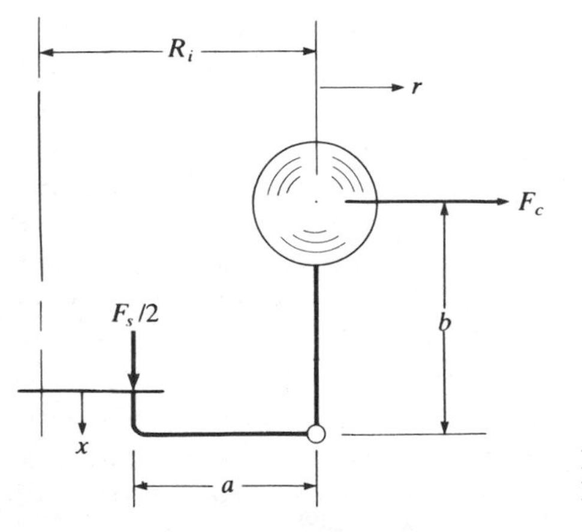

FIGURE 3.33
Free-body diagram of flyweight.

pivot to the line of action of F_c is b. Taking the moment of forces about the pivot point shows that

$$aF_s/2 = bF_c$$

or

$$F_s = 2\frac{b}{a}F_c = 2\frac{b}{a}C_g^2MRN_o^2 = C_fC_rRN_o^2 \tag{3.57}$$

where $C_f = 2C_g^2M$ and $C_r = b/a$. The two independent variables are R and N_o. Linearization gives

$$f_s = C_3r + C_4n_o \tag{3.58}$$

where

$$C_3 = \frac{\partial F_s}{\partial R}\bigg|_i = C_fC_rN_i^2$$

and

$$C_4 = \frac{\partial F_s}{\partial N_o}\bigg|_i = 2C_fC_rR_iN_i$$

The compression of the spring from its reference length is $z - x$. Thus, the variation in force exerted by the spring is

$$f_s = K_s(z - x) \tag{3.59}$$

where K_s is the spring constant. Setting Eqs. (3.58) and (3.59) equal, we have

$$K_s(z - x) = C_3r + C_4n_o$$

The geometry of Fig. 3.33 shows that the motions of r and x are related so that $r = -(b/a)x = -C_rx$. The reason for the minus sign is that as r increases, x decreases. Eliminating r from the preceding equation yields

$$K_sz - K_sx = -C_rC_3x + C_4n_o$$

or

$$x = \frac{K_sz - C_4n_o}{K_s - C_rC_3} \tag{3.60}$$

The block-diagram representation for Eqs. (3.55) and (3.60) is shown in Fig. 3.34, which is the comparator for the speed control system.

The operation of the servomotor was discussed in Sec. 3.3, and the block diagram was given in Fig. 3.12*c*.

The flow through the flow control valve is a function of position Y [that is, $Q = Q(Y)$]. Linearization gives

$$q = \frac{\partial Q}{\partial Y}\bigg|_i y = C_5y \tag{3.61}$$

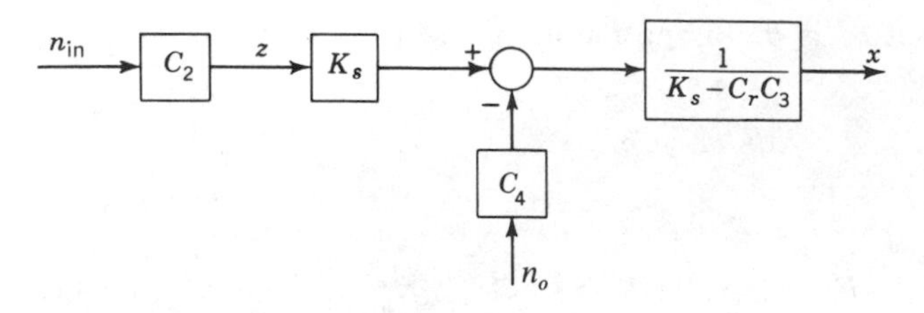

FIGURE 3.34
Block diagram for comparator.

where C_5 is the slope of the curve of Q versus Y evaluated at the reference position.

The speed N_o of an engine is a function of the fuel flow Q supplied to the engine and the engine torque T:

$$N_o = N_o(Q, T)$$

The linearized form of this expression is

$$n_o = \left.\frac{\partial N_o}{\partial Q}\right|_T q + \left.\frac{\partial N_o}{\partial T}\right|_Q t = C_6 q - C_7 t \tag{3.62}$$

where $C_6 = \partial N_o/\partial Q|_T$ is the change in speed per change in fuel flow with T maintained at the reference value and $C_7 = -\partial N_o/\partial T|_Q$ is the change in speed per change in torque with Q maintained constant at the reference value. For a constant T, the speed N_o increases as Q increases, so that $\partial N_o/\partial Q|_T$ is a positive number. For a given Q, the speed decreases as the torque increases, so that $\partial N_o/\partial T|_Q$ is a negative number. The minus sign $[C_7 = (-\partial N_o/\partial T)|_Q]$ converts C_7 to a positive constant. For convenience in analyzing control systems, it is desirable that all constants be positive numbers. The values of the preceding partial derivatives are obtained from the curve of operating characteristics of the particular engine under consideration. Typical operating characteristics for an engine are shown in Fig. 3.7.

The torque t produced by the engine minus the load torque t_L is the net torque available for acceleration. That is,

$$t - t_L = J\alpha = JDn_o \tag{3.63}$$

where $\alpha = dn_o/dt = Dn_o$ is the angular acceleration of the engine and J is the rotational inertia of the engine. Substituting t from Eq. (3.63) into Eq. (3.62) gives

$$n_o = C_6 q - C_7 t_L - C_7 JDn_o$$

or

$$n_o = \frac{C_6 q - C_7 t_L}{1 + C_7 JD} = \frac{C_6}{1 + \tau_2 D}(q - C_8 t_L) \tag{3.64}$$

where

$$\tau_2 = C_7 J \qquad \text{and} \qquad C_8 = \frac{C_7}{C_6} = \frac{-\partial N_o/\partial T|_i}{\partial N_o/\partial Q|_i} = \left.\frac{\partial Q}{\partial T}\right|_i$$

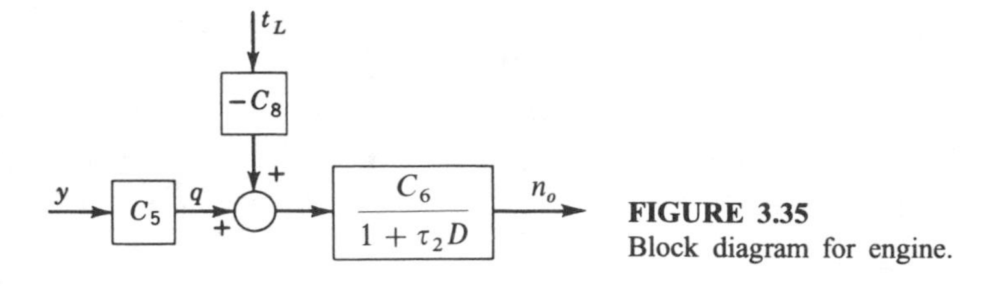

FIGURE 3.35
Block diagram for engine.

By writing $N_o = N_o(Q, T)$ in the implicit form $G(Q, T, N_o) = 0$, the product of the partial derivatives is

$$\frac{\partial Q}{\partial T}\frac{\partial T}{\partial N_o}\frac{\partial N_o}{\partial Q} = -1 \qquad \text{or} \qquad \frac{-\partial N_o/\partial T}{\partial N_o/\partial Q} = \frac{\partial Q}{\partial T} \tag{3.65}$$

The block-diagram representation for Eqs. (3.61) and (3.64) is shown in Fig. 3.35.

The overall block-diagram representation for this speed control system is obtained by combining Figs. 3.34, 3.12*c*, and 3.35, as shown in Fig. 3.36, in which $\tau_1 = 2A/C$.

By letting $K_1 = C_5/(K_s - C_r C_3)$ and eliminating the minor feedback loop, Fig. 3.36 may be represented as shown in Fig. 3.37. The operational form of the differential equation relating the output n_o to the input n_{in} and external disturbance t_L for the speed control system represented by the block diagram of Fig. 3.37 is obtained as follows: Subtract the feedback signal $C_4 n_o$ from the reference input $C_2 K_s n_{in}$, and perform the mathematical operations indicated by the feedforward portion of the block diagram until the output n_o is obtained. That is,

$$\left[(C_2 K_s n_{in} - C_4 n_o)\frac{K_1}{1+\tau_1 D} - C_8 t_L\right]\frac{C_6}{1+\tau_2 D} = n_o \tag{3.66}$$

Solving for n_o yields

$$n_o = \frac{C_2 C_6 K_1 K_s n_{in} - C_6 C_8 (1+\tau_1 D) t_L}{(1+\tau_1 D)(1+\tau_2 D) + C_4 C_6 K_1} \tag{3.67}$$

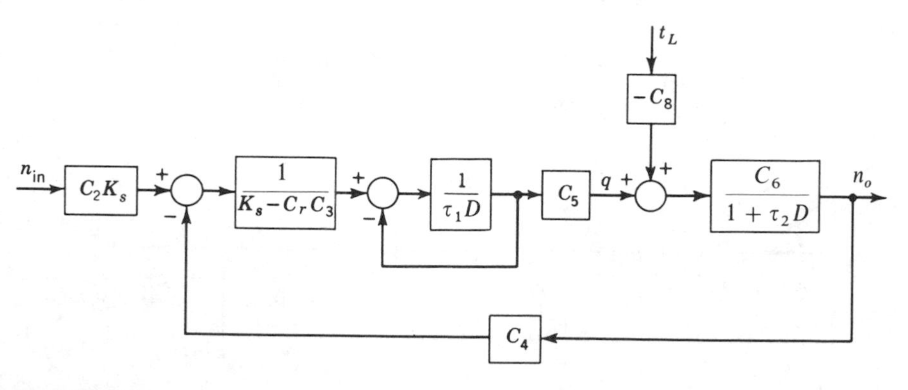

FIGURE 3.36
Combined block diagram.

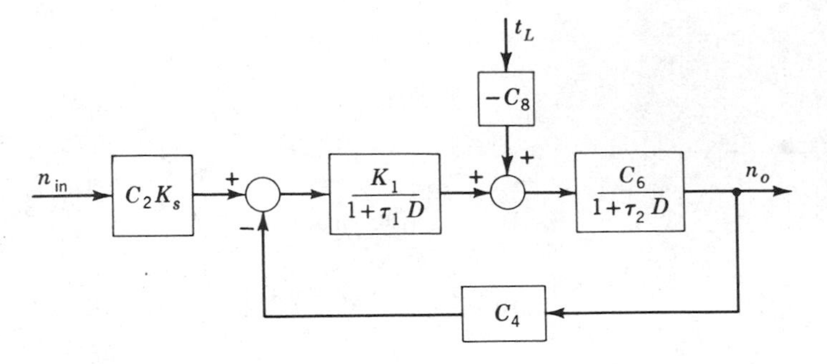

FIGURE 3.37
Overall block diagram for speed control system.

In determining the dynamic behavior of a system, one is interested in the variation of the system parameters from some reference condition. This is the type of information that is available from the block-diagram representation shown in Fig. 3.37. If absolute values are desired, it is an easy matter to convert from n_{in} to N_{in} or from n_o to N_o by merely adding the reference value.

In Fig. 3.38 is shown the block-diagram representation for a typical feedback control system. The overall equation of operation is

$$\{[r(t) - H(D)c(t)]G_1(D) + d(t)\}G_2(D) = c(t)$$

Solving for the output $c(t)$ gives

$$c(t) = \frac{G_1(D)G_2(D)r(t) + G_2(D)\,d(t)}{1 + G_1(D)G_2(D)H(D)} \tag{3.68}$$

The operators $G_1(D)$, $G_2(D)$, and $H(D)$ may be written in the form

$$G_1(D) = \frac{N_{G_1}}{D_{G_1}} \qquad G_2(D) = \frac{N_{G_2}}{D_{G_2}} \qquad H(D) = \frac{N_H}{D_H} \tag{3.69}$$

where N_{G_1} is the numerator of G_1 and D_{G_1} is the denominator of G_1, etc. Substitution of $G_1(D)$, $G_2(D)$, and $H(D)$ from Eq. (3.69) into Eq. (3.68) yields the general form

$$c(t) = \frac{N_{G_1}N_{G_2}D_H r(t) + N_{G_2}D_H D_{G_1}\,d(t)}{D_{G_1}D_{G_2}D_H + N_{G_1}N_{G_2}N_H} \tag{3.70}$$

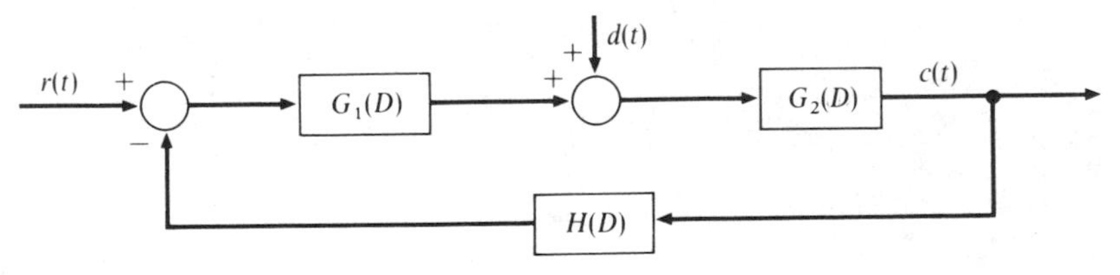

FIGURE 3.38
Typical feedback control system.

Note that the coefficient $N_{G_1}N_{G_2}D_H$ of $r(t)$ is the product of the numerator terms $N_{G_1}N_{G_2}$ from $r(t)$ to the output $c(t)$ and the denominator term D_H from the output $c(t)$ back to $r(t)$. Similarly, the coefficient $N_{G_2}D_HD_{G_1}$ of $d(t)$ is the product of the numerator term N_{G_2} from $d(t)$ to the output $c(t)$ and the denominator terms $D_HD_{G_1}$ from the output $c(t)$ back to $d(t)$. The denominator for Eq. (3.70) is seen to be the product of all the denominator terms in the loop $(D_{G_1}D_{G_2}D_H)$ plus the product of all the numerator terms $(N_{G_1}N_{G_2}N_H)$. In general, the coefficient for a signal entering any place in a loop is equal to the product of the numerator terms from the point where the signal enters the loop to the output times the product of the denominator terms from the output back to the point where the signal enters the loop. The denominator of the differential equation is always the product of all the denominator terms in the loop plus the product of all the numerator terms.

The form given by Eq. (3.70) saves considerable time and effort in obtaining the differential equation of operation for a control system. For example, from Fig. 3.37, note that $N_{G_1} = K_1$, $N_{G_2} = C_6$, $N_H = C_4$, $D_{G_1} = 1 + \tau_1 D$, $D_{G_2} = 1 + \tau_2 D$, $D_H = 1$, $r(t) = C_2K_sn_{\text{in}}$, $d(t) = -C_8t_L$, and $c(t) = n_o$. Substitution of these values into Eq. (3.70) yields directly Eq. (3.67).

3.9 GENERALIZED FEEDBACK CONTROL SYSTEM

A generalized representation for a feedback control system is shown in Fig. 3.39. Note that the command signal, or desired input, does not usually go directly to the comparator but must be converted to a suitable input for this device. Similarly, the controlled variable, or output, in the general case must also be changed by the feedback elements $H(D)$ before it can be measured by the comparator. The actuating signal e is amplified by the control elements $G_1(D)$ before it enters the system $G_2(D)$ being controlled. An external disturbance, as shown in Fig. 3.39, is a disturbance

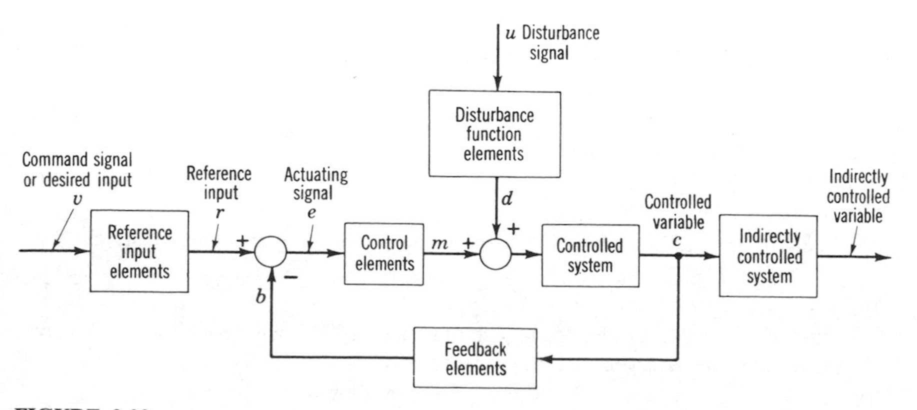

FIGURE 3.39
Generalized feedback control system.

which acts independently to affect the operation of the system. Although in Fig. 3.39 the external disturbance is shown entering the system between the control elements and the controlled system, in general, the external disturbance may enter the system at any point.

Also notice from this generalized representation of a control system that the controlled variable is not necessarily the quantity which it is desired to control. For example, a household thermostat controls the temperature of the air around the thermostat, and depending upon the circulation of air in the house, the temperatures of other areas may vary considerably. In addition, the idealized purpose of this control is to maintain the comfort of the persons of the household which depends upon humidity, their clothing, their amount of physical activity, etc. Thus, it is apparent that the controlled variable is not necessarily the ultimate quantity which it is desired to control.

PROBLEMS

3.1. Determine the linear approximation for each of the following functions.
(*a*) $Y = \sqrt{X}$
(*b*) $Y = 1\sqrt{X}$
For $X_i = 100$, use this approximation to evaluate Y when $X = 98$.

3.2. Determine the linear approximation for each of the following functions.
(*a*) $Y = \sin X$
(*b*) $Y = \tan X$
For $X_i = 30°$, use this approximation to evaluate Y when $X = 33°$.

3.3. Effect a linear approximation for $Y = X^{1/5}$. For $X_i = 32$, what is the approximate value of $\sqrt[5]{30}$?

3.4. Effect a linear approximation for $Y = X^{2/3}$. For $X_i = 1000$, what is the approximate value of $(997)^{2/3}$?

3.5. Effect a linear approximation for $Y = \log X$. For $X_i = 100$, what is the approximate value of log 101?

3.6. Determine the linear approximation for the equation $Z = \sin X \cos Y$. For $X_i = 60°$ and $Y_i = 30°$, what is the approximate value of Z when $X = 63°$ and $Y = 28°$?

3.7. Effect a linear approximation for the function $Z = X^2 \log Y$. For $X_i = 5$ and $Y_i = 100$, what is the approximate value of Z when $X = 5.1$ and $Y = 98$?

3.8. The volume V of a sphere is $V = \frac{4}{3}\pi R^3$. Determine the equation for the linear approximation for V.
(*a*) If $R_i = 10$, what percentage of error results from using this approximation for V when $R = 11$?
(*b*) What is the percentage of error for $R = 9$?
(*c*) The change in the volume due to a change in R is $S\,\Delta R$ where S is the surface area. Determine the equation for the surface area S.

3.9. The equation for the area of the parallelogram shown in Fig. P3.9 is

$$A = WL \sin\theta$$

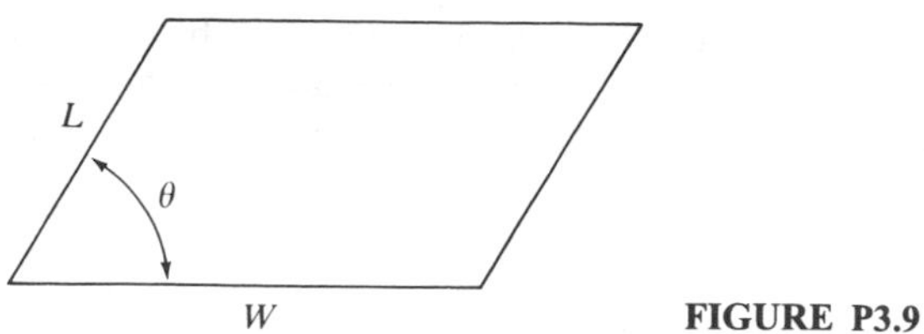

FIGURE P3.9

Determine the linear approximation for the area A. For $W_i = 12$, $L_i = 8$, and $\theta_i = 60°$, what is the approximate area when $W = 14$, $L = 9$, and $\theta = 58°$?

3.10. The equation for the area of the segment of a circle, as illustrated in Fig. P3.10, is

$$A = \tfrac{1}{2}R^2(\theta - \sin\theta)$$

Determine the linear approximation for the area A. The reference values are $R_i = 10$ and $\theta_i = 60°$. What is the approximate value of A when $R = 11$ and $\theta = 62°$?

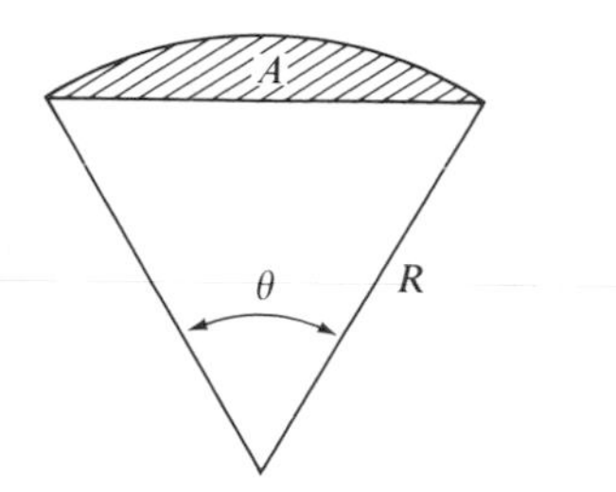

FIGURE P3.10

3.11. The equation for the volume of a cylinder is $V = \pi R^2 H$, where R is the radius of the base and H is the altitude. Determine the equation for the variation v in the volume due to a variation r in the radius and a variation h in the altitude.

3.12. The period of oscillation of a simple pendulum is

$$T = 2\pi\sqrt{\frac{L}{g}}$$

For $L_i = 100$ in and $g = 32.2$ ft/s^2, determine the change in the period due to (a) an increase in L of 1 in and (b) a decrease in g of 0.1 ft/s^2.

3.13. Linearize the equation

$$V = \frac{D}{T}$$

where V is velocity, D is distance, and T is time.

(a) Determine the linear approximation for v due to a change d in the distance and a change t in the time. Express this variation v in terms of the reference value V_i.

(b) With the speedometer indicating 60 mi/h, it is observed that 62 s is required to travel between successive mile indicator markers on the interstate highway. Approximately what is the velocity of the vehicle?

3.14. To linearize $V = D/T$ of Prob. 3.13 by the perturbation method, replace V by $V_i + v$, D by $D_i + d$, and T by $T_i + t$. Neglect higher-order variations (differential quantities), and then subtract the reference condition $V_i T_i = D_i$. Thus, verify the linearized form obtained in Prob. 3.13.

3.15. For subsonic flow of air through a restriction, the mass rate of flow is

$$M = 1.05A\sqrt{(P_1 - P_2)P_2/T}$$

The area of the restriction and the temperature T are constant. Determine the linear approximation for m due to a change $p_1 - p_2$ in the pressure drop across the restriction and due to a change p_2 in the downstream pressure. Express this variation m in terms of the reference value M_i.

3.16. Application of the law of cosines to the triangle shown in Fig. P3.16 shows that

$$C^2 = A^2 + B^2 - 2AB \cos \alpha$$

Effect a linear approximation for the change c in the length of side C. For $A_i = 8$, $B_i = 12$, and $\alpha_i = 120°$, determine the change c when $A = 9$, $B = 14$, and $\alpha = 123°$.

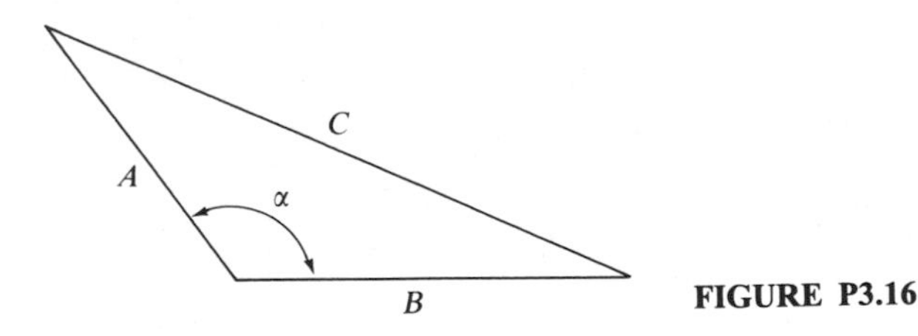

FIGURE P3.16

3.17. A chord of length L subtended by angle θ is illustrated in Fig. P3.17. The equation for the length is

$$L = 2R \sin \frac{\theta}{2}$$

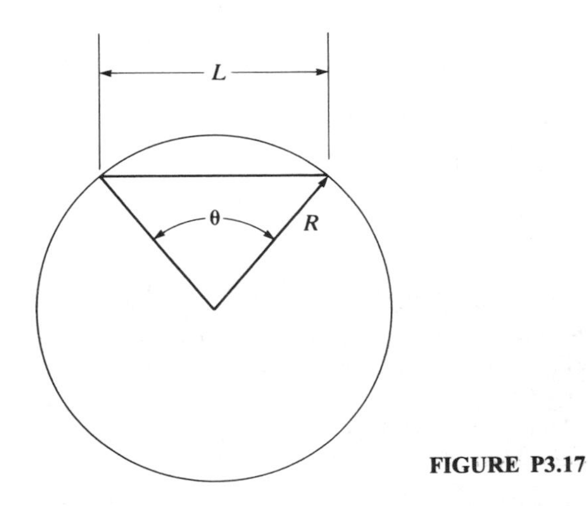

FIGURE P3.17

where R is the radius. Determine the equation for the variation l in the length due to a change r in the radius and change $\Delta\theta$ of the angle. For $R_i = 10$ and $\theta_i = 60°$, what is the approximate value of L when $R = 12$ and $\theta = 63°$?

3.18. The volume of a torus ("doughnut" shape, Fig. P3.18) is $V = 2\pi^2 R r^2$.

(*a*) Determine the equation for the variation ΔV in the volume due to a change ΔR in R and a change Δr in r.

(*b*) The change in volume due to a change in Δr only is $S\,\Delta r$, where S is the surface area. Determine the equation for the surface area S.

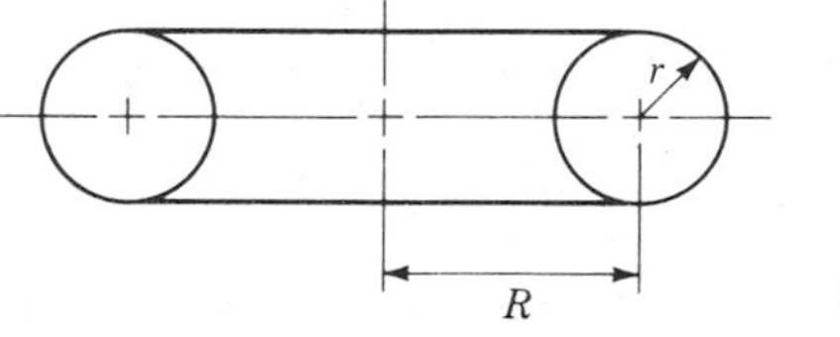

FIGURE P3.18

3.19. The error incurred in measuring the diameter D of a circle is d. Effect a linear approximation for the variation a of area A of the circle. How accurately must the diameter be measured so that the area will be correct to 1 percent?

3.20. The firing of a projectile on an inclined surface is illustrated in Fig. P3.20. The angle of inclination of the incline is β. The initial velocity of the projectile V is at an angle α from the horizontal. The equation for the range R is

$$R = \frac{V^2}{g\cos^2\beta}[\sin(2\alpha - \beta) - \sin\beta]$$

Effect a linear approximation for the change in range r due to a variation v in the initial velocity and a variation $\Delta\alpha$ of angle α. For $V_i = 100$ ft/s, $\alpha_i = 75°$, $\beta = 30°$, and $g = 32.2$ ft/s^2, what is the approximate value of R when $V = 110$ ft/s and $\alpha = 70°$?

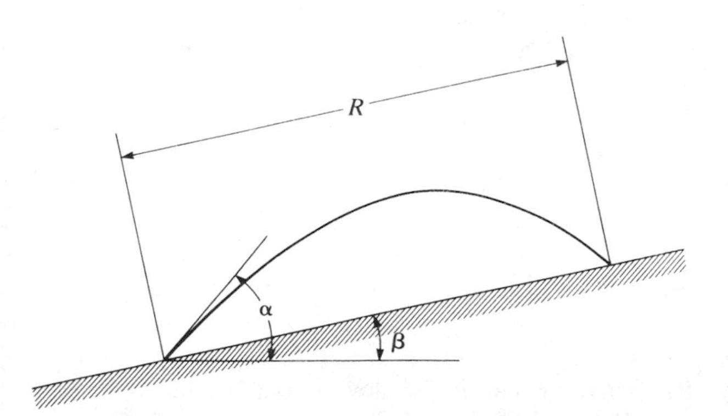

FIGURE P3.20

3.21. A certain motion is governed by the differential equation $\ddot{R} - R^{-3} + R^{-2} = 0$. Assuming that the variation of R from its reference value of unity remains small, linearize the differential equation. That is, determine the linear differential equation which describes the variation r of R.

3.22. The linearized equation of operation about a typical operating point for a hydraulic amplifier used in a machine tool is

$$f = 20{,}000e - 100\dot{y}$$

where $\dot{y} = Dy$ = velocity of power piston
f = change in force acting on piston
e = position of valve which controls piston

The load is due to the tool reactive force f_L and inertial force MD^2y; hence

$$f = f_L + MD^2y$$

The reference position x and controlled position y are connected by a walking-beam linkage such that

$$e = \frac{x - y}{2}$$

For $M = 1$, obtain the overall block-diagram representation for this system in which x is the reference input, y the controlled variable, and f_L the external disturbance.

3.23. Typical operating curves for a dc motor are shown in Fig. P3.23. These are curves of torque T versus operating speed N for constant values of voltage E applied to the motor. These curves are a plot of the function $N = N(T, E)$. Effect a linear approximation for N. Evaluate the partial derivatives in this approximation when T_i is 2 in · lb and E_i is 20 V.

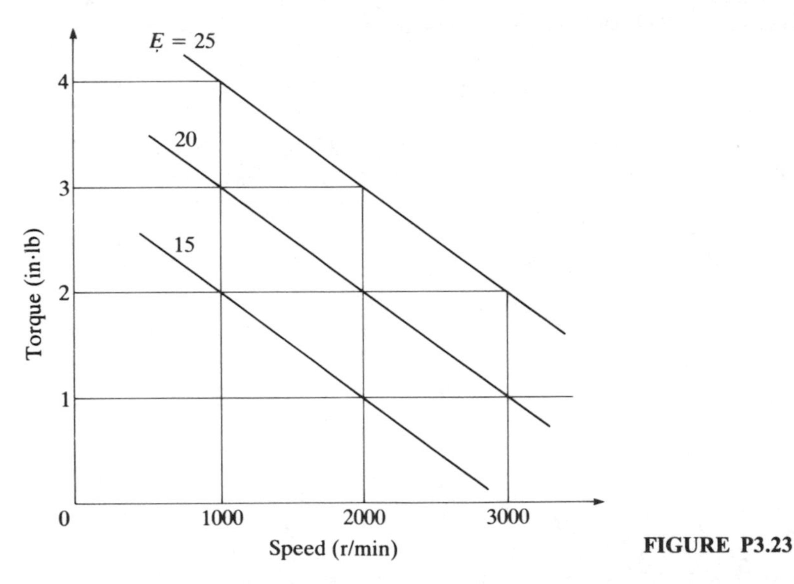

FIGURE P3.23

3.24. For the tank shown in Fig. P3.24, Q_{in} is constant. The flow out is a function of the head $Q_o = Q_o(H)$, where $H = H_i + h$. The equilibrium level is such that $Q_{in} = Q_o(H_i)$. The linear approximation for the flow out is $Q_o = Q_o(H_i) + (\partial Q_o/\partial H)_i h = Q_o(H_i) + Ch$, where $C = (\partial Q_o/\partial H)_i$. Show that the differential equation of operation for this tank is $dh/h = -(C/A)\, dt = -dt/\tau$, where $\tau = A/C$. Integrate both sides of this differential equation, and show that $h = h_0 e^{-t/\tau}$, where h_0 is the value of h at time $t = 0$. Note that as time approaches infinity, the level returns to its equilibrium value $h = 0$.

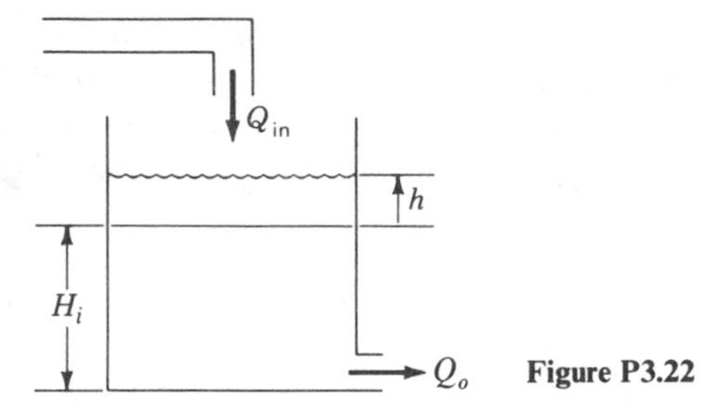

Figure P3.22

3.25. Figure P3.25 shows a hydraulic servomotor which is similar to the power-amplifying device used in power steering units. A movement in the x direction of the valve is seen to open passage 1 to the supply pressure, which in turn causes the big piston to move to the right. Because the sleeve is directly connected to this piston, the sleeve also moves to the right to close off flow from the valve. Construct the block diagram for this system. Determine the differential equation relating the input position x to the output y. Identify the time constant.

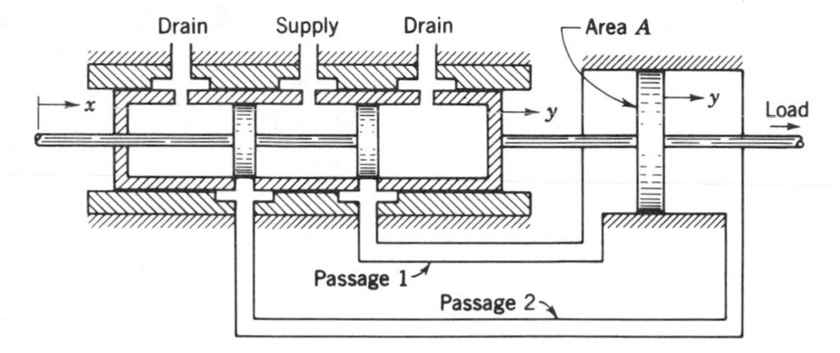

FIGURE P3.25

3.26. In Fig. P3.26 is shown a modification of the hydraulic power amplifier discussed in Prob. 3.25. Construct the block diagram for this system. Determine the differential equation relating input x to output y. Note that the position of the sleeve is $[a/(a + b)]y$.

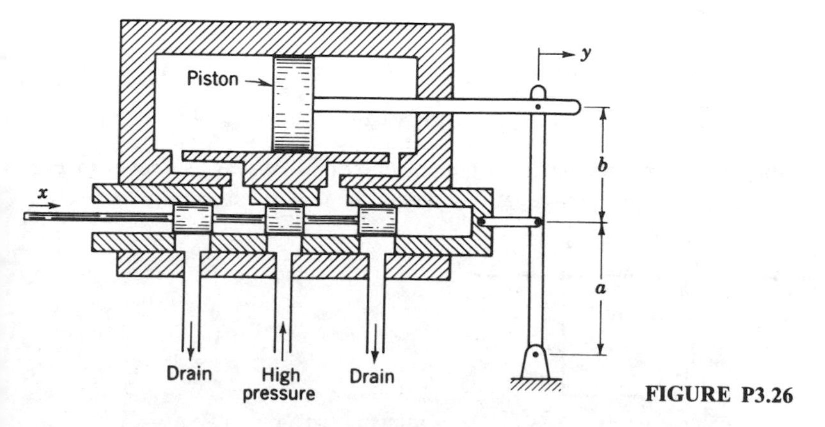

FIGURE P3.26

3.27. For the hydraulic amplifier shown in Fig. P3.27, determine the block diagram for the walking-beam linkage and the block diagrams relating e to y and y to w. Combine these diagrams to determine the overall block-diagram representation for the system.

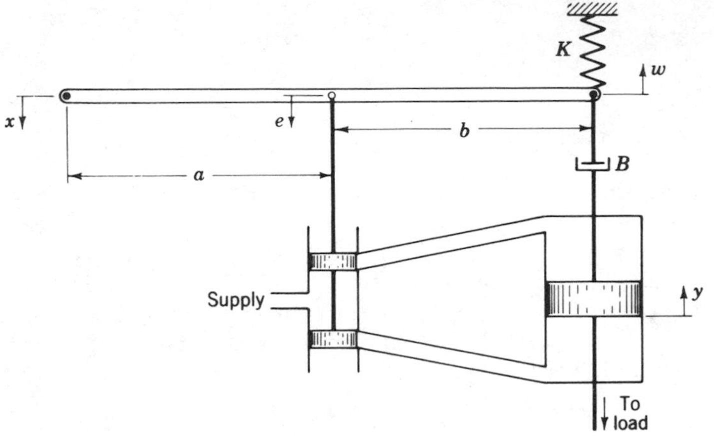

FIGURE P3.27

3.28. The block-diagram representation for a servomechanism is shown in Fig. P3.28. Determine the differential equation relating the output y to the input x. Determine the value of the steady-state gain K and the time constant τ.

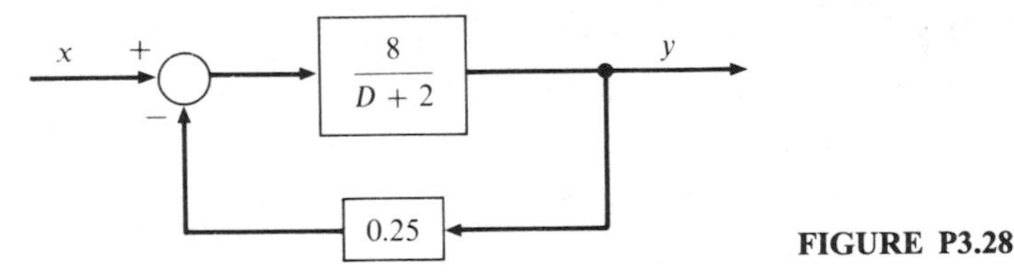

FIGURE P3.28

3.29. Determine the differential equation relating y and x for the hydraulic servomotor described by the block diagram shown in Fig. P3.29. Evaluate the time constant τ and steady-state gain K.

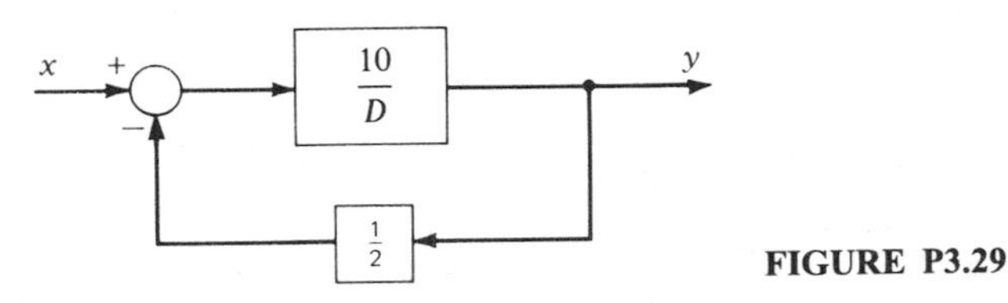

FIGURE P3.29

3.30. The block-diagram representation for a pneumatic control system is shown in Fig. P3.30. Determine the differential equation relating the controlled variable $c(t)$ to the

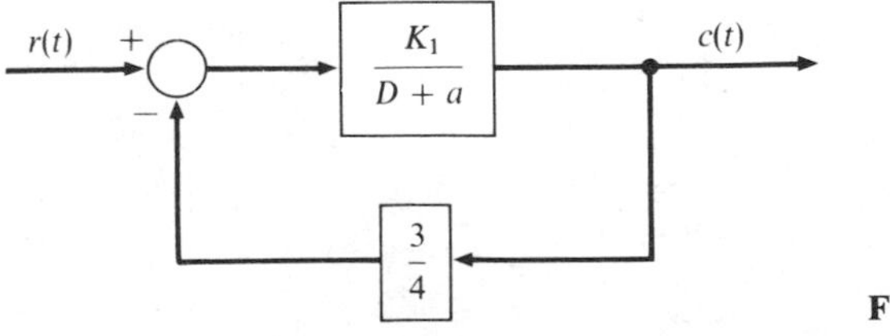

FIGURE P3.30

reference input $r(t)$. What are the values of the steady-state gain K and the time constant τ in terms of K_1 and a?

3.31. The block-diagram representation for a precision metering valve is shown in Fig. P3.31. Determine the differential equation relating the controlled variable $c(t)$ to the reference input $r(t)$. Determine the values of the steady-state gain K and the time constant τ in terms of K_1 and K_2.

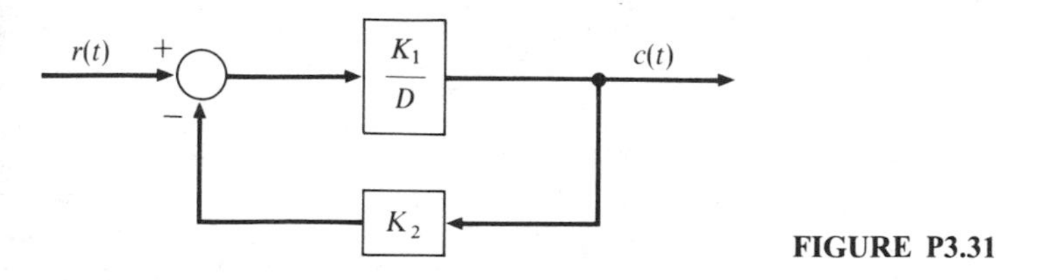

FIGURE P3.31

3.32. For the system shown in Fig. P3.32, determine K_1 and K_2 so that the system will have a steady-state gain of 1 and a time constant of 0.5 s.

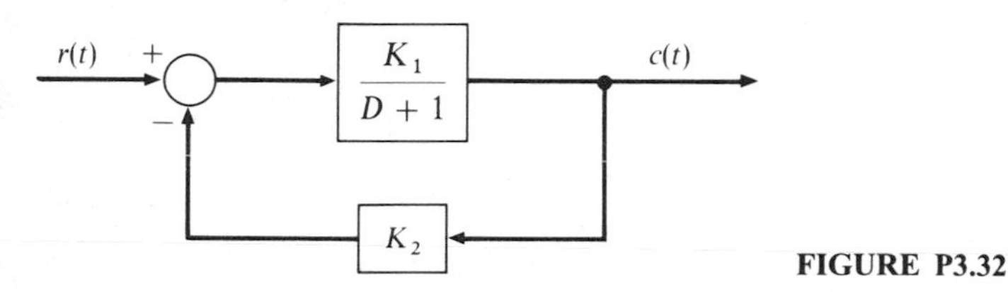

FIGURE P3.32

3.33. For the speed control system shown in Fig. P3.33, determine the differential equation relating the controlled variable $c(t)$ to the reference input $r(t)$ and the disturbance $d(t)$. What is the value of the time constant τ? What is the equation for steady-state operation?

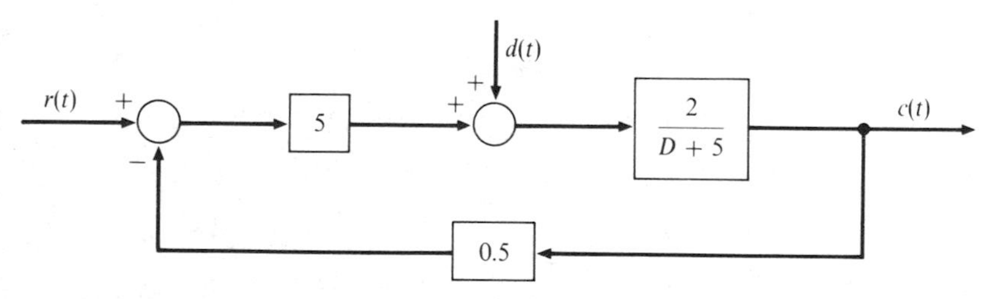

FIGURE P3.33

3.34. As shown in Fig. P3.34, a compressible fluid flows into a tank. The mass rate of flow into the tank is $M = M(P_{in} - P)$. The equation of state is $PV = WRT$, where W is the

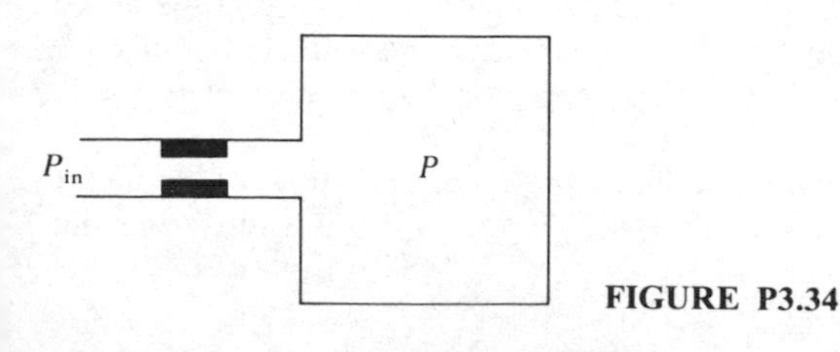

FIGURE P3.34

the mass of air in the tank. Linearization shows that $m = C_1(p_{in} - p)$ and $w = C_2 p$. Determine the differential equation relating the change in pressure p inside the tank to the change in inlet pressure p_{in}. What are the time constant τ and the steady-state gain?

3.35. Derive the equation relating p_2 and z for each of the pneumatic control elements shown in Fig. P3.35a and b. Explain the significance of each partial derivative which occurs in these equations.

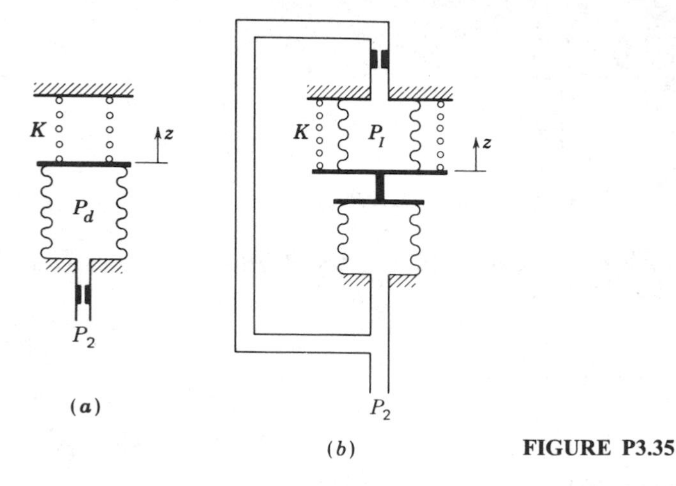

FIGURE P3.35

3.36. Figure P3.36 shows a flapper amplifier. Determine the overall block diagram which results when the controlled pressure P_2 is connected to each of the pneumatic control elements shown in Fig. P3.35. The position z is fed back to the walking-beam linkage as indicated. Note that for this overall block-diagram representation the input is the error e and the output is the position z. (The time constant for the flapper amplifier may be considered negligible so that $p_2 = -K_1 x$.)

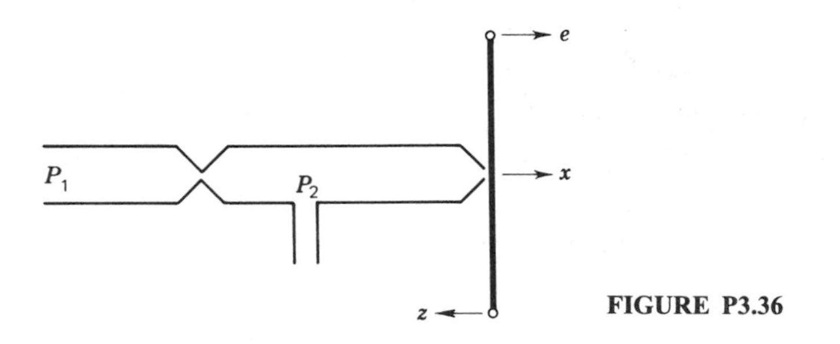

FIGURE P3.36

3.37. In Fig. P3.37 a tank is used as a mixer. Pure water flows in at a constant rate Q_w. A salt solution at a concentration C_s flows in at a constant rate Q_s. The mixed solution of concentration C_o leaves at the rate Q_o. Because the volume V of fluid in the tank is kept constant, the outflow Q_o is equal to $Q_w + Q_s$. The rate of accumulation of salt in the tank is $Q_s C_s - Q_o C_o = VDC_o$. The salt concentration C_s is controlled by a valve (not shown) such that $C_s = K\theta$.

Determine the differential equation relating the concentration C_o of the tank to the valve position θ in terms of the parameters Q_s, Q_w, and V. Identify the time constant.

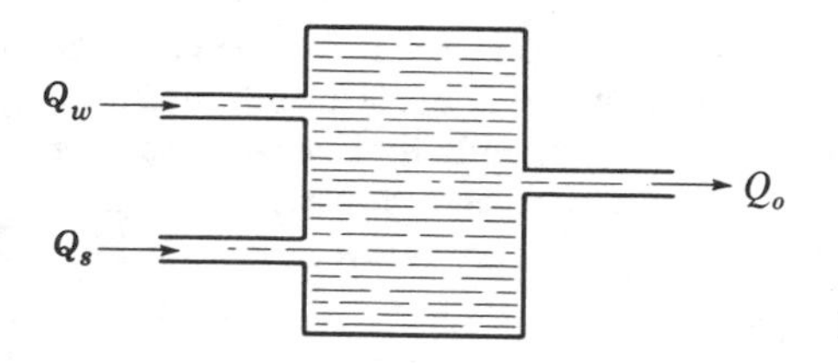

FIGURE P3.37

3.38. A block of material at temperature T_0 is placed in an oven which is held at the constant temperature T_1. The rate of heat flow q into the material is given by $q = hA(T_1 - T)$, where h is the coefficient of heat transfer of the surface, A is the area of the surface, and T is the temperature of the block. The rate of change of temperature of the block is $dT/dt = q/(Mc)$, where M is the mass and c is the average specific heat. Determine the differential equation which describes the temperature of the block as a function of time. Identify the time constant.

3.39. For the thermometer shown in Fig. P3.39, the ambient temperature is T_a and the temperature of the fluid is T. The rate of heat flow from the surrounding medium to the fluid is $Q = C_1(T_a - T)$. The rate of change of temperature of the fluid is $DT = C_2Q$. Construct the block-diagram representation for this system in which T_a is the input and T is the output. Determine the time constant τ.

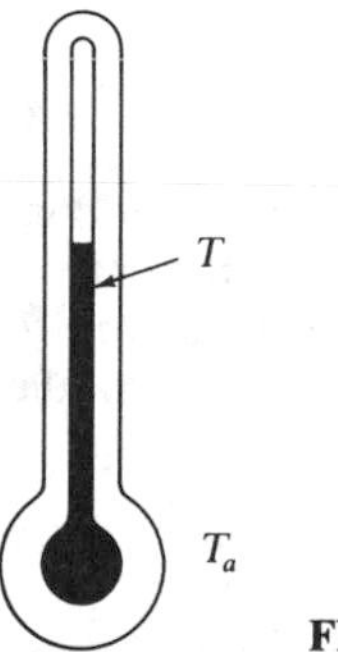

FIGURE P3.39

3.40. The oven shown in Fig. P3.40 is supplied with heat from an electric source. The rate of heat supplied Q_s is proportional to the voltage $Q_s = k_1E_s$. The rate at which heat is lost Q is proportional to the difference in the oven temperature T and the ambient temperature T_a, thus $Q = k_2(T - T_a)$. The rate of change of temperature of the oven is $DT = k_3(Q_s - Q)$. Construct the block-diagram representation for this oven. Determine the differential equation which relates the temperature T to the voltage E_s and the ambient temperature T_a. What is the time constant τ?

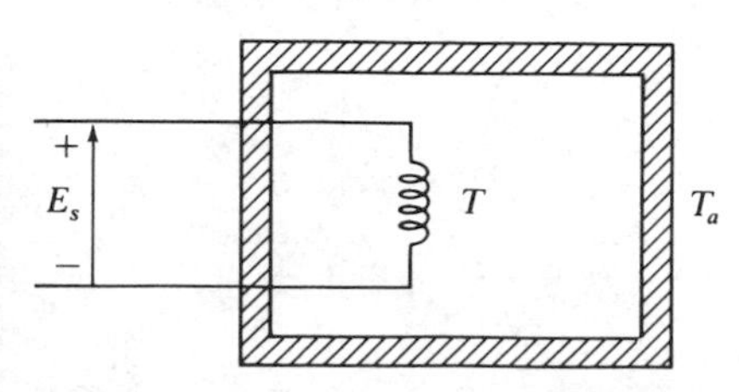

FIGURE P3.40

3.41. The speed torque curves for a dc motor are shown in Fig. P3.41, where V is the applied voltage, N is the speed, and T is the output torque. Determine the linear approximation for the change in torque t due to a change in speed n and a change in voltage v. The motor drives an inertial load such that $t = J\,dn/dt$, where J is the mass moment of inertia. For $J = 0.1$, determine the differential equation relating the change in speed n to the change in voltage v. Determine the time constant τ and the steady-state gain.

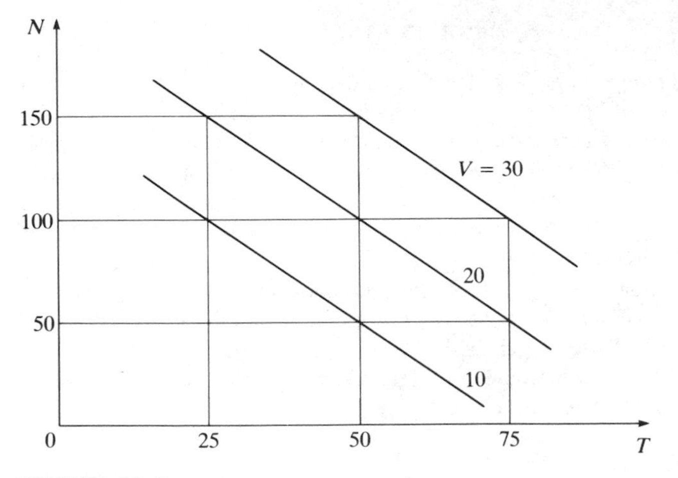

FIGURE P3.41

3.42. The characteristics of an engine are described by the family of curves shown in Fig. P3.42. Determine the linear approximation for the torque t delivered by the engine. When the torque t is greater than the load torque t_L, the excess $t - t_L$ is used to accelerate the engine so that $t - t_L = J\,dn/dt$, where J is the rotational inertia of the

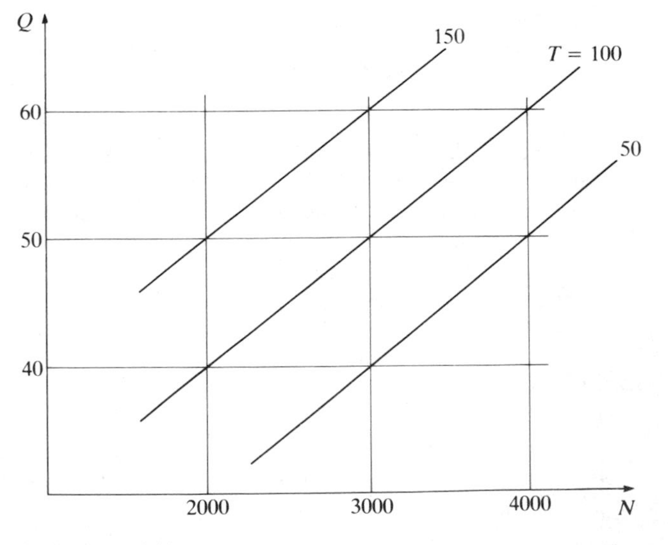

FIGURE P3.42

engine. For $J = 0.025$, determine the differential equation relating the change in speed n to the change in fuel flow q and the change in load torque t_L. What is the time constant τ?

3.43. The operating curves for a dc motor used to control the position of a robot are shown in Fig. P3.43, where T is the motor torque, N is the speed, and V is the applied voltage. The reference point of operation is $N_i = 1000$ and $T_i = 50$. Determine the linearized equation of operation for the change in the torque t about the reference point of operation. The net torque available for acceleration of the robot is $t - t_L = J\,dn/dt$, where t_L is the load torque and $J = 0.01$ is the mass moment of inertia. Determine the differential equation relating the output n to the input v and the load t_L. What is the time constant τ?

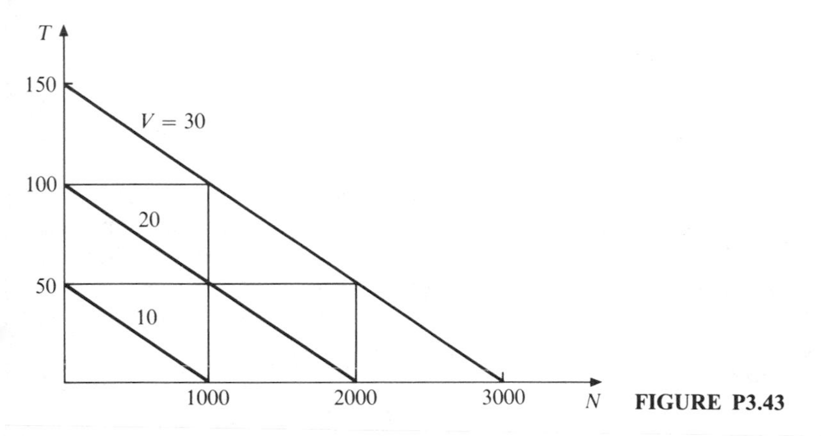

FIGURE P3.43

3.44. The characteristics of an engine are described by the family of curves shown in Fig. P3.44. Determine the linear approximation for the torque t delivered by the engine. The difference between the torque t produced by the engine and the load torque t_L is used to accelerate the engine $J\,dn/dt$ and to overcome viscous friction Bn. Thus, $t - t_L = J\,dn/dt + Bn$. For $J = 0.02$ and $B = 0.03$, determine the differential equation relating the change in speed n to the change in fuel flow q and the change in load torque t_L. Determine the time constant τ.

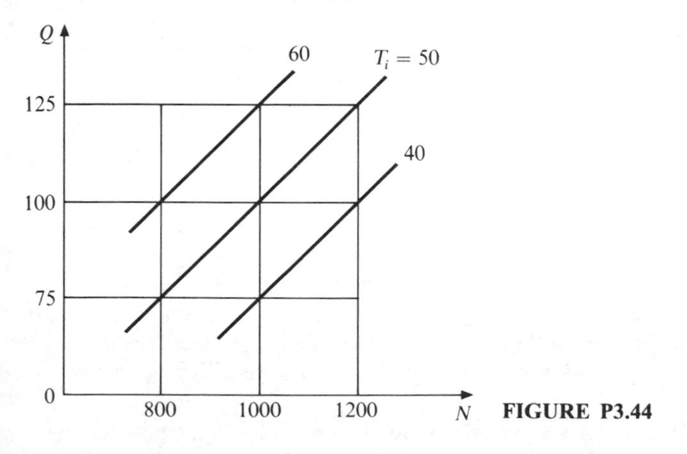

FIGURE P3.44

3.45. The tank shown in Fig. P3.45 is initially empty ($t < 0$). A constant rate of flow Q_{in} is added for $t > 0$. The rate at which flow leaves the tank is $Q_o = CH$. The cross-sectional area of the tank is A. Determine the differential equation for the head H. Identify the time constant. What is the final steady-state value of the head?

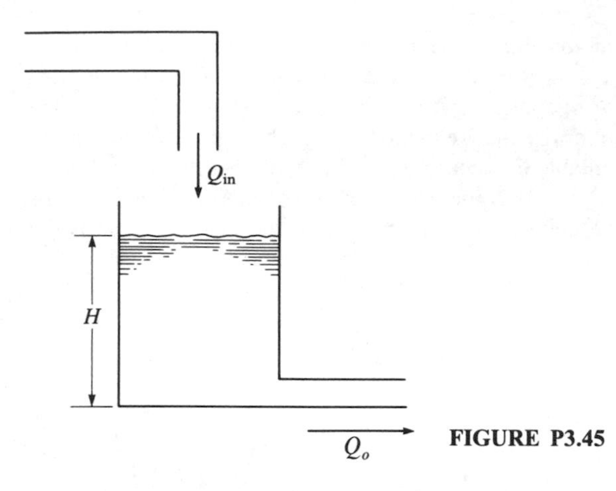

FIGURE P3.45

3.46. For the tank shown in Fig. P3.46, H_i represents the equilibrium level when $Q_{in} = Q_o = Q_i$. The rate of flow out is $Q_o = Q_i + Ch$, where h is the change in head from the equilibrium level. The rate of flow in is $Q_{in} = Q_i + q$, where q is the change in flow from the reference value. Determine the differential equation relating the change in head h due to the change in flow q. Identify the time constant and the steady-state gain.

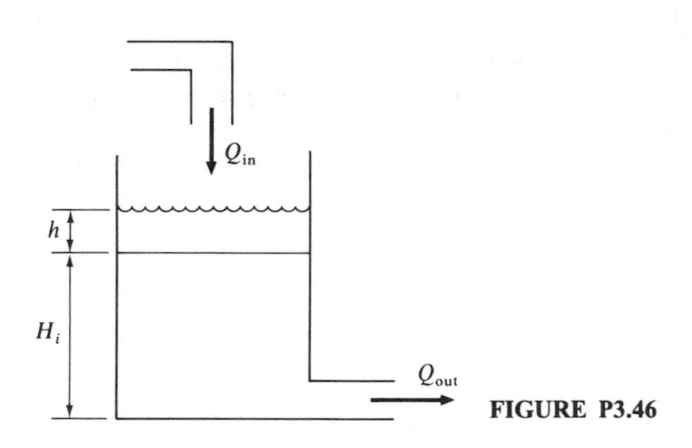

FIGURE P3.46

3.47. The dynamics of an electronic amplifier can be approximated by the differential equation

$$E_o = \frac{K}{1 + aD} E_{in}$$

where E_o is the output voltage and E_{in} is the input voltage. Explain the physical significance of parameters K and a. Describe the details of an experiment to determine the values for K and a. A power source, a switch, and a sensitive voltmeter are available.

3.48. In Fig. P3.48 is shown an electric speed control system. The input potentiometer provides a reference input voltage E_r proportional to the desired speed N_{in}, that is, $E_r = K_r N_{in}$. A voltage signal E_c which is proportional to the controlled output speed N_o is provided by the tachometer ($E_c = K_c N_o$). The error $E_r - E_c$ is amplified by an

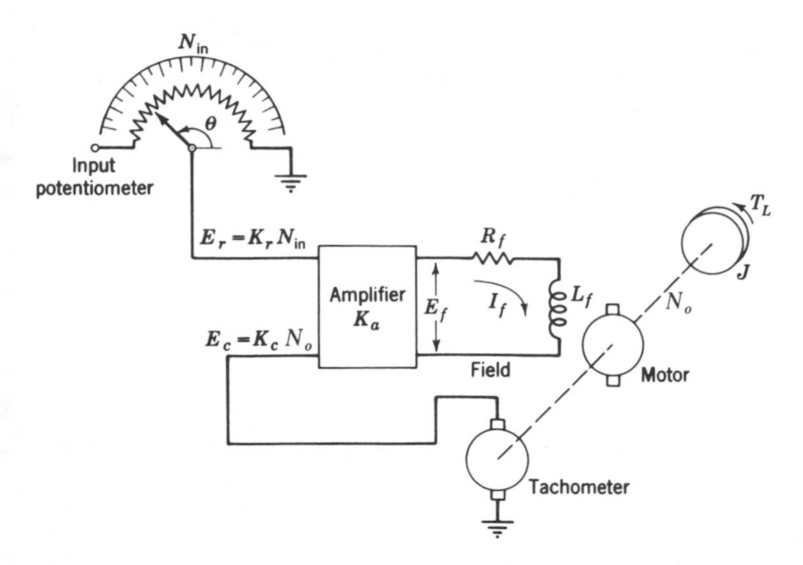

FIGURE P3.48

electronic amplifier whose output is $E_f = K_a(E_r - E_c)$. The voltage E_f is applied to the field of a field-controlled dc motor. The torque exerted on the shaft by the motor (air-gap) torque is proportional to the field current; that is, $T = K_m I_f$. Determine the overall block diagram for this speed control system for the case in which the load torque consists of an inertial load $JD^2\theta = JDN_o$ and an external torque T_L.

3.49. A field-controlled dc motor is shown in Fig. P3.49. The motor drives the load through a gearbox, so that $\omega_c = n\omega$, where n is the gear ratio, ω is the motor speed, and ω_c is the speed of the load (i.e., the controlled speed). The output shaft is connected to a tachometer, which produces a voltage proportional to the controlled speed ($E_c = K_c\omega_c$). An electronic amplifier is used to amplify the error signal by a factor K_a; that is, $E_f = K_a(E_r - E_c)$. Complete the overall block-diagram representation for this system.

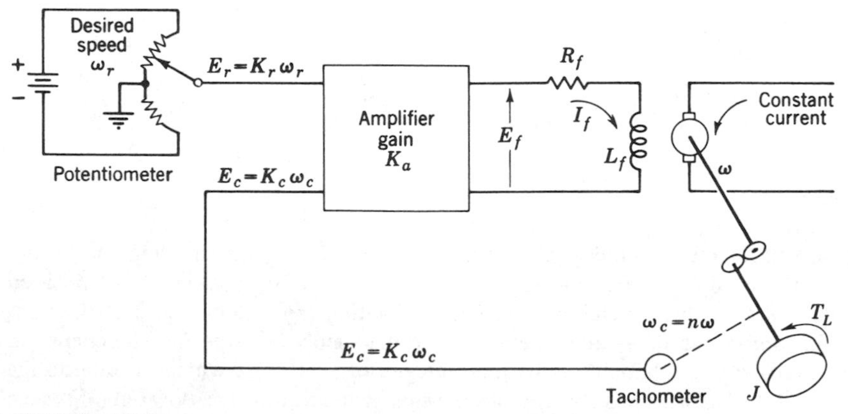

FIGURE P3.49

3.50. In Fig. P3.50 is shown a generator which is used as a voltage amplifier. The prime mover drives the generator at a constant speed. Determine the equation of operation for the amplification ratio E_2/E_1.

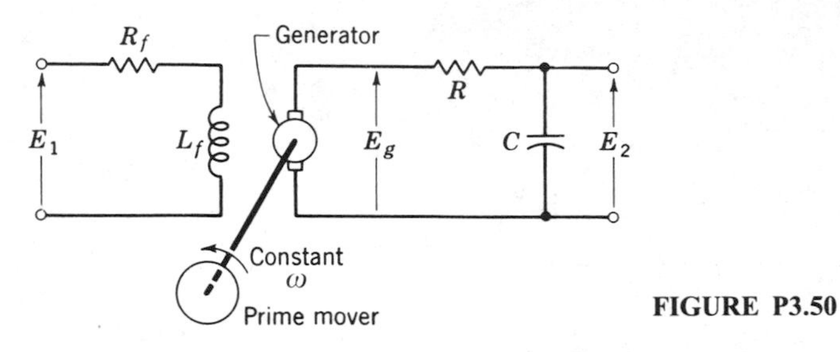

FIGURE P3.50

3.51. Figure P3.51 shows a tension-regulating apparatus such as is used in the paper industry. To ensure uniform winding, it is necessary to maintain a constant tension F_c as the sheet is being wound on the windup roll. To increase the tension in the paper, the tension control lever is raised. This raises the torque control arm of the motor, which increases the torque T_m applied by the motor to the windup roll. The change in torque provided by the motor is $t_m = K_m e/(1 + \tau D)$. For the windup roll, it follows that $F_c = T_m/R$, where R is the radius of the wheel. Determine the overall block diagram relating a variation f_r of the reference or desired tension to a variation of the controlled tension f_c.

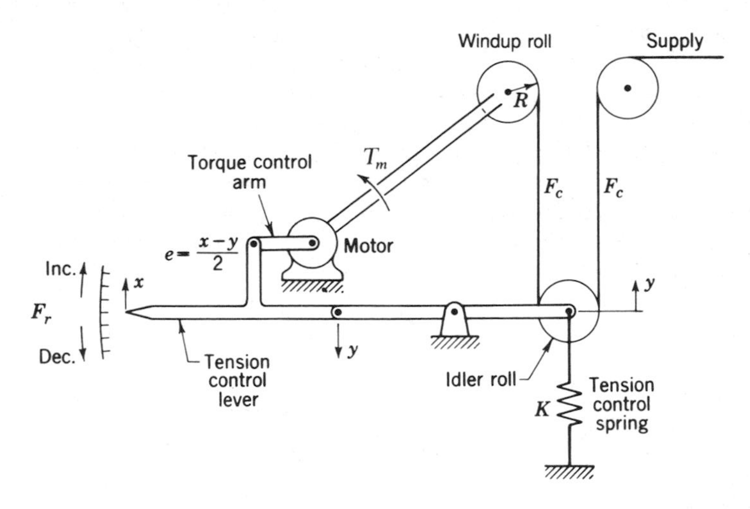

FIGURE P3.51

3.52. Figure P3.52 shows a liquid-level controller. To raise the level of the fluid, the control lever is moved up (i.e., position z is raised). This raises the valve position $e = (z - h_0)/2$, which increases y, thereby admitting more flow Q_{in}. The flow Q_{in} is a function of the flow valve opening Y and the supply pressure P_s. The change in volume of liquid in the tank is the time integral $(q_{in} - q_o)/D$, which is equal to the cross-sectional area of the tank A_T times the change in level h_0. The flow out Q_o is seen to depend on the pressure head H_0. Determine the overall block diagram for this system.

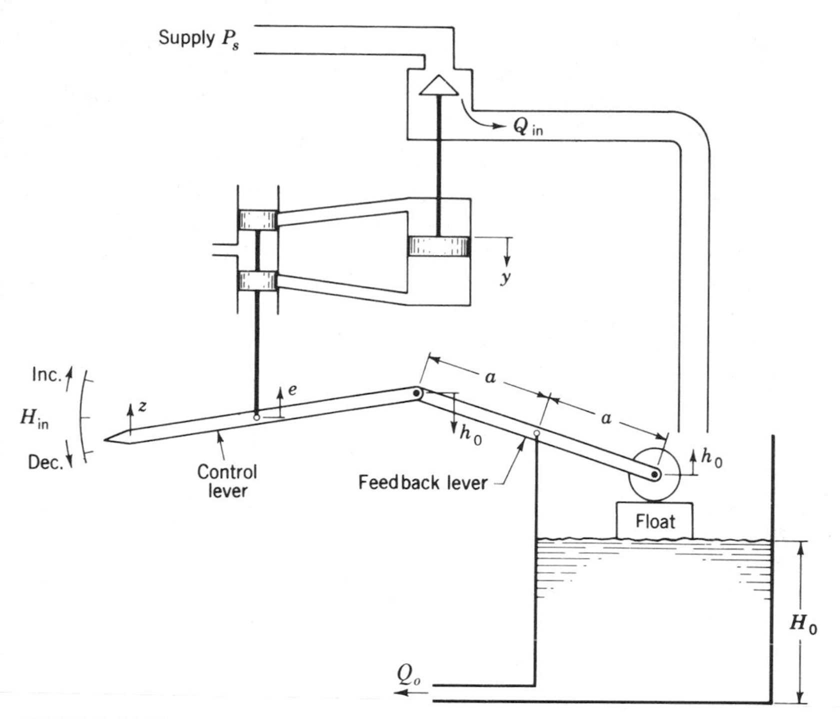

FIGURE P3.52

3.53. A system for controlling flow is shown in Fig. P3.53. Increasing the desired flow setting increases the compression of spring K_1, which causes x and the position e of the balanced valve to move up. This in turn causes the flow valve to move down, which increases the flow. The amount of flow out is measured by a venturi-type

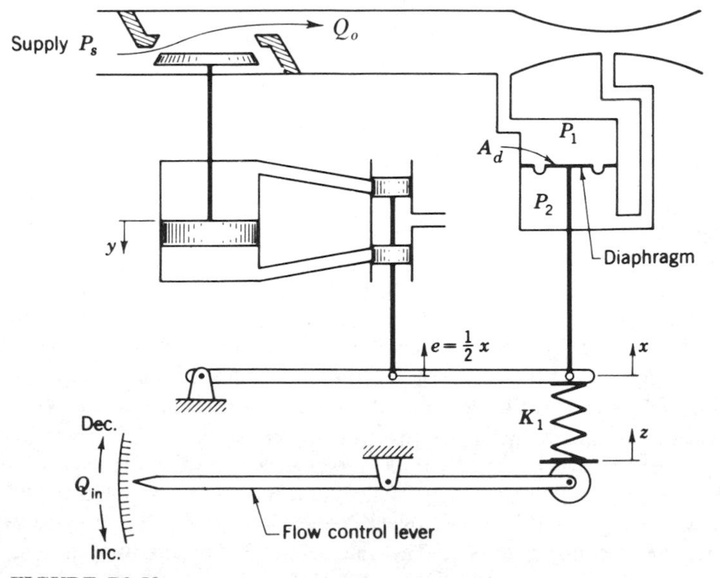

FIGURE P3.53

flowmeter, so that the pressure drop $P_1 - P_2$ is a function of Q_o. The diaphragm prevents leakage from the high pressure P_1 to the low pressure P_2, but it permits motion, just as a piston would. The effective area of the diaphragm is A_d. The flow Q_o is seen to be a function of the flow valve opening Y and the supply pressure P_s. Determine the overall block-diagram representation for this system.

3.54. The linearized equation of operation for the governor shown in Fig. 3.32 is given by Eq. (3.60). The governor shown in Fig. P3.54 is the same as that in Fig. 3.32 except $z = 0$. Thus, letting $z = 0$ in Eq. (3.60) yields the equation for the position x of the governor shown in Fig. P3.54, $x = -C_4 n_o/(K_s - C_r C_3)$. The rate of flow through the valve is $q = C_2 x$. The rate of change of volume in the power piston is $q = A\, dy/dt$. Because Z is fixed, N_{in} is constant and thus $n_{\text{in}} = 0$. The block diagram for this governor is also shown in Fig. P3.54. Show that $G_1(D) = K/D$. Determine K.

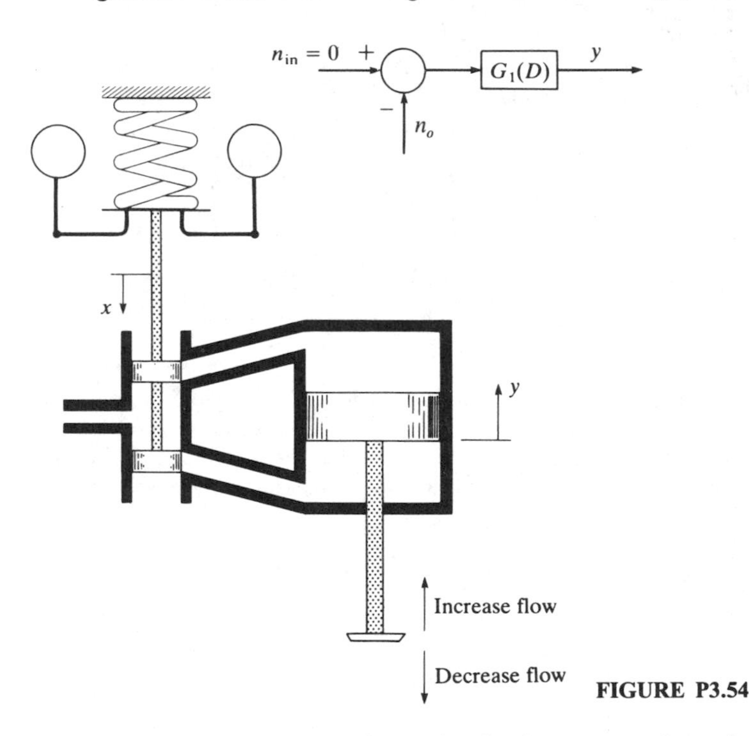

FIGURE P3.54

3.55. The linearized equation of operation for the governor shown in Fig. 3.32 is given by Eq. (3.60). The governor shown in Fig. P3.55 is the same as that in Fig. 3.32 except $z = -y/2$. Thus, letting $z = -y/2$ in Eq. (3.60) yields the equation for the position x of the governor shown in Fig. P3.55, $x = -(K_s y/2 + C_4 n_o)/(K_s - C_r C_3)$. The rate of flow through the valve is $q = C_2 x$. The rate of change of volume in the power piston is $q = A\, dy/dt$. Because N_{in} is constant, $n_{\text{in}} = 0$. The block diagram for this governor is also shown in Fig. P3.55. Show that $G_1(D) = K/(1 + \tau D)$. Determine K and τ.

3.56. The linearized equation of operation for the governor shown in Fig. 3.32 is given by Eq. (3.60). The governor shown in Fig. P3.56 is the same as that shown in Fig. 3.32 except $z = 0$. Thus, letting $z = 0$ in Eq. (3.60) yields the equation for the position x of the governor shown in Fig. P3.56, $x = -C_4 n_o/(K_s - C_r C_3)$. Because of the sleeve around the valve, the opening is $x - y$. The rate of flow through the valve is

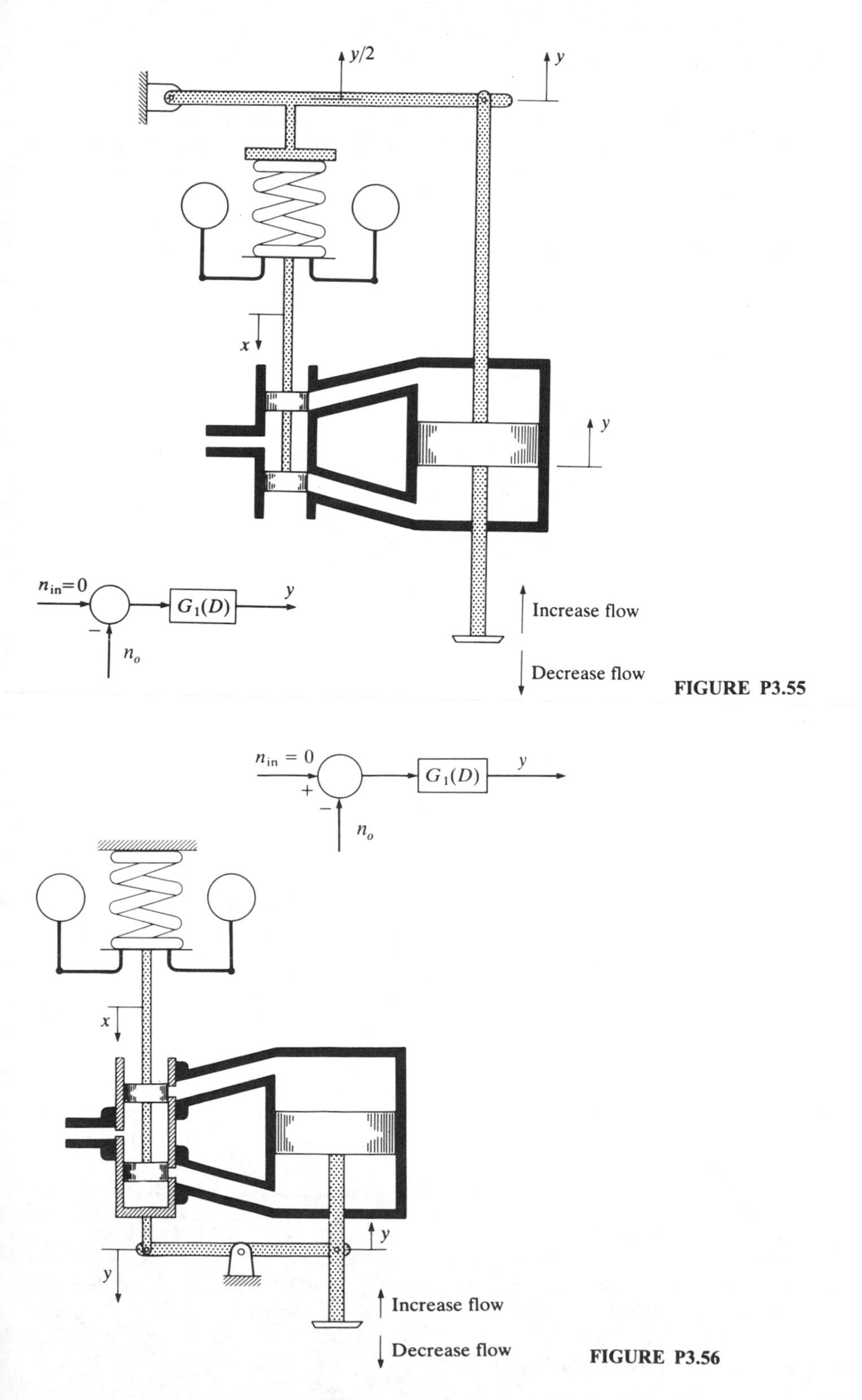

FIGURE P3.55

FIGURE P3.56

$q = C_2(x - y)$. The rate of change of volume in the power piston is $q = A\,dy/dt$. Because Z is fixed, N_{in} is constant and thus $n_{in} = 0$. The block-diagram representation for this governor is also shown in Fig. P3.56. Show that $G_1(D) = K/(1 + \tau D)$. Determine K and τ.

3.57. Figure P3.57 shows a speed-sensing device in which the position x is proportional to the angular velocity of the gear pump. The rate of flow q through the pump is proportional to the speed of rotation, or $q = C_1 n$. Because of the fixed orifice, the pressure p is proportional to the rate of flow q. Thus, $p = C_2 q$. The force balance of the bellows-spring combination is $K_s x = Ap$, where K_s is the spring rate and A is the area of the bellows. Determine the equation relating position x to speed n.

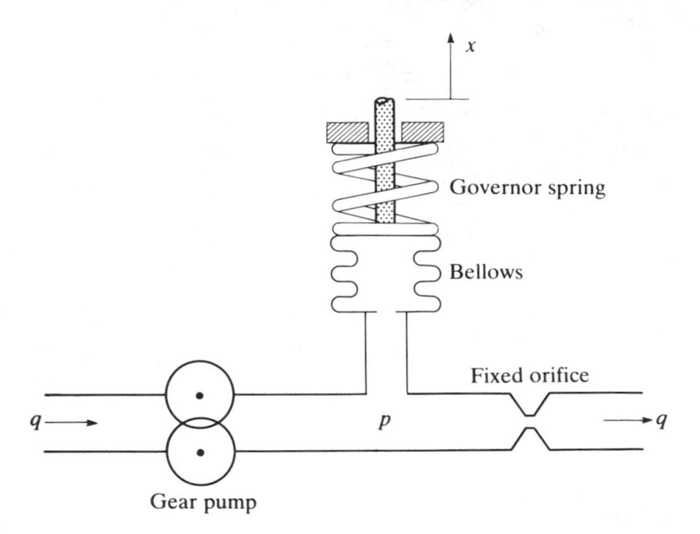

FIGURE P3.57

3.58. For the system shown in Fig. P3.58, $G_1(D) = K_1/(1 + \tau_1 D)$. Determine the differential equation of operation when $A = 1$, $B = -5$, $K_1 = 1$, $K_2 = 2$, $K_H = 0.5$, $\tau_1 = \frac{1}{6}$, and $\tau_2 = 1$.

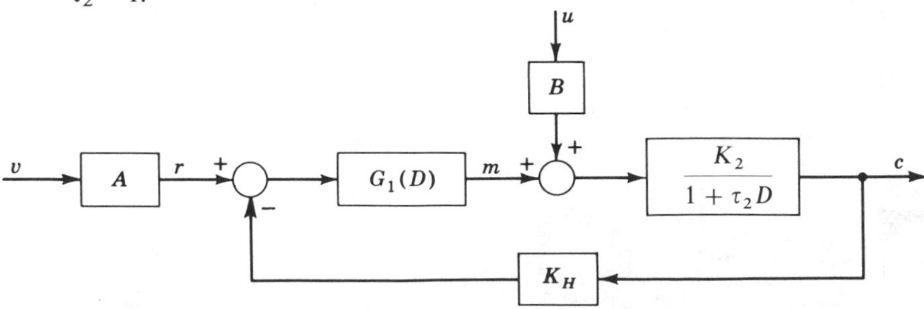

FIGURE P3.58

3.59. For the system shown in Fig. P3.58, $G_1(D) = K_1/D$. Determine the differential equation of operation when $A = 1$, $B = -1$, $K_1 = 1$, $K_2 = 0.75$, $K_H = 1.0$, and $\tau_2 = 0.25$.

3.60. Repeat Prob. 3.59 except $K_2 = 4$.

CHAPTER 4

STEADY-STATE OPERATION

By *steady-state operation* is meant the equilibrium state attained such that there is no change with respect to time of any of the system variables. The system remains at this equilibrium state of operation until it is excited by a change in the desired input or by an external disturbance. A transient condition is said to exist as long as any of the variables of the system is changing with time. In this chapter, we show that considerable information about the basic character of a system may be obtained from an analysis of its steady-state operation.

4.1 STEADY-STATE ANALYSIS

The general block-diagram representation for a feedback control system is shown in Fig. 4.1*a*. For steady-state operation c, v, and u will have constant values, and therefore terms resulting from powers of D operating on these constant quantities will be zero. The equation describing the steady-state operation of a control system is obtained by letting $D = 0$ in the differential equation of operation for the system. Similarly, the block diagram that describes the steady-state operation of a system is obtained by letting $D = 0$ in the general block-diagram representation for the system. The block diagram which describes the steady-state operation of the system of Fig. 4.1*a* is shown in Fig. 4.1*b*, in which

$$K_{G_1} = [G_1(D)]_{D=0} \qquad K_{G_2} = [G_2(D)]_{D=0} \qquad K_H = [H(D)]_{D=0}$$

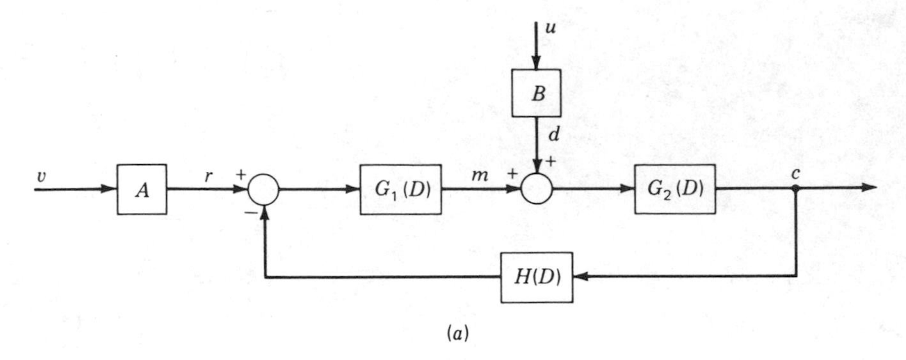

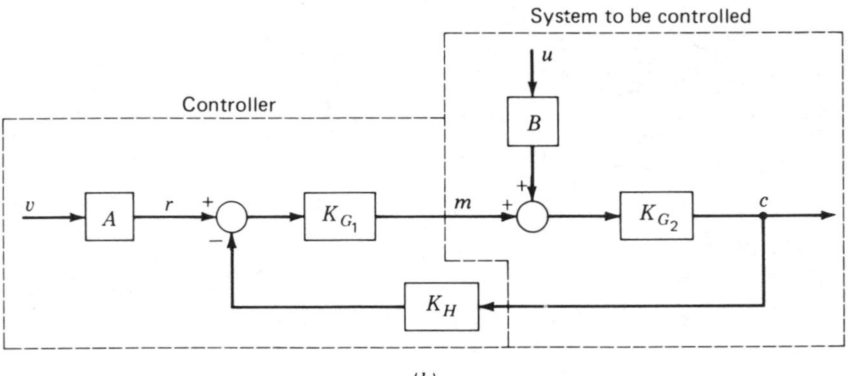

FIGURE 4.1
General block-diagram representation for a control system.

where K_{G_1} is obtained by letting $D = 0$ in the differential operator $G_1(D)$, etc. From Fig. 4.1*b*, the equation for steady-state operation is

$$[(Av - K_H c)K_{G_1} + Bu]K_{G_2} = c$$

or

$$c = \frac{AK_{G_1}K_{G_2}}{1 + K_{G_1}K_{G_2}K_H}v + \frac{BK_{G_2}}{1 + K_{G_1}K_{G_2}K_H}u \tag{4.1}$$

The constant A in Eq. (4.1) is, in effect, the scale factor for the input dial. To have the coefficient of the v term equal to unity, A must be selected such that

$$\frac{AK_{G_1}K_{G_2}}{1 + K_{G_1}K_{G_2}K_H} = 1$$

or

$$A = \frac{1 + K_{G_1}K_{G_2}K_H}{K_{G_1}K_{G_2}} = \frac{1}{K_{G_1}K_{G_2}} + K_H \tag{4.2}$$

When A is chosen in accordance with Eq. (4.2), the coefficient of the v term is unity, in which case Eq. (4.1) becomes

$$c = v + \frac{BK_{G_2}}{1 + K_{G_1}K_{G_2}K_H}u = v + \frac{B}{1/K_{G_2} + K_{G_1}K_H}u \tag{4.3}$$

To have the controlled variable c equal to the command signal v (that is, $c = v$), it is necessary that the coefficient of the u term be zero. This coefficient is zero if either K_{G_1} or K_H is infinite. From Eq. (4.1) it follows that if K_H is infinite, then c must be zero regardless of the value of v or u. In effect, no control is possible when K_H is infinite. Also note from Eq. (4.2) that an infinite value of K_H would necessitate A being infinite, which is physically impossible. Thus only K_{G_1} can be made infinite. This is accomplished by having an integrator in the control elements to yield a $1/D$ term, which gives the effect of an infinite constant during steady-state operation. This type of system is called an *integral control system.*

The left portion of the control system enclosed by the dotted lines in Fig. 4.1*b* is the controller. For the controller, it follows that

$$(Av - K_H c)K_{G_1} = m$$

or

$$c = -\frac{1}{K_{G_1}K_H}m + \frac{A}{K_H}v \tag{4.4}$$

Figure 4.2*a* shows a plot of the steady-state operating characteristics for a typical controller. Lines of constant values for the command signal V are plotted with the controlled variable C as the abscissa and the manipulated variable M as the ordinate. For $v = 0$, Eq. (4.4) shows that

$$-K_{G_1}K_H = \frac{m}{c}\bigg|_{v=0} = \frac{\Delta M}{\Delta C}\bigg|_{\Delta V=0} = \frac{\partial M}{\Delta C}\bigg|_V \tag{4.5}$$

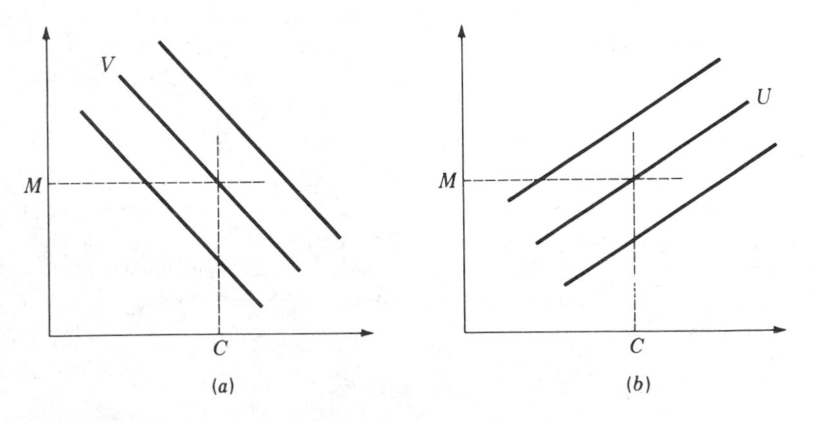

FIGURE 4.2
Steady-state operating characteristics for (*a*) controller and (*b*) system to be controlled.

where $v = \Delta V$ is the change in V from the reference value, $m = \Delta M$ is the change in M from the reference value, and $c = \Delta C$ is the change in C from the reference value. For $v = 0$, then V must be constant. The term $\partial M/\partial C|_V$ is the slope of the controller curves shown in Fig. 4.2*a*.

For the case in which $c = 0$, Eq. (4.4) becomes

$$AK_{G_1} = \frac{m}{v}\bigg|_{c=0} = \frac{\Delta M}{\Delta V}\bigg|_{\Delta C=0} = \frac{\partial M}{\partial V}\bigg|_C \tag{4.6}$$

For $c = 0$, then C must be constant. A line of constant C is a vertical line in Fig. 4.2*a*. The term $\partial M/\partial V|_C$ determines the vertical spacing between the lines of constant V.

Finally, for the case in which $m = 0$, Eq. (4.4) shows that

$$\frac{K_H}{A} = \frac{v}{c}\bigg|_{m=0} = \frac{\Delta V}{\Delta C}\bigg|_{\Delta M=0} = \frac{\partial V}{\partial C}\bigg|_M \tag{4.7}$$

For $m = 0$, then M must be constant. A line of constant M is a horizontal line. The term $\partial V/\partial C|_M$ determines the horizontal spacing between lines of constant V.

The right portion of Fig. 4.1*b* enclosed by the dotted lines represents the system to be controlled. The equation for the steady-state operation of the system to be controlled is

$$(m + Bu)K_{G_2} = c$$

or

$$c = K_{G_2}m + BK_{G_2}u \tag{4.8}$$

From an analysis similar to that for the controller, it follows that

$$\frac{\partial M}{\partial C}\bigg|_U = \frac{1}{K_{G_2}} \qquad \frac{\partial M}{\partial U}\bigg|_C = -B \qquad \frac{\partial C}{\partial U}\bigg|_M = BK_{G_2} \tag{4.9}$$

Typical steady-state operating curves for the system to be controlled are shown in Fig. 4.2*b*. The first partial $\partial M/\partial C|_U$ is the slope, the second partial $\partial M/\partial U|_C$ determines the vertical spacing, and the last partial $\partial C/\partial U|_M$ determines the horizontal spacing. In summary, the individual constants in the block diagram of Fig. 4.1*b* may be obtained from the steady-state operating curves, and vice versa.

Illustrative example 4.1. The block-diagram representation for an engine control system such as that shown in Figs. 3.30 and 3.32 is shown in Fig. 4.3. The reference operating point is $V_i = C_i = 100$, $M_i = 50$, and $U_i = 10$. Determine the steady-state constants for this system, and then sketch the steady-state operating curves. Select A in accordance with Eq. (4.2).

Solution. The steady-state constants are $B = -5$, $K_H = 0.5$, and

$$K_{G_1} = \frac{1}{1 + \tau_1 D}\bigg|_{D=0} = 1 \qquad K_{G_2} = \frac{2}{1 + \tau_2 D}\bigg|_{D=0} = 2$$

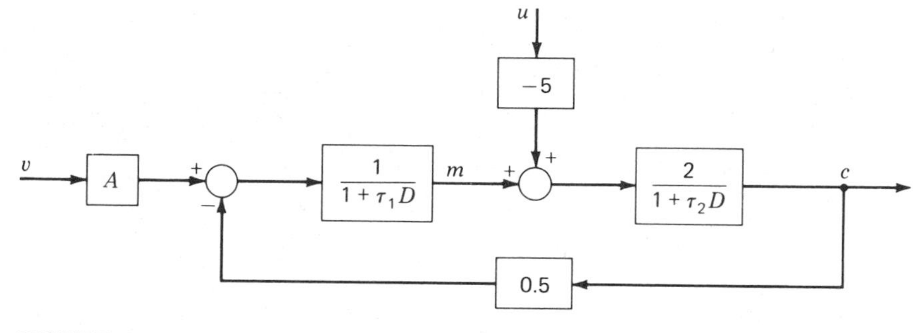

FIGURE 4.3
Engine control system.

The value of A is

$$A = \frac{1}{K_{G_1} K_{G_2}} + K_H = \frac{1}{(1)(2)} + 0.5 = 1.0$$

The slope of the operating curves for the controller is

$$\left.\frac{\partial M}{\partial C}\right|_V = -K_{G_1} K_H = -(1.0)(0.5) = -0.5$$

The line of $V_i = 100$ may now be drawn through the reference point ($M_i = 50$, $C_i = 100$) with a slope of -0.5, as shown in Fig. 4.4*a*. The horizontal spacing is

$$\left.\frac{\Delta V}{\Delta C}\right|_M = \frac{K_H}{A} = 0.5$$

For $\Delta C = 20$, then $\Delta V = 0.5\ \Delta C = 0.5(20) = 10$. Note in Fig. 4.4*a* that when M is constant (a horizontal line), the horizontal spacing is such that as V changes by 10 units, C changes by 20 units. The vertical spacing is

$$\left.\frac{\Delta M}{\Delta V}\right|_C = AK_{G_1} = (1.0)(1.0) = 1.0$$

For $\Delta V = 10$, then $\Delta M = 1.0\ \Delta V = 1.0(10) = 10.0$. Note in Fig. 4.4*a* that when C is constant (a vertical line), the vertical spacing is such that as V changes by 10 units, M changes by 10 units.

If any two of the three quantities of slope, horizontal spacing, or vertical spacing are known, then the third quantity is determined. Thus, only two of these three quantities are independent. This fact is proved by noting that for the controller enclosed by the dashed box in Fig. 4.1*b*, the output is the manipulated variable M, and the inputs are the command signal V and the controlled variable C. For a given controller the manipulated variable M is a function of V and C. That is,

$$M = M(V, C)$$

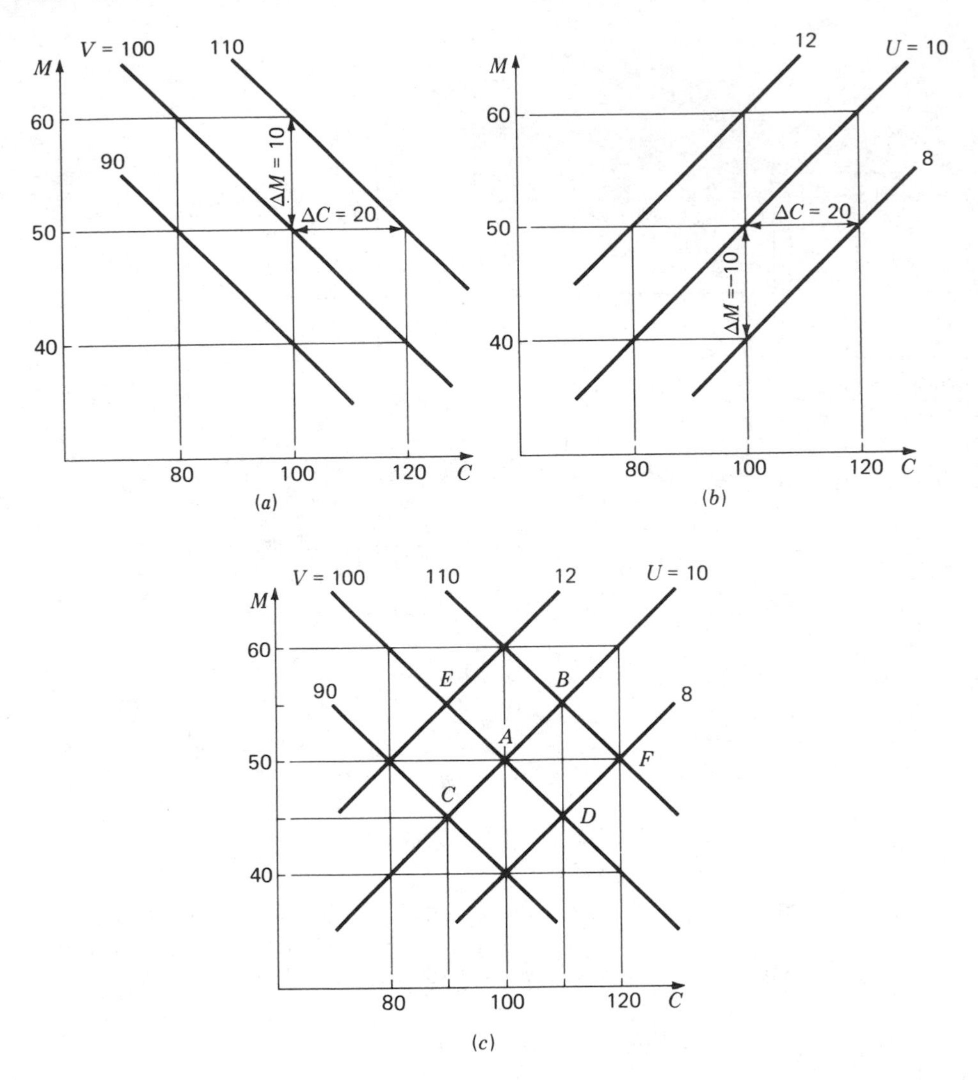

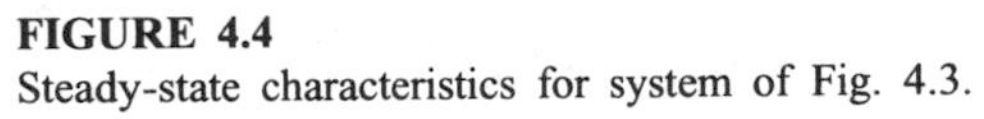

FIGURE 4.4
Steady-state characteristics for system of Fig. 4.3.

The implicit form for this relationship is

$$G(M, V, C) = 0$$

For an implicit function of n variables, the product of the n partial derivatives is $(-1)^n$. For the implicit function G of the three variables M, V, and C, it follows that

$$\left.\frac{\partial M}{\partial V}\right|_C \left.\frac{\partial V}{\partial C}\right|_M \left.\frac{\partial C}{\partial M}\right|_V = -1 \tag{4.10}$$

Solving for the slope $\partial M/\partial C|_V$ gives

$$\left.\frac{\partial M}{\partial C}\right|_V = -\left.\frac{\partial M}{\partial V}\right|_C \left.\frac{\partial V}{\partial C}\right|_M = -AK_{G_1}\frac{K_H}{A} = -K_{G_1}K_H$$

This illustrates the statement that if the vertical spacing $\partial M/\partial V|_C = AK_{G_1}$ and the horizontal spacing $\partial V/\partial C|_M = K_H/A$ are known, then the slope $\partial M/\partial C|_V = -K_{G_1}K_H$ is automatically determined.

The slope of the load lines for the system to be controlled is

$$\left.\frac{\partial M}{\partial C}\right|_U = \frac{1}{K_{G_2}} = \frac{1}{2} = 0.5$$

As shown in Fig. 4.4*b*, the load line for $U_i = 10$ is drawn through the reference point ($M_i = 50$, $C_i = 100$) with a slope of 0.5. The horizontal spacing is

$$\left.\frac{\Delta C}{\Delta U}\right|_M = BK_{G_2} = -5(2) = -10$$

For $\Delta U = -2$, then $\Delta C = -10\ \Delta U = -10(-2) = 20$. Note in Fig. 4.4*b* that when M is constant (a horizontal line), the horizontal spacing between the load lines is such that as C changes by 20 units, U changes by -2 units. The vertical spacing is

$$\left.\frac{\Delta M}{\Delta U}\right|_C = -B = -(-5) = 5$$

For $\Delta U = -2$, then $\Delta M = 5\ \Delta U = 5(-2) = -10$. Thus, when C is constant (a vertical line), the vertical spacing between load lines is such that as M changes by -10 units, U changes by -2 units.

As was the case for the lines of operation for the controller, the load lines can be completely determined if any two of the three quantities (slope, horizontal spacing, or vertical spacing) are known. As is shown in Fig. 4.1*b*, for the system to be controlled, the output is the controlled variable C and the inputs are the manipulated variable M and the disturbance U. Thus, the output C is a function of M and U. That is,

$$C = C(M, U)$$

The implicit form for this relationship is

$$G(C, M, U) = 0$$

The product of the partial derivatives is

$$\left.\frac{\partial C}{\partial M}\right|_U \left.\frac{\partial M}{\partial U}\right|_C \left.\frac{\partial U}{\partial C}\right|_M = -1 \tag{4.11}$$

Solving for the slope $\partial M/\partial C|_U$ gives

$$\left.\frac{\partial M}{\partial C}\right|_U = -\left.\frac{\partial M}{\partial U}\right|_C \left.\frac{\partial U}{\partial C}\right|_M = -(-B)\frac{1}{BK_{G_2}} = \frac{1}{K_{G_2}}$$

Thus, by knowing the vertical spacing $\partial M/\partial U|_C = -B$ and the horizontal spacing $\partial U/\partial C|_M = 1/(BK_{G_2})$, the slope $\partial M/\partial C|_U = 1/K_{G_2}$ can be determined.

Because Fig. 4.4*a* and Fig. 4.4*b* have the same coordinates (*C*, *M*), the two diagrams may be superimposed upon each other as shown in Fig. 4.4*c*. Substitution of the values of the constants into Eq. (4.1) yields the overall equation for steady-state operation

$$\begin{aligned} c &= \frac{AK_{G_1}K_{G_2}}{1+K_{G_1}K_{G_2}K_H}v + \frac{BK_{G_2}}{1+K_{G_1}K_{G_2}K_H}u \\ &= \frac{(1)(1)(2)}{1+(1)(2)(0.5)}v + \frac{(-5)(2)}{1+(1)(2)(0.5)}u = v - 5u \end{aligned} \tag{4.12}$$

The coefficient of the v term and the coefficient of the u term may be obtained directly from Fig. 4.4*c* without the need to evaluate all the steady-state constants. For the overall system shown in Fig. 4.1*b*, the output is the controlled variable C, and the inputs are the command signal V and the disturbance U. Thus, for any given system, C is a function of V and U. That is,

$$C = C(V, U)$$

Linearization gives

$$c = \left.\frac{\partial C}{\partial V}\right|_U v + \left.\frac{\partial C}{\partial U}\right|_V u \tag{4.13}$$

Comparison of this result with Eq. (4.1) shows that

$$\left.\frac{\partial C}{\partial V}\right|_U = \frac{AK_{G_1}K_{G_2}}{1+K_{G_1}K_{G_2}K_H} \tag{4.14}$$

and

$$\left.\frac{\partial C}{\partial U}\right|_V = \frac{BK_{G_2}}{1+K_{G_1}K_{G_2}K_H} \tag{4.15}$$

The term $\partial C/\partial V|_U$ is evaluated by dividing changes in C by corresponding changes in V with U fixed. In Fig. 4.4*c*, with $U = 10$, at point B the value of C is 110 and the value of V is 110. At point C, the value of C is 90 and the value of V is 90. Thus

$$\left.\frac{\Delta C}{\Delta V}\right|_U = \frac{C_B - C_C}{V_B - V_C} = \frac{110-90}{110-90} = 1$$

where C_B is the value of C at point B, V_B is the value of V at point B, etc. The term $\partial C/\partial U|_V$ is evaluated by dividing changes in C by corresponding changes in U with V fixed. In Fig. 4.4*c*, with $V = 100$, at point D the value of C is 110 and the value of U is 8. At point E, the value of C is 90 and the value of U is 12. Thus

$$\left.\frac{\Delta C}{\Delta U}\right|_V = \frac{110-90}{8-12} = \frac{20}{-4} = -5$$

Point A in Fig. 4.4c is the reference point of operation. In going from A to B the command signal changes from 100 to 110, so that $v = 10$. There is no change in the load, so that $u = 0$. Application of Eq. (4.12) gives $c = v - 5u = 10$. Note that in going from point A to point B, the controlled variable changes from 100 to 110, so that $c = 10$. In going from point A to point F, the command signal changes from 100 to 110 so that $v = 10$. The load U changes from 10 to 8, so that $u = -2$. Application of Eq. (4.12) gives $c = v - 5u = 10 - 5(-2) = 20$. In going from A to F the controlled variable changes from 100 to 120, so that $c = 20$.

Because the input scale factor has been selected in accordance with Eq. (4.2), at the reference load $U_i = 10$ $(u = 0)$ the controlled variable is equal to the command signal $(c = v)$. Note in Fig. 4.4c that $V = C = 100$ at point A; $V = C = 110$ at point B; and $V = C = 90$ at point C.

4.2 EQUILIBRIUM

Figure 4.5a shows a typical operating line for a controller in which V is the value of the command signal. When the value of the controlled variable is C_1, the value of the manipulated variable being supplied by the controller is M_1. Figure 4.5b shows a typical load line for a system to be controlled in which U is the load. When the value of the manipulated variable being supplied to the system is M_1, the output from the system is C_1. As indicated in Fig. 4.5c by point A, the intersection of the line of operation of the command signal V for the controller and the load line U for the system determines the equilibrium point of operation for the system. That is, at point A the amount of the manipulated variable M_1 being supplied by the controller is the same as that required to maintain the system output at C_1. If the system output were C_2, then point B would be the point of operation for the controller and point C would be the point of operation for the system to be controlled. The amount of the manipulated variable required to maintain the system at point C is M_C. Because the controller is supplying only the amount M_B, the system output C decreases until equilibrium is attained at point A.

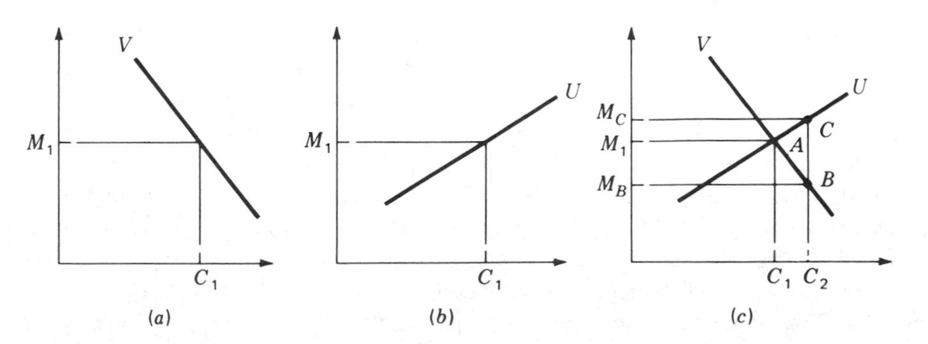

FIGURE 4.5
(a) Typical operating line for a controller, (b) typical load line for system to be controlled, and (c) superposition of (a) and (b).

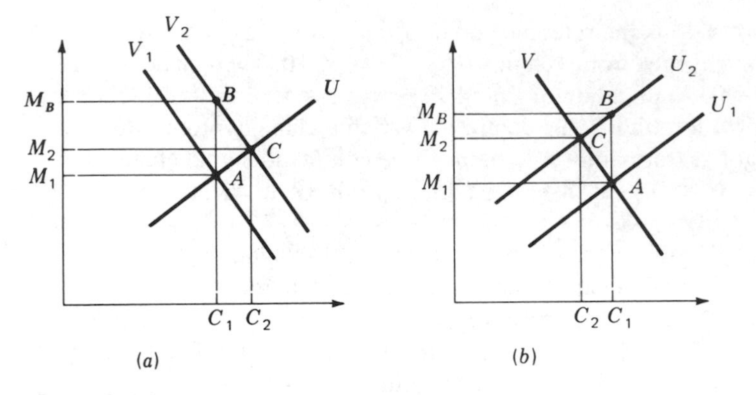

FIGURE 4.6
Effect of changing (*a*) the command signal and (*b*) the load.

The characteristics shown in Fig. 4.6*a* illustrate the effect of changing the command signal from V_1 to V_2 with the load U maintained constant. Initially the system is at equilibrium at point A, the intersection of the V_1 controller curve and the U load curve. When the command signal is changed from V_1 to V_2, the new operating point for the controller is point B, which is the intersection of the value of the output C_1 and the V_2 controller curves. The point of operation for the system to be controlled remains at point A which is the intersection of the output C_1 and the U load curves. Because the controller is supplying more of the manipulated variable than is required to maintain the system to be controlled at point A ($M_B > M_1$), output C increases. The new equilibrium point of operation will be at point C, where the value of the manipulated variable supplied by the controller M_2 is the value required to maintain the system output at C_2. If the input dial is calibrated in such a way that the value of the command signal V_1 is equal to the output C_1, the value of the command signal V_2 is equal to the output C_2, etc., then the system is calibrated so that the output will always equal the input for a given load condition. When this is so, Eq. (4.2) is automatically satisfied.

The characteristics shown in Fig. 4.6*b* illustrate the effect of changing the load from U_1 to U_2 with the command signal V maintained fixed. Initially the system is at equilibrium at point A, the intersection for the controller curve V and the load line U_1. When the load is changed to U_2, the new point of operation for the system to be controlled is point B, which is the intersection of the value of the output C_1 and the load line U_2. The point of operation of the controller remains at point A, which is the intersection of the value of the output C_1 and the controller line V. Because the amount of the manipulated variable M_1 being supplied by the controller is less than the amount M_B required to maintain the system to be controlled at point B, output C decreases. The new equilibrium point of operation will be point C, which is the intersection of the new load line U_2 and the controller line V. Note that with the command signal V kept fixed, the output has changed from C_1 to C_2 due to the change in load.

The slope of the controller lines V is $-K_{G_1}K_H$. For an integral-type controller, K_{G_1} is infinite. The resulting controller lines are vertical. For this case there is no change in the output due to a change in the load. Replacing the control element $G_1(D) = 1/(1 + \tau_1 D)$ in Fig. 4.3 by the element $G_1(D) = 1/D$ would yield such a system. The resulting steady-state characteristics are shown in Fig. 4.7*a*. Note that for steady-state operation, the value of the controlled variable C is always equal to the command signal V regardless of the value of the disturbance U. Using points B and C in Fig. 4.7*a* to evaluate the coefficient $\partial C/\partial V|_U$ and using points D and E to evaluate the coefficient $\partial C/\partial U|_V$ yield

$$\left.\frac{\Delta C}{\Delta V}\right|_U = \frac{C_B - C_C}{V_B - V_C} = \frac{110 - 90}{110 - 90} = 1$$

$$\left.\frac{\Delta C}{\Delta U}\right|_V = \frac{C_D - C_E}{U_D - U_E} = \frac{100 - 100}{8 - 12} = 0$$

Substitution of these results into Eq. (4.13) yields the equation for steady-state operation

$$c = v$$

Thus for an integral-type control system, the controlled variable c is equal to the command signal v regardless of the disturbance.

For an open-loop system, there is no feedback path. When $K_H = 0$, the feedback path of Fig. 4.1*b* is disconnected so that an open-loop system results. For this case the slope of the controller lines ($-K_{G_1}K_H = 0$) is zero. As shown in Fig. 4.7*b*, for an open-loop system the lines of operation for the controller are horizontal.

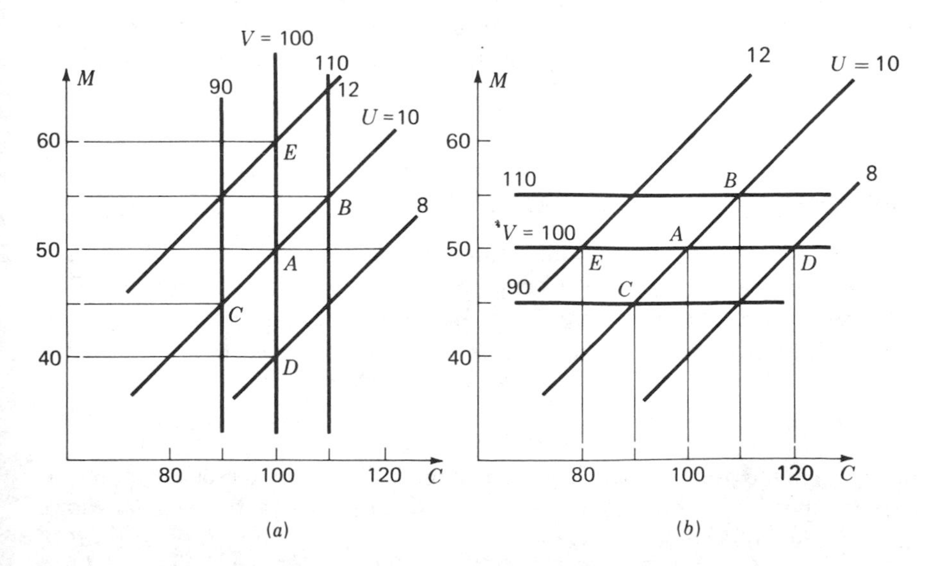

FIGURE 4.7
Steady-state characteristics for (*a*) an integral control system and (*b*) an open-loop system.

Using points B and C in Fig. 4.7b to evaluate the coefficient $\partial C/\partial V\,|_U$ and points D and E to evaluate the coefficient $\partial C/\partial U\,|_V$, we get

$$\left.\frac{\Delta C}{\Delta V}\right|_U = \frac{110-90}{110-90} = 1 \qquad \left.\frac{\Delta C}{\Delta U}\right|_V = \frac{120-80}{8-12} = \frac{40}{-4} = -10$$

Substitution of these results into Eq. (4.13) yields the equation for steady-state operation

$$c = v - 10u$$

The slope of the controller lines $\partial M/\partial C|_V = -K_{G_1}K_H$ varies from zero for an open-loop system ($K_H = 0$) to infinity for an integral control system ($K_{G_1} = \infty$). A proportional control system is one for which the controller lines have a finite slope, as shown in Fig. 4.4c. Usually the slope of the control lines for a proportional controller is very steep, so that they are almost vertical. The steeper the slope, the less the variation in the controlled variable C due to a change in the disturbance U. An open-loop system sees the greatest change in the controlled variable due to a disturbance, whereas an integral controller has no change in the controlled variable due to a disturbance.

Proportional control systems usually exhibit better transient characteristics than integral control systems. In addition, proportional control systems provide the operator with a "feel" as to what is going on. For example, a power steering system feeds back some of the torque applied to the steering wheel so that the driver has a measure or feel for the turning effort being applied to the wheels. Satisfactory performance may usually be achieved by making the control lines sufficiently steep; this makes the coefficient of the u term sufficiently small that variations in the external disturbance cause only slight errors. For integral control systems, the coefficient of the u term is zero. For proportional control systems, the coefficient of the u term is finite. The coefficient attains its maximum value for an open-loop system.

4.3 PROPORTIONAL CONTROL SYSTEMS

The differential equation relating the output n_o to the input n_{in} and external disturbance t_L for the speed control system represented by the block diagram of Fig. 3.37 is given by Eq. (3.67). Letting $D = 0$ in Eq. (3.67) yields, for the equation describing the steady-state operation of the speed control system,

$$n_o = \frac{C_2C_6K_1K_s n_{\text{in}} - C_6C_8t_L}{1 + C_4C_6K_1} \tag{4.16}$$

Letting $D = 0$ in the overall block diagram of Fig. 3.37 yields the block diagram for steady-state operation shown in Fig. 4.8a. Comparison with the general block diagram of Fig. 4.8b shows that $c = n_o$, $v = n_{\text{in}}$, $u = t_L$, $A = C_2K_s$, $B = -C_8$, $K_{G_1} = K_1$, $K_{G_2} = C_6$, and $K_H = C_4$. Substitution of these results into Eq. (4.1) verifies Eq. (4.16).

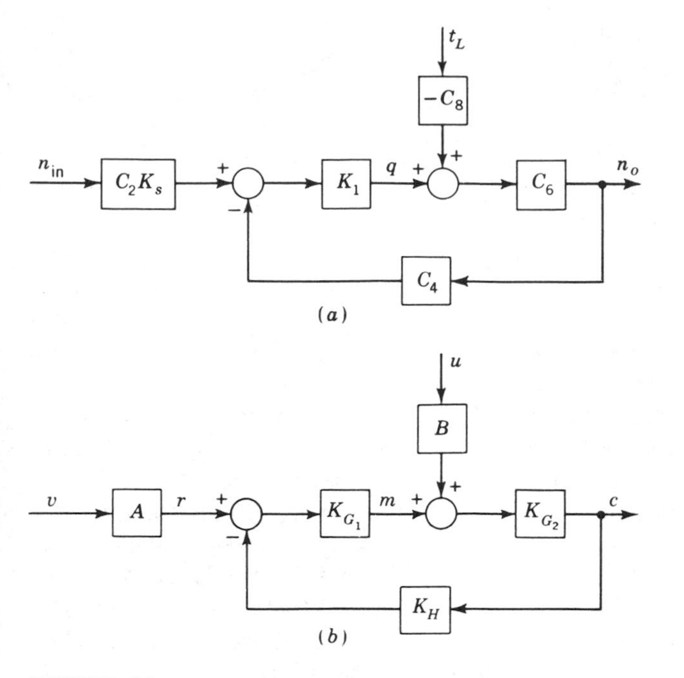

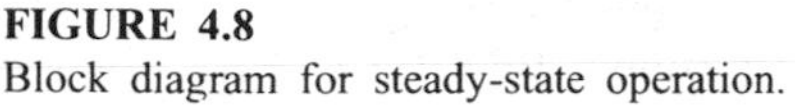

FIGURE 4.8
Block diagram for steady-state operation.

The speed control system shown in Fig. 3.37 is a proportional control system. Substitution of the corresponding values for this speed control system into Eq. (4.2) gives

$$A = C_2K_S = \frac{1}{C_6K_1} + C_4$$

or

$$C_2 = \frac{1}{K_S}\left(\frac{1}{C_6K_1} + C_4\right) \tag{4.17}$$

The term $C_2 = \partial Z/\partial N_{\text{in}}|_i$ is the scale factor for the speed-setting dial.

Because some of the terms in Eq. (4.17) are partial derivatives evaluated at the reference operating condition, the value of the scale factor C_2 is seen to vary for different reference points. This would result in a nonlinear scale for the input speed dial. The use of a nonlinear scale may be avoided by connecting the throttle lever to a cam which in turn sets the desired position of the top of the spring, as shown in Fig. 4.9. It is then a relatively easy matter to set up the speed control system so that Eq. (4.17) is satisfied for any reference condition. When this is so, Eq. (4.16) becomes

$$n_o = n_{\text{in}} - \frac{C_6C_8}{1 + C_4C_6K_1}t_L \tag{4.18}$$

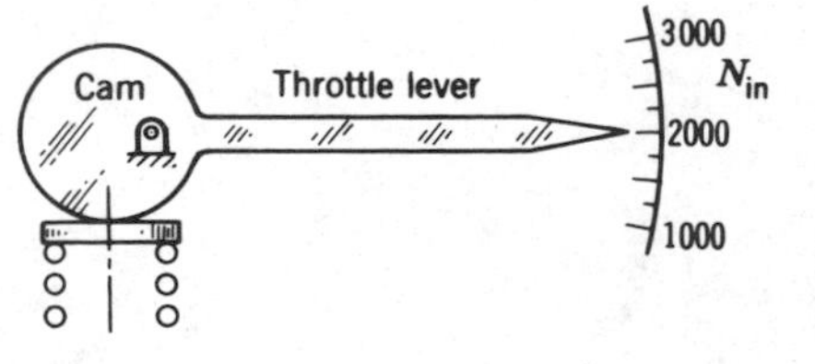

FIGURE 4.9
Cam to avoid nonlinear input scale.

Equation (4.18) is the typical form of the steady-state relationship that exists between the input, output, and external disturbance for a proportional control system. When the load torque T_L is not equal to the reference value (that is, $t_L \neq 0$), then n_o is not equal to n_{in}. For example, suppose that this is the speed control system for the gas turbine of a jet airplane and that T_{L_i} is the torque required for the airplane in level flight. When the airplane is inclined to gain altitude, a greater load torque T_L is required than that for level flight (that is, $t_L > 0$). Thus, Eq. (4.18) shows that the output speed is slightly less than the desired value for this flight condition.

The physical reason for this can be seen by looking at the schematic diagram in Fig. 3.32 for the speed control system. For level flight, the system is set up so that $N_o = N_{in}$. When the airplane is gaining altitude, the load torque is increased. This increased torque results in a decreased speed, which in turn causes a lower position for x. Because of the lower position for x, there is a greater flow of fuel. To have the airplane continue to gain altitude, more flow is required than for level flight. To maintain this increased flow, the engine speed must be slightly less than for level flight.

The steady-state operating curves for this speed control system are shown in Fig. 4.10. For an airplane in level flight, the curve of fuel flow Q required to maintain various speeds N_o is indicated by $T_{L_i} = T_2$. The curve marked T_3 would correspond to operation of the airplane at a certain angle of inclination. Similarly, curve T_1 would correspond to the airplane's losing altitude at a certain angle of declination.

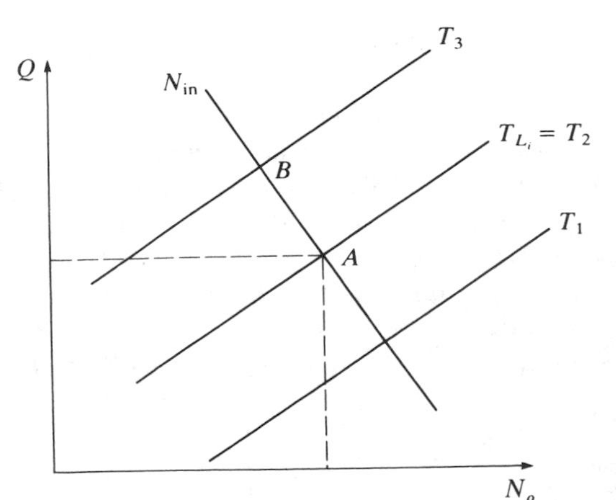

FIGURE 4.10
Steady-state operating curves for a speed control system.

The operating line AB for the controller is obtained by fixing the speed setting at some value N_{in} and then plotting corresponding values of fuel flow Q coming from the controller, for various speeds of rotation N_o. Note from Fig. 3.32 that an increased speed N_o of the flyweights increases the centrifugal force, which causes x to move up. This in turn causes y to move down, and this decreases the flow Q supplied by the controller.

Steady-state operation exists at the intersection of the line of operation of the controller and the torque line for the given flight condition because at this intersection just enough flow is being supplied to maintain the flight condition. For example, if the airplane is in level flight and the operating line for the desired speed setting is AB, then the intersection at point A of line T_2 and operating line AB is the steady-state operating point for the system. The speed-setting dial is calibrated by setting the value of speed N_{in} on the dial equal to the steady-state value of the output speed at the reference load T_2. For a given speed setting, such as that indicated by line AB, if the load is increased to T_3 while the desired speed is unchanged, then the new operating point must be on line T_3 at point B. Because AB is not a vertical line, variations in the load are seen to cause variations in the output speed. A proportional controller is sometimes called a *droop controller*, and line AB is referred to as the *droop line*.

Illustrative example 4.2. A typical family of steady-state operating curves for a unity-feedback ($K_H = 1$) speed control system is shown in Fig. 4.11. At the reference operating condition (point A), $N_{in} = N_o = 4000$, $Q_i = 1000$, and $T_i = 200$. Determine the steady-state constants and the equation for steady-state operation. With N_{in} held

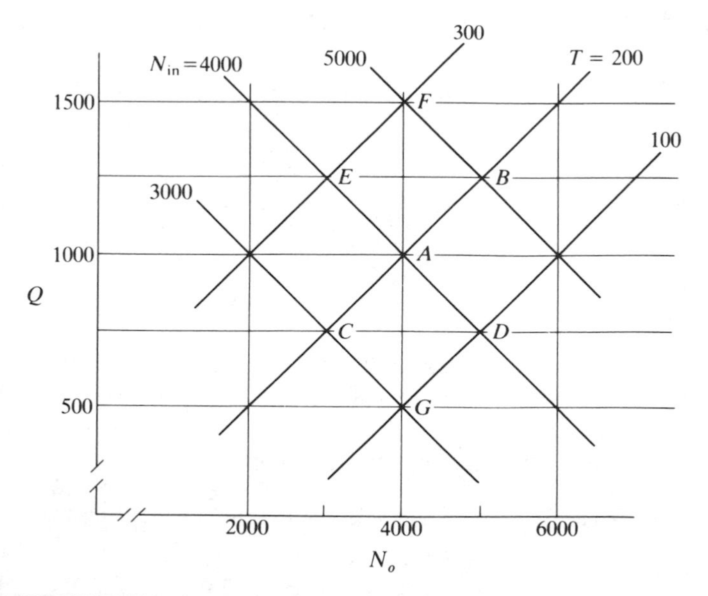

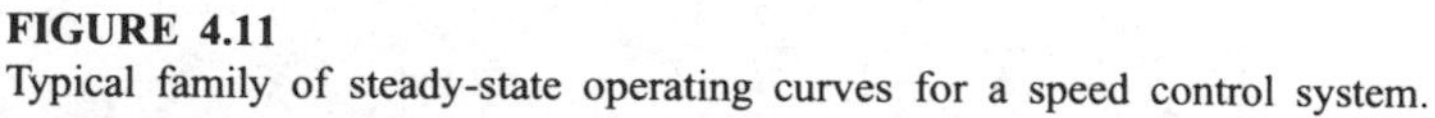
FIGURE 4.11
Typical family of steady-state operating curves for a speed control system.

fixed at its reference value, what is the change in speed N_o when the load T changes from the reference value $T_i = 200$ to $T = 300$? By what factor should the slope of the controller lines be changed so as to reduce this change by a factor of 50?

Solution. The constants for the system to be controlled are

$$\frac{1}{K_{G_2}} = \left.\frac{\partial M}{\partial C}\right|_U = \left.\frac{\Delta Q}{\Delta N_o}\right|_T = \frac{Q_B - Q_C}{N_{o_B} - N_{o_C}} = \frac{1250 - 750}{5000 - 3000} = \frac{1}{4}$$

$$B = -\left.\frac{\partial M}{\partial U}\right|_C = -\left.\frac{\Delta Q}{\Delta T}\right|_{N_o} = -\frac{Q_F - Q_G}{T_F - T_G} = -\frac{1500 - 500}{300 - 100} = -5$$

where Q_B is the value of Q at point B, N_{o_B} is the value of N_o at point B, etc.

The constants for the controller are

$$-K_{G_1}K_H = \left.\frac{\partial M}{\partial C}\right|_V = \left.\frac{\Delta Q}{\Delta N_o}\right|_{N_{\text{in}}} = \frac{Q_D - Q_E}{N_{o_D} - N_{o_E}} = \frac{750 - 1250}{5000 - 3000} = -\frac{1}{4}$$

$$AK_{G_1} = \left.\frac{\partial M}{\partial V}\right|_C = \left.\frac{\Delta Q}{\Delta N_{\text{in}}}\right|_{N_o} = \frac{Q_F - Q_G}{N_{\text{in}_F} - N_{\text{in}_G}} = \frac{1500 - 500}{5000 - 3000} = \frac{1}{2}$$

For $K_H = 1$, then $K_{G_1} = \frac{1}{4}$, and $A = 2$. The equation for steady-state operation is

$$n_o = \frac{AK_{G_1}K_{G_2}}{1 + K_{G_1}K_{G_2}K_H} n_{\text{in}} + \frac{BK_{G_2}}{1 + K_{G_1}K_{G_2}K_H} t$$

$$= \frac{(2)(\frac{1}{4})(4)}{1 + (\frac{1}{4})(4)(1)} n_{\text{in}} + \frac{(-5)(4)}{1 + (\frac{1}{4})(4)(1)} t = n_{\text{in}} - 10t$$

This result may be verified by application of Eq. (4.13). That is,

$$c = \left.\frac{\partial C}{\partial V}\right|_U v + \left.\frac{\partial C}{\partial U}\right|_V u \tag{4.13}$$

Thus

$$n_o = \left.\frac{\partial N_o}{\partial N_{\text{in}}}\right|_T n_{\text{in}} + \left.\frac{\partial N_o}{\partial T}\right|_{N_{\text{in}}} t = \frac{N_{o_B} - N_{o_C}}{N_{\text{in}_B} - N_{\text{in}_C}} n_{\text{in}} + \frac{N_{o_D} - N_{o_E}}{T_D - T_E} t$$

$$= \frac{5000 - 3000}{5000 - 3000} n_{\text{in}} + \frac{5000 - 3000}{100 - 300} t = n_{\text{in}} - 10t$$

With N_{in} fixed at the reference value, then $n_{\text{in}} = 0$. The change in speed n_o due to changing the load from the reference value $T_i = 200$ to $T = 300$ ($t = T - T_i = 100$) is

$$n_o = n_{\text{in}} - 10t = 0 - 10(100) = -1000$$

To decrease this change by a factor of 50, the coefficient of the t term must be changed from -10 to $-\frac{10}{50} = -\frac{1}{5}$ Thus

$$\frac{BK_{G_2}}{1 + K_{G_1}K_{G_2}K_H} = \frac{(-5)(4)}{1 + 4K_{G_1}K_H} = -\frac{1}{5}$$

Solving for $K_{G_1}K_H$ gives

$$K_{G_1}K_H = \frac{99}{4}$$

To decrease the speed error by a factor of 50, the slope of the controller lines must be increased by a factor of 99 (i.e., from $-K_{G_1}K_H = -\frac{1}{4}$ to $-99/4$).

4.4 INTEGRAL CONTROL SYSTEMS

By eliminating the linkage between x and y of Fig. 3.32 and using the hydraulic integrator shown in Fig. 4.12, the proportional control system is converted to an integral control system. The block-diagram representation for the integrator is also shown in Fig. 4.12. The substitution of this diagram for that of the servomotor which it replaces in Fig. 3.36 yields the block-diagram representation shown in Fig. 4.13*a*.

The value of K_{G_1} is computed as follows:

$$K_{G_1} = \left[\frac{C_1C_5}{(K_s - C_rC_3)A_1D}\right]_{D=0} = \left(\frac{K_I}{D}\right)_{D=0} = \infty \tag{4.19}$$

where $K_I = C_1C_5/[(K_s - C_rC_3)A_1]$ is the constant associated with the integrating portion of the system, as shown in Fig. 4.13*b*.

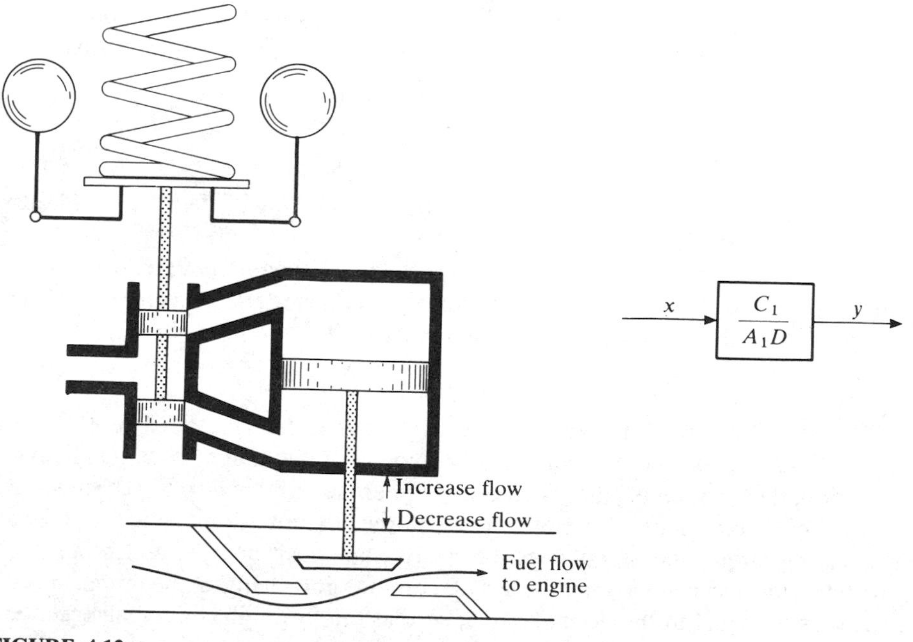

FIGURE 4.12
Integral control system.

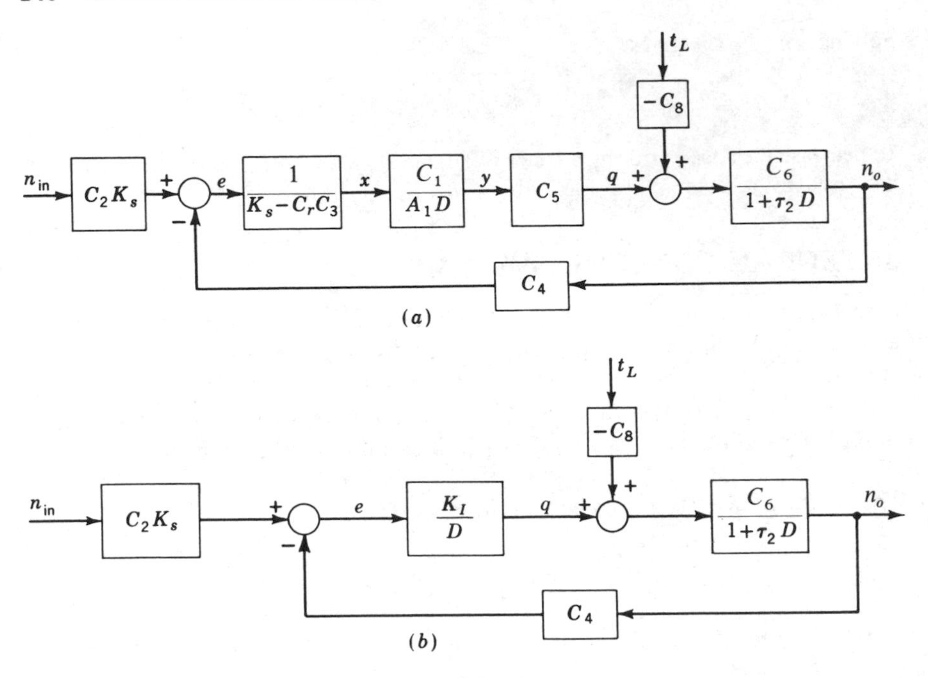

FIGURE 4.13
Block diagram for integral control system.

Because K_{G_1} is infinite, e must be zero for steady-state operation. Thus subtracting the feedback signal from the reference input in Fig. 4.13b gives

$$C_2K_sn_{\text{in}} - C_4n_o = e = 0$$

or

$$n_o = \frac{C_2K_s}{C_4}n_{\text{in}} \tag{4.20}$$

The preceding equation shows that the speed is independent of the load torque for an integral control system. It is an easy matter to adjust the scale factor C_2 for the input speed dial so that $C_2K_s/C_4 = 1$, in which case

$$n_o = n_{\text{in}} \tag{4.21}$$

The operation of an integral control system may be visualized as follows. From Fig. 4.12, we see that if x momentarily changes and then returns to its line-on-line position, the position of y has been changed permanently and so has the amount of flow going to the engine. Therefore, changing the amount of flow to account for a new operating torque does not change the steady-state position of x, which must be line on line. Because neither x nor the spring compression changes, the output speed must always be equal to the desired value, so that the flyweight force balances the spring force. (Note that, for the proportional control system, changing the fuel flow requires a permanent change in the position x.)

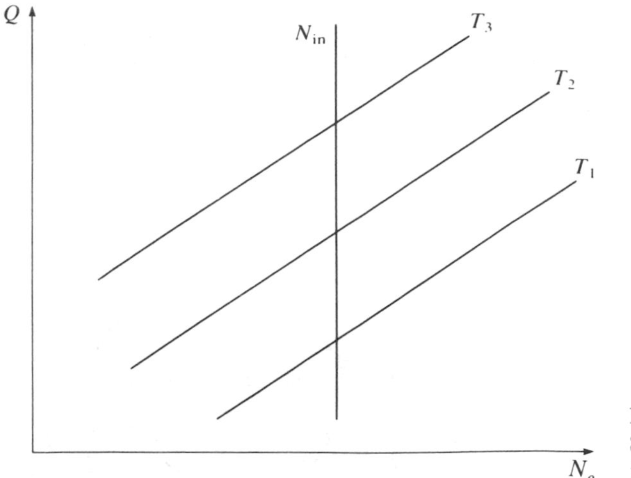

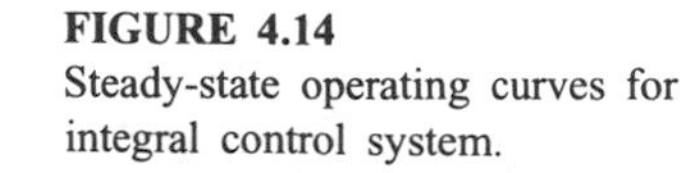

FIGURE 4.14
Steady-state operating curves for integral control system.

An integral control is easily recognized because there must be an integrating component yielding a $1/D$ term in the block diagram between the comparator and the point where the external disturbance enters the system. The line of operation of an integral controller is a vertical line. The operating characteristics of an integral control system are shown in Fig. 4.14.

An integral controller is also called a *floating controller* because of the floating action of the position y of the flow-setting valve. Two other terms used for an integral controller are *reset controller* and *isochronous controller.*

4.5 PROPORTIONAL PLUS INTEGRAL CONTROL SYSTEMS

From a consideration of steady-state operation only, integral control systems seem preferable to proportional systems. However, it is generally easier to achieve good transient behavior with a proportional system than with an integral system (techniques for determining the transient behavior of systems are presented in Chaps. 5 through 12). It is possible to combine the basic features of a proportional controller and an integral controller to form a proportional plus integral controller.

The action of a proportional plus integral controller in response to a change in the input or external disturbance is initially similar to that of a proportional controller, but as the new equilibrium point is reached, the control action becomes the same as that of an integral controller. (In effect, the slope of the controller line continually increases.)

A proportional plus integral controller combines the desirable transient characteristics of a proportional controller and the feature of no steady-state error of the integral controller.

A proportional plus integral controller is shown in Fig. 4.15. The proportional action is provided by unit 1, which is the same as that for the proportional controller

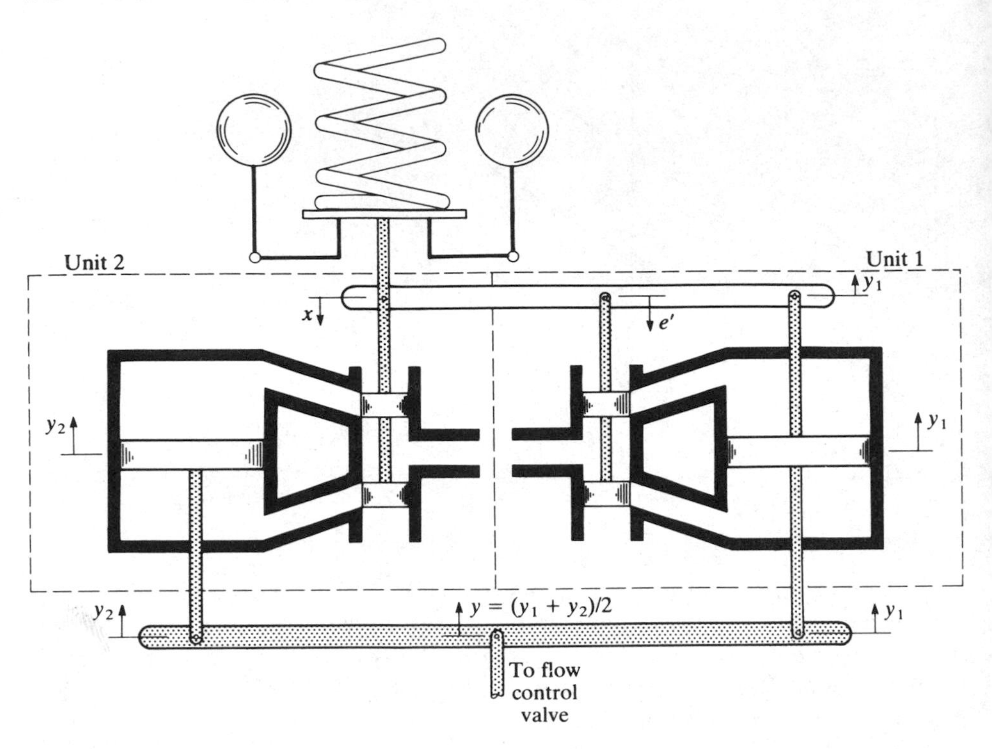

FIGURE 4.15
Proportional plus integral control system.

shown in Fig. 3.32. The equation for the proportional action is

$$y_1 = \frac{1}{1 + \tau_1 D} x \tag{4.22}$$

The integral action is provided by unit 2, which is the same as that for the integral controller shown in Fig. 4.12. The equation for this integral action is

$$y_2 = \frac{C}{AD} x \tag{4.23}$$

The proportional and integral actions are added by a walking-beam linkage such that

$$y = \frac{y_1 + y_2}{2} \tag{4.24}$$

Substitution of y_1 and y_2 into the preceding expression gives

$$y = \frac{1}{2}\left(\frac{1}{1 + \tau_1 D} + \frac{C}{AD}\right) x \tag{4.25}$$

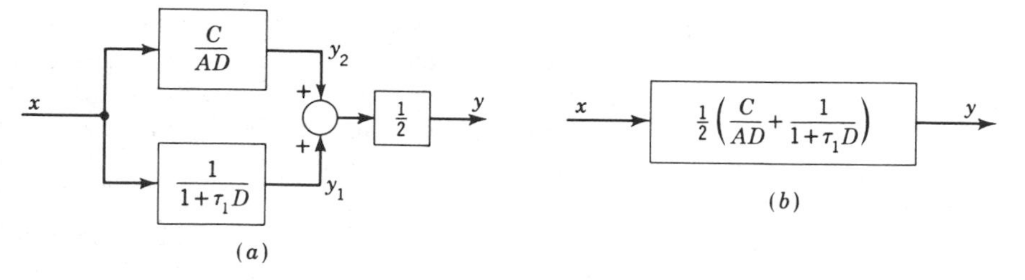

FIGURE 4.16
Block diagram for proportional plus integral action.

The individual block diagrams for Eqs. (4.22) to (4.24) are shown in Fig. 4.16*a*. The combined diagram is shown in Fig. 4.16*b*. Substitution of this combined diagram into its corresponding position between x and y of Fig. 4.13*a* yields the resulting representation for this proportional plus integral control system, shown in Fig. 4.17.

The value of K_{G_1} for this proportional plus integral control system is

$$K_{G_1} = \frac{C_5}{2(K_s - C_r C_3)} \left(\frac{1}{1 + \tau_1 D} + \frac{C}{AD} \right) \Bigg|_{D=0} = \infty \tag{4.26}$$

The proportional plus integral actions are clearly evidenced by Eq. (4.26). Because K_{G_1} is infinite for steady-state operation, it follows that e is zero during steady-state operation. Thus, from Fig. 4.17,

$$C_2 K_s n_{\text{in}} - C_4 n_o = e = 0$$

or

$$n_o = \frac{C_2 K_s n_{\text{in}}}{C_4} = n_{\text{in}} \tag{4.27}$$

Comparison of Eqs. (4.27) and (4.20) shows that the steady-state operation of a proportional plus integral control system is the same as that of an integral control system alone. A proportional plus integral control is sometimes referred to as a

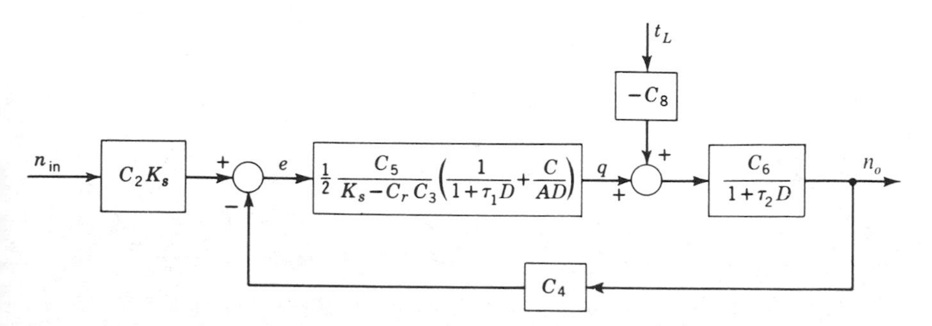

FIGURE 4.17
Overall block diagram for proportional plus integral control system.

compensated isochronous control. To better understand the action of this control, suppose that the throttle lever is moved to increase the speed. This causes the position x in Fig. 4.15 to move down, as does e'. The time constant τ_1 of the proportional unit is small, so that y_1 changes rapidly to increase the flow setting. The resulting motion of y_1 returns e' to its line-on-line position.

For the integrating unit, the quantity C/A is small, so that y_2 continues to move at a slower rate to provide corrective action. As the speed increases, the position x moves up. The integrating unit continues to provide corrective action until x is returned to its line-on-line position (that is, $x = 0$). In summary, for proportional plus integral control, the initial effect is provided primarily by the proportional action, and the final effect is provided by the integrator.

4.6 MODES OF CONTROL

For a proportional controller the manipulated variable is proportional to the error signal. That is,

$$m = K_p e \tag{4.28}$$

A plot of this relationship is shown in Fig. 4.18*a*.

For an integral controller, the manipulated variable is proportional to the integral of the error. Thus,

$$m = K_i \int e\,dt = \frac{K_i}{D} e \tag{4.29}$$

The horizontal line labeled e in Fig. 4.18*b* represents the case in which the error e is maintained constant. For a constant error e, the manipulated variable is $m = K_i et$. Thus, the manipulated variable is the ramp function labeled m.

A desirable feature of an integral controller is that there is no steady-state error due to variations in the load. The undesirable feature is that integral controllers tend to decrease stability with the result that a low system gain must be used to obtain a stable system. A high system gain tends to make the system more accurate and minimizes the effects of friction and other inaccuracies in the system.

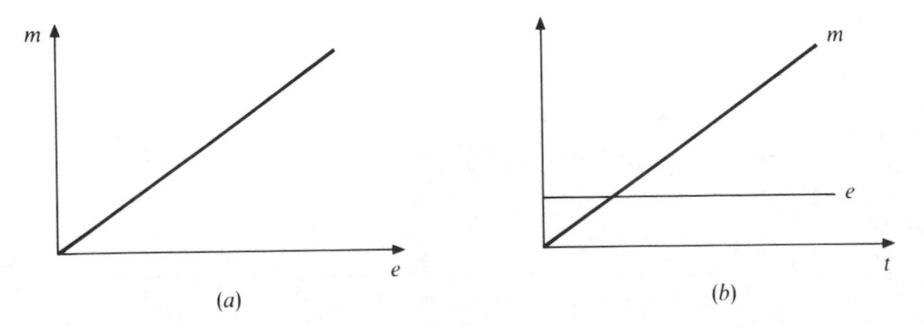

FIGURE 4.18
Corrective action for (*a*) proportional controller and (*b*) integral controller.

A proportional plus integral controller combines the desirable features of both proportional and integral controllers. That is, there is no steady-state error due to changes in the load, and the system gain can be made quite high before the transient response begins to deteriorate. That is, excessive oscillations begin to occur.

The equation relating the output m to the input e for a proportional plus integral controller is

$$m = K_p e + K_i \int e\,dt \tag{4.30}$$

Due to the proportional action, for a step change e in the error the manipulated variable undergoes a constant change $m_p = K_p e$. *Integral time* T_i is defined as the time required for the integral action to cause the same corrective action as the proportional action. That is,

$$m_i = K_i \int e\,dt = K_i e \int dt = K_i e T_i = K_p e$$

Thus

$$K_i = \frac{K_p}{T_i} \tag{4.31}$$

A plot of the proportional and integral corrective actions is shown in Fig. 4.19a. The equation for a proportional plus integral controller may be written in the form

$$m = K_p e + K_i \int e\,dt = K_p \left(1 + \frac{1}{T_i D}\right) e \tag{4.32}$$

When a change first occurs, the integral of $e\,dt$ is very small. Thus, initially most of the corrective action is due to the proportional controller. Later the integral of $e\,dt$ becomes significantly large so that most of the corrective action is provided by the integral controller. The transient response is predominantly determined by the initial response (proportional action), and the steady-state behavior is determined by the later response (integral action).

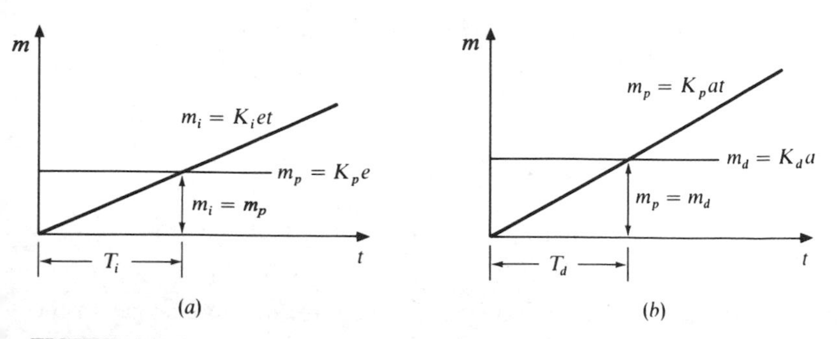

FIGURE 4.19
Corrective action for (a) proportional plus integral controller and (b) proportional plus derivative controller.

The equation for a derivative controller is

$$m = K_d \frac{d}{dt} e = K_d De \tag{4.33}$$

Derivative action is proportional to the rate of change of the error. Thus, it provides a corrective effect that tends to anticipate future values of the error.

For any steady-state value of the error, the rate of change of the error is zero. Derivative action is never used alone because for any steady-state operating condition (regardless of how big the error), it does not provide any corrective action. Consequently, derivative action is used only in combination with proportional or proportional plus integral action.

The equation of operation for a proportional plus derivative controller is

$$m = K_p e + K_d \frac{d}{dt} e \tag{4.34}$$

For a ramp input $e = at$, the derivative action is

$$m_d = K_d D(at) = K_d a$$

Derivative time T_d is defined as the time such that the corrective effect m_p due to the proportional action is equal to that due to the derivative action. That is,

$$m_p = K_p e = K_p at = K_p a T_d = K_d a$$

Thus

$$K_d = K_p T_d \tag{4.35}$$

A plot of the derivative and proportional corrective action is shown in Fig. 4.19*b*.

The equation for a proportional plus derivative controller may be written in the form

$$m = K_p e + K_d De = K_p(1 + T_d D)e \tag{4.36}$$

Because there is no integral action, there is a steady-state error due to variations in the load.

Proportional plus integral plus derivative (PID) controllers are the most commonly used controllers. They provide the greatest versatility and flexibility. The equation of operation for a PID controller is

$$\begin{aligned} m &= K_p e + \frac{K_i}{D} e + K_d De \\ &= K_p \left(1 + \frac{1}{T_i D} + T_d D\right) e \end{aligned} \tag{4.37}$$

Because of the integral action, there is no steady-state error due to changes in the load. Selecting values of the controller gain K_p, the integral time T_i, and the derivative time T_d to achieve good transient behavior is referred to as *tuning the controller.*

4.7 SUMMARY

The selection of the control elements $G_1(D)$ is seen to have a predominant effect upon the steady-state operation of a system. For more complex control systems, it becomes increasingly difficult, if not impossible, to distinguish the individual modes of control. However, regardless of the various modes that may be present, it is a relatively simple matter to determine whether K_{G_1} is finite or infinite. For an infinite value, the integral action predominates and there is no steady-state error due to variations in the external disturbance. For a finite value, the system behaves as a proportional control system.

A major problem in the design of control systems is the determination of the system parameters to obtain satisfactory transient performance. The transient behavior of a system is prescribed by the differential equation of operation for the system. In the next chapter, we show how such differential equations may be solved algebraically by the use of Laplace transforms. In Chap. 6, we show that the transient behavior is governed primarily by the roots of the characteristic equation for the system. Thus, the transient characteristics of a system may be ascertained directly from a knowledge of the roots of the characteristic equation.

PROBLEMS

4.1. For the control system shown in Fig. P4.1, determine the steady-state equation relating v, u, and c for each of the following cases:

(a) $G_1(D) = \dfrac{K_1}{1+\tau_1 D}$ (b) $G_1(D) = \dfrac{K_1}{D}$

(c) $G_1(D) = \dfrac{K_1}{D} + \dfrac{K_1}{1+\tau_1 D}$

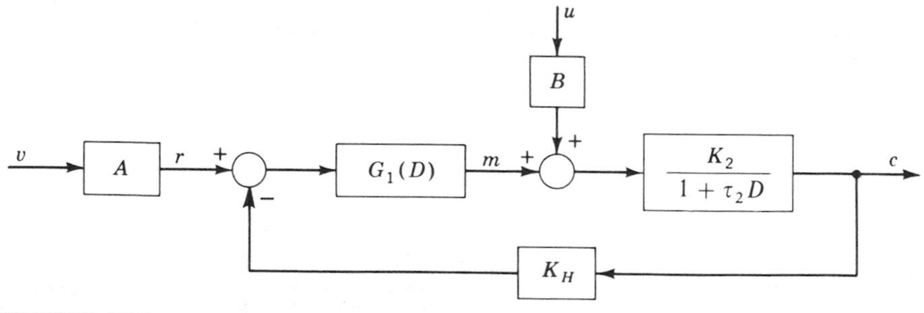

FIGURE P4.1

4.2. The steady-state operating curves for a proportional temperature control system are shown in Fig. P4.2.

(a) Determine the equation for steady-state operation about point A.

(b) If this were an open-loop rather than a closed-loop system, what would be the steady-state equation of operation?

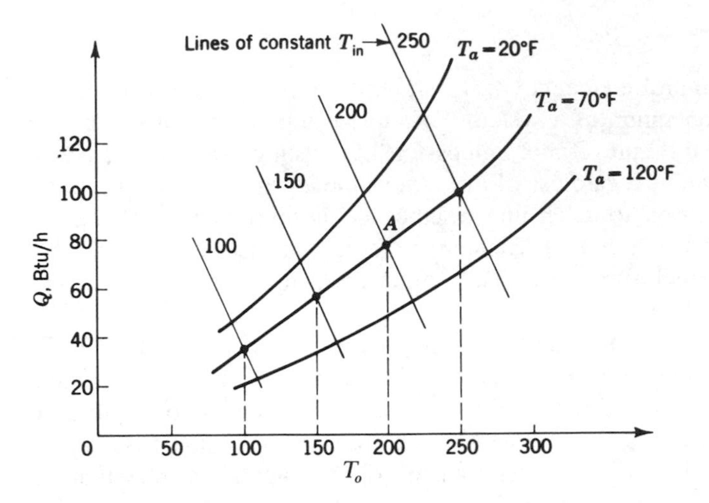

FIGURE P4.2

4.3. The block diagram representation for a speed control system is shown in Fig. P4.3*a*. Determine the differential equation of operation and the equation for steady-state operation. At the reference operating condition ($r = c = d = 0$), $R_i = C_i = 100$, $D_i = 40$, and $M_i = 60$. In Fig. P4.3*b*, sketch the load line ($D_i = 40$) and the controller line ($R_i = 100$). Also sketch the load line for $D = 60$ and the controller line for $R = 110$. Note that because $A = 1$, then $R = V$. Similarly, because $B = 1$, then $D = U$.

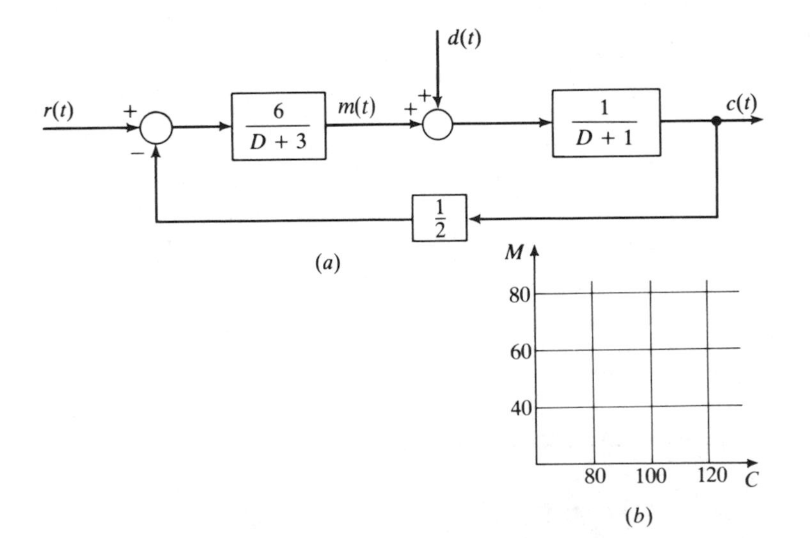

FIGURE P4.3

4.4. The block-diagram representation for a system that controls pressure is shown in Fig. P4.4*a*. Determine the differential equation of operation and the equation for steady-state operation. At the reference operating condition ($r = c = d = 0$), $R_i = C_i = 100$, $M_i = 250$, and $D_i = 150$. In Fig. P4.4*b*, sketch the load line ($D_i = 150$) and the

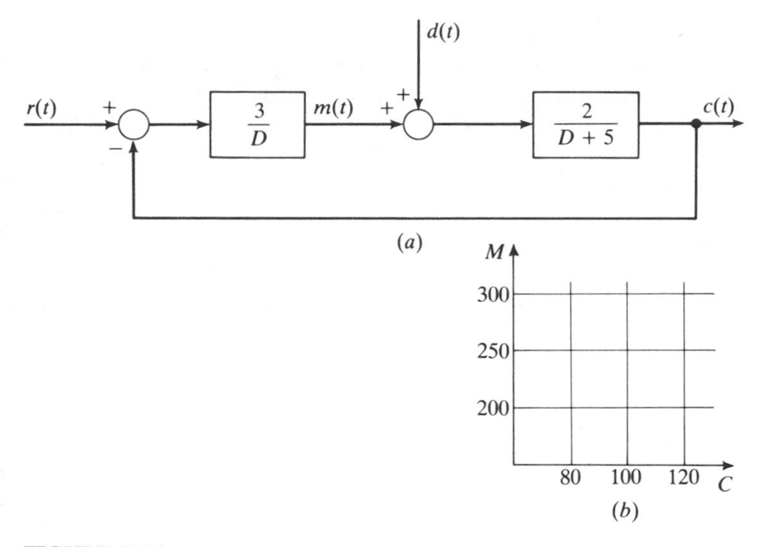

FIGURE P4.4

controller line ($R_i = 100$). Also sketch the load line for $D = 200$ and the controller line for $R = 120$. Note that for $A = 1$, then $R = V$. Similarly, for $B = 1$, then $D = U$.

4.5. The steady-state operating curves for a unity-feedback system ($K_H = 1$) are shown in Fig. P4.5. Construct the block diagram that describes the steady-state operation of this system.

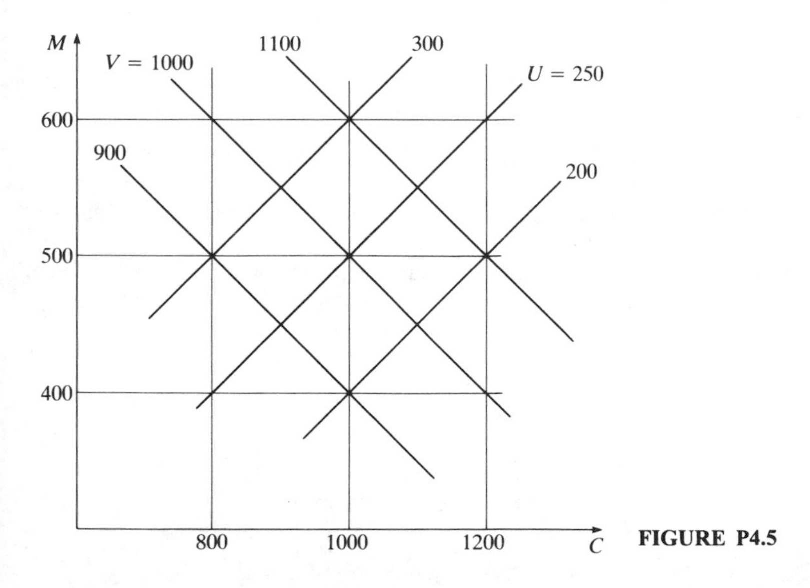

FIGURE P4.5

4.6. A controller is to be designed such that for no change in the command signal ($v = 0$), the output c will not change by more than 1 unit when the load u increases by 40 units. The steady-state operating curves for the system have a slope of 2. For $B = 0.3$, what is the required slope of the controller lines?

4.7. Repeat Prob. 4.6 except $B = -0.3$.

4.8. A controller is to be designed for a system such that the output C will not change by more than 1 unit when the load changes by 10 units. The steady-state operating curves for the system to be controlled have a slope of 5 units. What is the required slope of the controller lines when $B = 3$?

4.9. Repeat Prob. 4.8 except $B = -3$.

4.10. The steady-state operating curves for a system to be controlled are shown in Fig. P4.10. The reference operating condition is $V_i = C_i = 500$, $U_i = 100$, and $M_i = 200$. Determine the slope of the controller lines such that the controlled variable does not decrease by more than 4 percent of its reference value ($c = 0.04 \times 500 = 20$) when $U = 125$.

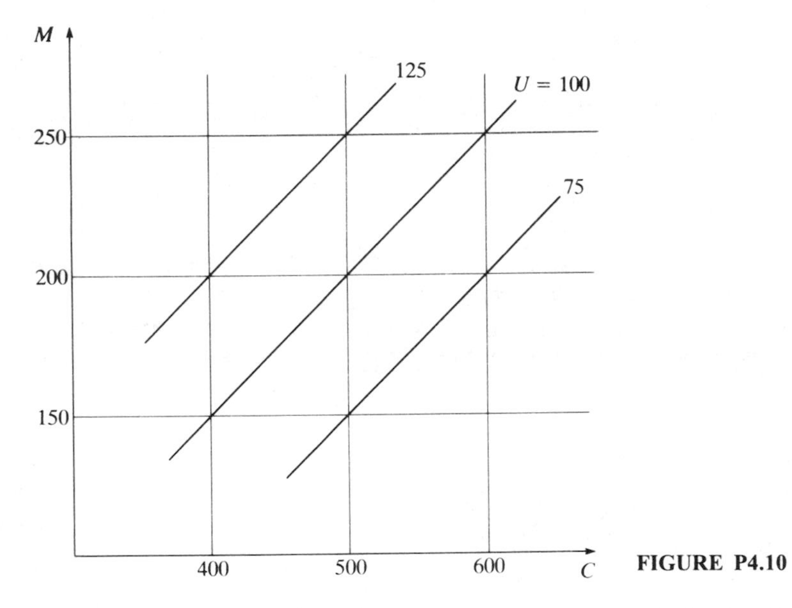

FIGURE P4.10

4.11. The block diagram for a flow control system is shown in Fig. P4.11. At the reference operating point, $V_i = C_i = 100$, $M_i = 25$, and $U_i = 50$. With V held fixed at its reference value ($V = V_i = 100$), calculate the change c in the controlled variable when U changes from its reference value $U_i = 50$ to $U = 60$. What is the new value of C?

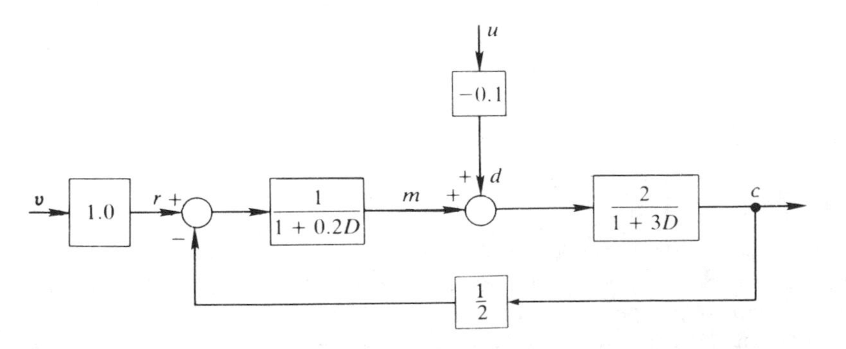

FIGURE P4.11

4.12. The block diagram for a pneumatic control system is shown in Fig. P4.12. At the reference operating point, $C_i = V_i = 500$, $M_i = 250$, and $U_i = 100$. Determine the value of A such that the coefficient of the v term in the equation for steady-state operation is unity. With the command signal V held constant at its reference value $V = V_i = 500$, determine the change c in the output (error) when the load changes from its reference value $U_i = 100$ to $U = 88$. Determine the required slope of the controller lines so as to reduce this error by a factor of 6. By what factor has the slope of the lines of constant V been changed?

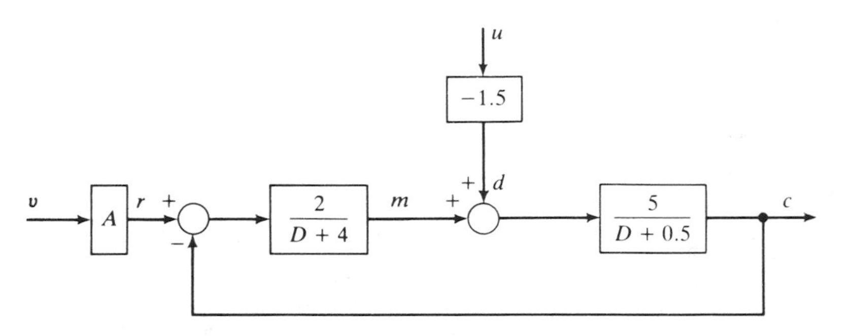

FIGURE P4.12

4.13. The block diagram of a feedback control system for operation about a reference operating point is shown in Fig. P4.13. The steady-state equation of operation is

$$c = \left.\frac{\partial C}{\partial V}\right|_U v + \left.\frac{\partial C}{\partial U}\right|_V u$$

(*a*) Determine the value of K_{G_1} such that $\partial C/\partial U|_V = 0.05$.
(*b*) Determine the value of A such that $\partial C/\partial V|_U = 1$.
(*c*) For the system to be controlled, what is the slope of the lines of constant U?
(*d*) For the system to be controlled, determine the horizontal spacing ΔC between the lines of constant U when $\Delta U = 10$ units.

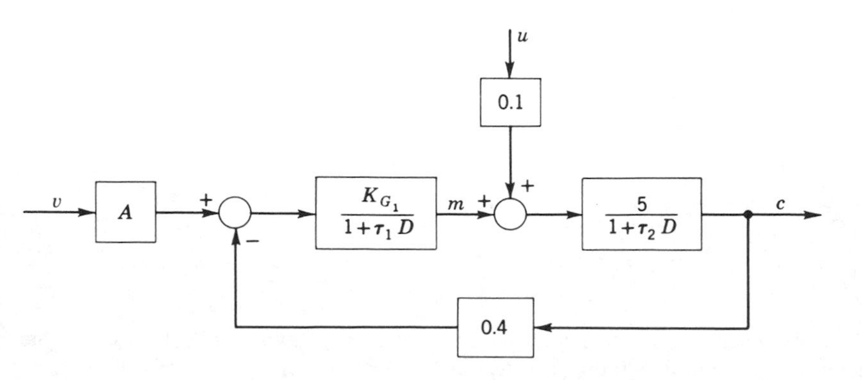

FIGURE P4.13

4.14. You are given the system shown in Fig. P4.14.
(*a*) Determine K_{G_1} such that $\partial C/\partial U|_V = 0.1$.
(*b*) Determine A such that $\partial C/\partial V|_U = 1$.

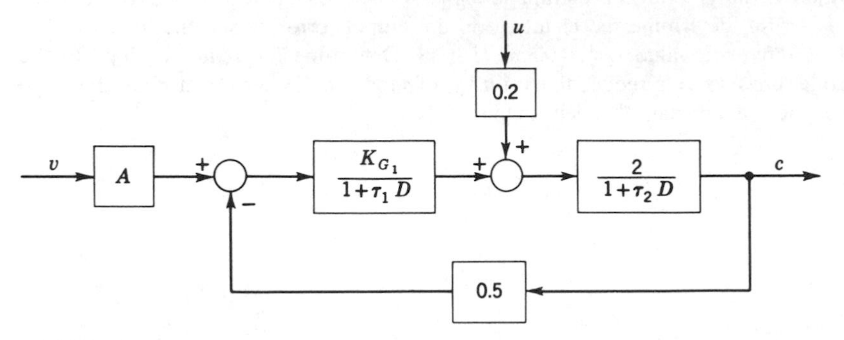

FIGURE P4.14

4.15. The reference point of operation for the system shown in Fig. P4.15*a* is $R_i = C_i = 1000$, $M_i = 200$, and $D_i = 100$. Because $A = B = 1$, then $V = R$ and $U = D$. Insert values on the M axis of Fig. P4.15*b* where the question marks appear such that the load line $D_i = 100$ has the correct slope. Draw in the load line for $D = 150$. Draw in the controller line for $R_i = 1000$.

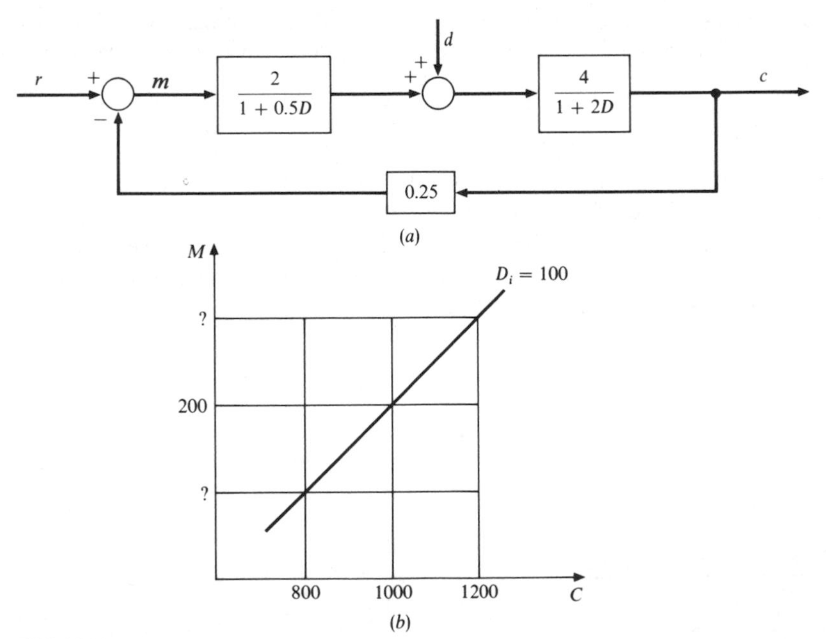

FIGURE P4.15

4.16. The block diagram for a hydraulic control system is shown in Fig. P4.16*a*. At the reference operating condition, $R_i = C_i = 100$, $D_i = 50$, and $M_i = 25$. Note because $A = B = 1$, then $V = R$ and $U = D$. The load line at the reference operating condition

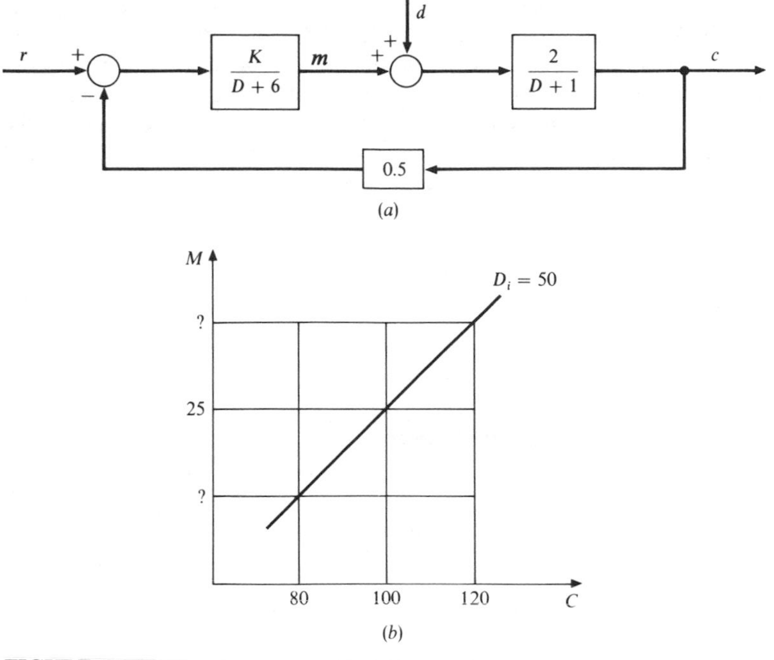

FIGURE P4.16

is drawn. Insert values on the M axis of Fig. P4.16b where the question marks appear such that the load line $D_i = 50$ has the correct slope. Next sketch the load line for $D = 40$. Determine the required value of K such that when the load changes from $D_i = 50$ to $D = 40$ (command signal $V = R$ remains constant), the controlled variable changes by 1 unit.

4.17. The load line $U_i = 25$ is shown in Fig. P4.17. Put in values on the C axis of Fig. P4.17 where the question marks appear such that the slope is 0.2. For $B = -0.2$, draw in the load line for $U = 30$. The controller is to be designed such that the change in the output c will be 1 unit when the load changes by 5 units. Determine the required slope of the controller lines.

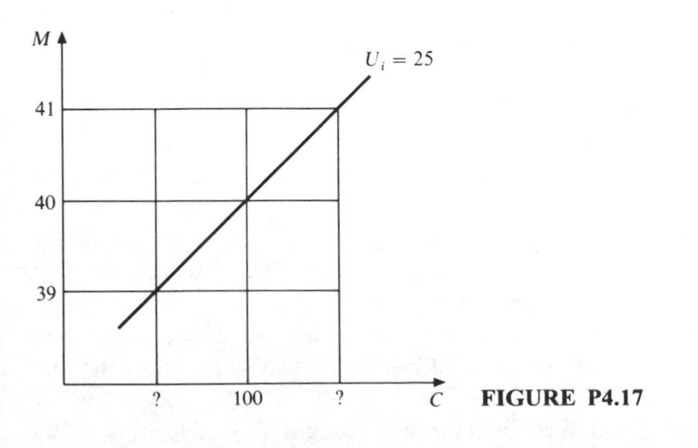

FIGURE P4.17

4.18. The block diagram for an industrial temperature control system is shown in Fig. P4.18*a*. At the reference operating point, $V_i = C_i = 100$, $M_i = 40$, and $U_i = 20$. The $U_i = 20$ load line is shown in Fig. P4.18*b*. Insert values on the C axis of Fig. P4.18*b* where the question marks appear such that it has the correct slope. Draw in the $U = 30$ load line. Determine the value of K_1 such that when $V = V_i$, the change of the controlled variable c will be 1 unit when U changes from its reference value $U_i = 20$ to $U = 30$. Determine A such that the coefficient of the v term in the equation for steady-state operation is unity.

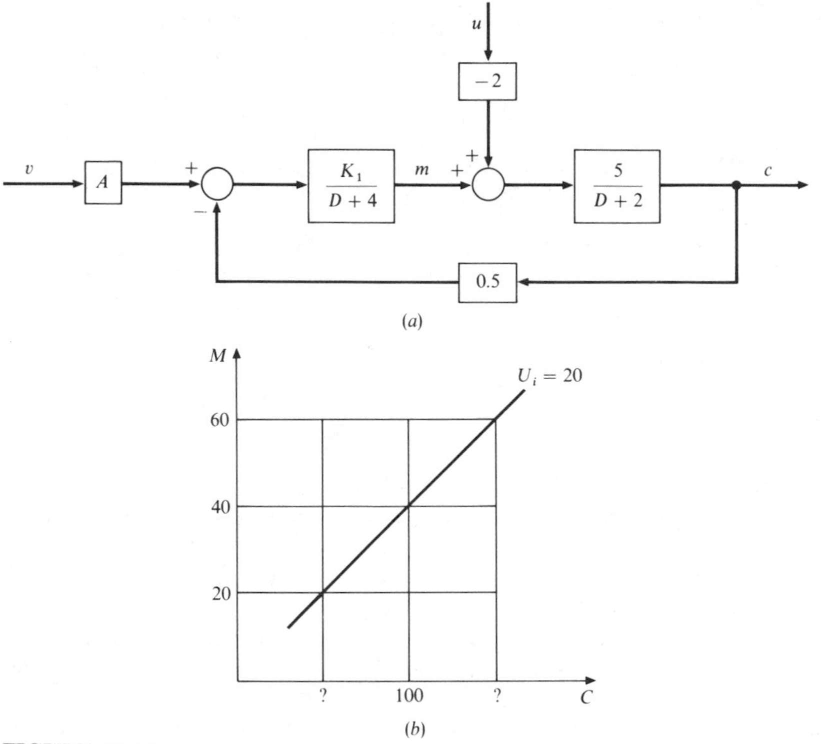

FIGURE P4.18

4.19. The steady-state operating curves for a controller are shown in Fig. P4.19. At the reference operating condition, $C_i = V_i = 100$, $M_i = 25$, and $U_i = 10$.

(*a*) What are the values of AK_{G_1} and $K_{G_1}K_H$? Determine K_{G_2} such that when $u = 0$, then $c = r$. Draw the reference load line $U_i = 10$.

(*b*) For $v = 0$, it is desired that when U changes from 10 to 15, the change in the controlled variable c will be $\frac{5}{3}$. Determine the required value of B, and draw in the load line for $U = 15$.

4.20. The block diagram for a system that controls the steering of a ship is shown in Fig. P4.20*a*. The reference point of operation is $V_i = C_i = 100$, $M_i = 50$, and $U_i = 10$. Write the equation for steady-state operation.

(*a*) Put the correct numbers on the M axis of the steady-state operating curves shown in Fig. P4.20*b* so that the load line $U_i = 10$ has the correct slope. Draw the load line for $U = 12$.

(*b*) Determine the value of K such that when the load changes from 10 to 12 units, the

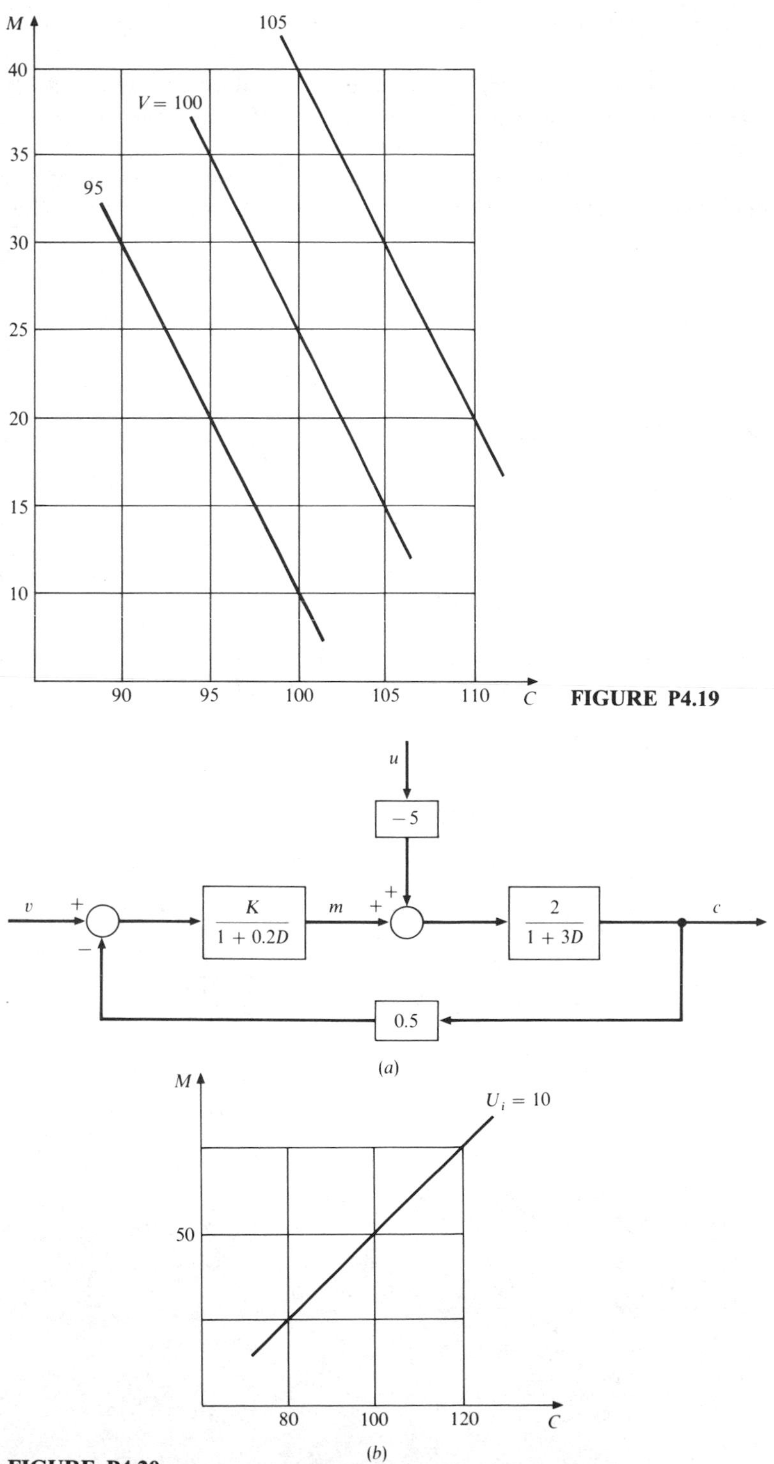

FIGURE P4.19

FIGURE P4.20

change in the controlled variable will not exceed 5 units. Draw the resulting controller line $V_i = 100$.

4.21. The steady-state operating curves for a system to be controlled are shown in Fig. P4.21. At the reference operating point, $C_i = V_i = 500$, $M_i = 20$, and $U_i = 100$.

(*a*) Determine B and K_{G_2} at the reference operating point.

(*b*) Determine the required slope of the lines of constant V such that the variation c will not exceed 2 percent of its reference value when U changes from 100 to 120.

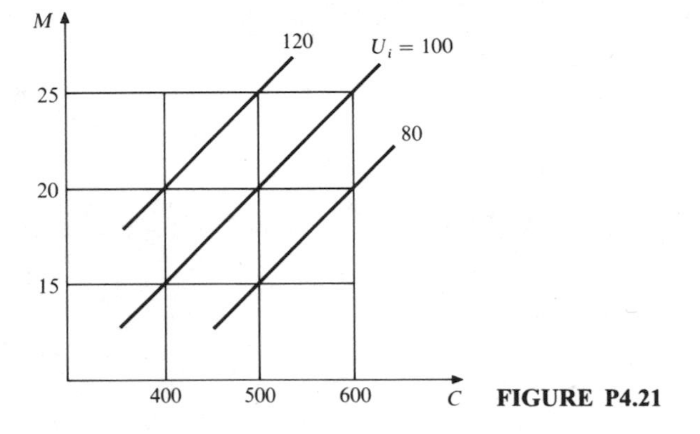

FIGURE P4.21

4.22. The block diagram and steady-state operating curves for a system that controls tension are shown in Fig. P4.22. At the reference operating condition, $V_i = C_i = 100$, $M_i = 50$,

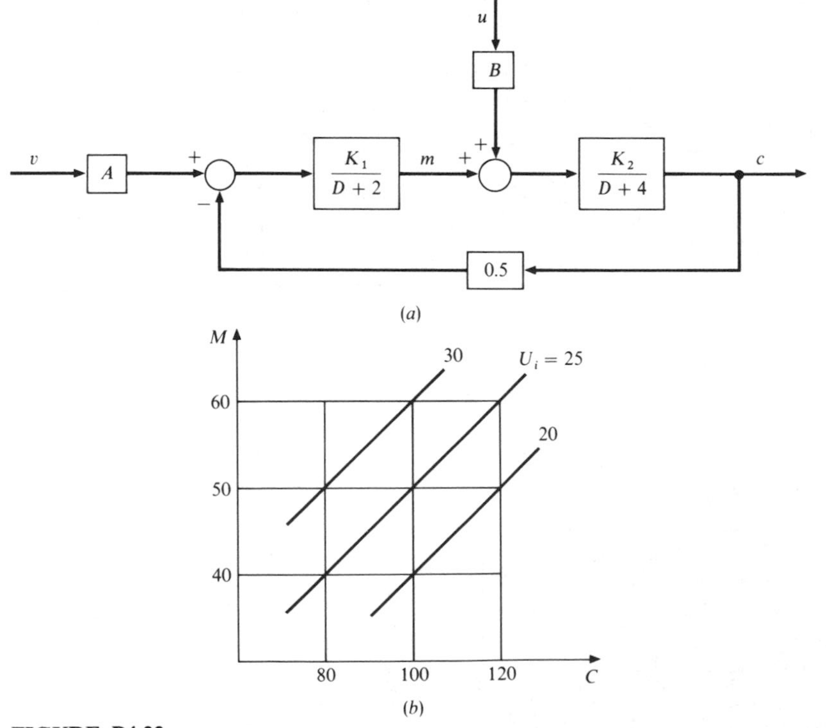

FIGURE P4.22

and $U_i = 25$. Determine K_2 and B. Determine K_1 such that when $V = V_i$ ($v = 0$), the change in the controlled variable will be 1 unit when the load changes by 5 units.

4.23. The steady-state curves for a system to be controlled are shown in Fig. P4.23. The reference operating condition is $V_i = C_i = 100$, $U_i = 40$, and $M_i = 60$. Determine the values of B and K_{G_2}. Determine the required slope of the controller lines such that when the load changes from 40 to 50, the output C will not change by more than 2 units. For $K_H = 1.6$, determine the value of A such that $c = v$ when $u = 0$.

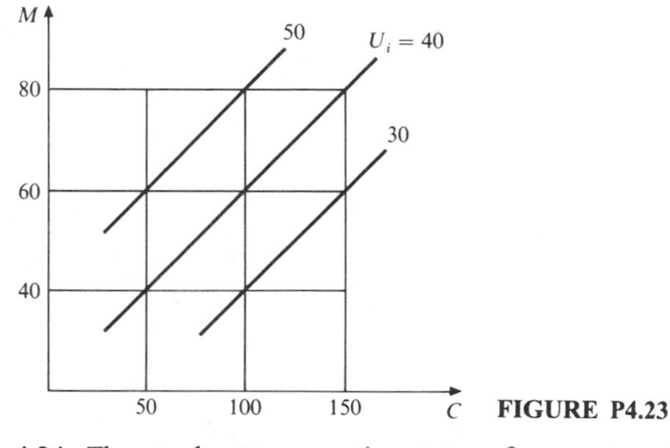

FIGURE P4.23

4.24. The steady-state operating curves for a system to be controlled are shown in Fig. P4.24.

(*a*) Determine B and K_{G_2} at the reference operating point P.

(*b*) Determine the required slope of the controller lines ($-K_{G_1}K_H$) such that the variation in C will not exceed 2 percent of its reference value [$c = 0.02(200) = 4$ units] when the external disturbance varies from its reference value to its maximum value ($u = U_{max} - U_i = 20$ units).

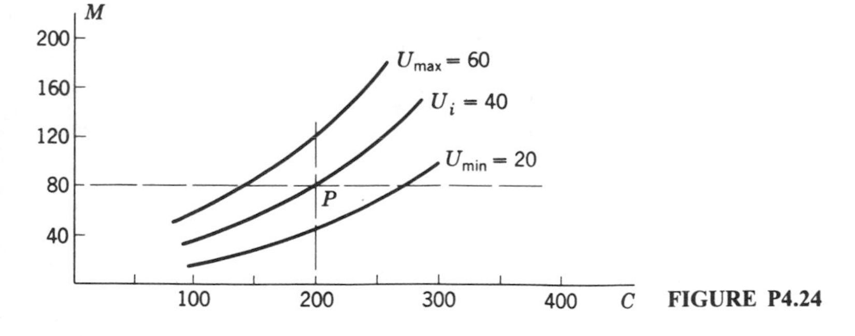

FIGURE P4.24

4.25. At the reference operating point, $C_i = 100$, $M_i = 5$, and $U_i = 10$, the characteristics of the system to be controlled are $B = 0.5$ and $K_{G_2} = 25$. Determine the slope of the load lines U.

(*a*) Draw the nominal load line $U = 10$ for the system to be controlled; also draw the $U = 6$ and the $U = 14$ load lines. Note: Straight lines will suffice because we are working with linearized information about the reference operating point.

(*b*) Calculate the required slope of the controller line such that the output C will not change by more than 1 unit when the load is changed from its nominal value $U_i = 10$ to $U = 14$.

4.26. Repeat Prob. 4.25 except $B = -0.5$.

4.27. The linearized equation of operation for an engine $N = N(Q, T)$ about the reference point $N_i = 2000$ r/min, $T_i = 400$ ft · lb, and $Q_i = 40$ lb/min is

$$n = 50q - 10t$$

Determine $\partial N/\partial Q\,|_i$, $\partial N/\partial T\,|_i$, and $\partial T/\partial Q\,|_i$. Sketch the operating curve of the engine for $T_i = 400$ ft · lb and the curve for $T = 500$ ft · lb.

4.28. The block diagram for an integral control system is shown in Fig. P4.28. At the reference operating condition $c = u = v = 0$.

(*a*) Determine the value of A such that $c = v$ for steady-state operation.

(*b*) For $u = \Delta U = 10$ units, what is the vertical distance ΔM between the lines of constant U?

(*c*) For $v = \Delta V = 10$ units, what is the horizontal distance ΔC between the lines of constant V?

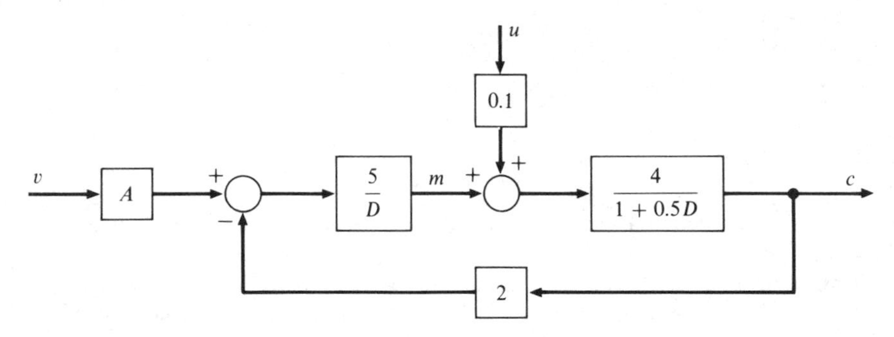

FIGURE P4.28

4.29. The steady-state operating curves for a unity-feedback system are shown in Fig. P4.29.

(*a*) Evaluate B, K_{G_2}, K_{G_1}, K_H, AK_{G_1}, $\partial C/\partial V\,|_U$, and $\partial C/\partial U\,|_V$.

(*b*) If the slope of the controller lines is increased by a factor of 10, what steady-state error results for $V = 400$ ($v = 0$) when U changes from 12 to 10 ($u = -2$)?

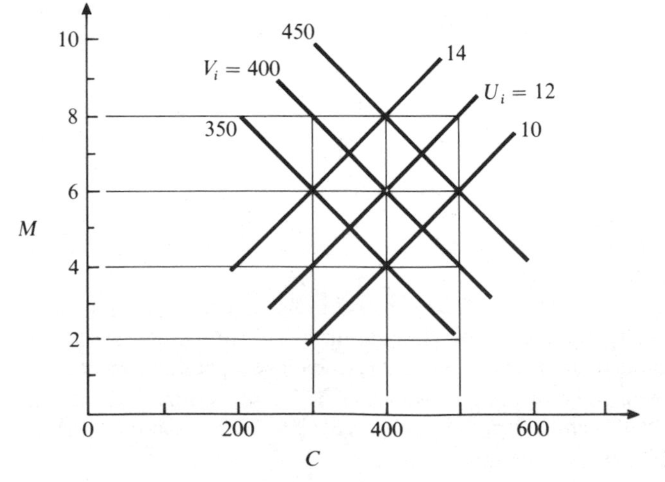

FIGURE P4.29

4.30. For the integral control system shown in Fig. P4.30, determine

(*a*) The slope of the controller lines and the slope of the load lines;

(*b*) The value of A such that $c = v$ for steady-state operation;

(*c*) The value of B such that the horizontal spacing between the load lines is $\Delta C = 0.5\ \Delta U$.

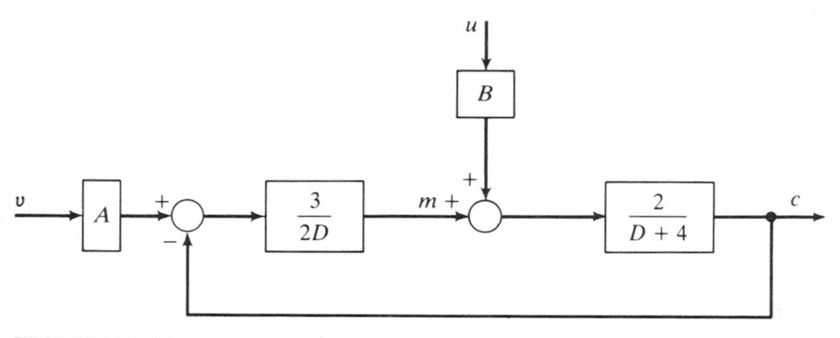

FIGURE P4.30

4.31. The block-diagram representation for a system to control the speed of an engine is shown in Fig. P4.31. The reference operating point is $C_i = V_i = 4000$ r/min, $M_i = 2000$ lb/h, and $U_i = 200$ ft · lb.

(*a*) Determine A such that $c = v$ when $u = 0$.

(*b*) For $V = V_i = 4000$ r/min ($v = \Delta V = 0$), determine the speed error $c = \Delta C$ when $U = 150$ ft · lb ($u = \Delta U = U - U_i = -50$).

(*c*) Determine the required slope of the controller lines such that this speed error is reduced to 25 r/min.

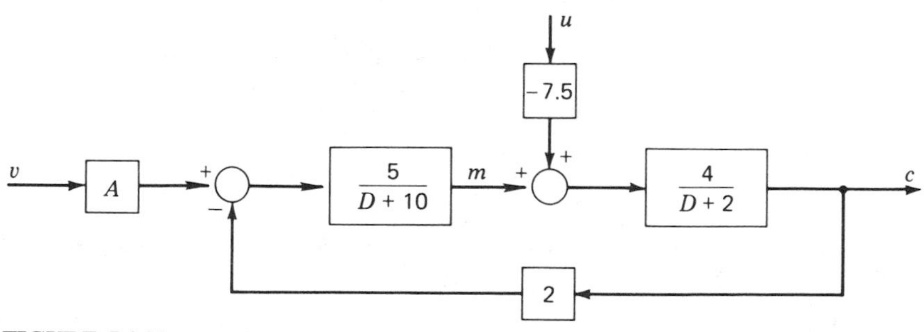

FIGURE P4.31

4.32. When the system to be controlled has an integrating element such as to make K_{G_2} infinite, the steady-state operating curves for the system become horizontal straight lines, as shown in Fig. P4.32. The steady-state operating curves for a remote-control positioning device are shown in Fig. P4.32. The controlled shaft position is designated by θ_o and the set position by θ_{in}. Variations are to be indicated by $\Delta\theta$. For this remote-control positioning system, the external disturbance is a load torque T_L, and the manipulated variable is a motor torque T_m.

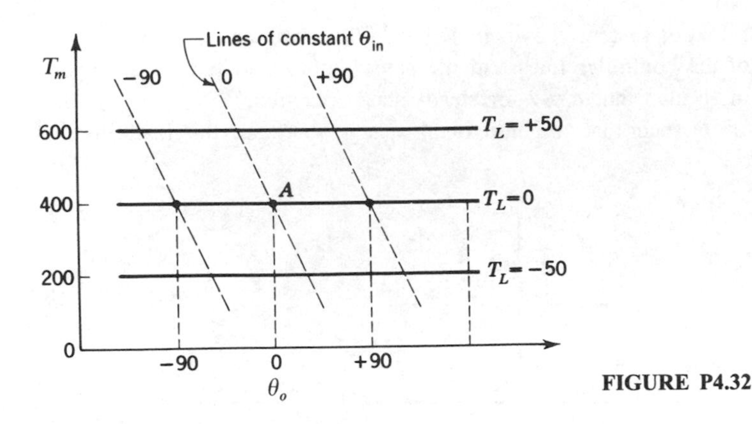

FIGURE P4.32

(a) Determine the steady-state equation of operation relating $\Delta\theta_o$, $\Delta\theta_{in}$, and t_L for operation about point A.

(b) To decrease the error caused by variations in the external disturbance by a factor of 5, what should be the new slope of the controller lines?

4.33. A controller is to be designed for a system such that the output C will not change by more than 1 unit when the load U increases by 40 units. The steady-state operating curves for the system to be controlled have a slope of 0.2 units. What is the required slope of the controller lines (a) when $B = 0.1$ and (b) when $B = -0.1$?

4.34. Figure P4.34 shows the steady-state block diagram for a system which is subjected to two external disturbances. Determine the steady-state equation relating c, v, u_1, and u_2. For which of the following cases will the steady-state operation be independent of u_1? For which cases will it be independent of variations in u_2? Note: Steady-state operation is independent of K_{G_3}.

(a) $K_{G_1} = \infty$, K_{G_2} and K_H are finite

(b) $K_{G_2} = \infty$, K_{G_1} and K_H are finite

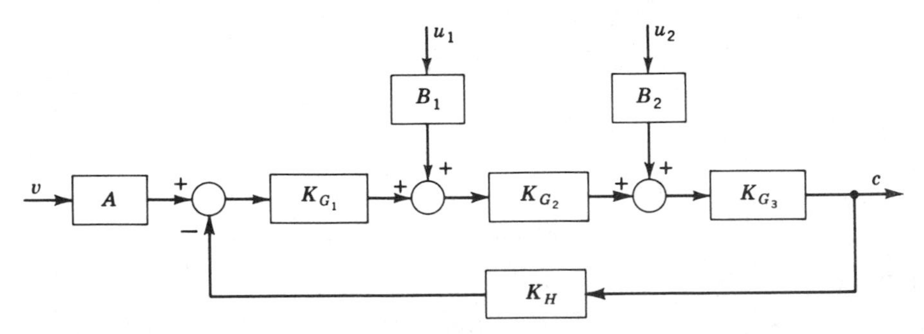

FIGURE P4.34

4.35. A reproducing shaper is shown in Fig. P4.35. The position y of the duplicating cutter is seen to follow the position x of the master cutter. Determine the mode of operation of the shaper. What modifications would be necessary to convert this to a proportional plus integral controller?

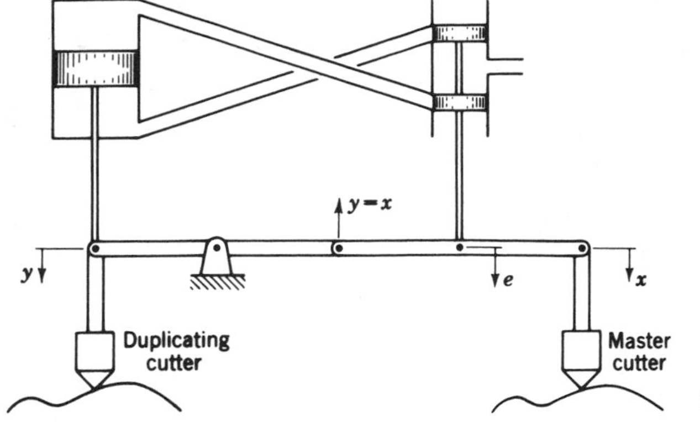

FIGURE P4.35

CHAPTER 5

LAPLACE TRANSFORMS

By *transient response* is meant the manner in which a system changes from some initial operating condition to some final condition. For example, in Fig. 5.1 we see that at some arbitrary time $t = 0$ the output is $y(0)$. The curve marked (a) represents the transient response of a system in which $y(t)$ slowly approaches its new operating condition. The curve marked (b) shows a system which successively overshoots and undershoots, but these oscillations gradually die out as the new operating condition is obtained.

A linear control system or component is one in which the operation is described by a linear differential equation, usually with constant coefficients. For a known input, classical methods could be used for determining the output. However, considerable time is saved by using the Laplace transform method of solving linear differential equations. In addition, as explained in later chapters, the Laplace transform analysis is closely related to other methods for evaluating system performance.

A brief review of classical methods for solving differential equations with constant coefficients is presented first. This is done so that when the Laplace transform method is explained, a clearer understanding of the similarities and the differences between the two methods results.

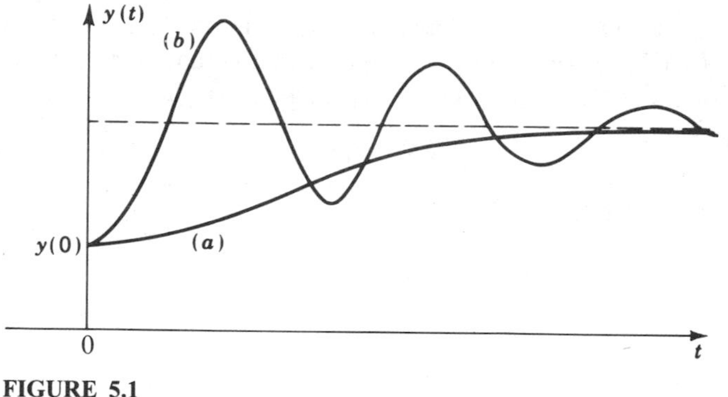

FIGURE 5.1
Transient response.

5.1 CLASSICAL METHODS

The transient response of a control system may be obtained by solving differential equations of the general form

$$(D^n + b_{n-1}D^{n-1} + \cdots + b_1 D + b_0)y(t)$$
$$= (a_m D^m + a_{m-1}D^{m-1} + \cdots + a_1 D + a_0)f(t)$$

Solving for $y(t)$ gives

$$y(t) = \frac{a_m D^m + a_{m-1}D^{m-1} + \cdots + a_1 D + a_0}{D^n + b_{n-1}D^{n-1} + \cdots + b_1 D + b_0} f(t) \tag{5.1}$$

where $a_0, a_1, \ldots, a_m$ and $b_0, b_1, \ldots, b_{n-1}$ are constants. The term $f(t)$ represents the excitation to the system. This is called the *forcing function* because it forces or excites the system. The output $y(t)$ is called the *response function* because it responds to the forcing function $f(t)$. The denominator of Eq. (5.1) is the characteristic function of the differential equation. The equation which results from setting the characteristic function equal to zero is called the *characteristic equation*. The value of the exponent n, the highest power of D in the characteristic function, is the order of the differential equation. We now show how Eq. (5.1), which is an nth-order differential equation, may be written as the sum of n first-order differential equations. First we rewrite Eq. (5.1) in the form

$$y(t) = \frac{L_m(D)}{L_n(D)} f(t) \tag{5.2}$$

where $L_n(D) = D^n + b_{n-1}D^{n-1} + \cdots + b_1 D + b_0$
$L_m(D) = a_m D^m + a_{m-1}D^{m-1} + \cdots + a_1 D + a_0$

The characteristic function $L_n(D)$ may be factored into the form

$$L_n(D) = (D - r_1)(D - r_2) \cdots (D - r_n) \tag{5.3}$$

where $r_1, r_2, \ldots, r_n$ are the roots of the characteristic equation $L_n(D) = 0$. In factoring the polynomial $L_n(D)$ as shown in Eq. (5.3), note that D may be treated as an algebraic quantity. For example, consider the function

$$L_n(D) = D^2 + 3D + 2 = (D - r_1)(D - r_2)$$

Setting $L_n(D)$ equal to zero yields

$$D^2 + 3D + 2 = 0$$

The roots of this equation are

$$r_{1,2} = \frac{-3 \pm \sqrt{9 - 8}}{2} = -1, \ -2$$

Therefore,

$$L_n(D) = [D - (-1)][D - (-2)] = (D + 1)(D + 2)$$

Because $r_1, r_2, \ldots, r_n$ are the values of D for which $L_n(D) = 0$, then $r_1, r_2, \ldots, r_n$ are also called the *zeros* of the function $L_n(D)$. Thus, the roots of the characteristic equation $L_n(D) = 0$ are the zeros of the characteristic function $L_n(D)$. The zeros of $L_n(D)$ are said to be *distinct* if each zero has a different value (that is, $r_1 \neq r_2 \neq r_3 \neq \cdots \neq r_n$). When two or more zeros are equal, the characteristic function is said to have *repeated zeros*.

For distinct zeros, $L_m(D)/L_n(D)$ in Eq. (5.2) may be written in the form

$$\frac{L_m(D)}{L_n(D)} = \frac{K_1}{D - r_1} + \frac{K_2}{D - r_2} + \cdots + \frac{K_i}{D - r_i} + \cdots + \frac{K_n}{D - r_n} \tag{5.4}$$

The procedure for obtaining any constant K_i is as follows: First multiply both sides of Eq. (5.4) by $D - r_i$:

$$(D - r_i)\frac{L_m(D)}{L_n(D)} = \frac{D - r_i}{D - r_1}K_1 + \frac{D - r_i}{D - r_2}K_2 + \cdots + K_i + \cdots + \frac{D - r_i}{D - r_n}K_n \tag{5.5}$$

The multiplication of the K_i term in Eq. (5.4) by $D - r_i$ cancels the denominator, thus leaving K_i alone, as shown in Eq. (5.5). By letting $D = r_i$ in Eq. (5.5), each term on the right side becomes zero except for K_i, which remains. Thus

$$K_i = \lim_{D \to r_i}\left[(D - r_i)\frac{L_m(D)}{L_n(D)}\right] \tag{5.6}$$

Successive application of Eq. (5.6), in which $i = 1, 2, \ldots, n$, yields each of the constants $K_1, K_2, \ldots, K_n$, respectively, in Eq. (5.4). As an example of the use of this partial-fraction expansion technique, suppose we want to expand $L_m(D)/L_n(D)$, where

$$\frac{L_m(D)}{L_n(D)} = \frac{5D + 8}{D^2 + 3D + 2} = \frac{K_1}{D + 2} + \frac{K_2}{D + 1}$$

Application of Eq. (5.6) yields

$$K_1 = \lim_{D \to -2}\left[(D+2)\frac{5D+8}{(D+2)(D+1)}\right] = \lim_{D \to -2}\frac{5D+8}{D+1} = \frac{-10+8}{-2+1} = 2$$

and

$$K_2 = \lim_{D \to -1}\left[(D+1)\frac{5D+8}{(D+2)(D+1)}\right] = \lim_{D \to -1}\frac{5D+8}{D+2} = \frac{-5+8}{-1+2} = 3$$

Thus

$$\frac{L_m(D)}{L_n(D)} = \frac{2}{D+2} + \frac{3}{D+1}$$

The general form for expressing an nth-order differential equation as the sum of n first-order differential equations is obtained by substitution of $L_m(D)/L_n(D)$ from Eq. (5.4) into Eq. (5.2). That is,

$$\begin{aligned} y(t) &= \frac{K_1}{D-r_1}f(t) + \cdots + \frac{K_i}{D-r_i}f(t) + \cdots + \frac{K_n}{D-r_n}f(t) \\ &= \sum_{i=1}^{n}\frac{K_i}{D-r_i}f(t) = \sum_{i=1}^{n}K_i y_i(t) \end{aligned} \tag{5.7}$$

where

$$y_i(t) = \frac{1}{D-r_i}f(t)$$

To solve this differential equation, first write it in the form

$$\frac{d}{dt}y_i(t) - r_i y_i(t) = f(t) \tag{5.8}$$

Next, multiply through by the integrating factor $e^{-r_i t}$.

$$e^{-r_i t}\frac{d}{dt}y_i(t) - r_i e^{-r_i t}y_i(t) = \frac{d}{dt}[e^{-r_i t}y_i(t)] = f(t)e^{-r_i t}$$

Note that the integrating factor makes it possible to write the left side as a perfect differential. Integration of the second and last terms on the right side shows that

$$e^{-r_i t}y_i(t) = \int f(t)e^{-r_i t}\,dt + c_i$$

Multiplying through by $e^{r_i t}$ yields

$$y_i(t) = e^{r_i t}\left[\int f(t)e^{-r_i t}\,dt + c_i\right]$$

or

$$y_i(t) = c_i e^{r_i t} + e^{r_i t}\int f(t)e^{-r_i t}\,dt \tag{5.9}$$

Because the constant of integration c_i is displayed separately in Eq. (5.9), it suffices to evaluate the integral at time t only. Substitution of Eq. (5.9) into Eq. (5.7) yields the following general solution:

$$y(t) = \sum_{i=1}^{n} k_i e^{r_i t} + \sum_{i=1}^{n} K_i e^{r_i t} \int f(t) e^{-r_i t}\, dt \tag{5.10}$$

where $k_i = c_i K_i$ is a constant. The first summation on the right side of Eq. (5.10) is the homogeneous solution, and the second summation is the particular solution. That is,

$$y_h(t) = \sum_{i=1}^{n} k_i e^{r_i t} \tag{5.11}$$

and

$$y_p(t) = \sum_{i=1}^{n} K_i e^{r_i t} \int f(t) e^{-r_i t}\, dt \tag{5.12}$$

where $y_h(t)$ is the homogeneous solution and $y_p(t)$ is the particular solution.

Illustrative example 5.1. Determine the solution of the following differential equation when all the initial conditions are zero and $f(t) = 2$.

$$(D^2 + 3D + 2)y(t) = f(t) \tag{5.13}$$

or

$$y(t) = \frac{1}{(D+1)(D+2)} f(t) = \left(\frac{K_1}{D+1} + \frac{K_2}{D+2}\right) f(t) = \left(\frac{1}{D+1} - \frac{1}{D+2}\right) f(t)$$

where

$$K_1 = \lim_{D \to -1} \left[(D+1)\frac{1}{(D+1)(D+2)}\right] = 1$$

and

$$K_2 = \lim_{D \to -2} \left[(D+2)\frac{1}{(D+1)(D+2)}\right] = -1$$

Solution. The substitution of $r_1 = -1$ and $r_2 = -2$ into Eq. (5.11) yields for the homogeneous solution

$$y_h = k_1 e^{r_1 t} + k_2 e^{r_2 t} = k_1 e^{-t} + k_2 e^{-2t}$$

For $K_1 = 1$ and $K_2 = -1$, the particular solution is evaluated from Eq. (5.12) as follows:

$$\begin{aligned} y_p &= {}_1 e^{-t} \int 2e^t\, dt + K_2 e^{-2t} \int 2e^{2t}\, dt \\ &= e^{-t}(2e^t) - e^{-2t}(e^{2t}) = 2 - 1 = 1 \end{aligned}$$

The general solution is the sum of the homogeneous and particular solutions.

$$y(t) = k_1 e^{-t} + k_2 e^{-2t} + 1$$

The derivative $dy/dt = y'(t)$ is

$$y'(t) = -k_1 e^{-t} - 2k_2 e^{-2t}$$

Because the initial conditions are zero, $y(0) = y'(0) = 0$. Evaluation of $y(t)$ and $y'(t)$ at time $t = 0$ gives

$$y(0) = k_1 + k_2 + 1 = 0$$
$$y'(0) = -k_1 - 2k_2 = 0$$

Solving these two equations simultaneously for k_1 and k_2 yields $k_1 = -2$ and $k_2 = 1$. Thus, the desired result is

$$y(t) = 1 - 2e^{-t} + e^{-2t} \tag{5.14}$$

Different initial conditions would yield different values for the constants k_1 and k_2.

Numerous techniques have been developed for solving linear differential equations with constant coefficients. However, the method of Laplace transforms, which is described next, is best suited for solving the types of problems of interest to control engineers. In many ways, the Laplace transform method is similar to the preceding method of using the partial-fraction expansion to reduce an nth-order equation to the sum of n first-order equations. A major difference is that, in the Laplace transform method, the response due to each term in the partial-fraction expansion is determined directly from the transform table. Thus, there is no need to perform the integrations indicated by Eq. (5.10). Because initial conditions are automatically incorporated into the Laplace transform, the resulting response expression yields directly the total solution (i.e., homogeneous and particular solutions). Thus, the constants arising from the initial conditions are automatically evaluated, so that the final desired result is obtained directly.

5.2 LAPLACE TRANSFORM METHOD

This method of solving differential equations is somewhat analogous to the process of multiplying or dividing by using logarithms. In the well-known transformation of logarithms, numbers are transformed to powers of base 10 or some other base. This process in effect makes it possible to multiply and divide by using the simpler operations of addition and subtraction. After the desired answer is found in logarithms, the transformation back to the real number system is accomplished by finding antilogarithms.

In the method of Laplace transforms, transformation of the terms of the differential equation yields an algebraic equation in another variable s. Thereafter the solution of the differential equation is effected by simple algebraic manipulations in the s domain (the new variable is s rather than time t). To obtain the desired time solution, it is necessary to invert the transform of the solution from the s domain back to the time domain. Actually, for much control work, information obtained in the s domain suffices, so that it may be unnecessary to invert back to the time domain.

The Laplace transformation $F(s)$ of a function of time $f(t)$ is*

$$F(s) = \mathscr{L}[f(t)] = \int_0^\infty f(t)e^{-st}\,dt \tag{5.15}$$

where $\mathscr{L}$ is the symbol for taking the Laplace transform. The symbol $\mathscr{L}$ is read "Laplace transform of," so that $\mathscr{L}[f(t)]$ means the Laplace transform of $f(t)$. For the integral on the right side of Eq. (5.15), the variable t vanishes after evaluation between the limits of integration. Thus, the resulting expression is a function of s only [that is, $F(s)$]. The mathematical derivation of the Laplace transform method including Eq. (5.15) is presented in App. B.

Transforming Functions from the Time Domain to the s Domain

Some input functions which are frequently used for investigating the characteristics of a control system are the step function, pulse function, impulse function, exponential function, and sinusoidal function.

The actual input to a system is usually unknown. Thus, a standard test input signal is normally chosen. There is good correlation between the response of a system to a standard input and the actual response of a system during normal operation. Also the use of a standard input allows the control engineer to compare several competing designs. The step function and the impulse function are commonly used standard input signals.

STEP FUNCTION. A graphical representation of a step function is shown in Fig. 5.2. Note that for $t < 0$ the value of the function is zero, and for $t > 0$ the value of the function is h. Thus, a step function represents a change in the input from one value to another. The manner in which a system responds to a step change in the input is a good indication of the manner in which the system goes from one operating condition to another. A step function is designated by the symbol $hu(t)$, where h is the height and $u(t)$ is the symbol for a unit step function whose height is 1. Application of Eq. (5.15) gives

$$\begin{aligned} F(s) &= \mathscr{L}[hu(t)] = \int_0^\infty he^{-st}\,dt \\ &= -\frac{he^{-st}}{s}\bigg|_0^\infty = \frac{h(-e^{-s(\infty)} + e^{-s(0)})}{s} = \frac{h}{s} \end{aligned} \tag{5.16}$$

* F. H. Raven, *Mathematics of Engineering Systems*, McGraw-Hill, New York, 1966.

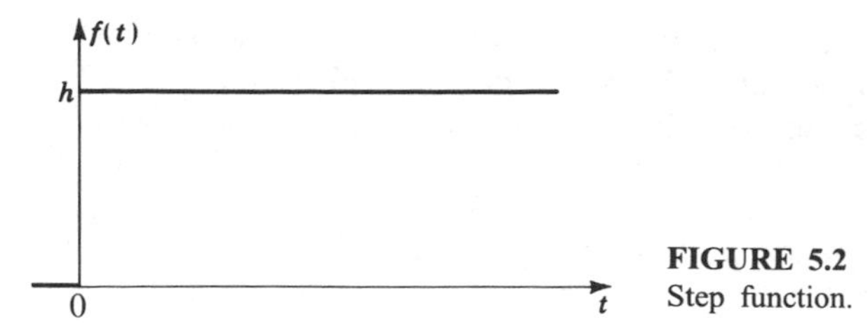

FIGURE 5.2
Step function.

In evaluating a transform, the term s is regarded as any constant which makes $F(s)$ convergent. As illustrated by Eq. (5.16), if s is any positive constant ($s > 0$), then $e^{-s(\infty)} = 0$ and $e^{-s(0)} = e^0 = 1$, so that the result follows. However, note that for negative values of s, then $e^{-s(\infty)} = e^{\infty} = \infty$, in which case $F(s)$ would be divergent. The operator s must be taken as any constant such that $F(s)$ is convergent. Although there are a range of values of s over which $F(s)$ is convergent, there is but one transform $F(s)$ corresponding to each time function $f(t)$. Table 5.1 shows a list of time functions $f(t)$ and their corresponding transforms $F(s)$. In solving problems by using Laplace transforms, the term s acts as a dummy operator, and thus there is no need to know the range of values over which $F(s)$ exists.

TABLE 5.1
Laplace transform pairs

$f(t)$	$F(s)$
$\delta(t)$	1
$u(t)$	$\frac{1}{s}$
t	$\frac{1}{s^2}$
e^{at}	$\frac{1}{s-a}$
t^n	$\frac{n!}{s^{n+1}}$
$t^n e^{at}$	$\frac{n!}{(s-a)^{n+1}}$
$\sin \omega t$	$\frac{\omega}{s^2+\omega^2}$
$\cos \omega t$	$\frac{s}{s^2+\omega^2}$
$e^{at} \sin \omega t$	$\frac{\omega}{(s-a)^2+\omega^2}$
$e^{at} \cos \omega t$	$\frac{s-a}{(s-a)^2+\omega^2}$

The list of transform pairs [i.e., corresponding values of $F(s)$ and $f(t)$] given in Table 5.1 is adequate for the solution of most problems which arise in control engineering. The derivation of most of these transform pairs is now explained.

PULSE FUNCTION. A pulse function is shown in Fig. 5.3. The height of the function is h, and the width is t_0, so that its area is ht_0. The Laplace transform is obtained by applying Eq. (5.15), in which $f(t) = h$ for $0 < t < t_0$ and $f(t) = 0$ for $t > t_0$.

$$F(s) = \int_0^{t_0} he^{-st}\,dt = h\left[\frac{-e^{-st}}{s}\right]_0^{t_0} = \frac{h}{s}(1 - e^{-t_0 s}) \tag{5.17}$$

IMPULSE FUNCTION. A special case of a pulse function is an impulse function. If the height is designated as $h = k/t_0$, it follows that the area is always equal to k. Now, as the width t_0 approaches zero, the height becomes infinite but the area remains equal to k. This limiting case of a pulse function is called an *impulse*. The symbol $k\delta(t)$ represents an impulse function whose area is k. Substitution of $h = k/t_0$ into Eq. (5.17) and taking the limit as t_0 approaches zero give the following transform for an impulse.

$$F(s) = \mathscr{L}[k\delta(t)] = \lim_{t_0 \to 0}\left[\frac{k}{t_0 s}(1 - e^{-t_0 s})\right] = \frac{0}{0}$$

Application of L'hopital's rule for evaluating the preceding indeterminant gives

$$F(s) = \lim_{t_0 \to 0} \frac{d/dt_0[k(1 - e^{-t_0 s})]}{d/dt_0(t_0 s)} = \lim_{t_0 \to 0} \frac{kse^{-t_0 s}}{s} = k \tag{5.18}$$

The transform of an impulse function thus equals the area of the function. The impulse function whose area is unity $\delta(t)$ is called a *unit impulse*. Much information about the transient behavior of a system may be obtained by determining the manner in which a system returns to its equilibrium state after the system has been excited by a momentary disturbance, such as a pulse or an impulse.

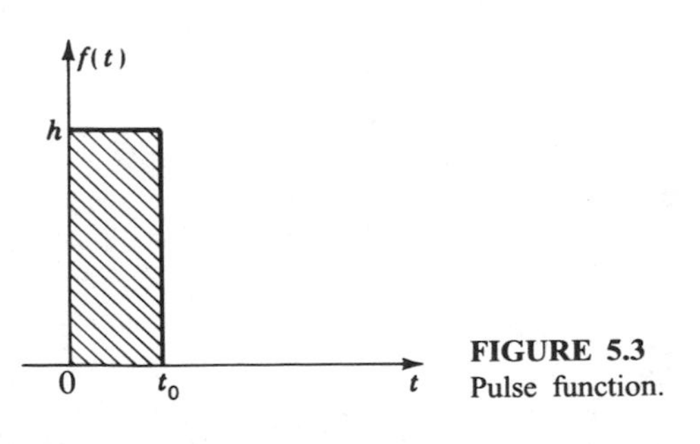

FIGURE 5.3
Pulse function.

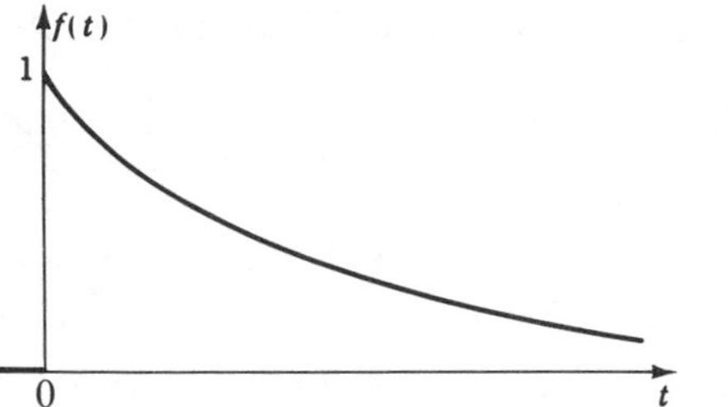

FIGURE 5.4
Exponentially decaying function.

EXPONENTIALLY DECAYING FUNCTION. The function $f(t) = e^{-at}$ is shown in Fig. 5.4. Applying Eq. (5.15) gives the transform of this exponentially decaying function:

$$F(s) = \mathscr{L}(e^{-at}) = \int_0^\infty e^{-(a+s)t}\,dt = -\frac{e^{-(a+s)t}}{s+a}\Bigg|_0^\infty = \frac{1}{s+a} \tag{5.19}$$

SINUSOIDAL FUNCTION. A sinusoidal function is shown in Fig. 5.5. The equation for this sinusoid is

$$f(t) = \sin \omega t$$

Thus

$$F(s) = \mathscr{L}(\sin \omega t) = \int_0^\infty e^{-st} \sin \omega t\, dt \tag{5.20}$$

The preceding integration is simplified by making use of Euler's equations.

$$\begin{aligned} e^{j\theta} &= \cos\theta + j\sin\theta \\ e^{-j\theta} &= \cos\theta - j\sin\theta \end{aligned} \tag{5.21}$$

Adding Euler's equations and dividing by 2 yield

$$\cos\theta = \frac{e^{j\theta} + e^{-j\theta}}{2} \tag{5.22}$$

Subtracting the second of Euler's equations from the first and dividing by $2j$ yield

$$\sin\theta = \frac{e^{j\theta} - e^{-j\theta}}{2j} \tag{5.23}$$

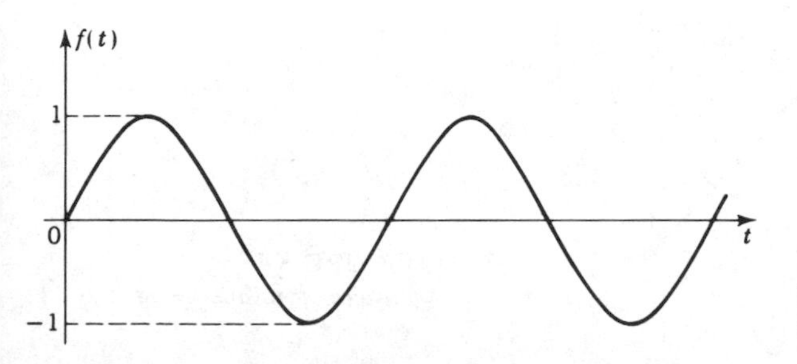

FIGURE 5.5
Sinusoidal function.

Substitution of the exponential form for $\sin\theta$ as given by Eq. (5.23) into Eq. (5.20) gives

$$F(s) = \int_0^\infty \frac{e^{j\omega t} - e^{-j\omega t}}{2j} e^{-st}\, dt$$
$$= \frac{1}{2j}\left(\frac{1}{s - j\omega} - \frac{1}{s + j\omega}\right) = \frac{\omega}{s^2 + \omega^2} \tag{5.24}$$

The response of a system to a sinusoidal forcing function forms the basis for appraising the performance of systems by frequency-response techniques, as is discussed in Chaps. 11 and 12.

The validity of Euler's equations is proved by expanding $e^{j\theta}$, $\cos\theta$, and $\sin\theta$ by the use of Maclaurin's series. Thus,

$$e^{j\theta} = 1 + j\theta + \frac{(j\theta)^2}{2!} + \frac{(j\theta)^3}{3!} + \frac{(j\theta)^4}{4!} + \frac{(j\theta)^5}{5!} + \cdots$$
$$= \left(1 - \frac{\theta^2}{2!} + \frac{\theta^4}{4!} - \cdots\right) + j\left(\theta - \frac{\theta^3}{3!} + \frac{\theta^5}{5!} - \cdots\right)$$

$$\cos\theta = 1 - \frac{\theta^2}{2!} + \frac{\theta^4}{4!} - \cdots$$

$$\sin\theta = \theta - \frac{\theta^3}{3!} + \frac{\theta^5}{5!} - \cdots$$

The results of Eqs. (5.21) follow directly from the preceding expansions.

Because the vector $e^{j\theta}$ is the vector sum of $\cos\theta + j\sin\theta$, the magnitude of $e^{j\theta}$ is always $\cos^2\theta + \sin^2\theta = 1$. The physical significance of the unit vector

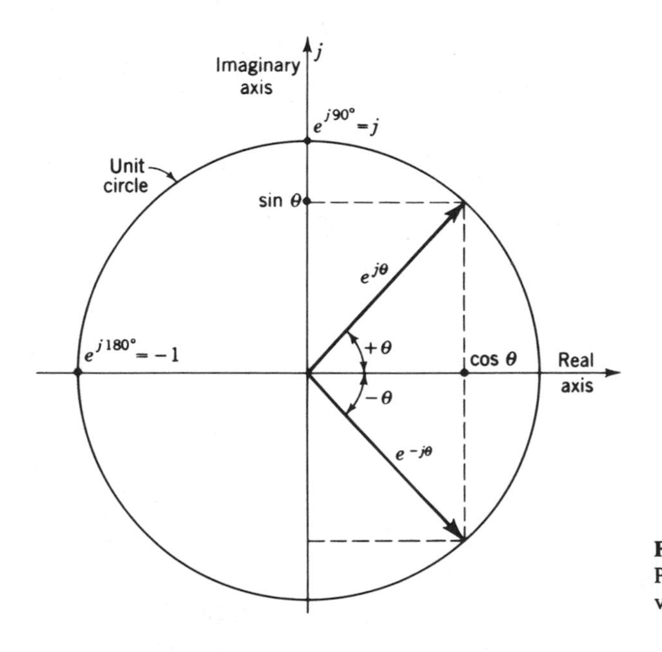

FIGURE 5.6
Physical significance of unit vector $e^{j\theta}$.

$e^{j\theta}$ is shown in Fig. 5.6. Here note that $e^{j\theta}$ is a unit vector which is rotated counterclockwise an angle θ from the real axis. For θ equal to 90°, we see from the unit circle that $e^{j90°} = j$. Squaring the preceding unit vector shows that $j^2 = e^{j90°}e^{j90°} = e^{j180°}$. From the unit circle, this vector is seen to be equal to -1. Thus, $j^2 = -1$, and $j = \sqrt{-1}$.

5.3 TRANSFORM PROPERTIES

Significant characteristics and theorems pertaining to Laplace transformations are developed in this section. A list of Laplace transform properties is given in Table 5.2. With the aid of these properties, the usefulness of the Laplace transform method is greatly extended. In addition, these theorems help us to obtain a much better understanding of the method.

Real Translation

The function $f(\tau)$ shown in Fig. 5.7 begins at time $t = t_0$ rather than at $t = 0$. Note that $\tau = 0$ when $t = t_0$. Because $\tau = t - t_0$, then $f(\tau) = f(t - t_0)$. From Eq. (5.15), the Laplace transform for $f(\tau)$ is

$$\mathscr{L}[f(\tau)] = \int_0^\infty f(\tau)e^{-st}\,dt = \int_{t_0}^\infty f(\tau)e^{-st}\,dt \tag{5.25}$$

TABLE 5.2
Laplace transform properties

Time function	Laplace transform
$kf(t)$	$kF(s)$
$f_1(t) \pm f_2(t)$	$F_1(s) \pm F_2(s)$
$f'(t)$	$sF(s) - f(0)$
$f''(t)$	$s^2F(s) - sf(0) - f'(0)$
$f^n(t)$	$s^nF(s) - s^{n-1}f(0) - \cdots - f^{n-1}(0)$
$f^{(-1)}(t)$	$\dfrac{F(s)}{s} + \dfrac{f^{(-1)}(0)}{s}$
$f^{(-n)}(t)$	$\dfrac{F(s)}{s^n} + \dfrac{f^{(-1)}(0)}{s^n} + \cdots + \dfrac{f^{(-n)}(0)}{s}$
$f(at)$	$\dfrac{1}{a}F\left(\dfrac{s}{a}\right)$
$e^{at}f(t)$	$F(s - a)$
$t^nf(t)$	$(-1)^n \dfrac{d^n}{ds^n}F(s)$
$f(\tau) = f(t - t_0)$	$e^{-t_0 s}F(s)$
$\int_0^t f(\lambda)g(t - \lambda)\,d\lambda$	$F(s)G(s)$

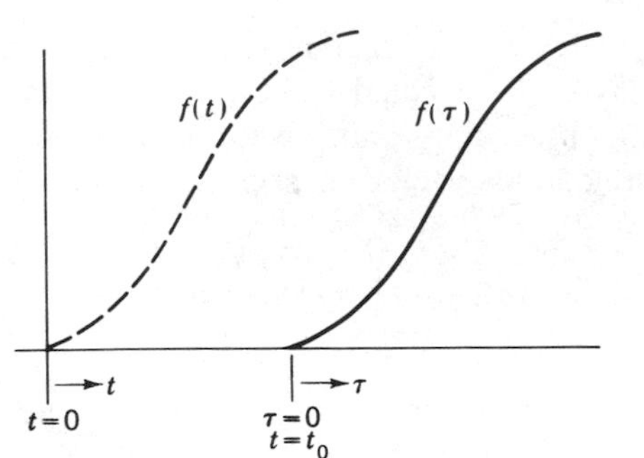

FIGURE 5.7
Delayed time function.

where $f(\tau) = f(t - t_0)$ is zero for $t < t_0$. In making the change of variable from t to τ, $dt = d\tau$, and the lower limit of integration $t = t_0$ corresponds to $\tau = 0$. Thus, the preceding integral becomes

$$\mathscr{L}[f(\tau)] = \int_0^\infty f(\tau)e^{-s(t_0+\tau)}\,d\tau = e^{-t_0 s}\int_0^\infty f(\tau)e^{-s\tau}\,d\tau \tag{5.26}$$

The right side may be written as

$$\int_0^\infty f(\tau)e^{-s\tau}\,d\tau = \int_0^\infty f(t)e^{-st}\,dt = F(s)$$

where $F(s)$ would be the transform of the function if it were not delayed. Substitution of the preceding result into Eq. (5.26) gives

$$\mathscr{L}[f(\tau)] = \mathscr{L}[f(t - t_0)] = e^{-t_0 s}F(s) \tag{5.27}$$

Thus, the transform of a function which is delayed a time t_0 is obtained by multiplying the transform of the function if it were not delayed [that is, $F(s)$] by $e^{-t_0 s}$.

An application of Eq. (5.27) is immediately evident by noting that the pulse function shown in Fig. 5.3 may be regarded as a step function of height h which begins at $t = 0$ minus a step function of height h which begins at $t = t_0$. The transform for the first step function is h/s, while that for the delayed step function is $(h/s)e^{-t_0 s}$. Subtracting the delayed step function from the first yields the following transform for the pulse:

$$\frac{h}{s}(1 - e^{-t_0 s})$$

This is the result given by Eq. (5.17).

Transform of a Derivative

Any linear differential equation will, of course, have derivatives of various orders. The order of each derivative is the same as the exponent of the operator D in the operational representation of the term. The general expression for transforming derivatives is obtained as follows. The form for integration by parts is

$$\int u\,dv = uv - \int v\,du$$

Let $u = f(t)$ and $v = -e^{-st}/s$; then $du = (d/dt)[f(t)]\,dt$ and $dv = e^{-st}\,dt$. Integration between the limits of zero and infinity gives

$$\int_0^\infty f(t)e^{-st}\,dt = -f(t)\,\frac{e^{-st}}{s}\Bigg|_0^\infty + \frac{1}{s}\int_0^\infty \frac{d}{dt}\,f(t)\,e^{-st}\,dt \tag{5.28}$$

The left side is $F(s)$. The upper limit of the first term on the right side is zero. Thus

$$\begin{aligned} F(s) &= \frac{f(0)}{s} + \frac{1}{s}\int_0^\infty \left[\frac{df(t)}{dt}\right] e^{-st}\,dt \\ &= \frac{f(0)}{s} + \frac{1}{s}\mathscr{L}\left[\frac{df(t)}{dt}\right] \end{aligned} \tag{5.29}$$

where $f(0)$ is the initial value of $f(t)$. Solving Eq. (5.29) for the transform of the derivative gives

$$\mathscr{L}\left[\frac{df(t)}{dt}\right] = \mathscr{L}[f'(t)] = sF(s) - f(0) \tag{5.30}$$

By extension of the preceding techniques to higher-order derivatives, the following equations for transforms of higher-order derivatives are obtained:

$$\begin{aligned} \mathscr{L}[f''(t)] &= s^2F(s) - sf(0) - f'(0) \\ \mathscr{L}[f'''(t)] &= s^3F(s) - s^2f(0) - sf'(0) - f''(0) \\ \mathscr{L}[f^n(t)] &= s^nF(s) - s^{n-1}f(0) - \cdots - f^{n-1}(0) \end{aligned} \tag{5.31}$$

where $f'(t) = df(t)/dt$, $f''(t) = d^2f(t)/dt^2, \ldots, f^n(t) = d^nf(t)/dt^n$, and $f'(0)$ is the initial value of $f'(t)$, etc. The initial conditions $f(0)$, $f'(0)$, $f''(0)$, ... associated with a particular differential equation must, of course, be given.

An interesting result is obtained when Eq. (5.30) is applied to the function shown in Fig. 5.8*a*. The initial value of this function is $f(0)$. Because of the step change of height h_c, the value of the function for $t > 0$ is $h = h_c + f(0)$. The transform of this function is

$$F(s) = \mathscr{L}[f(t)] = \mathscr{L}\{[h_c + f(0)]u(t)\} = \frac{h_c + f(0)}{s}$$

Application of Eq. (5.30) to obtain the transform of the derivative of this function gives

$$\mathscr{L}\left[\frac{d}{dt}f(t)\right] = sF(s) - f(0) = \frac{s[h_c + f(0)]}{s} - f(0) = h_c \tag{5.32}$$

The preceding transform is the same as that obtained for an impulse function. Thus, the derivative of a step change is an impulse function whose area is equal to the change in height h_c of the step.

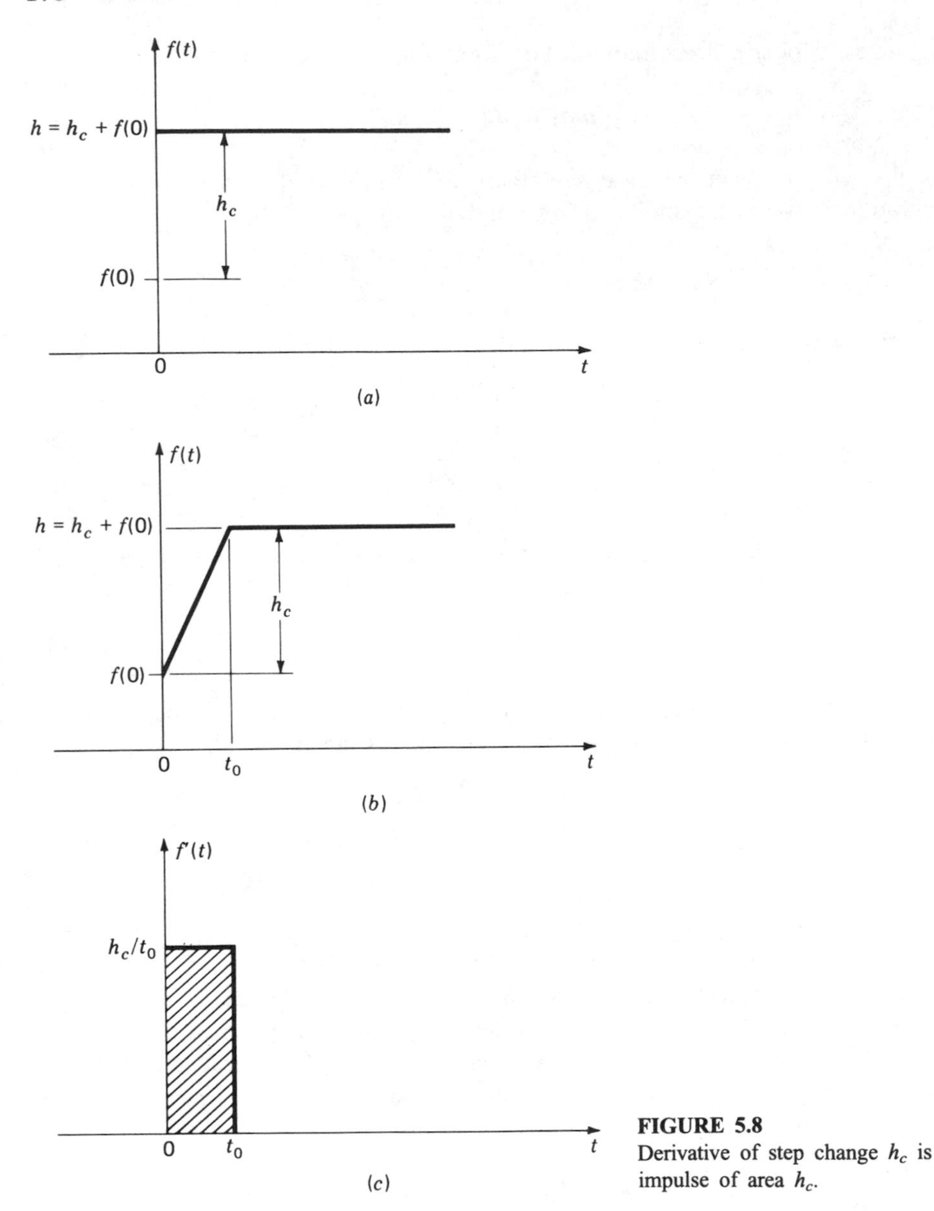

FIGURE 5.8
Derivative of step change h_c is impulse of area h_c.

This result may also be verified geometrically. Figure 5.8*b* shows the function $f(t)$ whose initial value is $f(0)$. The function increases linearly from its initial value $f(0)$ to the value $h = h_c + f(0)$ at time t_0. For $t > t_0$, the function maintains the value h. Note that as t_0 approaches zero, this function approaches the step change shown in Fig. 5.8*a*. The derivative $f'(t)$ is shown in Fig. 5.8*c*. For $0 < t < t_0$, the slope of $f(t)$ in Fig. 5.8*b* is h_c/t_0. For $t > t_0$, the slope is zero. The derivative $f'(t)$ is a pulse function of height h_c/t_0 and width t_0. As t_0 approaches zero, the derivative approaches an impulse of area h_c, where h_c is equal to the change in height of the step.

When a function $f(t)$ has a derivative which is a constant times the original function, then the transform $F(s)$ may be obtained by application of the equation for the transform of the derivative of the function. For example, for $f(t) = \sin \omega t$, it follows that $f'(t) = \omega \cos \omega t$ and $f''(t) = -\omega^2 \sin \omega t = -\omega^2 f(t)$. Thus, application of the general expression for the transform of the second derivative gives

$$\mathscr{L}[f''(t)] = \mathscr{L}[-\omega^2 f(t)] = -\omega^2 F(s) = s^2 F(s) - sf(0) - f'(0)$$

Substitution of $f(0) = [\sin \omega t]_{t=0} = 0$ and $f'(0) = [\omega \cos \omega t]_{t=0} = \omega$ into the preceding equation and solving for $F(s) = \mathscr{L}[f(t)] = \mathscr{L}(\sin \omega t)$ yield

$$F(s) = \frac{\omega}{s^2 + \omega^2}$$

This verifies the result previously established in Eq. (5.24).

Multiplication by t

The Laplace transform of $tf(t)$ is

$$\mathscr{L}[tf(t)] = -\frac{d}{ds} F(s) \tag{5.33}$$

To verify this theorem, note that

$$-\frac{d}{ds} F(s) = -\frac{d}{ds} \int_0^\infty f(t) e^{-st}\, dt = \int_0^\infty [tf(t)] e^{-st}\, dt$$

The right side is the Laplace transform of the term in brackets $tf(t)$. To illustrate this theorem, consider the unit step function $f(t) = u(t)$, for which $F(s) = 1/s = s^{-1}$. The Laplace transform of $tf(t) = tu(t) = t$ is

$$\mathscr{L}(t) = -\frac{d}{ds} s^{-1} = \frac{1}{s^2}$$

For $f(t) = t$ and $F(s) = 1/s^2$, the Laplace transform of $tf(t) = t^2$ is

$$\mathscr{L}(t^2) = -\frac{d}{ds}(s^{-2}) = \frac{2}{s^3}$$

In general, it follows that

$$\mathscr{L}(t^n) = \frac{n!}{s^{n+1}}$$

Multiplication by e^{at}

The Laplace transform of $e^{at} f(t)$ is

$$\mathscr{L}[e^{at} f(t)] = F(s - a) \tag{5.34}$$

This theorem is verified directly by replacing s by $s - a$ in the general expression for $F(s)$. That is,

$$F(s-a) = \int_0^\infty f(t)e^{-(s-a)t}\,dt = \int_0^\infty [e^{at}f(t)]e^{-st}\,dt$$

Thus, $F(s - a)$ is the Laplace transform of the term in brackets $e^{at}f(t)$. To illustrate the application of this theorem, let $f(t) = t^n$. For $F(s) = n!/s^{n+1}$, it follows that

$$\mathcal{L}(e^{at}t^n) = \frac{n!}{(s-a)^{n+1}}$$

As another application, consider the function $f(t) = \sin \omega t$. For $F(s) = \omega/(s^2 + \omega^2)$, it follows that

$$\mathcal{L}(e^{at}\sin \omega t) = \frac{\omega}{(s-a)^2 + \omega^2}$$

Change of Scale

The Laplace transform of $f(at)$ is

$$\mathcal{L}[f(at)] = \frac{1}{a}F\left(\frac{s}{a}\right) \tag{5.35}$$

To verify this theorem, first note that

$$\frac{1}{a}F\left(\frac{s}{a}\right) = \frac{1}{a}\int_0^\infty f(t)e^{-(s/a)t}\,dt$$

Next, replacing t by at, in which case dt is replaced by $a\,dt$, gives

$$\frac{1}{a}F\left(\frac{s}{a}\right) = \int_0^\infty f(at)e^{-st}\,dt$$

The right side is the transform of $f(at)$. To illustrate the application of this theorem, consider the function $f(t) = \sin t$, for which $F(s) = 1/(s^2 + 1)$. Thus,

$$\mathcal{L}(\sin \omega t) = \frac{1}{\omega}F\left(\frac{s}{\omega}\right) = \frac{1}{\omega}\frac{1}{(s/\omega)^2 + 1} = \frac{\omega}{s^2 + \omega^2}$$

Transform of an Integral

In using the Laplace transform method to solve integrodifferential equations, it is necessary to obtain the transform of an integral. The procedure for obtaining the equation for the transform of an integral is similar to that for a differential. In the general expression for integration by parts, let $u = \int f(t)\,dt$ and $v = -e^{-st}/s$; then $du = f(t)$ and $dv = e^{-st}\,dt$. This yields

$$\int_0^\infty \left[\int f(t)\,dt\right]e^{-st}\,dt = -\frac{e^{-st}}{s}\int f(t)\,dt\Big|_0^\infty + \frac{1}{s}\int_0^\infty f(t)e^{-st}\,dt$$

The left side is the Laplace transform of the integral of $f(t)$. The value of the first term on the right side at the upper limit is zero. Thus,

$$\mathscr{L}\left[\int f(t)\,dt\right] = \left.\frac{\int f(t)\,dt}{s}\right|_{t=0} + \frac{1}{s}F(s)$$

or

$$\mathscr{L}[f^{(-1)}(t)] = \frac{F(s)}{s} + \frac{f^{(-1)}(0)}{s} \tag{5.36}$$

where $f^{(-1)}(t) = \int f(t)\,dt$ and $f^{(-1)}(0) = \int f(t)\,dt|_{t=0}$ is the initial value of the integral.

By repeated application of this procedure, it is found that

$$\mathscr{L}[f^{(-2)}(t)] = \frac{F(s)}{s^2} + \frac{f^{(-1)}(0)}{s^2} + \frac{f^{(-2)}(0)}{s}$$

$$\mathscr{L}[f^{(-n)}(t)] = \frac{F(s)}{s^n} + \frac{f^{(-1)}(0)}{s^n} + \frac{f^{(-2)}(0)}{s^{n-1}} + \cdots + \frac{f^{(-n)}(0)}{s} \tag{5.37}$$

where $f^{(-n)}(t) = \int \cdots \int f(t)\,dt^n$.

Linearity Theorem

The linearity characteristic of Laplace transformations is a very useful property. If k is a constant or a variable which is independent of both t and s, then it follows from Eq. (5.15) that

$$\mathscr{L}[kf(t)] = k\mathscr{L}[f(t)] = kF(s) \tag{5.38}$$

Another important linearity property is

$$\mathscr{L}[f_1(t) \pm f_2(t)] = F_1(s) \pm F_2(s) \tag{5.39}$$

Final-Value Theorem

This theorem enables us to obtain the value $f(t)$ of a time function at $t = \infty$ directly from the Laplace transform $F(s)$. This is, in effect, the same type of information which is obtained from a steady-state analysis.

To develop the final-value theorem, first write the equation for the transform of a derivative in the form

$$\int_0^\infty f'(t)e^{-st}\,dt = sF(s) - f(0)$$

As s approaches zero, then $e^{-st} \approx 1$. Thus

$$\int_0^\infty f'(t)\,dt = \lim_{s\to 0} sF(s) - f(0)$$

The left side may be written in the form

$$\int_0^\infty f'(t)\,dt = f(t)\Big|_0^\infty = \lim_{t\to\infty} f(t) - f(0)$$

The desired final-value theorem is obtained by equating the right sides of the preceding expressions. Thus,

$$\lim_{t\to\infty} f(t) = \lim_{s\to 0} sF(s) \tag{5.40}$$

For functions such as sin t, t^n, or e^{at} where $a > 0$, then $\int_0^\infty f'(t)\,dt$ does not exist. When the poles of $sF(s)$ are located on or to the right of the imaginary axis, this integral does not exist. The final-value theorem is not applicable in such cases.

Initial-Value Theorem

With the aid of the initial-value theorem, the value $f(t)$ of a time function at $t = 0+$ may be computed directly from the transform $F(s)$ for the function. Note that $f(0+)$ is not the initial value $f(0)$, but rather the value of the function at a time slightly greater than zero.

The derivation of the initial-value theorem follows. For $t \approx 0$, then $e^{-st} \approx 1$; thus the equation for the transform of a derivative can be written as

$$\mathscr{L}[f'(t)] = \int_0^{0+} f'(t)(1)\,dt + \int_{0+}^\infty f'(t)e^{-st}\,dt = sF(s) - f(0)$$

As s approaches infinity, $e^{-st} \approx 0$. Thus, as $s \to \infty$, the second integral vanishes. Hence,

$$f(0+) - f(0) = \lim_{s\to\infty} sF(s) - f(0)$$

or

$$f(0+) = \lim_{s\to\infty} sF(s) \tag{5.41}$$

This is the mathematical formulation of the initial-value theorem.

Application of the initial-value theorem to the step function shown in Fig. 5.8*a* gives

$$f(0+) = [sF(s)]_{s=\infty} = \frac{s[h_c + f(0)]}{s}\Bigg|_{s=\infty} = h_c + f(0) \tag{5.42}$$

Thus, a step change of height h_c occurs at $t = 0$.

Illustrative example 5.2. Use the Laplace transform method to verify the solution of Illustrative example 5.1. That is,

$$(D^2 + 3D + 2)y(t) = f(t)$$

where $f(t) = 2u(t)$ and all initial conditions are zero.

Solution. Transforming each term yields for the transformed equation

$$[s^2Y(s) - sy(0) - y'(0)] + 3[sY(s) - y(0)] + 2Y(s) = F(s)$$

For $y(0) = y'(0) = 0$, the transformed equation becomes

$$(s^2 + 3s + 2)Y(s) = F(s)$$

Comparison of this transformed equation with the differential equation shows that when all the initial conditions are zero, the transformed equation is obtained by replacing D by s, $y(t)$ by $Y(s)$, and $f(t)$ by $F(s)$ in the differential equation. The transformed equation is now written in the form

$$\begin{aligned} Y(s) &= \frac{F(s)}{(s+1)(s+2)} = \frac{2}{s(s+1)(s+2)} \\ &= \frac{K_1}{s} + \frac{K_2}{s+1} + \frac{K_3}{s+2} \end{aligned}$$

where $F(s) = \mathscr{L}[f(t)] = \mathscr{L}[2u(t)] = 2/s$. The partial-fraction expansion constants are evaluated as follows:

$$K_1 = \lim_{s \to 0}\left[s\frac{2}{s(s+1)(s+2)}\right] = \left[\frac{2}{(s+1)(s+2)}\right]_{s=0} = 1$$

$$K_2 = \lim_{s \to -1}\left[(s+1)\frac{2}{s(s+1)(s+2)}\right] = \left[\frac{2}{s(s+2)}\right]_{s=-1} = -2$$

$$K_3 = \lim_{s \to -2}\left[(s+2)\frac{2}{s(s+1)(s+2)}\right] = \left[\frac{2}{s(s+1)}\right]_{s=-2} = 1$$

The resulting transformed equation is

$$Y(s) = \frac{1}{s} - \frac{2}{s+1} + \frac{1}{s+2}$$

Inverting yields

$$y(t) = 1 - 2e^{-t} + e^{-2t} \qquad t > 0$$

This verifies the result given by Eq. (5.14).

Illustrative example 5.3. Advanced aerodynamic design and extensive use of high-strength, low-weight composite materials contribute to the greater cruise efficiency and superior handling characteristics of the YF · 22 airplane shown in Fig. 5.9. The plane cruises at supersonic speeds without the use of afterburners. Very high-speed integrated circuits and advanced data processors maximize aircraft performance and minimize pilot workload. The block-diagram representation for a proposed proportional engine control system for this airplane is shown in Fig. 5.10. The command signal $v = n_{\text{in}}$ is the desired (input) speed, and the controlled variable $c = n_o$ is the actual (output) speed. For $A = 1$, $K_1 = 1$, $K_2 = 2$, $K_H = 0.5$, $B = -5$, $\tau_1 = \frac{1}{6}$, and $\tau_2 = 1.0$, this system is the same as that discussed in Illustrative example 4.1. Determine the response of this system when all initial conditions are zero for each of the following cases:

(*a*) n_{in} is a step function of constant value n_{in}, and $u = 0$.
(*b*) u is a step function of constant value u, and $n_{\text{in}} = 0$.

FIGURE 5.9
The YF · 22 airplane. (Courtesy Department of the Air Force.)

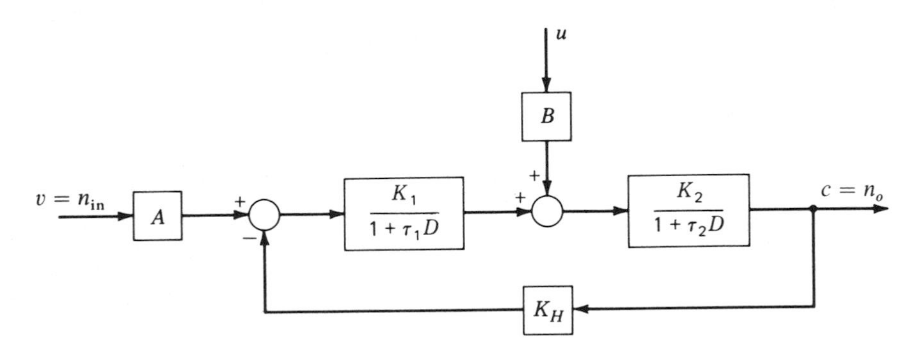

FIGURE 5.10
Proportional control system.

Solution. The differential equation of operation is

$$n_o = \frac{AK_1K_2n_{in} + BK_2(1+\tau_1 D)u}{(1+\tau_1 D)(1+\tau_2 D) + K_1K_2K_H} \tag{5.43}$$

Because the initial conditions are zero, the transformed equation is obtained by replacing D by s, n_o by $N_o(s)$, n_{in} by $N_{in}(s)$, and u by $U(s)$. Thus

$$N_o(s) = \frac{AK_1K_2N_{in}(s) + BK_2(1+\tau_1 s)U(s)}{(1+\tau_1 s)(1+\tau_2 s) + K_1K_2K_H}$$

$$= \frac{[AK_1K_2/(\tau_1\tau_2)]N_{in}(s) + [BK_2/(\tau_1\tau_2)](1+\tau_1 s)U(s)}{s^2 + (1/\tau_1 + 1/\tau_2)s + (K_1K_2K_H + 1)/(\tau_1\tau_2)}$$

Substitution of numerical values into this transformed equation gives

$$N_o(s) = \frac{12N_{\text{in}}(s) - 10(s+6)U(s)}{s^2 + 7s + 12} \tag{5.44}$$

(*a*) For $N_{\text{in}}(s) = n_{\text{in}}/s$ and $U(s) = 0$, the preceding becomes

$$N_o(s) = \frac{12n_{\text{in}}}{s(s+3)(s+4)} = \left(\frac{1}{s} - \frac{4}{s+3} + \frac{3}{s+4}\right)n_{\text{in}}$$

Inverting yields for the response

$$n_o(t) = (1 - 4e^{-3t} + 3e^{-4t})n_{\text{in}} \qquad t > 0$$

Note that as time approaches infinity, $n_o = n_{\text{in}}$. Because $n_0 = n_{\text{in}}$ for steady-state operation, the coefficient of the $v = n_{\text{in}}$ term in Eq. (4.1) must be unity. This fact is substantiated by Eq. (4.12), which describes the steady-state behavior of this system.

(*b*) For $N_{\text{in}}(s) = 0$ and $U(s) = u/s$, the transformed equation is

$$N_o(s) = \frac{-10(s+6)u}{s(s+3)(s+4)} = \left(\frac{-5}{s} + \frac{10}{s+3} - \frac{5}{s+4}\right)u$$

Inverting yields

$$n_o(t) = (-5 + 10e^{-3t} - 5e^{-4t})u \qquad t > 0$$

For steady-state operation, $n_o = -5u$. As is verified by Eq. (4.12), the coefficient of the u term is -5.

The steady-state behavior does not depend on the values of τ_1 and τ_2, but the transient behavior does. For example, if $\tau_1 = \frac{1}{15}$ and $\tau_2 = \frac{1}{2}$, the transformed equation becomes

$$N_o(s) = \frac{60N_{\text{in}}(s) - 20(s+15)U(s)}{(s+5)(s+12)} \tag{5.45}$$

For $N_{\text{in}}(s) = n_{\text{in}}/s$ and $U(s) = 0$, the response is

$$n_o(t) = (1 - \tfrac{12}{7}e^{-5t} + \tfrac{5}{7}e^{-12t})n_{\text{in}} \qquad t > 0$$

For $N_{\text{in}}(s) = 0$ and $U(s) = u/s$, the response is

$$c(t) = (-5 + \tfrac{40}{7}e^{-5t} - \tfrac{5}{7}e^{-12t})u \qquad t > 0$$

Illustrative example 5.4. The block-diagram representation for a proposed integral-type speed control system for the YF · 22 airplane shown in Fig. 5.9 is shown in Fig. 5.11. The output speed is n_o, the desired (input) speed is n_{in}, and the load torque is u.

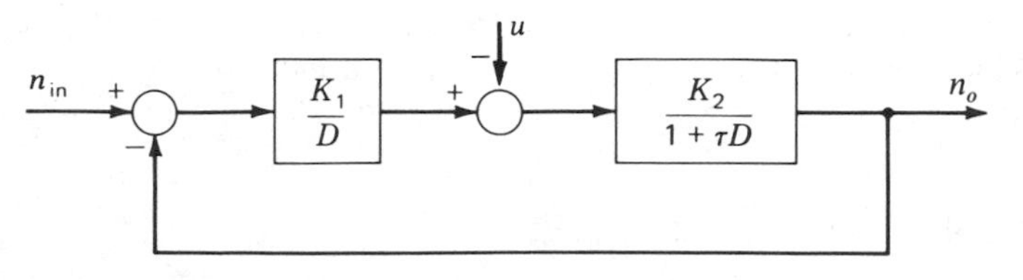

FIGURE 5.11
Integral control system.

For $\tau = 0.25$, $K_1 = 1.0$, $K_2 = 0.75$, and all initial conditions zero, determine the response of the system for each of the following cases:
(*a*) n_{in} is a step function of constant value n_{in}, and $u = 0$.
(*b*) u is a step function of constant value u, and $n_{in} = 0$.

Solution. The differential equation of operation is

$$n_o = \frac{K_1 K_2 n_{in} - K_2 D u}{D(1 + \tau D) + K_1 K_2} = \frac{(K_1 K_2/\tau) n_{in} - (K_2 D/\tau) u}{D^2 + (1/\tau) D + K_1 K_2/\tau} \tag{5.46}$$

Because the initial conditions are zero, the transformed equation is obtained by replacing D by s, n_o by $N_o(s)$, n_{in} by $N_{in}(s)$, and u by $U(s)$. Thus

$$N_o(s) = \frac{(K_1 K_2/\tau) N_{in}(s) + (K_2 s/\tau) U(s)}{s^2 + (1/\tau) s + K_1 K_2/\tau}$$

Substitution of $\tau = 0.25$, $K_1 = 1.0$, and $K_2 = 0.75$ gives

$$N_o(s) = \frac{3 N_{in}(s) - 3 s U(s)}{s^2 + 4s + 3} \tag{5.47}$$

(*a*) For $N_{in}(s) = n_{in}/s$ and $U(s) = 0$, the preceding becomes

$$N_o(s) = \frac{3 n_{in}}{s(s+1)(s+3)} = \left(\frac{1.0}{s} - \frac{1.5}{s+1} + \frac{0.5}{s+3}\right) n_{in}$$

Inverting yields for the response

$$n_o(t) = (1.0 - 1.5 e^{-t} + 0.5 e^{-3t}) n_{in} \qquad t > 0$$

Note that as time approaches infinity, $n_o = n_{in}$. Because $n_o = n_{in}$ for steady-state operation, the coefficient of the v term in Eq. (4.1) must be unity. Checking shows that

$$\frac{A K_{G_1} K_{G_2}}{1 + K_{G_1} K_{G_2} K_H} = \frac{A K_{G_2}}{1/K_{G_1} + K_{G_2} K_H} = \frac{A}{K_H} = 1$$

where $K_{G_1} = K_1/D|_{D=0} = \infty$, $A = 1$, and $K_H = 1$.
(*b*) For $N_{in}(s) = 0$ and $U(s) = u/s$, the transformed equation is

$$N_o(s) = \frac{-3u}{(s+1)(s+3)} = \left(\frac{1.5}{s+3} - \frac{1.5}{s+1}\right) u$$

Inverting yields

$$n_o(t) = 1.5(e^{-3t} - e^{-t}) u \qquad t > 0$$

Note that as time approaches infinity, $n_o = 0$. Because this is an integral-type control system, there is no steady-state error due to the disturbance u. For this case the coefficient of the u term in Eq. (4.1) must be zero. Checking shows that

$$\frac{B K_{G_2}}{1 + K_{G_1} K_{G_2} K_H} = \frac{(-1)(0.75)}{1 + (\infty)(0.75)(1)} = 0$$

where $B = -1$, $K_H = 1$, $K_{G_2} = [K_2/(1 + \tau D)]_{D=0} = K_2 = 0.75$, and $K_{G_1} = (K_1/D)_{D=0} = \infty$.

Principle of Superposition

The superposition principle is of fundamental importance. This principle states that if there is more than one excitation, the solution is the sum of the responses due to each excitation taken separately. Thus, if the step change in part *a* and the step change in part *b* of Illustrative examples 5.3 and 5.4 were to occur at the same time, the total response would be the sum of the solutions for parts *a* and *b*. It also follows from the principle of superposition that if the excitation is the sum of functions, such as a step function plus an impulse, then the response is the sum of the responses due to each taken separately.

5.4 INITIAL CONDITIONS

The initial state of a system is specified by the initial conditions. Because initial conditions are the state of the system as t approaches zero from the negative direction, to be more precise these initial-condition terms should be designated $f(0-)$, $f'(0-)$, $f''(0-)$, etc. However, for the sake of simplicity, initial conditions are designated as $f(0)$, $f'(0)$, $f''(0)$, etc., throughout this text.

The initial value at time $t = 0-$ for the function shown in Fig. 5.12*a* is indicated by $f(0)$. For $t > 0$, the excitation is he^{-t}. For $t = 0+$, the value of the function is h. Thus, a discontinuity or step change $h - f(0)$ occurs at the origin. The Laplace transform for $f(t)$ is

$$F(s) = \mathscr{L}[f(t)] = \int_0^\infty he^{-t}e^{-st}\,dt = h\int_0^\infty e^{-(s+1)t}\,dt = \frac{h}{s+1}$$

The inverse is

$$f(t) = \mathscr{L}^{-1}[F(s)] = he^{-t} \qquad t > 0$$

The symbol $\mathscr{L}^{-1}$ is used to indicate the *inverse Laplace transform*. Note that the inverse Laplace transform contains no information about the initial condition. Because the inverse yields the response for $t > 0$, the Laplace transform method is capable of solving problems in which there is a discontinuity at the origin, i.e., where $f(0-) \neq f(0+)$.

The derivative $f'(t)$ shown in Fig. 5.12*b* consists of the impulse at the origin whose area $h - f(0)$ is equal to the step change in $f(t)$ and the function $(d/dt)(he^{-t}) = -he^{-t}$. The Laplace transform for $f'(t)$ is

$$\mathscr{L}[f'(t)] = [h - f(0)] - \frac{h}{s+1}$$

This result may be verified by application of the formula for the transform of a derivative. That is,

$$\mathscr{L}[f'(t)] = sF(s) - f(0) = s\,\frac{h}{s+1} - f(0)$$

$$= \left(\frac{sh}{s+1} - h\right) + [h - f(0)] = [h - f(0)] - \frac{h}{s+1}$$

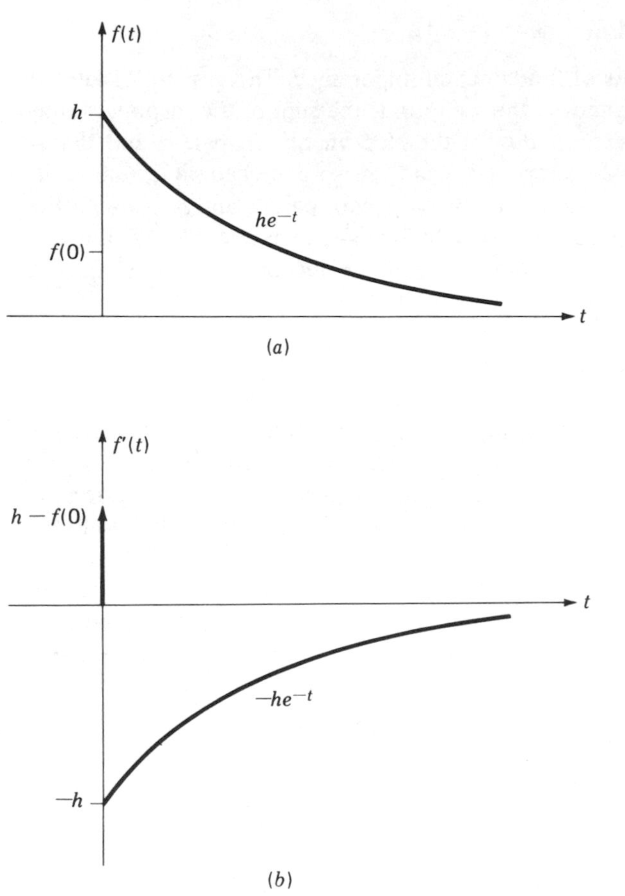

FIGURE 5.12
(*a*) Function with a discontinuity and (*b*) its derivative.

Note that the initial condition at time $t = 0-$ must be substituted into the equation for the transform of a derivative.

In taking the Laplace transform of a differential equation, initial-condition terms arise in transforming derivative terms. The initial conditions at time $t = 0-$ are substituted into the resulting transformed equation. Inverting this transformed equation yields the response for $t > 0$. Values at time $t = 0+$ are obtained by letting $t = 0+$ in the response equation. Because it is not necessary that the $0-$ values equal the $0+$ values, the transform method can be used for solving differential equations in which there are discontinuities. The following examples will help clarify this situation. For a more rigorous presentation of this distinction, see Raven.*

* F. H. Raven, *Mathematics of Engineering Systems*, McGraw-Hill, New York, 1966, chap. 2.

Illustrative example 5.5. Use the Laplace transform method to determine the solution of the following differential equation:

$$\frac{dy}{dt} + y = \delta(t) \tag{5.48}$$

The initial condition is $y(0) = 0$.

Solution. The transform of the differential equation is

$$[sY(s) - y(0)] + Y(s) = 1$$

For $y(0) = 0$, the transformed equation becomes

$$Y(s) = \frac{1}{s+1}$$

Inverting yields the response

$$y(t) = e^{-t} \qquad t > 0 \tag{5.49}$$

The value at time $t = 0+$ is

$$y(0+) = e^{-(0+)} = 1$$

This result may also be obtained by application of the initial-value theorem,

$$y(0+) = \lim_{s\to\infty} sY(s) = \left.\frac{s}{s+1}\right|_{s=\infty} = \left.\frac{1}{1+1/s}\right|_{s=\infty} = 1$$

As illustrated in Fig. 5.13, the response function $y(t)$ has a discontinuity at the origin. The initial value $y(0) = y(0-) = 0$ is indicated by the $\times$ at the left of the origin. The value of $y(0+)$ is 1.0. To solve such problems by classical methods, it is necessary to use independent methods to determine the state of the system at time $t = 0+$. Because the system is continuous for $t > 0$, the solution may then be found by classical procedures. To determine the state of the system at $t = 0+$, each term of the differential equation is integrated from $t = 0-$ to $t = 0+$. Thus

$$\int_{0-}^{0+} \frac{dy}{dt}\,dt + \int_{0-}^{0+} y\,dt = \int_{0-}^{0+} \delta(t)\,dt$$

The value of the first term is $y(0+) - y(0-)$. Because y is a bounded function, the integral from $0-$ to $0+$ vanishes. The right side is the area of the impulse at the origin,

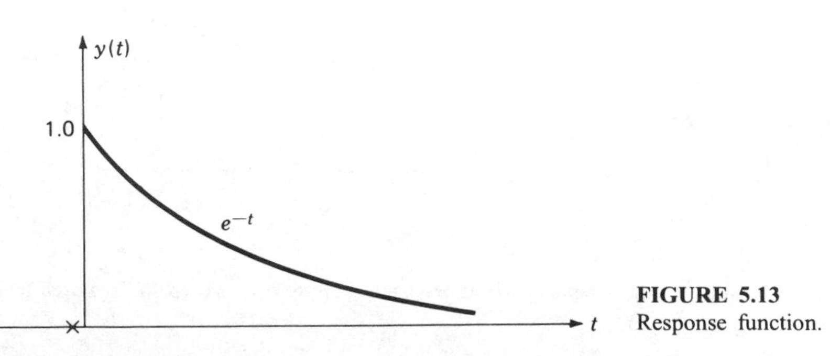

FIGURE 5.13
Response function.

which is unity. Thus

$$y(0+) - y(0-) = 1 \qquad \text{or} \qquad y(0+) = y(0-) + 1 = 1$$

For $t > 0$, the excitation $\delta(t)$ is zero. Thus, for $t > 0$ the differential equation is

$$\frac{dy}{dt} + y = 0 \qquad y(0+) = 1$$

Because $f(t) = 0$, the particular solution is zero. The general solution consists of the homogeneous solution only:

$$y(t) = k_1 e^{-t}$$

The constant k_1 is evaluated from the boundary condition that $y(0+) = 1$ when $t = 0+$. Thus

$$1 = k_1 e^{-(0+)} = k_1$$

The resulting response is

$$y(t) = e^{-t} \qquad t > 0$$

This verifies the solution obtained by the transform method, Eq. (5.49).

Illustrative example 5.6. The block-diagram representation for a precision positional control system operating as an open-loop system is shown in Fig. 5.14*a*. Because of the large mass of the load to be positioned, the value of the time constant τ is 5. The steady-state gain is $K = 1$. Determine the response of this system to a unit step change in the input x. All initial conditions are zero.

Solution. The equation relating the output y to the input x is

$$y = \frac{K}{1 + \tau D} x = \frac{1}{1 + 5D} x$$

or

$$Dy + \tfrac{1}{5} y = \tfrac{1}{5} x$$

Transforming each term yields for the Laplace transform

$$sY(s) - y(0) + 0.2Y(s) = 0.2X(s)$$

For $y(0) = 0$ and $X(s) = 1/s$, then

$$Y(s) = \frac{0.2X(s)}{s + 0.2} = \frac{0.2}{s(s + 0.2)} = \frac{1}{s} - \frac{1}{s + 0.2}$$

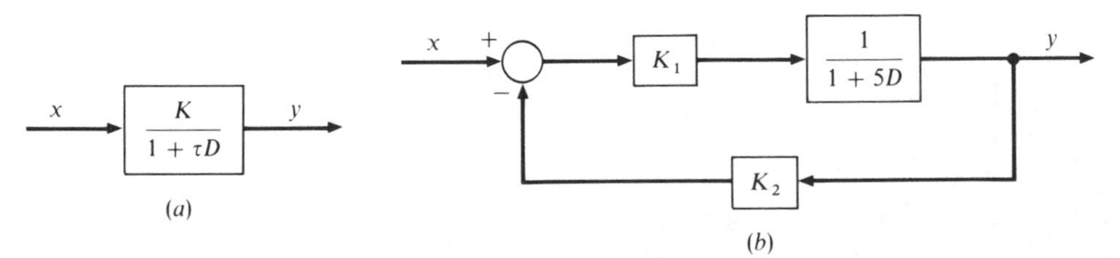

FIGURE 5.14
Precision positional control system operating (*a*) as an open-loop system and (*b*) as a closed-loop system.

Inverting yields for the response

$$y(t) = 1 - e^{-0.2t} = 1 - e^{-t/5} \qquad t > 0$$

This result shows that the time constant τ is 5 and the steady-state gain is 1.

Illustrative example 5.7. It is proposed to use feedback to speed up the response of the system of Illustrative example 5.6 by a factor of 10 while maintaining the same steady-state gain of 1. Figure 5.14*b* shows the block-diagram representation for the proposed system. Determine the required values of K_1 and K_2, and then determine the response of the resulting system to a unit step change in the input x.

Solution. The equation relating the output y to the input x is

$$y = \frac{K_1}{1 + K_1K_2 + 5D}x = \frac{K_1/(1 + K_1K_2)}{1 + [5/(1 + K_1K_2)]D}x$$

To decrease the time constant by a factor of 10, then $1 + K_1K_2 = 10$. To maintain the same steady-state gain of 1,

$$\frac{K_1}{1 + K_1K_2} = \frac{K_1}{10} = 1$$

Hence

$$K_1 = 10$$

From the relationship $1 + K_1K_2 = 1 + 10K_2 = 10$, it follows that $K_2 = 0.9$. The resulting input-output equation is

$$y = \frac{1}{1 + 0.5D}x$$

or

$$Dy + 2y = 2x$$

The Laplace transform is

$$sY(s) - y(0) + 2Y(s) = 2X(s)$$

For $y(0) = 0$ and $X(s) = 1/s$, then

$$Y(s) = \frac{2X(s)}{s + 2} = \frac{2}{s(s + 2)} = \frac{1}{s} - \frac{1}{s + 2}$$

Inverting yields for the response

$$y(t) = 1 - e^{-2t} = 1 - e^{-t/0.5} \qquad t > 0$$

This result shows that the time constant $\tau = 0.5$ has been reduced by a factor of 10 and the steady-state gain remains 1.

Illustrative example 5.8. Figure 5.15 shows NASA's shuttle-borne observatory Astro, which will explore the invisible ultraviolet and x-ray universe in greater depth and detail than ever before. The block-diagram representation for the attitude

FIGURE 5.15
NASA's shuttle-borne observatory, Astro. (Courtesy NASA.)

control system is shown in Fig. 5.16. Determine the response of this system for the case in which $r(t) = 2e^{-t}$ and all the initial conditions are zero.

Solution. From the block diagram, it follows that the input-output relationship is

$$c(t) = \frac{D+4}{D^2+5D+6} r(t) = \frac{D+4}{(D+2)(D+3)} r(t) \tag{5.50}$$

Because all the initial conditions are zero, substitution of s for D, $C(s)$ for $c(t)$, and $R(s)$ for $r(t)$ yields for the transformed equation

$$C(s) = \frac{s+4}{(s+2)(s+3)} R(s) = \frac{2(s+4)}{(s+1)(s+2)(s+3)} = \frac{K_1}{s+1} + \frac{K_2}{s+2} + \frac{K_3}{s+3}$$

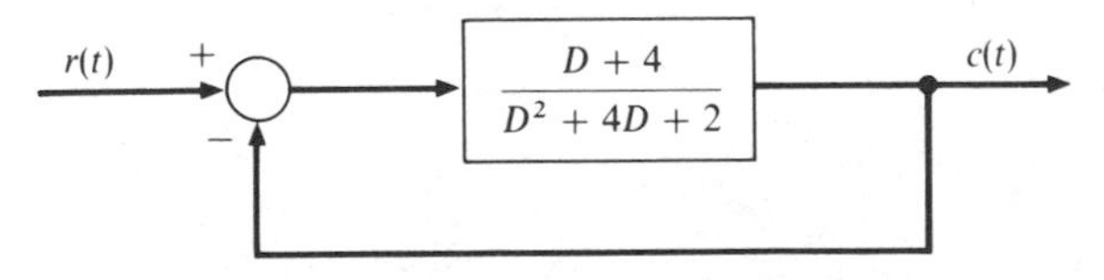

FIGURE 5.16
Attitude control system

where $R(s) = \mathscr{L}(2e^{-t}) = 2/(s+1)$. The partial-fraction expansion constants are $K_1 = 3$, $K_2 = -4$, and $K_3 = 1$. Thus

$$C(s) = \frac{3}{s+1} - \frac{4}{s+2} + \frac{1}{s+3}$$

Inverting yields for the response

$$c(t) = 3e^{-t} - 4e^{-2t} + e^{-3t} \qquad t > 0 \tag{5.51}$$

The derivative is

$$c'(t) = -3e^{-t} + 8e^{-2t} - 3e^{-3t}$$

Substitution of $t = 0+$ into the preceding expressions shows that $c(0+) = 0$ and $c'(0+) = 2$. These values may also be verified by application of the initial-value theorem.

$$c(0+) = \lim_{s\to\infty} s\mathscr{L}[c(t)] = sC(s)|_{s=\infty} = \left.\frac{2s(s+4)}{(s+1)(s+2)(s+3)}\right|_{s=\infty}$$

$$= \left.\frac{2(1/s + 4/s^2)}{(1+1/s)(1+2/s)(1+3/s)}\right|_{s=\infty} = 0$$

$$c'(0+) = \lim_{s\to\infty} s\mathscr{L}[c'(t)] = s[sC(s) - c(0)]_{s=\infty} = s^2C(s)|_{s=\infty} = 2$$

Note that the initial value $c'(0) = 0$ at time $t = 0-$ is different from the value $c'(0+) = 2$ at time $t = 0+$. Thus, this problem could not be solved by classical methods.

Illustrative example 5.9. Determine the solution of Eq. (5.50) for the case in which $r(t) = 0$, $c(0) = 2$, and $c'(0) = -5$.

Solution. For $r(t) = 0$, the differential equation becomes

$$(D^2 + 5D + 6)c(t) = 0 \tag{5.52}$$

Transforming each term gives

$$[s^2C(s) - sc(0) - c'(0)] + 5[sC(s) - c(0)] + 6C(s) = 0$$

$$(s^2 + 5s + 6)C(s) = (s+5)c(0) + c'(0) = 2s + 5$$

$$C(s) = \frac{2s+5}{(s+2)(s+3)} = \frac{1}{s+2} + \frac{1}{s+3}$$

Inverting yields for the response

$$c(t) = e^{-2t} + e^{-3t} \qquad t > 0 \tag{5.53}$$

Integrodifferential Equations

A resistor R and a capacitor C in series are shown in Fig. 5.17. The total voltage drop is the sum of the voltage drop $e_r = Ri$ across the resistor and the voltage drop $e_c = \left(\int i\,dt\right)/C$ across the capacitor.

$$e = Ri + \frac{1}{C}\int i\,dt \tag{5.54}$$

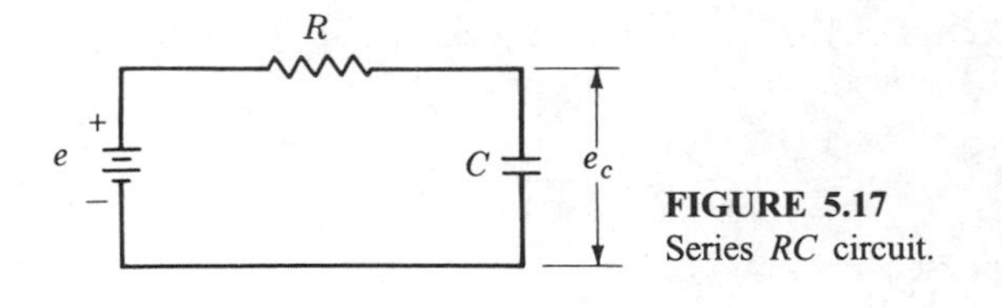

FIGURE 5.17
Series RC circuit.

The Laplace transform for this integrodifferential equation is

$$E(s) = RI(s) + \frac{1}{sC}\left[I(s) + \left(\int i\,dt\right)_{t=0}\right] = \frac{1}{sC}(1 + RCs)I(s) + \frac{e_c(0)}{s}$$

where $e_c(0) = \left(\int i\,dt\right)/C|_{t=0}$ is the initial voltage across the capacitor. Solving for $I(s)$ gives

$$I(s) = \frac{C[sE(s) - e_c(0)]}{1 + RCs} \tag{5.55}$$

The process of obtaining the transform of the integral term may be avoided by first differentiating Eq. (5.54) and then transforming this result. Thus

$$\frac{de}{dt} = R\,\frac{di}{dt} + \frac{1}{C}i$$

The corresponding transform is

$$sE(s) - e(0) = R[sI(s) - i(0)] + \frac{I(s)}{C}$$

or

$$\frac{1}{C}(1 + RCs)I(s) = sE(s) - [e(0) - Ri(0)] = sE(s) - e_c(0)$$

The voltage drop e_c across the capacitor is the applied voltage e minus the voltage drop Ri across the resistor. Solving for $I(s)$ yields the same result given by Eq. (5.55).

5.5 GENERAL PROCEDURES

Suppose we want to determine the solution of the following differential equation:

$$(1 + \tau D)y(t) = Kf(t) \tag{5.56}$$

For convenience in performing the partial-fraction expansion, it is desirable to have the coefficient of the highest power of D in $L_n(D)$ equal to 1. Thus

$$\left(D + \frac{1}{\tau}\right)y(t) = \frac{K}{\tau}f(t) \tag{5.57}$$

Transforming each term of this differential equation and solving for $Y(s)$ give

$$Y(s) = \frac{(K/\tau)F(s) + y(0)}{s + 1/\tau} \tag{5.58}$$

The symbol $F(s)$ is the transform of the input. The nature of the expression $F(s)$ depends upon the particular input to the system, such as a step function, exponential function, sinusoid, etc.

All Initial Conditions Zero

Suppose we want to determine the response of this system to a step-function input when all the initial conditions are zero. A plot of the input $f(t)$ is shown in Fig. 5.18*a*. The initial value is $f(0) = 0$, and then a step change h_c occurs so that the

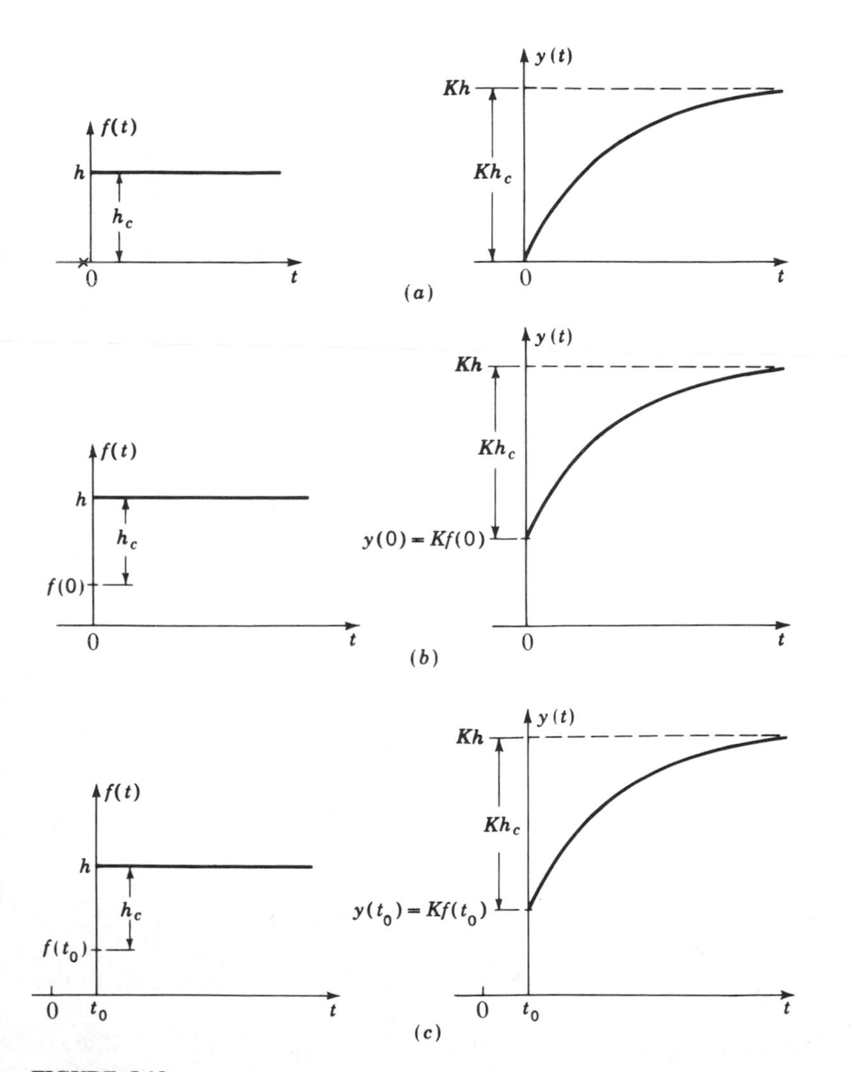

FIGURE 5.18
Response of a first-order system.

height of this function is $h = h_c$. The substitution of $F(s) = h_c/s$ and $y(0) = 0$ into Eq. (5.58) gives

$$Y(s) = \frac{Kh_c/\tau}{s(s+1/\tau)}$$

The partial-fraction expansion is

$$Y(s) = Kh_c\left(\frac{1}{s} - \frac{1}{s+1/\tau}\right)$$

Inverting yields for the response

$$y(t) = Kh_c(1 - e^{-t/\tau}) \qquad t > 0 \tag{5.59}$$

A graph of this response is also shown in Fig. 5.18*a*.

Initially at a Steady-State Operating Condition

A system is initially at a steady-state operating condition if the initial values of all the time derivatives are zero (that is, $D^n y = D^m f = 0$, $n = 1, 2, 3, \ldots$ and $m = 1, 2, 3, \ldots$). When this system is initially at a steady-state operating condition, then from Eq. (5.57) it follows that $y(0) = Kf(0)$.

Let us now determine the response for the case in which the initial value of the input is $f(0)$, as shown in Fig. 5.18*b*. The height h of the step input is the initial value $f(0)$ plus the change h_c. Thus

$$F(s) = \frac{h}{s} = \frac{h_c + f(0)}{s} = \frac{h_c + y(0)/K}{s}$$

The substitution of this value of $F(s)$ into Eq. (5.58) gives

$$Y(s) = \frac{Kh_c/\tau}{s(s+1/\tau)} + \frac{y(0)}{s}$$

The partial-fraction expansion is

$$Y(s) = Kh_c\left(\frac{1}{s} - \frac{1}{s+1/\tau}\right) + \frac{y(0)}{s} \tag{5.60}$$

Inverting yields

$$y(t) = Kh_c(1 - e^{-t/\tau}) + y(0) \qquad t > 0 \tag{5.61}$$

The first term on the right side is the response due to the step change h_c alone when all the initial conditions are zero [that is, Eq. (5.59)]. The response shown in Fig. 5.18*b* is the same as that in Fig. 5.18*a* except that it is raised by the initial value $y(0)$.

The vertical shifting theorem states that for a system which is initially at a steady-state operating condition, the response $y(t)$ is equal to the initial value $y(0)$ plus the response for the case in which all the initial conditions are zero.

Time Delay

To obtain the transform of the derivative of a delayed function, replace $f(\tau)$ by $df(\tau)/d\tau$ in Eq. (5.26). Thus

$$\mathscr{L}\left[\frac{df(\tau)}{d\tau}\right] = e^{-t_0 s}\int_0^\infty \frac{df(\tau)}{d\tau} e^{-s\tau}\, d\tau = e^{-t_0 s}[sF(s) - f(\tau)]_{\tau=0}$$
$$= [sF(s) - f(t_0)]e^{-t_0 s} \tag{5.62}$$

where $f(\tau)|_{\tau=0} = f(t)|_{t=t_0} = f(t_0)$. In general, it follows that

$$\mathscr{L}[f^n(\tau)] = [s^n F(s) - s^{n-1} f(t_0) - \cdots - f^{n-1}(t_0)]e^{-t_0 s} \tag{5.63}$$

Thus, to convert a transform to a delayed transform, multiply each term by $e^{-t_0 s}$ and evaluate the initial conditions at time t_0, which is the starting point for the delayed function.

Suppose we want to determine the response for the case in which the input of Fig. 5.18*b* is delayed by time t_0 as shown in Fig. 5.18*c*. Multiplying each term in Eq. (5.60) by $e^{-t_0 s}$ and evaluating the initial conditions at time t_0 give for the transformed equation

$$Y(s)e^{-t_0 s} = Kh_c\left(\frac{1}{s} - \frac{1}{s + 1/\tau}\right)e^{-t_0 s} + \frac{y(t_0)}{s}e^{-t_0 s} \tag{5.64}$$

The operator $e^{-t_0 s}$ merely means that t should be replaced by $t - t_0$ in the regular transform. Thus, the inverse of Eq. (5.64) is

$$y(t - t_0) = Kh_c(1 - e^{-(t-t_0)/\tau}) + y(t_0) \qquad t > t_0 \tag{5.65}$$

This is the same as Eq. (5.61) except that t has been replaced by $t - t_0$, and the starting point is at time t_0 rather than at $t = 0$. The response shown in Fig. 5.18*c* is the same as that in Fig. 5.18*b* except that it is delayed by a time t_0. It is not necessary that a system be initially at a steady-state operating condition to effect a time shift. For most problems in which a time shift occurs, it is more convenient to work the problem initially as though there were no time shift and then replace t by $t - t_0$ to obtain the desired result.

The general procedure used to solve differential equations by Laplace transforms may be summarized as follows:

1. Transform each term of the differential equation from the time domain to the s domain, and then solve for $Y(s)$.
2. Substitute the value of the initial conditions, and the transform of the input into the expression obtained in step 1.
3. Perform a partial-fraction expansion.
4. Invert each term back to the time domain to obtain the desired time response.

Much simplification in carrying out the algebraic manipulations of a Laplace transform solution is afforded for the following special cases:

1. *All initial conditions zero.* When all the initial conditions are zero, then the transform of a derivative is $\mathscr{L}[D^n f(t)] = s^n F(s)$. Thus, it follows that the transformed equation is obtained by substituting s for D, $Y(s)$ for $y(t)$, and $F(s)$ for $f(t)$ in the original differential equation.
2. *System initially at a steady-state operating condition.* When all the initial derivatives are zero, then evaluating Eq. (5.1) at $t = 0$ shows that $by(0) = af(0)$ or $y(0) = (a/b)f(0)$. For this case, application of the vertical shifting theorem states that the response $y(t)$ is obtained by adding the initial value $y(0)$ to the response for the case in which all the initial conditions are zero (i.e., case 1).
3. *Time shift.* A time shift is effected by substituting $t - t_0$ for t.

Illustrative example 5.10. Determine the response of the system described by the following differential equation:

$$(D^2 + 7D + 12)y(t) = 6(D + 2)f(t) \tag{5.66}$$

The forcing function $f(t)$ is a unit step function, and all initial conditions are zero.

Solution. Because the initial conditions are zero, substitution of s for D, $Y(s)$ for $y(t)$, and $F(s)$ for $f(t)$ yields

$$Y(s) = \frac{6(s+2)F(s)}{s^2 + 7s + 12} = \frac{6(s+2)}{s(s+3)(s+4)}$$

where $F(s) = 1/s$ for a unit step function. Performing a partial-fraction expansion gives

$$Y(s) = \frac{1}{s} + \frac{2}{s+3} - \frac{3}{s+4}$$

Inverting yields for the time response

$$y(t) = 1 + 2e^{-3t} - 3e^{-4t} \qquad t > 0 \tag{5.67}$$

Suppose now we desire to determine the response of this system for the case in which the system is initially at a steady-state operating condition (i.e., initially all derivative terms are zero). The initial value of $f(t)$ is $f(0) = 5$. The unit step function is superimposed upon the constant value $f(0) = 5$, so that for $t > 0$, the input is $f(0) + 1 = 5 + 1 = 6$. The initial steady-state value of y is obtained by letting all derivative terms be zero in the differential equation. Thus, $12y(0) = 6(2)f(0)$ or $y(0) = f(0) = 5$. Adding the initial value of y to the response for the case in which all initial conditions are zero yields for the response

$$y(t) = (1 + 2e^{-3t} - 3e^{-4t}) + y(0) = (1 + 2e^{-3t} - 3e^{-4t}) + 5 \qquad t > 0 \tag{5.68}$$

This response may also be verified by transforming the differential equation termwise:

$$[s^2Y(s) - sy(0) - y'(0)] + 7[sY(s) - y(0)] + 12Y(s) = 6[sF(s) - f(0)] + 12F(s)$$

Substituting $f(0) = y(0)$, $y'(0) = 0$, and $F(s) = [f(0) + 1]/s$ gives

$$(s^2 + 7s + 12)Y(s) = (s + 7)y(0) + 6 + \frac{12[y(0) + 1]}{s}$$

Thus,

$$Y(s) = \frac{6(s+2) + (s^2 + 7s + 12)y(0)}{s(s^2 + 7s + 12)} = \frac{6(s+2)}{s(s^2 + 7s + 12)} + \frac{y(0)}{s}$$

Inverting verifies the response obtained directly by application of the vertical shifting theorem, Eq. (5.68).

Illustrative example 5.11. The system shown in Fig. 5.19 is in a state of equilibrium, with $r = 1$ and $d = 0$. A step-function disturbance $d(t) = u(t)$ is then initiated at time $t = 0$. Determine the response $c(t)$ for $t > 0$.

Solution. The differential equation of operation for this system is

$$(D^2 + 5D + 6)c(t) = 6r(t) + 3Dd(t) \tag{5.69}$$

The expression for steady-state operation at $t = 0$ is obtained by letting $D = 0$ in the differential equation of operation. Thus $6c(0) = 6r(0)$ or $c(0) = r(0) = 1$. Transforming each term in the differential equation gives

$$[s^2C(s) - sc(0) - c'(0)] + 5[sC(s) - c(0)] + 6C(s) = 6R(s) + 3[sD(s) - d(0)]$$

For $c(0) = r(0) = 1$, $c'(0) = 0$, $d(0) = 0$, $R(s) = 1/s$, and $D(s) = 1/s$, the resulting transformed equation becomes

$$(s^2 + 5s + 6)C(s) = \frac{6}{s} + \frac{3}{s}s + s + 5 = \frac{s^2 + 8s + 6}{s}$$

Hence

$$C(s) = \frac{s^2 + 8s + 6}{s(s+2)(s+3)} = \frac{1}{s} + \frac{3}{s+2} - \frac{3}{s+3}$$

Inverting yields for the response

$$c(t) = 1 + 3e^{-2t} - 3e^{-3t} \qquad t > 0 \tag{5.70}$$

Because the system is initially at a steady-state operating condition, this result may be obtained directly by application of the vertical shifting theorem. That is, the resultant response is obtained by adding the initial value $c(0) = 1$ to the response of the system to the step change $d(t) = u(t)$ for the case in which all initial conditions are zero. The transformed equation, when all initial conditions are zero, is

$$C(s) = \frac{3sD(s)}{s^2 + 5s + 6} = \frac{3}{s^2 + 5s + 6} = \frac{3}{s+2} - \frac{3}{s+3}$$

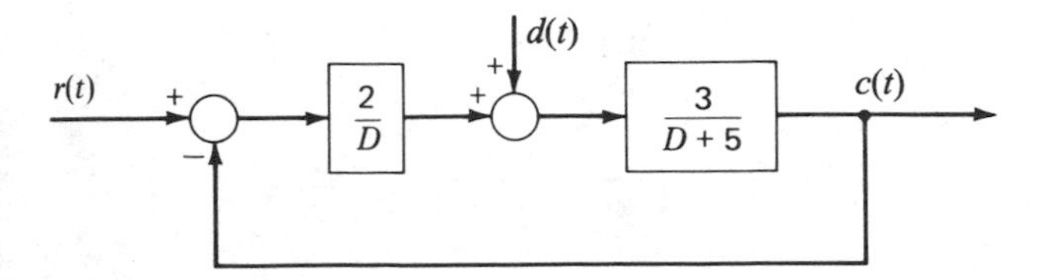

FIGURE 5.19
Feedback control system.

where $D(s) = 1/s$. Inverting yields for the response due to only the step change

$$c(t) = 3e^{-2t} - 3e^{-3t} \qquad t > 0$$

Thus, adding the initial value $c(0) = 1$ verifies the response previously obtained.

5.6 PIECEWISE CONTINUOUS FUNCTIONS

As illustrated in Fig. 5.20, a piecewise continuous function is characterized by the fact that the equation for the function changes from interval to interval. For example, in Fig. 5.20*a*, for $0 < t < t_0$, $f(t) = at$; and for $t > t_0$, $f(t) = at_0$. For the first interval $0 < t < t_0$, the input function is inclined at slope a. Such an inclined straight line is called a *ramp function*. For $t > t_0$, the input is a step function. Such problems are solved by starting with the first interval and successively solving for the response in each interval.

Illustrative example 5.12. Determine the solution of the differential equation

$$\frac{dy}{dt} + y = f(t) \tag{5.71}$$

where $f(t)$ is the function shown in Fig. 5.21*a* and all the initial conditions are zero.

Solution. The transform of the differential equation is

$$sY(s) - y(0) + Y(s) = F(s)$$

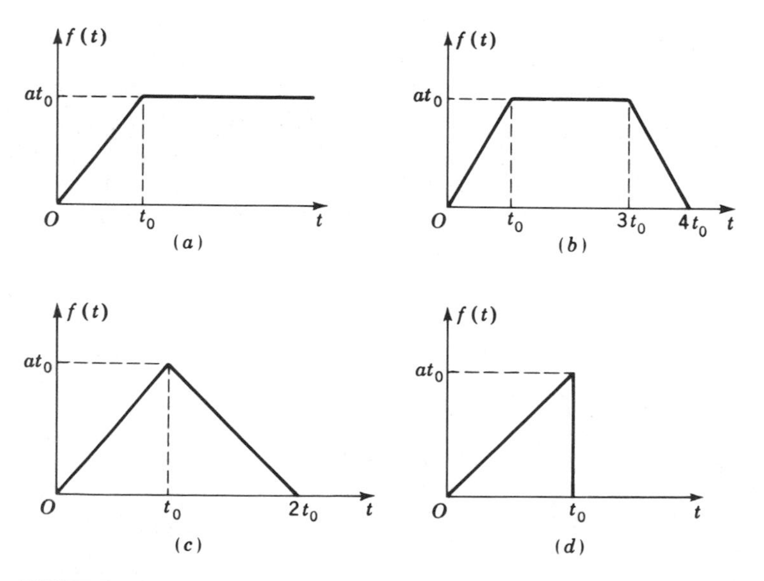

FIGURE 5.20
Piecewise continuous functions.

Solving for $Y(s)$ gives

$$Y(s) = \frac{F(s) + y(0)}{s+1} \tag{5.72}$$

For the interval $0 < t \leq 1$, the input is a unit step function. For $F(s) = 1/s$ and $y(0) = 0$, the transformed equation becomes

$$Y(s) = \frac{1}{s(s+1)} = \frac{1}{s} - \frac{1}{s+1} \tag{5.73}$$

Inverting yields for the response for the first interval

$$y(t) = 1 - e^{-t} \qquad 0 < t \leq 1 \tag{5.74}$$

For the second interval ($t > 1$), the input is zero. The initial condition for this second interval is obtained by evaluating the response for the preceding interval at the transition point $t = 1$. Thus

$$y(1) = 1 - e^{-1} = 0.632$$

It is more convenient to determine the response for the second interval as though there were no time delay and then to replace t by $t - t_0 = t - 1$, in the final answer. Substitution of $y(0) = 0.632$ and $F(s) = 0$ into Eq. (5.72), which is the transform of the differential equation, gives

$$Y(s) = \frac{0.632}{s+1}$$

Inverting yields

$$y(t) = 0.632e^{-t}$$

Replacing t by $t - t_0 = t - 1$ yields the response for $t > 1$

$$y(t-1) = 0.632e^{-(t-1)} \qquad t > 1 \tag{5.75}$$

The resulting response is indicated by the $y(t)$ curve in Fig. 5.21*a*.

In this method, the initial conditions for a new interval are obtained by evaluating the equation for the preceding interval at the time when the preceding interval ceases and the new one begins.

An alternate method for solving piecewise continuous problems is to regard the input as being the sum of separate functions, as illustrated in Fig. 5.21*b*. Subtracting the step function which begins at $t = 1$ from the step function which begins at $t = 0$ yields the pulse function shown in Fig. 5.21*a*. It is possible to represent any piecewise continuous function as the sum of other functions which are continuous.

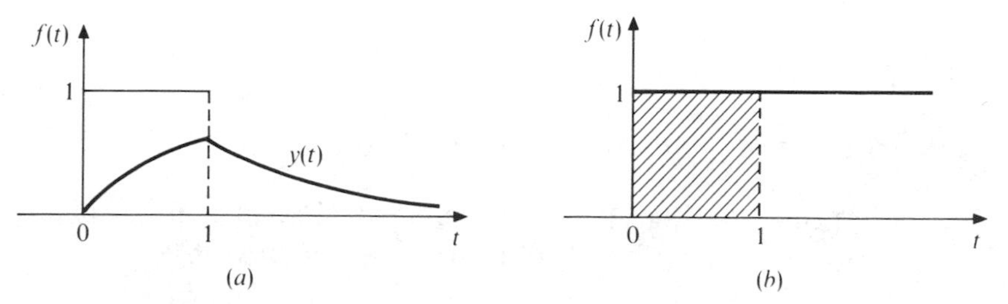

FIGURE 5.21
(*a*) System response to pulse function input and (*b*) alternate representation for a pulse function.

The transform for the first step function in Fig. 5.21*b* is $1/s$, and that for the second step function is $-(1/s)e^{-t_0 s}$, where $t_0 = 1$. Thus, the transform $F(s)$ of the input is

$$F(s) = \frac{1}{s} - \frac{1}{s}e^{-t_0 s}$$

Substitution of this result into Eq. (5.72) yields for the transformed equation

$$\begin{aligned} Y(s) &= \frac{1/s + y(0)}{s+1} - \frac{1/s}{s+1}e^{-t_0 s} \\ &= \frac{1/s}{s+1} - \frac{1/s}{s+1}e^{-t_0 s} \end{aligned} \tag{5.76}$$

Because of the delaying factor $e^{-t_0 s} = e^{-s}$, the second term on the right side should be ignored for $t < t_0 = 1$. Thus, for $0 < t \leq 1$, the transform is the same as that given by Eq. (5.73). The corresponding response for $0 < t \leq 1$ is given by Eq. (5.74). For $t > 1$, the response is the sum of that due to the second term of Eq. (5.76) and that already obtained for the first term. The inverse of $(1/s)/(s + 1)$ is given by Eq. (5.74). To take into account the time shift, replace t by $t - t_0 = t - 1$. Thus, for $t > 1$ the total response is

$$\begin{aligned} y(t) &= 1 - e^{-t} - [1 - e^{-(t-1)}] = (1 - e^{-1})e^{-(t-1)} \\ &= 0.632e^{-(t-1)} \qquad t > 1 \end{aligned}$$

where $e^{-t} = e^{-1}e^{-(t-1)}$. This verifies the result obtained by the first method, Eq. (5.75).

An advantage of this second method of solution is that initial conditions appear only in the transformed expression for the first interval. However, in this latter method, the amount of computational effort increases with the number of separate functions required to make up the overall piecewise continuous function. Thus, choice of the first or second method depends on the particular problem to be solved.

5.7 CONVOLUTION INTEGRAL

This integral may be used to determine the response of a system to any arbitrary input $f(t)$. Consider the transformed equation

$$Y(s) = G(s)F(s)$$

The inverse Laplace transform is

$$y(t) = \mathscr{L}^{-1}[G(s)F(s)] = \mathscr{L}^{-1}\left[G(s)\int_0^\infty f(\lambda)e^{-\lambda s}\,d\lambda\right] = \int_0^\infty f(\lambda)\mathscr{L}^{-1}[G(s)e^{-\lambda s}]\,d\lambda$$

where $F(s) = \int_0^\infty f(\lambda)e^{-\lambda s}\,d\lambda$. Recalling from the real translation theorem that $\mathscr{L}[f(t - t_0)] = F(s)e^{-t_0 s}$ where $f(t - t_0) = 0$ for $t_0 > t$, we see that

$$\mathscr{L}^{-1}[F(s)e^{-t_0 s}] = \begin{cases} f(t - t_0) & t_0 < t \\ 0 & t_0 > t \end{cases}$$

Similarly,

$$\mathcal{L}^{-1}[G(s)e^{-\lambda s}] = \begin{cases} g(t-\lambda) & \lambda < t \\ 0 & \lambda > t \end{cases}$$

Subdividing the integral from zero to infinity into the sum of the intervals from 0 to t and from t to infinity shows that

$$\int_0^\infty f(\lambda)\mathcal{L}^{-1}[G(s)e^{-\lambda s}]\, d\lambda = \int_0^t f(\lambda)g(t-\lambda)\, d\lambda + \int_t^\infty f(\lambda)(0)\, d\lambda$$

In the first integral on the right side, λ is less than t, so that the inverse of $G(s)e^{-\lambda s}$ is $g(t - \lambda)$. In the second integral, λ is greater than t, so the inverse is zero. Because this last integral vanishes

$$y(t) = \mathcal{L}^{-1}[G(s)F(s)] = \int_0^t f(\lambda)g(t-\lambda)\, d\lambda \tag{5.77}$$

This is the convolution integral.

The function $g(t - \lambda)$ is obtained by designating the inverse of $G(s)$ as $g(\lambda)$ and then replacing λ by $t - \lambda$. Similarly, $f(\lambda)$ is obtained by replacing t by λ in the equation for the forcing function $f(t)$. Note that the input $f(\lambda)$ is used directly (without transforming) in this integral procedure for determining the response $y(t)$.

Interchanging $F(s)$ and $G(s)$ in the preceding development yields the alternate form:

$$y(t) = \mathcal{L}^{-1}[F(s)G(s)] = \int_0^t g(\lambda)f(t-\lambda)\, d\lambda \tag{5.78}$$

The convolution integral may be used to obtain the inverse of the product of two transforms. For example, for $F(s) = 1/s^2$ and $G(s) = 1/(s + 1)$, then $f(\lambda) = \lambda$, $g(\lambda) = e^{-\lambda}$ and $g(t - \lambda) = e^{-(t-\lambda)}$. Thus,

$$\begin{aligned}\mathcal{L}^{-1}[G(s)F(s)] = \mathcal{L}^{-1}\left(\frac{1}{s^2}\,\frac{1}{s+1}\right) &= \int_0^t \lambda e^{-(t-\lambda)}\, d\lambda = e^{-t}\int_0^t \lambda e^{\lambda}\, d\lambda \\ &= e^{-t}[e^{\lambda}(\lambda - 1)]_0^t = e^{-t} + t - 1\end{aligned}$$

Using the alternative form $f(t - \lambda) = t - \lambda$ and $g(\lambda) = e^{-\lambda}$ yields the same result. It is important to note that $\mathcal{L}^{-1}[F(s)G(s)]$ is not equal to the product $\mathcal{L}^{-1}[F(s)]\mathcal{L}^{-1}[G(s)]$. For the preceding, $\mathcal{L}^{-1}[F(s)] = t$ and $\mathcal{L}^{-1}[G(s)] = e^{-t}$. This product te^{-t} is not the correct answer.

As another example, suppose we want to determine the inverse of $[\omega/(s^2 + \omega^2)]^2$. By letting

$$F(s) = G(s) = \frac{\omega}{s^2 + \omega^2}$$

then $f(\lambda) = \sin \omega\lambda$ and $g(t - \lambda) = \sin \omega(t - \lambda)$. Application of the convolution integral gives

$$\mathcal{L}^{-1}\left(\frac{\omega}{s^2+\omega^2}\right)^2 = \int_0^t [\sin \omega\lambda \sin \omega(t-\lambda)]\, d\lambda$$

Using the identity $\sin\alpha\sin\beta = [\cos(\alpha - \beta) - \cos(\alpha + \beta)]/2$ in which $\alpha = \omega\lambda$ and $\beta = \omega(t - \lambda)$ gives

$$\mathscr{L}^{-1}\left(\frac{\omega}{s^2+\omega^2}\right)^2 = \frac{1}{2}\int_0^t [\cos\omega(2\lambda - t) - \cos\omega t]\, d\lambda$$
$$= \frac{1}{2}\left(\frac{\sin\omega t}{\omega} - t\cos\omega t\right)$$

Illustrative example 5.13. Use the convolution integral to verify the solution of Illustrative example 5.1, i.e.,

$$(D^2 + 3D + 2)y(t) = f(t)$$

where $f(t) = 2u(t)$ and all initial conditions are zero.

Solution. The Laplace transform for $Y(s)$ is

$$Y(s) = \frac{1}{s^2+3s+2}F(s) + \frac{(s+3)y(0)+y'(0)}{s^2+3s+2} \tag{5.79}$$
$$= G(s)F(s) + \frac{(s+3)y(0)+y'(0)}{s^2+3s+2}$$

where $G(s) = 1/(s^2 + 3s + 2)$. This is the general form that is obtained in solving any differential equation by the convolution integral. The first term on the right side of Eq. (5.79) yields the response due to the forcing function, and the last term yields the response due to the initial conditions. Note that the initial-condition term is displayed separately and may be inverted directly. The physical significance of $g(t)$ is determined by noting that when $f(t)$ is a unit impulse and all the initial conditions are zero, then $Y(s) = G(s)$. Thus, $y(t) = g(t) = \mathscr{L}^{-1}[G(s)]$ is the response of a system to a unit impulse when all the initial conditions are zero. More simply, $g(t)$ is referred to as the *impulse response*.

For $y(0) = y'(0) = 0$, the inverse Laplace transform of Eq. (5.79) is

$$y(t) = \mathscr{L}^{-1}[G(s)F(s)] = \int_0^t f(\lambda)g(t-\lambda)\, d\lambda$$

where

$$g(t) = \mathscr{L}^{-1}[G(s)] = \mathscr{L}^{-1}\left(\frac{1}{s^2+3s+2}\right) = \mathscr{L}^{-1}\left(\frac{1}{s+1} - \frac{1}{s+2}\right) = e^{-t} - e^{-2t}$$

For $f(t) = 2u(t)$, then $f(\lambda) = 2$. Thus,

$$y(t) = \int_0^t (2)(e^{-(t-\lambda)} - e^{-2(t-\lambda)})\, d\lambda = 2e^{-t}\int_0^t e^{\lambda}\, d\lambda - 2e^{-2t}\int_0^t e^{2\lambda}\, d\lambda$$
$$= 1 - 2e^{-t} + e^{-2t} \tag{5.14}$$

This checks the result given by Eq. (5.14). With the aid of the convolution integral, the response of the system to any forcing function $f(t)$ may be obtained by knowing the response of the system to a unit impulse $g(t)$ when all the initial conditions are zero.

Illustrative example 5.14. It has been experimentally determined that the response of a system to a unit impulse when all the initial conditions are zero is e^{-t}. Use the convolution integral to determine the response of this system to a unit step function $f(t) = u(t)$ when all the initial conditions are zero.

Solution. For $g(t) = e^{-t}$ and $f(t) = 1$, then $g(\lambda) = e^{-\lambda}$ and $f(t - \lambda) = 1$. Thus,

$$y(t) = \int_0^t f(t-\lambda)g(\lambda)\, d\lambda = \int_0^t (1)e^{-\lambda}\, d\lambda = [-e^{-\lambda}]_0^t = 1 - e^{-t} \qquad t > 0$$

Illustrative example 5.15. Use the convolution integral to determine the solution of Illustrative example 5.12. The differential equation is

$$\frac{dy}{dt} + y = f(t) \qquad f(t) = \begin{cases} 1 & 0 < t < 1 \\ 0 & t > 1 \end{cases}$$

A plot of the forcing function $f(t)$ is shown in Fig. 5.21a.

Solution. The Laplace transform of the differential equation is

$$sY(s) - y(0) + Y(s) = F(s)$$

or

$$\begin{aligned} Y(s) &= \frac{1}{s+1}F(s) + \frac{y(0)}{s+1} \\ &= G(s)F(s) + \frac{y(0)}{s+1} \end{aligned} \tag{5.80}$$

where $G(s) = 1/(s + 1)$. Note that $G(s)$ is the Laplace transform that would result if all of the initial conditions were zero and $f(t)$ were a unit impulse. Thus, $g(t) = \mathscr{L}^{-1}[G(s)]$ is the impulse response of the system. For $y(0) = 0$, the inverse Laplace transform of Eq. (5.80) is

$$y(t) = \mathscr{L}^{-1}[G(s)F(s)] = \int_0^t f(\lambda)g(t-\lambda)\, d\lambda$$

where

$$g(t) = \mathscr{L}^{-1}[G(s)] = \mathscr{L}^{-1}\left(\frac{1}{s+1}\right) = e^{-t}$$

For $0 < t < 1$, $f(t) = 1$ in which case $f(\lambda) = 1$. Thus

$$\begin{aligned} y(t) &= \int_0^t (1)e^{-(t-\lambda)}\, d\lambda = e^{-t}\int_0^t e^{\lambda}\, d\lambda \\ &= 1 - e^{-t} \qquad 0 < t \le 1 \end{aligned}$$

This checks the result given by Eq. (5.74).

Because $f(\lambda) = 1$ for $0 < \lambda < 1$ and $f(\lambda) = 0$ for $\lambda > 1$, then for $t > 1$

$$y(t) = \int_0^1 (1)e^{-(t-\lambda)}\, d\lambda + \int_1^t (0)e^{-(t-\lambda)}\, d\lambda = e^{-t}(e-1) = 0.632e^{-(t-1)} \qquad t > 1$$

This checks the previously attained result given by Eq. (5.75).

5.8 ERROR COEFFICIENTS

For the system shown in Fig. 5.22, the actuating signal is $e(t) = r(t) - H(D)c(t) = r(t) - G(D)H(D)e(t)$. When all initial conditions are zero, the transformed equation is

$$E(s) = \frac{R(s)}{1 + G(s)H(s)}$$

Application of the final-value theorem yields for the steady-state error

$$e_{ss} = \lim_{t \to \infty} e(t) = \lim_{s \to 0} sE(s)$$

For a unit step input $[R(s) = 1/s]$, the steady-state error is

$$e_{ss} = \lim_{s \to 0} s\frac{1/s}{1 + G(s)H(s)} = \lim_{s \to 0} \frac{1}{1 + G(s)H(s)} = \frac{1}{1 + K_p} \tag{5.81}$$

where $K_p = \lim_{s \to 0} G(s)H(s)$ is called the *positional error constant.*

For a unit ramp input $[r(t) = t,\ R(s) = 1/s^2]$, the steady-state error is

$$e_{ss} = \lim_{s \to 0} s\frac{1/s^2}{1 + G(s)H(s)} = \lim_{s \to 0} \frac{1}{s + sG(s)H(s)} = \lim_{s \to 0} \frac{1}{sG(s)H(s)} = \frac{1}{K_v} \tag{5.82}$$

where $K_v = \lim_{s \to 0} sG(s)H(s)$ is the *velocity error constant.*

For a unit parabolic input $[r(t) = t^2/2,\ R(s) = 1/s^3]$, the steady-state error is

$$e_{ss} = \lim_{s \to 0} \frac{1}{s^2 + s^2G(s)H(s)} = \lim_{s \to 0} \frac{1}{s^2G(s)H(s)} = \frac{1}{K_a} \tag{5.83}$$

where $K_a = \lim_{s \to 0} s^2G(s)H(s)$ is the *acceleration error constant.* The error constant provides a convenient method for determining the steady-state error to a unit step, unit ramp, or unit parabolic input. Note that the unit ramp function is the integral of the unit step function $\int(1)\,dt = t$, and the unit parabolic function is the integral of the unit ramp function $\int t\,dt = t^2/2$. Because the response due to initial conditions dies out as time approaches infinity, the preceding results are valid even if the initial conditions are not zero.

Usually the command signal $v(t)$ is the desired value of the controlled variable $c(t)$ so that the error is $v(t) - c(t)$. By referring to Fig. 3.28, which is the general block-diagram representation for a feedback control system, note that the actuating signal $e(t)$ is a measure of the error but is not necessarily equal to the error. Thus, the

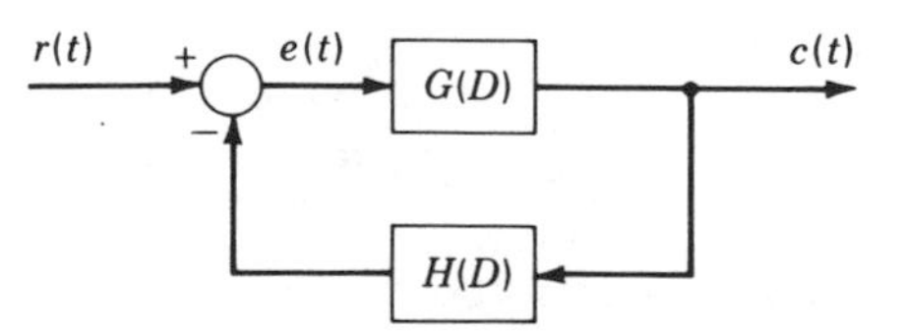

FIGURE 5.22
Feedback control system.

preceding criteria for steady-state error to a unit step input, or a unit ramp input, or to a unit parabolic input, are measures of these errors and not necessarily the errors themselves. This may cause erroneous results when these criteria are used as the basis for the design of control systems, as is done in many textbooks.

Illustrative example 5.16. The system shown in Fig. 5.10 is discussed in Illustrative example 5.3. For this system, determine the steady-state error to a unit step-function input.

Solution. From Fig. 5.10, it follows that

$$G(s) = \frac{K_1}{1+\tau_1 s}\,\frac{K_2}{1+\tau_2 s} \qquad H(s) = K_H$$

The positional error constant is

$$K_p = \lim_{s \to 0} G(s)H(s) = K_1 K_2 K_H = (1.0)(2.0)(0.5) = 1.0$$

Thus, the steady-state error due to a unit step input is

$$e_{ss} = \frac{1}{1+K_p} = 0.5$$

The corresponding steady-state values for c and v may be determined by noting from Fig. 5.10 that $e = r - K_H c = r - 0.5c$ and $r = Av = v$. For a unit step input, $r = 1$, and thus $c = 2(r - e_{ss}) = 2(1 - 0.5) = 1.0$ and $v = r = 1.0$. This agrees with the result of Illustrative example 5.3, in which it was found that for a constant input, $c = v$. As demonstrated by this example, the presence of a steady-state error signal does not necessarily imply that the controlled variable is not equal to the command signal.

Illustrative example 5.17. For the system shown in Fig. 5.23*a*, determine the response to a unit step function, a unit ramp function, and a unit parabolic function when all initial conditions are zero. What is the steady-state error to each of these inputs?

Solution. The differential equation of operation is

$$(D^2 + 3D + 2)c(t) = 2r(t)$$

The transformed equation is

$$C(s) = \frac{2R(s)}{(s+1)(s+2)} \tag{5.84}$$

For a unit step-function input, $R(s) = 1/s$. Carrying out the partial-fraction expansion and inverting yield for the time response

$$c(t) = 1 - 2e^{-t} + e^{-2t} \qquad t > 0$$

The corresponding input $r(t) = u(t)$ and response $c(t)$ are shown in Fig. 5.23*b*. The error $e(t) = r(t) - c(t)$ is $e(t) = 1 - (1 - 2e^{-t} + e^{-2t}) = 2e^{-t} - e^{-2t}$. Taking the limit as t approaches infinity yields for the steady-state error

$$e_{ss} = \lim_{t \to \infty} e(t) = \lim_{t \to \infty} (2e^{-t} - e^{-2t}) = 0$$

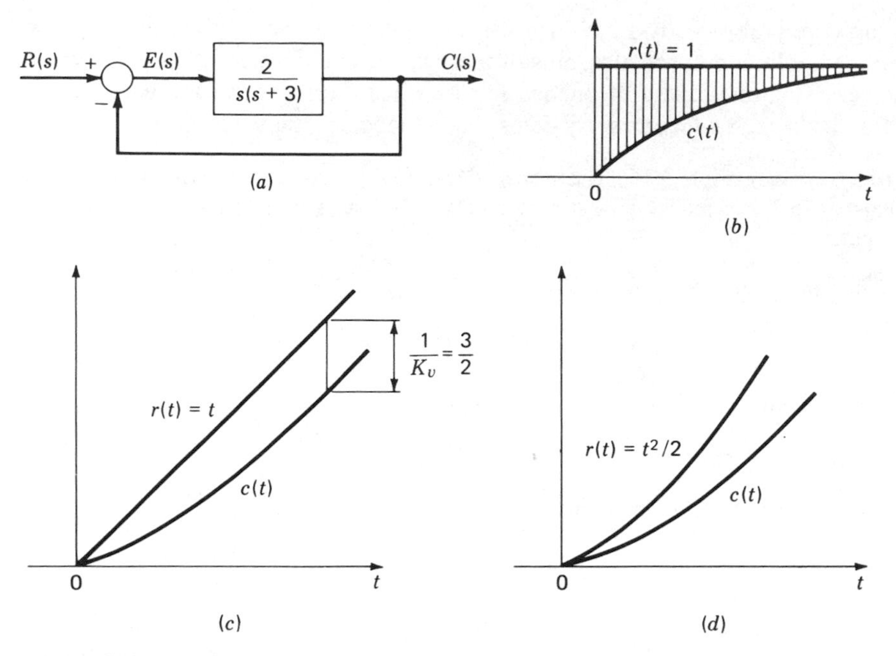

FIGURE 5.23

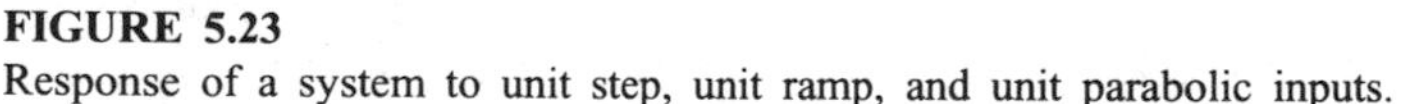
Response of a system to unit step, unit ramp, and unit parabolic inputs.

The positional error constant is

$$K_p = \lim_{s \to 0} G(s)H(s) = \left.\frac{2}{s(s+3)}\right|_{s=0} = \infty$$

Thus, the steady-state error to a unit step function is

$$e_{ss} = \frac{1}{1+K_p} = \frac{1}{1+\infty} = 0$$

For a unit ramp input, $r(t) = t$. Substitution of $R(s) = 1/s^2$ into Eq. (5.84) gives

$$C(s) = \frac{2}{s^2(s+1)(s+2)} = \frac{1}{s^2} - \frac{\frac{3}{2}}{s} + \frac{2}{s+1} - \frac{\frac{1}{2}}{s+2}$$

Inverting yields for the response

$$c(t) = t - \tfrac{3}{2} + 2e^{-t} - \tfrac{1}{2}e^{-2t}$$

A plot of the input $r(t) = t$ and the response $c(t)$ is shown in Fig. 5.23c. The error $e(t) = r(t) - c(t)$ is

$$e(t) = t - \left(t - \tfrac{3}{2} + 2e^{-t} - \tfrac{1}{2}e^{-2t}\right) = \tfrac{3}{2} - 2e^{-t} - \tfrac{1}{2}e^{-2t}$$

Taking the limit as t approaches infinity yields for the steady-state error

$$e_{ss} = \lim_{t \to \infty} e(t) = \tfrac{3}{2}$$

The velocity error constant is

$$K_v = \lim_{s \to 0} sG(s)H(s) = \frac{2s}{s(s+3)}\bigg|_{s=0} = \frac{2}{3}$$

Thus, the steady-state error to a unit ramp function is

$$e_{ss} = \frac{1}{K_v} = \frac{3}{2}$$

For a unit parabolic input, $r(t) = t^2/2$. Substitution of $R(s) = 1/s^3$ into Eq. (5.84) gives

$$C(s) = \frac{2}{s^3(s+1)(s+2)} = \frac{1}{s^3} - \frac{\frac{3}{2}}{s^2} + \frac{\frac{7}{4}}{s} - \frac{2}{s+1} + \frac{\frac{1}{4}}{s+2}$$

Inverting yields for the response

$$c(t) = \frac{t^2}{2} - \frac{3}{2}t + \frac{7}{4} - 2e^{-t} + \frac{1}{4}e^{-2t}$$

The error $e(t) = r(t) - c(t)$ is

$$e(t) = \tfrac{3}{2}t - \tfrac{7}{4} + 2e^{-t} - \tfrac{1}{4}e^{-2t}$$

In the limit as t approaches infinity, the error is

$$e_{ss} = \lim_{t \to \infty} e(t) = \left[\tfrac{3}{2}t - \tfrac{7}{4} + 2e^{-t} - \tfrac{1}{4}e^{-2t}\right]_{t=\infty} = \infty$$

A plot of the input $r(t) = t^2/2$ and the response $c(t)$ is shown in Fig. 5.23*d*. Note that the error keeps growing as t increases. The acceleration error constant is

$$K_a = \lim_{s \to 0} s^2G(s)H(s) = \frac{2s^2}{s(s+3)}\bigg|_{s=0} = \frac{2}{(1+3/s)}\bigg|_{s=0} = \frac{2}{1+\infty} = 0$$

Thus, the steady-state error to a unit parabolic input is

$$e_{ss} = \frac{1}{K_a} = \frac{1}{0} = \infty$$

Consider now the case in which the input is a unit step function $f(t) = u(t)$. The transform of the error signal is

$$E(s) = \frac{R(s)}{1 + G(s)H(s)} = \frac{1/s}{1 + G(s)H(s)}$$

Taking the limit as s approaches zero shows that

$$\lim_{s \to 0} E(s) = \lim_{s \to 0} \frac{1}{s + sG(s)H(s)} = \lim_{s \to 0} \frac{1}{sG(s)H(s)} = \frac{1}{K_v}$$

From the definition of the Laplace transform, it follows that

$$\lim_{s \to 0} E(s) = \lim_{s \to 0} \int_0^\infty e(t)e^{-st}\,dt = \int_0^\infty e(t)\,dt \tag{5.85}$$

The right side is the time integral of the error to a unit step input.

This integral shown shaded in Fig. 5.23*b* is

$$\int_0^\infty e(t)\,dt = \frac{1}{K_v} = \frac{3}{2}$$

This result may also be checked by integration. Thus

$$\int_0^\infty e(t)\,dt = \int_0^\infty (2e^{-t} - e^{-2t})\,dt = \left[-2e^{-t} + \tfrac{1}{2}e^{-2t}\right]_0^\infty = 0 - (-2 + \tfrac{1}{2}) = \tfrac{3}{2}$$

As demonstrated in the preceding, many interesting features and characteristics of systems may be obtained directly from the transformed equation without the need to invert and obtain the time response.

PROBLEMS

5.1. Determine the partial-fraction expansion for the following operators.

(*a*) $\dfrac{L_m(D)}{L_n(D)} = \dfrac{1}{(D+2)(D+3)}$ (*b*) $\dfrac{L_m(D)}{L_n(D)} = \dfrac{D+4}{(D+2)(D+3)}$

(*c*) $\dfrac{L_m(D)}{L_n(D)} = \dfrac{3D+7}{(D+1)(D+2)(D+3)}$ (*d*) $\dfrac{L_m(D)}{L_n(D)} = \dfrac{D^2+4D+5}{(D+1)(D+2)(D+3)}$

5.2. Determine the partial-fraction expansion for the following operators.

(*a*) $\dfrac{L_m(D)}{L_n(D)} = \dfrac{3}{(D+2)(D+5)}$ (*b*) $\dfrac{L_m(D)}{L_n(D)} = \dfrac{D+8}{(D+2)(D+5)}$

(*c*) $\dfrac{L_m(D)}{L_n(D)} = \dfrac{18D+30}{D(D+2)(D+5)}$ (*d*) $\dfrac{L_m(D)}{L_n(D)} = \dfrac{D^2+4D+10}{D(D+2)(D+5)}$

5.3. Consider the differential equation

$$\frac{dy}{dt} + y = f(t)$$

Use classical techniques to determine the solution for each of the following cases.
(*a*) $f(t) = 0$ and $y(0) = 1$
(*b*) $f(t) = 1$ and $y(0) = 0$

5.4. Consider the differential equation

$$\frac{dy}{dt} + 2y = 10f(t)$$

Using classical methods, determine the solution for each of the following cases.
(*a*) $f(t) = 1$ and $y(0) = 0$
(*b*) $f(t) = 0$ and $y(0) = 1$
(*c*) $f(t) = e^{-t}$ and $y(0) = 0$

5.5. Consider the differential equation

$$y(t) = \frac{12}{(D+1)(D+3)}f(t)$$

Using classical methods, determine the solution when $f(t) = 1$ for each of the following cases.

(a) $y(0) = y'(0) = 0$

(b) $y(0) = 2$ and $y'(0) = 0$

5.6. Repeat Prob. 5.5 except $f(t) = e^{-2t}$.

5.7. Use Eq. (5.15) to determine the Laplace transform of each of the following functions.

(a) $f(t) = t$

(b) $f(t) = t^2$

(c) $f(t) = \cos t$

5.8. Given that the Laplace transform of e^{-t} is $1/(s + 1)$, use appropriate theorems to determine the Laplace transform of each of the following functions.

(a) $f(t) = te^{-t}$

(b) $f(t) = t^2 e^{-t}$

(c) $f(t) = e^{-3t}$

5.9. Given that the Laplace transform of $\sin t$ is $1/(s^2 + 1)$, use appropriate theorems to determine the Laplace transform of each of the following functions.

(a) $f(t) = t \sin t$

(b) $f(t) = e^{-t} \sin t$

(c) $f(t) = \sin 3t$

5.10. Determine the Laplace transform for each of the following functions.

(a) $f(t) = 0$ for $t < 2$ and $f(t) = e^{-(t-2)}$ for $t > 2$

(b) $f(t) = 0$ for $t < \pi$ and $f(t) = \sin(t - \pi)$ for $t > \pi$

5.11. Use the equation for obtaining the Laplace transform of the second derivative of $f(t)$ to determine the transforms of

(a) $f(t) = \cos \omega t$

(b) $f(t) = \sinh \omega t$

5.12. Use the equation for obtaining the Laplace transform of the integral of $f(t)$ to determine the transforms of

(a) $t^2 = \int 2t\, dt$

(b) $\sin \omega t = \int (\omega \cos \omega t)\, dt$

(c) $\cos \omega t = \int (-\omega \sin \omega t)\, dt$

5.13. Use the Laplace transform method to determine the solution of Prob. 5.3.

5.14. Use the Laplace transform method to determine the solution of Prob. 5.4.

5.15. Use the Laplace transform method to determine the solution of Prob. 5.5.

5.16. The operation of an amplifier is described by the differential equation

$$\frac{dy}{dt} + 0.2y = f(t)$$

Determine the response $y(t)$ to a unit step function $f(t) = u(t)$. All the initial conditions are zero. Identify the time constant and steady-state gain.

5.17. Consider the differential equation

$$\frac{dy}{dt} + y = u(t) + e^{-t} \qquad y(0) = 0$$

where $u(t)$ is a unit step function. Determine the transform of the derivative $\mathscr{L}[y'(t)] = sY(s) - y(0)$, and then invert to obtain $y'(t)$.

5.18. The block diagram for a hydraulic servomotor is shown in Fig. P5.18. Determine K_1 and K_2 such that the servomotor will have a steady-state gain of 2 and a time constant of 0.4 s.

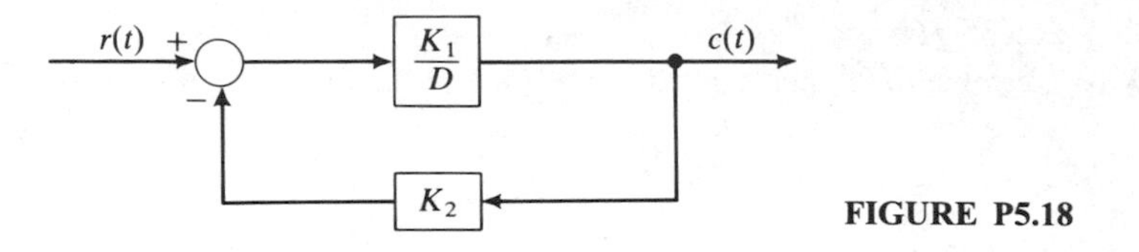

FIGURE P5.18

5.19. The block diagram for a jet pipe amplifier is shown in Fig. P5.19. Determine K and a, such that the amplifier will have a steady-state gain of 1 and a time constant of 0.25 s.

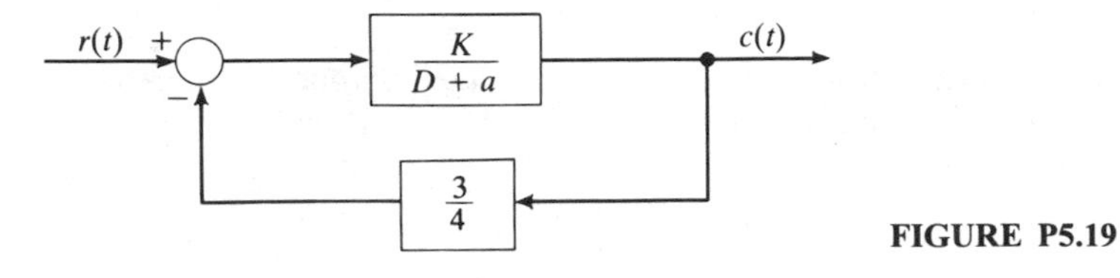

FIGURE P5.19

5.20. An armature-controlled dc motor is shown in Fig. P5.20. Determine K_1 and K_2 such that the system will have a steady-state gain of 2 and a time constant of 0.4 s.

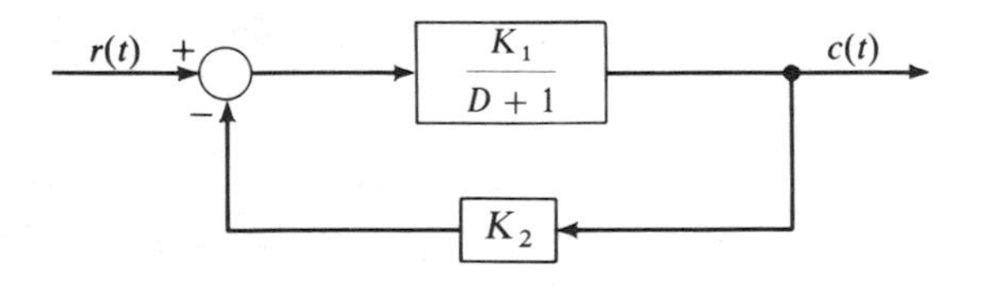

FIGURE P5.20

5.21. The differential equation of operation for a control system is

$$y(t) = \frac{12(D+1)}{(D+3)(D+4)} f(t)$$

Determine the response $y(t)$ when $f(t) = u(t)$ is a unit step function and all initial conditions are zero.

5.22. Determine the response $y(t)$ for the system of Prob. 5.21 when $f(t) = 0$, $y(0) = 1$, and $y'(0) = 0$.

5.23. Determine the response $y(t)$ for the system of Prob. 5.21 when $f(t) = 0$, $y(0) = 0$, and $y'(0) = 1$.

5.24. The differential equation of operation for a system is

$$c(t) = \frac{D+6}{D^2+5D+6} r(t)$$

Determine the response $c(t)$ for each of the following cases.

(*a*) $r(t) = u(t)$, $c(0) = \dot{c}(0) = 0$

(*b*) $r(t) = e^{-t}$, $c(0) = \dot{c}(0) = 0$

(*c*) $r(t) = 0$, $c(0) = 1$, and $\dot{c}(0) = 0$

5.25. A system for controlling the speed of a hypersonic plane is shown in Fig. P5.25. Determine the response $c(t)$ for each of the following cases.

(*a*) $r(t) = u(t)$ and $c(0) = \dot{c}(0) = 0$

(*b*) $r(t) = 0$, $c(0) = 1$, and $\dot{c}(0) = 1$

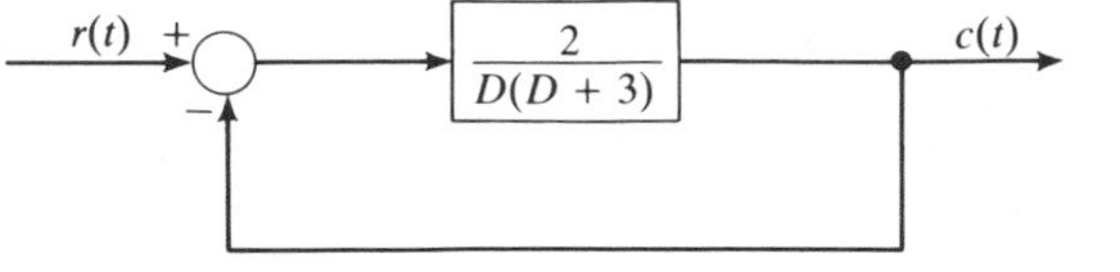

FIGURE P5.25

5.26. A system for controlling the steering of a modern ship is shown in Fig. P5.26. Determine the response $c(t)$ for each of the following cases.

(*a*) $r(t) = u(t)$ and $c(0) = \dot{c}(0) = 0$

(*b*) $r(t) = 0$, $c(0) = 1$, and $\dot{c}(0) = 1$

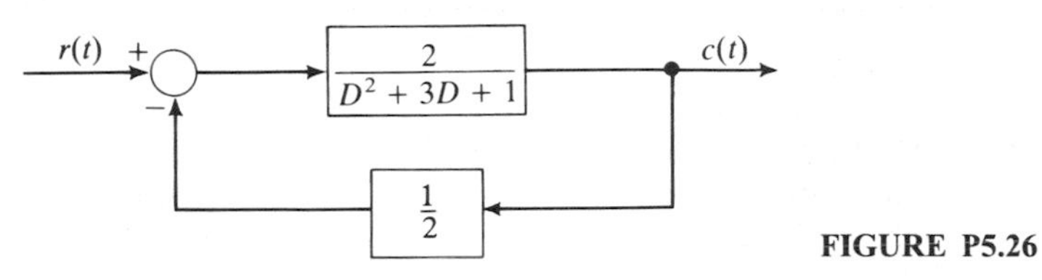

FIGURE P5.26

5.27. A system for controlling the temperature of a modern office building is shown in Fig. P5.27. Determine the response $c(t)$ for each of the following cases.

(*a*) $r(t) = u(t)$ and $c(0) = \dot{c}(0) = 0$

(*b*) $r(t) = 0$, $c(0) = 1$, and $\dot{c}(0) = -2$

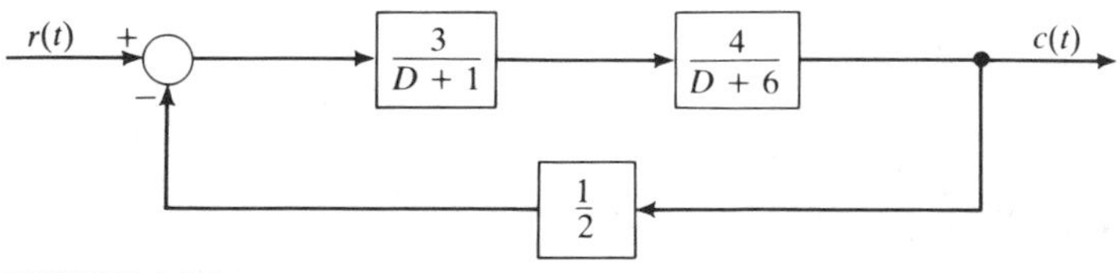

FIGURE P5.27

5.28. In the paper industry, it is important to maintain a constant tension on the continuous sheet of paper between the windoff and windup rolls. Such a system is shown in Fig. P5.28. Determine the response $c(t)$ when

(*a*) $r(t) = u(t)$, $d(t) = 0$, and $c(0) = \dot{c}(0) = 0$

(*b*) $r(t) = 0$, $d(t) = u(t)$, $c(0) = 1$, and $\dot{c}(0) = -1$

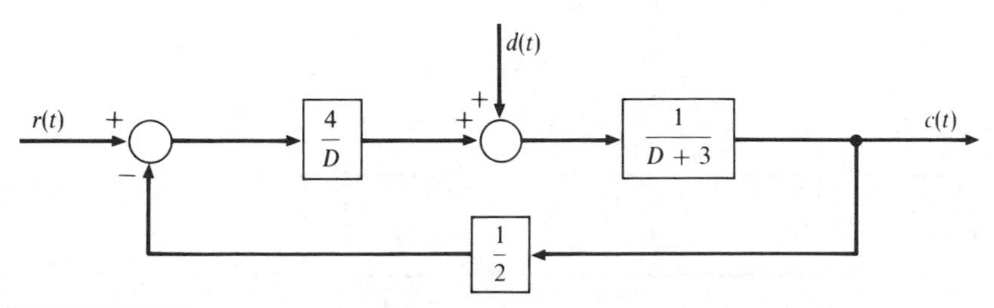

FIGURE P5.28

5.29. Robots can be programmed to move tools through a prescribed path very accurately. For the robotic control system shown in Fig. P5.29, all the initial conditions are zero and $K = 1.5$. Determine the response for each of the following cases.

(*a*) $r(t)$ is a unit step function and $d(t) = 0$.

(*b*) $d(t)$ is a unit step function and $r(t) = 0$.

(*c*) Both $r(t)$ and $d(t)$ are unit step functions.

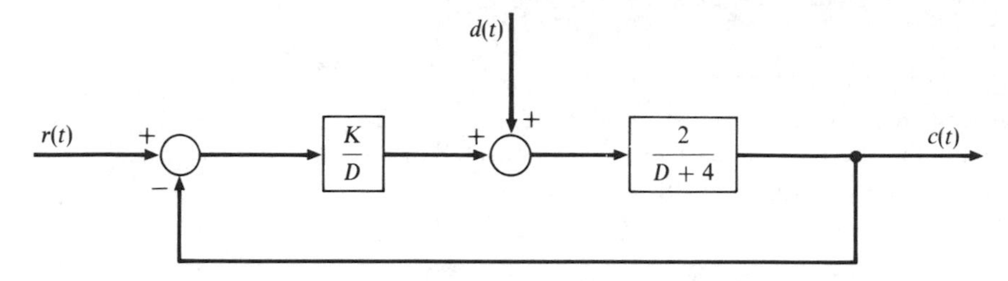

FIGURE P5.29

5.30. Solar cells generate a dc voltage that can be used to drive dc motors. A control system for maintaining the maximum available power as the angle of the sun changes during the day is shown in Fig. P5.30. Determine the response $c(t)$ for each of the following cases.

(*a*) $r(t) = u(t)$, $d(t) = 0$, and $c(0) = \dot{c}(0) = 0$

(*b*) $r(t) = 0$, $d(t) = \delta(t)$, and $c(0) = \dot{c}(0) = 0$

(*c*) $r(t) = d(t) = 0$, $c(0) = 1$, and $\dot{c}(0) = 0$

For each case use the initial-value theorem to check the result for $c(0+)$.

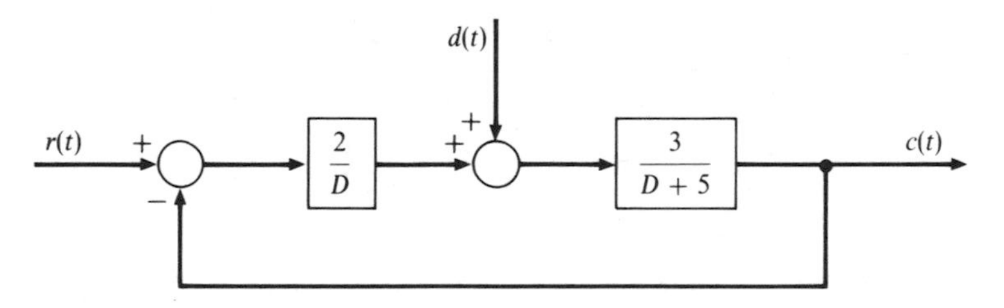

FIGURE P5.30

5.31. A system for controlling the speed of a computer-controlled lathe is shown in Fig. P5.31. Determine the response $c(t)$ for each of the following cases.

(*a*) $r(t) = u(t)$, $d(t) = 0$, and $c(0) = \dot{c}(0) = 0$

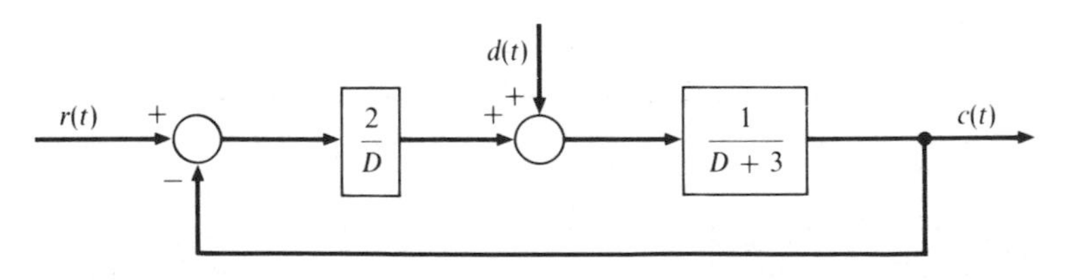

FIGURE P5.31

(*b*) $r(t) = 0$, $d(t) = u(t)$, and $c(0) = \dot{c}(0) = 0$
(*c*) $r(t) = d(t) = 0$, $c(0) = 1$, and $\dot{c}(0) = 2$

5.32. Using the Laplace transform method, determine the solution of the following differential equation for the case in which $f(t)$ is a step function of height $h = 6$; all initial conditions are zero.

$$y(t) = \frac{D+5}{(D+1)(D+3)} f(t)$$

5.33. A switch is inserted in series in the circuit of Fig. 5.17. For $t < 0$, the switch is open and all initial conditions are zero. For $t > 0$, the switch is closed and the voltage source $e = 100$ V is applied to the circuit. For $R = 10^6\ \Omega = 1\ \text{M}\Omega$ and $C = 10^{-6}\ \text{F} = 1\ \mu\text{F}$, determine the equation for the resulting current i that flows through the circuit and the voltage e_c across the capacitor.

5.34. Using the Laplace transform method, a student found the solution of the differential equation

$$\ddot{y} + 4\dot{y} + 3y = 4\delta(t) \qquad y(0) = \dot{y}(0) = 0$$

to be

$$y = 2e^{-t} - 2e^{-3t}$$

Being rather thorough, the student decided to check the answer by checking the initial conditions and checking whether the solution satisfied the differential equation. The results were $y(0) = 2 - 2 = 0$, $\dot{y}(0) = -2 + 6 = 4$, and $\ddot{y} + 4\dot{y} + 3y = 0$. Since this solution satisfied neither the differential equation nor the one initial condition, the student concluded that there had been an error in determining y. Is the error in the solution or in the conclusion? Explain.

5.35. A mass-spring damper system is shown in Fig. P5.35.
(*a*) Write the differential equation.
(*b*) Use Laplace transforms to solve this equation when $x(0) = 0$, $\dot{x}(0) = 1$, $f = 0$, $M = 1$, $B = 3$, and $K = 2$.

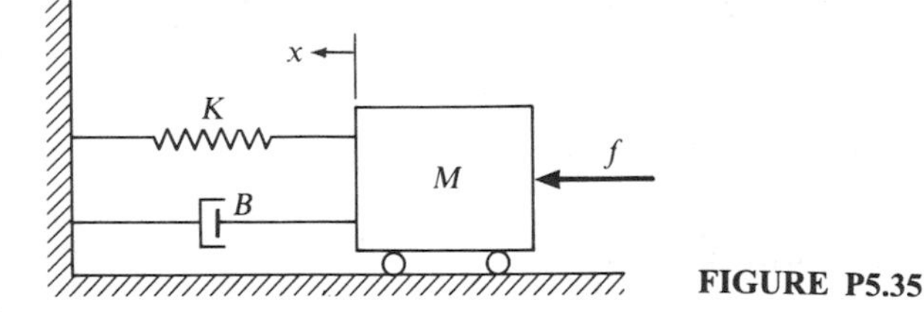

FIGURE P5.35

5.36. Using Laplace transforms, determine the response $y(t)$ for the following differential equations for the case in which all the initial conditions are zero and the input $f(t)$ is an impulse of area $k = 2$ which occurs at time $t = 0$.

(*a*) $$y(t) = \frac{2D+3}{(D+2)(D+3)} f(t)$$

(*b*) $$y(t) = \frac{2D+3}{D(D+2)(D+3)} f(t)$$

Use the initial-value theorem to determine $y(0+)$ and the final-value theorem to determine $y(\infty)$.

5.37. Repeat Prob. 5.36 except the system is initially at a steady-state operating condition, with $f(0) = 2$.

5.38. The dynamics of a liquid-level control system are described by the differential equation

$$y(t) = \frac{10(2D+1)}{(D+2)(D+5)} f(t)$$

Use the Laplace transform method to determine the response $y(t)$ when all the initial conditions are zero and the forcing function $f(t)$ is
(a) A unit impulse $\delta(t)$
(b) A unit step function $u(t)$
(c) An exponential e^{-t}

5.39. Repeat Prob. 5.38 except that the system is initially at a steady-state operating condition with $f(0) = 3$.

5.40. The proportional control system shown in Fig. P5.40 is in a state of equilibrium, with $r = 3$ and $d = 0$. A step-function disturbance $d(t) = 6u(t)$ is then initiated at time $t = 0$. Determine the response $c(t)$ for $t > 0$.

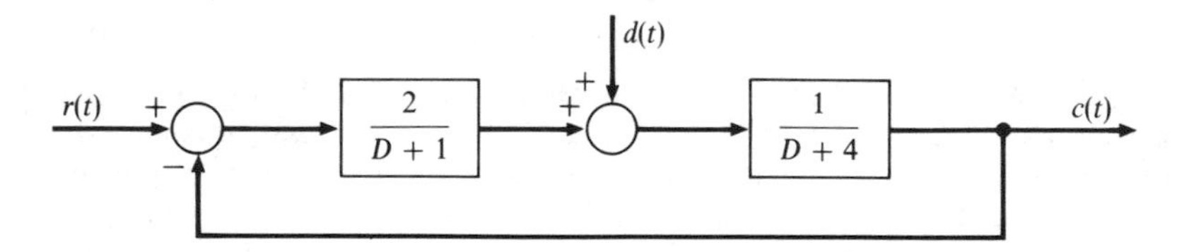

FIGURE P5.40

5.41. The proportional control system shown in Fig. P5.40 is in a state of equilibrium, with $r = 0$ and $d = 6$. A step-function change $r(t) = 3u(t)$ is then initiated at time $t = 0$. Determine the response $c(t)$ for $t > 0$.

5.42. The integral control system shown in Fig. P5.42 is initially at equilibrium, with $r = 1$ and $d = 0$. A step-function disturbance $d(t) = u(t)$ is then initiated at time $t = 0$. Determine the response $c(t)$ for $t > 0$.

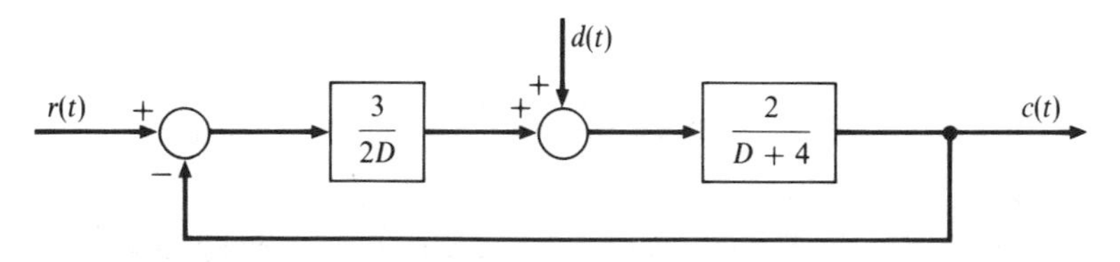

FIGURE P5.42

5.43. The following functions may be written as products of $1/s$, $1/s^2$, or $1/(s+1)$. Use the convolution integral to determine the inverse transforms of these functions.

(a) $Y(s) = \dfrac{1}{s^3}$ (b) $Y(s) = \dfrac{1}{s^4}$ (c) $Y(s) = \dfrac{1}{s^2(s+1)}$ (d) $Y(s) = \dfrac{1}{(s+1)^2}$

5.44. The following functions may be written as products of $1/s$, $1/(s^2+1)$, or $s/(s^2+1)$. Use the convolution integral to determine the inverse transforms of these functions.

$$(a)\quad Y(s)=\frac{1}{s(s^2+1)} \qquad (b)\quad Y(s)=\frac{1}{(s^2+1)^2}$$

$$(c)\quad Y(s)=\frac{s}{(s^2+1)^2} \qquad (d)\quad Y(s)=\frac{s^2}{(s^2+1)^2}$$

5.45. For the system shown in Fig. P5.45, use the Laplace transform method to determine the response $y(t)$ when $f(t)=\delta(t)$ is a unit impulse and all initial conditions are zero. Next use the convolution integral to determine the response for each of the following cases.
(a) $f(t)=u(t)$ and $y(0)=0$
(b) $f(t)=e^{-t}$ and $y(0)=1$

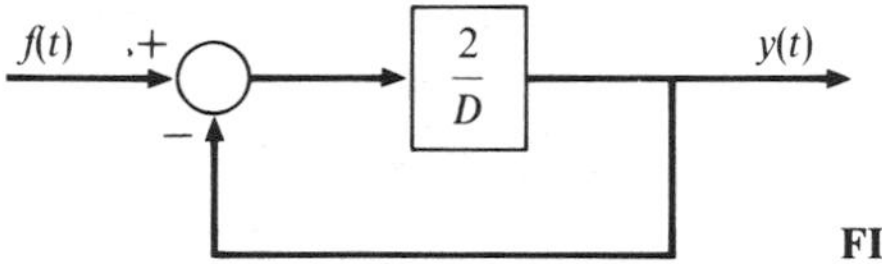

FIGURE P5.45

5.46. The operation of a control system is described by the differential equation

$$y(t)=\frac{20}{D^2+9D+20}f(t)$$

Use the convolution integral to determine the response for each of the following cases.
(a) $f(t)=u(t)$ and $y(0)=\dot{y}(0)=0$
(b) $f(t)=u(t)$ and $y(0)=\dot{y}(0)=1$

5.47. Repeat Prob. 5.46 except $f(t)=e^{-4t}$.

5.48. The operation of a control system is described by the differential equation

$$y(t)=\frac{10(D+2)}{D^2+9D+20}f(t)$$

Use the convolution integral to determine the response for each of the following cases.
(a) $f(t)=u(t)$ and $y(0)=\dot{y}(0)=0$
(b) $f(t)=u(t)$, $y(0)=0$, and $\dot{y}(0)=1$

5.49. It has been experimentally determined that the response of a system to a unit impulse when all the initial conditions are zero is $e^{-t}-e^{-2t}$. Use the convolution integral to determine the response of the system to the step function $f(t)=2u(t)$ when all the initial conditions are zero.

5.50. Determine the position, velocity, and acceleration error constants, and then determine the steady-state error to a unit step, a unit ramp, and a unit parabolic input for the system shown (a) in Fig. P5.40 and (b) in Fig. P5.42.

CHAPTER

6

TRANSIENT RESPONSE

Because an actual system may be subjected to all types and varieties of input excitations $f(t)$, it becomes impractical to calculate the system response for every possible excitation. In this chapter, we show that a very good measure of the transient behavior may be obtained directly from the zeros of the characteristic function (i.e., roots of the characteristic equation). This criterion for evaluating transient performance is obtained by considering the essential characteristics of a general system of order n.

The general operational representation for a differential equation of order n is

$$y(t) = \frac{a_m D^m + a_{m-1} D^{m-1} + \cdots + a_1 D + a_0}{D^n + b_{n-1} D^{n-1} + \cdots + b_1 D + b_0} f(t) \tag{6.1}$$

The transform of each term is

$$\begin{aligned} \mathscr{L}[D^n y(t)] &= s^n Y(s) - I(s)_n \\ b_{n-1}\mathscr{L}[D^{n-1} y(t)] &= b_{n-1} s^{n-1} Y(s) - I(s)_{n-1} \\ &\cdots\cdots\cdots \\ a_m \mathscr{L}[D^m f(t)] &= a_m s^m F(s) - I(s)_m \\ a_{m-1}\mathscr{L}[D^{m-1} f(t)] &= a_{m-1} s^{m-1} F(s) - I(s)_{m-1} \\ &\cdots\cdots\cdots \end{aligned}$$

where $I(s)_n$, $I(s)_{n-1}$, . . . , represent the initial conditions associated with each transform. Transforming each term of Eq. (6.1) accordingly and collecting terms yield

$$Y(s) = \frac{(a_m s^m + a_{m-1} s^{m-1} + \cdots + a_1 s + a_0)F(s) + I(s)}{s^n + b_{n-1} s^{n-1} + \cdots + b_1 s + b_0}$$
$$= \frac{L_m(s)F(s) + I(s)}{L_n(s)} \tag{6.2}$$

where $I(s) = I(s)_n + I(s)_{n-1} + \cdots - I(s)_m - I(s)_{m-1} - \cdots$ is the sum of all the initial-condition terms. By comparison of Eqs. (6.1) and (6.2), note that the characteristic function in the s domain $L_n(s)$ is the same as that in the D domain $L_n(D)$. The numerator also has the same form, except that the initial conditions $I(s)$ are added in the s domain. Comparison of Eqs. (6.1) and (6.2) shows that when all the initial conditions are zero, the transform is obtained by merely substituting s for D, $Y(s)$ for $y(t)$, and $F(s)$ for $f(t)$ in the operational form of the differential equation.

For this case,

$$Y(s) = \frac{L_m(s)}{L_n(s)} F(s) \tag{6.3}$$

where $L_m(s)/L_n(s)$ is called the *transfer function*. Note that $L_m(s)$ and $L_n(s)$ are obtained directly from the differential equation of operation for the system. Thus, the transfer function contains basic information concerning the essential characteristics of a system without regard to initial conditions or excitation.

Control engineers often write differential equations using the s notation rather than the operator D. This yields directly the form given by Eq. (6.3). This form suffices for most control work. When it is necessary to determine the effect of initial conditions, then s is replaced by D to obtain the differential equation. Transforming the differential equation yields the resulting transformed equation which contains the initial-condition terms.

The term $F(s)$ in Eq. (6.2) is the general representation for the transform of the input signal or forcing function. This term may be written as

$$F(s) = \frac{N_{F(s)}}{D_{F(s)}}$$

where $N_{F(s)}$ is the numerator of $F(s)$ and $D_{F(s)}$ is the denominator of $F(s)$. For example, for a unit step function, $F(s) = 1/s$, and thus $N_{F(s)} = 1$ and $D_{F(s)} = s$. Substitution of the preceding representation $F(s) = N_{F(s)}/D_{F(s)}$ into Eq. (6.2) yields the following general transformed form for $Y(s)$:

$$Y(s) = \frac{L_m(s)N_{F(s)} + I(s)D_{F(s)}}{L_n(s)D_{F(s)}} = \frac{A(s)}{B(s)} \tag{6.4}$$

where $A(s)$ and $B(s)$ are polynomials in s.

6.1 INVERSE TRANSFORMATIONS

By an *inverse transformation* is meant the process of inverting a function from the s domain back to the time domain. The inverse transform $\mathscr{L}^{-1}$ of a function $F(s)$ is defined by the equation

$$\mathscr{L}^{-1}[F(s)] = \frac{1}{2\pi j}\int_C F(s)e^{ts}\,ds = f(t) \tag{6.5}$$

where C is a suitably chosen contour in the s domain.* This integral method of evaluating the inverse transform is not employed when the much simpler process of entering a transform table with the given $F(s)$ and reading directly the desired $f(t)$ can be utilized, as is the case for ordinary control analysis. A partial listing of commonly used transforms is given in Table 5.1.

The transform table may be used to obtain the Laplace transform $F(s)$ of a given function of time or to obtain the inverse transform $f(t)$ for a given function of s. This process is analogous to using a logarithmic table for obtaining the logarithm of a number or to using the same table for the opposite process of obtaining antilogarithms.

At first it would appear that the list of transforms given in Table 5.1 would have to be extended considerably to be applicable to the wide range of problems encountered in the design of systems. However, this is not the case. The list in Table 5.1 is adequate for the solution of most ordinary problems that arise in control engineering. The reason for this is that relatively few different types of terms appear in the differential equation after it has been expanded by a partial-fraction expansion. In particular, the zeros of $B(s)$ are either distinct or repeated.

Distinct Zeros

The transformed function $B(s)$ is the denominator of Eq. (6.4). When the zeros of $B(s)$ are distinct, the denominator $B(s)$ can be factored in the form

$$B(s) = (s - r_1)(s - r_2)\cdots(s - r_n) \tag{6.6}$$

where $r_1, r_2, \ldots, r_n$ are n distinct zeros of $B(s)$.

The partial-fraction expansion of Eq. (6.4) is of the form

$$Y(s) = \frac{K_1}{s - r_1} + \frac{K_2}{s - r_2} + \cdots + \frac{K_i}{s - r_i} + \cdots + \frac{K_n}{s - r_n} \tag{6.7}$$

where $K_1, K_2, \ldots, K_n$ are constants. Each constant K_i may be evaluated by the method used to obtain Eq. (5.6). That is, first multiply both sides of Eq. (6.7) by $s - r_i$; then take the limit as s approaches r_i. After these operations are performed, the only term remaining on the right side of Eq. (6.7) is K_i. Thus

$$K_i = \lim_{s \to r_i} (s - r_i)Y(s) \tag{6.8}$$

* F. H. Raven, *Mathematics of Engineering Systems*, McGraw-Hill, New York, 1966, chap. 6.

The inverse transform of Eq. (6.7) is obtained directly from the transform table and is

$$y(t) = K_1 e^{r_1 t} + K_2 e^{r_2 t} + \cdots + K_n e^{r_n t} \tag{6.9}$$

Equation (6.9) shows that each distinct zero of $B(s) = L_n(s) D_{F(s)}$ yields an exponential-type term $K_i e^{r_i t}$ in the response function. The exponent r_i is the corresponding zero of $B(s)$. Each zero $r_1, r_2, \ldots, r_n$ must be negative so that each term $K_i e^{r_i t}$ in $y(t)$ will be a decaying function. If any zero of $B(s)$ is positive, $y(t)$ will increase without bound as t increases to infinity. A constant term results if $r_i = 0$, because $K_i e^{(0)t} = K_i$.

Repeated Zeros

For the case in which $B(s)$ has a multiple or repeated zero r which occurs q times, $B(s)$ may be factored in the form

$$B(s) = (s - r)^q (s - r_1)(s - r_2) \cdots (s - r_{n-q}) \tag{6.10}$$

The corresponding partial-fraction expansion for $Y(s)$ is

$$Y(s) = \frac{C_q}{(s-r)^q} + \frac{C_{q-1}}{(s-r)^{q-1}} + \cdots + \frac{C_1}{s-r} + \frac{K_1}{s-r_1} + \frac{K_2}{s-r_2} + \cdots + \frac{K_{n-q}}{s-r_{n-q}} \tag{6.11}$$

The constant coefficients for the multiple terms are evaluated as follows:

$$\begin{aligned} C_q &= \lim_{s \to r} \left[(s-r)^q Y(s)\right] \\ C_{q-1} &= \lim_{s \to r} \left\{\frac{d}{ds}\left[(s-r)^q Y(s)\right]\right\} \\ C_{q-k} &= \lim_{s \to r} \left\{\frac{1}{k!}\frac{d^k}{ds^k}\left[(s-r)^q Y(s)\right]\right\} \end{aligned} \tag{6.12}$$

From the transform table, the inverse transform of Eq. (6.11) is

$$y(t) = \left[\frac{C_q t^{q-1}}{(q-1)!} + \frac{C_{q-1} t^{q-2}}{(q-2)!} + \cdots + \frac{C_2 t}{1!} + C_1\right] e^{rt} + K_1 e^{r_1 t} + \cdots + K_{n-q} e^{r_{n-q} t} \tag{6.13}$$

Each response term associated with the repeated zero $(s - r)^q$ is multiplied by the exponential factor e^{rt}. If the value of r is positive, $y(t)$ will become infinite as time increases. For negative values of r, a decreasing exponential results, and thus the response term due to the repeated zero eventually vanishes.

Illustrative example 6.1. Suppose we want to determine the time response $y(t)$ for the transformed equation

$$Y(s) = \frac{11s + 28}{(s+2)^2(s+5)} = \frac{C_2}{(s+2)^2} + \frac{C_1}{s+2} + \frac{K_1}{s+5} \tag{6.14}$$

Solution. The constants are evaluated as follows:

$$C_2 = \lim_{s \to -2} \frac{11s + 28}{s + 5} = 2$$

$$C_1 = \lim_{s \to -2} \left[\frac{d}{ds}\frac{11s + 28}{s + 5}\right] = \lim_{s \to -2} \frac{(s + 5)(11) - (11s + 28)}{(s + 5)^2} = 3$$

$$K_1 = \lim_{s \to -5} \frac{11s + 28}{(s + 2)^2} = -3$$

Thus,

$$Y(s) = \frac{2}{(s + 2)^2} + \frac{3}{s + 2} - \frac{3}{s + 5}$$

By use of Table 5.1, the inverse transform of the preceding equation is

$$y(t) = (2t + 3)e^{-2t} - 3e^{-5t} \tag{6.15}$$

Differentiation was involved in the evaluation of C_1. An alternative technique which eliminates the need for differentiation is to evaluate all constants which do not require differentiation (for example, C_2 and K_1) and then write $Y(s)$ in terms of the unknown constant. Thus,

$$Y(s) = \frac{11s + 28}{(s + 2)^2(s + 5)} = \frac{2}{(s + 2)^2} + \frac{C_1}{s + 2} - \frac{3}{s + 5}$$

This result must hold for any value of s. For $s = -1$, it becomes

$$Y(-1) = \tfrac{17}{4} = 2 + C_1 - \tfrac{3}{4}$$

Hence

$$C_1 = 5 - 2 = 3$$

Illustrative example 6.2. Determine the solution of the differential equation

$$(D + 1)^2(D + 2)y(t) = (D^3 + 4D^2 + 5D + 2)y(t) = f(t)$$

where $f(t) = 2e^{-t}$ and all the initial conditions are zero.

Solution. Transforming each term in the differential equation yields for the transformed equation

$$\begin{aligned}&[(s^3Y(s) - s^2y(0) - sy'(0) - y''(0)]\\&+ 4[(s^2Y(s) - sy(0) - y'(0)] + 5[sY(s) - y(0)] + 2Y(s) = F(s)\end{aligned}$$

For $y(0) = y'(0) = y''(0) = 0$, it follows that

$$(s^3 + 4s^2 + 5s + 2)Y(s) = F(s)$$

The transformed equation may now be written in the form

$$Y(s) = \frac{F(s)}{(s+1)^2(s+2)} = \frac{2}{(s+1)^3(s+2)} = \frac{C_3}{(s+1)^3} + \frac{C_2}{(s+1)^2} + \frac{C_1}{s+1} + \frac{K_1}{s+2}$$

where $F(s) = \mathcal{L}(2e^{-t}) = 2/(s+1)$. The partial-fraction constants are

$$C_3 = \lim_{s \to -1} \left[(s+1)^3 \frac{2}{(s+1)^3(s+2)} \right] = \left[\frac{2}{s+2} \right]_{s=-1} = 2$$

$$C_2 = \lim_{s \to -1} \left[\frac{d}{ds} \frac{2}{s+2} \right] = \left[\frac{-2}{(s+2)^2} \right]_{s=-1} = -2$$

$$C_1 = \lim_{s \to -1} \left[\frac{1}{2!} \frac{d^2}{ds^2} \frac{2}{s+2} \right] = \left[\frac{2}{(s+2)^3} \right]_{s=-1} = 2$$

$$K_1 = \lim_{s \to -2} \left[(s+2) \frac{2}{(s+1)^3(s+2)} \right] = \left[\frac{2}{(s+1)^3} \right]_{s=-2} = -2$$

The resulting partial-fraction expansion is

$$Y(s) = \frac{2}{(s+1)^3} - \frac{2}{(s+1)^2} + \frac{2}{s+1} - \frac{2}{s+2}$$

Inverting yields

$$y(t) = t^2e^{-t} - 2te^{-t} + 2e^{-t} - 2e^{-2t} = (2 - 2t + t^2)e^{-t} - 2e^{-2t}$$

The alternative technique for evaluating C_1 and C_2 without the need for differentiation is to substitute the constants $C_3 = 2$ and $K_1 = -2$ which do not require differentiation into the expression for $Y(s)$. Thus

$$Y(s) = \frac{2}{(s+1)^3(s+2)} = \frac{2}{(s+1)^3} + \frac{C_2}{(s+1)^2} + \frac{C_1}{s+1} - \frac{2}{s+2}$$

This expression is valid for any value of s. For $s = 0$ and for $s = 1$, we obtain

$$Y(0) = 1 = 2 + C_2 + C_1 - 1 \qquad \text{or} \qquad C_1 + C_2 = 0$$

$$Y(1) = \frac{2}{2^3(3)} = \frac{1}{12} = \frac{1}{4} + \frac{C_2}{4} + \frac{C_1}{2} - \frac{2}{3} \qquad \text{or} \qquad 2C_1 + C_2 = 2$$

Solving these equations simultaneously verifies that $C_1 = 2$ and $C_2 = -2$.

6.2 COMPLEX CONJUGATE ZEROS

Complex zeros of $B(s)$ always occur in pairs, and furthermore these zeros are always conjugates of one another. That is, they have the same real part but equal and opposite imaginary parts. Thus, if the polynomial $B(s)$ has a complex zero $a + jb$, the complex conjugate $a - jb$ will also be a zero of $B(s)$. Although the preceding discussion of distinct zeros is also applicable to complex conjugate zeros, the following analysis brings out more clearly the fact that a pair of complex conjugate zeros in $B(s)$ combines to introduce an exponentially damped sinusoidal term in $y(t)$.

A pair of complex conjugate zeros when multiplied together yields the following quadratic:

$$[s-(a+jb)][s-(a-jb)] = s^2 - 2as + a^2 + b^2 \tag{6.16}$$

For any given quadratic term, the values of a and b may be computed by equating coefficients of like terms. Consider the expression

$$s^2 + 4s + 9$$

The coefficient 4 of the s term is equal to $-2a$, so that $-2a = 4$ or $a = -2$. Similarly, equating the constant terms gives $a^2 + b^2 = 9$ or $b = \sqrt{9-4} = \sqrt{5}$. Thus, the complex conjugate zeros are $a \pm jb = -2 \pm j\sqrt{5}$. If in the determination of b it is found that b is an imaginary number, then the two zeros are real and unequal rather than complex conjugates. For example, consider the quadratic

$$s^2 + 8s + 12$$

The value of a is equal to -4, so that $b = \sqrt{12-16} = j\sqrt{4} = j2$. For this case the zeros are $a \pm jb = -4 \pm j^2 2 = -4 \mp 2 = -6, -2$. Because the case of real zeros has been discussed previously, it is assumed in the following analysis that b is real, so that the zeros are complex conjugates.

For complex conjugate zeros $B(s)$ may be factored in the form

$$B(s) = [s-(a+jb)][s-(a-jb)](s-r_1)\cdots(s-r_{n-2}) \tag{6.17}$$

The partial-fraction expansion for $Y(s) = A(s)/B(s)$ is of the form

$$Y(s) = \frac{K_c}{s-(a+jb)} + \frac{K_{-c}}{s-(a-jb)} + \frac{K_1}{s-r_1} + \cdots + \frac{K_{n-2}}{s-r_{n-2}} \tag{6.18}$$

The inverse transform of Eq. (6.18) is

$$y(t) = K_c e^{(a+jb)t} + K_{-c} e^{(a-jb)t} + K_1 e^{r_1 t} + \cdots + K_{n-2} e^{r_{n-2} t} \tag{6.19}$$

The constants K_c and K_{-c} associated with the complex conjugate zeros are evaluated in the usual manner for distinct zeros. Thus, application of Eq. (6.8) yields

$$\begin{aligned} K_c &= \lim_{s \to a+jb} \left\{ [s-(a+jb)] \frac{A(s)}{[s-(a+jb)][s-(a-jb)](s-r_1)\cdots(s-r_{n-2})} \right\} \\ &= \lim_{s \to a+jb} \left[\frac{1}{2jb} \frac{A(s)}{(s-r_1)\cdots(s-r_{n-2})} \right] = \frac{1}{2jb} K(a+jb) \end{aligned} \tag{6.20}$$

where

$$K(a+jb) = \lim_{s \to a+jb} \frac{A(s)}{(s-r_1)\cdots(s-r_{n-2})} = \left[(s^2 - 2as + a^2 + b^2) \frac{A(s)}{B(s)} \right]_{s=a+jb}$$

Similarly, the constant K_{-c} is

$$K_{-c} = \lim_{s \to a - jb} \left\{ [s - (a - jb)] \frac{A(s)}{[s - (a + jb)][s - (a - jb)](s - r_1) \cdots (s - r_{n-2})} \right\}$$

$$= \lim_{s \to a - jb} \frac{A(s)}{(-2jb)(s - r_1) \cdots (s - r_{n-2})} = -\frac{1}{2jb} K(a - jb) \tag{6.21}$$

where

$$K(a - jb) = \lim_{s \to a - jb} \frac{A(s)}{(s - r_1) \cdots (s - r_{n-2})} = \left[(s^2 - 2as + a^2 + b^2) \frac{A(s)}{B(s)} \right]_{s = a - jb}$$

The constants $K(a + jb)$ and $K(a - jb)$ are complex conjugate numbers. These complex conjugate numbers may be represented as shown graphically in Fig. 6.1, whence

$$\begin{aligned} K(a + jb) &= |K(a + jb)| e^{j\alpha} \\ K(a - jb) &= |K(a - jb)| e^{-j\alpha} \end{aligned} \tag{6.22}$$

where $|K(a + jb)| = |K(a - jb)|$ is the length of either vector, α is the angle of vector $K(a + jb)$, and $-\alpha$ is the angle of vector $K(a - jb)$.

The constants K_c and K_{-c}, which are also complex conjugate numbers, may be written in the form

$$\begin{aligned} K_c &= \frac{1}{2jb} |K(a + jb)| e^{j\alpha} \\ K_{-c} &= -\frac{1}{2jb} |K(a + jb)| e^{-j\alpha} \end{aligned} \tag{6.23}$$

Substitution of K_c and K_{-c} from Eqs. (6.23) into Eq. (6.19) gives

$$y(t) = \frac{1}{b} |K(a + jb)| e^{at} \frac{e^{j(bt + \alpha)} - e^{-j(bt + \alpha)}}{2j} + K_1 e^{r_1 t} + \cdots + K_{n-2} e^{r_{n-2} t}$$

or

$$y(t) = \frac{1}{b} |K(a + jb)| e^{at} \sin(bt + \alpha) + K_1 e^{r_1 t} + \cdots + K_{n-2} e^{r_{n-2} t} \tag{6.24}$$

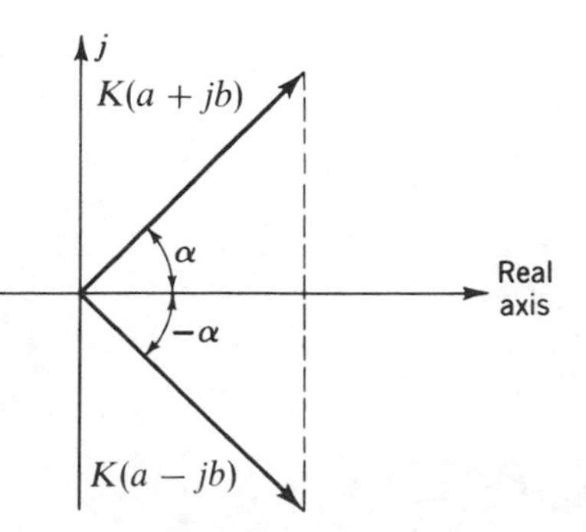

FIGURE 6.1
Vector representation for $K(a + jb)$ and $K(a - jb)$.

Illustrative example 6.3. Determine the inverse transformation of the following transformed equation:

$$Y(s) = \frac{75}{(s^2 + 4s + 13)(s + 6)}$$

Solution. Equating coefficients to obtain the values of a and b for the quadratic yields $-2a = 4$, or $a = -2$, and $a^2 + b^2 = 13$, or $b = \sqrt{13 - 4} = 3$. Evaluation of $K(a + jb)$ gives

$$K(a + jb) = \left[(s^2 - 2as + a^2 + b^2)\frac{A(s)}{B(s)}\right]_{s=a+jb}$$

$$= \left(\frac{75}{s + 6}\right)_{s=-2+j3} = \frac{75}{4 + j3} \tag{6.25}$$

As shown in Fig. 6.2, the vector whose real part is 4 and whose imaginary part is 3 may be expressed in polar form as

$$4 + j3 = 5\underline{/36.9^\circ}$$

Hence, Eq. (6.25) becomes

$$K(a + jb) = \frac{75}{5\underline{/36.9^\circ}} = 15\underline{/-36.9^\circ}$$

Thus

$$|K(a + jb)| = 15$$

and

$$\alpha = \measuredangle K(a + jb) = -36.9^\circ$$

The general form of the inverse transformation is

$$y(t) = \frac{1}{b}|K(a + jb)|e^{at} \sin(bt + \alpha) + K_1 e^{r_1 t}$$

Evaluation of K_1 gives

$$K_1 = \lim_{s \to -6} \frac{75}{s^2 + 4s + 13} = \frac{75}{25} = 3$$

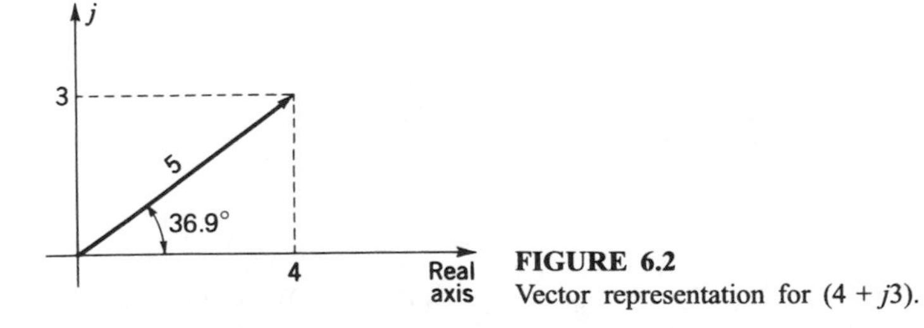

FIGURE 6.2
Vector representation for $(4 + j3)$.

Thus the desired result is

$$y(t) = 5e^{-2t} \sin(3t - 36.9°) + 3e^{-6t} \tag{6.26}$$

Application of the relationship $\sin(\alpha + \beta) = \sin\alpha \cos\beta + \cos\alpha \sin\beta$ in which $\alpha = 3t$ and $\beta = -36.9°$ yields the alternate form

$$y(t) = e^{-2t}(4 \sin 3t - 3 \cos 3t) + 3e^{-6t} \tag{6.27}$$

This form of the result may be obtained directly. The response term due to a pair of complex conjugate roots $a \pm jb$ may be written in the form

$$\begin{aligned} y(t) &= \frac{|K(a+jb)|}{b} e^{at} \sin(bt + \alpha) = \frac{e^{at}}{b} |K(a+jb)|(\cos\alpha \sin bt + \sin\alpha \cos bt) \\ &= \frac{e^{at}}{b}(A \sin bt + B \cos bt) \end{aligned} \tag{6.28}$$

where $A = |K(a+jb)| \cos\alpha$
$B = |K(a+jb)| \sin\alpha$

Figure 6.3 shows the vector $K(a + jb)$. Note that the horizontal component is $A = |K(a+jb)| \cos\alpha$, and the vertical component is $B = |K(a+jb)| \sin\alpha$. For the preceding example $|K(a+jb)| = 15$ and $\alpha = -36.9°$, hence $A = 15 \cos(-36.9°) = 12$ and $B = 15 \sin(-36.9°) = -9$. For $a = -2$ and $b = 3$, application of Eq. (6.28) yields for the response due to the complex conjugate roots

$$e^{-2t}(4 \sin 3t - 3 \cos 3t)$$

Another method for obtaining the response due to complex conjugate roots results from writing $Y(s)$ in the form

$$Y(s) = \frac{75}{(s^2 + 4s + 13)(s + 6)} = \frac{As + B}{(s+2)^2 + 3^2} + \frac{K_1}{s + 6}$$

Evaluation of the constants yields $A = -3$, $B = 6$, and $K_1 = 3$, whence

$$Y(s) = 4\frac{3}{(s+2)^2 + 3^2} - 3\frac{s+2}{(s+2)^2 + 3^2} + \frac{3}{s+6}$$

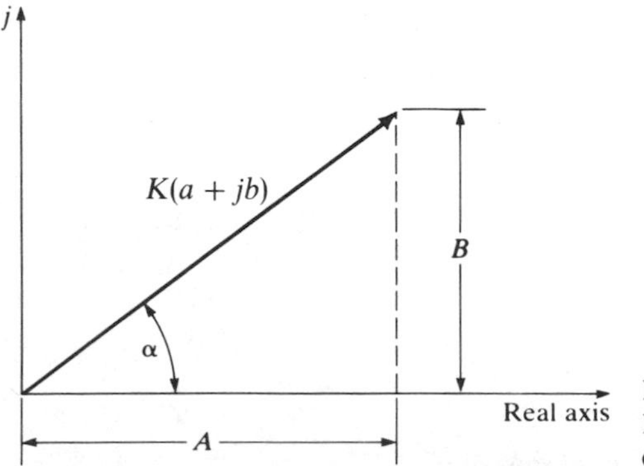

FIGURE 6.3
Horizontal component A and vertical component B of vector $K(a + jb)$.

Inverting yields the result for $y(t)$ given by Eq. (6.27). It should be pointed out that for higher-order systems the algebraic manipulations get very cumbersome with this latter method, whereas the algebraic manipulations required to evaluate $K(a + jb)$ do not appreciably increase in complexity for higher-order systems.

The general form of the response due to a pair of complex conjugate roots is

$$y(t) = \frac{|K(a+jb)|}{b} e^{at} \sin(bt + \alpha) \tag{6.29}$$

The exponential factor a is the real part of the complex conjugate roots. The imaginary part b is the frequency of oscillation of the exponentially damped sinusoid. Thus, b is referred to as the *damped frequency of oscillation* or *damped natural frequency.* The period of each oscillation is $2\pi/b$. The envelope of this sinusoid is $(1/b)|K(a+jb)|e^{at}$. To have the exponential term decreasing with time, it is necessary that a be negative. For the case in which $a = 0$, a sinusoid of constant amplitude $(1/b)|K(a+jb)|$ results. For $a = 0$, Eq. (6.29) becomes

$$y(t) = \frac{1}{b}|K(a+jb)| \sin(bt + \alpha) \tag{6.30}$$

Figure 6.4 graphically illustrates the types of time-response terms that result from complex conjugate roots. When the roots lie to the left of the imaginary axis ($a < 0$), a decreasing sinusoid results; when the roots are on the imaginary axis ($a = 0$), a sinusoid of constant amplitude results; and when the roots are to the right of the imaginary axis ($a > 0$), an increasing sinusoid results.

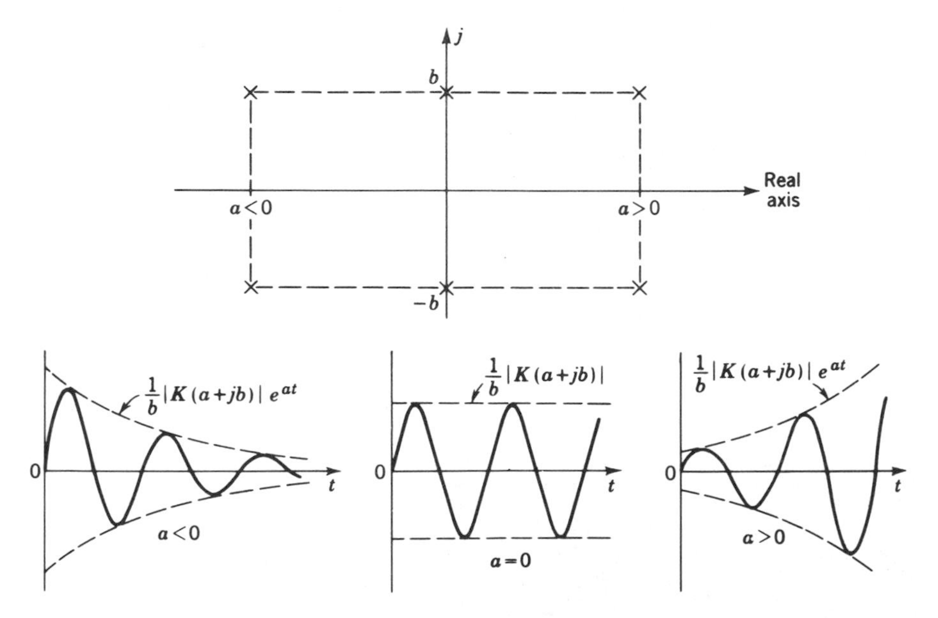

FIGURE 6.4
Response terms that result from complex conjugate zeros.

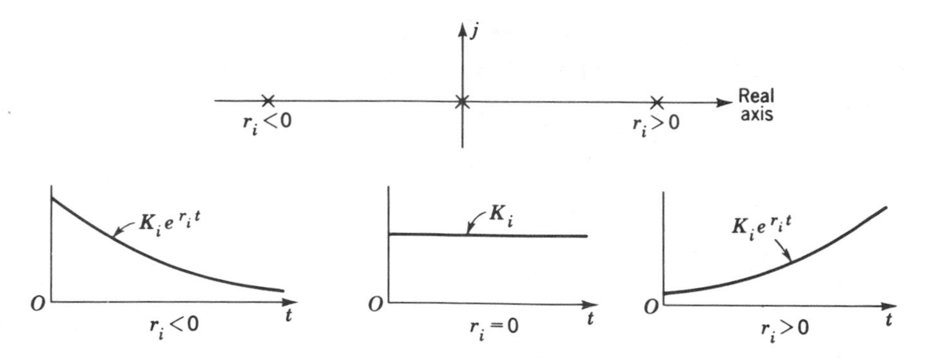

FIGURE 6.5
Response terms that result from real zeros.

Figure 6.5 shows a plot of the types of response terms that result from real roots. These results follow directly from Eq. (6.9). Because the exponential factor is the value of the root, a negative root ($r_i < 0$) yields an exponentially decreasing term, while a positive root ($r_i > 0$) yields an exponentially increasing term. A root at the origin ($r_i = 0$) results in a constant term. Even though a root may be repeated, the exponential term dominates the response.

6.3 DAMPING RATIO AND NATURAL FREQUENCY

Figure 6.6 shows a pair of complex conjugate roots. The root located at $a + jb$ is described by rectangular coordinates in which a is the horizontal component and b is the vertical component. This root is also described by polar coordinates in which

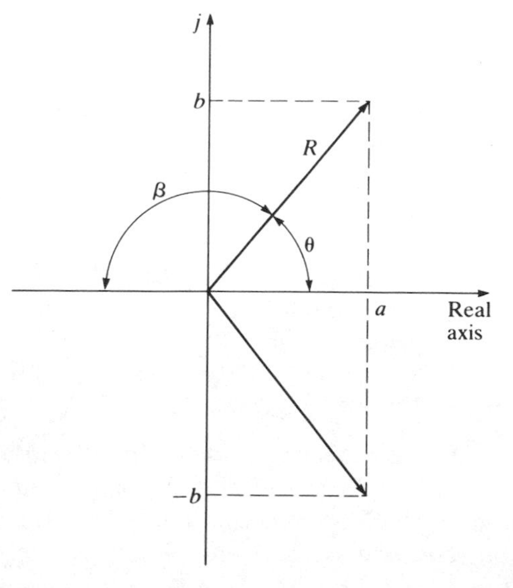

FIGURE 6.6
Polar representation $Re^{j\theta}$ of the vector $(a + jb)$.

R is the distance from the origin to the root and θ is the angle from the real axis to R. The rectangular and polar coordinates are related by the equations

$$\begin{aligned} a &= R\cos\theta \\ b &= R\sin\theta \\ a^2 + b^2 &= R^2 \end{aligned} \tag{6.31}$$

In terms of polar coordinates, the quadratic form is

$$s^2 - 2as + a^2 + b^2 = s^2 - (2R\cos\theta)s + R^2$$

The angle β which is measured from the negative real axis to R is such that $\theta = \pi - \beta$. Noting that $\cos\theta = \cos(\pi - \beta) = -\cos\beta$ yields for the quadratic form

$$s^2 + (2R\cos\beta)s + R^2 \tag{6.32}$$

When $\beta = 0$, the quadratic form becomes

$$s^2 + 2Rs + R^2 = (s + R)^2 \tag{6.33}$$

For this case the roots are located on the negative real axis at $-R$. The roots are no longer complex conjugate, but rather repeated and real. The response ceases to be sinusoidal. When $\beta = 0$, the system is said to be critically damped.

The coefficient of the s term ($2R\cos\beta$) in the quadratic form of Eq. (6.32) is a measure of the amount of damping in a system. The *damping ratio* ζ is defined as the ratio of the actual amount of damping in a system ($2R\cos\beta$) to the amount of damping when the system is critically damped ($2R$). Thus

$$\zeta = \frac{2R\cos\beta}{2R} = \cos\beta \tag{6.34}$$

When $\beta = 90°$, the quadratic form becomes

$$s^2 + R^2$$

For this case, the roots are located on the imaginary axis at $\pm jR$. Because $a = R\cos 90° = 0$ and $b = R$, the resulting response is

$$\frac{1}{b}|K(a + jb)|e^{at}\sin(bt + \alpha) = \frac{|K(a + jb)|}{R}\sin(Rt + \alpha)$$

This is a pure sinusoidal oscillation of constant amplitude. The frequency of oscillation is $R = \omega_n$, where ω_n is the natural frequency. Replacing R by ω_n and $\cos\beta$ by ζ in Eq. (6.32) yields the standard quadratic form

$$s^2 + 2\zeta\omega_n s + \omega_n^2 \tag{6.35}$$

The natural frequency ω_n is the frequency at which the system would oscillate if there were no damping. The natural frequency ω_n is also called the *undamped natural frequency*. The actual frequency at which the system does oscillate is b. The frequency b is called the *damped natural frequency*. Because $a^2 + b^2 = R^2 = \omega_n^2$

and $a = R\cos\theta = -R\cos\beta = -\omega_n\zeta$, it follows that

$$b = \sqrt{\omega_n^2 - a^2} = \omega_n\sqrt{1-\zeta^2} \tag{6.36}$$

Given the natural frequency ω_n and damping ratio ζ, then the damped natural frequency may be determined in accordance with Eq. (6.36).

For any given quadratic expression, numerical values of ζ and ω_n are computed by equating coefficients. For the quadratic $s^2 + 25$, it follows that $\omega_n^2 = 25$, or $\omega_n = 5$, and $2\zeta\omega_n = 0$ so that $\zeta = 0$. Similarly, it is found that $a = 0$ and $b = 5$. For the quadratic $s^2 + 6s + 25$, it is found that $\omega_n = 5$, $\zeta = 0.6$, $a = -3$, and $b = 4$. For the quadratic $s^2 + 10s + 25$, it follows that $\omega_n = 5$, $\zeta = 1.0$, $a = -5$, and $b = 0$. For the quadratic $s^2 + 15s + 25$, the values are $\omega_n = 5$, $\zeta = 1.5$, $a = -7.5$, and $b = j5.59$. Because b is imaginary, the root location is at $a \pm jb = -7.5 \pm j(j5.59) = -7.5 \pm 5.59 = -13.09, -1.91$. The root location for the four preceding quadratic terms is shown in Fig. 6.7. Note that when the damping ratio ζ is greater than 1, the roots are no longer complex conjugate but are real. The response is no longer sinusoidal, but rather exponential.

For the critically damped case $s^2 + 10s + 25$, the coefficient of the s term is 10. The damping ratio ζ is the ratio of the actual coefficient of the s term to the value when the system is critically damped. Thus, for the quadratic $s^2 + 6s + 25$, the damping ratio is $\zeta = 6/10 = 0.6$. Similarly, for the quadratic $s^2 + 15s + 25$, the damping ratio is $\zeta = 15/10 = 1.5$.

Figure 6.8 shows a general plot of a pair of complex conjugate roots. When a is negative, so that a decreasing sinusoid results, the roots are to the left of the imaginary axis, so that $0 < \beta < 90°$ in which case $1 > \zeta > 0$. For positive values of a, the roots are to the right of the imaginary axis, so that $90° < \beta < 180°$ and $0 > \zeta > -1$. A positive value of ζ yields a decreasing sinusoid, while a negative value results in an increasing sinusoid. For ζ equal to zero ($\beta = 90°$), a sinusoid of

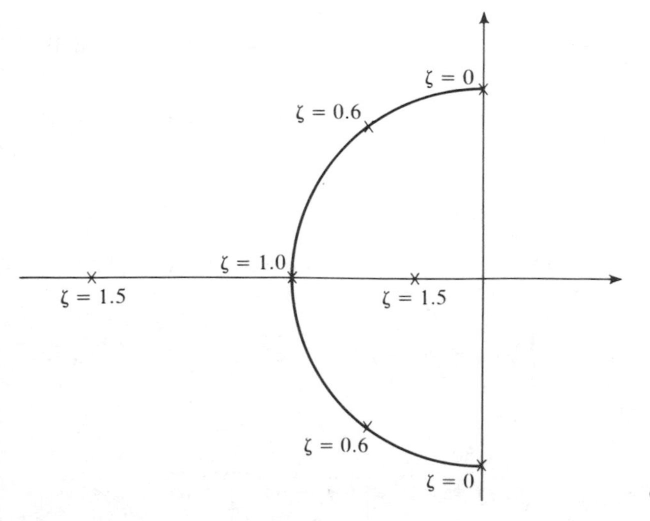

FIGURE 6.7
Root location for various damping ratios.

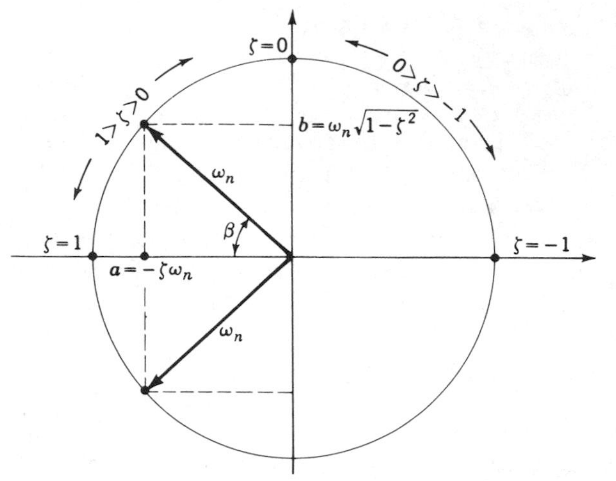

FIGURE 6.8
General plot of complex conjugate zeros.

constant amplitude is obtained. When the damping ratio ζ is greater than 1, then two negative real roots result. When ζ is less than -1, two positive real roots result.

Illustrative example 6.4. Suppose we want to determine the general equation for the transient response of a second-order system to a unit step-function change which occurs at $t = 0$. The operational form of the differential equation is

$$y(t) = \frac{\omega_n^2}{D^2 + 2\zeta\omega_n D + \omega_n^2} f(t) \tag{6.37}$$

Assume that all the initial conditions are zero.

Solution. The transform for this differential equation is

$$Y(s) = \frac{\omega_n^2 F(s)}{s^2 + 2\zeta\omega_n s + \omega_n^2} = \frac{\omega_n^2}{s(s^2 + 2\zeta\omega_n s + \omega_n^2)} \tag{6.38}$$

where $F(s) = 1/s$ is the transform for the step input. The partial-fraction expansion has the general form

$$Y(s) = \frac{K_1}{s} + \frac{K_c}{s-(a+jb)} + \frac{K_{-c}}{s-(a-jb)}$$

Because the complex conjugate zeros combine to yield a sinusoid, inverting yields the following general form for the response:

$$y(t) = K_1 + \frac{1}{b}|K(a+jb)|e^{at}\sin(bt+\alpha) \tag{6.39}$$

Evaluation of $K(a + jb)$ gives

$$K(a+jb) = \left.\frac{\omega_n^2}{s}\right|_{s=a+jb} = \frac{\omega_n^2}{a+jb} = \frac{\omega_n^2}{\sqrt{a^2+b^2}/\tan^{-1}(b/a)}$$

Thus,

$$|K(a+jb)| = \frac{\omega_n^2}{\sqrt{a^2+b^2}} = \omega_n \tag{6.40}$$

and

$$\begin{aligned} \alpha = \measuredangle K(a+jb) &= -\tan^{-1}\frac{b}{a} = \tan^{-1}\frac{-b}{a} \\ &= \tan^{-1}\frac{-\sqrt{1-\zeta^2}}{-\zeta} \end{aligned} \tag{6.41}$$

where

$$\begin{aligned} \sqrt{a^2+b^2} &= \omega_n \\ a &= -\zeta\omega_n \\ b &= \omega_n\sqrt{1-\zeta^2} \end{aligned}$$

The constant K_1 is evaluated as follows:

$$K_1 = \lim_{s\to 0} \frac{\omega_n^2}{s^2+2\zeta\omega_n s+\omega_n^2} = 1$$

Thus, the desired transient response is

$$y(t) = 1 + \frac{1}{\sqrt{1-\zeta^2}} e^{-\zeta\omega_n t} \sin\left[\left(\omega_n\sqrt{1-\zeta^2}\right)t + \alpha\right] \qquad -1 < \zeta < 1 \tag{6.42}$$

For the case in which $\zeta = 1$, the quadratic term in Eq. (6.38) is

$$s^2 + 2\omega_n s + \omega_n^2 = (s+\omega_n)^2$$

Thus the partial-fraction expansion for $Y(s)$ is

$$Y(s) = \frac{C_2}{(s+\omega_n)^2} + \frac{C_1}{s+\omega_n} + \frac{K_1}{s} \tag{6.43}$$

The constants C_2, C_1, and K_1 are evaluated as follows:

$$\begin{aligned} C_2 &= \lim_{s\to -\omega_n} \frac{\omega_n^2}{s} = -\omega_n \\ C_1 &= \lim_{s\to -\omega_n} \left(\frac{d}{ds}\frac{\omega_n^2}{s}\right) = \left.\frac{-\omega_n^2}{s^2}\right|_{s=-\omega_n} = -1 \\ K_1 &= \lim_{s\to 0} \frac{\omega_n^2}{s^2+2\omega_n s+\omega_n^2} = 1 \end{aligned}$$

Thus

$$y(t) = (C_2 t + C_1)e^{-\omega_n t} + K_1$$

or

$$y(t) = 1 - (\omega_n t + 1)e^{-\omega_n t} \tag{6.44}$$

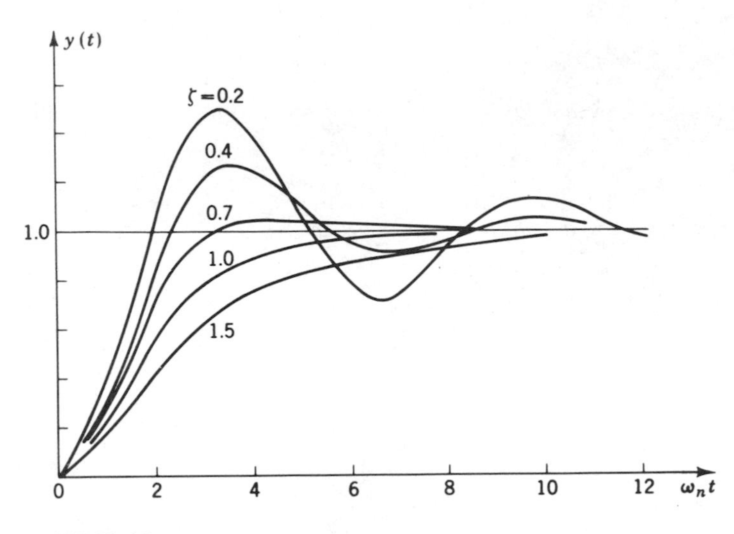

FIGURE 6.9
Response of a second-order system to a unit step input.

In a similar manner, the equation of the time response for $\zeta > 1$ or $\zeta < -1$ may also be derived. For these cases the zeros are real, and thus exponential terms are contributed to the response.

Figure 6.9 shows the response $y(t)$ to a unit step change in the input for various values of the damping ratio ζ. Note that for $\zeta < 0.4$, there is an excessive amount of overshooting and oscillations. For $\zeta > 1$, an excessive amount of time is required to reach the new operating condition. Thus, for most control work, it is desired to have $0.4 < \zeta < 1$.

As shown in Fig. 6.6, the form $a \pm jb$ is the rectangular representation for complex conjugate roots. That is, in rectangular coordinates, the real part a is the horizontal component, and the imaginary part b is the vertical component. Similarly, the polar form for complex conjugate roots is specified by the radius $\omega_n = \sqrt{a^2 + b^2}$ and the angle $\zeta = \cos \beta$. For computational purposes, the response is usually obtained most readily by using the rectangular form. It is then an easy matter to obtain the polar quantities ω_n and ζ. The polar form has the advantage of providing a more general insight into the actual behavior of the system. For example, Fig. 6.9 shows that the form of the response is determined by ζ. The speed of the response is governed by ω_n. That is, Fig. 6.9 shows that for $\zeta = 0.2$, the response $y(t)$ first crosses the value $y(t) = 1$ when $\omega_n t = 2$. Thus, if $\omega_n = 0.1$, the time is $t = 2/\omega_n = 2/0.1 = 20$ s. For $\omega_n = 10$, the time is $t = 2/\omega_n = 2/10 = 0.2$ s. Hence, the larger the value of ω_n, the faster the speed of response.

Logarithmic Decrement

For an exponentially damped sinusoid, as shown in Fig. 6.10, the amplitude of the sinusoid after each oscillation changes in a geometric series. At time t_1 the

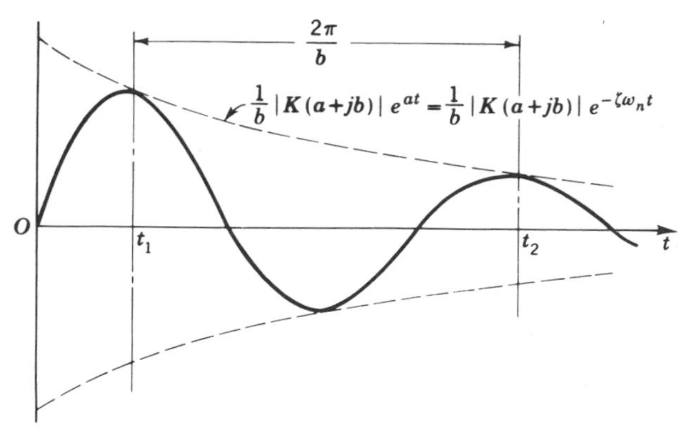

FIGURE 6.10
Logarithmic decrement.

amplitude is $(1/b)|K(a+jb)|e^{at_1}$. The time required to complete one period is $T = 2\pi/b$, and thus the time after one oscillation is completed is $t_2 = t_1 + T = t_1 + 2\pi/b$. The new amplitude is $(1/b)|K(a+jb)|e^{a(t_1 + 2\pi/b)}$. The ratio of amplitudes is

$$\frac{(1/b)|K(a+jb)|e^{at_1}}{(1/b)|K(a+jb)|e^{at_1}e^{2\pi a/b}} = e^{-2\pi a/b} = \exp\left[\frac{2\pi\zeta}{\sqrt{1-\zeta^2}}\right] \tag{6.45}$$

The natural logarithm of this amplitude ratio is $-2\pi a/b = 2\pi\zeta/\sqrt{1-\zeta^2}$, which is called the *logarithmic decrement*. The amplitude ratio after one oscillation is thus seen to be a function of the damping ratio only.

6.4 COMPUTER SOLUTION

The digital computer provides a very powerful tool for investigating the behavior of control systems. The solution of complex systems may be obtained on a small personal digital computer with rather simple programs. As shown in the following, to solve a differential equation it is only necessary to integrate numerically. Figure 6.11 shows a plot of the derivative $\dot{x}(t)$. The value of $x(t)$ at time t is

$$x(t) = \int \dot{x}(t)\,dt = \int_{t_0}^{t} \dot{x}(t)\,dt + C$$

Evaluation at time $t = t_0$ shows that $C = x(t_0)$. Thus

$$x(t) = \int_{t_0}^{t} \dot{x}(t)\,dt + x(t_0)$$

The preceding integral is the area under the curve of $\dot{x}(t)$ from t_0 to t. When the time increment $t - t_0$ is small, the area is closely approximated by the area of the trapezoid. Thus,

$$x(t) = 0.5(t - t_0)[\dot{x}(t_0) + \dot{x}(t)] + x(t_0)$$

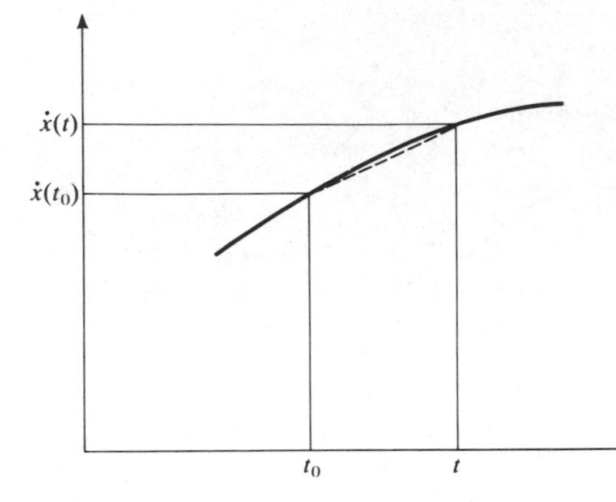

FIGURE 6.11
Trapezoidal approximation of area under a curve.

The computer statement for this integration is

```
X = 0.5*H*(DX0 + DX) + X0
```

where X $= x(t) =$ value of x at time t
X0 $= X(t_0) =$ value of x at time t_0
DX $= \dot{x}(t) =$ value of $\dot{x}$ at time t
DX0 $= \dot{x}(t_0) =$ value of $\dot{x}$ at time t_0
H $= t - t_0 =$ time increment

The computer programs in this text are written in Fortran. They can also be written in other languages such as Basic, Pascal, C, etc.

Illustrative example 6.5. Write the program for solving the differential equation

$$y(t) = \frac{D+5}{D^2 + 2D + 5} f(t)$$

The forcing function is $f(t) = u(t)$, and all initial conditions are zero.

Solution. When the numerator has a differential operator, the first step is to rewrite the differential equation so that the numerator operator is unity. That is,

$$x(t) = \frac{1}{D^2 + 2D + 5} f(t)$$

Hence

$$y(t) = (D+5)\frac{1}{D^2 + 2D + 5} f(t) = (D+5)x(t)$$

The differential equation for x is rewritten in the form

$$\ddot{x} + 2\dot{x} + 5x = f(t)$$

This differential equation for x is now solved for the highest-order derivative

$$\ddot{x} = f(t) - 2\dot{x} - 5x$$

Solving for the highest-order derivative is a fundamental step in the solution of a differential equation on a computer. The lower-order derivatives are then obtained by successive integration of the highest-order derivative. For this example

$$\dot{x} = \int \ddot{x}\, dt \qquad \text{and} \qquad x = \int \dot{x}\, dt$$

The computer statements for the differential equation are

```
D2X = F - 2.0*DX - 5.0*X
```

and

```
Y = 5.0*X + DX
```

where X = $x(t)$
DX = $\dot{x}(t)$
D2X = $\ddot{x}(t)$
F = $f(t)$
Y = $y(t)$

The computer statements for performing the integrations are

```
DX = 0.5*H*(D2X+D2X0)+DX0
 X = 0.5*H*(DX+DX0)+XO
```

The following program may now be written:

```
      T = 0.0
      X = 0.0
      DX = 0.0
      D2X = 0.0
      H = 0.01
      DO 25 K = 0,400
      F = 1.0
      Y = 5.0*X + DX
      IF (MOD(K,10).NE.0) GO TO 10
      WRITE (6,5) T,Y
 5    FORMAT (2F10.2)
10    T = T+H
      X0 = X
      DX0 = DX
      D2X0 = D2X
      D2X = F - 2.0*DX - 5.0*X
      DX = 0.5*H*(D2X+D2X0)+DX0
      X = 0.5*H*(DX+DX0)+X0
25    CONTINUE
```

The first four statements of this program initialize the system. That is, these statements set the initial values $t_0 = 0$, $x(t_0) = 0$, $\dot{x}(t_0) = 0$, and $\ddot{x}(t_0) = 0$. The next statement sets the integration interval $H = t - t_0 = 0.01$ s.

The following statement starts the DO loop. The process is repeated 400 times, so the final value will be the response when T = 400H = 400(0.01) = 4 s. The next statement is the forcing function. For a unit step function, F = 1.0. The statement Y = 5.0*X+DX computes the present value of the output $y(t)$. The first time through $y(t) = y(t_0)$ is the value of $y(t)$ at time t_0. The IF MOD statement causes corresponding values of T and Y to be printed every time the remainder of dividing K by 10 is zero (i.e., when K = 0, 10, 20, 30, etc.). The statement T = T+H sets the value of time at the end of the next integration interval. The statements X0 = X, DX0 = DX, and D2X0 = D2X set the values of $x(t)$, $\dot{x}(t)$, and $\ddot{x}(t)$ at the beginning of the new interval as the corresponding values $x(t)$, $\dot{x}(t)$, and $\ddot{x}(t)$ at the end of the preceding interval. The following statement computes the new value of the second derivative $\ddot{x}(t)$ = D2X at the end of this interval. The next statement integrates $\ddot{x}(t)$ over the interval. Similarly, the following statement integrates $\dot{x}(t)$ over the interval to obtain the value $x(t)$ = X at the end of the interval. The last statement sends the computer back to the beginning of the DO loop.

It is an easy matter to determine the response of this system to any forcing function by using the appropriate forcing function statement. For example, for $f(t) = 3e^t$,

```
F = 3.0*EXP(T)
```

When the forcing function is an impulse, then the impulse is approximated by a pulse. The unit impulse would be approximated by a pulse of width H = 0.01 and height F = 100. Thus

```
DO 25 K = 0,400
F = 0.0
IF(K.EQ.0)F = 100
```

Note that for the first interval (K = 0), F = 100. For all other intervals, F = 0. Trapezoidal integration is very accurate and yields good results in solving differential equations on digital computers.

One of the most accurate methods of numerical integration is the fourth-order Runge-Kutta. This program solves Illustrative example 6.5 by using the fourth-order Runge-Kutta method:

```
      EXTERNAL DERIV
      DIMENSION X(2), DX(2), WK1(2), WK2(2)
      COMMON F
      NDR = 2
      T = 0.0
      H = 0.01
      DT = H
      X(1)= 0.0
      X(2)= 0.0
      DO 20 K = 0,400
      F = 1.0
      Y = 5.0*X(1) + X(2)
      IF (MOD(K,10).NE.0) GO TO 10
      WRITE (6,5) T,Y
 5    FORMAT (2F10.2)
10    CALL RGKT(X, DX, T, DT, WK1, WK2, NDR, DERIV)
20    CONTINUE
      STOP
      END
```

```
      SUBROUTINE DERIV (X, DX, T)
      DIMENSION X(2), DX(2)
      COMMON F
      DX(1)= X(2)
      DX(2)= F - 2.0*X(2) - 5.0*X(1)
      RETURN
      END

      SUBROUTINE RGKT(Y, YD, T, DT, WK1, WK2, NDR, DERIV)
      DIMENSION Y(NDR), YD(NDR), WK1(NDR), WK2(NDR)
      DT2 = 0.5*DT
      M = 1
10    TP = T + REAL(M/2)*DT2
      CALL DERIV(Y, YD, TP)
      DO 20 I = 1, NDR
      GO TO (12, 14, 16, 18) M
12    WK1(I)= Y(I)
      E = YD(I)*DT2
      WK2(I)= E
      GO TO 20
14    E = YD(I)*DT2
      WK2(I)= WK2(I)+ 2.0*E
      GO TO 20
16    E = YD(I)*DT
      WK2(I)= WK2(I)+ E
      GO TO 20
18    E = (WK2(I)+ YD(I)*DT2)/3.0
20    Y(I) = WK1(I)+ E
      M = M+1
      IF (M.LE.4) GO TO 10
      T = TP
      RETURN
      END
```

In the Runge-Kutta method, the differential equation is expressed in a slightly different form than that used for the trapezoidal method. The differential equation for Illustrative example 6.5 is

$$\ddot{x} = f(t) - 2\dot{x} - 5x$$

By letting $x = x_1$, $\dot{x} = \dot{x}_1 = x_2$, and $\ddot{x} = \dot{x}_2$, this second-order differential equation may be written as two first-order differential equations

$$\dot{x}_1 = x_2 \qquad \dot{x}_2 = f(t) - 2x_2 - 5x_1$$

The Fortran form for these equations is

```
DX(1)= X(2)
DX(2)= F - 2.0*X(2) - 5.0*X(1)
```

This form is referred to as the *state-space representation*. The left side is always a first-order derivative, and no derivative terms appear on the right side. In general, in the state-space representation, an nth-order differential equation is written as n first-order differential equations. In the Runge-Kutta program, NDR is the order of the differential equation, which is 2 in this case. The statements before the DO loop initialize the system. Note that $x(t)$ is designated as X(1), and $\dot{x}(t)$ is designated at X(2). The forcing function statement is F = 1.0. The next statement computes the value of $y(t)$ from $x(t)$ and $\dot{x}(t)$. The IF MOD statement causes corresponding values of T and Y to be printed every time the remainder of dividing K by 10 is zero (i.e., when K = 0, 10, 20, . . .). The CALL RGKT statement transfers the program to the Runge-Kutta subroutine (i.e., SUBROUTINE RGKT). The differential equation to be solved is

$$\ddot{x} = f(t) - 2\dot{x} - 5x$$

The CALL DERIV statement causes SUBROUTINE DERIV to be entered. The following statements are the formulation of the differential equation to be solved:

```
DX(1)= X(2)
DX(2)= F - 2.0*X(2)- 5.0*X(1)
```

Here $x(t)$ = X(1), $\dot{x}(t)$ = DX(1) = X(2), and $\ddot{x}(t)$ = DX(2). Note that the left side is always a term of the form DX(1), DX(2), . . . , DX(N) and that the right side never has a term that appears on the left side. In summary, given the differential equation and the values at the beginning of the integration interval, the Runge-Kutta subroutine determines the values at the end of the interval.

Consider now the third-order differential equation

$$x(t) = \frac{f(t)}{D^3 + 3D^2 + 7D + 4}$$

whence

$$\dddot{x} = f(t) - 3\ddot{x} - 7\dot{x} - 4x$$

By letting $x = x_1$, $\dot{x} = \dot{x}_1 = x_2$, $\ddot{x} = \dot{x}_2 = x_3$, and $\dddot{x} = \dot{x}_3$, this third-order equation may be written as the three first-order equations

$$\dot{x}_1 = x_2$$

$$\dot{x}_2 = x_3$$

$$\dot{x}_3 = f(t) - 3x_3 - 7x_2 - 4x_1$$

The Fortran form for these equations is

```
DX(1)= X(2)
DX(2)= X(3)
DX(3)= F - 3.0*X(3) - 7.0*X(2) - 4.0*X(1)
```

In writing the computer program for a control system, first write the equation for the input and then proceed through the block diagram in the same manner in which the signals flow through the actual system. Thus, the computer program is

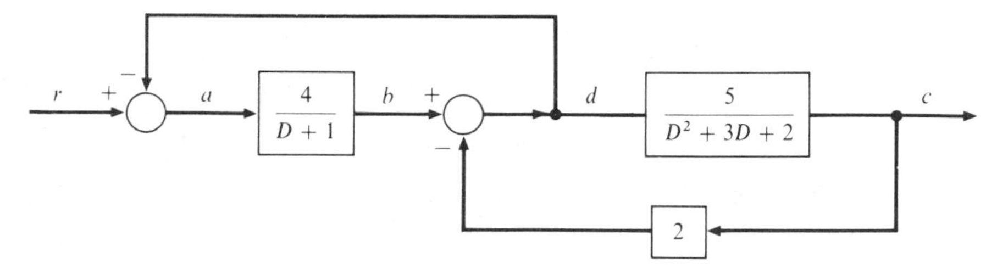

FIGURE 6.12
A multi-loop control system.

written for the input-output relationship for each block in the block-diagram representation for the system, and then these programs are hooked up in the same order in which the blocks are connected in the actual system. This technique is illustrated in the following example.

Illustrative example 6.6. Determine the computer program for the control system shown in Fig. 6.12. The input is $r(t) = u(t)$, a unit step function, and all initial conditions are zero.

Solution. The equation for the first summer is

$$a = r - d$$

The differential equation for the block following this summer is

$$b = \frac{4}{D+1}a \qquad \text{or} \qquad \dot{b} = 4a - b$$

Integration of $\dot{b}$ yields b. The corresponding computer statements are

```
A = R - D
DB = 4.0*A - B
B = 0.5*H*(DB0 + DB) + B0
```

The equation for the next summer is

$$d = b - 2c$$

The differential equation for the block following this summer is

$$c = \frac{5}{D^2 + 3D + 2}d \qquad \text{or} \qquad \ddot{c} = 5d - 3\dot{c} - 2c$$

Integration of $\ddot{c}$ yields $\dot{c}$, then integration of $\dot{c}$ yields c. The corresponding computer statements are

```
D = B - 2.0*C
D2C = 5.0*D - 3.0*DC - 2.0*C
DC = 0.5*H*(D2C+D2C0)+DC0
C = 0.5*H*(DC+DC0)+C0
```

The resulting computer program using trapezoidal integration is

```
      T = 0.0
      A = 0.0
      B = 0.0
      DB = 0.0
      C = 0.0
      DC = 0.0
      D2C = 0.0
      D = 0.0
      H = 0.01
      DO 25 K = 0,400
      R = 1.0
      IF (MOD(K,10).NE.0) GO TO 10
      WRITE(6,5) T,C
    5 FORMAT (2F10.2)
   10 T = T+H
      A = R-D
      B0 = B
      DB0 = DB
      DB = 4.0*A-B
      B = 0.5*H*(DB+DB0)+B0
      D = B-2.0*C
      C0 = C
      DC0 = DC
      D2C0 = D2C
      D2C = 5.0*D-3.0*DC - 2.0*C
      DC = 0.5*H*(D2C+D2C0)+DC0
      C = 0.5*H*(DC+DC0)+C0
   25 CONTINUE
      STOP
      END
```

When the fourth-order Runge-Kutta method is used to integrate, the differential equation for the first block is written in the form

```
DB(1)= 4.0*A - B(1)
```

where b = B(1) and $\dot{b}$ = DB(1). Similarly, the differential equation for the second block is expressed in the form

```
DC(1) = C(2)     DC(2) = 5.0*D - 3.0*C(2) - 2.0*C(1)
```

where c = C(1), $\dot{c}$ = DC(1) = C(2), and $\ddot{c}$ = DC(2). The resulting computer program is

```
      EXTERNAL DERVB, DERVC
      DIMENSION B(1),DB(1),C(2),DC(2)
      DIMENSION WK1B(1),WK2B(1), WK1C(2),WK2C(2)
      COMMON A,D
      NDRB = 1
      NDRC = 2
      T1 = 0.0
      T2 = 0.0
      H = 0.01
      DT = H
      B(1)= 0.0
      C(1)= 0.0
      C(2)= 0.0
      D = 0.0
      DO 25 K = 0,400
      IF (MOD(K,10).NE.0) GO TO 10
      WRITE (6,5) T1,C(1)
    5 FORMAT (2F10.2)
   10 R = 1.0
      A = R - D
      CALL RGKT (B,DB,T1,DT,WK1B,WK2B,NDRB,DERVB)
      D = B(1) - 2.0*C(1)
      CALL RGKT(C,DC,T2,DT,WK1C,WK2C,NDRC,DERVC)
   25 CONTINUE
      STOP
      END

      SUBROUTINE DERVB(B,DB,T1)
      DIMENSION B(1),DB(1)
      COMMON A,D
      DB(1)= 4.0*A - B(1)
      RETURN
      END

      SUBROUTINE DERVC(C,DC,T2)
      DIMENSION C(2),DC(2)
      COMMON A,D
      DC(1)= C(2)
      DC(2)= 5.0*D - 3.0*C(2) - 2.0*C(1)
      RETURN
      END
```

```
      SUBROUTINE RGKT(Y,YD,T,DT,WK1,WK2,NDR,DERIV)
      DIMENSION Y(NDR),YD(NDR),WK1(NDR),WK2(NDR)
      DT2 = 0.5*DT
      M = 1
10    TP = T + REAL(M/2)*DT2
      CALL DERIV(Y,YD,TP)
      DO 20 I = 1,NDR
      .
      .
      .
      RETURN
      END
```

The statements in the main program before the DO loop initialize the system. Following the WRITE statement is the R = 1 statement, which sets the input as a unit step function. The A = R − D statement is the equation for the first summer. The CALL RGKT statement transfers the program to the subroutine RGKT. Only the first few lines of this program are written because this subroutine is always the same. It was given in full in the Runge-Kutta program for Illustrative example 6.5. The arguments in the subroutine RGKT are replaced by the corresponding arguments in the calling statement. Thus, the arguments Y, YD, T, DT, WK1, WK2, NDR, and DERIV in subroutine RGKT are replaced by the arguments B, DB, T1, WK1B, WK2B, NDRB, and DERVB in the CALL RGKT statement. Since DERIV has been replaced by DERVB, when subroutine RGKT gets to the CALL DERIV statement, it will go to subroutine DERVB. Note that subroutine DERVB is the differential equation for the first block ($\dot{b} = 4a - b$). Subroutine RGKT starts with the value of b = B(1) at the beginning of the integration interval and integrates $\dot{b}$ = DB(1) to determine the value of b at the end of the interval. Next the statement for the summing point D = B(1) − 2.0*C(1) is executed. The next statement CALL RGKT again transfers the program to subroutine RGKT. The arguments Y, YD, T, DT, WK1, WK2, NDR, and DERIV in subroutine RGKT are now replaced by the arguments C, DC, T2, WK1C, WK2C, NDRC, and DERVC in the CALL RGKT statement. When the CALL DERIV statement is reached, the program will go to the subroutine DERVC, which is the differential equation for the second block ($\ddot{c} = 5d - 3\dot{c} - 2c$). Subroutine RGKT starts with the values c = C(1) and $\dot{c}$ = C(2) at the beginning of the integration interval and integrates $\ddot{c}$ = DC(2) and $\dot{c}$ = DC(1) to determine the values of $\dot{c}$ = C(2) and c = C(1) at the end of the interval.

The digital computer is easily programmed to handle nonlinear systems. For example, suppose that the equation relating the output y of a block to the input $f(t)$ is

$$\dddot{y} + \dot{y}\ddot{y} + y^2 = f(t)$$

Solving for the highest-order derivative yields

$$\dddot{y} = f(t) - y^2 - \dot{y}\ddot{y}$$

By letting y = X, $\dot{y}$ = DX, $\ddot{y}$ = D2X, $\dddot{y}$ = D3X, and $f(t)$ = F, this becomes

```
D3X = F - X**2 - DX*D2X
```

Trapezoidal integration may now be used to integrate $\dddot{y}$ = D3X to obtain $\ddot{y}$ = D2X. Integration of $\ddot{y}$ = D2X yields $\dot{y}$ = DX, and finally integration of $\dot{y}$ = DX yields y.

To use the Runge-Kutta method, let y = X(1), $\dot{y}$ = DX(1) = X(2), $\ddot{y}$ = DX(2) = X(3), and $\dddot{y}$ = DX(3). The resulting form of the differential equation to be used in subroutine DERIV is

```
DX(1)= X(2)
DX(2)= X(3)
DX(3)= F - X(1)**2 - X(2)*X(3)
```

6.5 TRANSIENT RESPONSE SPECIFICATIONS

A typical response of a second-order system to a step function input is shown in Fig. 6.13. Depending on the performance criterion for the particular system under investigation, various specifications may be employed. The *rise time* t_r is the time at which the response first attains its final steady-state value. The percentage of maximum overshoot is 100 times the maximum amount by which the response overshoots its final steady-state value divided by the final steady-state value. The *settling time* t_s is the time required before the response does not oscillate more than some small percentage, such as 2 or 5 percent, from the final steady-state value. For the case of a step input of height h, the right side of Eq. (6.42) is multiplied by h:

$$y(t) = h + \frac{h}{\sqrt{1-\zeta^2}} e^{-\zeta\omega_n t} \sin\left[\left(\omega_n\sqrt{1-\zeta^2}\right)t + \alpha\right] \tag{6.46}$$

where α is given by Eq. (6.41). The time t_r such that $y(t) = h$ occurs when the sine term vanishes, i.e., when

$$\left(\omega_n\sqrt{1-\zeta^2}\right)t_r + \alpha = 0$$

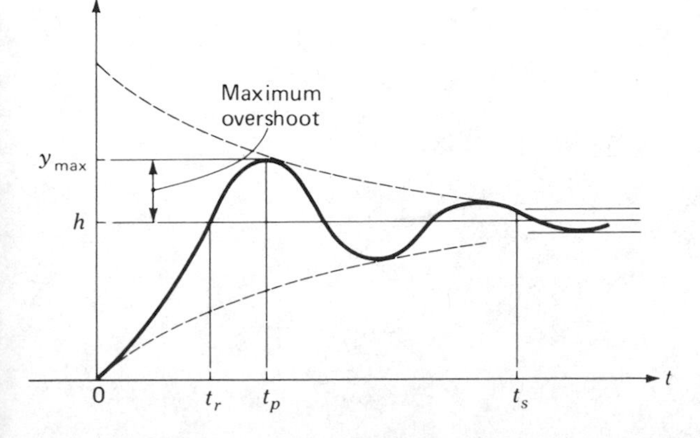

FIGURE 6.13
System response to a step input.

Thus,

$$t_r = \frac{-\alpha}{\omega_n\sqrt{1-\zeta^2}} = \frac{-\tan^{-1}\left[-\sqrt{1-\zeta^2}/(-\zeta)\right]}{\omega_n\sqrt{1-\zeta^2}} = \frac{\tan^{-1}\left[\sqrt{1-\zeta^2}/(-\zeta)\right]}{\omega_n\sqrt{1-\zeta^2}}$$

For a given value of ζ, the rise time is inversely proportional to the natural frequency ω_n. A plot of the normalized rise time $\omega_n t_r$ versus the damping ratio ζ is shown in Fig. 6.14*a*.

Because $\tan\alpha = -\sqrt{1-\zeta^2}/(-\zeta)$, it follows that $\cos\alpha = -\zeta$ and $\sin\alpha = -\sqrt{1-\zeta^2}$. Thus, Eq. (6.46) may be expressed in the form

$$y(t) = h - \frac{he^{-\zeta\omega_n t}}{\sqrt{1-\zeta^2}}\left[\zeta\sin\left(\omega_n\sqrt{1-\zeta^2}\right)t + \sqrt{1-\zeta^2}\cos\left(\omega_n\sqrt{1-\zeta^2}\right)t\right] \tag{6.47}$$

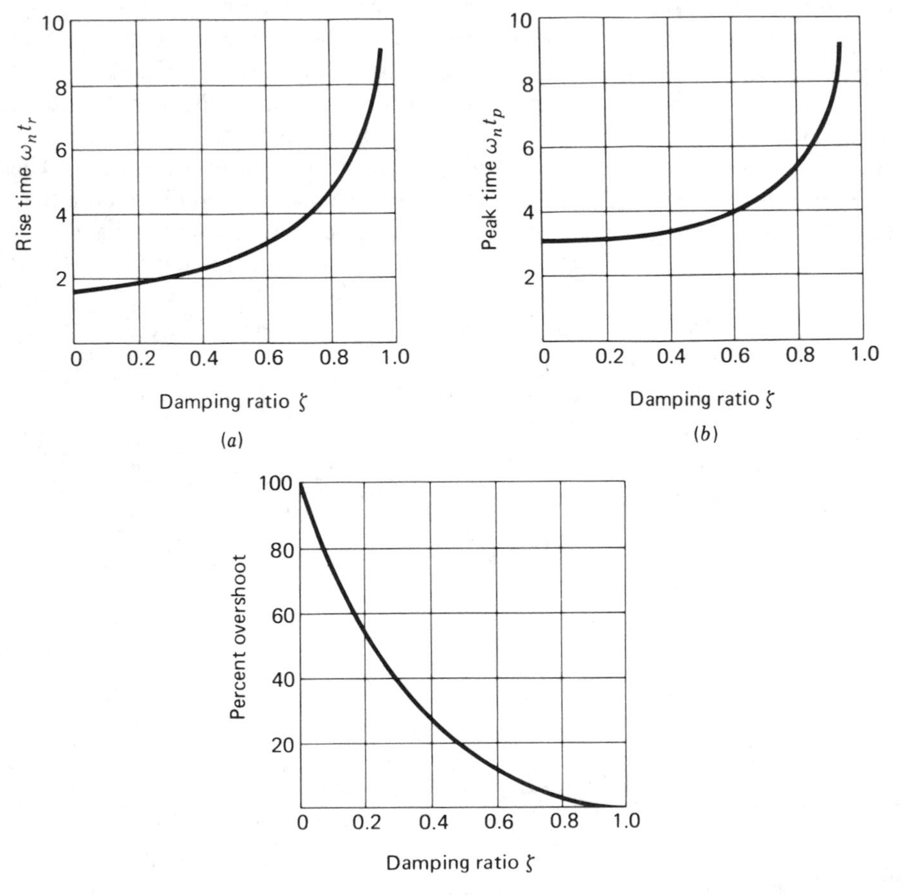

FIGURE 6.14
Performance criteria.

The derivative dy/dt is

$$\frac{dy}{dt} = \frac{\omega_n h e^{-\zeta\omega_n t}}{\sqrt{1-\zeta^2}} \sin\left(\omega_n \sqrt{1-\zeta^2}\right)t$$

Peak values of $y(t)$ occur when $dy/dt = 0$, that is, when

$$\left(\omega_n \sqrt{1-\zeta^2}\right)t_p = \pi,\ 2\pi,\ 3\pi,\ \ldots$$

The maximum value of $y(t)$ occurs when

$$t_p = \frac{\pi}{\omega_n \sqrt{1-\zeta^2}} \tag{6.48}$$

A plot of the normalized peak time $\omega_n t_p$ versus the damping ratio ζ is shown in Fig. 6.14*b*. Substitution of $(\omega_n \sqrt{1-\zeta^2})t = \pi$ into Eq. (6.47) yields for the maximum value

$$y_{\max} = h + h\exp(-\zeta\pi/\sqrt{1-\zeta^2}) \tag{6.49}$$

Subtracting the final steady-state value h, dividing by h, and multiplying by 100 yield for the maximum percentage of overshoot PO

$$\text{PO} = 100\exp\left(-\zeta\pi/\sqrt{1-\zeta^2}\right) \tag{6.50}$$

Note that the maximum value and percentage of overshoot depend on the damping ratio ζ only. A plot of percentage of overshoot versus damping ration ζ is shown in Fig. 6.14*c*.

The envelopes to the response curve are shown dashed in Fig. 6.13. The sine term in Eq. (6.46) causes the response to oscillate between these envelopes. The maximum value for the sine term is $+1$, and the minimum value is -1. The difference between each envelope and the final value is $he^{-\zeta\omega_n t}/\sqrt{1-\zeta^2}$. Each envelope is within 5 percent of the final value when

$$\frac{he^{-\zeta\omega_n t}}{h\sqrt{1-\zeta^2}} = \frac{e^{-\zeta\omega_n t}}{\sqrt{1-\zeta^2}} < 0.05$$

Because $e^{-3} = 0.05$, it is customary to approximate the settling time such that the response is within 5 percent of its final value when $\zeta\omega_n t > 3$. Thus,

$$t_s \geq \frac{3}{\zeta\omega_n} \tag{6.51}$$

Similarly, because $e^{-4} = 0.02$, the settling time such that the response is within 2 percent of the final value is approximated by

$$t_s \geq \frac{4}{\zeta\omega_n} \tag{6.52}$$

FIGURE 6.15
Solar thermal power generator. (Courtesy Sandia National Laboratories.)

Figure 6.15 shows a solar thermal power generator which has been developed by the Sandia National Laboratories. To power the engine, an array of 24 dish-shaped mirrors reflects the sunlight and focuses it onto the power generator. A control system automatically keeps the mirrors facing the sun so that the rays are concentrated on the power generator. An individual module can deliver 240 V, 60 Hz single-phase electricity to provide power for a remote village. When the sun does not shine, the engine can be heated by natural gas, propane, or heating oil. Individual modules can stand alone or can be connected to a grid to provide more power.

Illustrative example 6.7. The block-diagram representation of a control system to position the mirrors to track the sun is shown in Fig. 6.16. Determine the natural frequency, damping ratio, damped natural frequency, rise time, percentage of overshoot, and approximate 5 percent settling time.

Solution. The characteristic equation for this system is $s^2 + 5s + 25 = 0$. Comparison with the general form given by Eq. (6.35) shows that $\omega_n^2 = 25$, or $\omega_n = 5$, and $2\zeta\omega_n = 5$, or $\zeta = \frac{5}{10} = 0.5$. The damped natural frequency is $b = \omega_n\sqrt{1-\zeta^2} = 5\sqrt{\frac{3}{4}} = 4.33$ rad/s. The rise time is

$$t_r = \frac{\tan^{-1}\left[\sqrt{1-\zeta^2}/(-\zeta)\right]}{\omega_n\sqrt{1-\zeta^2}}$$

$$= \frac{\tan^{-1}\left[\left(\frac{\sqrt{3}}{2}\right)/\left(-\frac{1}{2}\right)\right]}{4.33} = \frac{\tan^{-1}(\sqrt{3}/-1)}{4.33} = \frac{2\pi/3}{4.33} = 0.483 \text{ s} \tag{6.53}$$

The angle whose tangent is $\sqrt{3}/(-1)$ is located in the second quadrant and is 120°, which is $2\pi/3$ rad. Application of Eq. (6.50) yields for the percentage of overshoot

$$\text{PO} = 100\exp(-\zeta\pi/\sqrt{1-\zeta^2}) = 100e^{-\pi/\sqrt{3}} = 16.3\,\% \tag{6.54}$$

The approximate 5 percent settling time is

$$t_s \approx \frac{3}{\zeta\omega_n} = \frac{3}{(0.5)(5)} = 1.2 \text{ s} \tag{6.55}$$

Illustrative example 6.8. The block diagram for the autopilot for a large commercial airliner is shown in Fig. 6.17. The system is to be designed such that the percentage of overshoot to a step change in the input does not exceed 10 percent, and the 5 percent settling time should be less than 1 s. Determine the required values of K and a.

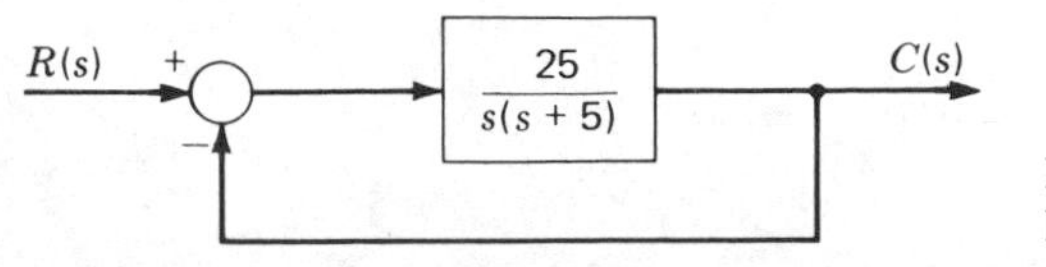

FIGURE 6.16
Feedback control system.

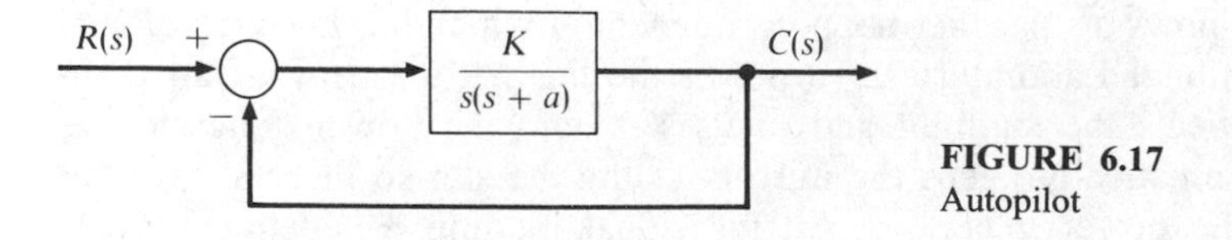

FIGURE 6.17
Autopilot

Solution. Taking the natural logarithm of Eq. (6.50) shows that

$$-\zeta\pi = \sqrt{1-\zeta^2}\ln\frac{\mathrm{PO}}{100} = \sqrt{1-\zeta^2}\ln\frac{10}{100} = -2.3\sqrt{1-\zeta^2}$$

Squaring and solving for ζ, we get

$$\zeta = \sqrt{\frac{5.3}{\pi^2 + 5.3}} = 0.59$$

Using a larger value of ζ results in less overshoot. For $\zeta = 0.6$ the percentage of overshoot is 9.5 percent. For a 5 percent settling time of 1 s

$$t_s \approx \frac{3}{\zeta\omega_n} = 1$$

To satisfy the setting time requirement, $\zeta\omega_n \geq 3$. Thus $\omega_n = 3/0.6 = 5$.

From Fig. 6.17, the closed-loop transfer function is

$$C(s) = \frac{K}{s^2 + as + K} R(s) = \frac{\omega_n^2}{s^2 + 2\zeta\omega_n s + \omega_n^2} R(s)$$

Hence

$$K = \omega_n^2 = 5^2 = 25 \qquad \text{and} \qquad a = 2\zeta\omega_n = 2(0.6)(5) = 6$$

Illustrative example 6.9. As shown in Fig. 6.18, the Chrysler Corporation uses highly efficient electric robot welders to assemble the unitized body structure of Dodge Spirit and Plymouth Acclaim four-door family sedans. Over 200 robots are used throughout this state-of-the-art assembly plant for welding, material handling, sealing, and painting. A closed-loop system for controlling the motion of such robots is shown in Fig. 6.19. The system is to be designed so that the percentage of overshoot will not exceed 5 percent and the 2 percent settling time should be less than 1 s. For steady-state operation, the controlled variable $c(t)$ is equal to the reference input $r(t)$. Determine the required values of a, b, and K.

Solution. The closed-loop transfer function is

$$C(s) = \frac{K}{s^2 + as + b + 0.5K} R(s)$$

The corresponding differential equation is

$$c(t) = \frac{K}{D^2 + aD + b + 0.5K} r(t)$$

FIGURE 6.18
Robot welders. (Courtesy the Chrysler Corporation.)

Letting the operator $D = 0$ shows that for steady-state

$$c_{ss} = \frac{K}{b + 0.5K} r_{ss}$$

Hence

$$K = b + 0.5K$$

or

$$b = 0.5K$$

$R(s)$ + − $\frac{K}{s^2 + as + b}$ $C(s)$ 0.5

FIGURE 6.19
System to control motion of robot.

The closed-loop transfer function may now be written in the form

$$C(s) = \frac{K}{s^2 + as + K} R(s) = \frac{\omega_n^2}{s^2 + 2\zeta\omega_n s + \omega_n^2} R(s)$$

Taking the natural logarithm of Eq. (6.50) shows that

$$-\zeta\pi = \sqrt{1-\zeta^2}\ln\frac{\text{PO}}{100} = \sqrt{1-\zeta^2}\ln\frac{5}{100} = -3.0\sqrt{1-\zeta^2}$$

Squaring and solving for ζ yield

$$\zeta = \sqrt{\frac{9}{\pi^2 + 9}} = 0.69$$

A larger value of ζ results in less overshoot. For $\zeta = \sqrt{2}/2 = 0.707$, the percentage of overshoot is 4.3 percent. For a 2 percent settling time of 1 s

$$t_s \approx \frac{4}{\zeta\omega_n} = 1$$

This settling time requirement is satisfied for $\zeta\omega_n = 4$. Thus

$$\omega_n = \frac{4}{\sqrt{2}/2} = 4\sqrt{2}$$

From the closed-loop transfer function, it follows that $K = \omega_n^2 = 32$, $a = 2\zeta\omega_n = 2(\sqrt{2}/2)(4\sqrt{2}) = 8$, and $b = 0.5K = 16$.

6.6 GENERAL FORM OF TRANSIENT RESPONSE

The general form of the transient response can be ascertained directly from the zeros of the transformed function $B(s)$. For example, suppose that the zeros of $B(s)$ are those plotted in Fig. 6.20. It follows that $B(s)$ may be factored in the form

$$B(s) = [s - (a_1 + jb_1)][s - (a_1 - jb_1)](s - jb_2)(s + jb_2)(s - 0)(s - r_2)(s - r)^2$$

Performing a partial-fraction expansion on $Y(s) = A(s)/B(s)$ yields

$$\begin{aligned} Y(s) = {} & \frac{K_{c_1}}{s - (a_1 + jb_1)} + \frac{K_{-c_1}}{s - (a_1 - jb_1)} + \frac{K_{c_2}}{s - jb_2} \\ & + \frac{K_{-c_2}}{s + jb_2} + \frac{K_1}{s} + \frac{K_2}{s - r_2} + \frac{C_2}{(s-r)^2} + \frac{C_1}{s - r} \end{aligned} \tag{6.56}$$

Taking the inverse transform of the preceding expression yields

$$\begin{aligned} y(t) = {} & \frac{1}{b_1}|K(a_1 + jb_1)|e^{a_1 t}\sin(b_1 t + \alpha_1) \\ & + \frac{1}{b_2}|K(jb_2)|\sin(b_2 t + \alpha_2) + K_1 + K_2 e^{r_2 t} + (C_2 t + C_1)e^{rt} \end{aligned} \tag{6.57}$$

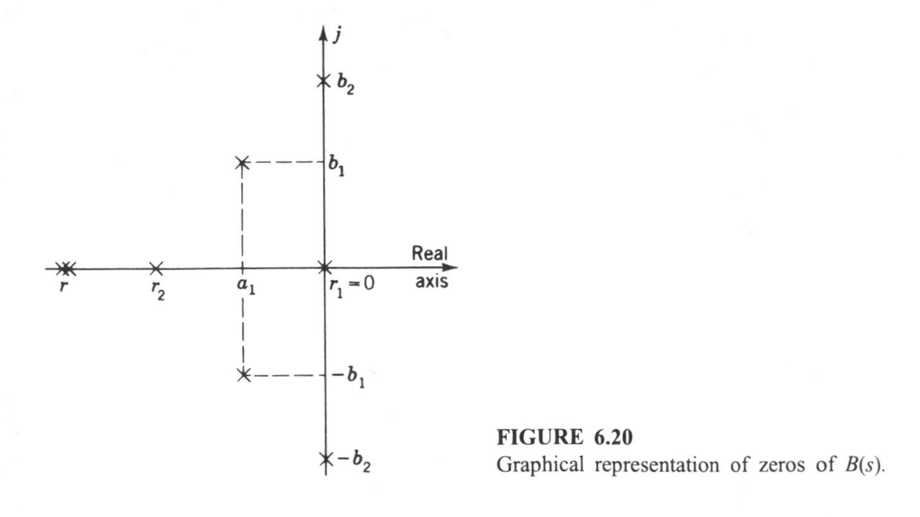

FIGURE 6.20
Graphical representation of zeros of $B(s)$.

Note that a pair of complex conjugate zeros yields an exponentially varying sinusoidal term. A pair of complex conjugate zeros on the imaginary axis yields a sinusoid with a constant amplitude. The zero at the origin contributes a constant term. Distinct or multiple zeros on the real axis yield exponential terms.

The term $B(s) = L_n(s)D_{F(s)}$ consists of the zeros of $L_n(s)$ of the characteristic function for the system plus the zeros of $D_{F(s)}$ corresponding to the denominator of the transform of the input excitation. If any zero of the characteristic function for the system $L_n(s)$ lies to the right of the imaginary axis, the response contains an increasing time function and will increase without bound. Thus, if any zero for $L_n(s)$ lies in the right half-plane (i.e., to the right of the imaginary axis), then the system is basically unstable. Whether a system is stable or unstable is a basic property of the system $L_n(s)$ itself and not of the particular input, or excitation, to the system.

The zeros of $D_{F(s)}$ yield response terms associated with the particular excitation to the system. For example, a ramp function $D_{F(s)} = s^2$ would give a response term of the form $C_1 t$. Note that the input which is a ramp function eventually becomes infinite, and thus the output of the system has been forced to infinity because of the particular input. As illustrated by the following example, the zeros of $D_{F(s)}$ do not affect the basic stability of a system but merely yield response terms appropriate to the particular excitation.

Illustrative example 6.10. The differential equation of operation for a control system is

$$y(t) = \frac{360(D^2 + D + 1)}{(D^2 + 2D + 5)(D^2 + 6D + 9)(D^2 + 6D + 8)} f(t) \tag{6.58}$$

Determine the general form of the response equation when the input excitation $f(t)$ is a unit step function.

Solution. Because of the unit step-function input, $N_{F(s)} = 1$, $D_{F(s)} = s$, so that $B(s) = L_n(s)D_{F(s)} = L_n(s)s$. Thus

$$Y(s) = \frac{360(s^2 + s + 1)(1) + sI(s)}{(s^2 + 2s + 5)(s^2 + 6s + 9)(s^2 + 6s + 8)s} = \frac{A(s)}{B(s)} \tag{6.59}$$

The partial-fraction expansion for $Y(s)$ gives

$$Y(s) = \frac{K_c}{s - (-1 + j2)} + \frac{K_{-c}}{s - (-1 - j2)} + \frac{C_2}{(s + 3)^2} + \frac{C_1}{s + 3} + \frac{K_1}{s + 2} + \frac{K_2}{s + 4} + \frac{K_3}{s}$$

Thus, the general form of the time solution is

$$y(t) = \tfrac{1}{2}|K(a + jb)|e^{-t} \sin (2t + \alpha) + (C_2 t + C_1)e^{-3t} + K_1 e^{-2t} + K_2 e^{-4t} + K_3 \tag{6.60}$$

Note in all cases that the exponent of each exponential term is equal to the horizontal distance from the imaginary axis to the zero of interest. That is, the exponential factor is equal to the numerical value of the real part of the zero. The terms due to zeros which are located far to the left of the imaginary axis have large exponential decaying factors and tend to decrease very rapidly to negligible quantities. Thus, zeros closer to the imaginary axis usually have greater effect on the transient behavior. Accordingly, the analysis of complicated control systems is often approximated by omitting from the characteristic function, the zeros which do not affect substantially the performance of the system.

6.7 RESPONSE TO AN EXTERNAL DISTURBANCE

In this section, we show that the characteristic function for the differential equation which relates the output of a system to a change in the external disturbance is the same as that for the differential equation which relates the output to a change in the desired input. Figure 6.21 shows the general representation for a feedback control system, in which $d(t)$ represents the external disturbance.

As previously discussed, the effect of the input $r(t)$ and external disturbance $d(t)$ on the output or controlled variable $c(t)$ may be considered individually and then each result added by superposition to obtain the total variation in $c(t)$. The block diagram which relates the input $r(t)$ to the output $c(t)$ without regard to the

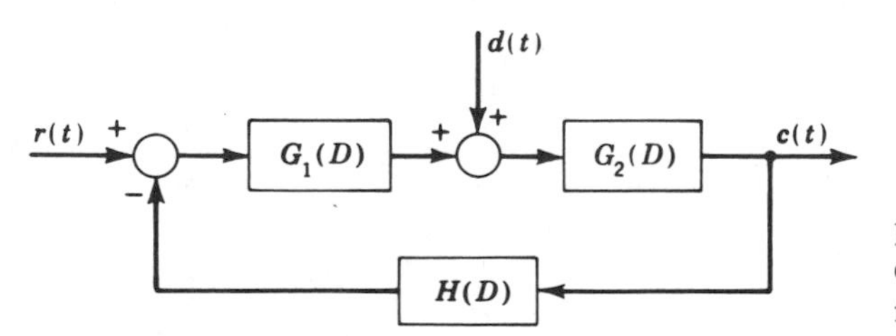

FIGURE 6.21
General representation for a feedback control system.

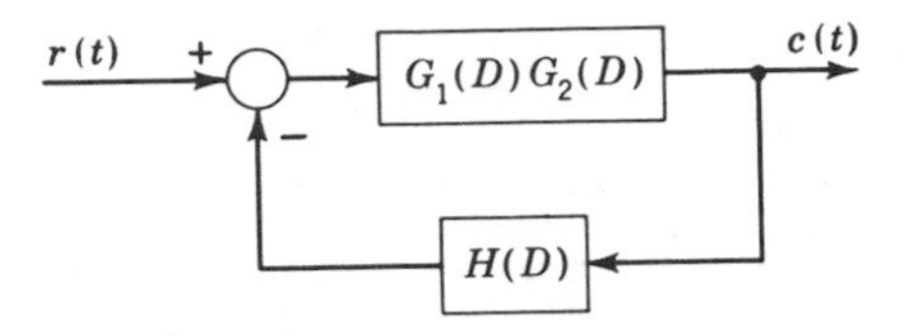

FIGURE 6.22
Block diagram for consideration of the input $r(t)$.

external disturbance is shown in Fig. 6.22. The equation relating $r(t)$ and $c(t)$ is

$$c(t) = \frac{G_1(D)G_2(D)}{1 + G_1(D)G_2(D)H(D)} r(t)$$

By using N_{G_1} to designate the numerator of $G_1(D)$ and D_{G_1} to designate the denominator of $G_1(D)$, etc., the preceding expression may be written in the form

$$c(t) = \frac{N_{G_1}N_{G_2}D_H}{D_{G_1}D_{G_2}D_H + N_{G_1}N_{G_2}N_H} r(t) \tag{6.61}$$

The characteristic function $L_n(D)$ for the system is

$$D_{G_1}D_{G_2}D_H + N_{G_1}N_{G_2}N_H$$

The block diagram which relates the external disturbance $d(t)$ to the output when $r(t)$ is considered zero is shown in Fig. 6.23. The equation relating $d(t)$ and $c(t)$ is

$$\begin{aligned} c(t) &= \frac{G_2(D)}{1 + G_1(D)G_2(D)H(D)} d(t) \\ &= \frac{N_{G_2}D_{G_1}D_H}{D_{G_1}D_{G_2}D_H + N_{G_1}N_{G_2}N_H} d(t) \end{aligned} \tag{6.62}$$

It can be shown that if an excitation enters any place in the loop, the differential equation relating the disturbance and the output will always have the same characteristic function $L_n(D)$, which is the product of all the denominator terms plus the product of all the numerator terms. The constants which appear in the partial-fraction expansion depend upon where the disturbance enters the system.

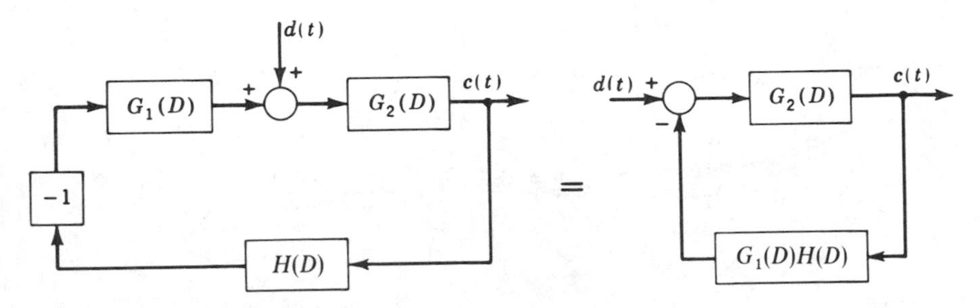

FIGURE 6.23
Block diagram for consideration of the external disturbance $d(t)$.

Thus, there is but one characteristic function for a system, and this function gives basic information as to the transient behavior of the system.

Impulse Response

The response of a system to a unit impulse excitation provides a good indication or measure of the general transient behavior of the system. The unit impulse is, in effect, a momentary disturbance which upsets the initial state of equilibrium of the system. In time, a stable system will return again to its equilibrium state.

Substitution of $F(s) = 1$ into Eq. (6.2) yields the transformed equation for a unit impulse excitation

$$Y(s) = \frac{L_m(s) + I(s)}{L_n(s)} \tag{6.63}$$

Note that the basic form of the response of a system to a unit impulse is determined entirely by the zeros of the characteristic function $L_n(s)$. If any zero lies to the right of the imaginary axis, the output increases without bound. Thus the system is basically unstable. For an unstable system, it is impossible to achieve any equilibrium state, because as soon as the power is turned on, the output continually increases with time.

Figure 6.24 shows the response to a unit impulse of the second-order system whose operational equation is

$$(D^2 + 2\zeta\omega_n D + \omega_n^2)y(t) = \omega_n^2 f(t) \tag{6.64}$$

We see from Fig. 6.24 that for $\zeta < 0.4$ there is a considerable amount of oscillation before the system again reaches equilibrium operation. Also, for $\zeta > 1.0$ a considerable amount of time is required for the system to return to its initial state. Thus, it seems desirable to have $0.4 < \zeta < 1.0$. This is the same result as was obtained from a consideration of the step function response. Because the scale of the

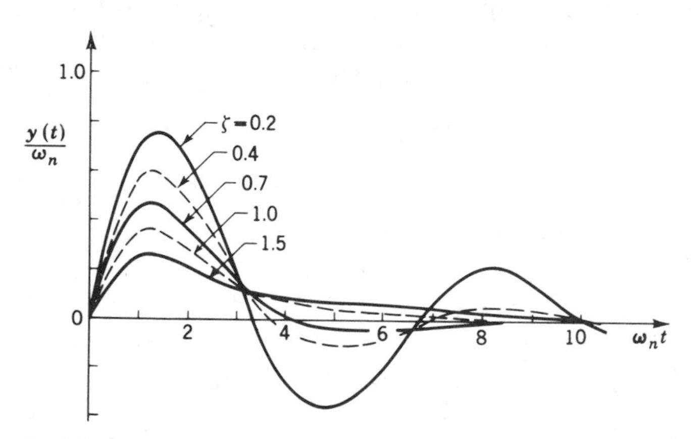

FIGURE 6.24
Response of a second-order system to a unit impulse.

horizontal axis is $\omega_n t$, the speed of response is governed by the natural frequency ω_n, as described in Sec. 6.3.

Controllers

Excellent control characteristics may be obtained by using controllers with various amounts of proportional, integral, and derivative action. A controller with proportional and integral action is referred to as a *PI controller*, while one with proportional and derivative action is a *PD controller*, and a controller having proportional plus integral plus derivative action is a *PID controller*. Controlling the amount of proportional or integral or derivative action makes it possible to design control systems to achieve desired transient performance characteristics.

Figure 6.25 shows a PUMA 500 robot (Programmable Universal Machine for Assembly) which is a very accurate 6 degree of freedom robot. This robot is checking tolerances of fabricated parts in an inspection cell. Parts that do not pass the inspection are automatically removed and then discarded.

Illustrative example 6.11. For most robotic control situations (as is typical of most positional control systems), the primary task is to drive the links so that their actual position corresponds to the desired input position. A system for controlling the position of a robot is shown in Fig. 6.26. Use a proportional plus integral controller to design a system whose characteristic equation has a damping ratio $\zeta = 1.0$ and a natural frequency $\omega_n = 4$.

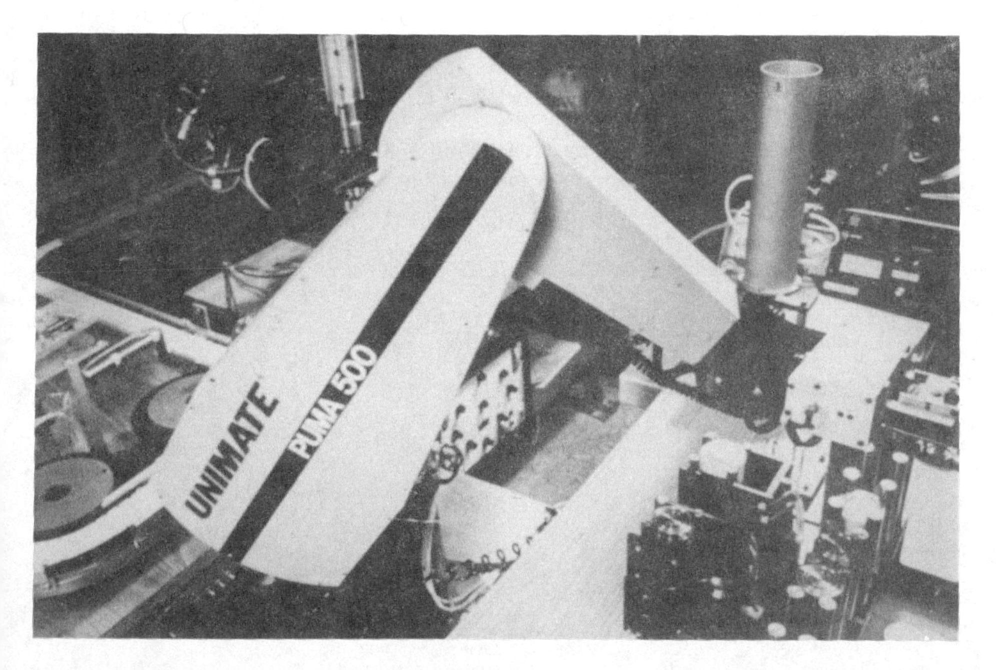

FIGURE 6.25
PUMA 500 robot. (Courtesy Stäubli Unimation Inc.)

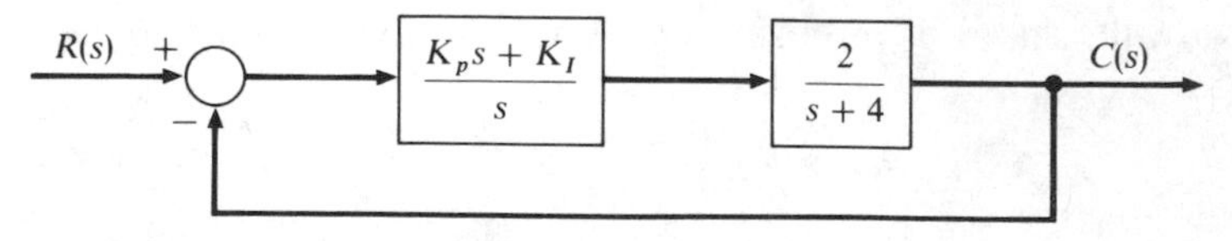

FIGURE 6.26
System for controlling the position of a robot.

Solution. The transfer function for a PI controller is

$$K_P + \frac{K_I}{s} = \frac{K_P s + K_I}{s}$$

The general form of the characteristic equation is

$$s(s+4) + 2(K_P s + K_I) = s^2 + (4 + 2K_P)s + 2K_I = 0$$

The desired form of the characteristic equation is

$$s^2 + 2\zeta\omega_n s + \omega_n^2$$

Comparing coefficients of the like powers of s shows that

$$K_I = \frac{\omega_n^2}{2} = \frac{16}{2} = 8$$

and

$$4 + 2K_P = 2\zeta\omega_n = 8$$

$$K_P = \frac{4}{2} = 2$$

Illustrative example 6.12. A positional control system for a robot is shown in Fig. 6.27. Use a PID controller to design a system whose characteristic equation has a real root at -10, a damping ratio $\zeta = 0.8$, and a natural frequency $\omega_n = 2$.

Solution. The transfer function for a PID controller is

$$K_P + \frac{K_I}{s} + K_D s = \frac{K_D s^2 + K_P s + K_I}{s}$$

The general form of the characteristic equation is

$$s(s^2 + 6s + 4) + 4(K_D s^2 + K_P s + K_I) = s^3 + (4K_D + 6)s^2 + 4(K_P + 1)s + 4K_I = 0$$

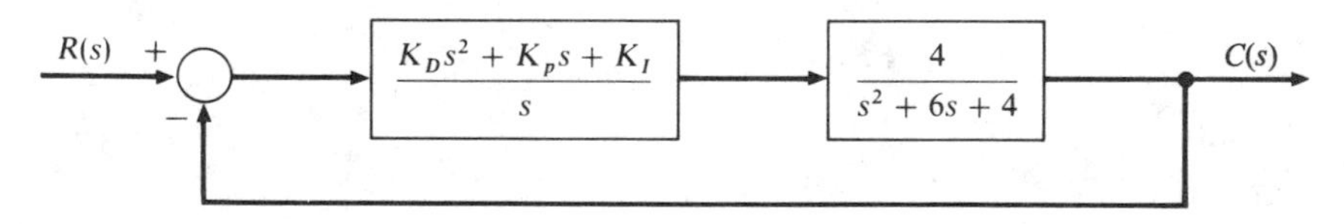

FIGURE 6.27
Positional control system for a robot.

The desired form of the characteristic equation is

$$(s+10)(s^2+2\zeta\omega_n s+\omega_n^2)=s^3+(10+2\zeta\omega_n)s^2+(20\zeta\omega_n+\omega_n^2)s+10\omega_n^2=0$$

Comparing coefficients of like powers of s yields for K_I, K_P, and K_D

$$K_I=\frac{10\omega_n^2}{4}=10$$

and

$$4(K_P+1)=20\zeta\omega_n+\omega_n^2=32+4=36$$
$$K_P=9-1=8$$

and

$$4K_D+6=10+2\zeta\omega_n=10+3.2=13.2$$
$$K_D=\frac{13.2-6}{4}=1.8$$

6.8 ROUTH'S STABILITY CRITERION

A major difficulty in using the Laplace transform method for determining the transient response of a feedback control system is that it necessitates determining the roots of the characteristic equation. The general form for the characteristic equation has previously been shown to be

$$D_{G_1}D_{G_2}D_H+N_{G_1}N_{G_2}N_H=0 \tag{6.65}$$

In determining the block diagram for a system, the terms N_{G_1}, N_{G_2}, N_H, D_{G_1}, D_{G_2}, and D_H are usually obtained in factored form. Because of the plus sign in Eq. (6.65), the roots of N_{G_1}, N_{G_2}, etc., are not the roots of the characteristic equation. Thus, it becomes necessary to determine the roots of the general polynomial represented by Eq. (6.65). This presents no difficulty for first- and second-order systems. The root of a first-order system is immediately obvious, and the two roots of a quadratic equation are readily solved. However, a third-order system requires determining the three roots of a cubic, a fourth-order system necessitates solving a quartic, etc.

*Routh's criterion** is a method for determining whether any of the roots of the characteristic equation are in the right half-plane. The application of this criterion is as follows: First write the characteristic equation in the general form

$$b_n s^n+b_{n-1}s^{n-1}+b_{n-2}s^{n-2}+\cdots+b_2 s^2+b_1 s+b_0 \tag{6.66}$$

* E. J. Routh, *Dynamics of a System of Rigid Bodies*, 3d ed., Macmillan, New York, 1877.

Next arrange the coefficients of the characteristic equation according to the following schedule:

$$
\begin{array}{lllll}
b_n & b_{n-2} & b_{n-4} & b_{n-6} & \cdots \\
b_{n-1} & b_{n-3} & b_{n-5} & b_{n-7} & \cdots \\
c_1 & c_2 & c_3 & c_4 & \cdots \\
d_1 & d_2 & d_3 & \cdots & \\
\cdots & \cdots & \cdots & \cdots & \cdots \\
e_1 & e_2 & 0 & & \\
f_1 & f_2 & 0 & & \\
g_1 & 0 & & & \\
h_1 & 0 & & &
\end{array}
\tag{6.67}
$$

After the b coefficients are arranged as shown, the row of c terms is evaluated as follows:

$$
\begin{aligned}
c_1 &= \frac{b_{n-1}b_{n-2} - b_n b_{n-3}}{b_{n-1}} \\
c_2 &= \frac{b_{n-1}b_{n-4} - b_n b_{n-5}}{b_{n-1}} \\
c_3 &= \frac{b_{n-1}b_{n-6} - b_n b_{n-7}}{b_{n-1}}
\end{aligned}
\tag{6.68}
$$

Examination of Eqs. (6.68) reveals the pattern for evaluating the c coefficients. By dropping down a row, the same pattern is used to obtain the d coefficients. That is,

$$
\begin{aligned}
d_1 &= \frac{c_1 b_{n-3} - b_{n-1} c_2}{c_1} \\
d_2 &= \frac{c_1 b_{n-5} - b_{n-1} c_3}{c_1}
\end{aligned}
\tag{6.69}
$$

This process is continued until one more row is obtained than the order of the differential equation. Thus, a third-order equation has four rows, a fourth-order equation has five rows, etc. To illustrate, consider the characteristic equation

$$s^4 + 3s^3 + s^2 + 6s + 2 = 0 \tag{6.70}$$

The first two rows of the following array are obtained directly from the coefficients of Eq. (6.70), and the remaining rows are computed as just described:

$$
\begin{array}{lrrrr}
s^4 & 1 & 1 & 2 & 0 \\
s^3 & 3 & 6 & 0 & \\
s^2 & -1 & 2 & 0 & \\
s^1 & 12 & 0 & & \\
s^0 & 2 & 0 & &
\end{array}
\tag{6.71}
$$

The left column identifies the row. The first row is indicated by s^4, the second by s^3, etc. The exponent in the first term s^4 is the order of the polynomial. Because there is one more row than the order of the polynomial, the last row is always indicated by s^0.

Routh's criterion states that the number of changes of sign of the coefficients in the first column of the array is equal to the number of roots of the characteristic equation located to the right of the imaginary axis. For the preceding example, the signs of the numbers in the first column are seen to go from plus to minus and then back to plus again, so that there are two changes of sign. Thus, there are two roots located to the right of the imaginary axis.

The number of coefficients in the first column is $n + 1$, where n is the order of the characteristic equation. If the Routh array is continued past the last row, only zeros result. The reason that it is necessary to keep track of the number of pertinent rows is that if the coefficient of the last row is zero, then the characteristic equation has a root at the origin. If the coefficients in the last two pertinent rows are zero, then the characteristic equation has a double root at the origin, etc.

A Zero in the First Column

When one of the coefficients in the first column is zero, it may be replaced by a very small number ε for the purpose of computing the remaining coefficients in the array. Thus, consider the characteristic equation

$$s^5 + 2s^4 + 4s^3 + 8s^2 + 10s + 6 = 0 \tag{6.72}$$

The Routh array is

$$\begin{array}{lcccc} s^5 & 1 & 4 & 10 & 0 \\ s^4 & 2 & 8 & 6 & 0 \\ s^3 & \varepsilon \approx 0 & 7 & 0 & \\ s^2 & 8 - \dfrac{14}{\varepsilon} & 6 & 0 & \\ s^1 & 7 & 0 & & \\ s^0 & 6 & 0 & & \end{array}$$

where $d_1 = (8\varepsilon - 14)/\varepsilon \approx 8 - 14/\varepsilon$ is a very large negative number when ε is positive and is a very large positive number when ε is negative. Regardless of whether ε is taken as a small positive number or a small negative number, there are two sign changes in the first column. Thus, the characteristic equation has two roots located to the right of the imaginary axis.

A Row of Zeros

When two or more of the roots of the characteristic equation are symmetrically located about the origin, then a row of Routh's array will contain all zeros. This situation occurs when the characteristic equation has a pair of real roots with opposite signs ($\pm r$), complex conjugate roots on the imaginary axis ($\pm j\omega$), or a pair of complex conjugate roots with opposite real parts ($a \pm jb$, $-a \pm jb$). The row preceding the row of zeros is the polynomial of such factors. This polynomial is called the *auxiliary equation*.

Thus consider the characteristic equation

$$s^6 + 6s^5 + 10s^4 + 12s^3 + 13s^2 - 18s - 24 = 0 \tag{6.73}$$

The Routh array is

$$\begin{array}{llllll} s^6 & 1 & 10 & 13 & -24 & 0 \\ s^5 & 6 & 12 & -18 & 0 & \\ s^4 & 8 & 16 & -24 & 0 & \\ s^3 & 0 & 0 & 0 & & \end{array}$$

The s^3 row contains all zeros. The auxiliary equation $A(s)$ is obtained from the coefficients of the preceding row as follows:

$$A(s) = 8s^4 + 16s^2 - 24 = 8(s^4 + 2s^2 - 3)$$

The auxiliary equation is always an even order. The power of s for the first coefficient (8) is s^4, the power of s for the next coefficient (16) is s^2, and the power of s for the last coefficient (-24) is $s^0 = 1$. In general, the power of s for the first coefficient is the order of the row, and the power decreases by 2 for each succeeding coefficient. Dividing the polynomial by the auxiliary equation yields

$$\begin{array}{r} s^2 + 6s + 8 \\ s^4 + 2s^2 - 3 \,\overline{)\, s^6 + 6s^5 + 10s^4 + 12s^3 + 13s^2 - 18s - 24} \\ \underline{s^6 \qquad + 2s^4 \qquad\qquad - 3s^2 \qquad\qquad} \\ 6s^5 + 8s^4 + 12s^3 + 16s^2 - 18s - 24 \\ \underline{6s^5 \qquad\qquad + 12s^3 \qquad\qquad - 18s \qquad} \\ 8s^4 \qquad + 16s^2 \qquad - 24 \\ \underline{8s^4 \qquad + 16s^2 \qquad - 24} \end{array}$$

Thus, the original characteristic equation may be written in the form

$$s^6 + 6s^5 + 10s^4 + 12s^3 + 13s^2 - 18s - 24 = (s^4 + 2s^2 - 3)(s^2 + 6s + 8) = 0$$

The roots of $s^4 + 2s^2 - 3 = (s^2 - 1)(s^2 + 3)$ consist of the pair of real and opposite roots $s^2 - 1 = (s - 1)(s + 1)$ and the pair of complex conjugate roots on the imaginary axis $s^2 + 3 = (s + j\sqrt{3})(s - j\sqrt{3})$.

As another example, consider the characteristic equation

$$s^6 + 3s^5 + 2s^4 + 4s^2 + 12s + 8 = 0 \tag{6.74}$$

The Routh array is

$$\begin{array}{lllll} s^6 & 1 & 2 & 4 & 8 \\ s^5 & 3 & 0 & 12 & 0 \\ s^4 & 2 & 0 & 8 & 0 \\ s^3 & 0 & 0 & 0 & \end{array}$$

The auxiliary equation is

$$A(s) = 2s^4 + 8 = 2(s^4 + 4)$$

Dividing the characteristic equation by $s^4 + 4$ yields

$$\begin{array}{r|l}
 & s^2 + 3s + 2 \\
\hline
s^4 + 4 & s^6 + 3s^5 + 2s^4 + 0s^3 + 4s^2 + 12s + 8 \\
 & s^6 \qquad\qquad\qquad\quad + 4s^2 \\
\hline
 & \quad 3s^5 + 2s^4 \qquad\qquad\qquad + 12s + 8 \\
 & \quad 3s^5 \qquad\qquad\qquad\qquad\; + 12s \\
\hline
 & \qquad\quad 2s^4 \qquad\qquad\qquad\quad + 8 \\
 & \qquad\quad 2s^4 \qquad\qquad\qquad\quad + 8 \\
\hline
\end{array}$$

The factored form for the original characteristic equation is

$$s^6 + 3s^5 + 2s^4 + 4s^2 + 12s + 8 = (s^4 + 4)(s^2 + 3s + 2) = 0$$

The roots of $s^4 + 4 = (s^2 + 2s + 2)(s^2 - 2s + 2)$ consist of the complex pair located at $-1 \pm j$ and the pair located at $+1 \pm j$. Such a set of complex conjugate roots is called a *quadripole*.

It is not necessary to use Routh's criterion if any of the coefficients of the characteristic equation are zero or negative. When this is so, it can be shown that at least one root is located on, or to the right of, the imaginary axis.

To determine how many roots of the characteristic equation lie to the right of some vertical line a distance σ from the imaginary axis (i.e., the number of roots that have a real part greater than σ), transform the characteristic equation by substituting $s + \sigma$ for s, and then apply Routh's criterion as just described. The number of changes of sign in the first column for this new function is equal to the number of roots which are located to the right of the vertical line through σ.

Illustrative example 6.13. A system for controlling the output torque from an engine is shown in Fig. 6.28. Apply Routh's criterion to determine the range of values of a such that the system is stable.

Solution. The characteristic equation for this system is

$$(s + 1)(s + 2)(s + 5) + 8(s + a) = s^3 + 8s^2 + 25s + 8a$$

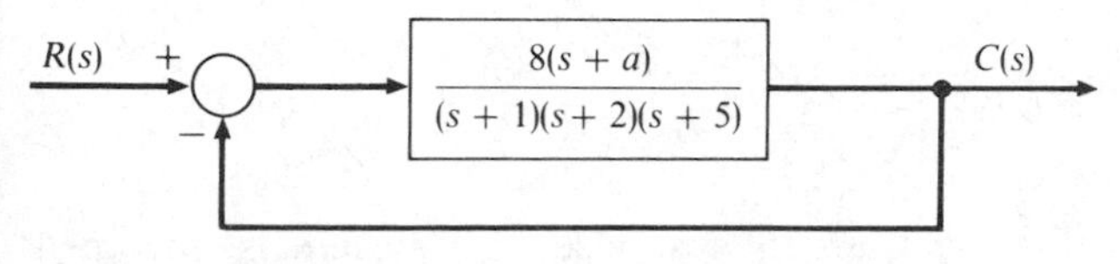

FIGURE 6.28
System for controlling torque.

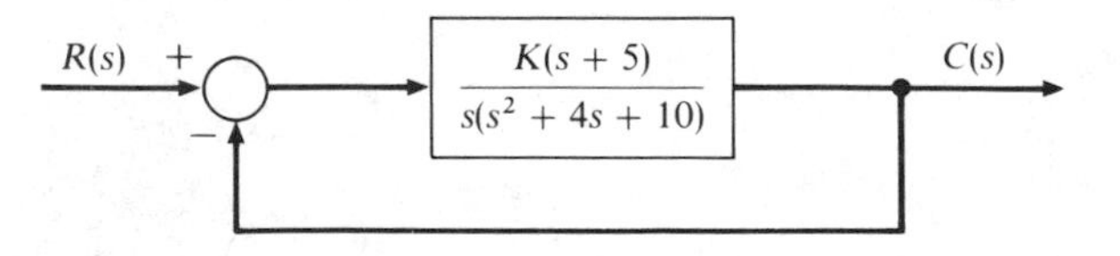

FIGURE 6.29
Unity-feedback control system.

The Routh's array is

$$\begin{array}{lcc} s^3 & 1 & 25 \\ s^2 & 8 & 8a \\ s^1 & \dfrac{200-8a}{8} & 0 \\ s^0 & 8a & \end{array}$$

For the system to be stable, all the coefficients in the first column must be positive:

$$200 - 8a = 8(25 - a) > 0 \qquad \text{and} \qquad 8a > 0$$

Thus, for a stable system

$$0 < a < 25$$

Illustrative example 6.14. For the unity-feedback control system shown in Fig. 6.29, determine the range of values for K such that the system is stable.

Solution. The characteristic equation is

$$s(s^2 + 4s + 10) + K(s + 5) = s^3 + 4s^2 + (K + 10)s + 5K$$

The Routh's array is

$$\begin{array}{lcc} s^3 & 1 & K+10 \\ s^2 & 4 & 5K \\ s^1 & \dfrac{4(K+10)-5K}{4} & 0 \\ s^0 & 5K & \end{array}$$

To be stable, all the coefficients in the first column must be positive.

$$40 - K > 0 \qquad \text{and} \qquad 5K > 0$$

Thus, for a stable system

$$0 < K < 40$$

6.9 SUMMARY

In this chapter, we showed that the transient response of a system is governed primarily by the location of the zeros of $B(s) = L_n(s)D_{F(s)}$. The zeros of $D_{F(s)}$ yield response terms appropriate to the particular excitation to the system. The

TABLE 6.1
Location of zeros and corresponding response functions

Zeros of $B(s) = L_n(s)D_{F(s)}$	Type of response
Left half-plane (distinct or repeated)	Decaying exponential and/or decaying sinusoid
Right half-plane (distinct or repeated)	Increasing exponential and/or increasing sinusoid
Imaginary axis:	
Distinct	Constant and/or constant sinusoid
Repeated	Increasing time function and/or increasing sinusoid

characteristic function $L_n(s)$ is a basic property of the system itself. When all the zeros of $L_n(s)$ are located in the left half-plane, the system is stable (i.e., for any bounded input, the response is also bounded). If any zero of $L_n(s)$ is located in the right half-plane, the system is unstable (i.e., the response is always unbounded). The imaginary axis is the borderline between stable and unstable systems.

Complex imaginary zeros are undesirable because they yield constant sinusoids. A zero of $L_n(s)$ at the origin is also undesirable because it indicates an integration of the input. Note that the integrator in the feedforward elements (which integrates the error signal) does not yield a zero of $L_n(s)$ at the origin because the zeros of $D_{G(s)}$ are not the zeros of $L_n(s)$. When the characteristic function $L_n(s)$ has a zero at the origin, a constant input (i.e., a step function) yields an unbounded time term $C_1 t$ in the output, which is the integral of the input.

The basic form of the response due to repeated zeros is the same as that for distinct zeros, except that repeated zeros on the imaginary axis yield increasing time terms rather than time terms with constant amplitudes. Because $B(s) = L_n(s)D_{F(s)}$, the zeros of $L_n(s)$ and $D_{F(s)}$ act independently to yield the time response unless one or more of the zeros of $L_n(s)$ and $D_{F(s)}$ are the same. This introduces repeated zeros in $B(s)$, which affects the basic form of the response equation only if the zeros are on the imaginary axis. For example, a repeated complex imaginary zero results in an increasing sinusoid rather than a constant sinusoid. Similarly, a repeated zero at the origin yields an increasing time function rather than a constant. To ensure stability, zeros of $L_n(s)$ should be excluded not only from the right half-plane but also from the imaginary axis.

Table 6.1 summarizes the types of response terms associated with the zeıos of $B(s)$.

PROBLEMS

6.1. Determine the time response $y(t)$ for each of the following transformed equations.

(*a*) $Y(s) = \dfrac{4}{s(s+1)(s+2)}$ (*b*) $Y(s) = \dfrac{s+4}{s(s+1)(s+2)}$

(*c*) $Y(s) = \dfrac{4}{s(s-1)(s-2)}$ (*d*) $Y(s) = \dfrac{s+4}{s(s-1)(s-2)}$

6.2. Determine the time response $y(t)$ for each of the following transformed equations.

$$(a)\quad Y(s)=\frac{s+3}{(s+2)^2}\qquad (b)\quad Y(s)=\frac{s+1}{(s+2)^3}$$

$$(c)\quad Y(s)=\frac{s^2+2}{(s+2)^3}\qquad (d)\quad Y(s)=\frac{s^3+2s^2+4s+2}{(s+2)^4}$$

6.3. Invert each of the following to determine the response $y(t)$.

$$(a)\quad Y(s)=\frac{2s+1}{(s+2)^2(s+5)}\qquad (b)\quad Y(s)=\frac{5s+1}{(s+2)(s+5)^2}$$

$$(c)\quad Y(s)=\frac{3(s-1)}{(s+2)^2(s+5)^2}\qquad (d)\quad Y(s)=\frac{3(2s+1)}{(s+2)^3(s+5)}$$

6.4. Use the Laplace transform method to determine the solution of each of the following differential equations when all the initial conditions are zero.

$(a)\quad (D+1)(D+2)y=e^{-t}$ $\qquad (b)\quad (D+1)(D+2)y=e^{-2t}$

$(c)\quad (D+1)(D+2)y=e^{-t}-e^{-2t}$

6.5. Use the Laplace transform method to determine the solution of the following differential equation when $f(t)=\delta(t)$ is a unit impulse, $y(0)=1$, and $\dot{y}(0)=0$.

$$(D+1)^2y(t)=f(t)$$

6.6. Use the Laplace transform method to determine the solution of the following differential equation for the case in which $f(t)=u(t)$ is a unit step function and all initial conditions are zero.

$$y(t)=\frac{4(D+1)}{(D+2)^2}f(t)$$

6.7. Use the Laplace transform method to determine the solution of the following differential equation when $f(t)=e^{-t}$ and all initial conditions are zero.

$$y(t)=\frac{D+4}{(D+1)(D+2)}f(t)$$

6.8. Repeat Prob. 6.7 except $f(t)=e^{-2t}$.

6.9. Repeat Prob. 6.7 except $f(t)=e^{-t}+e^{-2t}$.

6.10. A unity-feedback system with an integral controller is shown in Fig. P6.10. Determine the response $c(t)$ for each of the following cases:

$(a)\quad r(t)=u(t)$ and $c(0)=\dot{c}(0)=0$

$(b)\quad r(t)=e^{-t}$ and $c(0)=\dot{c}(0)=1$

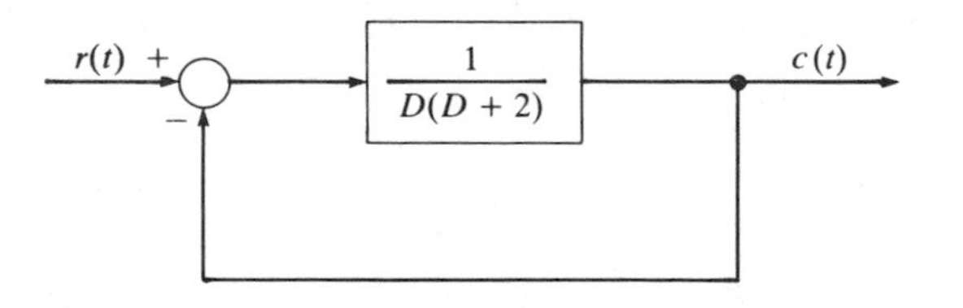

FIGURE P6.10

6.11. For the system shown in Fig. P6.11, determine K and a such that the response to a unit impulse $[r(t) = \delta(t)]$ has the form

$$c(t) = C_1 e^{-t} + C_2 e^{-4t}$$

Evaluate C_1 and C_2 when all the initial conditions are zero.

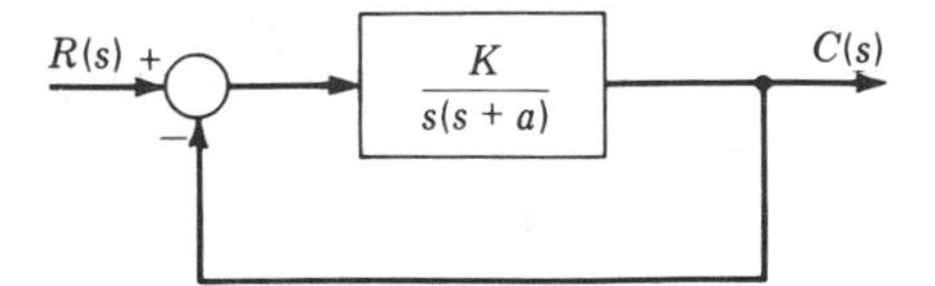

FIGURE P6.11

6.12. Repeat Prob. 6.11, except that it is desired that the impulse response have the form

$$c(t) = (C_0 + C_1 t)e^{-2t}$$

Evaluate C_0 and C_1 when all the initial conditions are zero.

6.13. Repeat Prob. 6.11, except that the impulse response should have the form

$$c(t) = Ce^{-t} \sin (2t + \alpha)$$

Evaluate C and α when all the initial conditions are zero.

6.14. The differential equation describing the operation of a control system is

$$(D^2 + 2D + 1)c(t) = r(t)$$

Determine the response $c(t)$ when $r(t) = 2e^{-t}$, $c(0) = 1$, and $\dot{c}(0) = 0$.

6.15. Use the Laplace transform method to solve the following differential equation:

$$y(t) = \frac{25(D + 1)}{D^2 + 10D + 25} f(t)$$

All the initial conditions are zero, and $f(t)$ is a unit step function.

6.16. Use the Laplace transform method to solve the differential equation:

$$\frac{dy}{dt} + y = f(t) \qquad y(0) = 0$$

The forcing function is $f(t) = 4 \sin t + 2 \cos t$.

6.17. The main servosystem for a computer-controlled milling machine is shown in Fig. P6.17. Determine the response $c(t)$ when

(*a*) $K = 1$, $r(t) = e^{-2t}$, $r(0) = 0$, $c(0) = 1$, and $\dot{c}(0) = 0$

(*b*) $K = 5$, $r(t) = u(t)$, $r(0) = c(0) = 0$, and $\dot{c}(0) = 1$

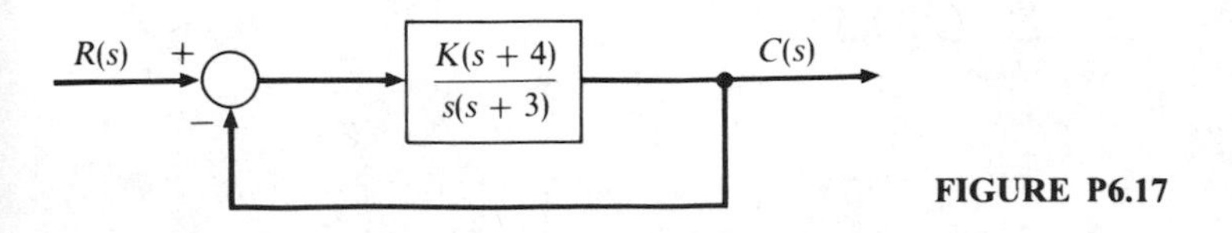

FIGURE P6.17

6.18. A system is described by the differential equation

$$(D^2 + 1)y = f(t)$$

All the initial conditions are zero. A step input $f(t) = u(t)$ is applied for π seconds [$f(t) = 1$ for $0 < t < \pi$ and $f(t) = 0$ for $t > \pi$]. Determine the response for $0 < t < \pi$ and for $t > \pi$.

6.19. Use the Laplace transform method to solve the following differential equation:

$$y(t) = \frac{25(D+1)}{D^2 + 8D + 25} f(t)$$

All the initial conditions are zero, and $f(t)$ is a unit step function.

6.20. The differential equation describing the operation of a control system is

$$c(t) = \frac{3(D+5)}{D^2 + 2D + 10} r(t)$$

Determine the response $c(t)$ for the case in which $r(t) = \delta(t)$ and all the initial conditions are zero. What are the damping ratio, natural frequency, and damped natural frequency?

6.21. Repeat Prob. 5.29 except $K = 4$.

6.22. Consider the differential equation

$$y(t) = \frac{5(D+1)}{D^2 + 2D + 5} f(t)$$

Determine the response $y(t)$ for each of the following cases.

(*a*) $f(t) = u(t)$ and $y(0) = \dot{y}(0) = 0$

(*b*) $f(t) = 0$ and $y(0) = \dot{y}(0) = 1$

What are the damping ratio, the natural frequency, and the damped natural frequency?

6.23. A control system for accurately controlling the position of a laser beam is shown in Fig. P6.23. Determine the response $c(t)$ for each of the following cases.

(*a*) $r(t) = u(t)$, $d(t) = 0$, and $c(0) = \dot{c}(0) = 0$

(*b*) $r(t) = 0$, $d(t) = u(t)$, and $c(0) = \dot{c}(0) = d(0) = 0$

What are the damping ratio, natural frequency, and damped natural frequency?

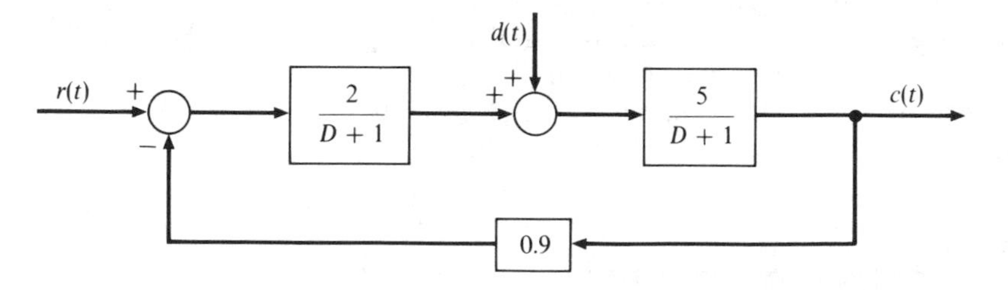

FIGURE P6.23

6.24. For the feedback control system shown in Fig. P6.24, determine K_1, K_2, and a such that the system will have a steady-state gain of 1.0, a damping ratio $\zeta = 0.5$, and a natural frequency $\omega_n = 4.0$.

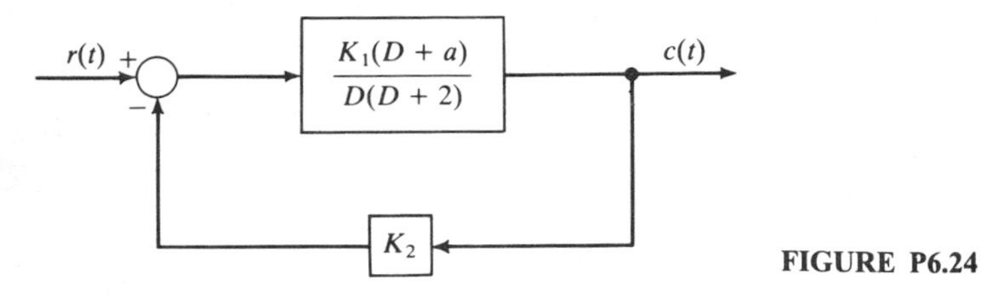

FIGURE P6.24

6.25. For the control system shown in Fig. P6.25, determine K_1, K_2, and a such that the system will have a steady-state gain of 1.0, a damping ratio $\zeta = 0.6$, and a natural frequency $\omega_n = 5.0$.

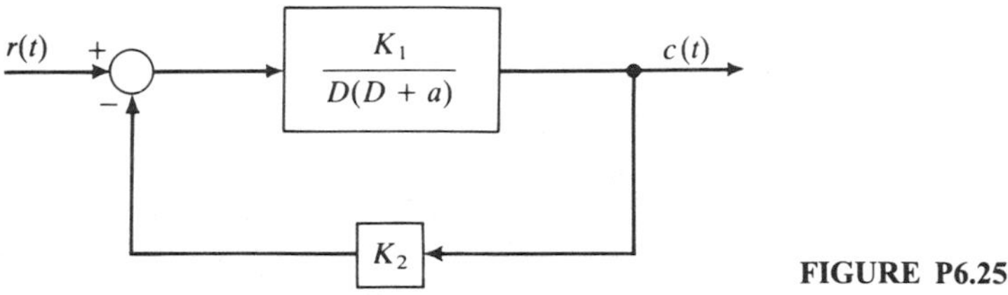

FIGURE P6.25

6.26. A servomechanism is shown in Fig. P6.26. Determine K_1, K_2, and a such that the system will have a natural frequency of 2, a steady-state gain of 1, and a damping ratio $\zeta = 0.5$.

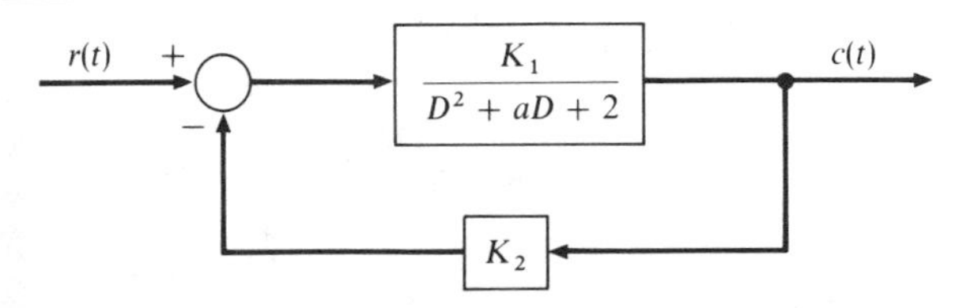

FIGURE P6.26

6.27. For the proportional control system shown in Fig. P6.27, determine K_1, K_2, and a such that the system will have a steady-state gain of 1, a natural frequency $\omega_n = 2$, and a damping ratio $\zeta = 0.5$.

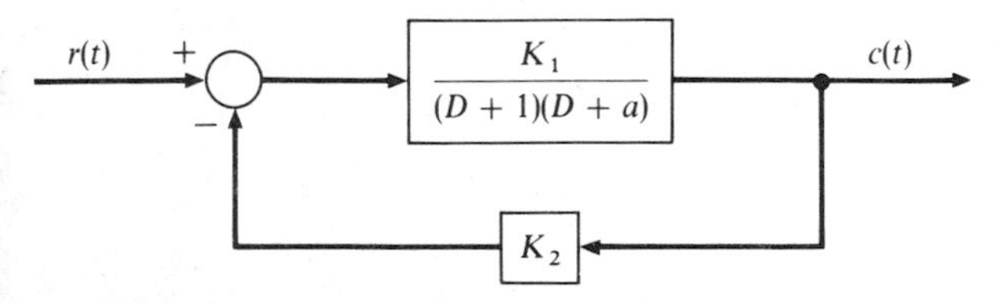

FIGURE P6.27

6.28. An aircraft attitude control system is shown in Fig. P6.28. Determine the response $c(t)$ for each of the following cases.

(*a*) $K = 1$, $r(t) = 2e^{-2t}$, and $c(0) = \dot{c}(0) = 1$

(*b*) $K = 10$, $r(t) = 0$, and $c(0) = \dot{c}(0) = 1$

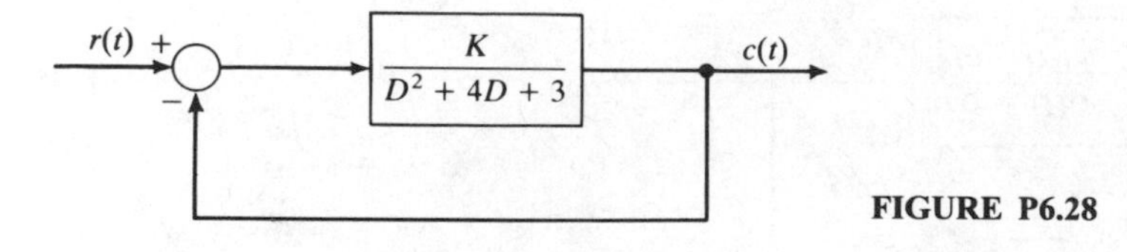

FIGURE P6.28

6.29. For the integral control system shown in Fig. P6.29, determine the value of K and a such that the characteristic equation has

(a) A real root at -1 and a real root at -5

(b) Repeated roots at -3

(c) Complex conjugate roots at $-3 \pm j4$

For each case determine the response $c(t)$ when $r(t) = 0$, $c(0) = 4$, and $\dot{c}(0) = 0$. What is the damping ratio for each case?

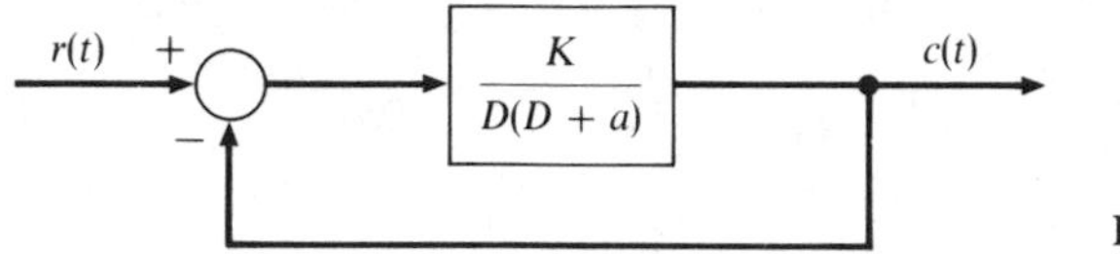

FIGURE P6.29

6.30. A system for controlling the temperature of an electric furnace is shown in Fig. P6.30. Determine the response $c(t)$ for each of the following cases.

(a) $K = 1$, $r(t) = 2e^{-3t}$, $c(0) = 1$, and $\dot{c}(0) = 0$

(b) $K = 5$, $r(t) = u(t)$, $c(0) = 0$, and $\dot{c}(0) = 1$

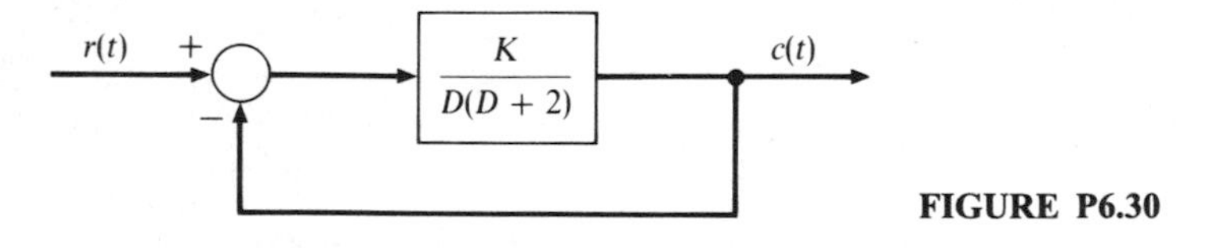

FIGURE P6.30

6.31. A control system with an external disturbance is shown in Fig. P6.31. Determine the response $c(t)$ when

(a) $K = 5$, $r(t) = d(t) = 0$, $c(0) = 4$, and $\dot{c}(0) = 0$

(b) $K = 9$, $r(t) = u(t)$, $d(t) = 0$, and $c(0) = \dot{c}(0) = 0$

(c) $K = 25$, $r(t) = 0$, $d(t) = \delta(t)$, $c(0) = 0$, and $\dot{c}(0) = 6$

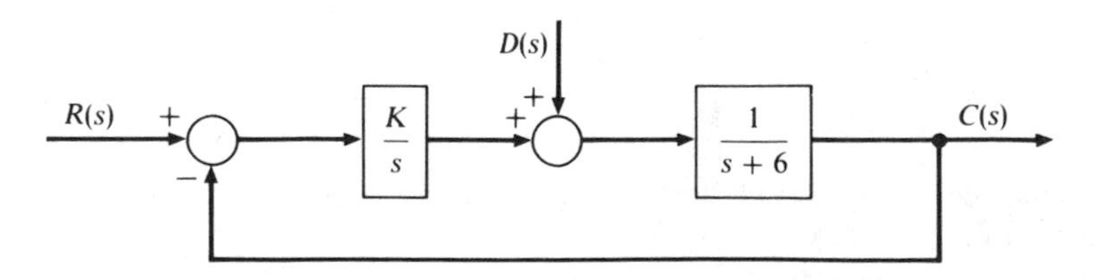

FIGURE P6.31

6.32. A system for controlling the air-fuel ratio of an automobile engine is shown in Fig. P6.32. Determine the response $c(t)$ when $r(t) = u(t)$, $c(0) = 0$, and $\dot{c}(0) = 1$ for each of the following cases.

(a) $K = 4$
(b) $K = 5$

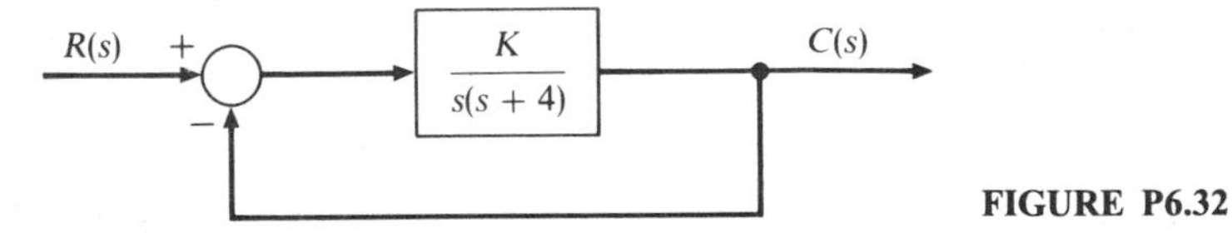

FIGURE P6.32

6.33. For the control system shown in Fig. P6.33, the forcing function is $r(t) = e^{-t}$, and all the initial conditions are zero. Determine the response $c(t)$ for each of the following cases.

(a) $K = 1$
(b) $K = 2$

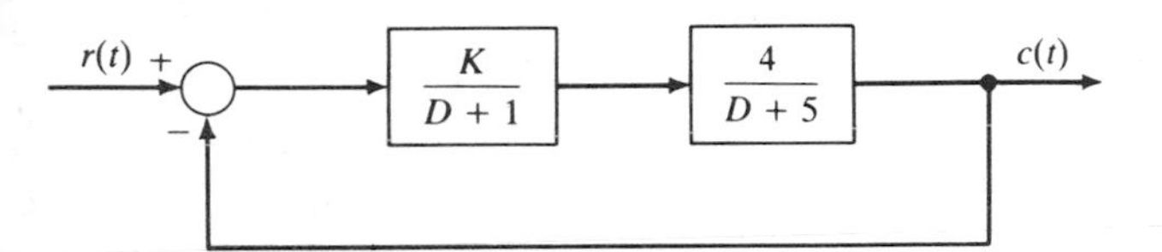

FIGURE P6.33

6.34. A hydraulic system which provides the muscle for a numerically controlled machine tool is shown in Fig. P6.34. The forcing function is $r(t) = u(t)$, and all the initial conditions are zero. Determine the response $c(t)$ when

(a) $K = 5$
(b) $K = 9$
(c) $K = 25$

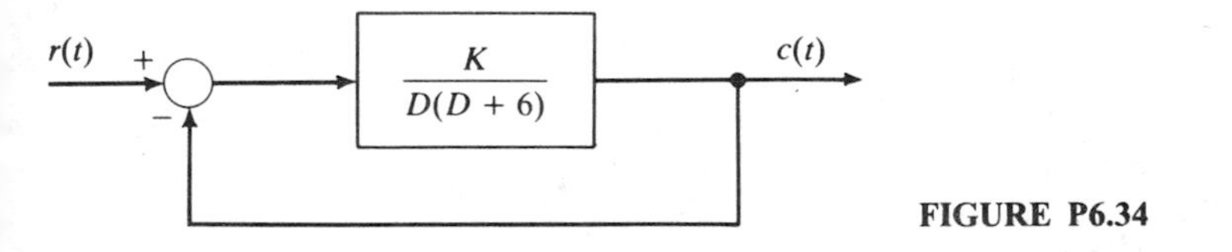

FIGURE P6.34

6.35. A flight control system provides stabilizing action on the aircraft. For the flight control system shown in Fig. P6.35, determine the response $c(t)$ for each of the following cases.

(a) $r(t) = 4u(t)$, $d(t) = 0$, and $c(0) = \dot{c}(0) = 0$
(b) $r(t) = 0$, $d(t) = 4u(t)$, and $c(0) = \dot{c}(0) = 0$
(c) $r(t) = d(t) = 0$, $c(0) = 4$, and $\dot{c}(0) = 0$

What are the damping ratio, natural frequency, and damped natural frequency for this system?

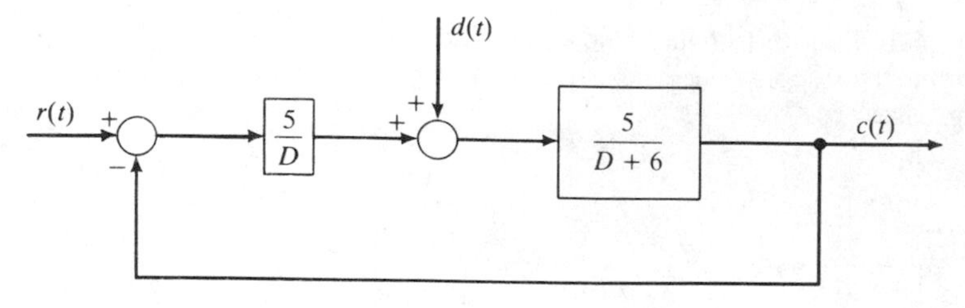

FIGURE P6.35

6.36. A dc motor position control system is shown in Fig. P6.36. Determine the response $c(t)$ for each of the following cases.
(*a*) $r(t) = 5u(t)$, $d(t) = 0$, and $c(0) = \dot{c}(0) = 0$
(*b*) $r(t) = 0$, $d(t) = 5u(t)$, and $c(0) = \dot{c}(0) = 0$
(*c*) $r(t) = d(t) = 0$, $c(0) = 1$, and $\dot{c}(0) = 0$

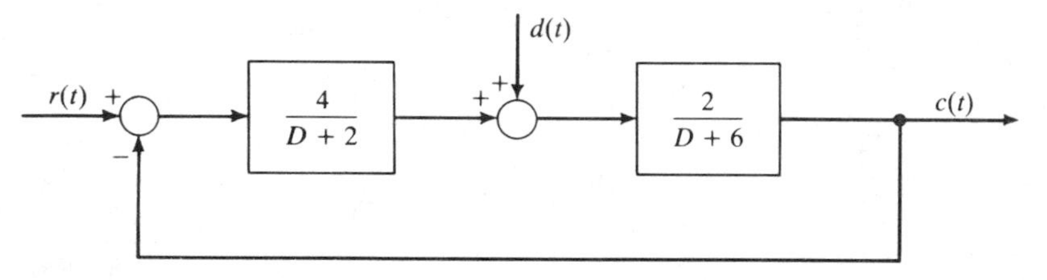

FIGURE P6.36

6.37. For the system shown in Fig. 5.11, $K_1 = 1.0$, $K_2 = 0.75$, $\tau = 0.25$, and all initial conditions are zero. Determine the response for each of the following cases.
(*a*) n_{in} is a unit impulse, and $u = 0$
(*b*) u is a unit impulse, and $n_{in} = 0$

6.38. Repeat Prob. 6.37 except that $K_2 = 1.25$.

6.39. Determine the damping ratio, undamped natural frequency, and damped natural frequency for the speed control system shown in Fig. P6.39. What is the response $c(t)$ of this system to a unit step function excitation $r(t) = u(t)$ when all initial conditions are zero?

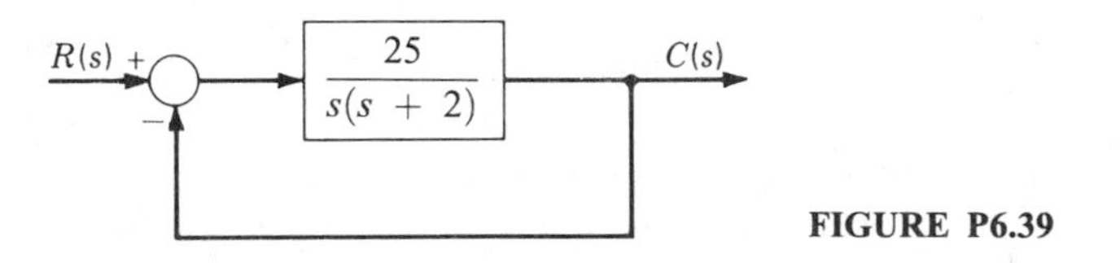

FIGURE P6.39

6.40. To improve the transient behavior of the speed control system of Fig. P6.39, a controller with proportional and derivative action is added, as shown in Fig. P6.40. Determine the value of K such that the resulting system will have a damping ratio of 0.5. What is the response $c(t)$ of this resulting system to a unit step function excitation $r(t) = u(t)$ when all initial conditions are zero?

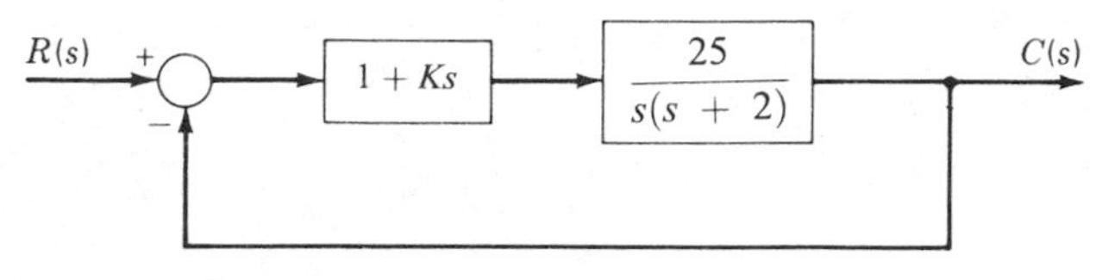

FIGURE P6.40

6.41. To improve the dynamic performance of the speed control system of Fig. P6.39, derivative action (Ks) is added to the feedback path, as shown in Fig. P6.41. Determine the value of K such that the resulting system will have a damping ratio of 0.5. What is the response $c(t)$ of this resulting system to a unit step function excitation $r(t) = u(t)$ when all initial conditions are zero?

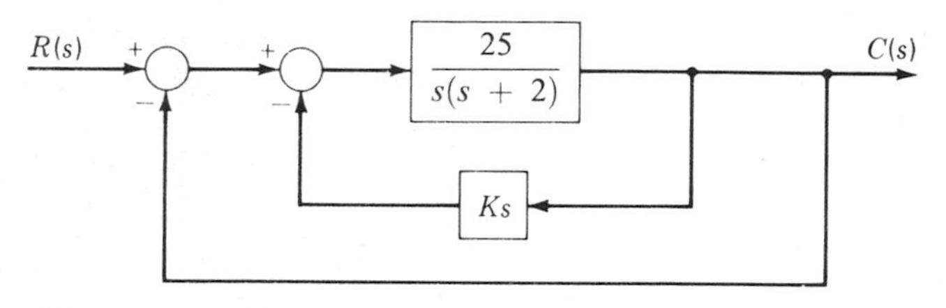

FIGURE P6.41

6.42. For the speed control system shown in Fig. P6.39, determine the rise time t_r, the peak time t_p, the 2 percent settling time t_s, and the percentage of overshoot.

6.43. Change the term $s + 2$ in Fig. P6.39 to $s + 10$, and then determine the rise time t_r, the peak time t_p, the 2 percent settling time t_s, and the percentage of overshoot.

6.44. Write a computer program to determine the response $c(t)$ for each of the systems shown in Fig. P6.44. The forcing function is $r(t) = u(t)$, and all initial conditions are zero. Use

(*a*) Trapezoidal integration

(*b*) Fourth-order Runge-Kutta integration

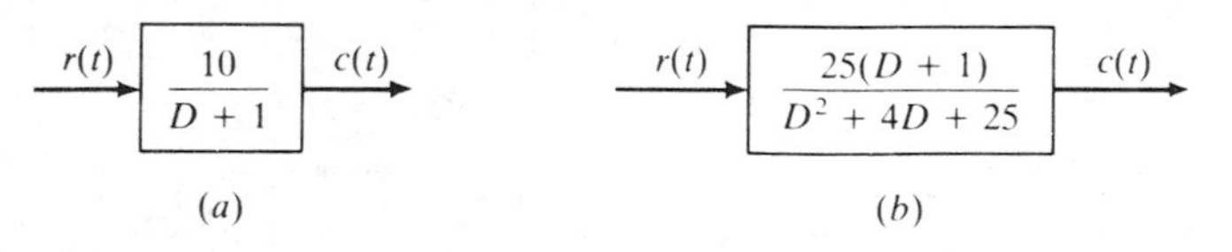

FIGURE P6.44

6.45. Write a computer program to evaluate the integral of e^t. Use

(*a*) Trapezoidal integration

(*b*) Fourth-order Runge-Kutta integration

Let t vary from 0 to 10 in increments of $\Delta t = 0.1$. Print out corresponding values of t, e^t, and the integral of e^t when $t = 0, 1, 2, \ldots, 10$.

6.46. Write a computer program for determining the response $c(t)$ for the feedback control system shown in Fig. P6.46. Incorporate into this program the differential equation relating $c(t)$ to $e(t)$. Show where the number 2 in the $D + 2$ term and the number 5 in

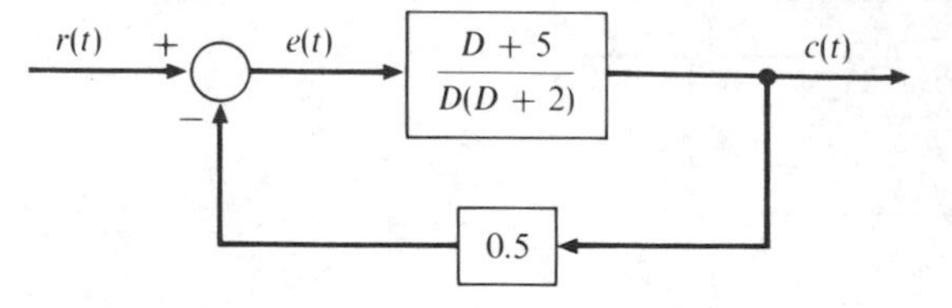

FIGURE P6.46

the $D + 5$ term appear in your program. The input is $r(t) = u(t)$, and all initial conditions are zero. Use

(*a*) Trapezoidal integration

(*b*) Fourth-order Runge-Kutta integration

6.47. Write a computer program for determining the response $c(t)$ for the control system shown in Fig. P6.47. Incorporate the differential equation relating $c(t)$ to $e(t)$ into this program. The input is $r(t) = u(t)$, and all initial conditions are zero. Use

(*a*) Trapezoidal integration

(*b*) Fourth-order Runge-Kutta integration

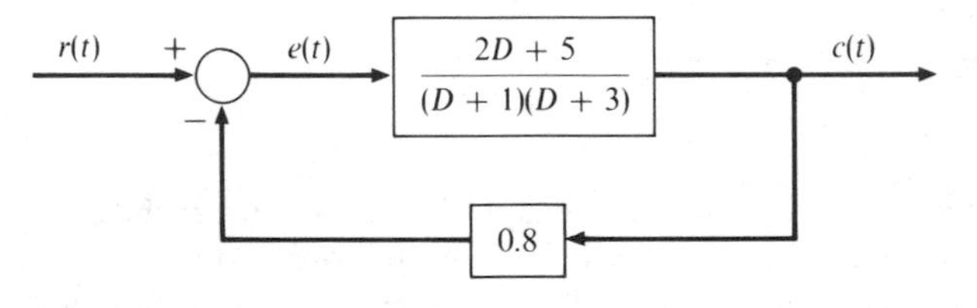

FIGURE P6.47

6.48. Write a computer program to determine the response $c(t)$ for each of the systems shown in Fig. P6.48. The forcing function is $r(t) = u(t)$, and all initial conditions are zero. Use

(*a*) Trapezoidal integration

(*b*) Fourth-order Runge-Kutta integration

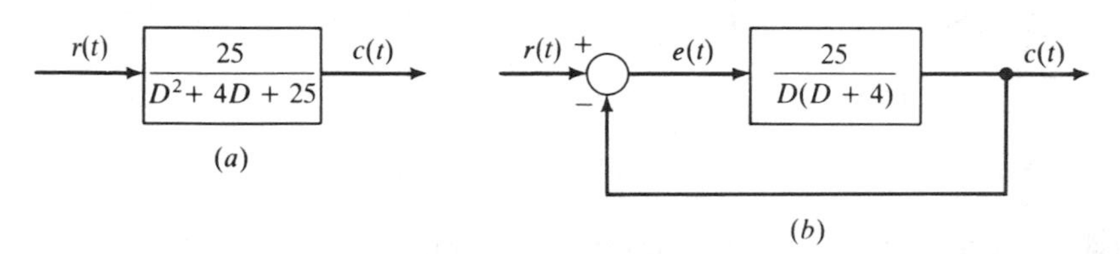

FIGURE P6.48

6.49. Write a computer program to determine the response $c(t)$ for each of the systems shown in Fig. P6.49. The forcing function is $r(t) = u(t)$, and all initial conditions are zero. Use

(*a*) Trapezoidal integration

(*b*) Fourth-order Runge-Kutta integration

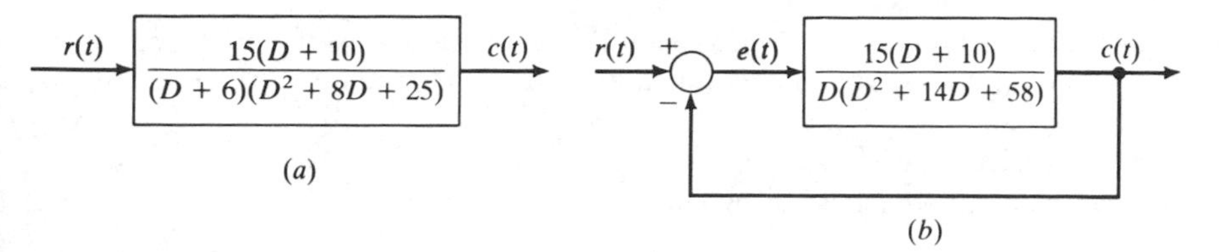

FIGURE P6.49

6.50. Write a computer program for determining the response $c(t)$ for the control system with a disturbance signal shown in Fig. P6.50. Incorporate into this program the differential equation relating $c(t)$ to $b(t)$ and the differential equation relating $m(t)$ to $e(t)$. The input is $r(t) = 5u(t)$, the disturbance is $d(t) = u(t)$ and all initial conditions are zero. Use

(*a*) Trapezoidal integration
(*b*) Fourth-order Runge-Kutta integration

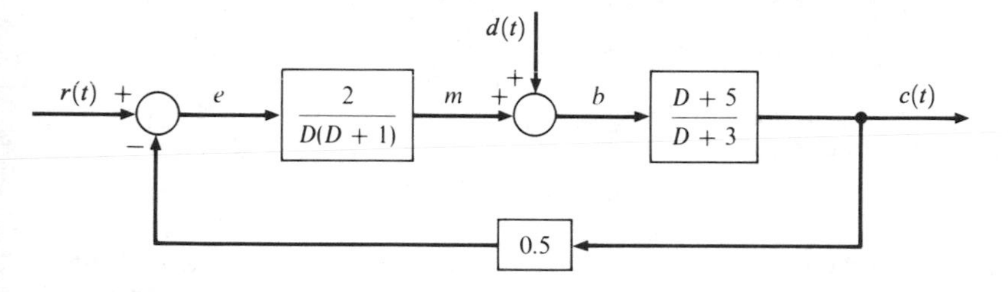

FIGURE P6.50

6.51. Write a computer program to determine the response of the integral control system with a disturbance signal shown in Fig. P6.51. Determine the response when $r(t) = d(t) = u(t)$ and all initial conditions are zero. To check that superposition holds, determine the response to $r(t)$ and $d(t)$ separately, and then add the results. Use

(*a*) Trapezoidal integration
(*b*) Fourth-order Runge-Kutta integration

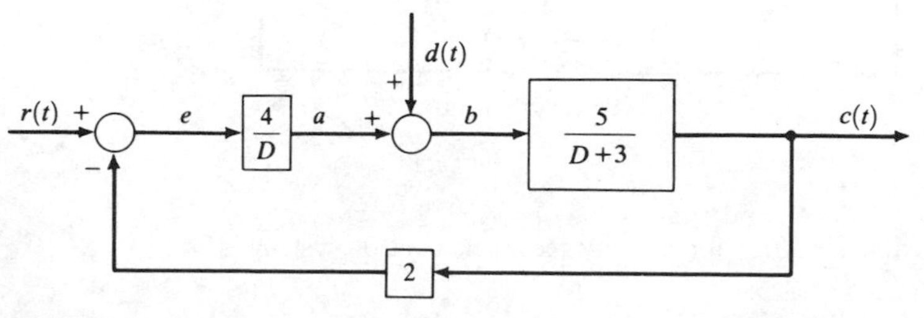

FIGURE P6.51

6.52. Write a computer program to determine the response $c(t)$ for the multiloop system shown in Fig. P6.52. The forcing function is $r(t) = u(t)$, and all initial conditions are zero. Use

(a) Trapezoidal integration

(b) Fourth-order Runge-Kutta integration

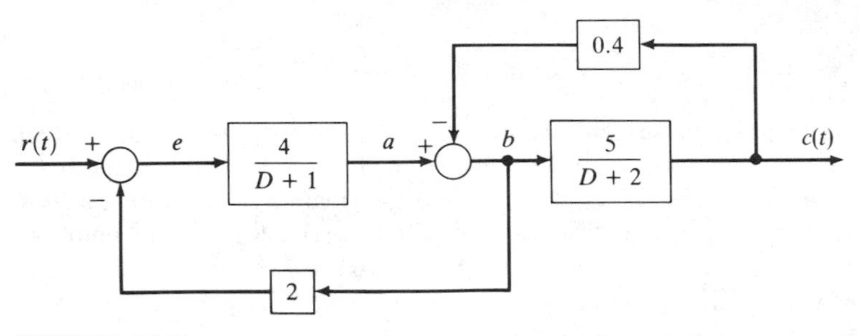

FIGURE P6.52

6.53. The characteristic function for a control system is known to be

$$L_n(D) = (D+5)(D^2+4D+13)$$

Determine the general form of the equation for the response $y(t)$ of this system when the input excitation $f(t)$ is

(a) An impulse

(b) A step function

Is the general form of these response expressions affected by the initial conditions?

6.54. Determine the characteristic function for the control system shown in Fig. P6.54. Use block-diagram algebra to move constants A and B into the main loop. Does this affect the characteristic function?

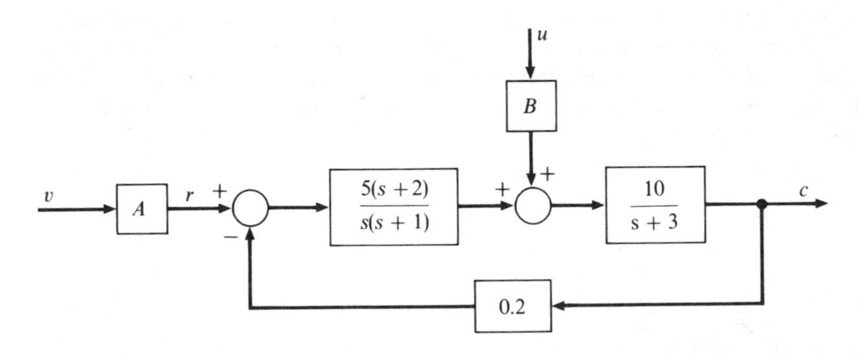

FIGURE P6.54

6.55. The transfer function $G(s)$ for a unity-feedback control system is

$$G(s) = \frac{K}{s(s^2+12s+45)}$$

When $K = 50$, the roots of the characteristic equation are -5, -5, and -2. Thus,

$$s(s^2 + 12s + 45) + 50 = (s + 5)^2(s + 2)$$

Determine the response $c(t)$ of this system for the case in which $r(t) = 0$, $c(0) = 0$, $\dot{c}(0) = 0$, and $\ddot{c}(0) = 45$.

6.56. Repeat Prob. 6.55 except $K = 104$. For this case the roots of the characteristic equation are -8, $-2 + j3$, and $-2 - j3$. Thus,

$$s(s^2 + 12s + 45) + 104 = (s + 8)(s^2 + 4s + 13)$$

6.57. (*a*) Apply Routh's criterion to determine an equation for K in terms of a, b, and c such that the cubic equation given below will have no roots with positive real parts.

$$as^3 + bs^2 + cs + K = 0$$

(*b*) Repeat (*a*) except use the equation

$$as^3 + bs^2 + Ks + c = 0$$

6.58. The characteristic equation for a feedback control system is

$$(s + 2)(s^2 + 4s + 8) + K = 0$$

Use Routh's criterion to determine the range of values of K such that the system is stable.

6.59. The characteristic equation for a control system for controlling the pressure in a chamber is

$$s(s^2 + 4s + 10) + K(s + 5) = 0$$

Use Routh's criterion to determine the range of values of K such that the system is stable.

6.60. The characteristic equation for a control system to control the output torque of an engine is

$$s(s^2 + 8s + a) + 4(s + 8) = 0$$

Use Routh's criterion to obtain the range of values of a for which the system is stable.

6.61. For each of the characteristic functions given below:

$$s^3 + 2s^2 + 5s + 24$$

$$s^4 + 2s^3 + 6s^2 + 2s + 5$$

$$s^4 + 3s^3 + 4s^2 + 6s$$

(*a*) Determine the number of zeros that lie on or to the right of the imaginary axis.
(*b*) Determine the number of zeros that have a real part greater than or equal to -4.

6.62. The characteristic equation for a certain system is

$$s(s^3 + 4s^2 + 2s + 3) + K(s + 1) = 0$$

Using Routh's criterion, determine the range of values of K for which the system is stable.

6.63. Apply Routh's criterion to determine the number of roots that lie in the right half-plane for each of the following characteristic equations.
(*a*) $s^3 + 5s^2 + 6s = 0$
(*b*) $s^3 + s^2 - s - 1 = 0$
(*c*) $s^3 + 2s^2 + 4s + 8 = 0$
(*d*) $s^4 + 5s^3 + 6s^2 = 0$
(*e*) $s^4 + 5s^3 + 5s^2 - 5s - 6 = 0$
(*f*) $s^4 + 5s^3 + 7s^2 + 5s + 6 = 0$
(*g*) $s^5 + s^4 + 5s^3 + 5s^2 + 4s + 4 = 0$
(*h*) $s^5 + 4s^4 + 6s^3 + 24s^2 + 25s + 100 = 0$
Identify the roots that are symmetrically located about the origin.

6.64. Determine the range of values of K such that the system whose characteristic equation is given below is stable:

$$s(s^2 + 2s + 5) + K(s + 4) = 0$$

Determine the factored form of the characteristic equation when K is such that the characteristic equation has a pair of roots on the imaginary axis.

6.65. For the system shown in Fig. P6.65, determine the range of values of K such that the system is stable. Determine the factored form of the characteristic equation when K is such that the characteristic equation has a pair of roots on the imaginary axis.

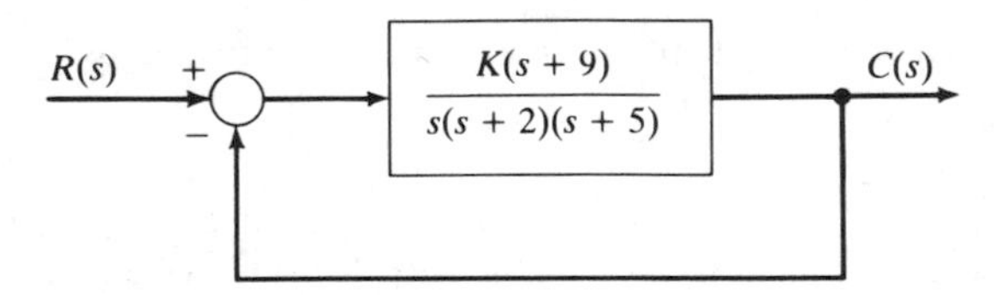

FIGURE P6.65

6.66. For the system shown in Fig. P6.66, determine the range of values of a such that the system is stable. Determine the factored form of the characteristic equation when a is such that the characteristic equation has a pair of roots on the imaginary axis.

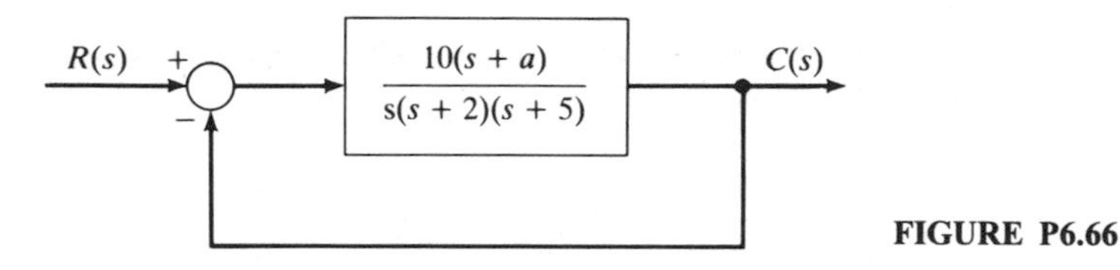

FIGURE P6.66

6.67. The characteristic equation for a system is

$$s(s^2 + 6s + 13) + K = 0$$

(*a*) Determine the value of K such that the characteristic equation has a pair of complex conjugate roots on the imaginary axis. Write the factored form of the characteristic equation for this case.
(*b*) Determine the value of K such that the characteristic equation has a pair of complex conjugate roots whose real part is -1. Write the factored form of the characteristic equation for this case.

6.68. For the system shown in Fig. P6.68, determine the range of values of K such that the system is stable. Write the factored form of the characteristic equation for the case in which the characteristic equation has a pair of complex conjugate roots on the imaginary axis.

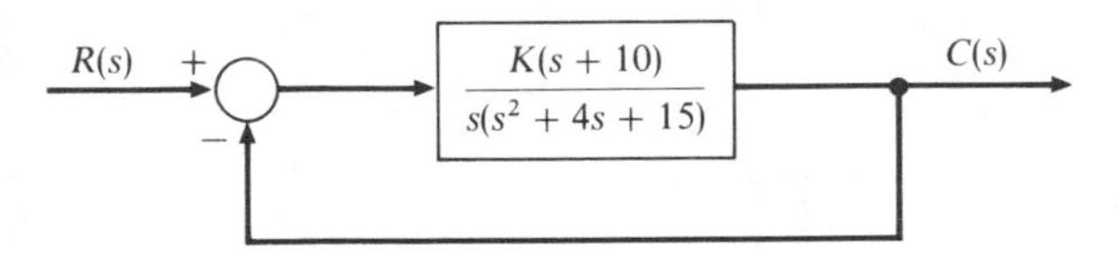

FIGURE P6.68

6.69. For the system shown in Fig. P6.69, determine the range of values of K such that the system is stable. Write the factored form of the characteristic equation for the case in which the characteristic equation has a pair of complex conjugate roots on the imaginary axis.

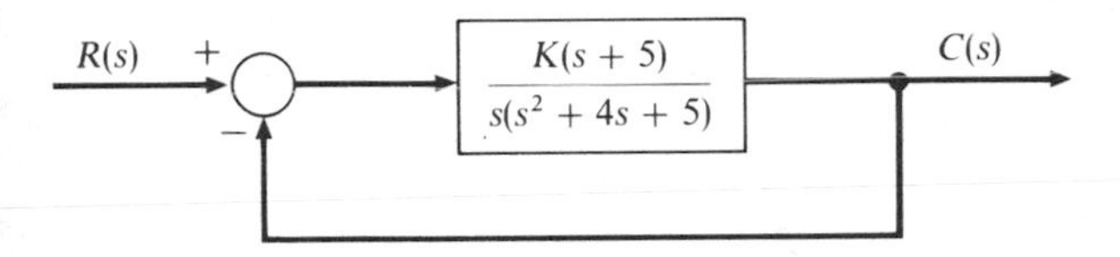

FIGURE P6.69

6.70. The characteristic equation for a feedback control system is

$$s(s^2 + 8s + 20) + K = 0$$

(*a*) Determine the value of the gain K such that the characteristic equation has a pair of roots on the imaginary axis. Write the factored form of the characteristic equation for this case.

(*b*) Determine the value of the gain K such that the characteristic equation has a pair of roots on the vertical axis which passes through -1. Write the factored form of the characteristic equation for this case.

6.71. For the system shown in Fig. P6.71, determine the range of values of K such that the system is stable.

(*a*) Write the factored form of the characteristic equation when K is such that the characteristic equation has a pair of roots on the imaginary axis.

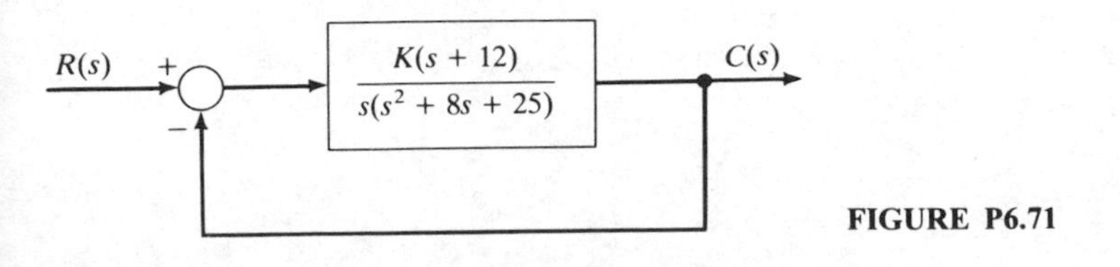

FIGURE P6.71

(*b*) Determine the value of K such that the characteristic equation has a pair of complex conjugate roots whose real part is equal to -1. What is the factored form of the characteristic equation for this case?

6.72. For the system shown in Fig. P6.72, determine the range of values of a such that the system is stable. Determine the factored form of the characteristic equation when a is such that the characteristic equation has a pair of roots on the imaginary axis.

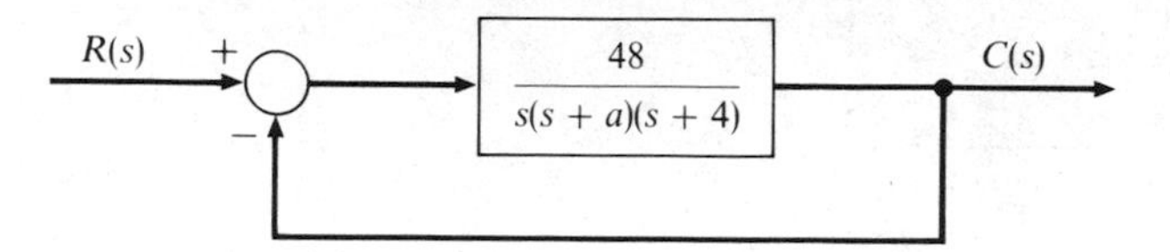

FIGURE P6.72

CHAPTER

7

THE ROOT-LOCUS METHOD

The root-locus method was developed by W. R. Evans.* This method enables one to determine the roots of the characteristic equation (i.e., the zeros of the characteristic function) by knowing the factored form of the feedforward and feedback elements of a control system. As discussed in Chap. 6, the transient behavior of a system is governed primarily by the roots of the characteristic equation for the system. Neither the initial conditions nor the particular excitation affects the basic operation of a system.

7.1 SIGNIFICANCE OF ROOT LOCI

The characteristic equation for the control system shown in Fig. 7.1*a* is

$$s(s+4)+K=s^2+4s+K=0 \tag{7.1}$$

The roots of the characteristic equation depend upon the value of K, which is the static loop sensitivity. As shown in Fig. 7.1*a*, the static loop sensitivity K is the product of all the constant terms in the control loop when the coefficient of each s term is unity.

---

* W. R. Evans, "Graphical Analysis of Control Systems," *Trans. AIEE*, 67: 547–551 (1948); *Control-System Dynamics*, McGraw-Hill, New York, 1954.

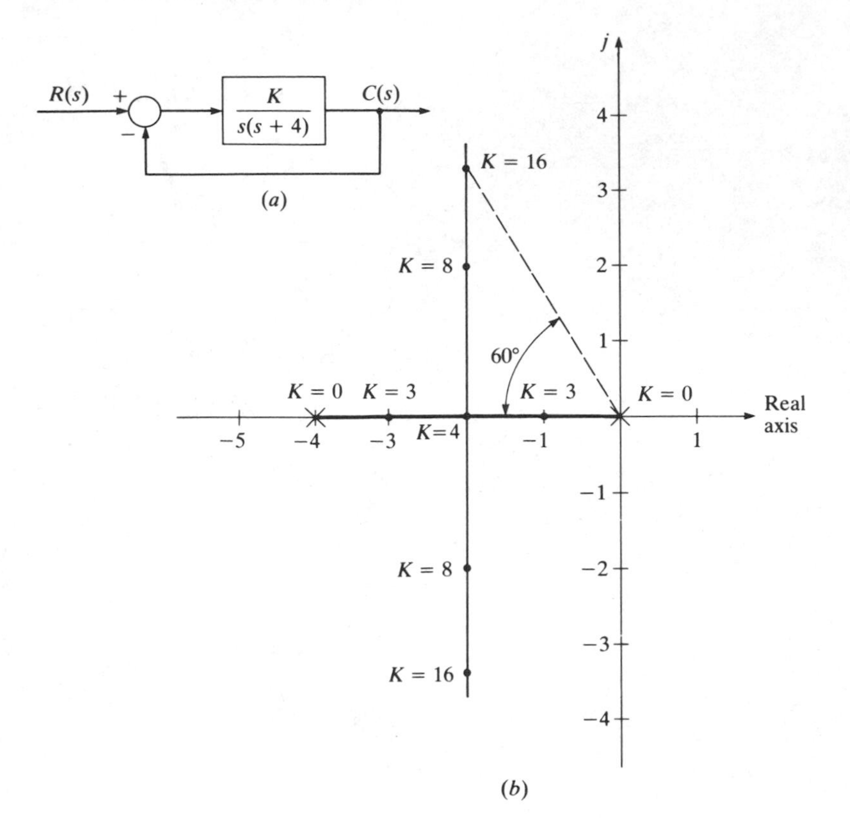

FIGURE 7.1
Root-locus plot for $s(s + 4) + K = 0$.

Because Eq. (7.1) is a quadratic, the roots are $r_{1,2} = a \pm jb$, in which $-2a = 4$, or $a = -2$, and $a^2 + b^2 = K$, or $b = \sqrt{K - 4}$. Thus

$$\begin{aligned} &\text{For } K > 4 \qquad r_{1,2} = -2 \pm j\sqrt{K - 4} \\ &\text{For } K = 4 \qquad r_1 \;= r_2 = -2 \\ &\text{For } K < 4 \qquad r_{1,2} = -2 \pm \sqrt{4 - K} \end{aligned} \tag{7.2}$$

In this last case, the roots are real and unequal. The heavy lines in Fig. 7.1*b* are a plot of the roots of this characteristic equation for various values of K. When $K = 0$, the roots are $r_{1,2} = 0, -4$; when $K = 3$, the roots are -1 and -3; when $K = 4$, the roots are $r_1 = r_2 = -2$; when $K = 8$, the roots are $r_{1,2} = -2 \pm j2$; etc.

Such a plot of the roots of the characteristic equation for each value of K as K varies from zero to infinity is called a *root-locus plot*. From such a root-locus plot, it is an easy matter to select the value of K to yield the desired roots of the characteristic equation. For example, suppose we want to have a damping ratio $\zeta = 0.5$. As discussed in the preceding chapter, $\beta = \cos^{-1}\zeta = \cos^{-1} 0.5 = 60°$. As is shown in Fig. 7.1*b*, the line inclined at angle $\beta = 60°$ intersects the root-locus plot at a value of $K = 16$. From this plot, the corresponding roots are

$$r_{1,2} = -2 \pm j\sqrt{12} = -2 \pm j2\sqrt{3}$$

Thus, the factored form of the characteristic equation is

$$(s - r_1)(s - r_2) = [s - (-2 + j2\sqrt{3})][s - (-2 - j2\sqrt{3})] = s^2 + 4s + 16 \quad (7.3)$$

It follows that $\omega_n^2 = 16$, or $\omega_n = 4$, and $2\zeta\omega_n = 4$, or $\zeta = 4/(2\omega_n) = \frac{1}{2}$.

Illustrative example 7.1. The space station Freedom will be the first permanently occupied research facility in space. As depicted in Fig. 7.2, the space shuttle will dock with Freedom. For an operating space station in orbit, it is necessary to maintain the orientation of the station relative to the sun for generating power. As shown, solar power arrays and utility systems are mounted on the station's truss, along with collapsible radiators which store solar energy until it is needed. The block-diagram representation for a proposed orientation control system is shown in Fig. 7.1*a* in which $c(t)$ is the actual orientation and $r(t)$ is the desired orientation. Determine the response $c(t)$ when $r(t) = 0$ and $c(0) = \dot{c}(0) = 1$ for each of the following cases.

(*a*) $K = 3$
(*b*) $K = 4$
(*c*) $K = 8$

FIGURE 7.2
Space Station *Freedom*—Gateway to the Future. (Space Art by Harold Smelcer; National Aeronautics and Space Administration.)

Solution. From Fig. 7.1a, it follows that the Laplace transform for $C(s)$ is

$$C(s) = \frac{K}{s(s+4)+K}R(s) \tag{7.4}$$

This is the transformed equation when all initial conditions are zero. Replacing s by D, $C(s)$ by $c(t)$, and $R(s)$ by $r(t)$ yields the differential equation for the system

$$c(t) = \frac{K}{D(D+4)+K}r(t)$$

or

$$(D^2 + 4D + K)c(t) = Kr(t)$$

For $r(t) = 0$, the Laplace transform is

$$s^2C(s) - sc(0) - \dot{c}(0) + 4[sC(s) - c(0)] + KC(s) = 0$$

Hence

$$(s^2 + 4s + K)C(s) = (s+4)c(0) + \dot{c}(0)$$

For $c(0) = \dot{c}(0) = 1$, this becomes

$$(s^2 + 4s + K)C(s) = s + 5$$

(a) For $K = 3$, the transformed equation for $C(s)$ is

$$C(s) = \frac{s+5}{s^2+4s+3} = \frac{s+5}{(s+1)(s+3)} = \frac{2}{s+1} - \frac{1}{s+3}$$

Inverting yields for the response

$$c(t) = 2e^{-t} - e^{-3t} \qquad t > 0$$

For $K = 3$, the factored form of the characteristic equation is

$$s^2 + 4s + 3 = (s+1)(s+3) = 0$$

Note from the root-locus plot shown in Fig. 7.1b that when $K = 3$, one root is located at -1 and the other root is located at -3.

(b) For $K = 4$, the transform for $C(s)$ is

$$C(s) = \frac{s+5}{s^2+4s+4} = \frac{s+5}{(s+2)^2} = \frac{3}{(s+2)^2} + \frac{1}{s+2}$$

Inverting yields for the response

$$c(t) = (1 + 3t)e^{-2t} \qquad t > 0$$

For $K = 4$, the characteristic equation has repeated roots at -2. That is,

$$s^2 + 4s + 4 = (s+2)^2$$

Note from the root-locus plot that when $K = 4$, the characteristic equation has a repeated root at -2.

(c) For $K = 8$, the transform for $C(s)$ is

$$C(s) = \frac{s+5}{s^2+4s+8}$$

This characteristic equation has complex conjugate roots located at $-2 \pm j2$. Evaluating $K(a + jb)$ yields

$$K(a+jb) = (s+5)|_{s=-2+j2} = 3 + j2 = \sqrt{13}\underline{/33.7^\circ}$$

Hence

$$c(t) = \frac{|K(a+jb)|}{b} e^{at} \sin(bt + \alpha) = \frac{\sqrt{13}}{2} e^{-2t} \sin(2t + 33.7^\circ)$$

When $K(a + jb)$ is written in the form $A + jB$, then the result may be expressed in the equivalent form

$$c(t) = \frac{e^{at}}{b}(A \sin bt + B \cos bt) = \frac{e^{-2t}}{2}(3 \sin 2t + 2 \cos 2t)$$

where A is the real part of $K(a + jb)$ and B is the imaginary part. Note from the root-locus plot that when $K = 8$, the characteristic equation has complex conjugate roots at $-2 + j2$ and $-2 - j2$.

Note from Fig. 7.1*b* that for $0 < K < 4$ the roots of the characteristic equation are real. For example, when $K = 3$, the roots are located at -1 and -3. For real roots the system response is the sum of the exponential terms associated with each root. For $K > 4$, the roots of the characteristic equation are complex conjugates. For the characteristic equation $s^2 + 4s + K$, it follows that $\omega_n = \sqrt{K}$ and $2\zeta\omega_n = 4$, or $\zeta = 2/\sqrt{K}$. The larger the value of K, the smaller is the damping ratio and the more oscillatory is the response. Many systems are such that they become unstable for large values of K. The larger the value of K, the better the steady-state accuracy. Thus, K must be sufficiently large to ensure good steady-state accuracy but not so large that the system will be too oscillatory or, worse yet, unstable.

Figure 7.3 shows a general feedback control system. The transfer function $G(s)$ for the feedforward elements may be written in the form $G(s) = N_G/D_G$, where N_G is the numerator of $G(s)$ and D_G is the denominator. Similarly, the transfer function for the feedback elements $H(s)$ may be written in the form $H(s) = N_H/D_H$. The characteristic equation is

$$D_G D_H + N_G N_H = 0$$

Dividing through by $D_G D_H$ yields

$$1 + G(s)H(s) = 1 + \frac{N_G N_H}{D_G D_H} = 1 + \frac{K(s-z_1)(s-z_2)\cdots(s-z_m)}{(s-p_1)(s-p_2)\cdots(s-p_n)} = 0 \tag{7.5}$$

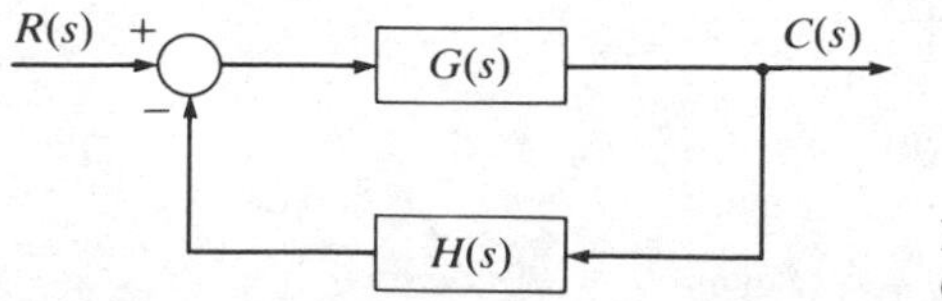

FIGURE 7.3
General feedback control system.

where $G(s)H(s) = N_GN_H/(D_GD_H)$ is the open-loop transfer function. The zeros of the open-loop transfer function $z_1, z_2, \ldots, z_m$ are the roots of $N_GN_H = 0$. The poles of the open-loop transfer function $p_1, p_2, \ldots, p_n$ are the roots of $D_GD_H = 0$.

The preceding equation may be written in the form

$$\frac{(s-p_1)(s-p_2)\cdots(s-p_n)}{(s-z_1)(s-z_2)\cdots(s-z_m)} = -K \tag{7.6}$$

In constructing root-locus plots, the roots $p_1, p_2, \ldots, p_n$ of $D_GD_H = 0$ are plotted as $\times$'s and the roots $z_1, z_2, \ldots, z_m$ of $N_GN_H = 0$ are plotted as circles, $\bigcirc$'s. In Fig. 7.4*a* is shown a typical $\times$ located at p_i. The vector from the origin to the $\times$ is p_i. The vector from the origin to any point s is s. The vector from the $\times$ at p_i to s is $s - p_i$. Similarly, in Fig. 7.4*b* is shown a typical circle located at z_i. The vector from the origin to the circle is z_i. The vector from z_i to s is $s - z_i$. The numerator terms in Eq. (7.6) are the vectors from all the $\times$'s to a trial point s. Similarly, the denominator terms in Eq. (7.6) are the vectors from all the $\bigcirc$'s to the trial point s. The magnitude of Eq. (7.6) is

$$\frac{|s-p_1||s-p_2|\cdots|s-p_n|}{|s-z_1||s-z_2|\cdots|s-z_m|} = \frac{\prod_{i=1}^{n}|s-p_i|}{\prod_{i=1}^{m}|s-z_i|} = |-K| \tag{7.7}$$

This is referred to as the *magnitude condition*. The magnitude condition is used to obtain the value of the gain K at any point on a root-locus plot. From Eq. (7.7), it follows that the magnitude of the gain K at any point on a root-locus plot is equal to the product of the distances from all the $\times$'s to the point divided by the product of distances from all the $\bigcirc$'s to the point. When there are no $\bigcirc$'s, the magnitude of the gain is simply the product of the distances from all the $\times$'s to the point. For example, at the point $s = -1$ in Fig. 7.1*b*, the distance from the $\times$ at the origin to the point at -1 is 1. The distance from the $\times$ at -4 to the point at -1 is 3. Because there are no $\bigcirc$'s, the gain is the product of the distances from all the $\times$'s to the point, which in this case is $K = (1)(3) = 3$. At the point $s = -2 + j2$, the distance from the $\times$ at the origin to the point is $\sqrt{(-2)^2 + 2^2} = \sqrt{8}$. The distance from the $\times$ at -4 to the

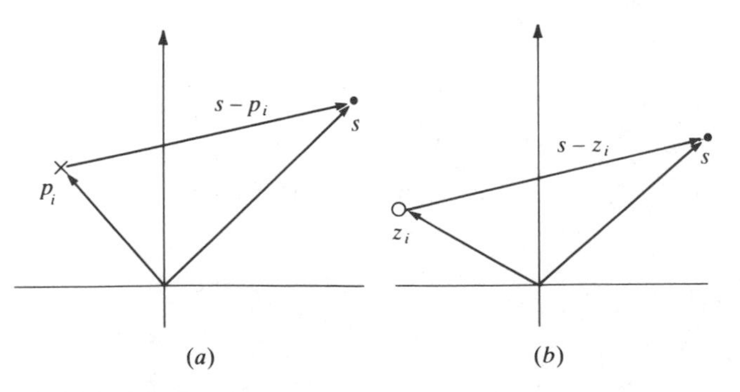

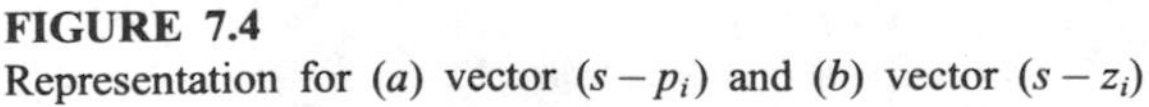

FIGURE 7.4
Representation for (*a*) vector $(s - p_i)$ and (*b*) vector $(s - z_i)$

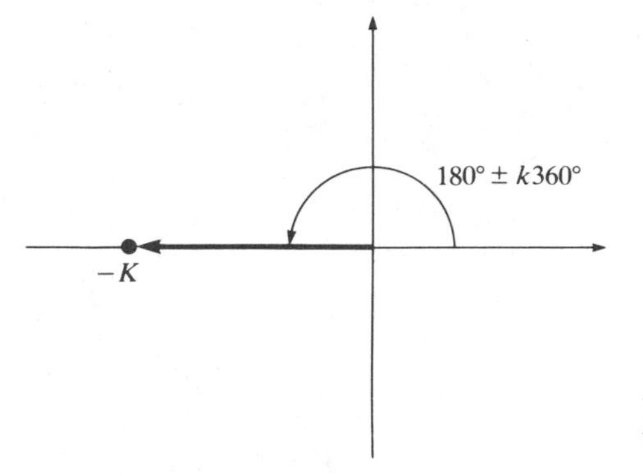

FIGURE 7.5
Vector representation for $-K$.

point $-2 + j2$ is $\sqrt{2^2 + 2^2} = \sqrt{8}$. The gain is the product of these distances, which is $K = \sqrt{8}\,\sqrt{8} = 8$.

Taking the angle of each vector represented in Eq. (7.6) shows that

$$[\measuredangle(s - p_1) + \measuredangle(s - p_2) + \cdots + \measuredangle(s - p_n)] - [\measuredangle(s - z_1) + \measuredangle(s - z_2) + \cdots + \measuredangle(s - z_m)] = \measuredangle -K$$

For positive values of K, the point $-K$ lies on the negative axis, as shown in Fig. 7.5. The vector $-K$ is the vector from the origin to the point $-K$ on the negative axis. The angle of this vector is 180°, 180° ± 360°, 180° ± 720°, etc. Thus, the preceding equation becomes

$$\sum_{i=1}^{n} \measuredangle(s - p_i) - \sum_{i=1}^{m} \measuredangle(s - z_i) = 180^\circ \pm k360^\circ \qquad k = 0,\ 1,\ 2,\ \ldots \tag{7.8}$$

This is called the *angle condition*. The left side is the sum of the angles from all the ×'s to a point s minus the sum of the angles from all the ○'s to point s. In order that a point s be a point on a root-locus plot, it is necessary that the sum of the angles from all the ×'s to the point minus the sum of the angles from all the ○'s to the point equal 180° ± k360°.

7.2 CONSTRUCTION OF LOCI

By determining certain points and asymptotes, the loci may be sketched in quite readily. From Eq. (7.6), it follows that when $s = p_i$, then $K = 0$. Thus, one locus begins or originates at each ×. The number of ×'s is equal to the order n of the characteristic equation. A second-order equation has two loci, a third-order equation has three loci, etc. It also follows from Eq. (7.6) that when $s = z_i$, then $K = \infty$. Thus, one locus ends or terminates at each of the m ○'s. Usually there are more ×'s than ○'s. The remaining $n - m$ loci terminate along asymptotes at infinity.

The first place to start investigating the location of loci is along the real axis. Figure 7.6a shows an × which is located on the real axis and a trial point s which

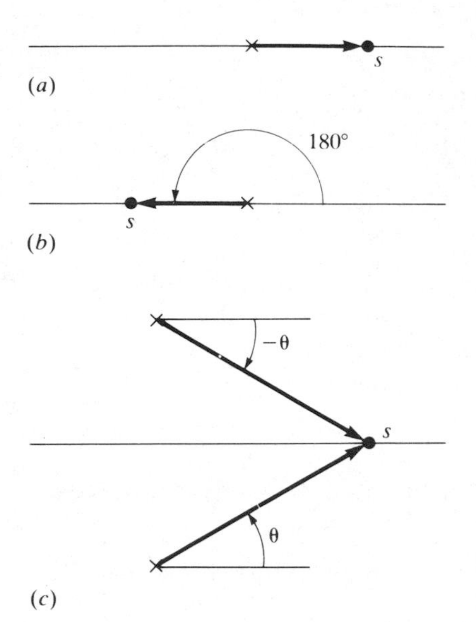

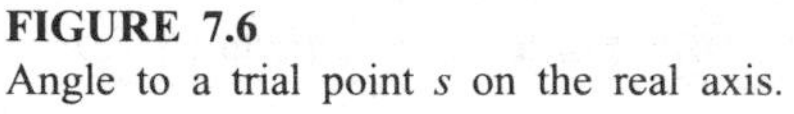
FIGURE 7.6
Angle to a trial point s on the real axis.

lies to the right of the $\times$. The angle of the vector from the $\times$ to the point s is 0°. Figure 7.6*b* is the same as Fig. 7.6*a*, except that point s is located to the left of the $\times$. The angle of the vector from the $\times$ to point s is now 180°. As the trial point s moves from the right side of an $\times$ or $\bigcirc$ to the left side, the angle of the vector changes from 0° to 180°. Figure 7.6*c* shows a pair of complex conjugate $\times$'s and a trial point s on the real axis. The angle from the lower $\times$ to the trial point is θ. The angle from the upper $\times$ to the trial point is $-\theta$. The sum $\theta + (-\theta)$ is zero. Thus, complex conjugate $\times$'s or $\bigcirc$'s do not affect the location of the loci on the real axis.

Figure 7.7 shows the $\times$'s and $\bigcirc$'s which are located on the real axis for a characteristic equation. To the far right of all the $\times$'s and $\bigcirc$'s, the angles from each $\times$ and from each $\bigcirc$ to the trial point are 0°. There is no locus in this region because the sum of the angles from all the $\times$'s to the trial point minus the sum of the angles from all the $\bigcirc$'s to the trial point is zero. In the region between the origin and -1, the angle of the vector from the $\times$ at the origin to the trial point has changed by 180°. All other angles are still zero. There is a locus in the region from the origin to -1 because the sum of the angles from all the $\times$'s to the trial point minus the sum of

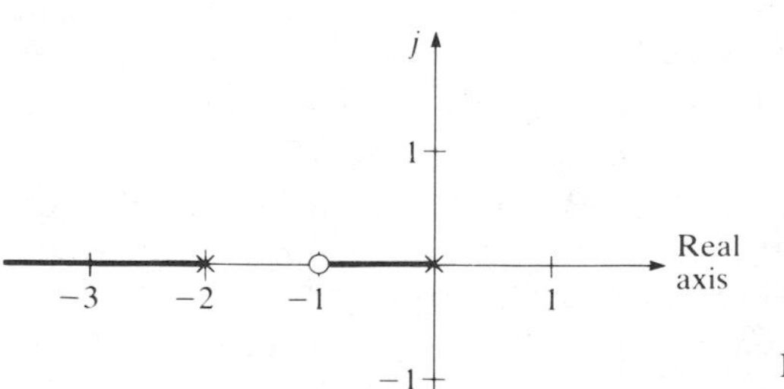

FIGURE 7.7
Determination of loci on the real axis.

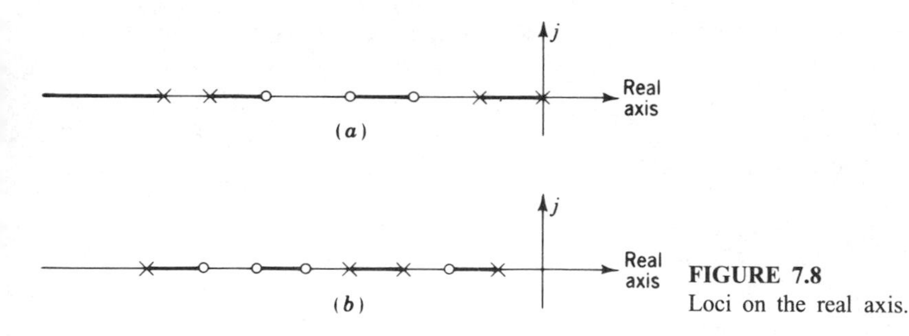

FIGURE 7.8
Loci on the real axis.

the angles from all the ○'s to the trial point is 180°. In the region between −1 and −2, the angle of the vector from the ○ at −1 to the trial point has changed by 180°. The angle condition is no longer satisfied, so there is no locus in this region. Note that every time a trial point moves from the right of an × or ○ to the left, the angle changes by 180°. As shown in Fig. 7.7, the location of the loci on the real axis alternates between ×'s and ○'s. As illustrated in Fig. 7.8, there is never a locus to the right of the first × or ○ on the real axis, but there is always a locus to the left of the first × or ○; there is never a locus to the left of the second × or ○, there is always a locus to the left of the third × or ○; never left of the fourth, always left of the fifth; and so on, alternating. Note in Fig. 7.1*b* that there is an × at the origin and an × at −4. The locus on the real axis lies between these ×'s.

The next step in the construction of the loci is to determine the asymptotes as s approaches infinity, i.e., the location of the loci for large values of s. Figure 7.9 shows an × located at point p_i. The point s is near infinity and cannot be seen on this plot. However, following the vector which is marked "to s" leads to the point s. In order that the vector $s - p_i$ meet point s at infinity, it is necessary that these two

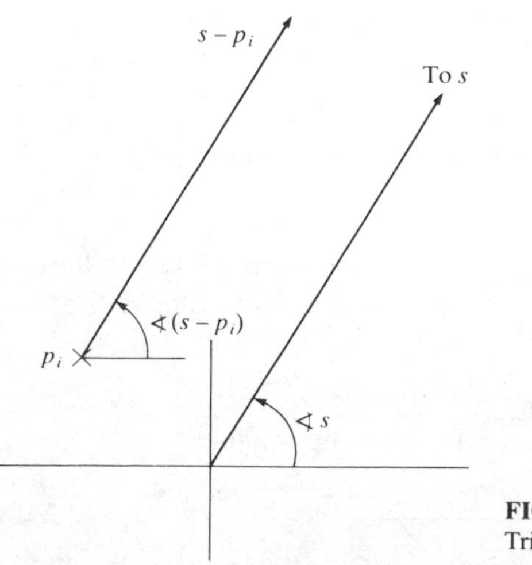

FIGURE 7.9
Trial point s located at infinity.

vectors be parallel. Thus, the angle of vector $s - p_i$ must be equal to the angle of vector s. That is,

$$\measuredangle (s - p_i) \approx \measuredangle s$$

Similarly, it follows that for very large values of s,

$$\measuredangle (s - z_i) \approx \measuredangle s$$

For very large values of s, the angle condition [Eq. (7.8)] becomes

$$\lim_{s \to \infty} \sum_{i=1}^{n} \measuredangle (s - p_i) - \lim_{s \to \infty} \sum_{i=1}^{m} \measuredangle (s - z_i) = \sum_{i=1}^{n} \measuredangle s - \sum_{i=1}^{m} \measuredangle s$$

$$= (n - m) \measuredangle s = 180° \pm k360°$$

Thus, the angles of the asymptotes are

$$\measuredangle s = \frac{180° \pm k360°}{n - m} \qquad k = 0,\ 1,\ 2,\ 3,\ \ldots \tag{7.9}$$

where n = highest power of s in $D_G D_H$ = no. of ×'s
m = highest power of s in $N_G N_H$ = no. of O's.

The number of distinct asymptotes is equal to $n - m$. Although it would appear from Eq. (7.9) that there are more asymptotes, the angles repeat for values of k after $n - m$ distinct angles have been determined. For $n - m = 1, 2, 3$, and 4, Eq. (7.9) becomes

$$\measuredangle s = 180° \pm k360° \qquad n - m = 1$$

$$\measuredangle s = \frac{180° \pm k360°}{2} = 90° \pm k180° \qquad n - m = 2$$

$$\measuredangle s = \frac{180° \pm k360°}{3} = 60° \pm k120° \qquad n - m = 3$$

$$\measuredangle s = \frac{180° \pm k360°}{4} = 45° \pm k90° \qquad n - m = 4$$

Figure 7.10 shows a plot of the asymptotes for the cases in which $n - m = 1$, 2, 3, and 4. For $n - m = 1$, the angle is 180°; for $n - m = 2$ the angles are ±90°; for $n - m = 3$ the angles are ±60° and 180°; and for $n - m = 4$ the angles are ±45° and ±135°.

To locate the asymptotes, it is necessary to know where they intersect the real axis. The point σ_c where the asymptotes cross the real axis is determined from the general equation.

$$\sigma_c = \frac{\Sigma \text{ roots of } D_G D_H - \Sigma \text{ roots of } N_G N_H}{(\text{no. roots of } D_G D_H) - (\text{no. roots of } N_G N_H)}$$

$$= \frac{(p_1 + p_2 + \cdots + p_n) - (z_1 + z_2 + \cdots + z_m)}{n - m} \tag{7.10}$$

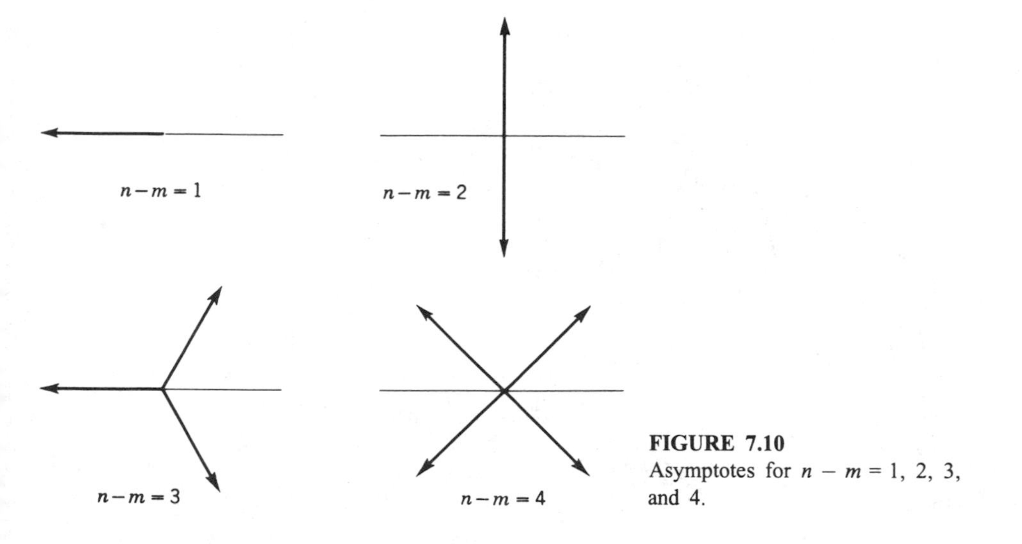

FIGURE 7.10
Asymptotes for $n - m = 1$, 2, 3, and 4.

To verify Eq. (7.10), first write the characteristic equation [Eq. (7.6)] in the form

$$(s-p_1)(s-p_2)\cdots(s-p_n)+K(s-z_1)(s-z_2)\cdots(s-z_m)=0$$

Solving for $-K$ gives

$$-K=\frac{(s-p_1)(s-p_2)\cdots(s-p_n)}{(s-z_1)(s-z_2)\cdots(s-z_m)}=\frac{s^n-(p_1+p_2+\cdots+p_n)s^{n-1}+\cdots}{s^m-(z_1+z_2+\cdots+z_m)s^{m-1}+\cdots} \tag{7.11}$$

Note that the coefficient of the second term in a polynomial is the negative of the sum of the roots.

Each numerator factor and each denominator factor may be written in the following form:

$$s-p_i=(s-\sigma_c)-(p_i-\sigma_c)$$
$$s-z_i=(s-\sigma_c)-(z_i-\sigma_c)$$

In Fig. 7.11, point s is indicated by a dot, point p_i by an $\times$, point z_i by a $\bigcirc$, and σ_c by a plus sign. The various vector quantities are indicated on this diagram. For a given value of p_i, the vector $p_i - \sigma_c$ remains constant. Hence, as s becomes infinite, $s - \sigma_c \gg p_i - \sigma_c$. Thus, for very large s

$$s-p_i\approx s-\sigma_c$$

Similarly, as s becomes infinite, it follows that

$$s-z_i\approx s-\sigma_c$$

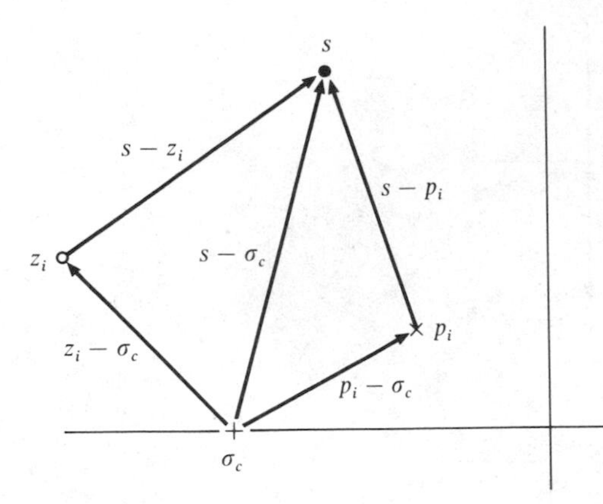

FIGURE 7.11
Asymptote crossing σ_c.

Thus,

$$-K \approx \frac{(s-\sigma_c)^n}{(s-\sigma_c)^m} = (s-\sigma_c)^{n-m} = s^{n-m} - (n-m)\sigma_c s^{n-m-1} + \cdots$$

Performing the division indicated in Eq. (7.11) gives

$$-K = s^{n-m} - [(p_1 + p_2 + \cdots + p_n) - (z_1 + z_2 + \cdots + z_n)]s^{n-m-1} + \cdots$$

Equating the coefficients of the second terms in the two preceding equations for $-K$ and solving for σ_c verify the result given by Eq. (7.10). For the case of Fig. 7.1*b*, there is an × at the origin and an × at −4, and there are no ○'s. For $n = 2$ and $m = 0$, then $n - m = 2$. From Fig. 7.9, for $n - m = 2$ the angle of the asymptotes is ±90°. Application of Eq. (7.10) in which $p_1 = 0$ and $p_2 = -4$ yields

$$\sigma_c = \frac{0 + (-4) - 0}{2} = -2$$

For the case of Fig. 7.1*b*, the root-locus plot lies on the asymptotes. In Fig. 7.1*b*, note that there is a locus on the real axis between the origin and −4. Because there are no ○'s, the two loci must terminate at infinity along the asymptotes. Thus, it is necessary that the loci break away from the real axis. The location of point σ_b at which the locus breaks away from the real axis is distinguished by the fact that the gain K has its maximum value on the real axis at the *breakaway point*. Note that at the origin and at −4 the value of K is 0, at −1 and at −3 the value of K is 3, and at −2 the value of K is 4. Differentiating the equation for K with respect to s and then setting dK/ds equal to zero yield the value of s at which K is a maximum. The characteristic equation for Fig. 7.1*b* is

$$s(s+4) + K = 0$$

Solving for $-K$ shows that

$$-K = s^2 + 4s$$

Differentiating and setting $-dK/ds$ equal to zero yield

$$-\frac{dK}{ds} = 2(s+2) = 0$$

Thus, the value of s at which the locus breaks away from the real axis is $s = \sigma_b = -2$. As is illustrated in Fig. 7.1b, two loci always break away from the real axis in a direction perpendicular to the axis.

Figure 7.12a shows the block diagram for a feedback control system, and Fig. 7.12b shows the corresponding root-locus plot. The characteristic equation for this system is

$$(s^2 + 8s + 25) + K(s+1)(s+5) = 0$$

This system has a pair of complex conjugate ×'s at $-4 \pm j3$, a ○ at -1, and a ○ at -5. Because there is a locus on the real axis between -1 and -5 and because one locus terminates at each of the ○'s, it is necessary that the loci break into the real axis. A *break-in point* is similar to a breakaway point except the locus breaks into the real axis at a break-in point whereas it breaks away from the real axis at a breakaway point. At a break-in point, K has its minimum value on the real axis.

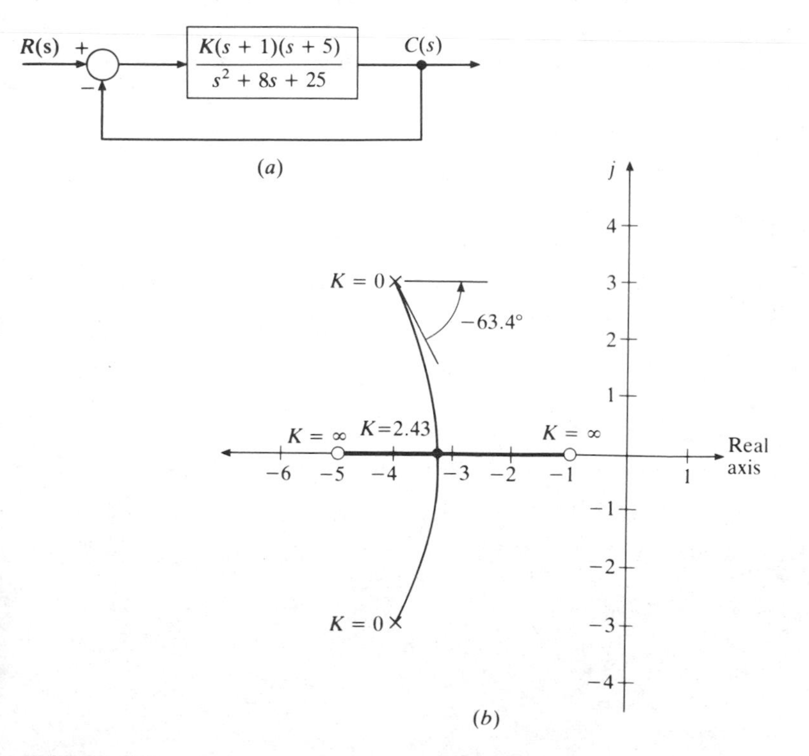

FIGURE 7.12
Root-locus plot for $(s^2 + 8s + 25) + K(s + 1)(s + 5) = 0$.

Setting dK/ds equal to zero yields the value $s = \sigma_b$ at which the break-in point occurs. Thus

$$-K = \frac{s^2 + 8s + 25}{s^2 + 6s + 5}$$

$$-\frac{dK}{ds} = \frac{(s^2 + 6s + 5)(2s + 8) - (s^2 + 8s + 25)(2s + 6)}{(s^2 + 6s + 5)^2}$$

$$= -\frac{2(s^2 + 20s + 55)}{(s^2 + 6s + 5)^2} = 0$$

The value of s which makes the numerator go to zero is the location of the break-in point, $s = \sigma_b = -3.29$. Substitution of this value of s into the equation for K shows that the value of K at the break-in point is $K = 2.43$.

As shown in Fig. 7.12b, the angle at which the locus departs or leaves the $\times$ at $-4 + j3$ is $-63.4°$. This angle is called the *angle of departure*. The value of the angle of departure is obtained by taking a trial point s which is located close to the $\times$ at $-4 + j3$ and then applying the angle condition. In Fig. 7.13a, angle ϕ_1 is the angle from the $\times$ at $-4 - j3$ to the point indicated by the dot which is close to the $\times$ at $-4 + j3$. The unknown angle of departure ϕ_d is the angle from the $\times$ at $-4 + j3$ to the nearby point. In Fig. 7.13b, angle ϕ_2 is the angle from the $\bigcirc$ at -1 to the point near the $\times$ at $-4 + j3$, and angle ϕ_3 is the angle from the $\bigcirc$ at -5 to this

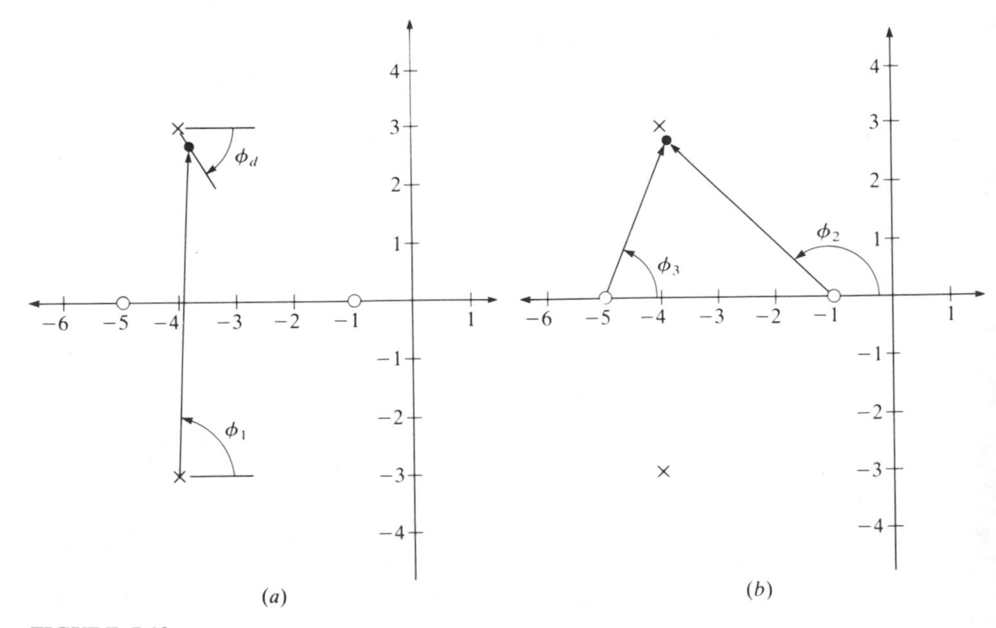

FIGURE 7.13
Determination of the angle of departure ϕ_d.

point. Application of the angle condition shows that

$$(\phi_1 + \phi_d) - (\phi_2 + \phi_3) = 180^\circ \pm k360^\circ$$

where $\phi_1 + \phi_d$ is the sum of the angles from the $\times$'s to the trial point and $\phi_2 + \phi_3$ is the sum of the angles from the $\bigcirc$'s to the trial point.

As the trial point approaches the $\times$ at $-4 + j3$, then ϕ_1, ϕ_2, and ϕ_3 become

$$\phi_1 = \tan^{-1}\frac{6}{0} = 90^\circ$$

$$\phi_2 = \tan^{-1}\frac{3}{-3} = 135^\circ$$

$$\phi_3 = \tan^{-1}\frac{3}{1} = 71.6^\circ$$

Different values of k merely add or subtract 360° from the angle. For $k = 0$, the preceding angle condition becomes

$$(90^\circ + \phi_d) - (135^\circ + 71.6^\circ) = 180^\circ$$

Hence

$$\phi_d = 296.6^\circ = -63.4^\circ$$

Because complex roots always occur as conjugate pairs, root-locus plots are always symmetric about the real axis. Thus, there is no need to obtain the angle of departure from the $\times$ at $-3 - j4$. The angle of departure must be 63.4°.

Figure 7.14*a* shows the block diagram for the same system as in Fig. 7.12 except that the $\times$'s and $\bigcirc$'s have been interchanged. Figure 7.14*b* shows the corresponding root-locus plot. Because there are two $\times$'s and two $\bigcirc$'s, one locus begins at each $\times$ and one locus terminates at each $\bigcirc$. The breakaway point is obtained as follows:

$$-K = \frac{s^2 + 6s + 5}{s^2 + 8s + 25}$$

$$-\frac{dK}{ds} = \frac{2(s^2 + 20s + 55)}{(s^2 + 8s + 25)^2} = 0$$

The value of s for which $dK/ds = 0$ is the breakaway point $s = \sigma_b = -3.29$. As illustrated in Fig. 7.14*b*, the angle at which the locus approaches the $\bigcirc$ at $-4 + j3$ is $-63.4°$. This angle is called the *angle of arrival*. In Fig. 7.15*a*, the angle from the $\times$ at -1 to a point near the circle at $-4 + j3$ is ϕ_1. The angle from the $\times$ at -5 to this point is ϕ_2. In Fig. 7.15*b*, the angle from the $\bigcirc$ at $-4 - j3$ to the trial point is ϕ_3. The unknown angle of arrival ϕ_a is the angle from the circle at $-4 + j3$ to the trial point which is nearby. Application of the angle condition shows that

$$(\phi_1 + \phi_2) - (\phi_3 + \phi_a) = 180^\circ \pm k360^\circ$$

where $\phi_1 + \phi_2$ is the sum of the angles from the $\times$'s to the trial point and $\phi_3 + \phi_a$ is the sum of the angles from the $\bigcirc$'s to the trial point.

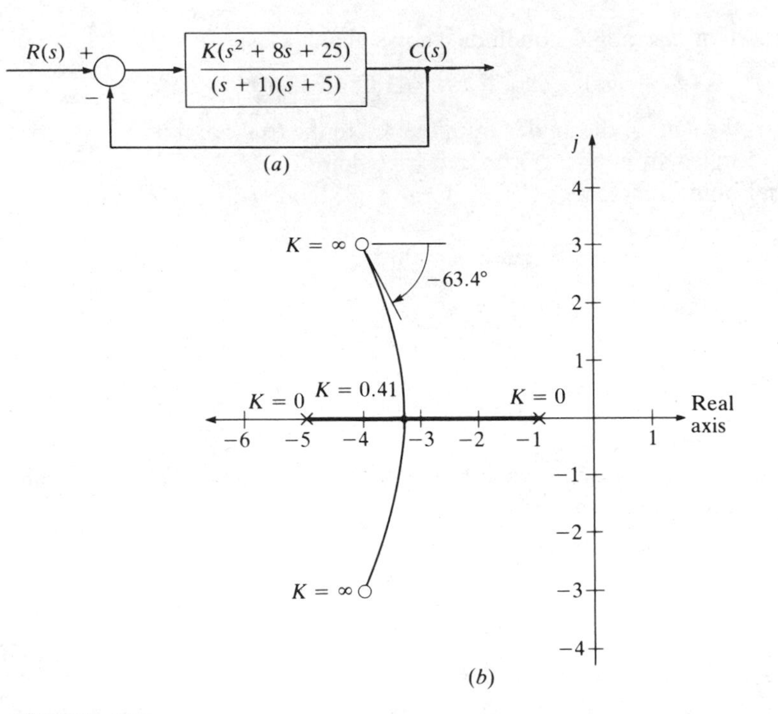

FIGURE 7.14
Root-locus plot for $(s + 1)(s + 5) + K(s^2 + 8s + 25) = 0$.

As the trial point s approaches the circle at $-4 + j3$, the values of ϕ_1, ϕ_2, and ϕ_3 are

$$\phi_1 = \tan^{-1}\frac{3}{-3} = 135^\circ$$

$$\phi_2 = \tan^{-1}\frac{3}{1} = 71.6^\circ$$

$$\phi_3 = \tan^{-1}\frac{6}{0} = 90^\circ$$

For $k = 0$, the preceding angle condition becomes

$$135^\circ + 71.6^\circ - (90^\circ + \phi_a) = 180^\circ$$

Hence

$$\phi_a = -63.4^\circ$$

For constructing the path of the loci in the complex plane, a trial point is selected in the region, and then the angle condition is applied. It is a good idea to start by choosing all trial points on the same horizontal line. For a trial point on one side of the locus, the angle condition will yield an angle greater than 180°, and for a trial point on the other side, an angle of less than 180° is obtained. This information

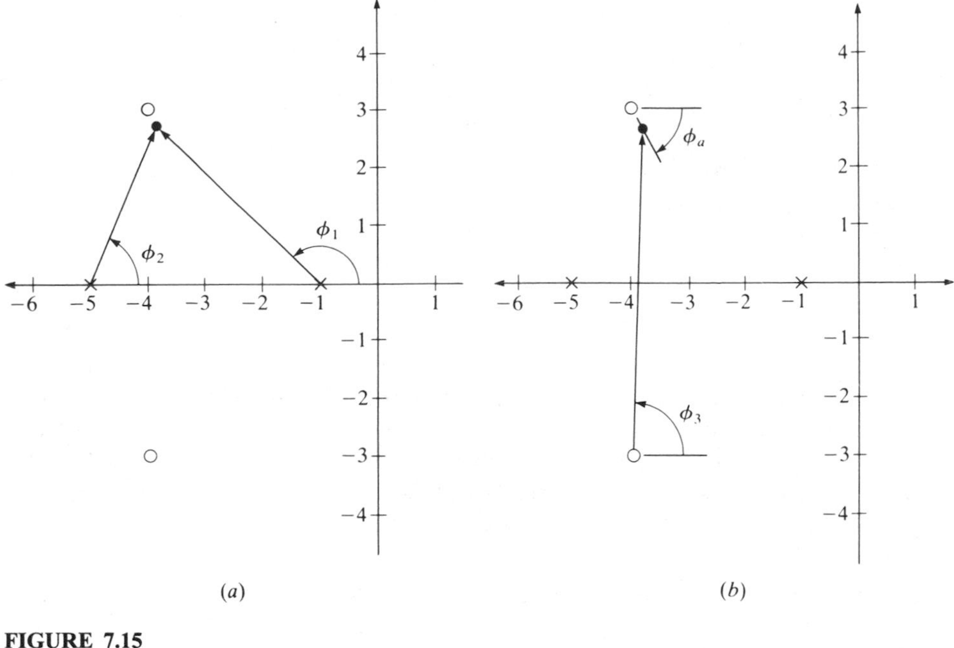

FIGURE 7.15

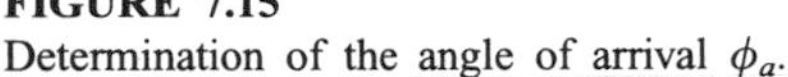
Determination of the angle of arrival ϕ_a.

indicates in what direction a new trial point should be taken. The use of a spirule, which is a commercially available device, saves much effort in finding the angle of a trial point. As is explained in Sec. 7.7, the digital computer can be used to obtain the root-locus plot for a given characteristic equation. Points on the locus corresponding to various values of the gain K are automatically obtained by the digital computer.

7.3 GENERAL PROCEDURE

In the preceding section, we showed that the loci could be sketched in quite accurately by knowing a few critical points such as breakaway or break-in points, location on the real axis, asymptotes, angle of departure from complex conjugate roots, etc. The general procedure for constructing root loci is summarized as follows:

1. *Origin.* When K is zero, the roots of the characteristic equation are the roots of $D_G D_H$. Thus, each locus originates at a root of $D_G D_H$ (designated by $\times$'s) and the number of individual loci is equal to n, the number of roots of $D_G D_H$.
2. *Terminus.* As K becomes very large, m loci (m is the number of roots of $N_G N_H$) will approach the m roots of $N_G N_H$. That is, one locus will terminate at each of the m roots of $N_G N_H$ designated as $\bigcirc$'s. The remaining $n - m$ loci will approach infinity along asymptotes.

3. *Asymptotes.* The angle at which each of the $n - m$ loci approaches infinity is determined from Eq. (7.9):

$$\measuredangle s = \frac{180° \pm k360°}{n - m} \tag{7.9}$$

The point σ_c at which the asymptotes intersect or cross the real axis is computed by Eq. (7.10):

$$\sigma_c = \frac{(p_1 + p_2 + \cdots + p_n) - (z_1 + z_2 + \cdots + z_m)}{n - m} \tag{7.10}$$

4. *Loci on real axis.* Complex conjugate roots of $D_G D_H$ or $N_G N_H$ have no effect on the location of loci on the real axis. The place where the loci are located along the real axis is determined by considering only roots of $D_G D_H$ and $N_G N_H$ which lie on the real axis. As illustrated in Fig. 7.8, there is never a locus to the right of the first × or ○ on the real axis, but there is always a locus to the left of the first × or ○; there is never a locus to the left of the second × or ○, but there is always a locus to the left of the third × or ○; never left of the fourth, always left of the fifth; and so on, alternating.
5. *Angle of departure.* The angle of departure of a locus from a complex conjugate root of $D_G D_H$ is obtained by selecting a trial point very close to this root and applying the angle condition.
6. *Angle of arrival.* The angle at which a locus will terminate at a complex conjugate root of $N_G N_H$ is determined by taking a trial point which is very close to this root and applying the angle condition. This process is similar to that used to obtain the angle of departure.
7. *Breakaway or break-in points.* The point σ_b at which the locus breaks away from or breaks into the real axis is determined by finding the real values $s = \sigma_b$ at which $dK/ds = 0$.

The occurrence of a breakaway or break-in point can be recognized from a consideration of the ×'s and ○'s which lie on the real axis. Every locus begins at an × and terminates at a ○ or along an asymptote at infinity. Thus, there must be a breakaway point between any two adjacent ×'s on the real axis which are connected by a locus. Similarly, a break-in point is required if a ○ on the real axis is not connected to an adjacent × on the real axis by a locus. Thus, if the locus is not located entirely on the real axis between an adjacent × or ○ , it must come into the real axis from elsewhere. The preceding rules may be verified for the root-locus plots shown in Fig. 7.16.

Illustrative example 7.2. Figure 7.17 shows a mobile laser ranging system which transmits short, intense laser pulses to an orbiting retroreflection-equipped satellite. Scientists compare the round-trip pulse time from this station with that of other stations to determine their movement and the corresponding shifting in the earth's tectonic plates. A laser (light amplification by stimulated emission of radiation) is a source of extremely intense, monochromatic, coherent light. A system for controlling

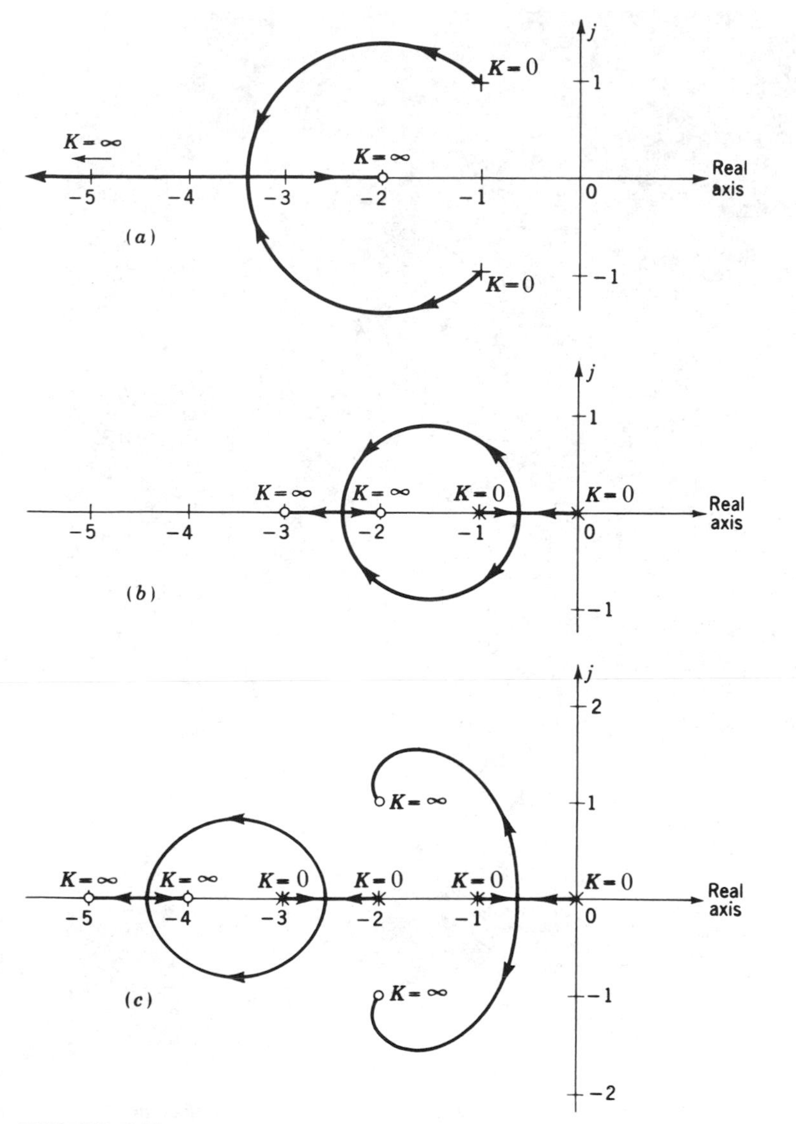

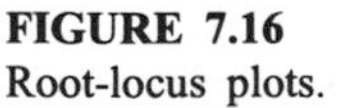
FIGURE 7.16
Root-locus plots.

the intensity of the light is shown in Fig. 7.18a. Construct the root-locus plot for this control system.

Solution. The characteristic equation is

$$s(s+4)(s+6)+K=0 \tag{7.12}$$

As is shown in Fig. 7.18b, there is an $\times$ at the origin, an $\times$ at -4, and an $\times$ at -6. There are no ○'s. On the real axis, there is a locus between the origin and -4, and one

FIGURE 7.17
Mobile laser ranging system. (Courtesy National Aeronautics and Space Administration.)

from -6 to $-\infty$. Because there are no ○'s, all three loci must terminate at infinity along asymptotes. Application of Eq. (7.10) yields for the point σ_c at which the asymptotes intersect the real axis

$$\sigma_c = \frac{0 + (-4) + (-6)}{3} = -\frac{10}{3} = -3\frac{1}{3}$$

For $n - m = 3$, the angles of the asymptotes are $\pm 60^\circ$ and 180°. The information obtained thus far is shown in Fig. 7.19. To determine the breakaway point, note that

$$-K = s(s+4)(s+6) = s^3 + 10s^2 + 24s$$

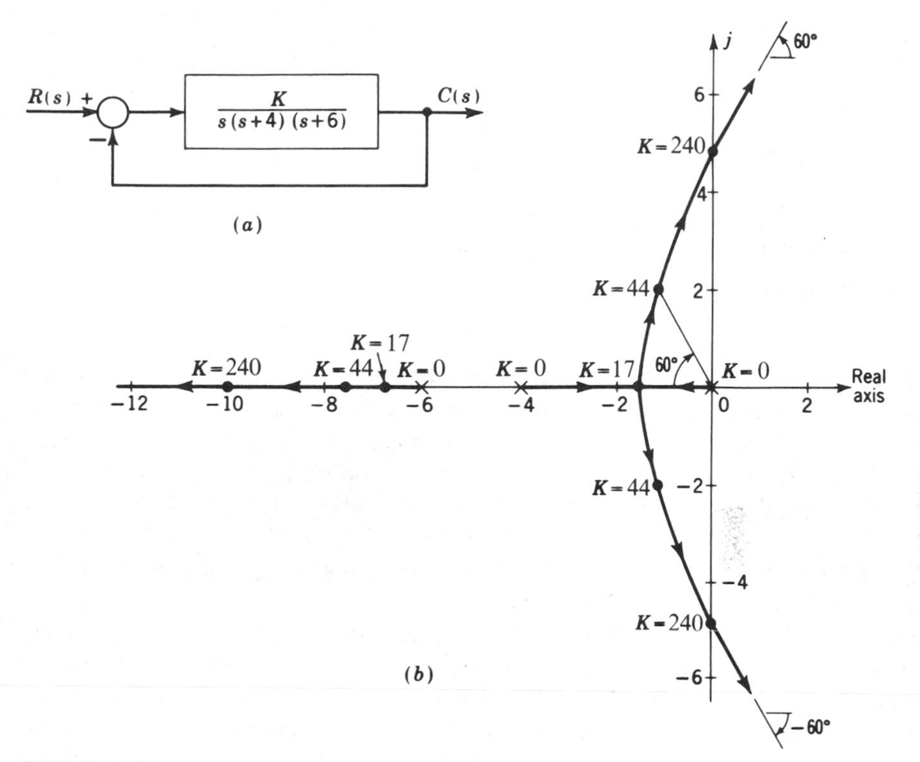

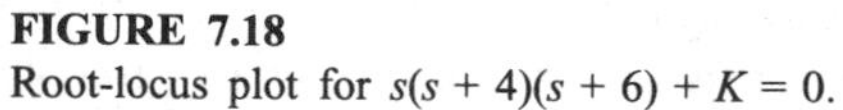

FIGURE 7.18
Root-locus plot for $s(s + 4)(s + 6) + K = 0$.

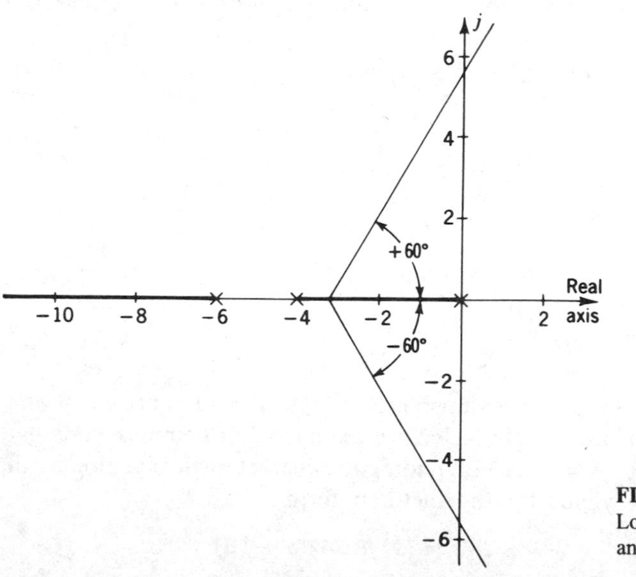

FIGURE 7.19
Location of loci on the real axis and the asymptotes.

Differentiating and setting $-dK/ds$ equal to zero give

$$\frac{-dK}{ds} = 3s^2 + 20s + 24 = 0 \tag{7.13}$$

whence $\sigma_b = -1.57$.

The resulting root-locus plot may now be drawn as shown in Fig. 7.18*b*.

A plot of the roots of the characteristic equation as K varies from 0 to ∞ yields very valuable information. In particular, it permits the characteristic equation to be written in factored form as

$$s(s+4)(s+6) + K = (s - r_1)(s - r_2)(s - r_3) \tag{7.14}$$

The right side is the factored form of the characteristic equation. Because the number of roots $r_1, r_2, \ldots, r_n$ is equal to the order of the equation, the number of loci is also equal to the order of the equation, which in this case is 3. For each value of K, there corresponds a particular value of r_1, r_2, and r_3. Thus, for each value of K from 0 to ∞, the root-locus plot is the plot of values of r_1, r_2, and r_3.

The three loci for Eq. (7.14) are drawn in Fig. 7.18*b*. One locus starts at $r_1 = -6$ for $K = 0$ and proceeds out the negative real axis as K increases. Another locus starts at $r_2 = -4$ and goes to the right along the real axis to the breakaway point $\sigma_b = -1.57$, and then it leaves the real axis and proceeds out along the 60° asymptote toward infinity. The third locus starts at $r_3 = 0$ and moves along the negative real axis to the breakaway point $\sigma_b = -1.57$. This locus then leaves the real axis and proceeds toward infinity along the $-60°$ asymptote. In the construction of the loci, the three loci are determined without regard to which is considered the r_1, the r_2, or the r_3 locus. That is, from the similarity of terms in Eq. (7.14), we see that the particular subscripts 1, 2, and 3 may be used interchangeably.

The value K at which a root-locus plot crosses the imaginary axis may be obtained by application of Routh's criterion. The application of Routh's criterion to the characteristic equation

$$s(s+4)(s+6) + K = s^3 + 10s^2 + 24s + K = 0$$

gives the following array:

$$\begin{array}{llll} s^3 & 1 & 24 & 0 \\ s^2 & 10 & K & 0 \\ s^1 & \dfrac{240 - K}{10} \approx \varepsilon & 0 & \\ s^0 & K & & \end{array}$$

The value of K which makes the s^1 row vanish is $K = 240$. The auxiliary equation is $10s^2 + K = 10(s^2 + 24)$. Thus, the characteristic equation has complex conjugate roots located at $r_{1,2} = \pm j\sqrt{24} = \pm j4.9$. Dividing the characteristic equation by the auxiliary function $s^2 + 24$ yields for the factored form

$$s^3 + 10s^2 + 24s + 240 = (s^2 + 24)(s + 10)$$

For each value of K the corresponding roots of the characteristic equation may be determined directly from the root-locus plot. These roots in turn govern the transient behavior. From the root-locus plot, the designer may select the value of K such that the system will have a desired transient response. For example, suppose we want to have a damping ratio of 0.5. As shown in Fig. 7.18b, the intersection of the line drawn at the angle $\beta = \cos^{-1} 0.5 = 60°$ with the locus is at the point $s = -1.2 + j2.1$. As illustrated in Fig. 7.20, application of the magnitude condition yields for the value of K at this point

$$K = |s||s+4||s+6| = \sqrt{(-1.2)^2 + 2.1^2}\sqrt{2.8^2 + 2.1^2}\sqrt{4.8^2 + 2.1^2} = 44$$

For $a = -1.2$ and $b = 2.1$, the corresponding quadratic term is $s^2 + 2.4s + 5.85$. Dividing the characteristic equation by this quadratic shows that when $K = 44$, the factored form of the characteristic equation is

$$s^3 + 10s^2 + 24s + 44 = (s^2 + 2.4s + 5.85)(s + 7.6)$$

To illustrate the application of the preceding concepts, let us determine the response $c(t)$ of the system shown in Fig. 7.18a for the case in which $K = 44$, the input $r(t)$ is a unit impulse, and all the initial conditions are zero.

Replacing s by D in the block diagram yields for the differential equation

$$c(t) = \frac{Kr(t)}{D(D+4)(D+6)+K} = \frac{Kr(t)}{D^3 + 10D^2 + 24D + K}$$

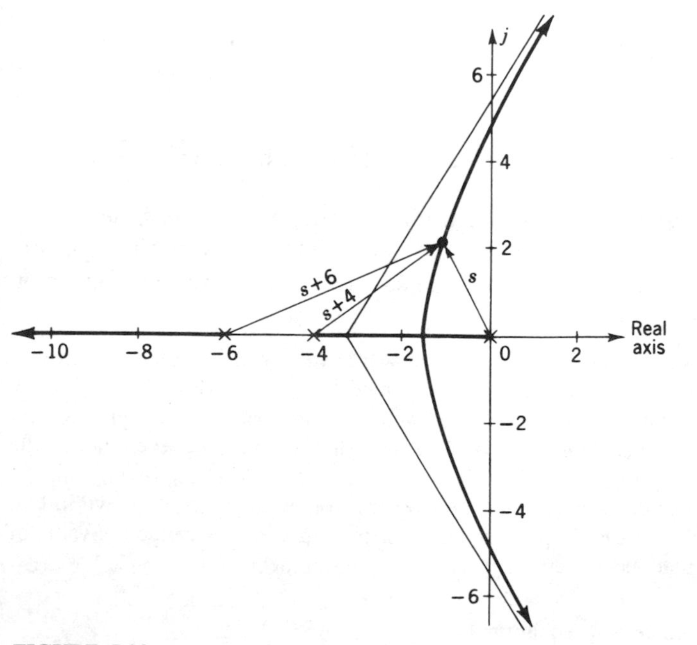

FIGURE 7.20
Application of the magnitude condition.

The transform is

$$C(s) = \frac{KR(s) + [s^2c(0) + s\dot{c}(0) + \ddot{c}(0)] + 10[sc(0) + \dot{c}(0)] + 24c(0)}{s^3 + 10s^2 + 24s + K}$$

For $K = 44$, $R(s) = 1$, and all initial conditions zero,

$$C(s) = \frac{44}{(s + 7.6)(s^2 + 2.4s + 5.85)}$$

The general form of the response is

$$c(t) = K_1 e^{-7.6t} + \frac{|K(a + jb)|}{b} e^{at} \sin(bt + \alpha)$$

The value of K_1 is

$$K_1 = \lim_{s \to -7.6} \frac{44}{s^2 + 2.4s + 5.85} = \frac{44}{81.85} = 0.54$$

The value of $K(a + jb)$ is

$$K(a + jb) = \lim_{s \to -1.2 + j2.1} \frac{44}{s + 7.6} = \frac{44}{6.4 + j2.1}$$

$$= \frac{44}{45.37}(6.4 - j2.1) = 6.2 - j2.0 = 6.532\underline{/ - 18.2^\circ}$$

Thus the desired solution is

$$c(t) = 0.54e^{-7.6t} + 3.11e^{-1.2t}(\sin 2.1t - 18.2^\circ)$$

or

$$c(t) = 0.54e^{-7.6t} + \frac{e^{-1.2t}}{2.1}(6.2 \sin 2.1t - 2.0 \cos 2.1t)$$

Note that regardless of the particular problem being solved, the general form of the transient response is determined by the roots of the characteristic equation. For a given value of K, these roots are ascertained directly from the root-locus plot.

Illustrative example 7.3. Helicopters perform a wide variety of important services. In addition to their well-known use as emergency medical services (EMS) aircraft, they are used for search and rescue operations as well as numerous other law enforcement and civilian applications. Helicopters are extraordinarily reliable even under the harshest conditions. Figure 7.21 shows a Bell helicopter. Unlike fixed-wing aircraft, helicopters tend to be quite unstable. A helicopter stabilizing system is shown in Fig. 7.22*a*. Construct the root-locus plot for this system, and then determine the value of the gain K such that the system will be critically damped.

Solution. The characteristic equation is

$$s(s^2 + 12s + 45) + K = 0$$

FIGURE 7.21
A Bell 412 series helicopter. (Courtesy Bell Helicopter Textron, Inc.)

As shown in Fig. 7.22*b*, there is an × at the origin, and there is a complex conjugate pair of ×'s at $-6 \pm j3$. On the real axis there is a locus from the origin to minus infinity. For three ×'s and no ○'s, all three loci terminate along asymptotes at infinity. The angles of the asymptotes are +60°, −60°, and 180°. Application of Eq. (7.10) yields for the point σ_c, where the asymptotes intersect the real axis,

$$\sigma_c = \frac{(-6+j3)+(-6-j3)+(0)-0}{3\ -\ 0} = \frac{-12}{3} = -4$$

As shown in Fig. 7.23, to determine the angle of departure from the × at the point $-6+j3$, a trial point is selected close to this ×.

Application of the angle condition gives

$$\phi_1 + \phi_2 + \phi_d = 180^\circ$$

In the limit as the trial point approaches the × at $-6+j3$, the values of ϕ_1 and ϕ_2 are

$$\phi_1 = \tan^{-1}\frac{3}{-6} = 153.4^\circ$$

$$\phi_2 = \tan^{-1}\frac{6}{0} = 90^\circ$$

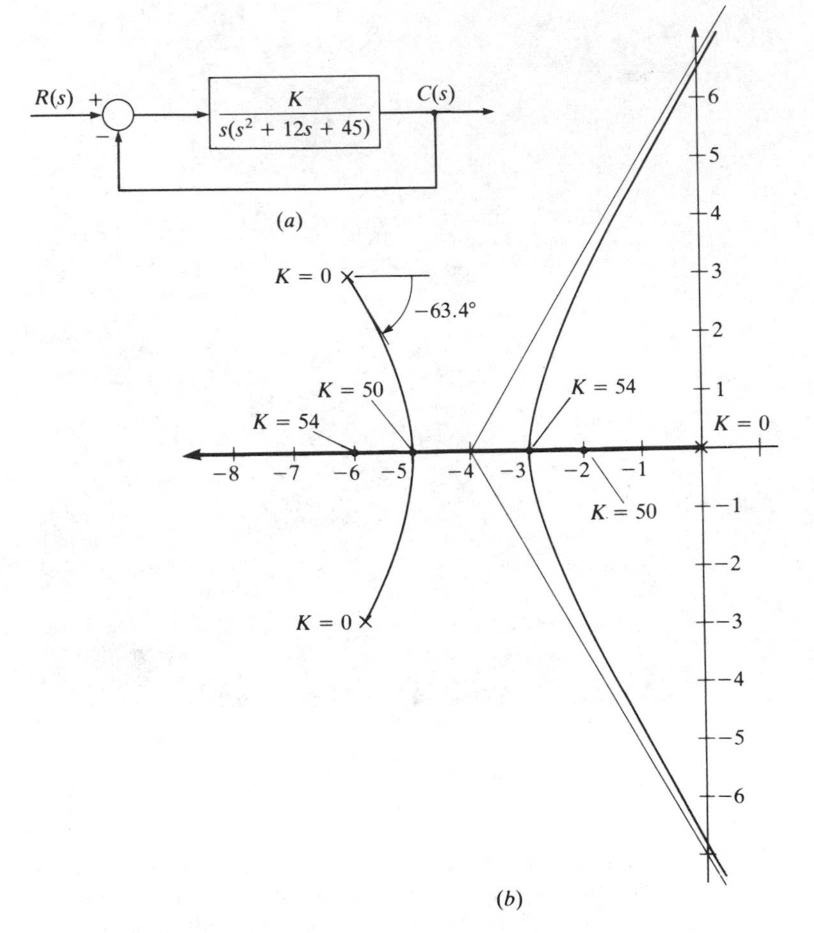

FIGURE 7.22
Root-locus plot for $s(s^2 + 12s + 45) + K = 0$.

Substitution of these values into the angle condition gives

$$153.4 + 90^\circ + \phi_d = 180^\circ$$

Thus

$$\phi_d = -63.4^\circ$$

This angle of departure is such as to lead the loci toward the real axis where there is a locus. There is a tendency for loci in the complex plane to be drawn to loci on the real axis. To check for break-in and breakaway points, first write

$$-K = s^3 + 12s^2 + 45s$$

Differentiation shows that

$$\frac{-dK}{ds} = 3(s^2 + 8s + 15) = 0$$

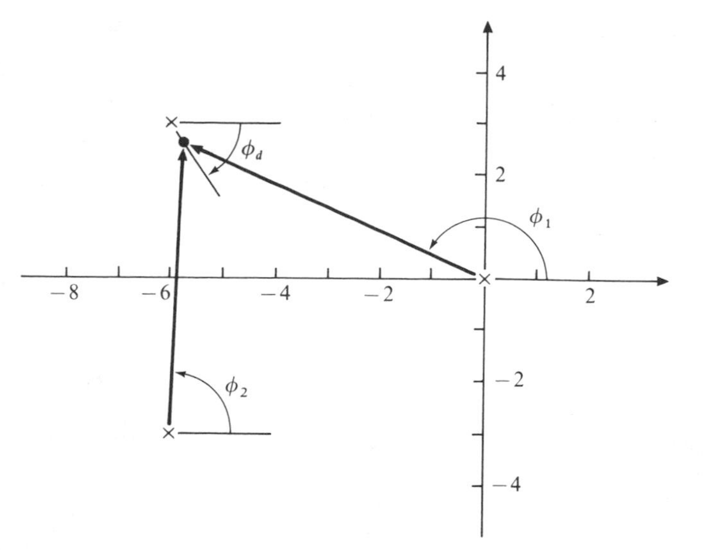

FIGURE 7.23
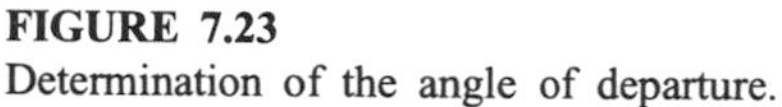
Determination of the angle of departure.

The values of s for which $dK/ds = 0$ are

$$\sigma_b = \frac{-8 \pm \sqrt{64 - 60}}{2} = -4 \pm 1 = -5, \ -3$$

As shown in Fig. 7.22*b*, the point $\sigma_b = -5$ is a break-in point, and the point $\sigma_b = -3$ is a breakaway point. The value of K at which the locus breaks into the real axis is $K = 50$. The value of K at which the locus breaks away from the real axis is $K = 54$. For a given value of K there are always three roots of the characteristic equation. The factored form is

$$s(s^2 + 12s + 45) + K = (s - r_1)(s - r_2)(s - r_3)$$

For $0 < K < 50$, one root lies on the locus between the $\times$ at $-6 + j3$ and the break-in point. The conjugate root lies on the locus between the $\times$ at $-6 - j3$ and the break-in point. The third root lies on the real axis between the $\times$ at the origin and the -2 point. For $50 < K < 54$, one root lies on the real axis between the break-in point at -5 and the point -6. The second root lies between the break-in point and the breakaway point. The third root lies between the -2 point and the breakaway point. For $K > 54$, one root lies on the real axis between -6 and $-\infty$. Another root lies on the locus which leaves the breakaway point and approaches the 60° asymptote. The third root lies on the conjugate locus.

From the root-locus plot, we see that for $K = 50$ there is a repeated root at -5. The factored form of the characteristic equation for this case is

$$(s + 5)^2(s + 2) = s(s^2 + 12s + 45) + 50$$

There is also a repeated root at -3 when $K = 54$. The factored form for this case is

$$(s+3)^2(s+6) = s(s^2+12s+45)+54$$

Either of these values of K (50 or 54) will yield a critically damped system. For $50 < K < 54$, the three roots are real and the system will be slightly overdamped. For $K < 50$ or $K > 54$, the system is underdamped. Application of Routh's criterion shows that

$$\begin{array}{llll} s^3 & 1 & 45 & 0 \\ s^2 & 12 & K & 0 \\ s^1 & 45 - K/12 & & \\ s^0 & K & & \end{array}$$

Thus, the range of values of K for a stable system is $0 < K < 540$. Using a gain K of 52 results in a system which is slightly overdamped. If the gain should change somewhat due to aging or temperature variations, etc., the system will remain close to being critically damped.

Illustrative example 7.4. The characteristic equation for the control system for an ac induction motor is

$$(s+2)(s+4)(s+a)+K=0$$

To achieve good dynamic behavior, it is desired that the damping ratio $\zeta = 0.5$ and that the natural frequency $\omega_n = 4$. Determine a and K.

Solution. When $\zeta = 0.5$ and $\omega_n = 4$, the characteristic equation has the quadratic form

$$s^2 + 2\zeta\omega_n s + \omega_n^2 = s^2 + 4s + 16$$

The complex conjugate roots are located at $-2a = 4$, or $a = -2$, and $a^2 + b^2 = 16$, or $b = \sqrt{16-4} = 2\sqrt{3}$, so that $a \pm jb = -2 \pm j2\sqrt{3}$.

In Fig. 7.24, the $\times$'s are located at -2 and -4, and the unknown point is $-a$. In order that the point indicated by the dot $(-2 + j2\sqrt{3})$ be on the root-locus plot, the angle condition must be satisfied:

$$\phi_1 + \phi_2 + \phi_3 = 180^\circ$$

where

$$\phi_1 = \tan^{-1}\frac{2\sqrt{3}}{0} = 90^\circ$$

$$\phi_2 = \tan^{-1}\frac{2\sqrt{3}}{2} = 60^\circ$$

Hence,

$$\phi_3 = \tan^{-1}\frac{2\sqrt{3}}{a-2} = 180^\circ - \phi_1 - \phi_2 = 30^\circ$$

Taking the tangent of both sides shows that

$$\frac{2\sqrt{3}}{a-2} = \tan 30^\circ = \frac{1}{\sqrt{3}}$$

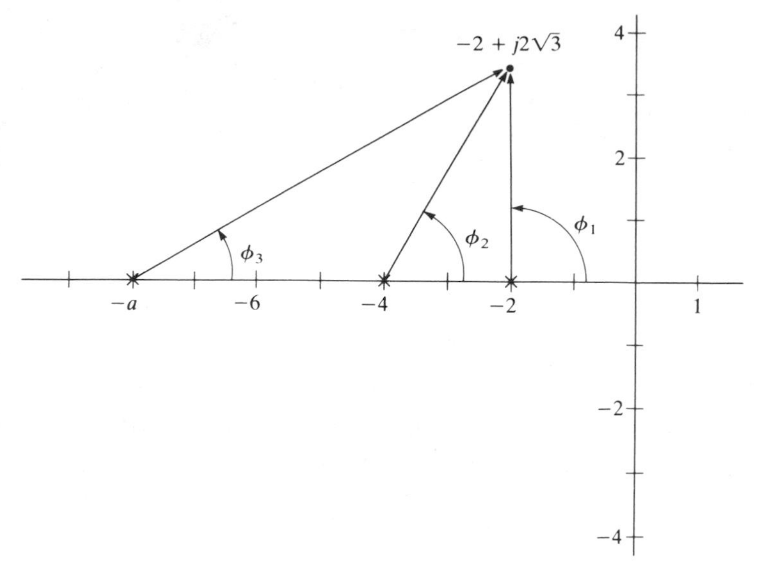

FIGURE 7.24
Determination of a such that the point $-2 + j2\sqrt{3}$ lies on the root-locus plot.

Solving for a gives

$$a - 2 = 2(3) = 6$$
$$a = 8$$

Application of the magnitude condition yields for the value of K at the point $-2 + j2\sqrt{3}$

$$K = |s+2|\,|s+4|\,|s+8| = 2\sqrt{3}\sqrt{(2\sqrt{3})^2 + 2^2}\sqrt{(2\sqrt{3})^2 + 6^2}$$
$$= 2\sqrt{3}\sqrt{12+4}\sqrt{12+36} = 96$$

Dividing the characteristic function $(s + 2)(s + 4)(s + 8) + 96$ by the quadratic factor $s^2 + 4s + 16$ yields the third factor $s + 10$. Thus, the factored form of the characteristic equation is

$$(s+2)(s+4)(s+8) + 96 = (s+10)(s^2 + 4s + 16)$$

7.4 NEWTON'S METHOD AND THE REMAINDER THEOREM

When the equation for dK/ds is a cubic, quartic, or higher-degree polynomial, solving for the break-in or breakaway point is a very laborious process. Newton's method is a technique for obtaining a close approximation to the break-in or breakaway point for such cases.

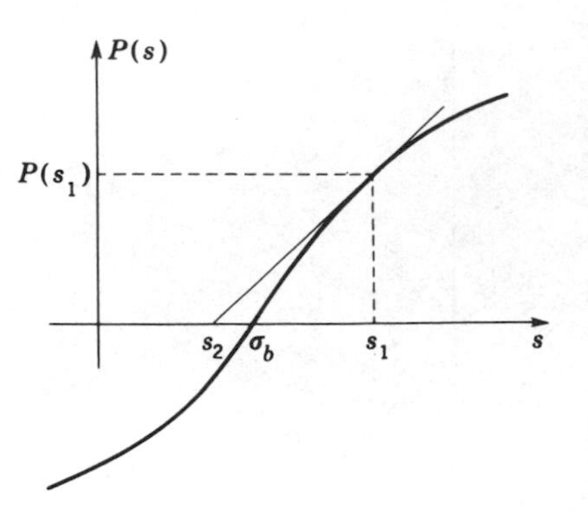

FIGURE 7.25
Typical plot of $P(s)$ versus s.

Let $P(s)$ be the numerator of dK/ds. Thus when $P(s) = 0$, then $dK/ds = 0$. Figure 7.25 shows a plot of $P(s)$ versus s. The tangent to this curve at $s = s_1$ is

$$P'(s_1) = \frac{P(s_1)}{s_1 - s_2}$$

Solving for s_2 gives

$$s_2 = s_1 - \frac{P(s_1)}{P'(s_1)} \tag{7.15}$$

The point s_2, which is the intersection of the tangent and the horizontal axis, is a closer approximation to σ_b than s_1 is. After s_2 is evaluated, applying the same process yields a closer approximation s_3, etc. Usually one trial yields sufficient accuracy for constructing the root-locus plot. We now show how $P(s_1)/P'(s_1)$ may be obtained by use of the remainder theorem.

Dividing $P(s)$ by $s - s_1$ gives

$$\frac{P(s)}{s - s_1} = Q(s) + \frac{R_1}{s - s_1}$$

where R_1 is the remainder that results when $P(s)$ is divided by $s - s_1$. Multiplying through by $s - s_1$ gives

$$P(s) = (s - s_1)Q(s) + R_1 \tag{7.16}$$

Taking the limit as s approaches s_1 shows that

$$P(s_1) = R_1 \tag{7.17}$$

By similarly dividing $Q(s)$ by $s - s_1$, it follows that

$$\frac{Q(s)}{s - s_1} = T(s) + \frac{R_2}{s - s_1} \tag{7.18}$$

where R_2 is the remainder that results when $Q(s)$ is divided by $s - s_1$. Multiplying through Eq. (7.18) by $s - s_1$ and then taking the limit as s approaches s_1 reveal that

$$Q(s_1) = R_2$$

To show that $Q(s_1) = P'(s_1)$, first differentiate Eq. (7.16) with respect to s:

$$P'(s) = (s - s_1)Q'(s) + Q(s)$$

Taking the limit as s approaches s_1 shows that

$$P'(s_1) = Q(s_1) = R_2 \tag{7.19}$$

Equation (7.15) may now be expressed in the form

$$s_2 = s_1 - \frac{R_1}{R_2} \tag{7.20}$$

To illustrate the application of this method, let us assume as a first trial that the breakaway point for Fig. 7.18 occurs at $s_1 = -2.0$. From Eq. (7.13), the polynomial $P(s)$ is

$$P(s) = 3s^2 + 20s + 24$$

Division by $s - s_1 = s + 2$ yields

$$\begin{array}{r|l} & 3s + 14 \\ s + 2 & 3s^2 + 20s + 24 \\ & 3s^2 + 6s \\ \hline & 14s + 24 \\ & 14s + 28 \\ \hline & -4 \end{array}$$

Hence, $Q(s) = 3s + 14$ and $R_1 = -4$. Dividing $Q(s)$ by $s - s_1$ gives

$$\begin{array}{r|l} & 3 \\ s + 2 & 3s + 14 \\ & 3s + 6 \\ \hline & 8 \end{array}$$

Hence, $T(s) = 3$ and $R_2 = 8$. Application of Eq. (7.20) yields the closer approximation

$$s_2 = -2 - \frac{-4}{8} = -1.5$$

The preceding divisions are simplified by the use of synthetic division, as illustrated below.

$$\begin{array}{ccc|c} 3 & 20 & 24 & -2 \\ & -6 & -28 & \\ \hline 3 & 14 & \boxed{-4} & \end{array}$$

The first row consists of the coefficients of the polynomial $P(s)$ and the root -2. The first coefficient 3 is brought below the line; then multiplying 3 by -2 yields -6, as shown. Adding 20 and -6 gives 14, as shown. Multiplying 14 by -2 yields the -28 term. Finally, adding 24 and -28 yields the remainder $R_1 = -4$, which is in the box. Note that the coefficients of $Q(s)$ appear to the left of the remainder. Remainder R_2 is now obtained by similarly applying synthetic division to the $Q(s)$ polynomial:

$$\begin{array}{cc|l} 3 & 14 & \underline{)\,-2} \\ & -6 & \\ \hline 3 & \boxed{8} & \end{array}$$

Thus, remainder R_2 is 8 and $T(s)$ is simply 3.

Let us now take -1.5 as the new trial breakaway point and then employ synthetic division to obtain the next closer approximation. Thus

$$\begin{array}{cccl} 3.00 & 20.00 & 24.00 & \underline{)\,-1.50} \\ & -4.50 & -23.25 & \\ \hline 3.00 & 15.50 & \boxed{0.75} & \\ & -4.50 & & \\ \hline 3.00 & \boxed{11.00} & & \end{array}$$

Thus, $R_1 = 0.75$, $R_2 = 11.00$, $Q(s) = 3s + 15.50$, and $T(s) = 3$. The closer approximation is

$$s_3 = s_2 - \frac{R_1}{R_2} = -1.50 - \frac{0.75}{11.00} = -1.57$$

7.5 LOCI EQUATIONS

In studying control systems, one frequently encounters the case in which there are two $\times$'s and one $\circ$, as shown in Fig. 7.26. We now show that the locus follows a circular path in the complex plane. The characteristic equation for this case is

$$(s - p_1)(s - p_2) + K(s - z) = 0 \tag{7.21}$$

As shown in Fig. 7.26, a trial point s may be expressed in the form $s = x + jy$. Substitution of $s = x + jy$ into the preceding equation gives

$$(x^2 + 2jxy - y^2) - (p_1 + p_2 - K)(x + jy) + p_1 p_2 - Kz = 0$$

In order that this equation equal zero, both the real parts and the imaginary parts (that is, j components) must be zero. Thus,

$$x^2 - y^2 - (p_1 + p_2 - K)x + p_1 p_2 - Kz = 0$$
$$2xy - (p_1 + p_2 - K)y = 0$$

To eliminate K between the two preceding equations, first solve each for K.

$$K = \frac{(x^2 - y^2) - (p_1 + p_2)x + p_1 p_2}{z - x} \tag{7.22}$$

$$K = (p_1 + p_2) - 2x \tag{7.23}$$

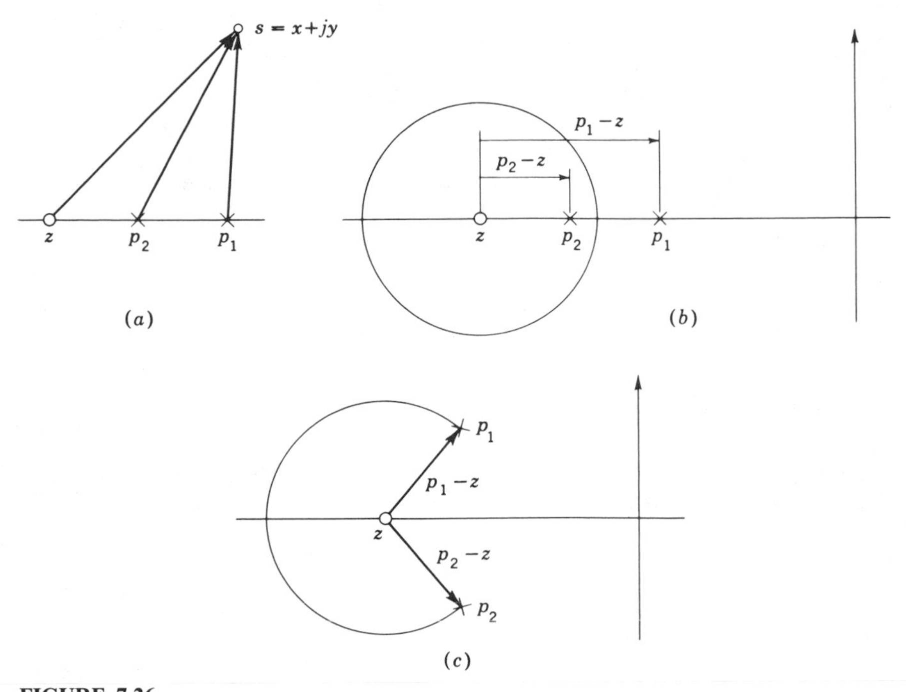

FIGURE 7.26
Circular loci.

The term K is now eliminated by equating the right sides of the preceding expressions. Thus

$$x^2 + y^2 = p_1 p_2 + [2x - (p_1 + p_2)]z$$

Completing the square gives

$$(x - z)^2 + y^2 = (p_1 - z)(p_2 - z) \tag{7.24}$$

This is recognized as the equation of a circle with center at z. The radius R is the square root of the product of the distance from z to p_1 and the distance from z to p_2. That is,

$$R = \sqrt{(p_1 - z)(p_2 - z)} \tag{7.25}$$

A typical root-locus plot for the case in which p_1 and p_2 are real is shown in Fig. 7.26*b*. For the case in which p_1 and p_2 are complex conjugate roots, the distance from z to p_1 equals the distance from z to p_2. Thus, the radius is simply the distance from z to either of the complex conjugate roots, as shown in Fig. 7.26*c*.

Equation (7.23) may be used to determine the gain K for points on the loci in the complex plane (i.e., off the real axis). For loci on the real axis, $y = 0$. Thus, letting $y = 0$ in Eq. (7.22) yields an equation for evaluating the gain K for loci on the real axis.

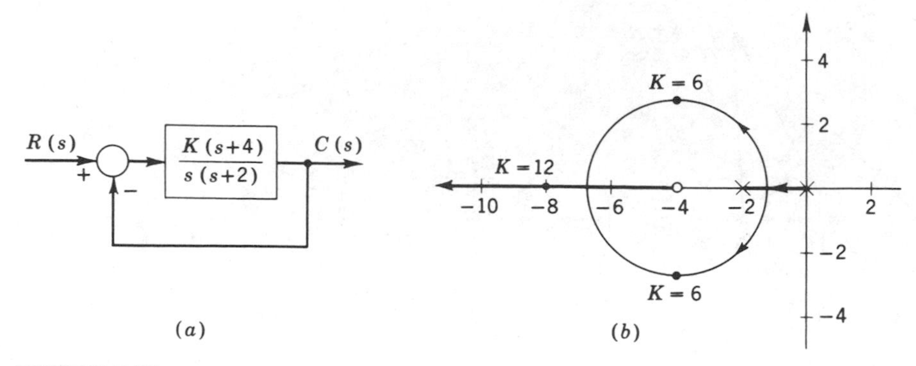

FIGURE 7.27
Root-locus plot for $s(s + 2) + K(s + 4) = 0$.

Illustrative example 7.5. Construct the root-locus plot for the system shown in Fig. 7.27*a*, and then determine the gain K at $x = -4$ and at $x = -8$.

Solution. The characteristic equation for this system is

$$s(s+2) + K(s+4) = 0$$

Thus, $z = -4$, $p_1 = 0$, and $p_2 = -2$.

The locus on the real axis may be drawn as shown in Fig. 7.27*b*. Because there are two ×'s and one ○, in the complex plane (i.e., off the real axis) the locus is a circle with center at $z = -4$ and radius

$$R = \sqrt{(p_1 - z)(p_2 - z)} = \sqrt{(4)(2)} = \sqrt{8} = 2\sqrt{2}$$

Application of Eq. (7.23) to determine the gain K at $x = -4$ gives

$$K = p_1 + p_2 - 2x = 0 - 2 - 2(-4) = 6$$

For the locus on the real axis at $x = -8$, application of Eq. (7.22) in which $y = 0$ gives

$$K = \frac{x^2 - (p_1 + p_2)x + p_1 p_2}{z - x} = \frac{64 - (-2)(-8)}{-4 - (-8)} = \frac{48}{4} = 12$$

The preceding method for obtaining the equation for a circular locus may be applied to any locus. The method is to replace s by $x + jy$ in the characteristic equation. Because both the real and imaginary parts must be zero, two equations are obtained. One equation gives values of gain K for the locus on the real axis, and the other equation gives values of gain K for the locus in the complex plane. Eliminating K between these two equations yields a general equation in x and y for the loci. Corresponding values of x and y that satisfy this equation are points on the loci. In the preceding example the equation was recognized as that of a circle.

7.6 VARIATION OF PARAMETERS

Thus far the discussion of root locus has been concerned with the case in which the gain K is the variable parameter. By algebraically rearranging the characteristic equation, the effect of the change of any parameter can be investigated.

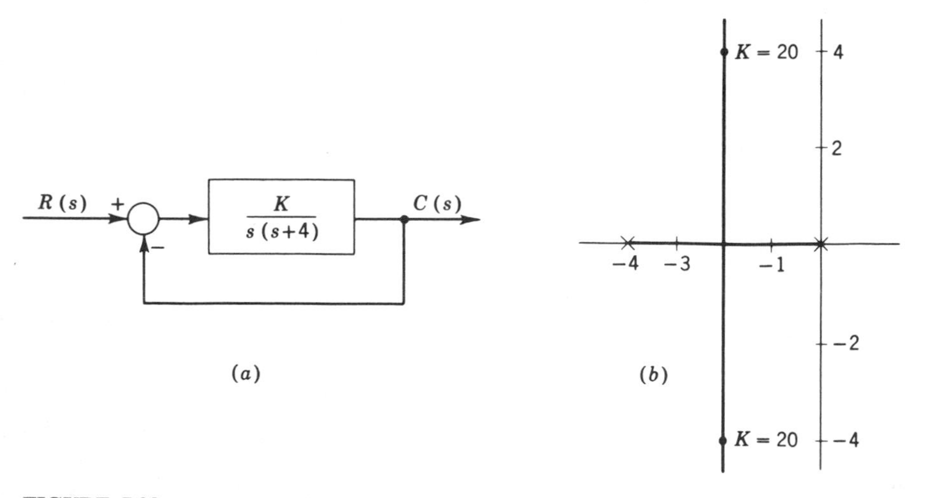

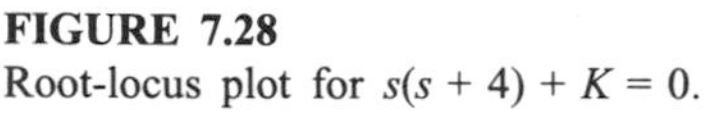

FIGURE 7.28
Root-locus plot for $s(s + 4) + K = 0$.

To illustrate this procedure, consider the system shown in Fig. 7.28*a*. The characteristic equation for this system is

$$s(s + 4) + K = 0$$

The basic root-locus plot is shown in Fig. 7.28*b*. For $K = 20$, note that the roots of the characteristic equation are $-2 \pm j4$. Thus

$$s(s + 4) + 20 = [s - (-2 - j4)][s - (-2 + j4)] = 0$$

Now let us investigate the effect of varying the parameter 4 when $K = 20$. Replacing the parameter by 4 by $4 + \Delta$, where Δ is the change in the parameter, yields for the new characteristic equation

$$s[s + (4 + \Delta)] + 20 = [s(s + 4) + 20] + \Delta s$$
$$= [s - (-2 - j4)][s - (-2 + j4)] + \Delta s = 0 \qquad (7.26)$$

This equation has the familiar form except that Δ is now the variable rather than K. The new root-locus plot begins at the value of the roots on the basic plot for $K = 20$. Equation (7.26) shows that the $\times$'s for the new plot are at $s = -2 \pm j4$ and there is a $\bigcirc$ at the origin. The corresponding root-locus plot for positive values of Δ is shown in Fig. 7.29*a*.

In investigating the effect of a change in parameter, we want to know the effect of decreasing as well as increasing the parameter. Thus far, root-locus plots have been constructed for positive values of K only. From Eqs. (7.6) and (7.8), it follows that when K (or Δ) is positive, the summation of the angles is $180° \pm k360°$. When K (or Δ) is a negative number, then $-K$ is a positive number, which may be represented by a point on the positive real axis. For negative K the angle is

$$\measuredangle(-K) = 0° \pm k360°$$

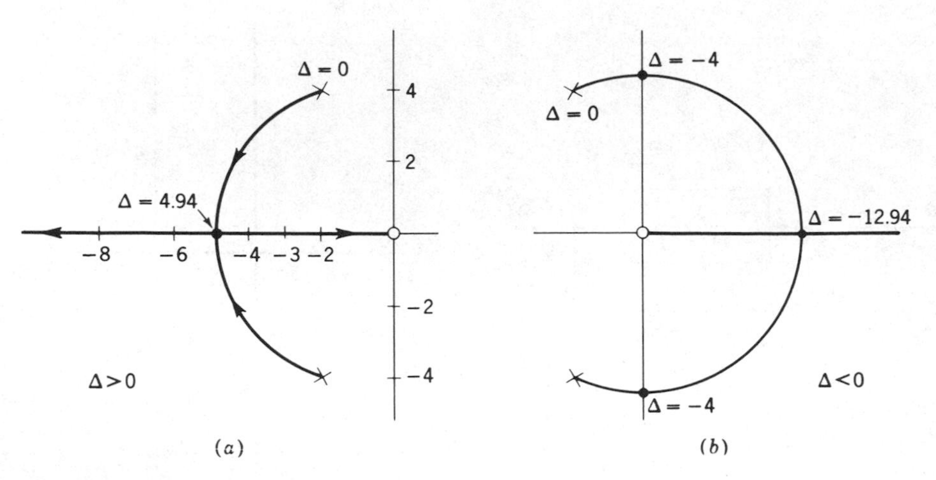

FIGURE 7.29
Root-locus plot for $[s(s+4)+20]+\Delta s=0$.

Thus, for negative values of K the angle condition becomes

$$\sum_{i=1}^{n} \measuredangle (s-p_1) - \sum_{i=1}^{m} \measuredangle (s-z_1) = 0° \pm k360° \qquad k = 0,\ 1,\ 2,\ 3,\ \ldots \tag{7.27}$$

The angles for the asymptotes are

$$\measuredangle\, s = \frac{0° \pm k360°}{n-m} \qquad k = 0,\ 1,\ 2,\ 3,\ \ldots \tag{7.28}$$

For $n - m = 1$, 2, 3, and 4, Eq. (7.28) becomes

$$\begin{aligned} \measuredangle\, s &= 0° \pm k360° & n-m&=1 \\ \measuredangle\, s &= 0° \pm k180° & n-m&=2 \\ \measuredangle\, s &= 0° \pm k120° & n-m&=3 \\ \measuredangle\, s &= 0° \pm k90° & n-m&=4 \end{aligned}$$

The corresponding asymptotes for $n - m = 1$, 2, 3, and 4 are shown in Fig. 7.30. For $n - m = 1$, the angle is 0°; for $n - m = 2$ the angles are 0° and 180°; for $n - m = 3$ the angles are 0° and ±120°; and for $n - m = 4$ the angles are 0°, ±90°, and 180°. The point σ_c at which the asymptotes intersect the real axis is obtained by Eq. (7.10) in the same manner as for positive K. That is,

$$\sigma_c = \frac{\sum_{i=1}^{n} p_i - \sum_{i=1}^{m} z_i}{n-m} \tag{7.10}$$

The following rules for constructing a root-locus plot for negative values of K (or Δ) are very similar to those given in Sec. 7.3 for positive values of K.

1. *Origin.* One locus begins at each ×, as was the case for positive K.
2. *Terminus.* One locus ends at each ○, as was the case for positive K. The remaining $n - m$ loci terminate at infinity along asymptotes.
3. *Asymptotes.* The angle at which each of the $n - m$ loci approaches infinity is

$$\measuredangle s = \frac{0^\circ \pm k360^\circ}{n - m}$$

The corresponding asymptotes for $n - m = 1$, 2, 3, and 4 are shown in Fig. 7.30. The point σ_c at which the asymptotes intersect the real axis is obtained by Eq. (7.10) in the same manner as for positive K.
4. *Loci on the real axis.* The location of loci on the real axis is determined by the ×'s and ○'s that lie on the real axis. The summation of angles at any point on the real axis is either 0° or 180°. For positive K the loci are regions on the real axis where the angle is 180°. The remaining regions on the real axis are regions where the angle is 0°, and thus these are the regions of the loci for negative values of K. Figure 7.31 shows the location of loci on the real axis for negative K for the same × and ○ pattern as Fig. 7.8. The loci now appear at all regions on the real axis that did not have loci in Fig. 7.8. For the case of negative K, a locus always

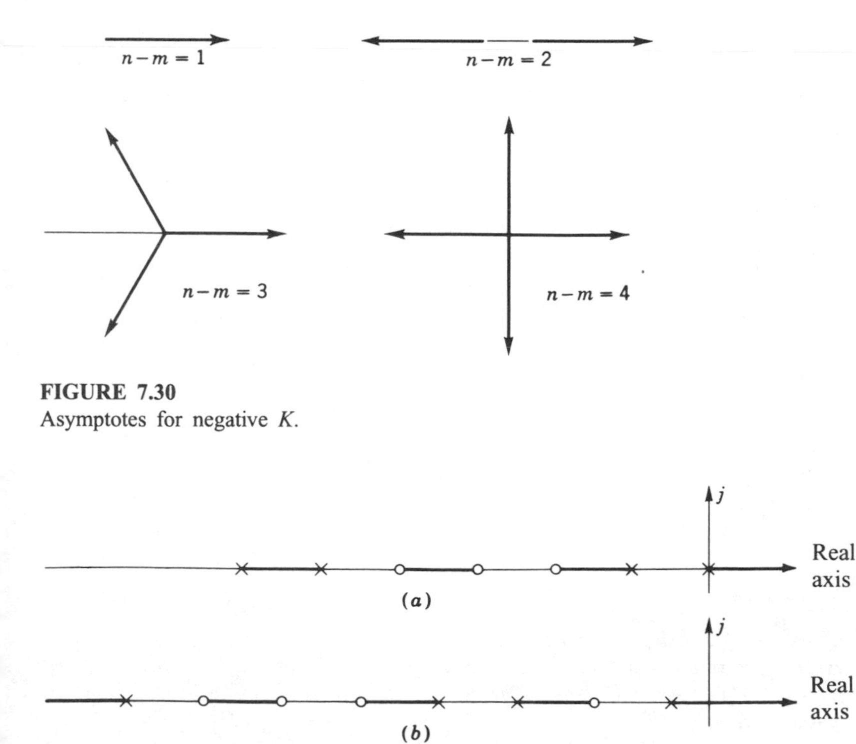

FIGURE 7.30
Asymptotes for negative K.

FIGURE 7.31
Location of loci on the real axis for negative K.

appears to the right of the first × or ○ on the real axis, there is never a locus to the left of the first × or ○; there is always a locus to the left of the second × or ○, there is never a locus to the left of the third × or ○; and so on, alternating.

5. *Angle of departure.* Because the summation of the angles is now 0° rather than 180°, the angle of departure for negative K is always 180° different from that obtained for positive K.
6. *Angle of arrival.* The angle of arrival for negative K is always 180° different from that obtained for positive K.
7. *Breakaway or break-in points.* These are determined in the same way as for positive K. In determining the breakaway or break-in points for positive K, often extra or extraneous values were obtained. Actually, these values are the ones for the case of negative K.

Figure 7.29*b* shows the root-locus plot for negative values of Δ. Because there are two ×'s and one ○, the loci form a circle in the complex plane. The system becomes unstable when the locus crosses the imaginary axis, which occurs at $\Delta = -4$. The preceding concepts are further illustrated in the following example.

Illustrative example 7.6. For the system for controlling the intensity of light for the laser system shown in Fig. 7.17, suppose that the gain K must be maintained at 240 to achieve satisfactory steady-state operation. For $K = 240$ the roots of the characteristic equation are $s = \pm j\sqrt{24}$ and $s = -10$. Hence, the factored form of the characteristic equation is

$$s(s+4)(s+6) + 240 = (s+10)(s^2+24)$$

Determine whether this system can be stabilized and good dynamic behavior achieved by changing the parameter 4 in the term $s + 4$.

Solution. Replacing 4 by $4 + \Delta$ yields for the new characteristic equation

$$\begin{aligned} s[(s+4)+\Delta](s+6)+240 &= [s(s+4)(s+6)+240] + \Delta s(s+6) \\ &= (s+10)(s^2+24) + \Delta s(s+6) = 0 \end{aligned}$$

To construct the root-locus plot for positive values of Δ, first draw the loci on the real axis as shown in Fig. 7.32*a*. Next the angle of departure from × at $s = j\sqrt{24}$ is

$$\left(\phi_d + 90° + \tan^{-1}\frac{\sqrt{24}}{10}\right) - \left(90° + \tan^{-1}\frac{\sqrt{24}}{6}\right) = 180°$$

Solving for ϕ_d gives $\phi_d = 193.1°$.

A break-in point is seen to occur between 0 and -6 on the real axis. As a first trial, let s_1 be the midpoint -3. To apply Newton's method, first solve the characteristic equation for $-\Delta$:

$$-\Delta = \frac{(s+10)(s^2+24)}{s(s+6)}$$

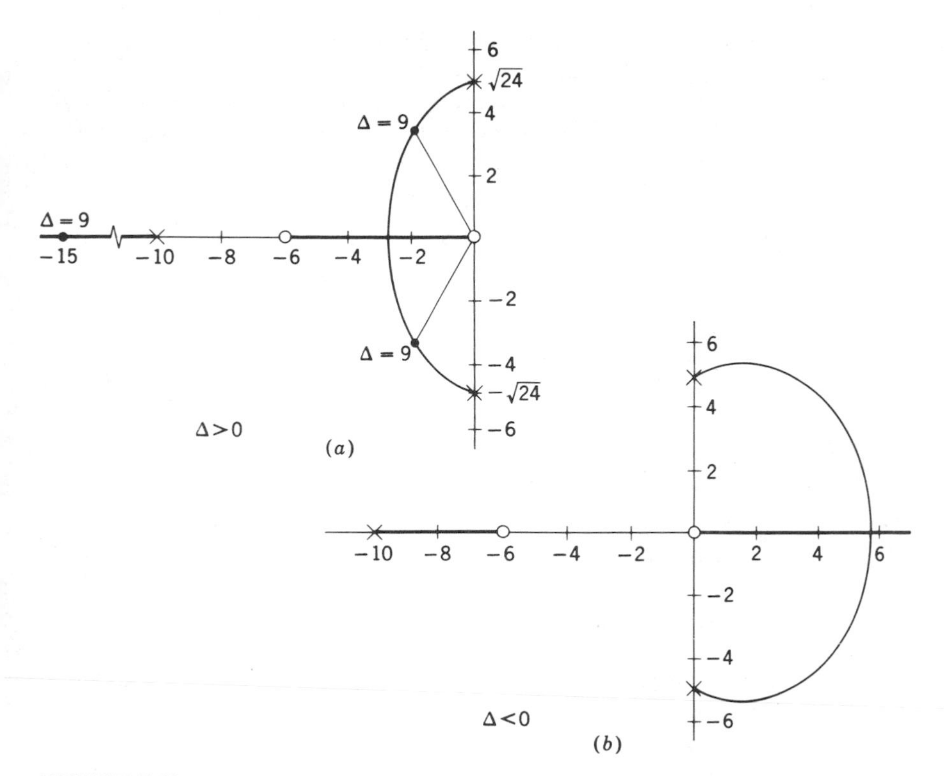

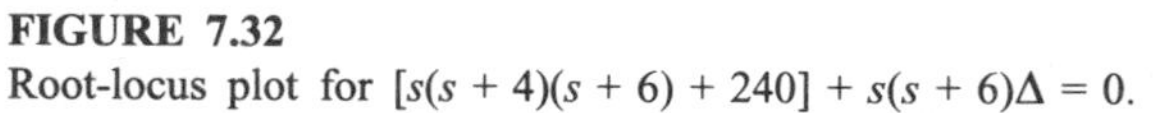

FIGURE 7.32
Root-locus plot for $[s(s+4)(s+6)+240]+s(s+6)\Delta=0$.

Differentiation to obtain $-d\Delta/ds$ gives

$$-\frac{d\Delta}{ds}=\frac{s^4+12s^3+36s^2-480s-1440}{(s^2+6s)^2}$$

The polynomial $P(s)$ is taken as the numerator of $-d\Delta/ds$. Thus, application of double synthetic division to obtain the remainders yields

$$\begin{array}{rrrrrl}
1 & 12 & 36 & -480 & -1440 & \underline{)\ -3} \\
 & -3 & -27 & -27 & 1521 & \\
\hline
1 & 9 & 9 & -507 & \boxed{81} & \\
 & -3 & -18 & 27 & & \\
\hline
1 & 6 & -9 & \boxed{-480} & &
\end{array}$$

Thus, $R_1 = 81$, $R_2 = -480$, $Q(s) = s^3 + 9s^2 + 9s - 507$, and $T(s) = s^2 + 6s - 9$. The value of s_2 is

$$s_2 = -3 - \frac{81}{-480} = -2.83$$

The remaining portion of the root-locus plot for positive values of Δ may now be sketched as shown in Fig. 7.32*a*.

To construct the root-locus plot for negative Δ, first draw the loci on the real axis as shown in Fig. 7.32*b*. The angle of departure for negative values of Δ is 180° different from that for positive Δ; hence

$$\phi_d = 193.1° - 180.0° = 13.1°$$

We see that a break-in point must occur on the positive real axis. Taking $s_1 = 5$ as the first trial and then applying double synthetic division give

$$\begin{array}{rrrrrl} 1 & 12 & 36 & -480 & -1440 & \underline{)5} \\ & 5 & 85 & 605 & 625 & \\ \hline 1 & 17 & 121 & 125 & \boxed{-815} & \\ & 5 & 110 & 1155 & & \\ \hline 1 & 22 & 231 & \boxed{1280} & & \end{array}$$

Thus, $R_1 = -815$, $R_2 = 1280$, $Q(s) = s^3 + 17s^2 + 121s + 125$, and $T(s) = s^2 + 22s + 231$. Application of Eq. (7.24) to obtain the closer approximation to the break-in point gives

$$s_2 = 5.0 - \frac{-815}{1280} = 5.64$$

From the root-locus plot we see that good dynamic behavior is obtained for $\Delta = 9$, in which case the three roots are $-2 \pm j\sqrt{12}$ and -15. For the complex conjugate roots, $\zeta = 0.5$ and $\omega_n = 4$.

7.7 COMPUTER SOLUTION

Many digital computers have subprograms for obtaining the roots of a polynomial. On such computers it is an easy matter to obtain all the roots of a characteristic equation for various values of K. To illustrate this procedure, consider the characteristic equation

$$s(s^2 + 8s + 25) + K(s + 2)(s + 4) = s^3 + (K + 8)s^2 + (6K + 25)s + 8K = 0$$

This is a third-order system. Thus, for each value of K there are three roots. Although the format for calling up the subprogram for obtaining the roots will vary from computer to computer, a typical program is as follows:

```
DO 10 K = 0,25
RK = REAL (K)
COEF (1) = 1.0
COEF (2) = RK+8.0
COEF (3) = 6.0*RK+25.0
COEF (4) = 8.0*RK
CALL ROOTS (COEF,3,ROOTR,ROOTI)
WRITE (6,*)
   K,ROOTR(1),ROOTI(1),ROOTR(2),ROOTI(2),ROOTR(3),ROOTI(3)
10 CONTINUE
```

The DO statement specifies the value of K for which the roots will be obtained. For this example, the roots are obtained for $K = 0, 1, 2, 3, \ldots, 25$. Because K is an integer, the REAL(K) statement is needed so that RK will be a real number having the value of K. The next four statements input the values of the coefficients of the polynomial: COEF(1) is the coefficient of s^3, COEF(2) is the coefficient of s^2, COEF(3) is the coefficient of s, and COEF(4) is the constant. The CALL ROOTS statement calls up the subprogram for obtaining the roots of the polynomial. ROOTR is the real part, and ROOTI is the imaginary part of the root. The write statement prints out the value of K and the real and imaginary parts of all three roots for that particular value of K. ROOTR(1) is the real part and ROOTI(1) is the imaginary part of the first root; ROOTR(2) is the real part and ROOTI(2) is the imaginary part of the second root; etc.

To obtain a root-locus plot, the real and imaginary parts of each root are stored for each value of K. The real part is the x component of the root location, and the imaginary part is the y component. The real and imaginary parts may be stored by putting a $J = 1$ statement before the DO loop and then adding the following statements immediately before the CONTINUE statement:

```
XP(J)= ROOTR(1)
YP(J)= ROOTI(1)
XP(J+1)= ROOTR(2)
YP(J+1)= ROOTI(2)
XP(J+2)= ROOTR(3)
YP(J+2)= ROOTI(3)
J = J+3
```

These points may now be plotted to obtain the root-locus plot.

7.8 SENSITIVITY

The parameters used in the design of control systems vary due to factors such as wear, aging, variations in the operating point, temperature, etc. It is thus desired to know the effect of small variations in these parameters upon the dynamic response of the system. In Sec. 7.6, we showed how to determine the effect of such variations upon the locations of the roots of the characteristic equation. The sensitivity method is described in the following.

The *sensitivity* S is defined as the percentage change in the system transmittance T compared to the percentage change in parameter K; that is,

$$S_K^T = \frac{dT/T}{dK/K} = \frac{K}{T}\frac{dT}{dK} = \frac{d \ln T}{d \ln K} \tag{7.29}$$

where $T(s) = C(s)/R(s)$ is the closed-loop transfer function for the system. The term K may represent any parameter, such as the gain, time constant, damping ratio, natural frequency, etc. To have the transmittance T which represents the system dynamics insensitive to variations in a parameter, the sensitivity S should be zero, or as small as possible.

The transfer function for the system of Fig. 7.28 may be expressed in the general form

$$G(s) = \frac{K}{s(s+a)} = \frac{K}{s(s+1/\tau)}$$

where $a = 1/\tau$ and τ is the time constant. As indicated in the following, for computational purposes it is easier to work with a. It is an easy matter to convert the resulting answer from a to τ if so desired.

The transmittance T is

$$T(s) = \frac{C(s)}{R(s)} = \frac{G(s)}{1+G(s)} = \frac{K}{s(s+a)+K}$$

The sensitivity with respect to the gain K is

$$S_K^T = \frac{K}{T}\frac{dT}{dK} = \frac{K[s(s+a)+K]}{K}\frac{[s(s+a)+K-K]}{[s(s+a)+K]^2}$$

$$= \frac{s(s+a)}{s(s+a)+K} \tag{7.30}$$

The sensitivity with respect to parameter a is

$$S_a^T = \frac{a}{T}\frac{dT}{da} = \frac{a[s(s+a)+K]}{K}\frac{-sK}{[s(s+a)+K]^2}$$

$$= \frac{-as}{s(s+a)+K} \tag{7.31}$$

The steady-state sensitivity is obtained by letting $s = 0$. For both Eqs. (7.30) and (7.31), the steady-state sensitivity obtained by letting $s = 0$ is $S_K^T = S_a^T = 0$. The dynamic sensitivity is obtained by replacing s by $j\omega$ in the sensitivity equations and then making plots of the sensitivity as a function of the frequency ω. The significance of replacing s by $j\omega$ is explained in Chaps. 11 and 12.

By defining sensitivity in a slightly different manner, it is possible to determine the change in the roots of the characteristic equation for small variations of a parameter. Thus, sensitivity is now defined as the rate of change of the root location with respect to the parameter, i.e.,

$$S_K^s = \frac{ds}{dK} \tag{7.32}$$

where ds/dK is the rate of change of the root location s in the characteristic equation with respect to K. For the system of Fig. 7.28, the characteristic equation is

$$s(s+a)+K = (s-r_1)(s-r_2) = 0$$

The derivative with respect to K is

$$2s\frac{ds}{dK} + a\frac{ds}{dK} + 1 = 0$$

Thus, the sensitivity with respect to parameter K is

$$S_K^s = \frac{ds}{dK} = \frac{-1}{a + 2s}$$

At the reference operating condition $a = 4$ and $K = 20$, the roots are $r_1 = -2 + j4$ and $r_2 = -2 - j4$. The sensitivity of the location of root r_1 is obtained by evaluating S_K^s at $s = r_1$. Thus,

$$S_K^{r_1} = \frac{dr_1}{dK} = \frac{-1}{4 + 2(-2 + j4)} = \frac{-1}{j8} = \frac{j}{8}$$

Because $dr_1 = (dr_1/dK)\, dK = S_K^{r_1}\, dK$, then

$$\Delta r_1 \approx S_K^{r_1}\, \Delta K = \frac{j}{8}\, \Delta K \tag{7.33}$$

If ΔK is 4, then $\Delta r_1 \approx 0.5j$. Hence, for $K = 20 + \Delta K = 24$, then $r_1 \approx (-2 + j4) + 0.5j = -2 + j4.50$. From the characteristic equation, the exact value of r_1 is found to be $-2 + j\sqrt{20} = -2 + j4.47$. In a similar manner, the change in the location of root r_2 may also be found.

The sensitivity of the location of the roots with respect to parameter a is

$$S_a^s = \frac{ds}{da}$$

The derivative of the characteristic equation with respect to a is

$$2s\,\frac{ds}{da} + a\,\frac{ds}{da} + s\,\frac{da}{da} = 0$$

Thus, the sensitivity ds/da is

$$S_a^s = \frac{ds}{da} = \frac{-s}{a + 2s}$$

The root sensitivity at $s = r_1$ is

$$S_a^{r_1} = \frac{dr_1}{da} = \frac{-(-2 + j4)}{4 + 2(-2 + j4)} = -\frac{2 + j}{4}$$

Thus

$$\Delta r_1 \approx -\frac{2 + j}{4}\, \Delta a \tag{7.34}$$

Because $a = 1/\tau$, if the time constant τ is changed from $\frac{1}{4}$ to $\frac{1}{6}$, then $\Delta a = 6 - 4 = 2$. Thus

$$r_1 \approx -2 + j4 - \frac{2 + j}{4}2 = -3 + j3.50$$

From the characteristic equation (for $a = 4 + 2 = 6$ and $K = 20$) the exact value of r_1 is found to be $-3 + j\sqrt{11} = -3 + j3.32$.

When the sensitivity is expressed in polar form, Eqs. (7.33) and (7.34) become

$$\Delta r_1 \approx 0.125\,\Delta K\underline{/90^\circ}$$
$$\Delta r_1 \approx 0.59\,\Delta a\underline{/26.6^\circ}$$

The angles indicate the direction in which the roots leave the reference value for positive variations in ΔK or Δa. The direction is reversed by 180° for negative variations.

If both K and a change simultaneously, then

$$\Delta s \approx \frac{\partial s}{\partial K}\,\Delta K + \frac{\partial s}{\partial a}\,\Delta a = S_K^s\,\Delta K + S_a^s\,\Delta a$$

Thus, the effects add when more than one parameter varies. For example, if $\Delta K = 4$ and $\Delta a = 2$, then the change Δr_1 is

$$\Delta r_1 \approx \frac{j}{8}(4) - \frac{2+j}{4}(2) = -1$$

The new value of r_1 is $(-2 + j4) - 1 = -3 + j4$. From the characteristic equation for $K = 20 + 4 = 24$ and $a = 4 + 2 = 6$, the exact value is found to be $-3 + j\sqrt{15} = -3 + j3.87$.

PROBLEMS

7.1. The root-locus plot for the system of Fig. 7.1*a* is given in Fig. 7.1*b*. Determine the response $c(t)$ for the case in which $r(t)$ is a unit step function and $K = 8$. All the initial conditions are zero.

7.2. The location of the ×'s and ○'s of the characteristic equation for various systems is shown in Fig. P7.2. Determine the characteristic equation for each system, and then sketch the root-locus plot. For each case, determine the value of K when the characteristic equation has a root located at -2.

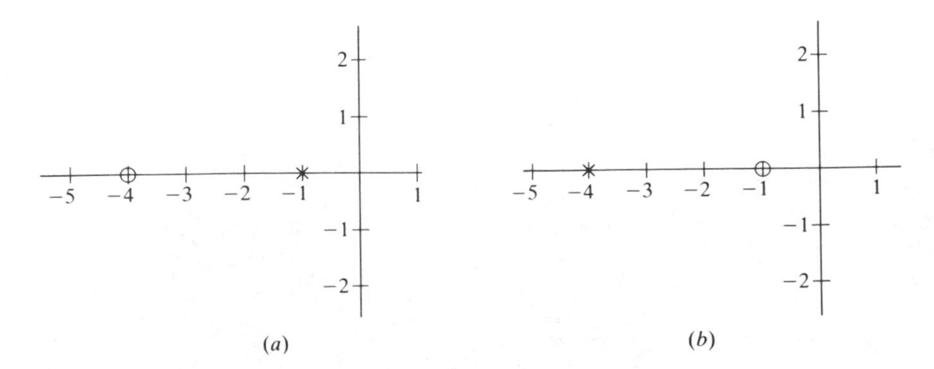

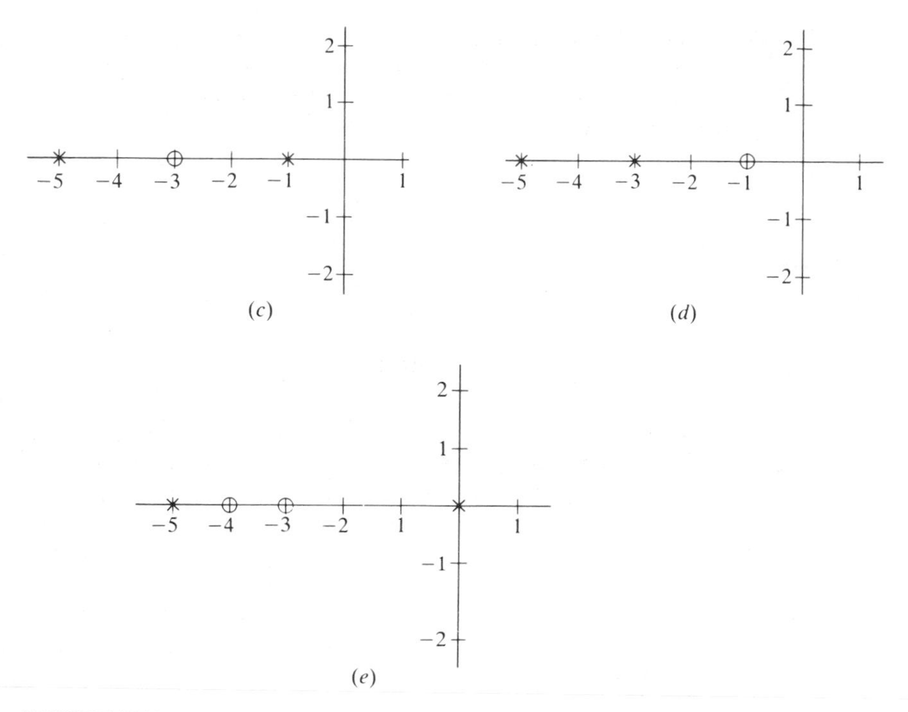

FIGURE P7.2

7.3. For the system shown in Fig. P7.3, write the characteristic equation and then determine the roots for $K = 5$, $K = 9$, and $K = 25$. Sketch the root-locus plot. Show on this plot where $K = 5$, 9, and 25. For each case, determine the response $c(t)$ when $r(t) = 0$, $c(0) = 0$, and $\dot{c}(0) = 1$.

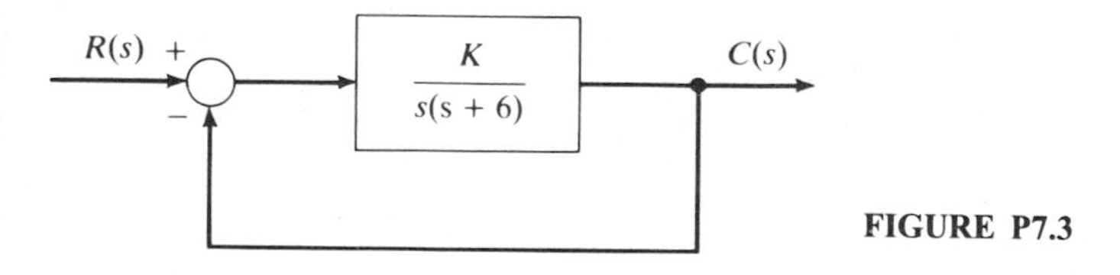

FIGURE P7.3

7.4. Figure P7.4 shows a system for controlling the air-fuel ratio for an automobile engine. Write the characteristic equation and then determine the roots for $K = 3$, $K = 4$, and $K = 20$. For each case, determine the response $c(t)$ when $r(t) = 0$, $c(0) = 1$, and $\dot{c}(0) = 0$.

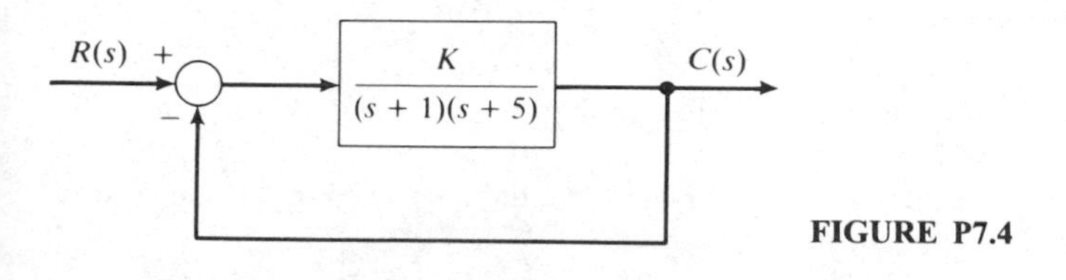

FIGURE P7.4

7.5. Figure P7.5 shows a dc motor positional control system. From the root-locus plot, it is found that when $K = 3$, the roots of the characteristic equation are located at $-4 \pm j3$. Determine the response $c(t)$ of this system for each of the following cases.
(*a*) $r(t) = 0$, $c(0) = 0$, and $\dot{c}(0) = 3$
(*b*) $r(t) = e^{-t}$ and $c(0) = \dot{c}(0) = 0$

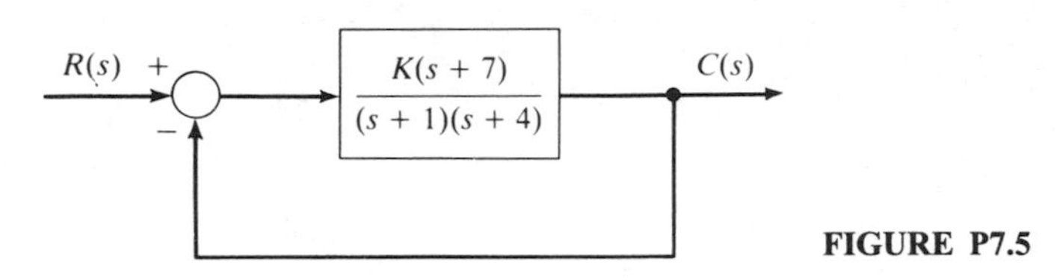

FIGURE P7.5

7.6. To decelerate modern aircraft, the braking action must be controlled to achieve a minimum stopping distance while maintaining passenger comfort. Figure P7.6 shows a control system for an aircraft brake. From the root-locus plot, it is found that when $K = 6$, the roots of the characteristic equation are located at -1, -3, and -4. Thus,

$$s(s^2 + 8s + 13) + 6(s + 2) = (s + 1)(s + 3)(s + 4)$$

Determine the response $c(t)$ when $r(t) = \delta(t)$ is a unit impulse and all initial conditions are zero.

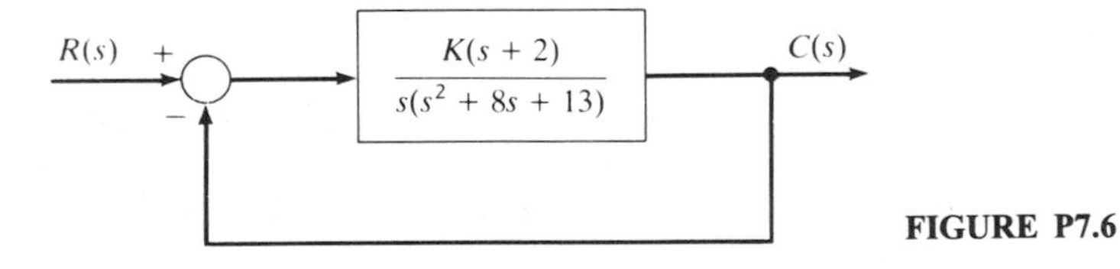

FIGURE P7.6

7.7. When $K = 10$, the roots of the characteristic equation for the system shown in Fig. P7.7 are located at -2, $-1 + j2$, and $-1 - j2$. Thus

$$s(s^2 + 4s + 9) + 10 = (s + 2)(s^2 + 2s + 5)$$

Determine the response $c(t)$ for the case in which $K = 10$, $r(t) = 0$, $c(0) = \dot{c}(0) = 0$, and $\ddot{c}(0) = 10$.

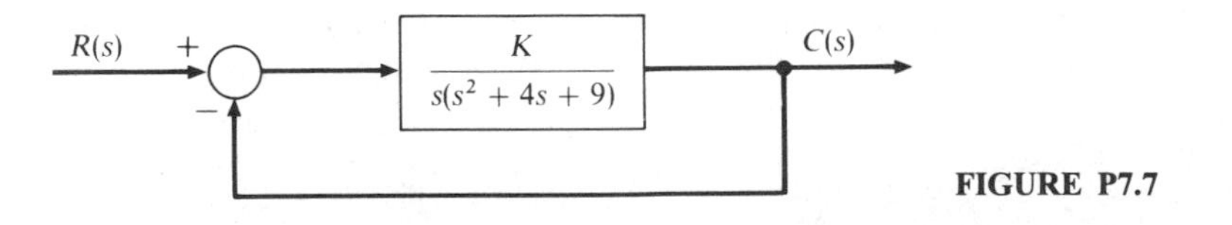

FIGURE P7.7

7.8. The control system for a computer-controlled milling machine is shown in Fig. P7.8. From the root-locus plot, it is found that when $K = 40$, the roots of the characteristic equation are located at -4 and $-1 \pm j3$. Determine the response $c(t)$ for the case in which $r(t) = 0$, $c(0) = \dot{c}(0) = 0$, and $\ddot{c}(0) = 18$.

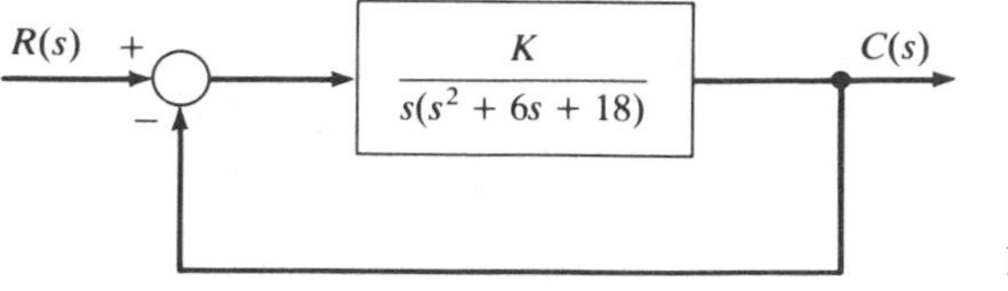

FIGURE P7.8

7.9. Apply the angle condition to determine whether the point $s = -2 + j2$ is on the root-locus plot for $s(s+2) + K(s+4) = 0$. If it is, determine the value of K at this point and the factored form of the characteristic equation.

7.10. Apply the angle condition to determine whether the point $s = -4 + j3$ is on the root-locus plot for $s(s+4)(s+7)(s+8) + K(s+1) = 0$. If it is, determine the value of K at this point and the factored form of the characteristic equation.

7.11. The characteristic equation for a system is

$$s(s+a)(s+8) + K = 0$$

Apply the angle condition to determine the value of a such that the characteristic equation will have a pair of complex conjugate roots at $-2 \pm j2$. Determine the value of the gain K at this point. Write the factored form of the characteristic equation for this case.

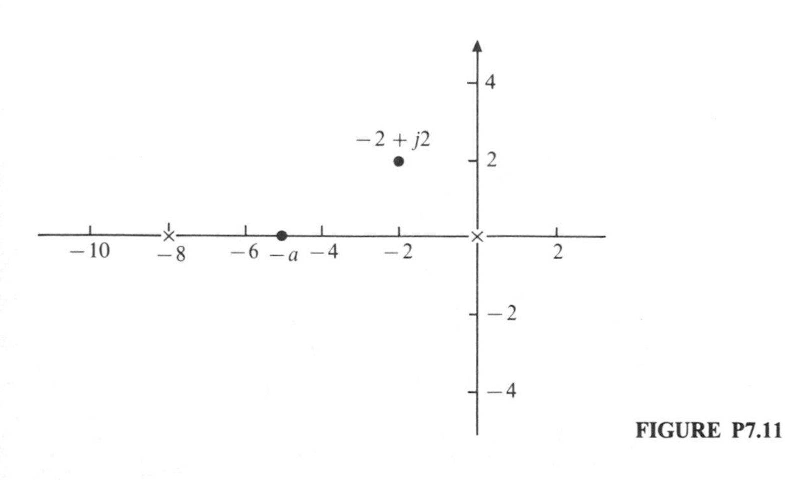

FIGURE P7.11

7.12. It is desired that the root-locus plot for the system shown in Fig. P7.12 go through the point $-4 + j3$. Apply the angle condition to determine the location of the $\times$ at $-a$ such that the plot does go through the point $-4 + j3$. Determine the value of the gain K at this point. Write the factored form of the characteristic equation for this case.

7.13. For the system shown in Fig. P7.13, it is desired that the root-locus plot go through the point $-2 + j2\sqrt{3}$. Apply the angle condition to determine the value of a such that the plot does go through this point. Determine the value of the gain K at this point. Write the factored form of the characteristic equation for this case.

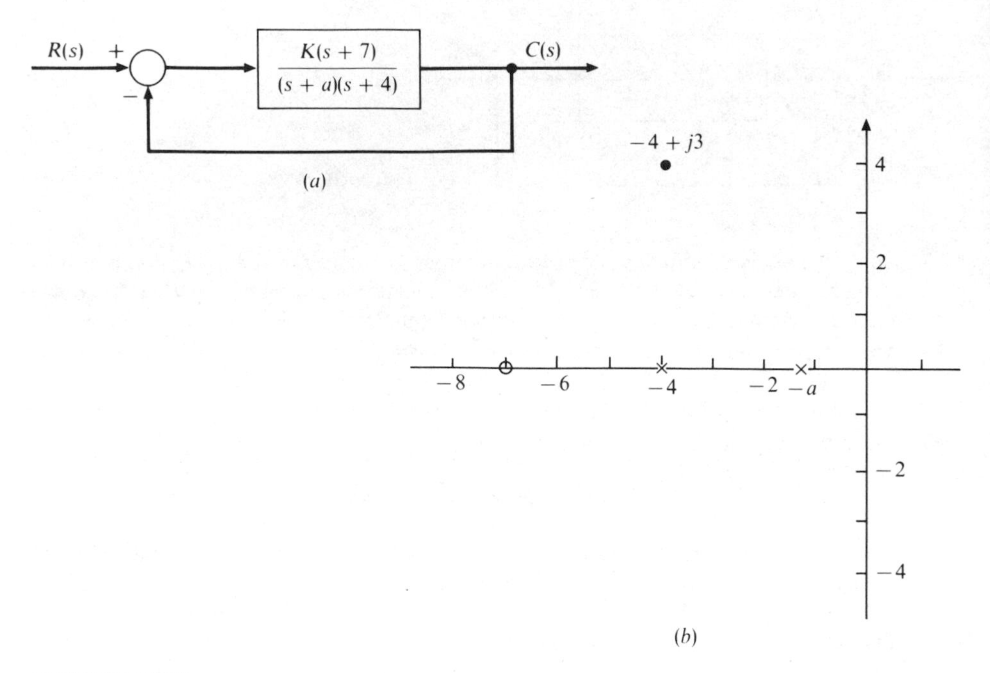

FIGURE P7.12

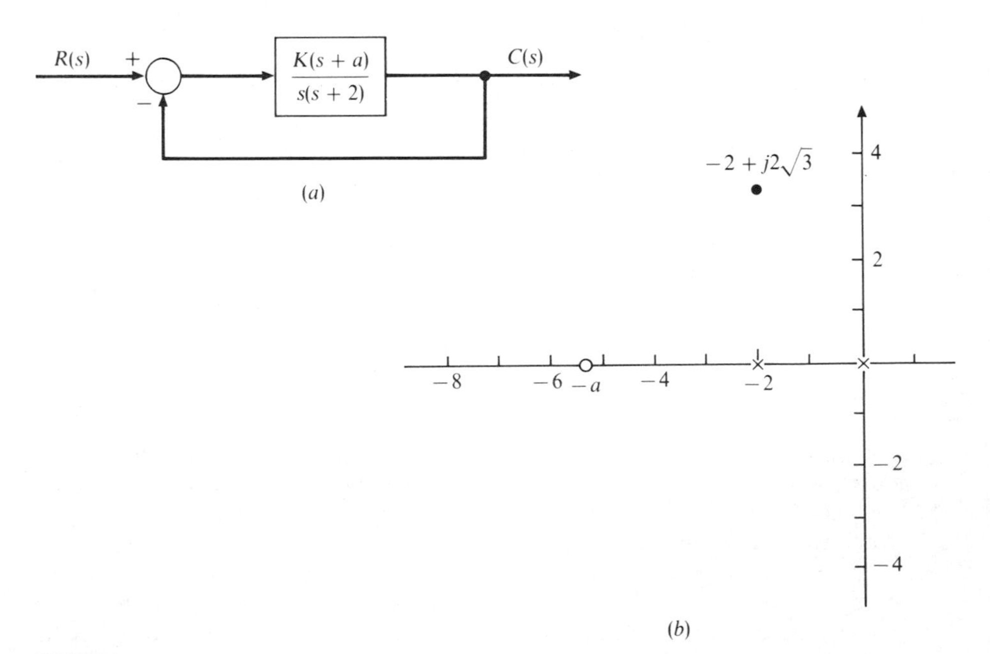

FIGURE P7.13

7.14. The location of the $\times$'s and $\bigcirc$'s of the characteristic equation for various systems is shown in Fig. P7.14. For each system, determine the characteristic equation and then sketch the root-locus plot.

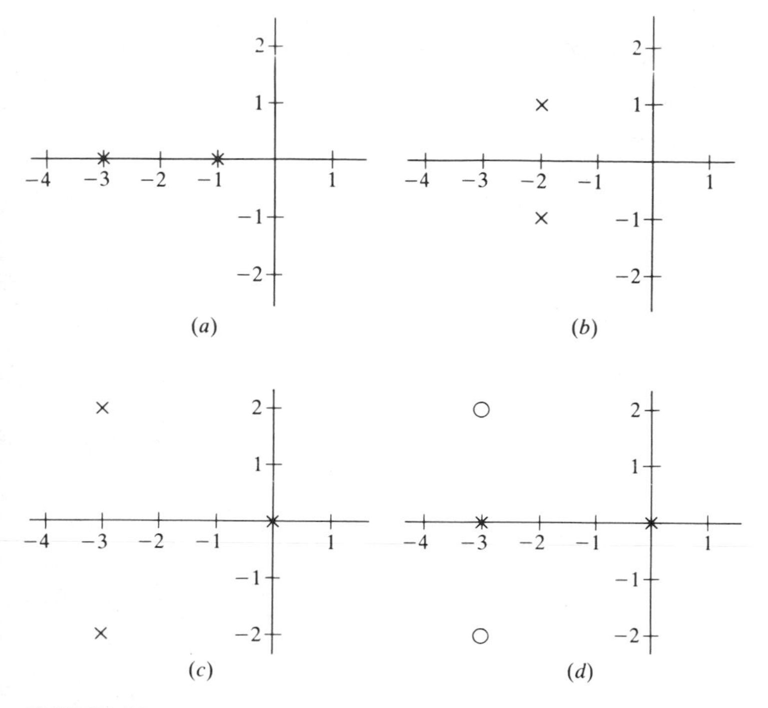

FIGURE P7.14

7.15. Sketch the root-locus plot for the system shown in Fig. P7.15. Determine the value of K to yield a damping ratio of 0.5.

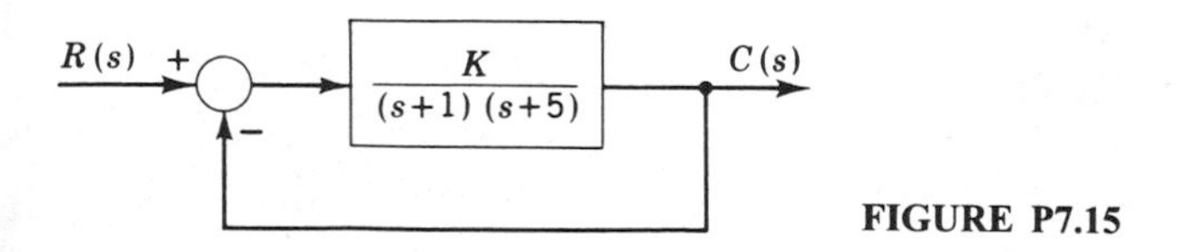

FIGURE P7.15

7.16. For the system shown in Fig. P7.16, determine the roots of the characteristic equation when $K = 1$, 5, 9, and 10. Sketch the root-locus plot, and then identify the points on this plot where $K = 1$, 5, 9, and 10.

(*a*) For $K = 1$, determine the response $c(t)$ when all the initial conditions are zero and $r(t) = \delta(t)$.

(*b*) For $K = 5$, determine the response $c(t)$ when all the initial conditions are zero and $r(t) = \delta(t)$.

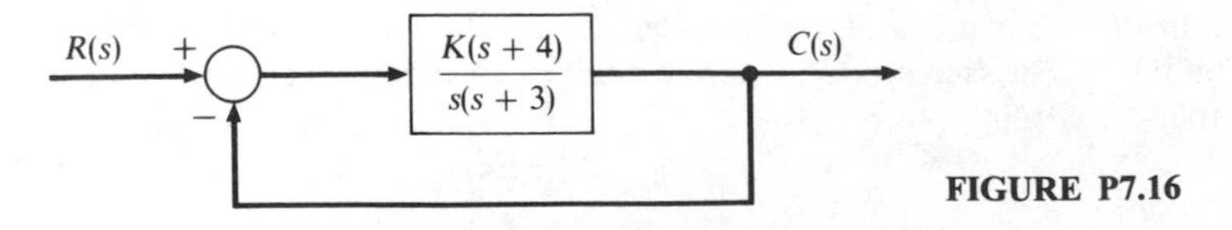

FIGURE P7.16

7.17. The control system and corresponding root-locus plot for a robot with flexibility are shown in Fig. P7.17.

(*a*) Determine the differential equation relating the output $c(t)$ and the input $r(t)$.

(*b*) Determine the response $c(t)$ for the case in which all the initial conditions are zero, $K = 20$, and the input excitation $r(t)$ is a unit impulse.

(*c*) Use Routh's criterion to determine the value of the gain K at which the system becomes unstable.

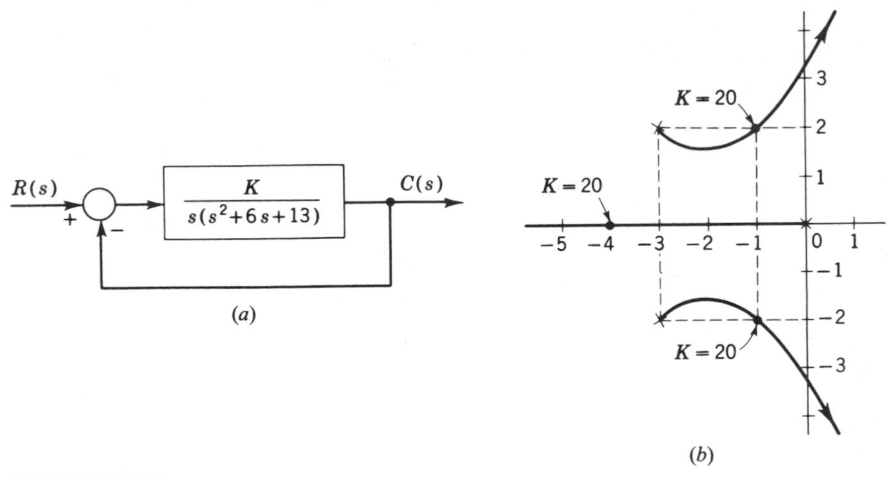

FIGURE P7.17

7.18. The block diagram for a control system is shown in Fig. P7.18. The controller is represented by the block $G_c(s)$. Determine the characteristic equation, and then sketch the corresponding root-locus plot for each of the following cases.

(*a*) Proportional controller, $G_c(s) = K_1$, and $H(s) = 1$

(*b*) Integral controller, $G_c(s) = K_1/s$, and $H(s) = 1$

(*c*) Proportional plus integral controller, $G_c(s) = K_1(1 + 1/s)$, and $H(s) = 1$

(*d*) Proportional plus derivative feedback, $G_c(s) = K_1$, and $H(s) = K_3(1 + s)$

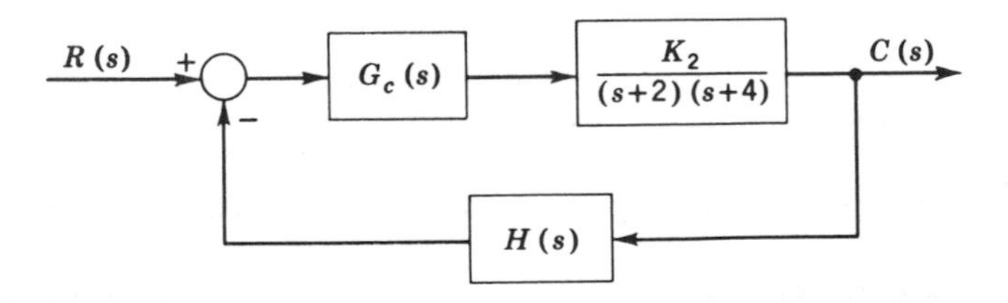

FIGURE P7.18

7.19. For the control system shown in Fig. P7.19, the box $G_c(s)$ represents the characteristics of the controller. Construct the root-locus plot for each of the following cases.

(*a*) Proportional controller, $G_c(s) = K_1$

(*b*) Integral controller, $G_c(s) = K_1/s$

(*c*) Integral plus proportional controller, $G_c(s) = K_1(1 + 1/s)$

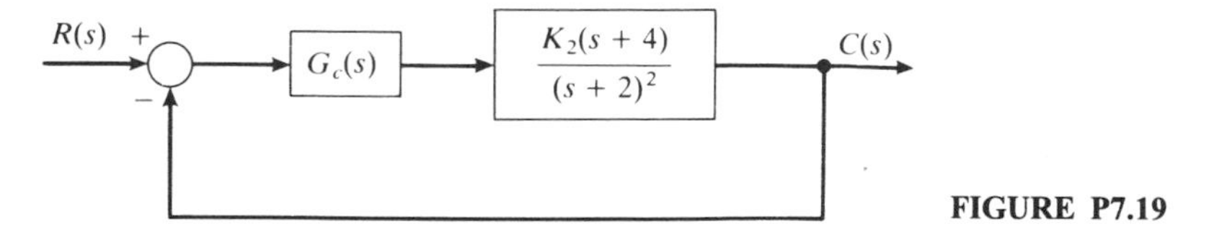

FIGURE P7.19

7.20. Sketch the root-locus plot for the system shown in Fig. P7.20 for each of the following cases.

(*a*) $G_1(s) = \dfrac{K_1}{s}$ (*b*) $G_1(s) = \dfrac{K_1}{s+4}$

For each case, determine the value of $K_1K_2K_3$ to yield a damping ratio of 0.5 for the dominant roots (i.e., the ones located nearest the imaginary axis).

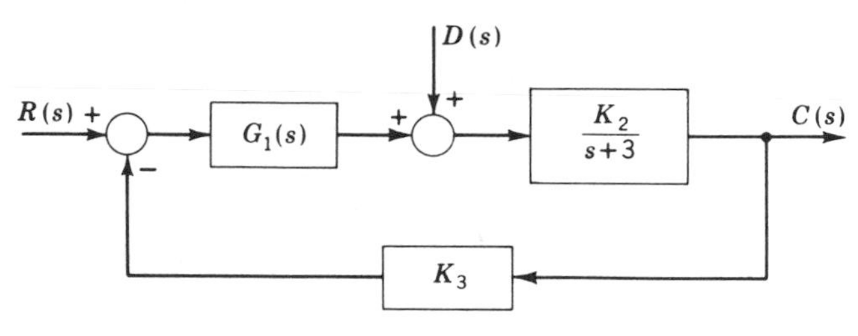

FIGURE P7.20

7.21. Sketch the root-locus plot for each of the two systems shown in Fig. P7.21. Determine the value of K at which each system becomes unstable. Comment on the effect of adding an integrating element, as is done in (*b*).

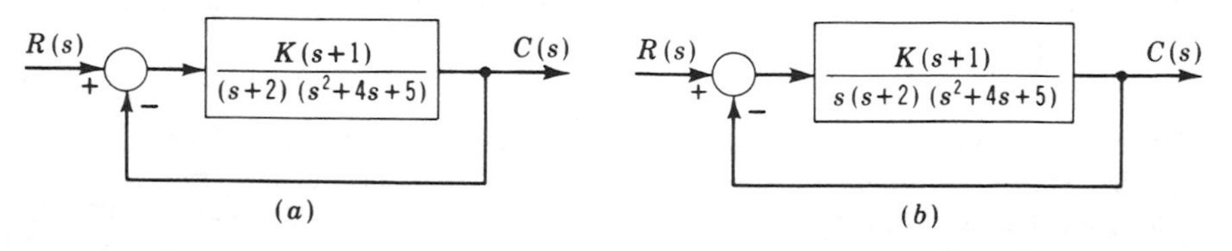

FIGURE P7.21

7.22. The block diagram and root-locus plot for a control system are shown in Fig. P7.22. Repeated roots occur at $s = -6$.

(*a*) Determine the value of the gain K such that the characteristic equation has repeated roots.

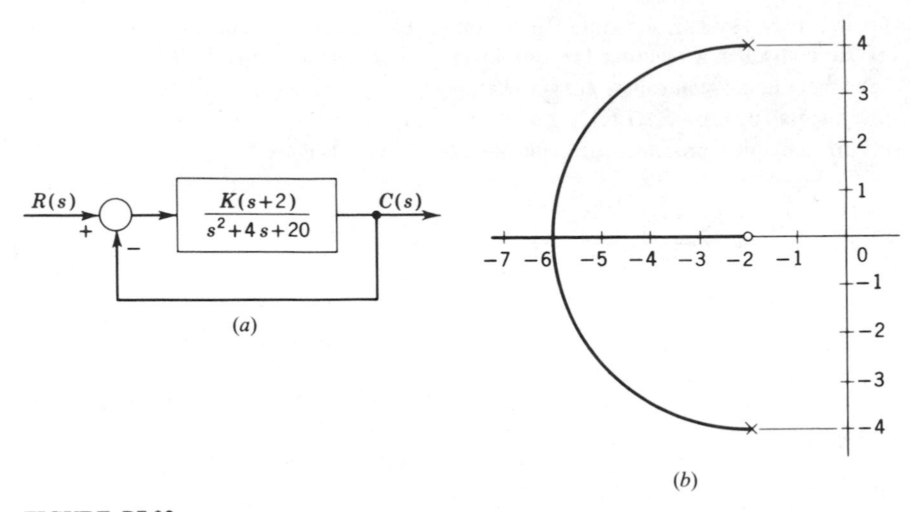

FIGURE P7.22

(*b*) For the case in which the characteristic equation has repeated roots, determine the response $c(t)$ when $r(t)$ is a unit impulse and all the initial conditions are zero.

7.23. For the system shown in Fig. P7.23, sketch the root-locus plot for each of the following cases.

(*a*) $H(s) = 1$ (*b*) $H(s) = s + 2$

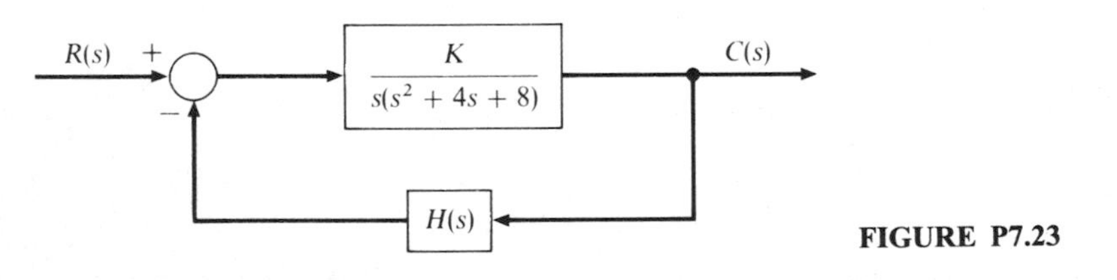

FIGURE P7.23

7.24. Sketch the root-locus plot for each of the following characteristic equations.

(*a*) $(s^2 + 8s + 25) + Ks(s + 4) = 0$

(*b*) $s(s + 4) + K(s^2 + 8s + 25) = 0$

7.25. Sketch the root-locus plot for each of the following characteristic equations.

(*a*) $s(s^2 + 4s + 8) + K = 0$

(*b*) $s(s^2 + 4s + 5) + K = 0$

7.26. The root-locus plot for the system of Fig. 7.18*a* is given in Fig. 7.18*b*. When $K = 17$, the characteristic equation has a repeated root at -1.57 and a distinct root at -6.9. Determine the response $c(t)$ for the case in which $r(t)$ is a unit impulse and $K = 17$. All the initial conditions are zero.

7.27. The characteristic equation for the system shown in Fig. 7.18*a* is given by Eq. (7.12). If a zero is added to the system at -2, the characteristic equation becomes

$$s(s + 4)(s + 6) + K(s + 2) = 0$$

Sketch the root-locus plot for this new system. Comment on the effect of adding the zero.

7.28. Sketch the root-locus plot for each of the characteristic equations given.
(*a*) $s^2(s + 8) + K = 0$ (*b*) $s^2(s + 8) + K(s + 2) = 0$

7.29. Construct the root-locus plot for the system shown in Fig. P6.40. Determine the value of K so that the resulting system will have a damping ratio of 0.6.

7.30. Sketch the root-locus plot for each of the following characteristic equations.
(*a*) $(s + 1)(s^2 + 2s + 2) + K = 0$ (*b*) $(s + 1)(s^2 + 2s + 2) + K(s + 2) = 0$
(*c*) $(s + 1)(s^2 + 2s + 2) + K(s + 2)(s + 4) = 0$

7.31. Sketch the root-locus plot for each of the following characteristic equations.
(*a*) $s(s + 3) + K(s + 4) = 0$ (*b*) $(s + 1)^2 + K(s + 3) = 0$

7.32. Sketch the root-locus plot for each of the following characteristic equations.
(*a*) $s(s^2 + 8s + 25) + K = 0$ (*b*) $s(s^2 + 8s + 25) + K(s + 2) = 0$
(*c*) $s(s^2 + 8s + 25) + K(s + 2)(s + 4) = 0$

7.33. Sketch the root-locus plot for each of the following characteristic equations.
(*a*) $(s + 2)^2 + K(s + 4) = 0$ (*b*) $(s + 2)^3 + K(s + 4) = 0$
(*c*) $s(s^2 + 10s + 26) + K = 0$ (*d*) $(s + 3)^2 + K(s^2 + 10s + 26) = 0$

7.34. A pneumatic actuator has the characteristic equation

$$s(s^2 + 8s + 20) + K(s + 8) = 0$$

Construct the root-locus plot for this system. Determine the value of the gain K when the characteristic equation has a pair of complex conjugate roots whose real part is -2.

7.35. Sketch the root-locus plot for the system shown in Fig. P7.35 for each of the following cases. For each case, determine the value of K at which each system becomes unstable. Comment on the effect of adding derivative action in the feedback path (i.e., case *b*).
(*a*) $H(s) = 1$ (*b*) $H(s) = s + 1$

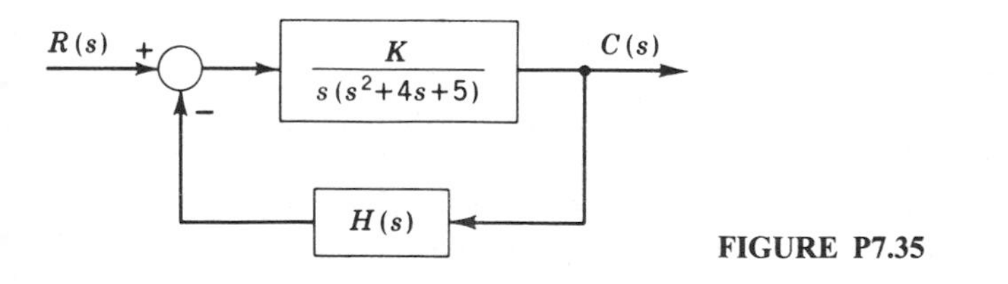

FIGURE P7.35

7.36. The root-locus method may be used to factor a polynomial. Note that the polynomial $s^2 + 4s + 20$ may be written as $s(s + 4) + K_1 = 0$ or as $s^2 + K_2(s + 5) = 0$. The intersection of these root-locus plots are the roots of $s^2 + 4s + 20 = 0$. Construct the root-locus plots and then determine the roots.

7.37. For each of the characteristic equations given below, sketch the root-locus plot for both positive and negative K.
(*a*) $(s + 1)(s + 3)(s + 5) + K = 0$
(*b*) $s(s + 4) + K(s^2 + 8s + 25) = 0$

7.38. For each of the characteristic equations given below, sketch the root-locus plot for both positive and negative K.
(*a*) $s(s + 3)^2 + K = 0$ (*b*) $s(s + 2)(s + 4)(s + 6) + K = 0$
(*c*) $s(s + 6)^2 + K(s^2 + 6s + 18) = 0$ (*d*) $s(s + 2)(s + 4) + K(s^2 + 4s + 8) = 0$

7.39. The control system for a dynamic vibration absorber is shown in Fig. P7.39. Sketch the root-locus plot for
(*a*) $K > 0$
(*b*) $K < 0$

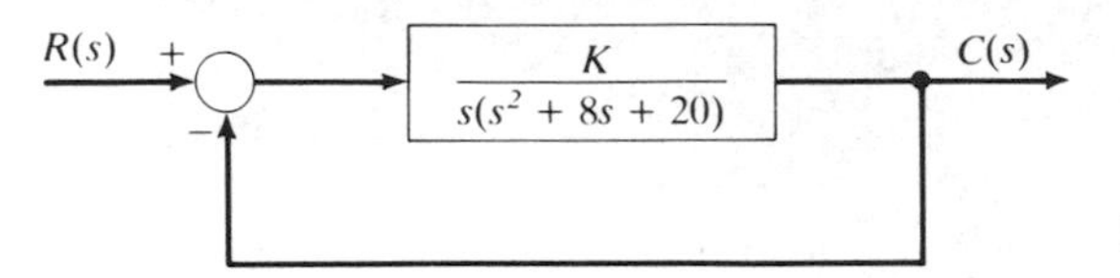

FIGURE P7.39

7.40. The root-locus plot for the system whose characteristic equation is $s(s+2) + K(s+4) = 0$ is shown in Fig. 7.27. When $K = 6$, the roots are $-4 \pm j\sqrt{8}$. Hence, for $K = 6$, the factored form of the characteristic equation is

$$s(s+2) + 6(s+4) = s^2 + 8s + 24$$

To investigate the effect of changing the parameter 4, replace 4 by $4 + \Delta$.
(*a*) Sketch the root-locus plot for positive Δ.
(*b*) Sketch the root-locus plot for negative Δ.

7.41. Repeat Prob. 7.40 except vary the parameter 2 instead of 4.

7.42. From the root-locus plot for the system shown in Fig. P7.42, it is known that when $K = 8$, the roots of the characteristic equation are located at $-2 \pm j2$. Determine the effect of varying the parameter 4 when $K = 8$. Replace 4 by $4 + \Delta$ in the characteristic equation, and then proceed to construct the root-locus plot for $\Delta > 0$ and the plot for $\Delta < 0$.

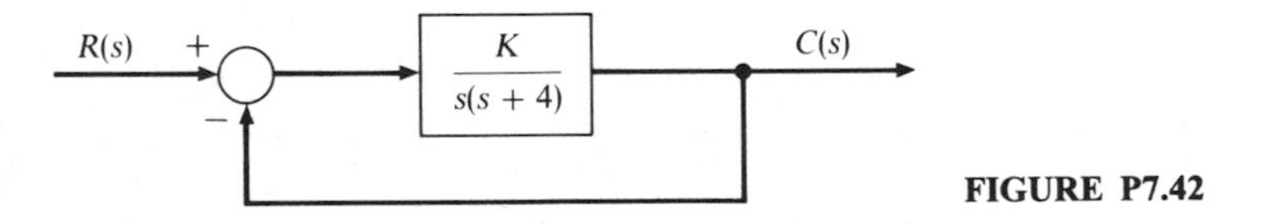

FIGURE P7.42

7.43. A positional control system for an industrial robot is shown in Fig. P7.43. It is known that when $K = 96$, the roots of the characteristic equation are located at -10, $-2 \pm j2\sqrt{3}$. Thus,

$$(s+2)(s+4)(s+8) + 96 = (s+10)(s^2 + 4s + 16)$$

Determine the effect of varying the parameter 8 on the location of the roots of the characteristic equation. Thus, replace the parameter 8 by $8 + \Delta$, and then proceed to construct the root-locus plot for $\Delta > 0$ and for $\Delta < 0$.

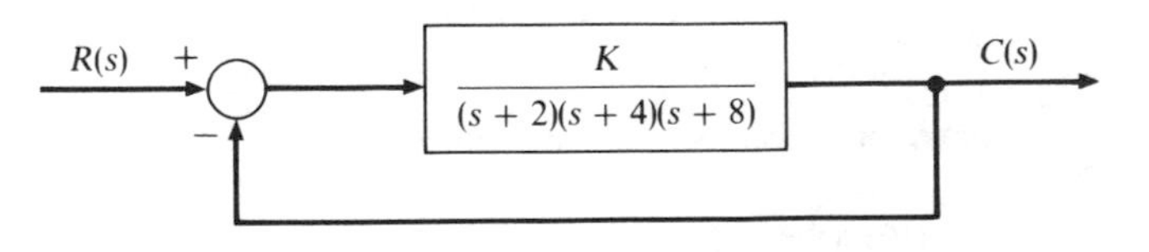

FIGURE P7.43

7.44. A system for minimizing the effect of disturbances and noise in an audio system is shown in Fig. P7.44. It is determined that when $K = 16$, the roots of the characteristic equation are located at -2, -4, and -6. Thus

$$s(s^2 + 12s + 28) + 16(s + 3) = (s + 2)(s + 4)(s + 6)$$

Determine the effect of varying the parameter 3 on the location of the roots of the characteristic equation. Thus, replace the parameter 3 by $3 + \Delta$ in the characteristic equation, and proceed to construct the root-locus plot for $\Delta > 0$ and for $\Delta < 0$.

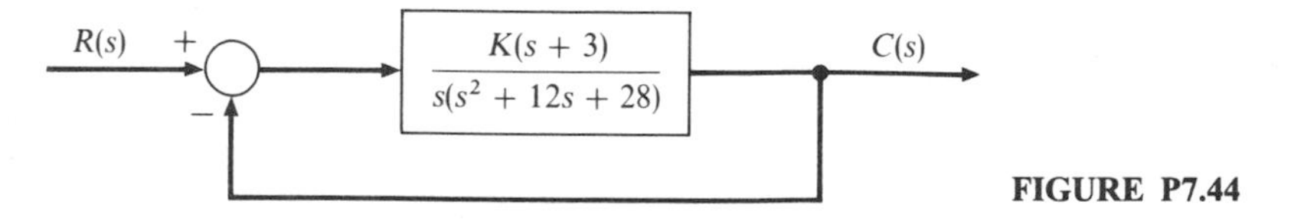

FIGURE P7.44

7.45. The characteristic equation for a system for automatically controlling the focus of a camera is

$$s(s^2 + 10s + 16) + K(s + 2) = 0$$

From the root-locus plot, it is found that when $K = 16$, the roots are located at -2, -4, and -4. Determine the effect of varying the parameter 16. Thus, replace 16 by $16 + \Delta$, and then construct the root-locus plot for $\Delta > 0$ and for $\Delta < 0$.

7.46. For the system of Prob. 7.17, investigate the effect of varying the parameter 13 when $K = 20$. Thus replace the parameter 13 by $13 + \Delta$ in the characteristic equation, and then sketch the resulting root-locus plot for $\Delta > 0$ and for $\Delta < 0$.

7.47. For the system of Prob. 7.17, investigate the effect of varying the parameter 6 when $K = 20$. Thus replace the parameter 6 by $6 + \Delta$, and then sketch the resulting root-locus plot for $\Delta > 0$ and for $\Delta < 0$.

7.48. A feedback control system and its root-locus plot are shown in Fig. P7.48. This is the same system shown in Fig. 7.22.

(*a*) When $K = 50$, the roots of the characteristic equation are located at -5, -5, and -2. Replace the parameter 45 in the characteristic equation by $45 + \Delta$, and proceed to construct the root-locus plot for $\Delta > 0$ and for $\Delta < 0$.

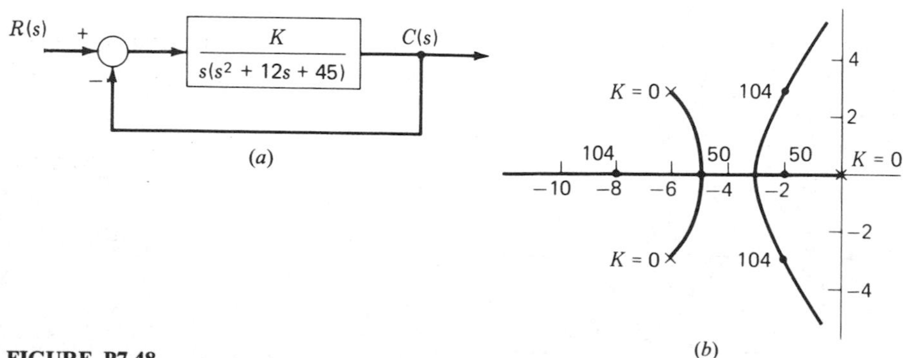

FIGURE P7.48

(*b*) When $K = 104$, the roots of the characteristic equation are located at -8, $-2 + j3$, and $-2 - j3$. Replace the parameter 45 by $45 + \Delta$, and proceed to construct the root-locus plot for $\Delta > 0$ and for $\Delta < 0$.

7.49. The feedforward transfer function for a unity-feedback control system is

$$G(s) = \frac{K(s+8)}{s(s^2 + 8s + 20)}$$

The root-locus plot for this system is shown in Fig. P7.49. When the gain K is 4, the roots of the characteristic equation are -4, $-2 + j2$, and $-2 - j2$. For $K = 4$, investigate the effect of varying the parameter 20 in the transfer function $G(s)$. Sketch the resulting root-locus plot for both positive and negative values of Δ.

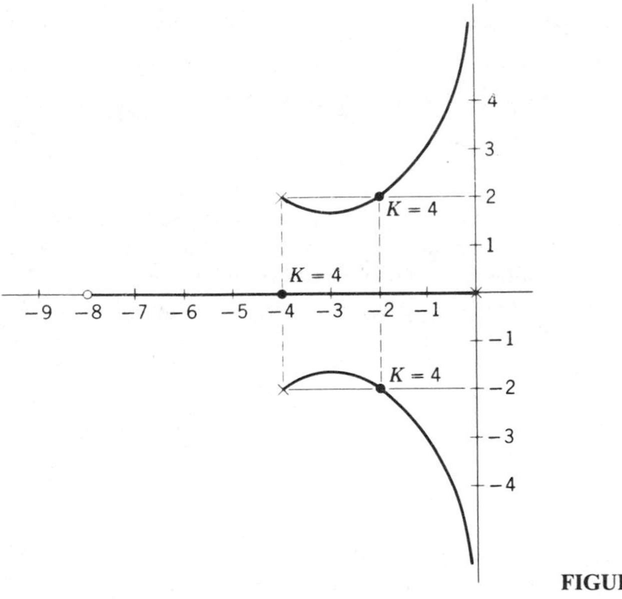

FIGURE P7.49

7.50. It is desired to design a compensator to stabilize the system shown in Fig. P7.50. The only means available is a constant multiplier K which can be inserted in the feedback loop as shown. Sketch the root-locus plot, and then use Routh's criterion to determine the gain K such that the roots of the characteristic equation lie on

(*a*) The imaginary axis

(*b*) The vertical axis at -3

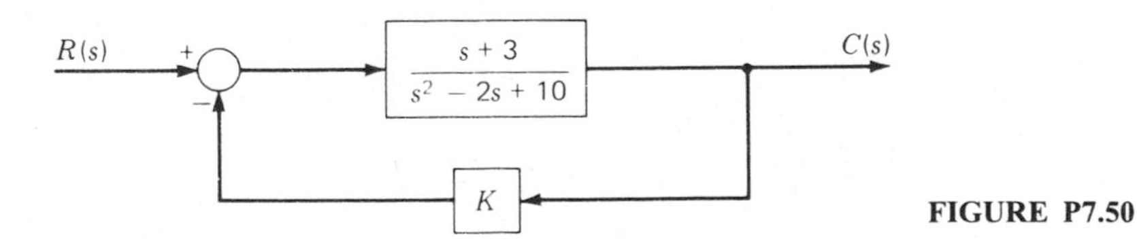

FIGURE P7.50

7.51. A feedback control system with internal feedback is shown in Fig. P7.51. Determine the overall equation relating the output $C(s)$ to the input $R(s)$.

(*a*) Construct the root-locus plot for the basic system without internal feedback ($\beta = 0$).

(*b*) Construct the root-locus plot for the actual system in which β is the variable parameter, $K_1 = 2.5$, and $K_2 = 10.0$.

(*c*) Determine the value of β such that the resulting system will have a damping ratio $\zeta = 0.6$.

(*d*) Determine the value of β such that the resulting system will have a damping ratio $\zeta = 1.0$.

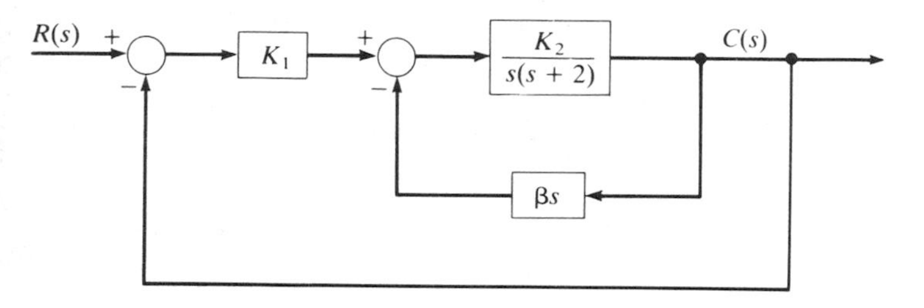

FIGURE P7.51

7.52. The portion of the system enclosed by the dashed box in Fig. P7.52 represents the plant to be controlled. The root-locus plot for this plant shows that for $K_1 = 2$, one root is located at the origin and the other root is located at -2. For $K_1 = 2$, construct the root-locus plot for the entire system, $0 < K_2 < \infty$. For what values of K_2 does the system have an oscillatory response? What is the smallest possible damping ratio?

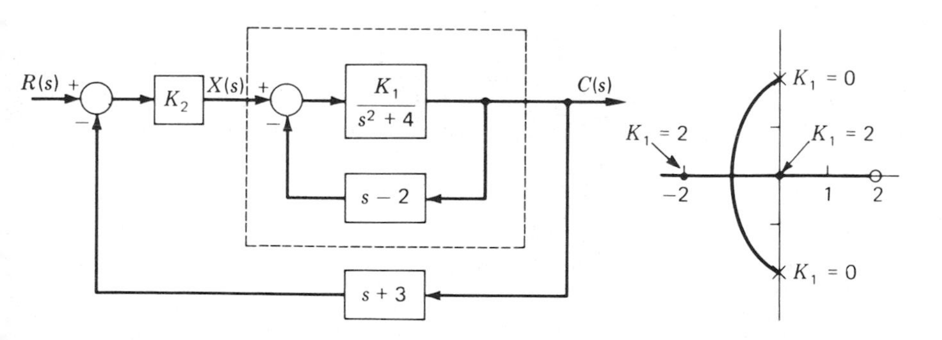

FIGURE P7.52

7.53. Figure P7.53 shows a feedback control system with internal feedback. Construct the root-locus plot for the basic system without the internal feedback ($\beta = 0$), and then construct the root-locus plot for the modified system in which β is the variable parameter, $K_1 = 2$, and $K_2 = 5$. Determine the value of β such that the resulting system will have a damping ratio of (*a*) $\zeta = 0.5$ and (*b*) $\zeta = 1.0$.

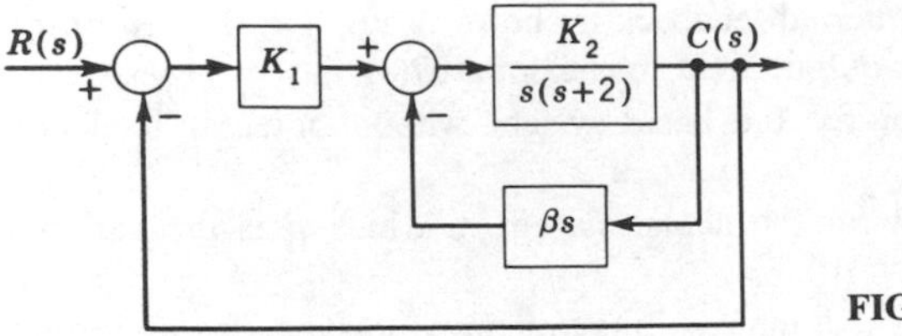

FIGURE P7.53

7.54. The block diagram for controlling the azimuth angle ϕ of the stable platform of an inertial guidance system is shown in Fig. P7.54*a*. Construct the root-locus plot, and determine the roots of the characteristic equation when $K = 4$.

The use of internal feedback with derivative action (βs) to stabilize this system is shown in Fig. P7.54*b*. For $K = 4$, construct the root-locus plot in which β is the variable parameter, and then determine the value of β such that the new characteristic equation will have a damping ratio of (*a*) $\zeta = 0.5$ and (*b*) $\zeta = 1.0$.

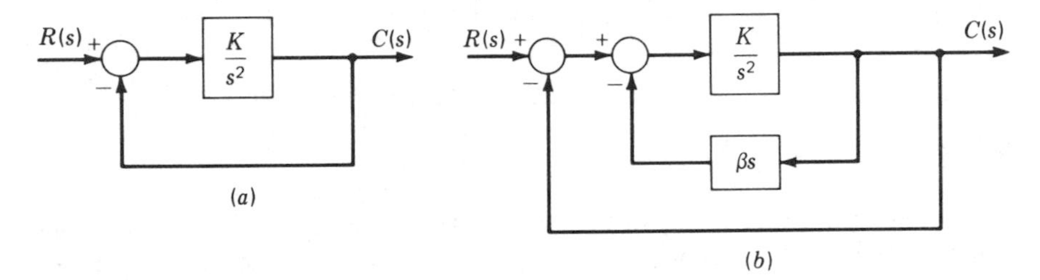

FIGURE P7.54

7.55. The transfer function for a unity-feedback control system is

$$G(s) = \frac{K}{s^2 + 2\zeta\omega_n s + \omega_n^2} = \frac{15}{s^2 + 4s + 5}$$

The roots of the characteristic equation are $r_{1,2} = -2 \pm j4$. Determine the equation for the change Δr_1 in root r_1 due to a change in (*a*) K, (*b*) ω_n, and (*c*) ζ.

CHAPTER 8

ANALOG COMPUTERS

The use of computers has played a major role in recent advances in the design of automatic control systems. These computers may be divided into two types, analog computers and digital computers.

An analog computer is one in which the equation describing the operation of the computer is analogous to that for the actual system. The most commonly used analog computer is the electronic analog computer, in which voltages at various places within the computer are proportional to the variable terms in the actual system. As is shown in this chapter, the operation of a control system can be simulated by the use of an analog computer.

Basically a digital computer can only add and subtract. Thus, it is necessary to reduce all problems to rather elementary mathematical manipulations. This process is called *programming*. The programming of a problem for solution on a digital computer makes extensive use of the methods of numerical analysis to convert the problem to the numerical operations which the computer can perform. It may require weeks or even months to program a problem for a computer, which in turn completes the solution in a few minutes or seconds. A digital computer has been referred to as an "energetic moron" in that it is capable of performing thousands of simple additions and subtractions in a second. A digital computer must store information for use in later computations. This is usually done by means of a magnetic disk, which acts as a memory device.

The input to a digital computer consists of numbers and instructions for the operation of the machine on these numbers. These numbers and instructions may be typed directly into the machine at the computer terminal, or they may be read from a magnetic tape.

Because of the ability of both digital and analog computers to solve complicated mathematical equations almost instantaneously, they are often incorporated as part of a control system to compute desired information. This information may then be used immediately to improve the control of the particular system. For example, in an inertial guidance system, the output of three mutually perpendicular accelerometers is fed into a computer which in turn calculates the position of the vehicle. Thus, the output of this computer is the actual position of the vehicle, which is compared with the desired position to yield an error signal for actuating the steering mechanism.

The electronic analog computer is a very powerful tool for investigating the performance of control systems. For more complex systems, the advantages of the analog computer become more apparent. Analog computers are used for many purposes besides investigating linear and nonlinear control systems. For example, they are used to solve nonlinear differential equations, partial differential equations, systems of differential or partial differential equations, matrix and eigenvalue problems, operations research problems, etc. New applications and uses for this versatile computing device are continually being discovered.

This chapter is primarily concerned with the use of these computers for simulating control systems. For this purpose, the equation describing the operation of the analog computer is analogous to that representing the actual physical system. The variable quantities of the actual system, such as the output, input, error, etc., are represented by voltages at various places within the analog computer. Permanent records of the voltages may be obtained by using recording equipment. By using potentiometers or variable capacitors to vary the resistance or capacitance at various places within the computer, the effect upon system performance of changing the corresponding parameters in the actual system (e.g., gain, time constants, damping ratio, etc.) may be determined immediately.

8.1 COMPUTER OPERATIONS

To solve any linear differential equation with constant coefficients, it is necessary only to make use of the processes of integration, summation, and multiplication by a constant. This is illustrated by the block diagram of Fig. 8.1 for the equation

$$M\ddot{y} + B\dot{y} + Ky = f(t) \tag{8.1}$$

To set up the block-diagram representation for a differential equation, first solve for the highest-order differential appearing in the original equation. The highest-order term appearing in Eq. (8.1) is $M\ddot{y}$. Solving Eq. (8.1) for $M\ddot{y}$ yields

$$M\ddot{y} = f(t) - B\dot{y} - Ky \tag{8.2}$$

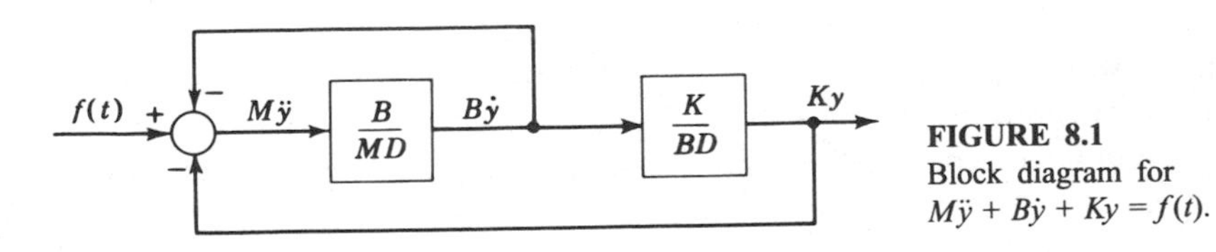

FIGURE 8.1
Block diagram for $M\ddot{y} + B\dot{y} + Ky = f(t)$.

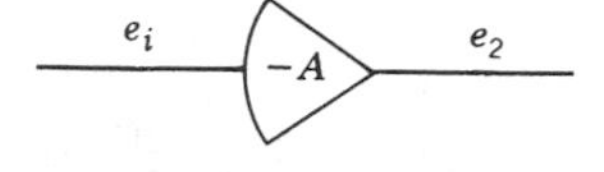

FIGURE 8.2
Schematic representation for an operational amplifier.

Successive integration of the highest-order differential and multiplication by appropriate constants yield the other lower-order terms

$$\frac{B}{MD}(M\ddot{y}) = B\dot{y}$$
$$\frac{K}{BD}(B\dot{y}) = Ky \tag{8.3}$$

Each term on the right side of Eq. (8.2) goes into the summer of Fig. 8.1, so that the output of the summer is proportional to the acceleration. Successive integration of this acceleration yields the velocity and displacement.

The heart of the electronic analog computer is the operational amplifier, which is a very high-gain dc amplifier. This device may be used as an integrator, summer, or multiplier. The particular mathematical operation depends upon the network of resistors and capacitors which are placed around it.

Figure 8.2 shows the schematic representation of an operational amplifier. The input voltage is e_i, the output is e_2, and the amplification is $-A$. Thus

$$e_2 = -Ae_i \tag{8.4}$$

The reason for the minus sign is that the amplifier reverses the phase of the input. For most operational amplifiers, the value of A is very large. Values of A may range from 100×10^3 to 100×10^6.

Multiplication by a Constant

By feeding the input voltage through a resistor R_1 and by putting a resistor R_2 in parallel with the amplifier as shown in Fig. 8.3, a circuit for multiplication by a constant is obtained. Because the input grid of the amplifier draws little or no current (a typical value is 10^{-9} A),

$$i_1 \approx i_2 \tag{8.5}$$

The values of i_1 and i_2 are

$$i_1 = \frac{e_1 - e_i}{R_1} \qquad i_2 = \frac{e_i - e_2}{R_2} \tag{8.6}$$

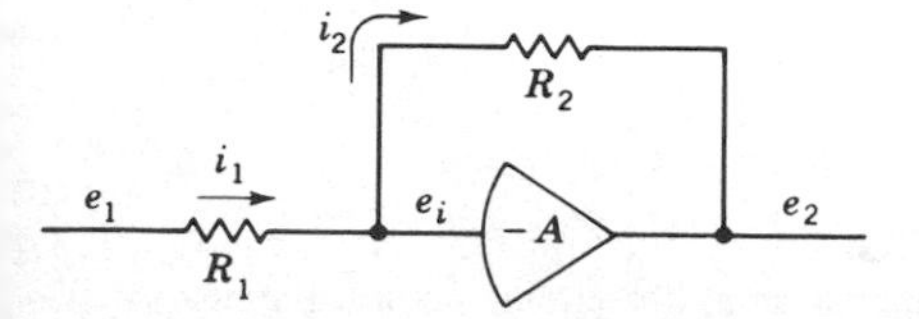

FIGURE 8.3
Operational amplifier circuit for multiplying by a constant.

Equating i_1 and i_2 yields

$$\frac{e_1 - e_i}{R_1} = \frac{e_i - e_2}{R_2} \tag{8.7}$$

Usually e_2 is less than 10 V; so for the very large values of A, it follows from Eq. (8.4) that $e_i = -e_2/A \approx 0$. The substitution of $e_i \approx 0$ into Eq. (8.7) gives

$$e_2 = -\frac{R_2}{R_1} e_1 \tag{8.8}$$

The value of the multiplication constant is $-R_2/R_1$. When $R_2 = R_1$, the constant is -1, which means simply that the phase of the input signal has been inverted.

Integration

Replacing resistor R_2 of Fig. 8.3 with a capacitor, as shown in Fig. 8.4, yields an integration circuit. The current i_2 flowing through this capacitor is

$$i_2 = C_2 D(e_i - e_2) \tag{8.9}$$

By equating i_1 and i_2 as before, it follows that

$$\frac{e_1 - e_i}{R_1} = C_2 D(e_i - e_2) \tag{8.10}$$

Because $e_i \approx 0$, the preceding reduces to

$$e_2 = \frac{-e_1}{R_1 C_2 D} = \frac{-1}{R_1 C_2} \int_0^t e_1 \, dt + e_2(0) \tag{8.11}$$

where $e_2(0)$ is the value of e_2 when $t = 0$. In addition to integration, the circuit of Fig. 8.4 multiplies by the constant $-1/(R_1 C_2)$.

Summation

The effect of summation is obtained by placing the desired quantities to be added in parallel at the input to the computer circuit. The output will then be the summation of the effect due to each individual input. A general summing circuit is shown in Fig. 8.5*a*, and a circuit to integrate more than one quantity is shown in Fig. 8.5*b*. A simplified notation for these various computer elements is also shown in Fig. 8.5*a* and *b*. This schematic notation saves much effort in drawing the computer diagram for a circuit.

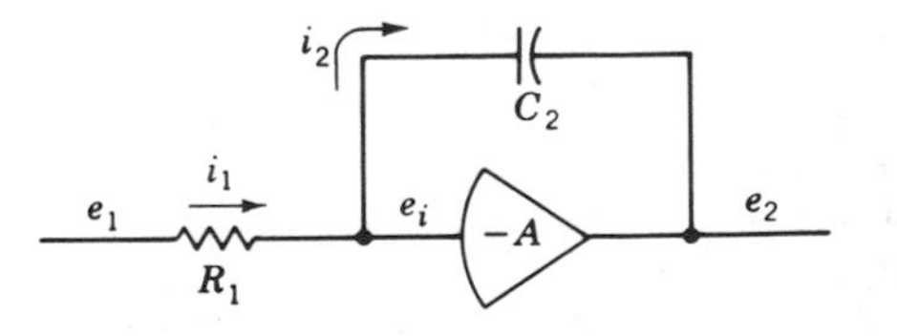

FIGURE 8.4
Operational amplifier circuit for integrating.

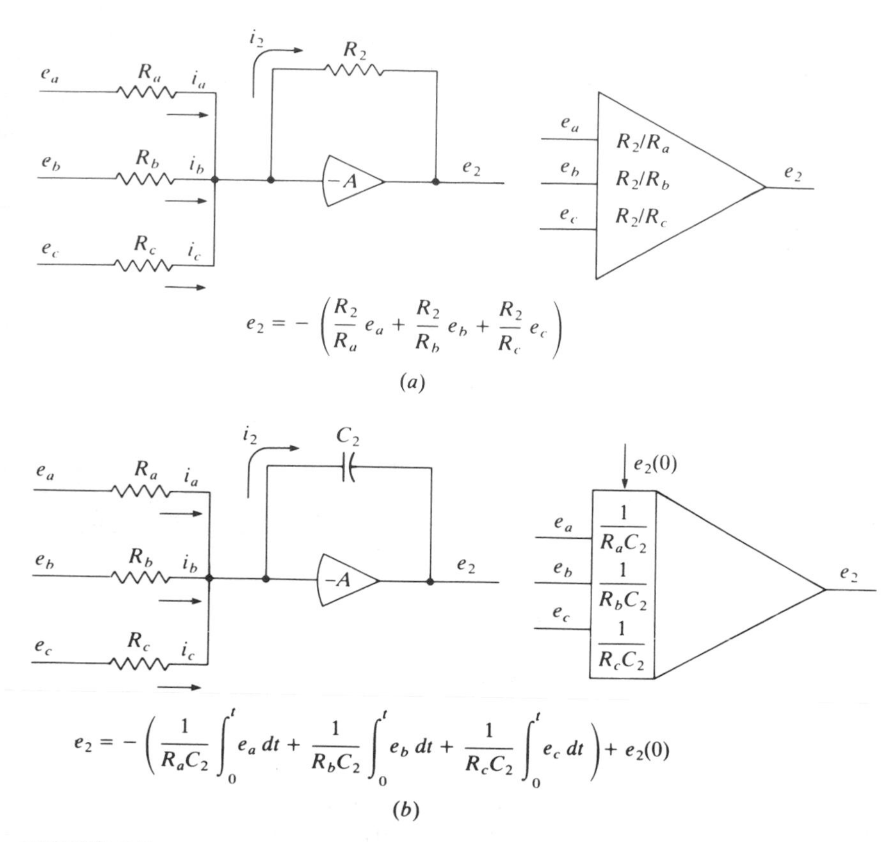

FIGURE 8.5
(*a*) General summing circuit, (*b*) general integrating circuit.

For the summing circuit of Fig. 8.5*a*, the sum of currents i_a plus i_b plus i_c is

$$i_a + i_b + i_c = \frac{e_a - e_i}{R_a} + \frac{e_b - e_i}{R_b} + \frac{e_c - e_i}{R_c} \approx \frac{e_a}{R_a} + \frac{e_b}{R_b} + \frac{e_c}{R_c}$$

Current i_2 is

$$i_2 = \frac{e_i - e_2}{R_2} \approx -\frac{e_2}{R_2}$$

Equating the right sides of these equations and solving for e_2 give

$$e_2 = -\left(\frac{R_2}{R_a}e_a + \frac{R_2}{R_b}e_b + \frac{R_2}{R_c}e_c\right) \tag{8.12}$$

The minus sign results because an amplifier always reverses the sign of the input voltage. Note that the equation for voltage e_2 may be obtained directly from the schematic diagram by multiplying each input voltage by the value of the ratio of the resistors for that particular input. That is, e_a is multiplied by R_2/R_a, e_b by R_2/R_b, and e_c by R_2/R_c.

For the integrating circuit in Fig. 8.5*b*, the sum of currents i_a plus i_b plus i_c is

$$i_a + i_b + i_c = \frac{e_a - e_i}{R_a} + \frac{e_b - e_i}{R_b} + \frac{e_c - e_i}{R_c} \approx \frac{e_a}{R_a} + \frac{e_b}{R_b} + \frac{e_c}{R_c}$$

Current i_2 is

$$i_2 = C_2 D(e_i - e_2) \approx -C_2 D e_2$$

Equating the right sides of these equations and solving for e_2 show that

$$\begin{aligned} e_2 &= -\left(\frac{1}{R_a C_2 D} e_a + \frac{1}{R_b C_2 D} e_b + \frac{1}{R_c C_2 D} e_c\right) \\ &= -\left(\frac{1}{R_a C_2}\int_0^t e_a\,dt + \frac{1}{R_b C_2}\int_0^t e_b\,dt + \frac{1}{R_c C_2}\int_0^t e_c\,dt\right) + e_2(0) \end{aligned} \tag{8.13}$$

Note that the schematic representation reveals directly this equation. That is, the equation for voltage e_2 is obtained by multiplying the integral of each input voltage by the corresponding $1/(RC)$ factor for that input and adding the initial value $e_2(0)$.

8.2 DIRECT PROGRAMMING

Suppose we want to determine the computer diagram for solving the following differential equation:

$$\ddot{y} + b\dot{y} + ay = f(t) \tag{8.14}$$

The initial conditions are designated as $y(0)$ and $\dot{y}(0)$. The expected maximum values for y, $\dot{y}$, $\ddot{y}$, and $f(t)$ are designated as y_m, $\dot{y}_m$, $\ddot{y}_m$, and $f(t)_m$.

Solving Eq. (8.14) for the highest-order derivative $\ddot{y}$ gives

$$\ddot{y} = f(t) - ay - b\dot{y} \tag{8.15}$$

Dividing through by $\ddot{y}_m$ and expressing $f(t)$ as $f(t)_m\,[f(t)/f(t)_m]$, y as $y_m(y/y_m)$, and $\dot{y}$ as $\dot{y}_m(\dot{y}/\dot{y}_m)$ yield

$$\frac{\ddot{y}}{\ddot{y}_m} = \frac{f(t)_m}{\ddot{y}_m}\left[\frac{f(t)}{f(t)_m}\right] - a\frac{y_m}{\ddot{y}_m}\left[\frac{y}{y_m}\right] - b\frac{\dot{y}_m}{\ddot{y}_m}\left[\frac{\dot{y}}{\dot{y}_m}\right] \tag{8.16}$$

Note that each term in brackets is the ratio of the variable compared to the maximum value [that is, $f(t)/f(t)_m$, y/y_m, and $\dot{y}/\dot{y}_m$]. Depending on the type of operational amplifiers used in a particular computer, the manufacturer suggests limiting the output voltage to a certain value such as 10 V. If the output voltage exceeds the suggested maximum value, the amplifier begins to saturate. Each amplifier has a light that goes on when it saturates.

The general computer diagram is shown in Fig. 8.6*a*. Because an amplifier reverses the sign from the input to the output, the sign in front of each scale factor alternates in going from one amplifier to the next. This fact is illustrated in Fig. 8.6*a* in which the output of the first amplifier is $e_1 = -(\ddot{y}/\ddot{y}_m)e_m$, the output of the second is $e_2 = (\dot{y}/\dot{y}_m)e_m$, and the output of the third is $e_3 = -(y/y_m)e_m$.

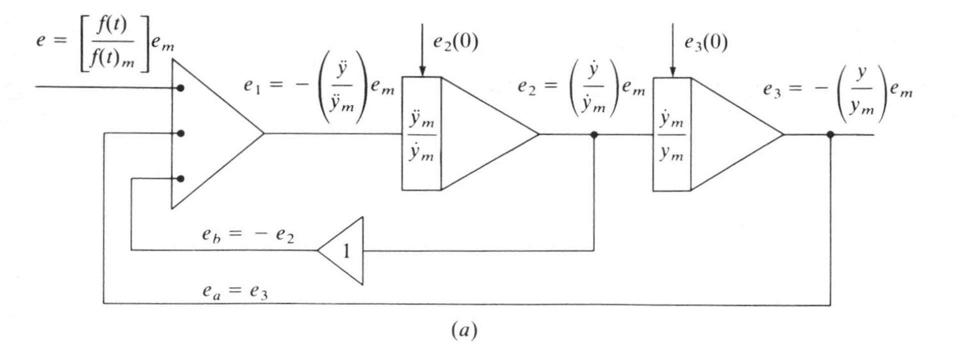

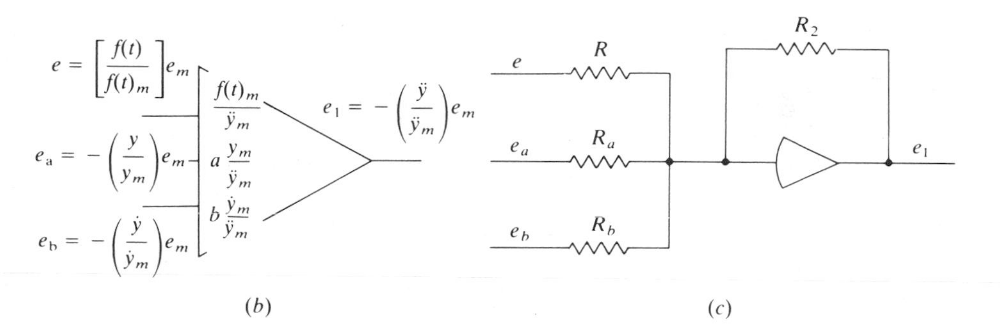

FIGURE 8.6
Analog computer diagram for $\ddot{y} + b\dot{y} + ay = f(t)$: (*a*) overall diagram, (*b*) first amplifier, (*c*) circuit diagram.

A more detailed diagram of the first amplifier is shown in Fig. 8.6*b*. Multiplying through Eq. (8.16) by e_m, where e_m is the suggested maximum voltage, gives

$$\frac{\ddot{y}}{\ddot{y}_m}e_m = \frac{f(t)_m}{\ddot{y}_m}\left[\frac{f(t)}{f(t)_m}\right]e_m - a\,\frac{y_m}{\ddot{y}_m}\left[\frac{y}{y_m}\right]e_m - b\,\frac{\dot{y}_m}{\ddot{y}_m}\left[\frac{\dot{y}}{\dot{y}_m}\right]e_m \tag{8.17}$$

As shown in Fig. 8.6*b*, each term on the right side is inputted to the first amplifier to generate $e_1 = (\ddot{y}/\ddot{y}_m)e_m$. Because the amplifier reverses sign, the output of the first amplifier is $e_1 = -(\ddot{y}/\ddot{y}_m)e_m$. Note that Eq. (8.17) is merely an algebraic rearrangement of Eq. (8.15). That is, the ay and $b\dot{y}$ terms in Eq. (8.15) have been taken to the right side, multiplied, and divided by y_m and $\dot{y}_m$, respectively, and then each term has been divided by $\ddot{y}_m$ and multiplied by e_m. The circuit diagram for the first amplifier is shown in Fig. 8.6*c*. The voltage relationship is

$$e_1 = -\frac{R_2}{R}e - \frac{R_2}{R_a}e_a - \frac{R_2}{R_b}e_b \tag{8.18}$$

Because the output of the second amplifier is $e_2 = (\dot{y}/\dot{y}_m)e_m$, it is necessary to use a sign changer to obtain $e_b = -e_2 = (-\dot{y}/\dot{y}_m)e_m$. The voltage $e_a = e_3 = (-y/y_m)e_m$ may be fed back directly. Multiplying through Eq. (8.17) by -1 gives

$$-\frac{\ddot{y}}{\ddot{y}_m}e_m = -\frac{f(t)_m}{\ddot{y}_m}\left[\frac{f(t)}{f(t)_m}\right]e_m - a\frac{y_m}{\ddot{y}_m}\left[-\frac{y}{y_m}\right]e_m - b\frac{\dot{y}_m}{\ddot{y}_m}\left[-\frac{\dot{y}}{\dot{y}_m}\right]e_m \tag{8.19}$$

Termwise comparison of Eq. (8.18), which is the voltage relationship for the first amplifier, with Eq. (8.19), which is the physical relationship to be solved by the first amplifier, shows that to have $e_1 = -(\ddot{y}/\ddot{y}_m)e_m$, $e = [f(t)/f(t)_m]e_m$, $e_a = (-y/y_m)e_m$, and $e_b = (-\dot{y}/\dot{y}_m)e_m$, then

$$\frac{R_2}{R} = \frac{f(t)_m}{\ddot{y}_m} \qquad \frac{R_2}{R_a} = a\frac{y_m}{\ddot{y}_m} \qquad \frac{R_2}{R_b} = b\frac{\dot{y}_m}{\ddot{y}_m} \tag{8.20}$$

Note that the value of $1/(R_1C_2)_2$ is equal to the ratio of the maximum value of the input $\ddot{y}_m$ to the maximum value of the output $\dot{y}_m$. The initial condition expression could also have been obtained by the evaluation of $e_2 = (\dot{y}/\dot{y}_m)e_m$ at time $t = 0$.

$$e_2 = -\frac{1}{(R_1C_2)_2}\int_0^t e_1\,dt + e_2(0) \tag{8.21}$$

where $(R_1C_2)_2$ is the product of the resistance and capacitance for amplifier 2. Because the integral of acceleration is velocity, it follows that

$$\dot{y} = \int_0^t \ddot{y}\,dt + \dot{y}(0)$$

This may be rewritten in the form

$$\frac{\dot{y}}{\dot{y}_m}e_m = -\frac{\ddot{y}_m}{\dot{y}_m}\int_0^t\left(-\frac{\ddot{y}}{\ddot{y}_m}\right)e_m\,dt + \frac{\dot{y}(0)}{\dot{y}_m}e_m \tag{8.22}$$

Equation (8.21) is the voltage relationship, and Eq. (8.22) is the corresponding physical relationship. To have $e_2 = (\dot{y}/\dot{y}_m)e_m$ and $e_1 = -(\ddot{y}/\ddot{y}_m)e_m$, termwise comparison of Eqs. (8.21) and (8.22) shows that

$$\frac{1}{(R_1C_2)_2} = \frac{\ddot{y}_m}{\dot{y}_m} \qquad \text{and} \qquad e_2(0) = \frac{\dot{y}(0)}{\dot{y}_m}e_m \tag{8.23}$$

Note that the value of $1/(R_1C_2)_2$ is equal to the ratio of the maximum value of the input $\ddot{y}_m$ to the maximum value of the output $\dot{y}_m$. The initial condition expression could also have been obtained by the evaluation of $e_2 = (\dot{y}/\dot{y}_m)e_m$ at time $t = 0$.

The voltage equation for the third amplifier is

$$e_3 = -\frac{1}{(R_1C_2)_3}\int_0^t e_2\,dt + e_3(0) \tag{8.24}$$

Because the integral of velocity is displacement,

$$y = \int_0^t \dot{y}\, dt + y(0)$$

This may be rewritten in the form

$$-\frac{y}{y_m} e_m = -\frac{\dot{y}_m}{y_m} \int_0^t \left(\frac{\dot{y}}{\dot{y}_m}\right) e_m\, dt - \frac{y(0)}{y_m} e_m \tag{8.25}$$

Termwise comparison of the voltage relationship [Eq. (8.24)] with the physical relationship [Eq. (8.25)] shows that to have $e_3 = -(y/y_m)e_m$ and $e_2 = (\dot{y}/\dot{y}_m)e_m$, then

$$\frac{1}{(R_1C_2)_3} = \frac{\dot{y}_m}{y_m} \qquad \text{and} \qquad e_3(0) = -\frac{y(0)}{y_m} e_m \tag{8.26}$$

The value of $1/(R_1C_2)_3$ is equal to the ratio of the maximum value of the input $\dot{y}_m$ to the maximum value of the output y_m. In general, the value of $1/(R_1C_2)$ for an integrator is equal to the ratio of the maximum value of the input (physical quantity to be integrated) to the maximum value of the output. Evaluating the voltage relationship $e_3 = -(y/y_m)e_m$ at time $t = 0$ verifies the result shown for the preceding initial-condition relationship.

Each integrator may be initially biased by a dc voltage to give the effect of initial conditions, as is represented diagrammatically at the top of each integrator in Fig. 8.6*a*. In practice, the initial condition is obtained by placing a source of constant potential such as a battery of potential $e_2(0)$ in parallel with the capacitor, as is shown in Fig. 8.7. For $t < 0$, the switch is up so that the battery can charge the capacitor to the desired initial value. At $t = 0$, the switch moves down to disconnect the battery so that the integrator circuit functions as previously discussed. There is never any need to apply an initial voltage $e(0)$ to account for initial conditions when an amplifier is used to multiply by a constant. Only integration requires initial conditions.

The relationships $e_1 = -(\ddot{y}/\ddot{y}_m)e_m$, $e_2 = (\dot{y}/\dot{y}_m)e_m$, and $e_3 = -(y/y_m)e_m$ may be rewritten in the form

$$\frac{e_1}{e_m} = -\frac{\ddot{y}}{\ddot{y}_m} \qquad \frac{e_2}{e_m} = \frac{\dot{y}}{\dot{y}_m} \qquad \frac{e_3}{e_m} = -\frac{y}{y_m} \tag{8.27}$$

For numerical purposes, suppose $\ddot{y}_m = 20$, $\dot{y}_m = 10$, and $y_m = 5$. If $e_m = 10$ V for the analog computer, then when $e_1 = 7.5$ V, the corresponding value of $\ddot{y} = -15$.

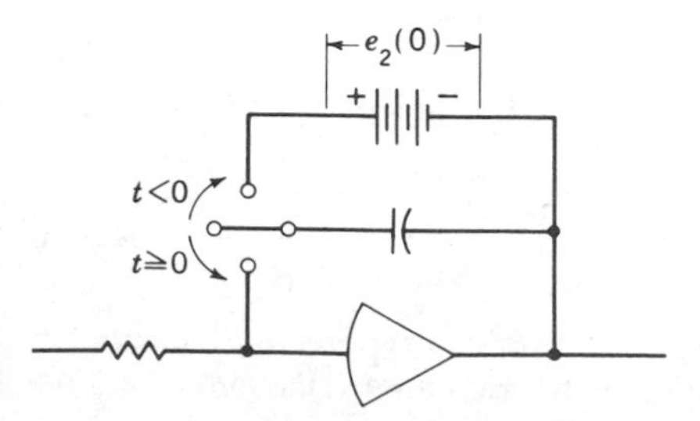

FIGURE 8.7
Circuit for obtaining initial conditions.

For $e_2 = 7.5$ V, then $\dot{y} = 7.5$. For $e_3 = 7.5$ V, then $y = -3.75$. Consider now the case of an analog computer for which $e_m = 100$ V. When $e_1 = 75$ V, the corresponding value of $\ddot{y}$ is -15. For $e_2 = 75$ V, then $\dot{y} = 7.5$. For $e_3 = 75$ V, then $y = -3.75$. From Eq. (8.27) we see that the ratio of the actual voltage to e_m is equal to the ratio of the actual value of the parameter to the maximum value of the parameter. With this method, the computer diagram is independent of the value of e_m for the particular computer being used.

It is not necessary to rearrange algebraically the differential equation and then compare it to the voltage relationship in order to obtain the value of the resistors for the first amplifier. A much simpler technique is now described. Consider the differential equation

$$\ddot{y} + b\dot{y} + ay = f(t) \tag{8.14}$$

Solving for $\ddot{y}$, dividing through by $\ddot{y}_m$, multiplying through by e_m, and then expressing each variable as a ratio compared to its maximum value give

$$\frac{\ddot{y}}{\ddot{y}_m} e_m = \frac{f(t)_m}{\ddot{y}_m}\left[\frac{f(t)}{f(t)_m}\right] e_m + a\frac{y_m}{\ddot{y}_m}\left(-\frac{y}{y_m}\right) e_m + b\frac{\dot{y}_m}{\ddot{y}_m}\left(-\frac{\dot{y}}{\dot{y}_m}\right) e_m \tag{8.17}$$

Because an amplifier reverses sign, when each term on the right side is fed into the first amplifier, the output is the negative of the left side, that is, $e_1 = -(\ddot{y}/\ddot{y}_m)e_m$. Referring to Fig. 8.6$b$, we note that the ratio of resistors R_2/R for the input voltage $e = [f(t)/f(t)_m]e_m$ is the coefficient $f(t)_m/\ddot{y}_m$ of the term $[f(t)/f(t)_m]e_m$ in Eq. (8.17). Similarly, the ratio of resistors R_2/R_a for the input voltage $e_a = (-y/y_m)e_m$ is the coefficient $ay_m/\ddot{y}_m$ of the $(-y/y_m)e_m$ term. Finally, the ratio of resistors R_2/R_b for the input voltage $e_b = -e_2 = (-\dot{y}/\dot{y}_m)e_m$ is the coefficient $b\dot{y}_m/\ddot{y}_m$ of the $(-\dot{y}/\dot{y}_m)e_m$ term. The need for the sign changer is easily detected because in the computer diagram of Fig. 8.6a, the voltage $e_2 = (\dot{y}/\dot{y}_m)e_m$ is generated whereas in Eq. (8.17) the term $(-\dot{y}/\dot{y}_m)e_m$ is obtained. The value of $1/(R_1C_2)_2$ for the second integrator is $\ddot{y}_m/\dot{y}_m$ and the value of $1/(R_1C_2)_3$ for the third integrator is $\dot{y}_m/y_m$.

Illustrative example 8.1. Suppose we want to determine the computer diagram for solving the following differential equation:

$$\ddot{y} + 2\dot{y} + 10y = f(t) \tag{8.28}$$

The initial conditions are $y(0) = 3$ and $\dot{y}(0) = 5$. The maximum values are $y_m = 5$, $\dot{y}_m = 10$, $\ddot{y}_m = 20$, and $f(t)_m = 8$.

Solution. Solving Eq. (8.28) for the highest-order derivative yields

$$\ddot{y} = f(t) - 10y - 2\dot{y} \tag{8.29}$$

Dividing through by $\ddot{y}_m$ and expressing $f(t)$, y, and $\dot{y}$ in terms of ratios compared to their respective maximum values give

$$\frac{\ddot{y}}{\ddot{y}_m} = \frac{f(t)_m}{\ddot{y}_m}\left[\frac{f(t)}{f(t)_m}\right] + 10\frac{y_m}{\ddot{y}_m}\left[-\frac{y}{y_m}\right] + 2\frac{\dot{y}_m}{\ddot{y}_m}\left[-\frac{\dot{y}}{\dot{y}_m}\right] \tag{8.30}$$

Figure 8.8a shows the analog computer diagram for this differential equation. It suffices to write the ratio of the physical quantity to its maximum value throughout the

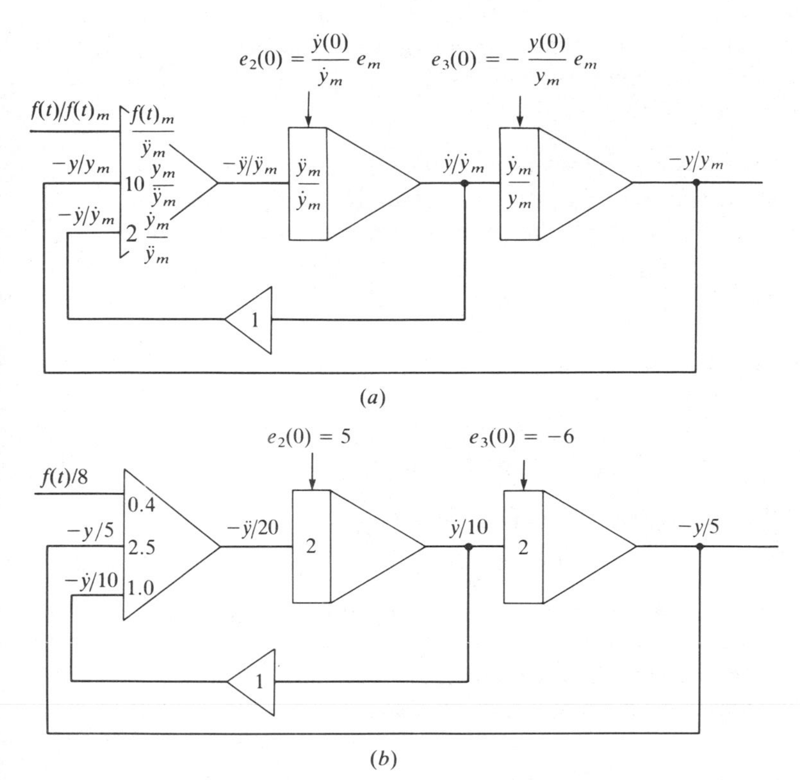

FIGURE 8.8
Analog computer diagram for $\ddot{y} + 2\dot{y} + 10y = f(t)$.

computer diagram. That is, although the output from the first amplifier is $e_1 = -(\ddot{y}/\ddot{y}_m)e_m$, only $-\ddot{y}/\ddot{y}_m$ is written. Similarly, the output from the second amplifier is $e_2 = (\dot{y}/\dot{y}_m)e_m$, but only $\dot{y}/\dot{y}_m$ is written, etc.

Each term in brackets on the right side of Eq. (8.30) represents a voltage which is inputted to the first amplifier. The coefficient for each term is the ratio of resistors for that signal. That is, the ratio of resistors R_2/R for the $f(t)/f(t)_m$ signal is $f(t)_m/\ddot{y}_m$, the ratio R_2/R_a for the $-y/y_m$ signal is $10y_m/\ddot{y}_m$, and the ratio R_2/R_b for the $\dot{y}/\dot{y}_m$ signal is $2\dot{y}_m/\ddot{y}_m$. The need for the sign changer is apparent because the signal $\dot{y}/\dot{y}_m$ occurs as the output of the second amplifier in the computer diagram whereas $-\dot{y}/\dot{y}_m$ occurs in Eq. (8.30). With the right side of Eq. (8.30) as the input to the first amplifier, the output is the negative of the left side, $-\ddot{y}/\ddot{y}_m$. The value of $1/(R_1C_2)$ for the second amplifier is $\ddot{y}_m/\dot{y}_m$, and that for the third is $\dot{y}_m/y_m$. Because $e_2 = (\dot{y}/\dot{y}_m)e_m$ and $e_3 = -(y/y_m)e_m$, the initial biasing voltages are $e_2(0) = (\dot{y}(0)/\dot{y}_m)e_m$ and $e_3(0) = -(y(0)/y_m)e_m$.

Substituting numerical values for y_m, $\dot{y}_m$, $\ddot{y}_m$, and $f(t)_m$ into Eq. (8.30) gives

$$\frac{\ddot{y}}{20} = \frac{8}{20}\left[\frac{f(t)}{8}\right] + 10\left(\frac{5}{20}\right)\left(-\frac{y}{5}\right) + 2\left(\frac{10}{20}\right)\left(-\frac{\dot{y}}{10}\right)$$

$$= 0.4\left[\frac{f(t)}{8}\right] + 2.5\left(-\frac{y}{5}\right) + 1.0\left(-\frac{\dot{y}}{10}\right) \tag{8.31}$$

The resulting computer diagram is shown in Fig. 8.8*b*. Note that $\ddot{y}_m/\dot{y}_m = 20/10 = 2$ and $\dot{y}_m/y_m = 10/5 = 2$. Similarly, $e_2(0) = (\dot{y}(0)/\dot{y}_m)e_m = (5/10)10 = 5$ and $e_3(0) = -[y(0)/y_m]e_m = -(3/5)10 = -6$.

It is a relatively easy matter to determine the differential equation that is being solved by a particular computer diagram. For example, multiplying each input for the first amplifier of Fig. 8.8*b* by the ratio of resistors for each input gives

$$0.4\left[\frac{f(t)}{8}\right] + 1.0\left(-\frac{\dot{y}}{10}\right) + 2.5\left(-\frac{y}{5}\right) \tag{8.32}$$

Because the amplifier reverses sign, the output $-\ddot{y}/20$ is the negative of the preceding input. That is,

$$-\frac{\ddot{y}}{20} = -\frac{f(t)}{20} + \frac{2\dot{y}}{20} + \frac{10y}{20} \tag{8.33}$$

Rearranging yields the original differential equation

$$\ddot{y} + 2\dot{y} + 10y = f(t) \tag{8.28}$$

Because $e_2(0)/e_m = \dot{y}(0)/\dot{y}_m = \dot{y}(0)/10$ and $e_3(0)/e_m = -y(0)/y_m = -y(0)/5$, then for $e_2(0) = 5$, $\dot{y}(0) = 5$ and for $e_3(0) = -6$, $y(0) = 3$. The input to the second amplifier of Fig. 8.8*b* is $-\ddot{y}/20$. Because this is an integrator, the output $\dot{y}/10$ is the negative of the integral of the input multiplied by the factor $[1/(R_1C_2)]_2 = 2$. That is,

$$\frac{\dot{y}}{10} = -2\int\left(-\frac{\ddot{y}}{20}\right)dt$$

Hence

$$\dot{y} = \int \ddot{y}\, dt$$

Similarly, for the third amplifier the output $-y/5$ is the negative of the integral of the input $\dot{y}/10$ multiplied by the factor $[1/(R_1C_2)]_3 = 2$. That is,

$$-\frac{y}{5} = -2\int \frac{\dot{y}}{10}\, dt$$

or

$$y = \int \dot{y}\, dt$$

For each system being studied, usually sufficient information is available to make a reasonable estimate of the maximum value of each term. If an error is made in predicting the maximum value of the term, then the maximum value of the voltage corresponding to that term will not be e_m. Such a situation is easily detected and corrected as follows: Suppose $e_m = 10$ V and the maximum value of voltage e_1 is found to be 7.5 V. The maximum value of $\ddot{y}$ is $7.5/10 = 3/4$ of the originally estimated maximum value of 20, hence the actual maximum value is 15. The computer diagram may be revised by using $\ddot{y}_m = 15$, in which case the maximum value of voltage e_1 will be 10 V. Notice that the original solution for which the maximum value of e_1 was 7.5 V and the corresponding $\ddot{y}_m$ is 15 is correct. The reason for revising the computer diagram is to obtain better accuracy by using the full range of 10 V rather than 7.5 V.

The general procedure for setting up an electronic analog computer to solve a differential equation of order n may be summarized as follows.

1. Solve the differential equation for the highest-order derivative. For example, solving Eq. (8.14) for $\ddot{y}$ gives

$$\ddot{y} = f(t) - ay - b\dot{y}$$

2. Divide through the preceding by $\ddot{y}_m$, and then express $f(t)$, y, and $\dot{y}$ in terms of ratios compared to their maximum values.

$$\frac{\ddot{y}}{\ddot{y}_m} = \frac{f(t)_m}{\ddot{y}_m}\left[\frac{f(t)}{f(t)_m}\right] + a\frac{y_m}{\ddot{y}_m}\left[-\frac{y}{y_m}\right] + b\frac{\dot{y}_m}{\ddot{y}_m}\left[-\frac{\dot{y}}{\dot{y}_m}\right] \tag{8.34}$$

The input to the first amplifier is the right side. Because an amplifier reverses sign, the output voltage is $e_1 = -(\ddot{y}/\ddot{y}_m)e_m$, but only $-\ddot{y}/\ddot{y}_m$ is written on the computer diagram. The output voltages of successive integrators are $e_2 = (\dot{y}/\dot{y}_m)e_m$ and $e_3 = (-y/y_m)e_m$, but only $\dot{y}/\dot{y}_m$ and $-y/y_m$ are written on the diagram. The coefficients in front of the brackets are the ratios of resistors to be used at the input to the first amplifier [that is, $f(t)_m/\ddot{y}_m = R_2/R$, $ay_m/\ddot{y}_m = R_2/R_a$, and $b\dot{y}_m/\ddot{y}_m = R_2/R_b$]. When the sign of the term in the computer $\dot{y}/\dot{y}_m$ is opposite to that of the corresponding term in the brackets $-\dot{y}/\dot{y}_m$, a sign changer must be used to multiply the feedback signal by -1. The value of $1/(R_1C_2)$ for each integrator is equal to the ratio of the maximum value for the input divided by the maximum value for the output [that is, $1/(R_1C_2)_2 = \ddot{y}_m/\dot{y}_m$ and $1/(R_1C_2)_3 = \dot{y}_m/y_m$].

The equation that is being solved by a summing amplifier is obtained by noting that the output is the negative of the sum of each input multiplied by its ratio of resistors. For an integrating amplifier, the output is the negative of the integral of the sum of each input multiplied by its $1/(RC)$ value.

Consider now the differential equation

$$z = \frac{D+4}{D^2+2D+10}f(t) \tag{8.35}$$

This may be expressed in the form

$$z = (D+4)y = \dot{y} + 4y \tag{8.36}$$

where

$$y = \frac{f(t)}{D^2+2D+10}$$

or

$$\ddot{y} + 2\dot{y} + 10y = f(t) \tag{8.28}$$

The differential equation for y is the same as Eq. (8.28), and the corresponding computer diagram is shown in Fig. 8.8. Dividing through Eq. (8.36) by z_m and then expressing y and $\dot{y}$ in terms of ratios compared to their maximum values give

$$\frac{z}{z_m} = \frac{\dot{y}_m}{z_m}\frac{\dot{y}}{\dot{y}_m} + 4\frac{y_m}{z_m}\frac{y}{y_m} \tag{8.37}$$

For $y_m = 5$, $\dot{y}_m = 10$, and $z_m = 25$ this becomes

$$\frac{z}{25} = \frac{10}{25}\left(\frac{\dot{y}}{10}\right) + 4\left(\frac{5}{25}\right)\left(\frac{y}{5}\right) = 0.4\left(\frac{\dot{y}}{10}\right) + 0.8\left(\frac{y}{5}\right) \tag{8.38}$$

The final computer diagram is shown in Fig. 8.9. The portion for solving Eq. (8.28) is the same as Fig. 8.8*b*. To obtain $z/25$, the input to the last amplifier is the negative of the right side of Eq. (8.38), that is,

$$0.4\left(-\frac{\dot{y}}{10}\right) + 0.8\left(-\frac{y}{5}\right)$$

The coefficient 0.4 is the ratio of resistors for the input $-\dot{y}/10$, and 0.8 is the ratio of resistors for the input $-y/5$. Because $\dot{y}/10$ is obtained as an output in the portion of the computer diagram for solving Eq. (8.28), it is necessary to use a sign changer to obtain $-\dot{y}/10$. The term $-y/5$ is obtained directly.

A feedback control system is shown in Fig. 8.10*a*. The maximum expected values are $c_m = 2$, $\dot{c}_m = 4$, $\ddot{c}_m = 10$, $f(t)_m = 2$, and $r_m = 1$. The initial conditions are $c(0) = 1$ and $\dot{c}(0) = 2$. To construct the analog computer diagram for this control system, first construct the diagram relating $c(t)$ and $f(t)$. Thus,

$$c(t) = \frac{10}{D(D+2)} f(t)$$

or

$$\ddot{c} + 2\dot{c} = 10f(t)$$

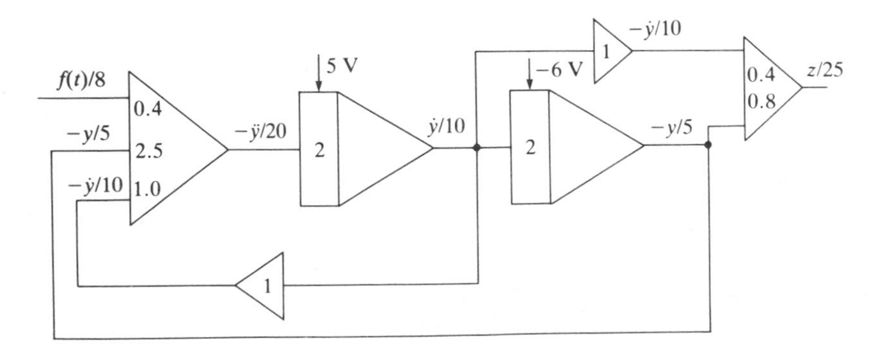

FIGURE 8.9
Analog computer diagram for $(D^2 + 2D + 10)z = (D + 4)f(t)$.

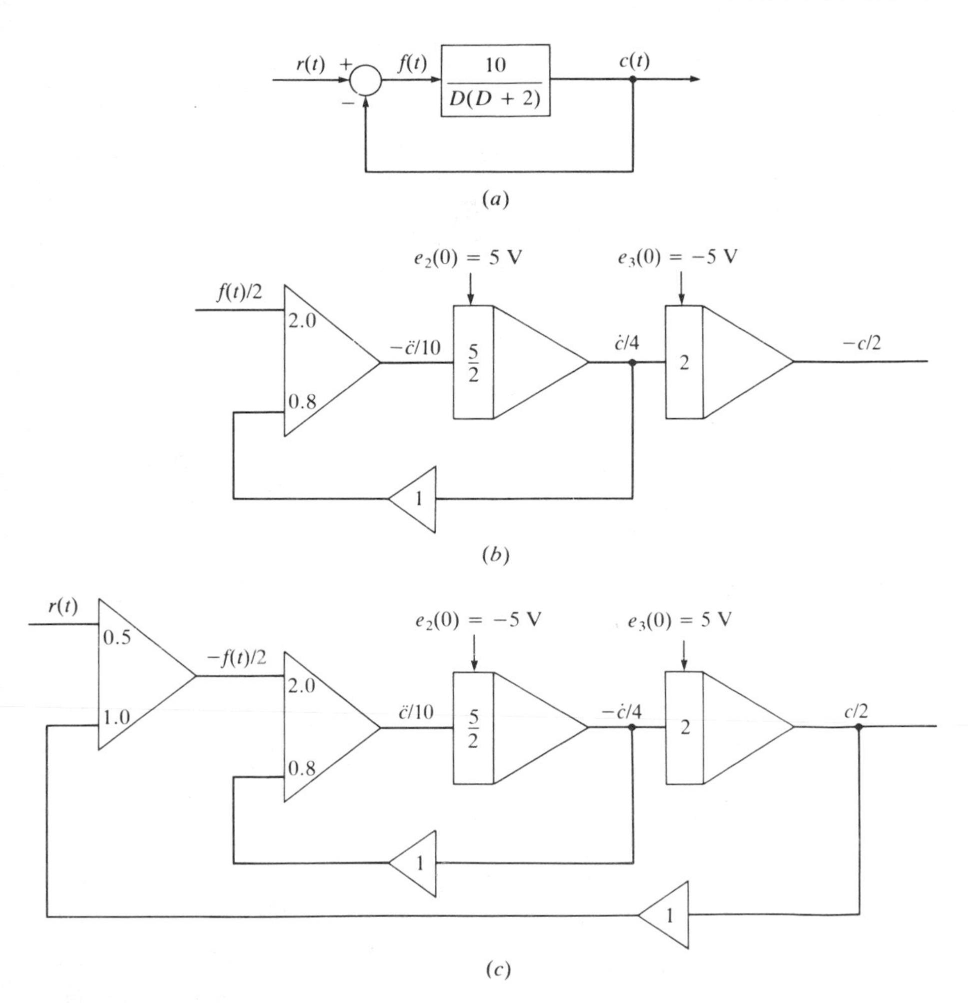

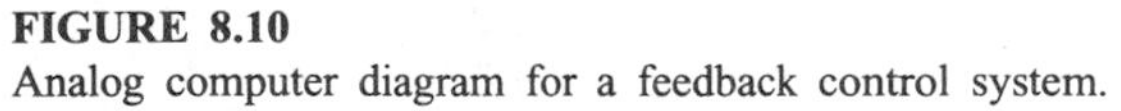

FIGURE 8.10
Analog computer diagram for a feedback control system.

Solving for $\ddot{c}$, the highest-order derivative, gives

$$\ddot{c} = 10f(t) - 2\dot{c}$$

Dividing through by $\ddot{c}_m$ and expressing each term as a ratio compared to its maximum value give

$$\frac{\ddot{c}}{\ddot{c}_m} = 10\,\frac{f(t)_m}{\ddot{c}_m}\left[\frac{f(t)}{f(t)_m}\right] - 2\frac{\dot{c}_m}{\ddot{c}_m}\left[\frac{\dot{c}}{\dot{c}_m}\right]$$

For $\ddot{c}_m = 10$, $\dot{c}_m = 4$, and $f(t)_m = 2$, this becomes

$$\frac{\ddot{c}}{10} = 2\left[\frac{f(t)}{2}\right] + 0.8\left[-\frac{\dot{c}}{4}\right]$$

The resulting computer diagram is shown in Fig. 8.10*b*. The input to the first amplifier is the right side. The ratio of resistors for the $f(t)/2$ signal is 2, and that for the $-\dot{c}/4$ signal is 0.8. Because $\dot{c}/4$ appears in the computer diagram, it is necessary to use a sign changer to obtain $-\dot{c}/4$. The output of this amplifier is $-\ddot{c}/10$. The first integration yields the $\dot{c}/\dot{c}_m = \dot{c}/4$ signal, and the second integration yields the $-c/c_m = -c/2$ signal. The value of $1/(RC)$ for the first integration is $\ddot{c}_m/\dot{c}_m = 10/4 = 2.5$, and that for the second integration is $\dot{c}_m/c_m = 4/2 = 2$. The biasing voltages are $e_2(0) = [\dot{c}(0)/4]e_m = (2/4)10 = 5$ V and $e_3(0) = [-c(0)/2]e_m = (-\frac{1}{2})10 = -5$ V.

The equation for the comparator is

$$f(t) = r(t) - c(t)$$

In terms of maximum values, this becomes

$$\frac{f(t)}{f(t)_m} = \frac{r(t)_m}{f(t)_m}\left[\frac{r(t)}{r(t)_m}\right] - \frac{c(t)_m}{f(t)_m}\left[\frac{c(t)}{c(t)_m}\right]$$

For $f(t)_m = 2$, $r(t)_m = 1$, and $c(t)_m = 2$, this becomes

$$\frac{f(t)}{2} = \frac{1}{2}\left[\frac{r(t)}{1}\right] + \left[-\frac{c(t)}{2}\right] \tag{8.39}$$

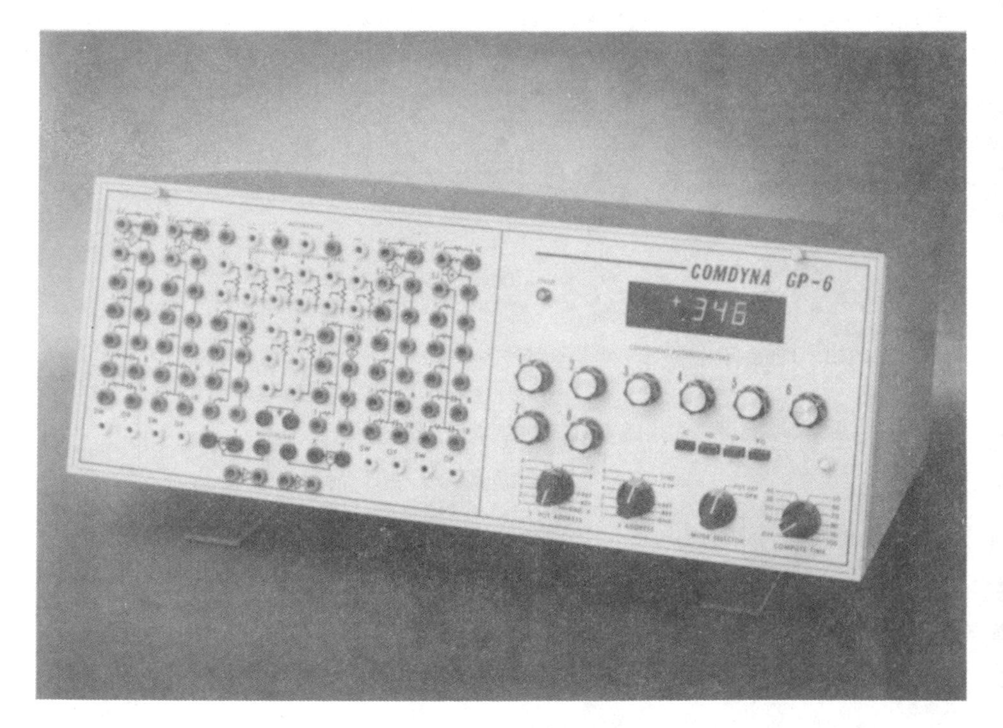

FIGURE 8.11
GP-6 Analog Computer. (Courtesy Comdyna, Inc.)

The resulting computer diagram for simulating the system shown in Fig. 8.10a may now be completed as shown in Fig. 8.10c. The preceding equation for the comparator is solved by the first amplifier. The input to this amplifier is the right side of Eq. (8.39). The ratio of resistors for the $r(t)/1 = r(t)$ signal is $\frac{1}{2}$, and that for the $-c(t)/2$ signal is 1. The output is $-f(t)/2$. The input to the diagram shown in Fig. 8.10b is $f(t)/2$. To avoid a sign changer, the signs of all the signals are changed in going from Fig. 8.10b to c. Thus $f(t)/2$ becomes $-f(t)/2$, $-\ddot{c}/10$ becomes $\ddot{c}/10$, $\dot{c}/4$ becomes $-\dot{c}/4$, $-c/2$ becomes $c/2$, and the signs of the biaising voltages are changed.

Figure 8.11 shows a picture of the GP-6 analog computer which is manufactured by Comdyna, Inc. Wires with banana plug tips are used to connect the various operational amplifiers, potentiometers, etc., in the same manner in which they are drawn on the analog computer diagram. Desired voltages are easily selected to be displayed on an oscilloscope or may be sent to a plotter to obtain a permanent record. This analog computer is capable of simulating nonlinear as well as linear systems.

8.3 TIME SCALE

For many problems, it is desired that the speed at which the analog computer solves the problem be different from the speed at which the phenomena actually occur. For example, various phenomena of astronomy require years, so obviously it is desirable to increase the speed at which such problems are solved in the computer. For other phenomena which take place very rapidly, it is necessary to slow down the speed at which such problems are simulated by the computer. Letting t represent the time at which a phenomenon actually occurs and τ represent the time required for this phenomenon to occur on the computer, we see that $\tau = at$ relates the actual time t to the computer or machine time τ. For $a < 1$, the phenomenon occurs faster in the computer than it does in nature. For example, if $a = 0.1$, something which actually requires 10 s to complete is completed in the computer in $\tau = 0.1t = 0.1 \times 10 = 1$ s. Similarly, if $a > 1$, the phenomenon is slowed down by the computer.

Illustrative example 8.2. Slow down the computer solution of Eq. (8.28) by a factor of 5.

Solution. The first step is to transform the original equation from a function of actual time t to a function of machine time τ. This is accomplished by noting that

$$\tau = at \tag{8.40}$$

and

$$\frac{d\tau}{dt} = a \tag{8.41}$$

Thus

$$\dot{y} = \frac{dy}{dt} = \frac{d\tau}{dt}\frac{dy}{d\tau} = a\,\frac{dy}{d\tau}$$
$$\ddot{y} = \frac{d^2y}{dt^2} = \frac{d}{dt}\frac{dy}{dt} = \frac{d\tau}{dt}\frac{d}{d\tau}a\,\frac{dy}{d\tau} = a^2\,\frac{d^2y}{d\tau^2} \tag{8.42}$$

Similarly, it may be shown that in general

$$\frac{d^ny}{dt^n} = a^n\,\frac{d^ny}{d\tau^n} \tag{8.43}$$

Application of the preceding rules to convert the original time expression given by Eq. (8.28) from a function of t to a function of τ gives

$$a^2\,\frac{d^2y}{d\tau^2} + 2a\,\frac{dy}{d\tau} + 10y = f\left(\frac{\tau}{a}\right) \tag{8.44}$$

where $f(\tau/a)$ is obtained by substitution of τ/a for t in the original function $f(t)$. Because of the change of variable, the term y in Eq. (8.44) is now a function of τ rather than t. To slow down the solution by a factor of 5, the value of a is 5, so that Eq. (8.44) becomes

$$25\,\frac{d^2y}{d\tau^2} + 10\,\frac{dy}{d\tau} + 10y = f\left(\frac{\tau}{5}\right) \tag{8.45}$$

The transformed initial conditions are

$$\left.\frac{dy}{d\tau}\right|_{\tau=0} = \frac{1}{a}\left.\frac{dy}{dt}\right|_{t=0} = \frac{1}{5}(5) = 1.0$$

Similarly, the transformed maximum values are

$$\left.\frac{d^2y}{d\tau^2}\right|_m = \frac{1}{a^2}\left.\frac{d^2y}{dt^2}\right|_m = \frac{20}{25} = 0.8$$

$$\left.\frac{dy}{d\tau}\right|_m = \frac{1}{a}\left.\frac{dy}{dt}\right|_m = \frac{10}{5} = 2.0$$

As shown in Fig. 8.12, the function $f(\tau/5)$ is obtained by multiplying the original time scale by the factor $a = 5$. Thus, a time-scale change does not effect the initial and maximum values of $f(\tau/a)$. Similarly, the initial and maximum values of y

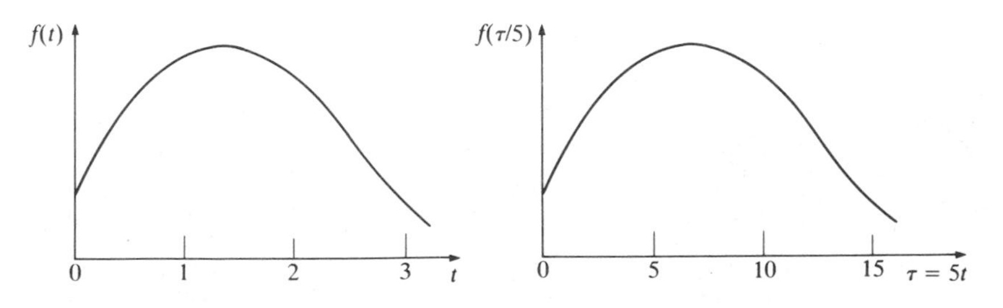

FIGURE 8.12
Graph of $f(t)$ versus t and $f(\tau/5)$ versus τ.

are unaffected by a change in time scale. By using the symbolism $\ddot{y}_\tau = d^2y/d\tau^2$, $\dot{y}_\tau = dy/d\tau$, etc., the maximum values are $\ddot{y}_{\tau_m} = 0.8$, $\dot{y}_{\tau_m} = 2.0$, $y_{\tau_m} = 5$, and $f_{\tau_m} = 8$. The initial values are $y_\tau(0) = 3$ and $\dot{y}_\tau(0) = 1$.

The computer diagram is now obtained by application of the general procedure given in the preceding section. Solving Eq. (8.45) for $\ddot{y}_\tau = d^2y/d\tau^2$, dividing through by $\ddot{y}_{\tau_m}$, and then expressing f_τ, y_τ, and $\dot{y}_\tau$ as ratios compared to their maximum values yield

$$\frac{\ddot{y}_\tau}{\ddot{y}_{\tau_m}} = \frac{f_{\tau_m}}{25\ddot{y}_{\tau_m}}\left(\frac{f_\tau}{f_{\tau_m}}\right) + \frac{10}{25}\frac{y_{\tau_m}}{\ddot{y}_{\tau_m}}\left(-\frac{y_\tau}{y_{\tau_m}}\right) + \frac{10}{25}\frac{\dot{y}_{\tau_m}}{\ddot{y}_{\tau_m}}\left(-\frac{\dot{y}_\tau}{\dot{y}_{\tau_m}}\right) \tag{8.46}$$

Substitution of numerical values gives

$$\begin{aligned}\frac{\ddot{y}_\tau}{0.8} &= \frac{8}{20}\left(\frac{f_\tau}{8}\right) + 10\left(\frac{5}{20}\right)\left(-\frac{y_\tau}{5}\right) + 10\left(\frac{2}{20}\right)\left(-\frac{\dot{y}_\tau}{2}\right)\\ &= 0.4\left(\frac{f_\tau}{8}\right) + 2.5\left(-\frac{y_\tau}{5}\right) + 1.0\left(-\frac{\dot{y}_\tau}{2}\right)\end{aligned} \tag{8.47}$$

The resulting computer diagram is shown in Fig. 8.13*a*.

With the right side of Eq. (8.47) as the input to the first amplifier, the output is the negative of the left side, or $-\ddot{y}_\tau/0.8$. The coefficient 0.4 is the ratio of resistors for $f_\tau/8$, the coefficient 2.5 is the ratio of resistors for $-y_\tau/5$, and the coefficient 1.0 is the ratio of resistors for $-\dot{y}_\tau/2$. Because $\dot{y}_\tau/2$ appears as the output of the second amplifier in Fig. 8.13*a*, it is necessary to use a sign changer to obtain the $-\dot{y}_\tau/2$ term which appears in Eq. (8.47). The value of $1/(R_1C_2)$ for the second amplifier is $\ddot{y}_{\tau_m}/\dot{y}_{\tau_m} = 0.8/2 = 0.4$, and the initial voltage is $e_2(0) = [\dot{y}_\tau(0)/\dot{y}_{\tau_m}]e_m = (\frac{1}{2})10 = 5$ V. For the third amplifier the value of $1/(R_1C_2)$ is $\dot{y}_{\tau_m}/y_{\tau_m} = \frac{2}{5} = 0.4$, and the initial voltage is $e_3(0) = -[y_\tau(0)/y_{\tau_m}]e_m = -(\frac{3}{5})10 = -6$ V.

Comparison of Fig. 8.8*b* which is the computer diagram when there is no time-scale change with Fig. 8.13*a* shows that the gains (0.4, 2.5, 1.0) in the first amplifier are unaffected by the time-scale change. Only the gains in the amplifiers which integrate are affected. Note that the value of the gain in the integrators of Fig. 8.13*a* is equal to that in Fig. 8.8*b* divided by the time-scale factor *a* (that is, $2/a = \frac{2}{5} = 0.4$). In general, the values of the gains of a summing amplifier are unaffected by a time-scale change, and the new value of the gain in an integrating amplifier is equal to the value for the case of no time-scale change divided by the value of *a*. Also note that the values of the initial biasing voltages for the integrators are unaffected by the time-scale change. Thus, the easiest way to obtain the computer diagram for a time-scale change is to first obtain the diagram for the case of no time-scale change and then divide the gain of each integrator by *a*. The maximum values are most readily obtained by noting that the maximum value ($y_m = y_{\tau_m} = 5$) for the y term is unaffected by the time-scale change [that is, $e_3 = (-y/5)e_m = (-y_\tau/5)e_m$]. The ratio $\dot{y}_{\tau_m}/y_{\tau_m} = 1/(R_1C_2)_3 = 0.4$ is the gain of the third amplifier. Thus, $\dot{y}_{\tau_m}$ is equal to the gain of the third amplifier times y_{τ_m} [that is, $\dot{y}_{\tau_m} = 0.4(5) = 2$]. Hence, the output of the second amplifier is $e_2 = (\dot{y}_\tau/2)e_m$. Similarly, $\ddot{y}_{\tau_m}$ is equal to the gain of the second amplifier times $\dot{y}_{\tau_m}$ [that is,

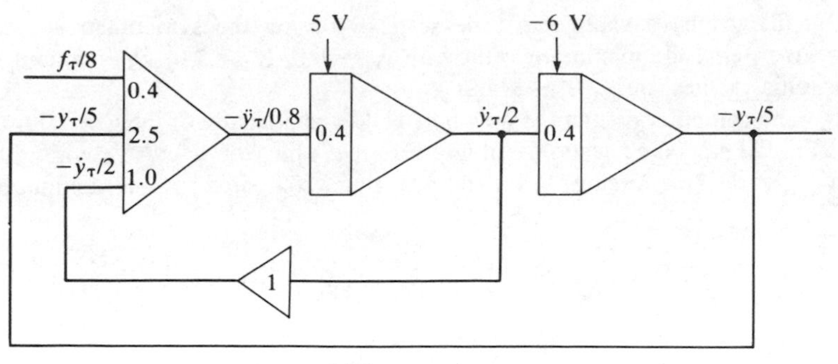

(*a*)

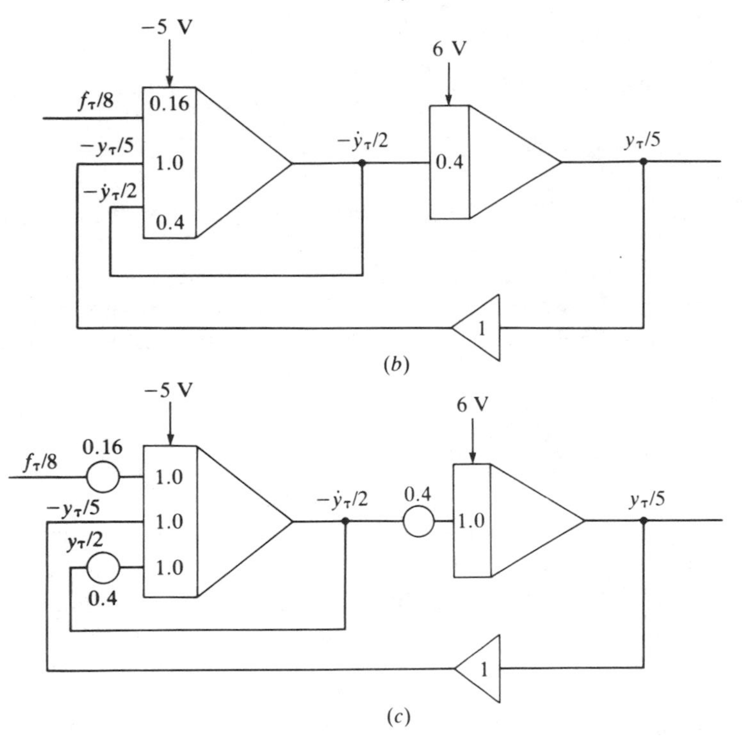

FIGURE 8.13
Analog computer diagram for $\ddot{y} + 2\dot{y} + 10y = f(t)$ with a time-scale change $\tau = 5t$.

$\ddot{y}_{\tau_m} = 0.4(2) = 0.8$]. Thus, the output of the first amplifier is $e_1 = (-\ddot{y}_\tau/0.8)e_m$. The values of the initial biasing voltages [$e_2(0) = 5$ and $e_3(0) = -6$] are unaffected by the time-scale change.

If it is not necessary to measure $d^2y/d\tau^2$, then the first amplifier of Fig. 8.13*a* which sums and the second amplifier which integrates may be combined into one amplifier which both sums and integrates. The new computer diagram in which the first two amplifiers of Fig. 8.13*a* have been combined into one amplifier is shown in

Fig. 8.13*b*. The gains (0.16, 1.0, and 0.4) for the new first amplifier are equal to the product of the gains (0.4, 2.5, and 1.0) for the summing amplifier and the gain (0.4) for the integrating amplifier [that is, (0.4) (0.4) = 0.16, (2.5) (0.4) = 1.0, and (1.0) (0.4) = 0.4]. Because one amplifier has been eliminated in going from Fig. 8.13*a* to Fig. 8.13*b*, the signals $\dot{y}_\tau/2$ and $-y_\tau/5$ of Fig. 8.13*a* have opposite signs in Fig. 8.13*b*. The feedback path from $\dot{y}_\tau/2$ of Fig. 8.13*a* to the first amplifier has a sign changer, but the path from $y_\tau/5$ to the first amplifier does not. Because the signs of these terms have been reversed in Fig. 8.13*b*, there is no sign changer in the feedback path from $-\dot{y}_\tau/2$ to the new first amplifier but there is a sign changer in the feedback path from $y_\tau/5$ to the new first amplifier. Note also that the signs of all the initial biasing voltages have been switched in going from Fig. 8.13*a* to *b*. To summarize, the elimination of one amplifier changes the signs of all the resulting output voltages, and thus the sign of each feedback quantity must be reversed. Similarly, the sign of each initial-condition voltage must be changed.

Some operational amplifiers use standard resistors 1×10^6, 0.25×10^6, and 0.10×10^6 Ω and a standard capacitor of 1×10^{-6} F. Thus, only gains of $1/(RC)$ equal to 1, 4, or 10 are readily available. It is possible to put two resistors in parallel or in series at the input to obtain some other effective value of resistance. For example, two 1×10^6 Ω resistors in series yield a 2×10^6 Ω resistance, while two 1×10^6 Ω resistors in parallel yield a 0.5×10^6 Ω resistance.

In general, it is necessary to use a potentiometer to obtain the desired effective resistance. Figure 8.14 shows a schematic diagram of a potentiometer. The voltage relationship is

$$e_o = \frac{R_b}{R_a} e_{\text{in}} = k e_{\text{in}} \tag{8.48}$$

where

$$k = \frac{R_b}{R_a} \qquad 0 < k < 1$$

The computer diagram of Fig. 8.13*b* may be modified by the addition of three potentiometers as shown in Fig. 8.13*c* such that only gains of 1, 4, or 10 are

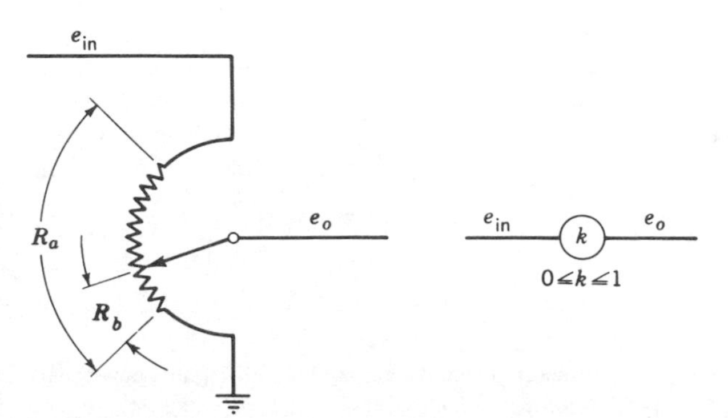

FIGURE 8.14
Potentiometer.

necessary at each amplifier. By comparison of Fig. 8.13*b* and *c*, note that the effective gain of an amplifier is the product of the gain of the amplifier and the value of k for the potentiometer in front of the amplifier.

8.4 SIMULATION

A very important application of the analog computer is the simulation of automatic control systems. One method that could be used to simulate a control system would be to determine the overall differential equation and solve this on the computer. Usually, however, one is interested in determining the effect on the system performance when certain parameters are varied. Using the preceding technique would mean solving a new differential equation for each change.

Because of the similarity between a block diagram and a computer diagram, it is customary to simulate each portion of the system and then interconnect these elements. Thus, the effect of changing one of the terms in the original block diagram may be achieved by changing the corresponding quantity in the computer diagram.

Figure 8.15 shows a schematic representation of an operational amplifier in which the input impedance is Z_1 and the parallel impedance is Z_2. The equation for this amplifier is

$$\frac{e_2}{e_1} = -\frac{Z_2}{Z_1} \tag{8.49}$$

When the input impedance is a resistor R_1 and the parallel impedance is a resistor R_2, the preceding expression reduces to the result given by Eq. (8.8). For the case in which $Z_1 = R_1$ and $Z_2 = 1/(C_2D)$, the result given by Eq. (8.11) is verified.

Table 8.1 shows a number of computer circuits for simulating various transfer functions. For the first circuit, $Z_1 = R_1$ and $Z_2 = 1/(1/R_2 + C_2D) = R_2/(1 + R_2C_2D)$. Substitution of the values of these impedances into Eq. (8.49) yields for the equation of operation for this computer circuit

$$e_2 = -\frac{R_2}{R_1(1 + R_2C_2D)}e_1 \tag{8.50}$$

Figure 8.16*a* shows a typical block diagram, and the corresponding computer diagram is shown in Fig. 8.16*b*.

Illustrative example 8.3. The system shown in Fig. 8.16*a* is used to control the position of a 6 degree-of-freedom robot. For this system, it is known that $K_1 = 10$, $K_2 = 5$, and $\tau = 1.0$. The initial values are $m(0) = 2.0$ and $c(0) = 1.0$. The maximum values have been estimated to be $c(t)_m = 2$ rad, $r(t)_m = 1$ rad, $m(t)_m = 5$ N · m, and

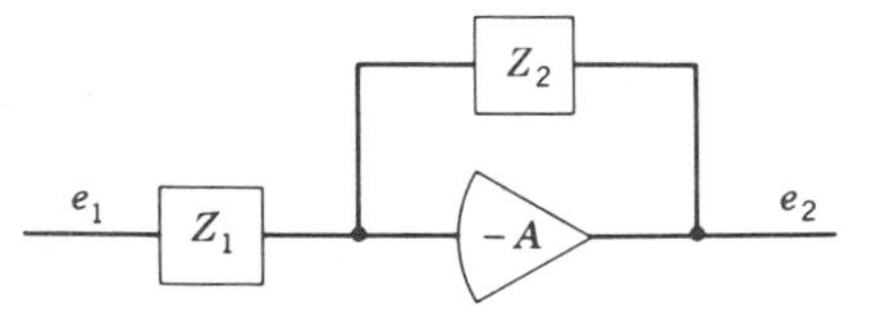

FIGURE 8.15
General schematic representation for an operational amplifier.

Table 8.1

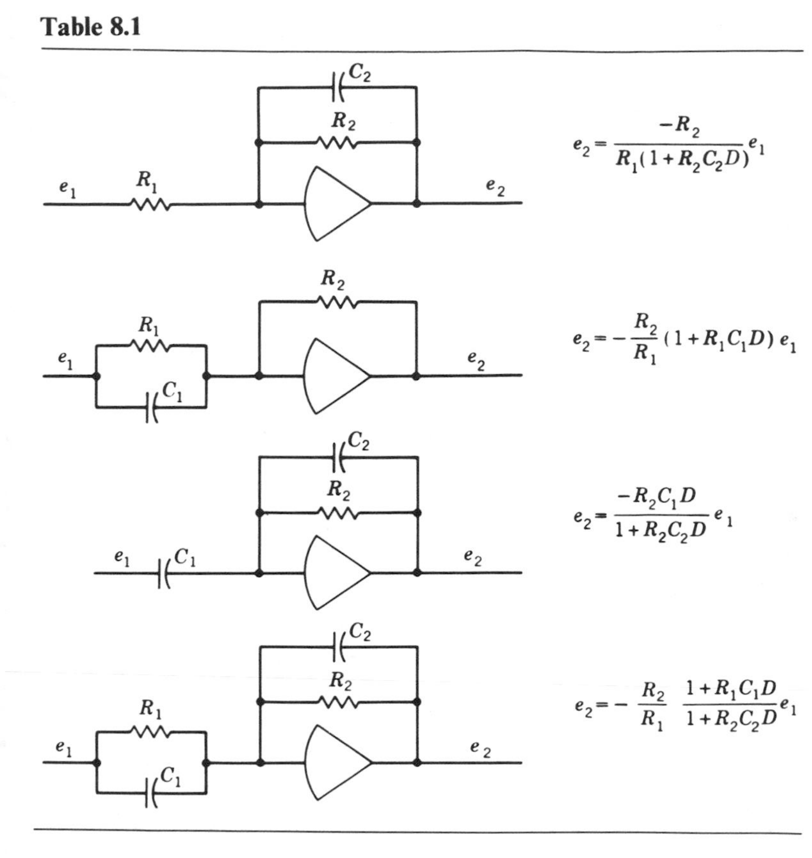

$d(t)_m = 10$ N · m. Determine the values of the resistors and capacitors for the computer diagram of Fig. 8.16*b*.

Solution. The voltage equation for the first amplifier is

$$e = -\frac{1}{R_r C_e}\int_0^t e_r\, dt - \frac{1}{R_c C_e}\int_0^t e_b\, dt + e(0) \tag{8.51}$$

The equation which describes the operation of the corresponding portion of the actual system is

$$\begin{aligned} m(t) &= K_1 \int_0^t [r(t) - c(t)]\, dt + m(0) \\ &= K_1 \int_0^t r(t)\, dt - K_1 \int_0^t c(t)\, dt + m(0) \end{aligned} \tag{8.52}$$

Multiplying through by the maximum voltage e_m, dividing through by $-m(t)_m$, and then expressing $r(t)$ and $c(t)$ in terms of ratios compared to their maximum values yield

$$-\left(\frac{m}{m_m}\right) e_m = -K_1 \frac{r_m}{m_m}\int_0^t \left(\frac{r}{r_m}\right) e_m\, dt - K_1 \frac{c_m}{m_m}\int_0^t \left(-\frac{c}{c_m}\right) e_m\, dt - \left[\frac{m(0)}{m_m}\right] e_m \tag{8.53}$$

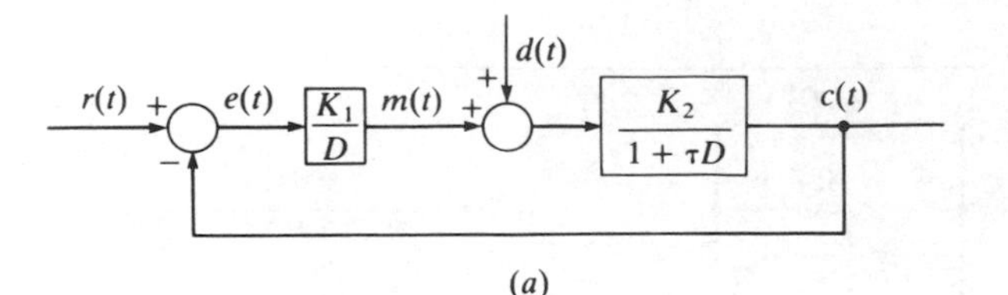

(a)

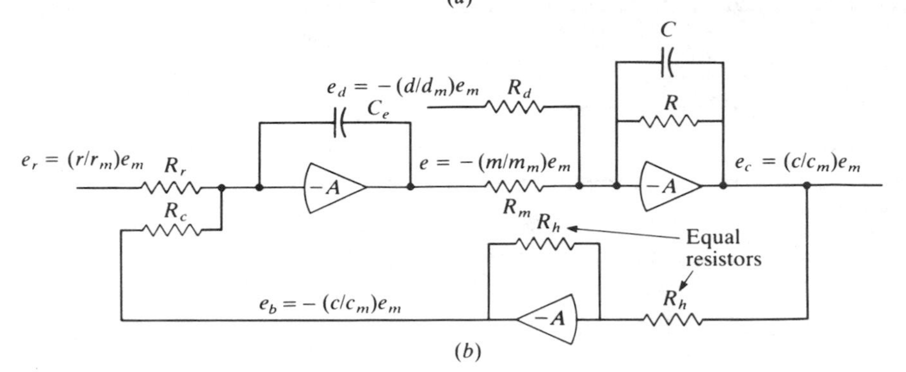

(b)

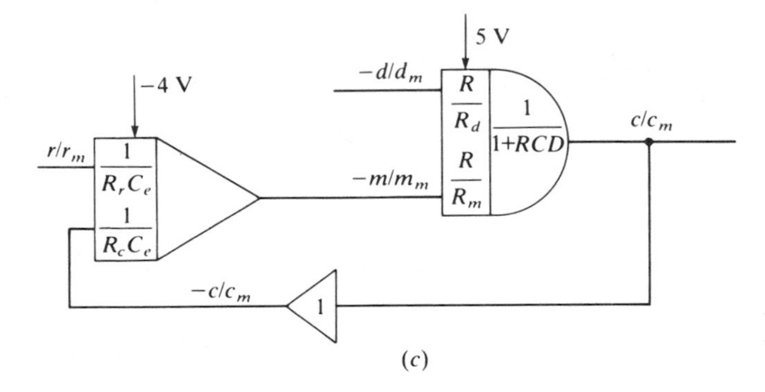

(c)

FIGURE 8.16
Analog computer diagram for a feedback control system.

In order that $e = -(m/m_m)e_m$, $e_r = (r/r_m)e_m$ and $e_b = -(c/c_m)e_m$, comparison of corresponding terms in Eqs. (8.51) and (8.53) shows that

$$\frac{1}{R_rC_e} = K_1\frac{r_m}{m_m} = 10\left(\frac{1}{5}\right) = 2 \qquad \frac{1}{R_cC_e} = K_1\frac{c_m}{m_m} = 10\left(\frac{2}{5}\right) = 4 \tag{8.54}$$

Because $e = -(m/m_m)e_m$, the initial biasing voltage is $e(0) = -[m(0)/m_m]e_m = -(\frac{2}{5})10 = -4$ V. For any system r_m, m_m, c_m, R_r, and R_c are constant. Thus the gain K_1 may be varied independently by varying the capacitance C_e. If a 1-μF capacitor is used for C_e, then

$$R_r = \frac{1}{2C_e} = \frac{1}{2 \times 10^{-6}} = 500{,}000\ \Omega$$

$$R_c = \frac{1}{4C_e} = \frac{1}{4 \times 10^{-6}} = 250{,}000\ \Omega$$

From Table 8.1, it follows that the voltage equation for the second amplifier is

$$e_c = -\frac{R}{R_m(1+RCD)}e - \frac{R}{R_d(1+RCD)}e_d \tag{8.55}$$

The equation for the corresponding portion of the actual system is

$$c(t) = \frac{K_2}{1+\tau D}[m(t) + d(t)] \tag{8.56}$$

Expressing this in terms of ratios compared to maximum values gives

$$\left(\frac{c}{c_m}\right)e_m = \frac{m_m K_2}{c_m(1+\tau D)}\left(\frac{m}{m_m}\right)e_m + \frac{d_m K_2}{c_m(1+\tau D)}\left(\frac{d}{d_m}\right)e_m \tag{8.57}$$

Comparison of Eqs. (8.55) and (8.57) shows that to have $e_c = (c/c_m)e_m$, $e = -(m/m_m)e_m$, and $e_d = -(d/d_m)e_m$, then

$$\frac{R}{R_m} = K_2\frac{m_m}{c_m} = 5\left(\frac{5}{2}\right) = 12.5 \qquad \frac{R}{R_d} = K_2\frac{d_m}{c_m} = 5\left(\frac{10}{2}\right) = 25 \qquad \tau = RC = 1 \tag{8.58}$$

Because $e_c = (c/c_m)e_m$, the initial biasing voltage is $e_c(0) = [c(0)/c_m]e_m = (\frac{1}{2})10 = 5$ V. If a 1-μF capacitor is used for C, then

$$R = \frac{1}{C} = 1{,}000{,}000\ \Omega$$

$$R_m = \frac{R}{12.5} = 80{,}000\ \Omega$$

$$R_d = \frac{R}{25} = 40{,}000\ \Omega$$

Figure 8.16*c* shows the schematic diagram for the analog computer circuit shown in Fig. 8.16*b*. The first amplifier is an integrator. The input to this integrator is $[1/(R_rC_e)]\ (r/r_m)$ plus $[1/(R_cC_e)]\ (-c/c_m)$. Because an amplifier changes sign from input to output, the negative of the output is equal to the integral of the input. Thus,

$$\frac{m}{m_m} = \frac{1}{R_rC_e}\int_0^t \frac{r}{r_m}\,dt - \frac{1}{R_cC_e}\int_0^t \frac{c}{c_m}\,dt + \frac{m(0)}{m_m} \tag{8.59}$$

The second amplifier of Fig. 8.16*c* multiplies the input by $1/(1 + RCD)$. The input is $(R/R_d)(-d/d_m) + (R/R_m)(-m/m_m)$. Because of the sign reversal from input to output, the output c/c_m is the negative of the input. Thus,

$$\frac{c}{c_m} = \frac{1}{1+RCD}\left[\frac{R}{R_m}\left(\frac{m}{m_m}\right) + \frac{R}{R_d}\left(\frac{d}{d_m}\right)\right] \tag{8.60}$$

To obtain the analog computer diagram for simulating a control system, it is not necessary to write both the voltage equation for the circuit and the physical relationship for the system. It suffices to write the physical relationship only. The equation for the first part of the system shown in Fig. 8.16*a* is

$$m = \frac{K_1}{D}(r - c) = K_1\int_0^t r\,dt - K_1\int_0^t c\,dt + m(0) \tag{8.61}$$

Expressing this in terms of ratios compared to maximum values gives

$$\frac{m}{m_m} = K_1 \frac{r_m}{m_m} \int_0^t \frac{r}{r_m}\, dt + K_1 \frac{c_m}{m_m} \int_0^t \frac{c}{c_m}\, dt + \frac{m(0)}{m_m} \tag{8.62}$$

With the right side as the input, the output will be $-m/m_m$. Referring to Fig. 8.16*c*, we notice that the coefficient of the first integral $K_1 r_m/m_m$ is the value $1/(R_r C_e)$ for the signal r/r_m. Similarly, the coefficient $K_1 c_m/m_m$ of the second integral is the value $1/(R_c C_e)$ for the signal $-c/c_m$. Thus

$$\frac{1}{R_r C_e} = K_1 \frac{r_m}{m_m} \qquad \frac{1}{R_e C_e} = K_1 \frac{c_m}{m_m} \tag{8.54}$$

These are the same results given by Eq. (8.54). The equation for the second portion of the system shown in Fig. 8.16*a* is

$$c = \frac{K_2}{1 + \tau D}(m + d) \tag{8.63}$$

Expressing this in terms of ratios compared to maximum values gives

$$\frac{c}{c_m} = \frac{1}{1 + \tau D}\left[K_2 \frac{m_m}{c_m}\left(\frac{m}{m_m}\right) + K_2 \frac{d_m}{c_m}\left(\frac{d}{d_m}\right)\right] \tag{8.64}$$

The $1/(1 + RCD)$ term in the schematic diagram for this second amplifier indicates that the input is multiplied by this factor. With the negative of the right side as the input to this amplifier, then the output is c/c_m. The coefficient $K_2 m_m/c_m$ of the m/m_m term is the value of R/R_m for the $-m/m_m$ signal in Fig. 8.16*c*. Similarly, the coefficient $K_2 d_m/c_m$ is the value of R/R_d for the $-d/d_m$ signal. Thus

$$\frac{R}{R_m} = K_2 \frac{m_m}{c_m} \qquad \frac{R}{R_d} = K_2 \frac{d_m}{c_m} \qquad \tau = RC \tag{8.58}$$

These are the same results given by Eq. (8.58). Because C appears only in the equation $\tau = RC$, τ may be varied independently by varying C. A variable capacitor provides a convenient means for varying C. To change the value of K_2, both R_m and R_d must be changed accordingly. Because output e_c is to be subtracted from input e_r, it is necessary to multiply the output by -1, as shown in the feedback path of Fig. 8.16*b*.

If, in testing, it is found that $e_{c_m} \neq 10$ V, this is evidence that the originally estimated value for c_m is not correct. Because $e_c = (c/c_m)e_m$, the actual value of c_m is now easily determined. When the maximum value of c is near 10 V, there is no need to change the originally estimated value of c_m. However, if the maximum voltage is quite small, then it is desirable to increase the accuracy with which the output voltage can be read by using the new value of c_m. Similarly, if c_m were enough greater than 10 V to cause overloading, then it would be necessary to use the new value of c_m. For most amplifiers, an overload light turns on when the maximum voltage is high enough to overload an amplifier.

A major application of analog computers is found in the design of systems with nonlinear components. Standard electronic circuits are available for simulating commonly encountered nonlinear effects such as coulomb friction, backlash, dead zone, saturation, continuous nonlinear functions, etc.

PROBLEMS

8.1. (*a*) Figure P8.1*a* shows the analog computer diagram for solving the differential equation

$$\dot{y} + 4y = f(t)$$

Complete this diagram for the case in which $y_m = 1$, $\dot{y}_m = 2$, and $f(t)_m = 1$.

(*b*) Figure P8.1*b* shows the wiring diagram corresponding to the analog computer diagram shown in Fig. P8.1*a*. Determine the values of R_a, R_b, R_1, and R_2 for this diagram.

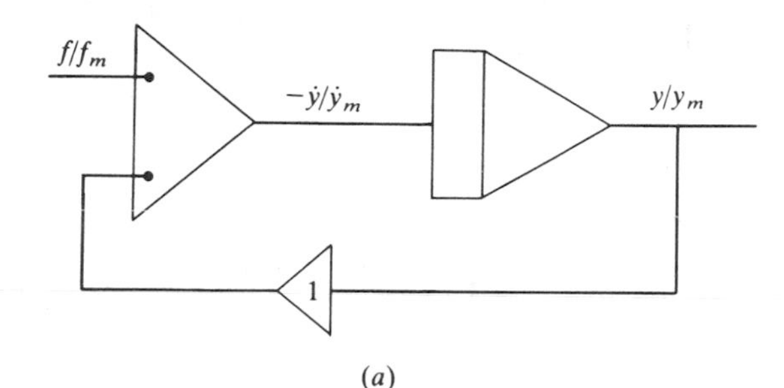

(*a*)

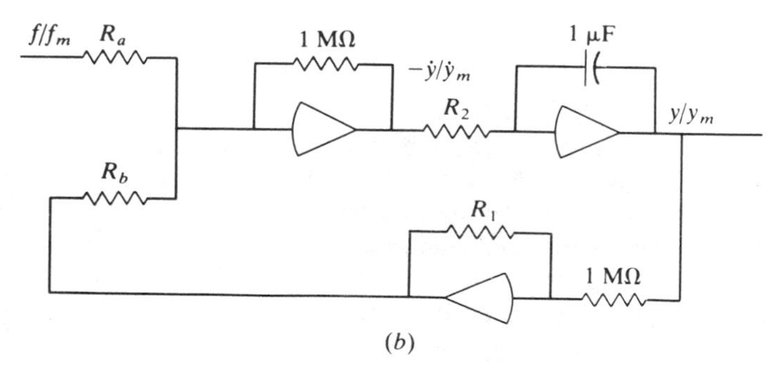

(*b*)

FIGURE P8.1

8.2. (*b*) Figure P8.2*a* shows the analog computer diagram for solving the differential equation

$$\ddot{y} + 2\dot{y} + 5y = f(t)$$

Complete this diagram for the case in which $y_m = 1$, $\dot{y}_m = 2$, $\ddot{y}_m = 4$, and $f(t)_m = 0.5$.

(*b*) Figure P8.2*b* shows the wiring diagram corresponding to the analog computer diagram shown in Fig. P8.2*a*. Determine the values of R_a, R_b, R_c, R_1, R_2, and R_3 for this diagram.

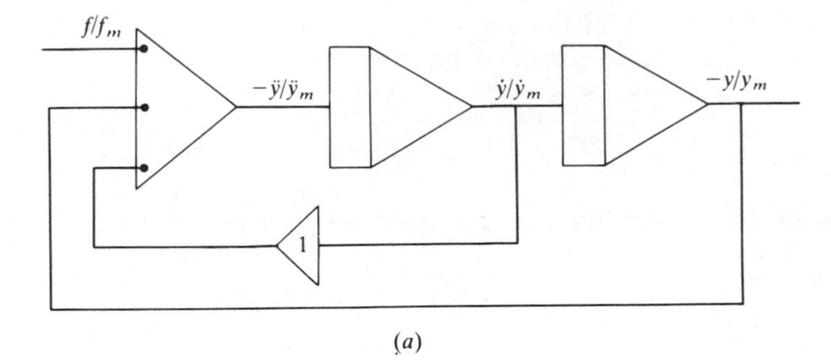

(a)

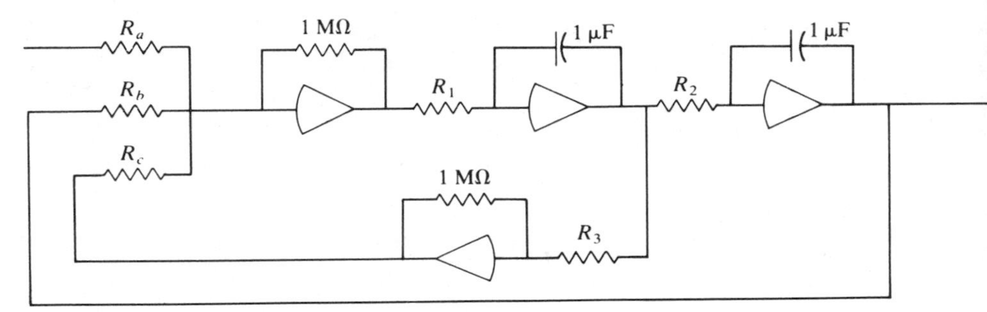

(b)

FIGURE P8.2

8.3. In running Prob. 8.2, it is found that the maximum voltage obtained for the $-y/y_m$ signal is 5 rather than 10 V. What is the actual value of y_m? Revise the analog computer diagram so that the maximum voltage corresponding to $-y/y_m$ will be 10 V.

8.4. In running Prob. 8.2, it is found that the maximum voltage obtained for the $\dot{y}/\dot{y}_m$ signal is 5 rather than 10 V. What is the actual value of $\dot{y}_m$? Revise the analog computer diagram so that the maximum voltage corresponding to $\dot{y}/\dot{y}_m$ will be 10 V.

8.5. In running Prob. 8.2, it is found that the maximum voltage obtained for the $-\ddot{y}/\ddot{y}_m$ signal is 5 rather than 10 V. What is the actual value of $\ddot{y}_m$? Revise the analog computer diagram so that the maximum voltage corresponding to $-\ddot{y}/\ddot{y}_m$ will be 10 V.

8.6. Determine the computer diagram for the following first-order equation:

$$(0.5D + 1)y = f(t)$$

The initial condition is $y(0) = 0.2$, and the maximum expected values are $f(t)_m = 1$, $y_m = 0.5$, and $\dot{y}_m = 2$.

8.7. Determine the computer diagram for simulating the following differential equation:

$$\ddot{y} + 2\dot{y} + 5y = f(t)$$

The maximum expected values are $y_m = 0.5$, $\dot{y}_m = 1.0$, $\ddot{y}_m = 2$, and $f(t)_m = 5.0$. The initial conditions are $y(0) = 0.2$ and $\dot{y}(0) = 0.5$.

8.8. Determine the computer diagram for simulating the following differential equation:

$$\ddot{y} + 6\dot{y} + 5y = f(t)$$

The maximum expected values are $y_m = 0.5$, $\dot{y}_m = 1.0$, $\ddot{y}_m = 2.0$, and $f(t)_m = 5.0$. The initial conditions are $y(0) = 0.2$ and $\dot{y}(0) = -0.2$.

8.9. Determine the computer diagram for simulating the following differential equation:

$$\ddot{y} + 25y = f(t)$$

The maximum expected values are $y_m = 0.1$, $\dot{y}_m = 0.5$, $\ddot{y}_m = 2.5$, and $f(t)_m = 1.0$. The initial conditions are $y(0) = 0$ and $\dot{y}(0) = 0.5$.

8.10. Determine the computer diagram to solve the following differential equation:

$$\ddot{y} + 2y = 5f(t)$$

The initial conditions are $y(0) = 3$ and $\dot{y}(0) = 2$. The maximum expected values are $y_m = 10$, $\dot{y}_m = 20$, $\ddot{y}_m = 50$, and $f(t)_m = 2$.

8.11. Construct the analog computer diagram for the differential equation

$$\ddot{y} + 10y = 2f(t)$$

The initial conditions are $y(0) = \dot{y}(0) = 2$. The maximum expected values are $y_m = 2$, $\dot{y}_m = 4$, $\ddot{y}_m = 10$, and $f(t)_m = 5$.

8.12. Determine the differential equation that is being solved by the computer diagram shown in Fig. P8.12.

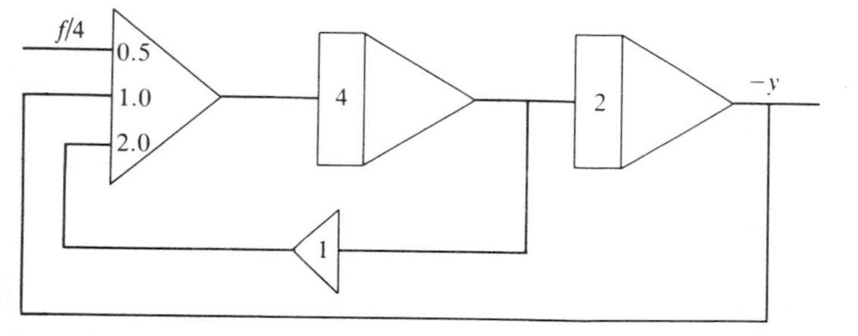

FIGURE P8.12

8.13. (*a*) Determine the differential equation that is being solved by the analog computer diagram shown in Fig. P8.13*a*.

(*b*) Figure P8.13*b* shows the wiring diagram for the first amplifier. Determine the equations for R_a, R_b, and R_c.

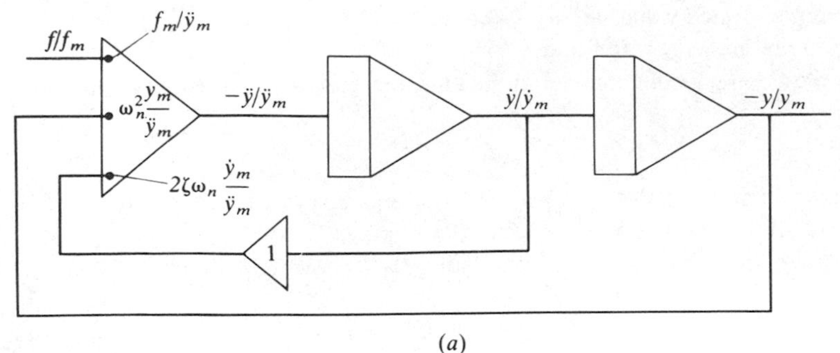

(a)

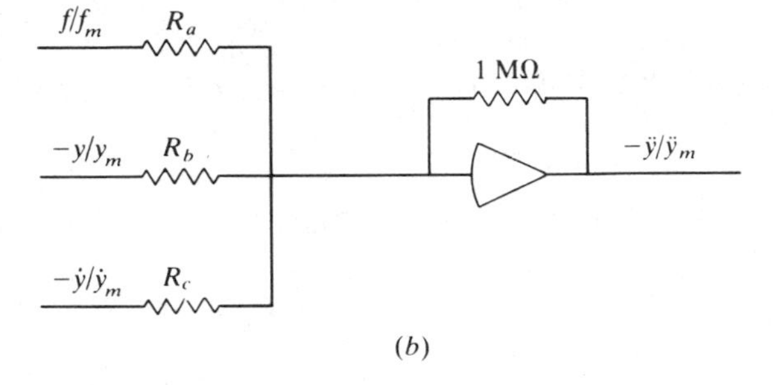

(b)

FIGURE P8.13

8.14. Determine the differential equation that is being solved by the computer diagram shown in Fig. P8.14. What are the initial conditions $y(0)$ and $\dot{y}(0)$?

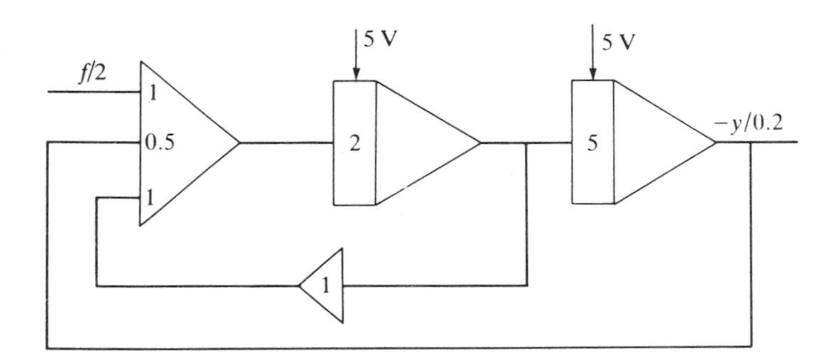

FIGURE P8.14

8.15. Determine the differential equation that is being solved by the computer diagram shown in Fig. P8.15. Specify the initial conditions for the differential equation.

8.16. Determine the differential equation that is being solved by the computer diagram shown in Fig. P8.16. What are the initial values $x(0)$ and $\dot{x}(0)$? What are the expected

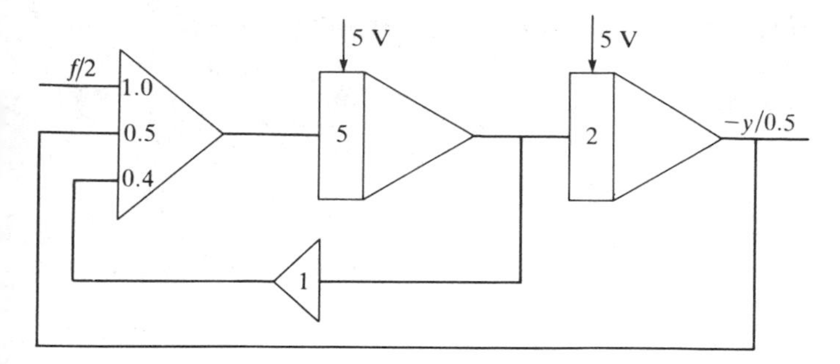

FIGURE P8.15

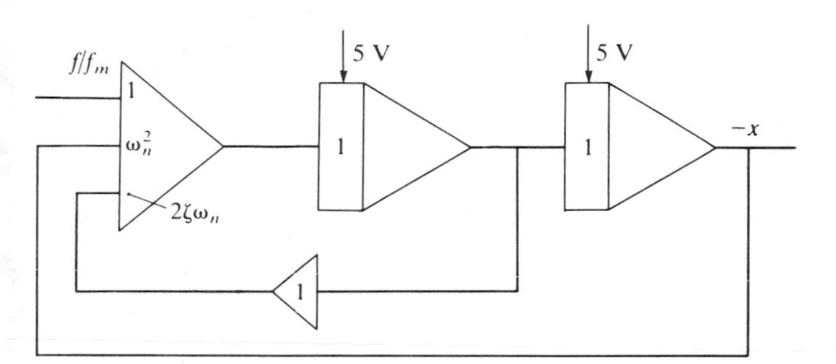

FIGURE P8.16

maximum values for x, $\dot{x}$, and $\ddot{x}$? Revise the computer diagram for each of the following cases.

(*a*) The maximum value of x is found to be twice its expected value.

(*b*) The maximum value of $\dot{x}$ is found to be twice its expected value.

(*c*) The maximum value of $\ddot{x}$ is found to be twice its expected value.

8.17. Determine the differential equation that is being solved by the computer diagram shown in Fig. P8.17.

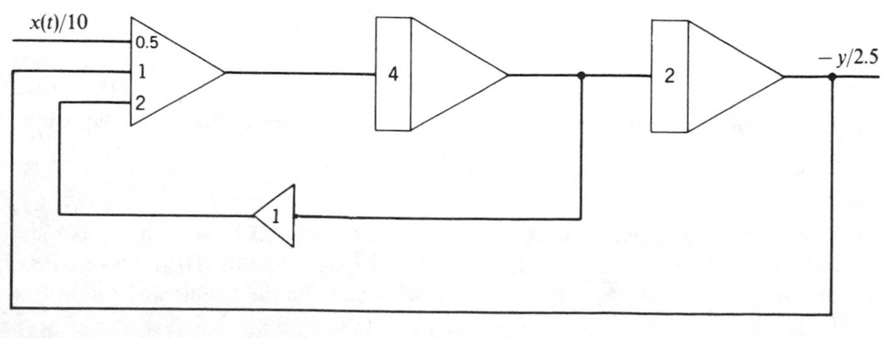

FIGURE P8.17

8.18. (*a*) Determine the differential equation that is being solved by the analog computer diagram shown in Fig. P8.18.
(*b*) For the first amplifier,

$$\frac{by_m}{\ddot{y}_m} = \frac{R_2}{R_b} \qquad \frac{a\dot{y}_m}{\ddot{y}_m} = \frac{R_2}{R_a}$$

Let $a = 2\zeta\omega_n$ and $b = \omega_n^2$. Solve for ζ and ω_n in terms of these electrical constants and scale factors. Can either ζ or ω_n be varied independently?

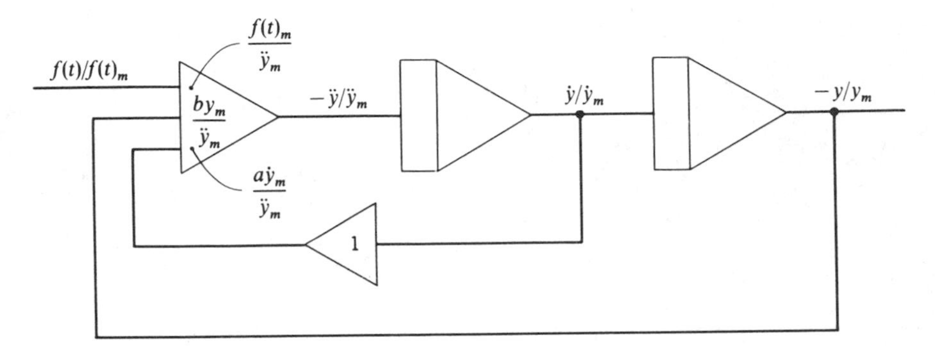

FIGURE P8.18

8.19. Speed up the computer solution for Prob. 8.6 by a factor of 5. Determine the computer diagram for this case.

8.20. Let it be desired to speed up the solution for Prob. 8.7 by a factor of 2. Determine the computer diagram for this case.

8.21. Determine the computer diagram which speeds up the solution of Prob. 8.8 by a factor of 5.

8.22. Determine the computer diagram for the following third-order differential equation:

$$(D^3 + 2D^2 + 5D + 10)y = f(t)$$

The initial conditions are $y(0) = 0.5$, $\dot{y}(0) = 0$, and $\ddot{y}(0) = 1.0$. The maximum expected values are $f(t)_m = 2.5$, $y_m = 1.0$, $\dot{y}_m = 2.0$, $\ddot{y}_m = 5.0$, and $\dddot{y}_m = 10$.

8.23. Determine the computer diagram which slows down the solution of Prob. 8.22 by a factor of 2.

8.24. (*a*) Determine the computer diagram to solve the following differential equation:

$$\dddot{y} + 8\ddot{y} + 25\dot{y} + 20y = f(t)$$

The initial conditions are $\ddot{y}(0) = 3$, $\dot{y}(0) = 1$, and $y(0) = 1$. The maximum expected values are $\dddot{y}_m = 10$, $\ddot{y}_m = 5$, $\dot{y}_m = 2$, $y_m = 1$, and $f(t)_m = 40$.

(*b*) Determine the computer diagram which will speed up the solution of part *a* by a factor of 2.

8.25. (*a*) Construct the analog computer diagram for simulating the differential equation

$$\ddot{x} + 4\dot{x} + 10x = f(t)$$

The maximum expected values are $x_m = 1$, $\dot{x}_m = 2$, $\ddot{x}_m = 8$, and $f(t)_m = 4$. The initial conditions are $x(0) = \dot{x}(0) = 1$.

(*b*) Show the necessary additions to your answer for part *a* in order to obtain the analog computer diagram for simulating the following differential equation:

$$\ddot{y} + 4\dot{y} + 10y = (2D + 5)f(t)$$

The maximum expected value for y is $y_m = 5$.

8.26. (*a*) Construct the analog computer diagram for simulating the differential equation

$$\ddot{x} + 5\dot{x} + 20x = f(t)$$

The maximum expected values are $x_m = 1$, $\dot{x}_m = 2$, $\ddot{x}_m = 5$, and $f(t)_m = 10$. The initial conditions are $x(0) = \dot{x}(0) = 1$.

(*b*) Show the necessary additions to your answer for part *a* in order to obtain the analog computer diagram for simulating the following differential equation:

$$\ddot{y} + 5\dot{y} + 20y = (D + 2)f(t)$$

The maximum expected value for y is $y_m = 5$.

8.27. (*a*) Construct the analog computer diagram for simulating the differential equation

$$\ddot{x} + 6\dot{x} + 25x = f(t)$$

The maximum expected values are $x_m = 1$, $\dot{x}_m = 2$, $\ddot{x}_m = 5$, and $f(t)_m = 10$. The initial conditions are $x(0) = \dot{x}(0) = 1$.

(*b*) Show the necessary additions to your answer for part *a* in order to obtain the analog computer diagram for simulating the following differential equation:

$$\ddot{y} + 6\dot{y} + 25y = (D + 5)f(t)$$

The maximum expected value for y is $y_m = 4$.

8.28. (*a*) Determine the analog computer diagram for simulating the differential equation

$$\ddot{x} + 3\dot{x} + 2x = f(t)$$

The maximum expected values are $x_m = 0.2$, $\dot{x}_m = 0.5$, $\ddot{x}_m = 1.0$, and $f(t)_m = 0.4$. The initial conditions are $x(0) = \dot{x}(0) = 0.1$.

(*b*) Show the necessary additions to your answer for part *a* in order to obtain the analog computer diagram for simulating the differential equation

$$\ddot{y} + 3\dot{y} + 2y = (D + 4)f(t)$$

The maximum expected value for y is $y_m = 1.0$.

8.29. The block-diagram representation for a control system is shown in Fig. P8.29*a*. The analog computer diagram for simulating this system is shown in Fig. P8.29*b*. The portion enclosed by the dashed box simulates the feedforward portion of the block diagram from e to c. Complete the analog computer diagram for the case in which $c_m = 2$, $\dot{c}_m = 4$, $e_m = 1$, $r_m = 1$, and $c(0) = 1$.

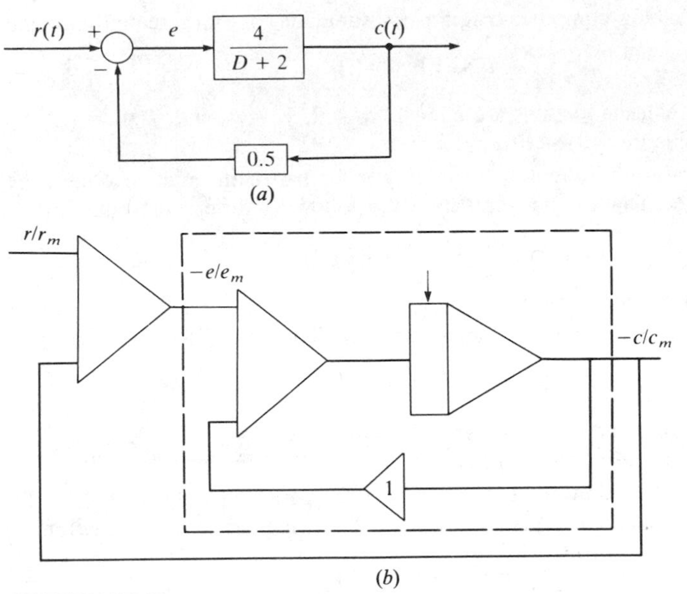

FIGURE P8.29

8.30. A feedback control system is shown in Fig. P8.30*a*. The analog computer diagram for simulating this system is shown in Fig. P8.30*b*. The portion enclosed by the dashed box simulates the feedforward portion of the system from e to c. The initial conditions are $c(0) = 1$ and $\dot{c}(0) = 2$. The maximum expected values are $c_m = 2$, $\dot{c}_m = 4$, $\ddot{c}_m = 10$, $e_m = 1$, and $r_m = 2$. Complete the analog computer diagram.

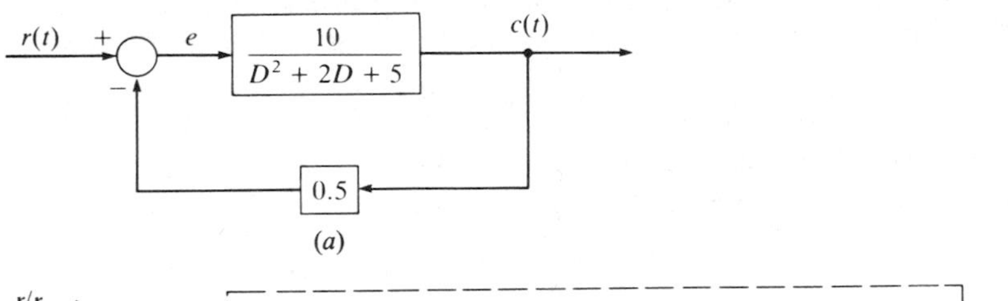

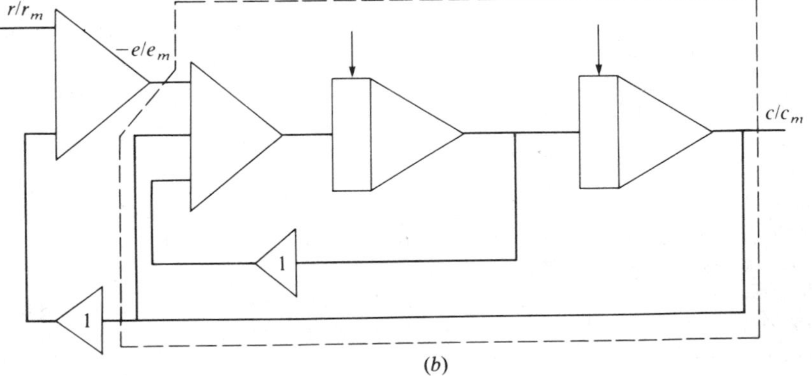

FIGURE P8.30

8.31. Determine the analog computer diagram to simulate the system shown in Fig. P8.31. The maximum expected values are $c_m = 1$, $\dot{c}_m = 2$, $\ddot{c}_m = 4$, $e(t)_m = 4$, and $r_m = 2$. The initial conditions are $c(0) = \dot{c}(0) = 1$. First obtain the diagram relating $c(t)$ to $e(t)$, and then hook up the comparator $e(t) = r(t) - 0.5c(t)$.

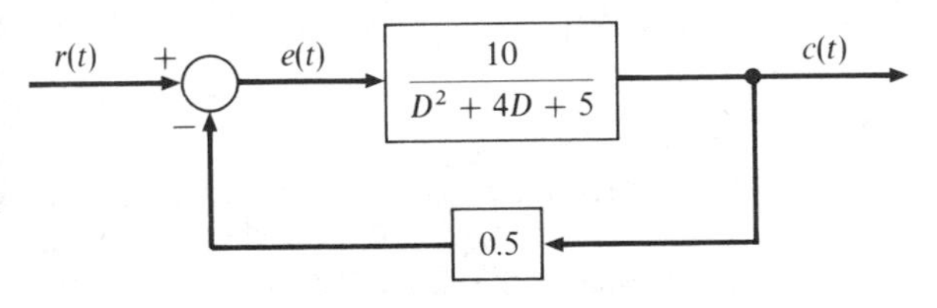

FIGURE P8.31

8.32. Determine the analog computer diagram for simulating the system shown in Fig. P8.32. The maximum expected values are $c_m = 1$, $\dot{c}_m = 2$, $\ddot{c}_m = 4$, $e_m = 2$, and $r_m = 1$. The initial conditions are $c(0) = \dot{c}(0) = 1$. First obtain the diagram relating $c(t)$ to $e(t)$, and then hook up the comparator $e(t) = r(t) - c(t)$.

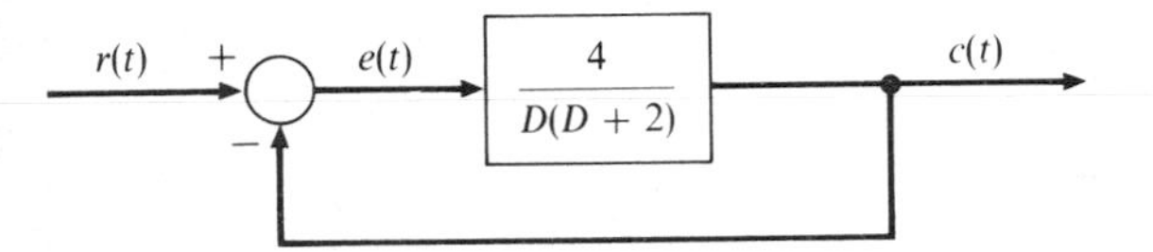

FIGURE P8.32

8.33. Determine the analog computer diagram for simulating the system shown in Fig. P8.33. The maximum expected values are $c_m = 0.5$, $\dot{c}_m = 1.0$, $\ddot{c}_m = 2.0$, $e_m = 5.0$, and $r_m = 4.0$. The initial conditions are $c(0) = 0.2$ and $\dot{c}(0) = -0.2$. First obtain the diagram relating $c(t)$ to $e(t)$, and then hook up the comparator $e(t) = r(t) - c(t)$.

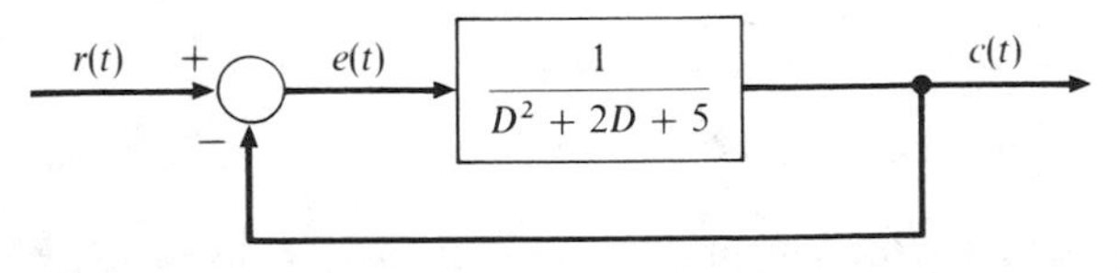

FIGURE P8.33

8.34. Determine the analog computer diagram for simulating the system shown in Fig. P8.34. The maximum expected values are $c_m = 2$, $\dot{c}_m = 4$, $\ddot{c}_m = 10$, $e_m = 5$, and $r_m = 2$. The initial conditions are $c(0) = 1$ and $\dot{c}(0) = 2$. First obtain the diagram relating $c(t)$ to $e(t)$, and then hook up the comparator $e(t) = r(t) - 2c(t)$.

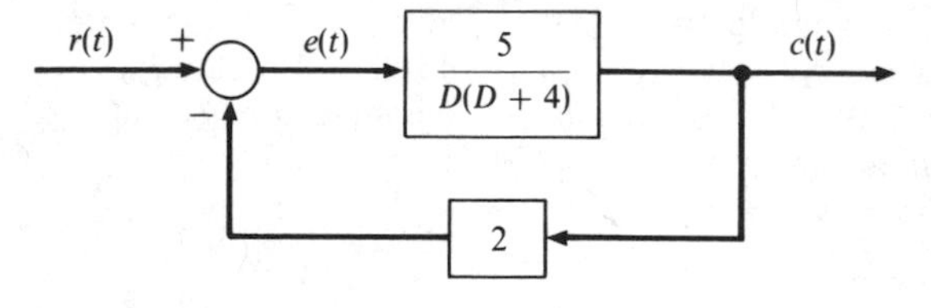

FIGURE P8.34

8.35. (*a*) Determine the block diagram for the system which is being simulated by the analog computer diagram shown in Fig. P8.35. (*Hint*: First obtain the block diagram for the portion of the circuit enclosed by the dashed lines, i.e., the block diagram relating input e to output c.)

(*b*) Draw directly the new computer diagram for speeding up the solution of the above problem by a factor of 2.

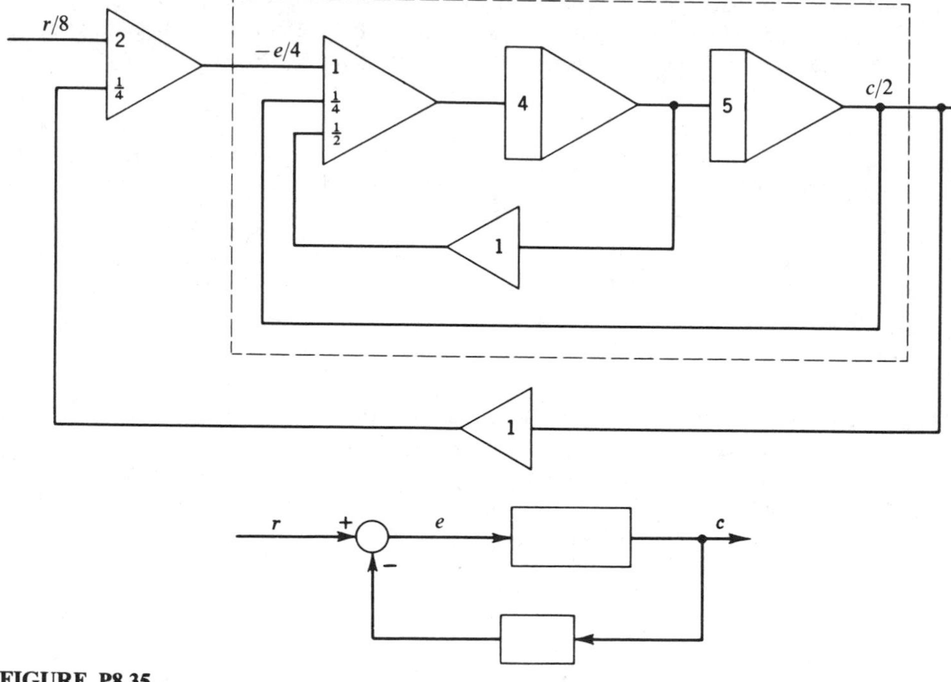

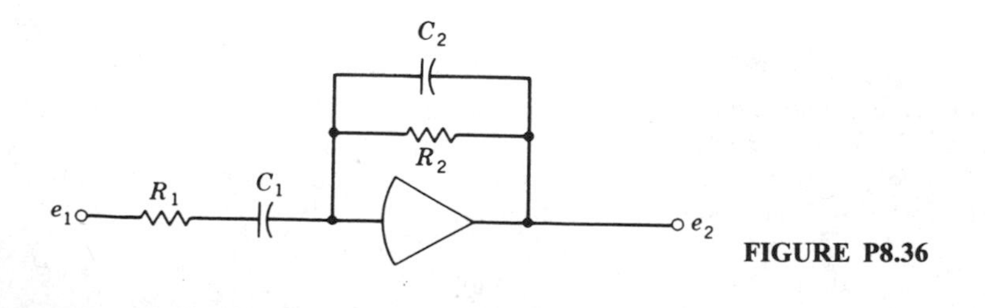

FIGURE P8.35

8.36. Derive the equation of operation for the amplifier circuit shown in Fig. P8.36.

FIGURE P8.36

8.37. Construct the computer diagram for simulating the control system shown in Fig. P8.37. The maximum expected values are $r(t)_m = 2.0$ and $c(t)_m = 1.0$.

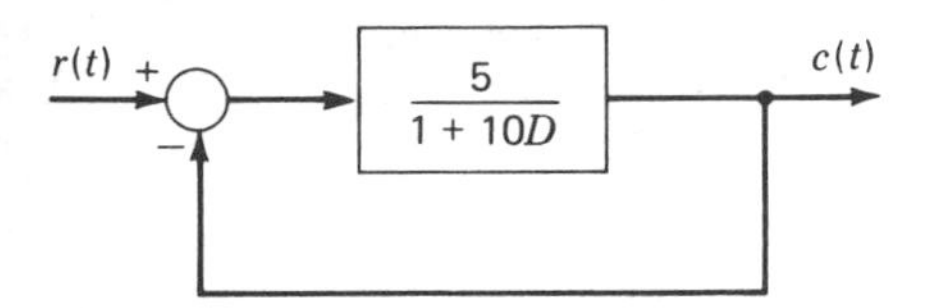

FIGURE P8.37

8.38. Construct the computer diagram for simulating the system shown in Fig. P8.38 for the case in which $r(t)_m = 2.0$, $c(t)_m = 1.0$, $K = 20$, and $\tau = 0.5$. (Use a 1-μF capacitor.)

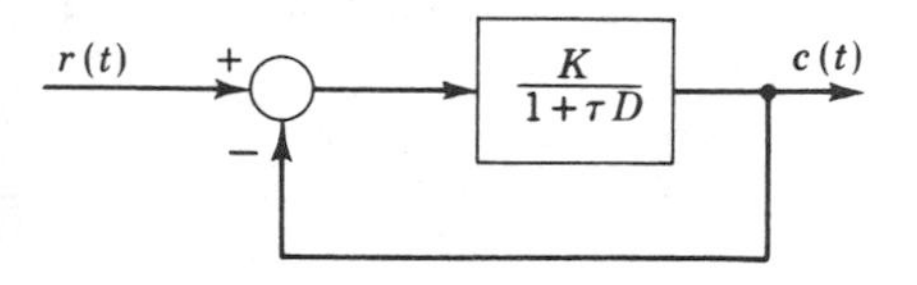

FIGURE P8.38

8.39. A control system and the corresponding computer diagram for simulating the control system are shown in Fig. P8.39. For $r_m = 2$, $c_m = 1$, $C_2 = 1\ \mu F$, $R_2 = 1\ M\Omega$, $R_r = 0.1\ M\Omega$, and $R_c = 0.2\ M\Omega$, determine the corresponding values of K_1, K_2, and τ.

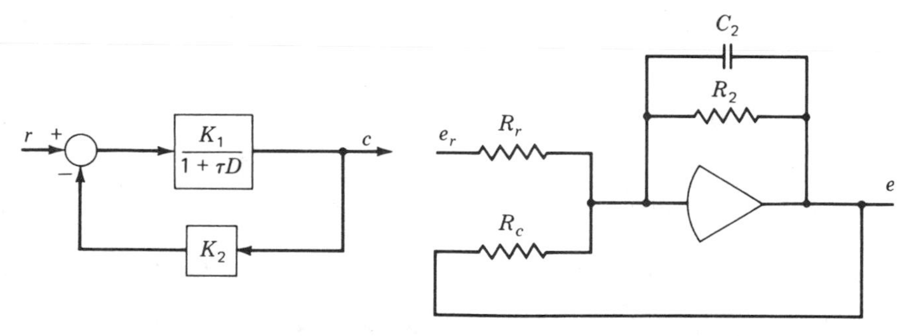

FIGURE P8.39

8.40. Determine the block-diagram representation for the control system which is being simulated by the analog computer diagram shown in Fig. P8.40.

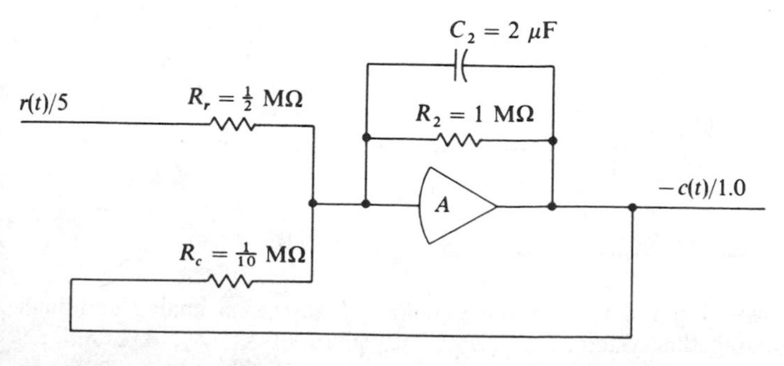

FIGURE P8.40

8.41. Construct the computer diagram for simulating the control system shown in Fig. P8.41. For this system $K_1 = 5$, $K_2 = 2$, and $\tau = 1$. The maximum expected values are $r(t)_m = 1.0$ and $c(t)_m = 2.0$. (Let $C_2 = 1\ \mu F$.)

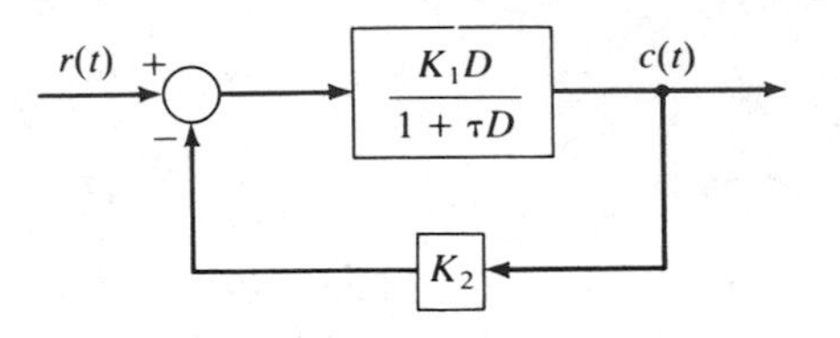

FIGURE P8.41

8.42. Both the analog computer diagram and the system it is simulating are shown in Fig. P8.42. Determine the values of K_1, K_2, and τ for the system.

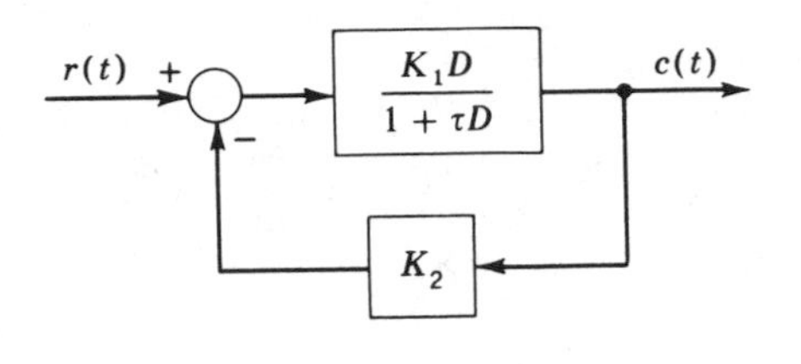

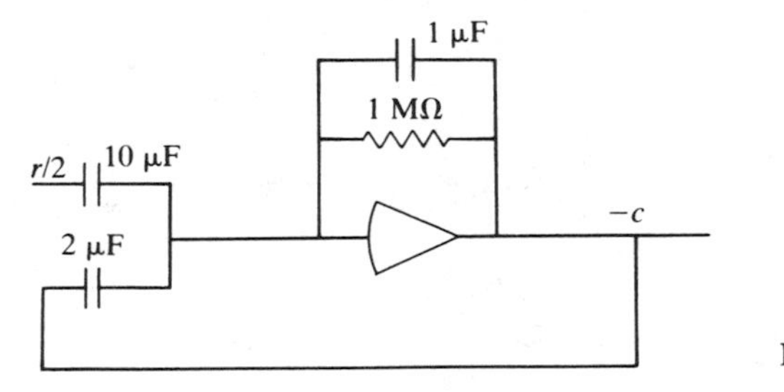

FIGURE P8.42

8.43. Determine the computer diagram for simulating the control system shown in Fig. P8.43. The maximum expected values are $r(t)_m = 0.5$, $d(t)_m = 1.0$, $m(t)_m = 1.0$, and $c(t)_m = 0.2$.

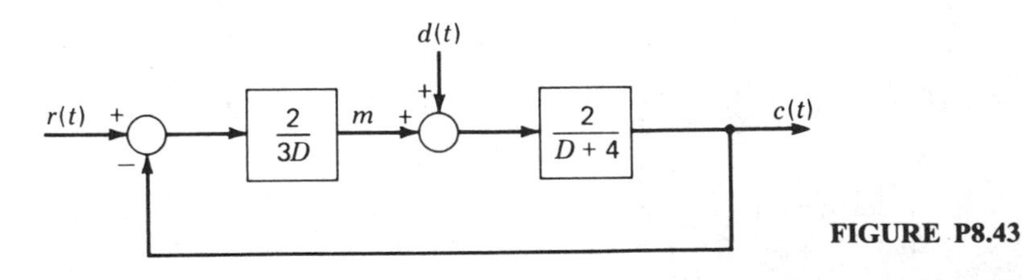

FIGURE P8.43

8.44. Figure P8.44*a* shows a control system and Fig. P8.44*b* shows an analog computer diagram for simulating the system. Determine the values of K_1, K_2, K_H, and τ.

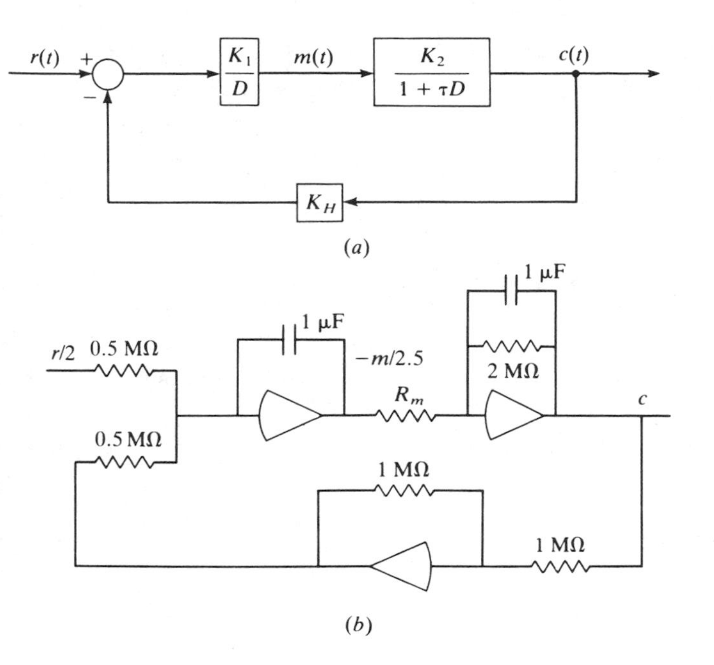

FIGURE P8.44

8.45. Determine the computer diagram for simulating the control system shown in Fig. P8.45. The maximum expected values are $r(t)_m = 0.5$, $c(t)_m = 1.0$, $d(t)_m = 2.5$, $m(t)_m = 2.0$. The values of the constants are $K_1 = 10$, $K_2 = 20$, $\tau_1 = 0.2$, and $\tau_2 = 1.0$. (Use 1-μF capacitors.)

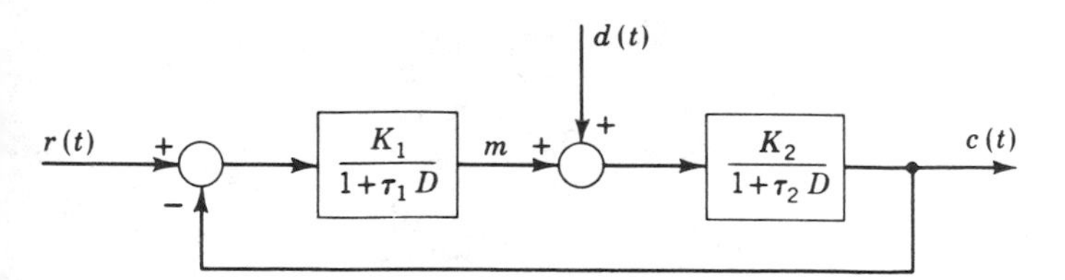

FIGURE P8.45

8.46. A system is described by the two simultaneous equations

$$\dot{x} + 4x + 2y = f(t)$$
$$2x + 4y + \dot{y} = 0$$

Solve the first differential equation for $\dot{x}$ and the second for $\dot{y}$; then proceed to complete the computer diagram shown in Fig. P8.46. That is, indicate the various amplifier gains, and hook up the appropriate interconnecting and feedback signals. The maximum expected values are $f_m = 2.0$, $x_m = 0.5$, $\dot{x}_m = 1.0$, $y_m = 1.0$, and $\dot{y}_m = 2.0$. The initial conditions are $x(0) = 0.2$ and $y(0) = 0.1$.

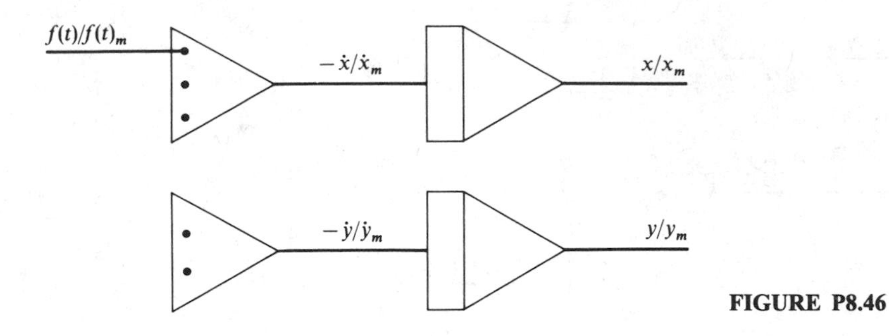

FIGURE P8.46

8.47. Construct the computer diagram for solving the two simultaneous differential equations

$$\dot{x} + 5x = y + f(t)$$
$$\ddot{y} + 10y = 2x$$

The initial conditions are $x(0) = 1$, $\dot{y}(0) = 0$, and $y(0) = 0.2$. The maximum expected values are $x_m = 4$, $\dot{x}_m = 10$, $y_m = 1$, $\dot{y}_m = 2$, $\ddot{y}_m = 4$, and $f(t)_m = 10$.

CHAPTER 9

STATE-SPACE METHODS

State-space methods provide a generalized technique for investigating the behavior of control systems. This approach is especially well suited to the analysis of more complex and nonlinear control systems. A feature of state-space methods is that they lend themselves to solution by either digital or analog computers.

9.1 SYSTEM REPRESENTATION

Common techniques employed for obtaining the state-space representation for a control system are direct programming, parallel programming, series programming, and general programming. The choice of the state-space representation depends on the nature of the particular problem to be solved.

Direct Programming

To illustrate this method, consider the differential equation

$$y(t) = \frac{D+3}{(D+1)(D+2)} f(t) \tag{9.1}$$

In direct programming, x_1 is taken as the differential equation in which the numerator operator is 1, that is,

$$x_1 = \frac{1}{(D+1)(D+2)} f(t) = \frac{1}{D^2 + 3D + 2} f(t)$$

Thus,

$$y(t) = (D+3)x_1 = \dot{x}_1 + 3x_1$$

By letting $\dot{x}_1 = x_2$, then

$$y(t) = 3x_1 + x_2 \tag{9.2}$$

The differential equation for x_1 may be written in the form

$$(D^2 + 3D + 2)x_1 = \ddot{x}_1 + 3\dot{x}_1 + 2x_1 = f(t)$$

Solving for $\ddot{x}_1 = \dot{x}_2$ gives

$$\dot{x}_2 = -2x_1 - 3x_2 + f(t) \tag{9.3}$$

The block diagram for obtaining the computer solution of Eq. (9.3) is shown in Fig. 9.1*a*. Note that the output of the summer (indicated by the circle) is $\dot{x}_2$. Successive integration of $\dot{x}_2$ yields in turn $x_2 = \dot{x}_1$ and x_1. The final computer diagram for obtaining $y(t)$ from x_1 and x_2 by using Eq. (9.2) is shown in Fig. 9.1*b*.

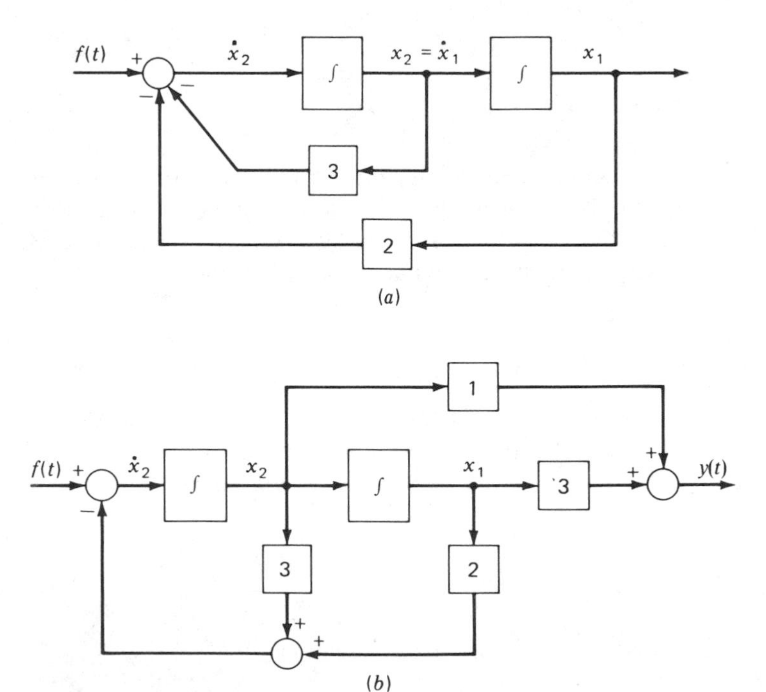

FIGURE 9.1
Computer diagram for direct programming.

The state-space representation for the differential equation, Eq. (9.1), is given by Eqs. (9.2) and (9.3). These system equations may be expressed in matrix form:

$$\begin{bmatrix} \dot{x}_1 \\ \dot{x}_2 \end{bmatrix} = \begin{bmatrix} 0 & 1 \\ -2 & -3 \end{bmatrix} \begin{bmatrix} x_1 \\ x_2 \end{bmatrix} + \begin{bmatrix} 0 \\ 1 \end{bmatrix} f(t)$$
$$y = \begin{bmatrix} 3 & 1 \end{bmatrix} \begin{bmatrix} x_1 \\ x_2 \end{bmatrix} \tag{9.4}$$

The x's are referred to as the *state variables*. Note that the output of each integrator in Fig. 9.1 is a state variable.

The general procedure for obtaining the state-space representation by direct programming is to let x_1 be the original differential equation in which the numerator operator (that is, $D + 3$) is taken as 1. Each derivative of x_1 is then assigned a state variable (for example, $\dot{x}_1 = x_2$, $\ddot{x}_1 = \dot{x}_2 = x_3$, etc.). The number of state variables is equal to the order of the differential equation. For the case in which the operator is 1, then $y = x_1$, $\dot{y} = \dot{x}_1 = x_2$, etc.

Because of the simplicity of the state-space relationships $\dot{x}_1 = x_2$, $\dot{x}_2 = x_3, \ldots,$ direct programming is the most commonly used method for obtaining the state-space representation.

Parallel Programming

In parallel programming the differential equation is expressed in the form of a partial-fraction expansion. Thus,

$$y(t) = \frac{D+3}{(D+1)(D+2)} f(t) = \left(\frac{2}{D+1} - \frac{1}{D+2} \right) f(t) \tag{9.5}$$

This result may be rewritten as

$$y(t) = 2x_1 - x_2 \tag{9.6}$$

where

$$x_1 = \frac{f(t)}{D+1} \qquad x_2 = \frac{f(t)}{D+2} \tag{9.7}$$

Figure 9.2*a* shows the computer diagram for generating the basic term

$$x = \frac{f(t)}{D+a}$$

or

$$\dot{x} = f(t) - ax$$

The computer diagram for generating the state variables as defined in Eqs. (9.7) is shown to the left of the dotted partition in Fig. 9.2*b*. To the right of the dotted partition is indicated the manner in which the state variables are combined in

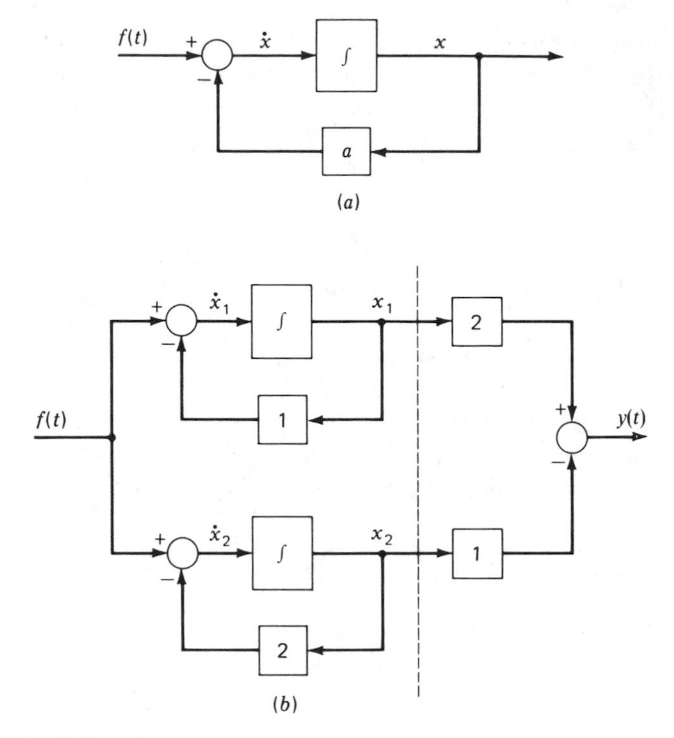

FIGURE 9.2
Computer diagram for parallel programming.

Eq. (9.6) to yield the desired solution $y(t)$. Equations (9.6) and (9.7) make up a state-space representation for the system, Eq. (9.1). In terms of matrix notation, Eqs. (9.6) and (9.7) become

$$\begin{bmatrix} \dot{x}_1 \\ \dot{x}_2 \end{bmatrix} = \begin{bmatrix} -1 & 0 \\ 0 & -2 \end{bmatrix} \begin{bmatrix} x_1 \\ x_2 \end{bmatrix} + \begin{bmatrix} 1 \\ 1 \end{bmatrix} f(t)$$
$$y(t) = [2 \quad -1] \begin{bmatrix} x_1 \\ x_2 \end{bmatrix} \tag{9.8}$$

The characteristic function for Eq. (9.5) is $(D + 1)(D + 2)$. Note in Eq. (9.8) that -1 and -2, which are the roots of the characteristic equation, are displayed on the principal diagonal of the first matrix to the right of the equals sign. All other elements of this matrix are zero. In general, with parallel programming the roots of the characteristic equation are displayed on the principal diagonal.

Series Programming

Consider the general nth-order differential equation

$$(D^n + a_1 D^{n-1} + \cdots + a_{n-1} D + a_n) y(t)$$
$$= (b_1 D^{n-1} + b_2 D^{n-2} + \cdots + b_{n-1} D + b_n) f(t) \tag{9.9}$$

This differential equation may be written in matrix form as

$$\begin{bmatrix} \dot{x}_1 \\ \dot{x}_2 \\ \vdots \\ \dot{x}_n \end{bmatrix} = \begin{bmatrix} -a_1 & 1 & 0 & \cdots & 0 \\ -a_2 & 0 & 1 & \cdots & 0 \\ \cdots & \cdots & \cdots & \cdots & \cdots \\ -a_n & 0 & 0 & \cdots & 0 \end{bmatrix} \begin{bmatrix} x_1 \\ x_2 \\ \vdots \\ x_n \end{bmatrix} + \begin{bmatrix} b_1 \\ b_2 \\ \vdots \\ b_n \end{bmatrix} f(t) \tag{9.10}$$

where $y(t) = x_1$.

For Eq. (9.1), it follows that $a_1 = 3$, $a_2 = 2$, $b_1 = 1$, and $b_2 = 3$. The resulting matrix representation is

$$\begin{aligned} \begin{bmatrix} \dot{x}_1 \\ \dot{x}_2 \end{bmatrix} &= \begin{bmatrix} -3 & 1 \\ -2 & 0 \end{bmatrix} \begin{bmatrix} x_1 \\ x_2 \end{bmatrix} + \begin{bmatrix} 1 \\ 3 \end{bmatrix} f(t) \\ y(t) &= [1 \quad 0] \begin{bmatrix} x_1 \\ x_2 \end{bmatrix} \end{aligned} \tag{9.11}$$

To prove that this matrix representation satisfies Eq. (9.1), first write the matrix equations in the form

$$\dot{x}_1 = -3x_1 + x_2 + f(t)$$
$$\dot{x}_2 = -2x_1 + 3f(t)$$

Differentiating the first equation and then substituting the resulting expression for $\dot{x}_2$ into the second give

$$\dot{x}_2 = \ddot{x}_1 + 3\dot{x}_1 - f'(t) = -2x_1 + 3f(t)$$

or

$$\ddot{x}_1 + 3\dot{x}_1 + 2x_1 = f'(t) + 3f(t)$$

This is identical to Eq. (9.1), with $x_1 = y(t)$. The block-diagram representation for the matrix equations is shown in Fig. 9.3.

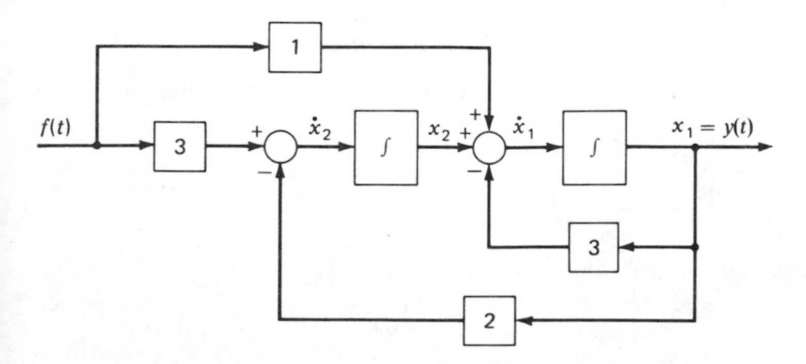

FIGURE 9.3
Computer diagram for series programming.

Series programming has the feature that y is always equal to x_1. However, the other state-space relationships tend to be somewhat awkward. In the case of the preceding example $\dot{x}_1 = -3x_1 + x_2 + f(t)$. Note that $\dot{x}_1$ is not equal to x_2 as was the case for direct programming.

General Programming

Consider the differential equation

$$(D^2 + a_1 D + a_2)y(t) = (b_0 D^2 + b_1 D + b_2)f(t) \tag{9.12}$$

The $b_0 D^2 f(t)$ term is eliminated by letting $y(t) = x_1 + b_0 f(t)$. Thus,

$$\ddot{x}_1 + a_1\dot{x}_1 + a_2 x_1 = (b_1 - a_1 b_0)f'(t) + (b_2 - a_2 b_0)f(t) \tag{9.13}$$

Now let us transform this to the form

$$\begin{bmatrix} \dot{x}_1 \\ \dot{x}_2 \end{bmatrix} = \begin{bmatrix} 0 & 1 \\ -a_2 & -a_1 \end{bmatrix} \begin{bmatrix} x_1 \\ x_2 \end{bmatrix} + \begin{bmatrix} h_1 \\ h_2 \end{bmatrix} f(t) \tag{9.14}$$

The corresponding matrix relationships are

$$\dot{x}_1 = x_2 + h_1 f(t)$$
$$\dot{x}_2 = -a_2 x_1 - a_1 x_2 + h_2 f(t)$$

Solving the first matrix relationship for x_2 shows that

$$x_2 = \dot{x}_1 - h_1 f(t)$$

Differentiation gives

$$\dot{x}_2 = \ddot{x}_1 - h_1 f'(t)$$

Substitution of the preceding results for x_2 and $\dot{x}_2$ into the second matrix relationship yields

$$\ddot{x}_1 - h_1 f'(t) = -a_2 x_1 - a_1[\dot{x}_1 - h_1 f(t)] + h_2 f(t)$$

or

$$\ddot{x}_1 + a_1\dot{x}_1 + a_2 x_1 = h_1 f'(t) + (h_2 + a_1 h_1) f(t)$$

This has the same form as Eq. (9.13). Equating coefficients of the $f'(t)$ and $f(t)$ terms shows that

$$h_1 = b_1 - a_1 b_0$$
$$h_2 = b_2 - a_2 b_0 - a_1 h_1$$

For the case of the differential equation given by Eq. (9.1), $a_1 = 3$, $a_2 = 2$, $b_0 = 0$, $b_1 = 1$, and $b_2 = 3$. Hence, $h_1 = 1$ and $h_2 = 0$. The resulting state-space representation is

$$\begin{bmatrix} \dot{x}_1 \\ \dot{x}_2 \end{bmatrix} = \begin{bmatrix} 0 & 1 \\ -2 & -3 \end{bmatrix} \begin{bmatrix} x_1 \\ x_2 \end{bmatrix} + \begin{bmatrix} 1 \\ 0 \end{bmatrix} f(t) \tag{9.15}$$

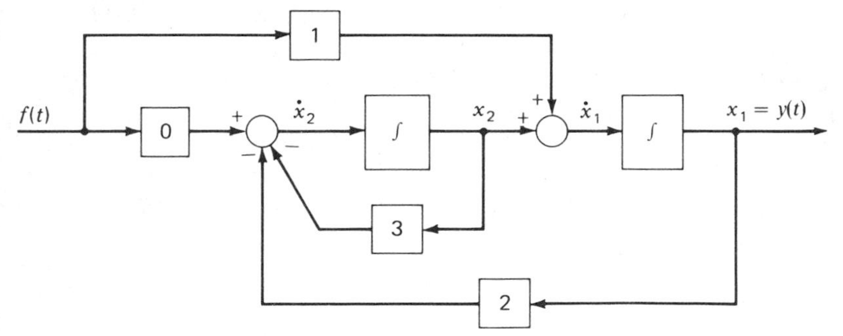

FIGURE 9.4
Computer diagram for general programming.

where $y(t) = x_1 + b_0 f(t) = x_1$. The block-diagram representation is shown in Fig. 9.4.

In general, the differential equation

$$(D^n + a_1 D^{n-1} + \cdots + a_{n-1} D + a_n) y(t) = (b_0 D^n + b_1 D^{n-1} + \cdots + b_{n-1} D + b_n) f(t) \tag{9.16}$$

is transformed to the form

$$\begin{bmatrix} \dot{x}_1 \\ \dot{x}_2 \\ \vdots \\ \dot{x}_n \end{bmatrix} = \begin{bmatrix} 0 & 1 & 0 & \cdots & 0 \\ 0 & 0 & 1 & \cdots & 0 \\ \cdots & \cdots & \cdots & \cdots & \cdots \\ -a_n & -a_{n-1} & -a_{n-2} & \cdots & -a_1 \end{bmatrix} \begin{bmatrix} x_1 \\ x_2 \\ \vdots \\ x_n \end{bmatrix} + \begin{bmatrix} h_1 \\ h_2 \\ \vdots \\ h_n \end{bmatrix} f(t) \tag{9.17}$$

where

$$\begin{aligned} h_1 &= b_1 - a_1 b_0 \\ h_2 &= b_2 - a_2 b_0 - a_1 h_1 \\ h_3 &= b_3 - a_3 b_0 - a_2 h_1 - a_1 h_2 \\ &\vdots \\ h_n &= b_n - a_n b_0 - a_{n-1} h_1 - a_{n-2} h_2 - \cdots - a_1 h_{n-1} \end{aligned} \tag{9.18}$$

and the state variables are

$$\begin{aligned} x_1 &= y(t) - b_0 f(t) \\ x_2 &= \dot{x}_1 - h_1 f(t) \\ x_3 &= \dot{x}_2 - h_2 f(t) \\ &\vdots \\ x_n &= \dot{x}_{n-1} - h_{n-1} f(t) \end{aligned} \tag{9.19}$$

In this method, the state-space representation is obtained by substituting the coefficients of the differential equation into a general form, Eq. (9.17). Note that the highest power of D on both sides of Eq. (9.16) is n. Other methods of obtaining the state-space representation are limited to the case in which the highest power of D on the right side is 1 less $(n - 1)$ than that on the left side (n).

There is no unique state-space representation for a system. Comparison of Eqs. (9.4), (9.8), (9.11), and (9.15) shows that any system whose operation is described by a linear differential equation may be written in the general matrix form

$$\dot{\mathbf{x}} = A\mathbf{x} + \mathbf{b}f(t) \tag{9.20}$$

$$y = \mathbf{c}^T\mathbf{x} \tag{9.21}$$

In the following section, we show that the preceding matrix form is also applicable to closed-loop systems.

Closed-Loop Systems

The state-space representation may be obtained directly from the block diagram for a system. The system shown in Fig. 9.5 represents the dynamics of a motor control system for an electric (battery-powered) car. The input is $u(t)$, and the output is $y(t)$. One state variable $x_1, x_2, \ldots$ is assigned for each order of D in the denominator.

For the $1/(D + 5)$ term, the state variable is x_1. For the $2/(D^2 + 3D + 4)$ term, the state variables are x_2 and x_3, where $x_3 = \dot{x}_2$. The equation for each block is

$$\frac{x_1}{x_2} = \frac{1}{D+5} \qquad \frac{x_2}{e} = \frac{2}{D^2 + 3D + 4}$$

Thus,

$$\begin{aligned}
\dot{x}_1 &= -5x_1 + x_2 \\
\dot{x}_2 &= x_3 \\
\dot{x}_3 &= \ddot{x}_2 = -3\dot{x}_2 - 4x_2 + 2e
\end{aligned}$$

The equation for e is

$$e = u - y = u - x_1$$

Hence

$$\dot{x}_3 = -2x_1 - 4x_2 - 3x_3 + 2u$$

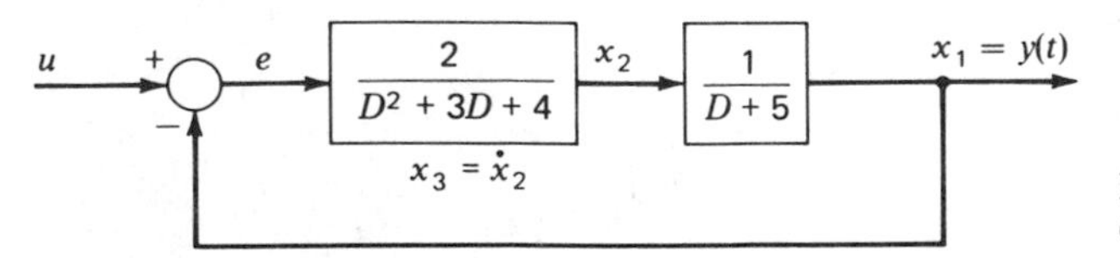

FIGURE 9.5
Closed-loop system.

The resulting state-space representation for this system is

$$\begin{bmatrix} \dot{x}_1 \\ \dot{x}_2 \\ \dot{x}_3 \end{bmatrix} = \begin{bmatrix} -5 & 1 & 0 \\ 0 & 0 & 1 \\ -2 & -4 & -3 \end{bmatrix} \begin{bmatrix} x_1 \\ x_2 \\ x_3 \end{bmatrix} + \begin{bmatrix} 0 \\ 0 \\ 2 \end{bmatrix} u$$

The output equation is

$$y = x_1$$

As illustrated in Fig. 9.5, the output from any block which contains the operator D in the denominator is assigned a state variable. Thus, the output of the $1/(D + 5)$ block is assigned x_1, and the output of the $2/(D^2 + 3D + 4)$ block is assigned x_2. When the order of the denominator is greater than 1, additional state variables are assigned ($x_3 = \dot{x}_2$, etc.) such that the number of state variables is equal to the order of D in the denominator.

Figure 9.6*a* shows a modified version of the motor control system shown in Fig. 9.5. This new system contains the numerator term $D + 2$. Numerator terms do not require additional state variables. To show this, the numerator term is displayed separately in Fig. 9.6*b*. The output z serves as a dummy variable. The general procedure when there is a numerator term is to put it in a separate block and place it ahead of the block containing the denominator term. The output from the block containing the numerator term is the dummy variable z. The output from the block containing the denominator term which is the input to the block containing the numerator is a state variable. The equation for each block is

$$\frac{x_1}{z} = \frac{1}{D + 5} \qquad \frac{x_2}{e} = \frac{1}{D^2 + 3D + 4}$$

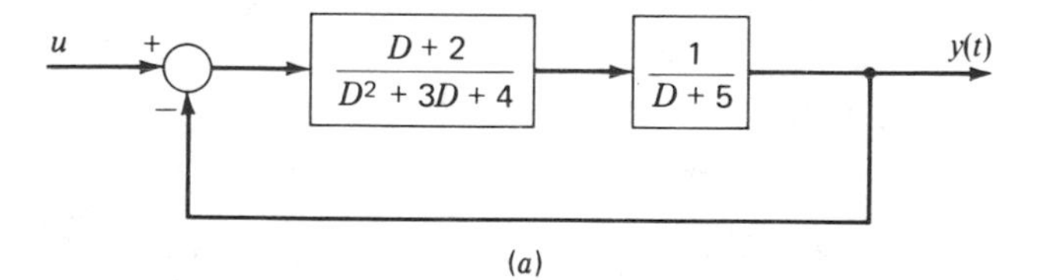

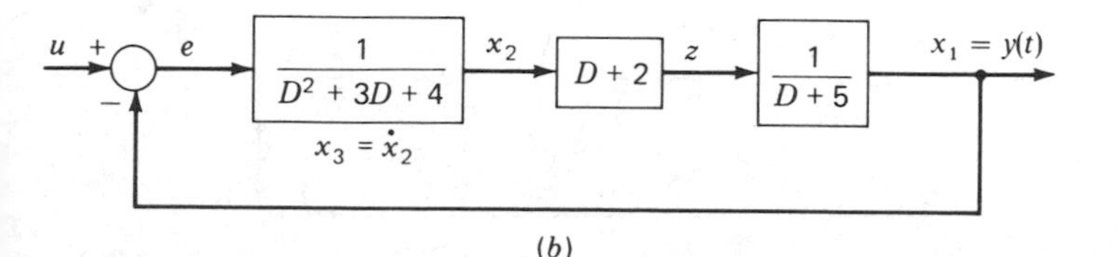

FIGURE 9.6
Closed-loop system.

Thus,

$$\dot{x}_1 = -5x_1 + z$$
$$\dot{x}_2 = x_3$$
$$\dot{x}_3 = \ddot{x}_2 = -3\dot{x}_2 - 4x_2 + e$$

Using the relationship $e = u - x_1$ to eliminate e and the relationship $z = \dot{x}_2 + 2x_2 = x_3 + 2x_2$ to eliminate z yields for the state-space representation for this system

$$\begin{bmatrix} \dot{x}_1 \\ \dot{x}_2 \\ \dot{x}_3 \end{bmatrix} = \begin{bmatrix} -5 & 2 & 1 \\ 0 & 0 & 1 \\ -1 & -4 & -3 \end{bmatrix} \begin{bmatrix} x_1 \\ x_2 \\ x_3 \end{bmatrix} + \begin{bmatrix} 0 \\ 0 \\ 1 \end{bmatrix} u$$

The output relationship is

$$y = x_1$$

To obtain the state-space representation for the system shown in Fig. 9.7*a*, the numerator term is placed in a separate block and placed ahead of the denominator term, as shown in Fig. 9.7*b*. The output of the numerator term is z, which is the system output y. The output of the denominator term $1/(D^2 + 3D + 4)$ is x_1. Because this denominator is second-order in D, the additional state variable $x_2 = \dot{x}_1$ is assigned. The output of the $1/D$ block is x_3.

The equation for the numerator block is

$$z = (D^2 + 2D + 5)x_1 = \ddot{x}_1 + 2\dot{x}_1 + 5x_1 = \dot{x}_2 + 2x_2 + 5x_1$$

The equation for the denominator block is

$$(D^2 + 3D + 4)x_1 = \dot{x}_2 + 3x_2 + 4x_1 = x_3$$

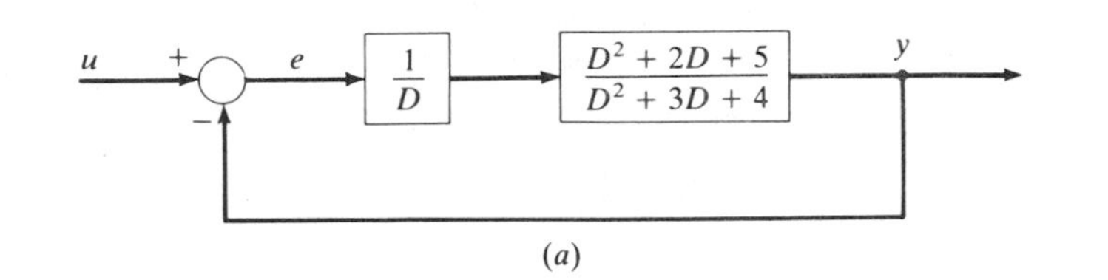

(*a*)

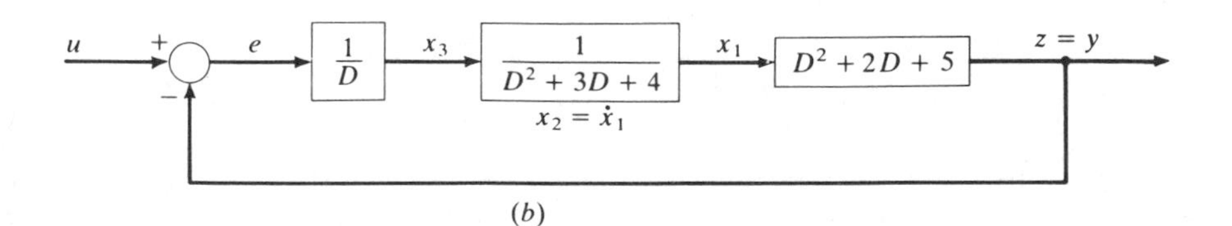

(*b*)

FIGURE 9.7
Closed-loop system.

Solving for $\dot{x}_2$ in terms of the state variables gives

$$\dot{x}_2 = -4x_1 - 3x_2 + x_3$$

Substituting this result into the expression for z yields

$$z = x_1 - x_2 + x_3$$

The equation for the $1/D$ block is

$$Dx_3 = \dot{x}_3 = e = u - z = -x_1 + x_2 - x_3 + u$$

Thus, the state-space representation for this system is

$$\begin{bmatrix} \dot{x}_1 \\ \dot{x}_2 \\ \dot{x}_3 \end{bmatrix} = \begin{bmatrix} 0 & 1 & 0 \\ -4 & -3 & 1 \\ -1 & 1 & -1 \end{bmatrix} \begin{bmatrix} x_1 \\ x_2 \\ x_3 \end{bmatrix} + \begin{bmatrix} 0 \\ 0 \\ 1 \end{bmatrix} u$$

Because $y = z$, the output relationship is

$$y = [1 \quad -1 \quad 1] \begin{bmatrix} x_1 \\ x_2 \\ x_3 \end{bmatrix}$$

Note that the preceding equations have the general form given by Eqs. (9.20) and (9.21).

A system which has one input u and one output y is referred to as a *single-input single-output* system. A system with more than one input is a multiple-input system, and a system with more than one output is a multiple-output system. A system with more than one input and more than one output is a multiple-input multiple-output system. Matrix methods lend themselves to the analysis of such systems. The system shown in Fig. 9.8 has two inputs and two outputs. The equation for each of the state variables is

$$\frac{x_1}{e_1} = \frac{1}{D+1} \qquad \frac{x_2}{x_3} = \frac{1}{D+2} \qquad \frac{x_3}{e_2} = \frac{1}{D}$$

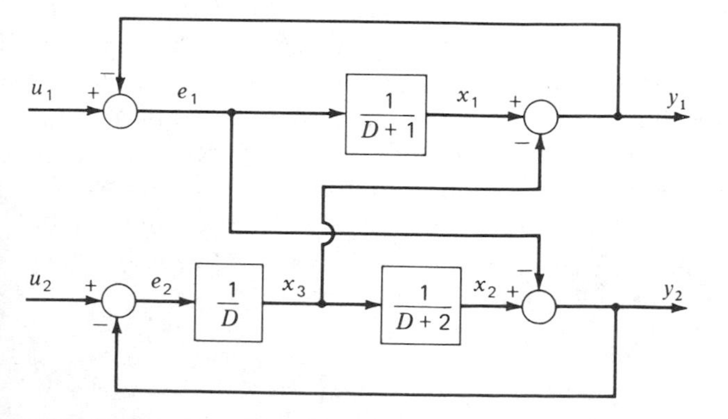

FIGURE 9.8
Multiple-input multiple-output system.

Thus,

$$\dot{x}_1 = -x_1 + e_1$$
$$\dot{x}_2 = -2x_2 + x_3$$
$$\dot{x}_3 = e_2$$

The equations for e_1 and e_2 are

$$\begin{aligned} e_1 &= u_1 - y_1 = u_1 - (x_1 - x_3) = -x_1 + x_3 + u_1 \\ e_2 &= u_2 - y_2 = u_2 - (x_2 - e_1) = u_2 - x_2 + (-x_1 + x_3 + u_1) \\ &= -x_1 - x_2 + x_3 + u_1 + u_2 \end{aligned}$$

Substituting e_1 and e_2 into the system equations and expressing the result in matrix form yield

$$\begin{bmatrix} \dot{x}_1 \\ \dot{x}_2 \\ \dot{x}_3 \end{bmatrix} = \begin{bmatrix} -2 & 0 & 1 \\ 0 & -2 & 1 \\ -1 & -1 & 1 \end{bmatrix} \begin{bmatrix} x_1 \\ x_2 \\ x_3 \end{bmatrix} + \begin{bmatrix} 1 & 0 \\ 0 & 0 \\ 1 & 1 \end{bmatrix} \begin{bmatrix} u_1 \\ u_2 \end{bmatrix}$$

The output relationships are

$$y_1 = x_1 - x_3$$
$$y_2 = x_2 - e_1 = x_1 + x_2 - x_3 - u_1$$

The corresponding matrix form is

$$\begin{bmatrix} y_1 \\ y_2 \end{bmatrix} = \begin{bmatrix} 1 & 0 & -1 \\ 1 & 1 & -1 \end{bmatrix} \begin{bmatrix} x_1 \\ x_2 \\ x_3 \end{bmatrix} + \begin{bmatrix} 0 & 0 \\ -1 & 1 \end{bmatrix} \begin{bmatrix} u_1 \\ u_2 \end{bmatrix}$$

The general matrix representation for a multiple-input multiple-output system is

$$\dot{\mathbf{x}} = A\mathbf{x} + B\mathbf{u} \tag{9.22}$$

The general output relationship is

$$\mathbf{y} = C\mathbf{x} + D\mathbf{u} \tag{9.23}$$

The block-diagram representation for these matrix relationships is shown in Fig. 9.9. The portion enclosed by the dashed box is the system equation. The remaining portion is the output relationship. The double-width arrows represent more than one variable. For the case of a single input u, B becomes the column matrix $\mathbf{b}$ and D becomes the column matrix $\mathbf{d}$. For the case of a single output y, C becomes $\mathbf{c}^T$. For the usual case in which there is no feedforward path from input u to the output, the D term in Eq. (9.23) vanishes. Thus, most systems can be represented in the form given by Eqs. (9.20) and (9.21), in which $\mathbf{b}$ is replaced by B for multiple-input systems and $\mathbf{c}^T$ is replaced by C for multiple-output systems.

Computer Solution

The state-space representation for a differential equation or for a control system lends itself very well to solution on a digital computer. The solution is obtained by

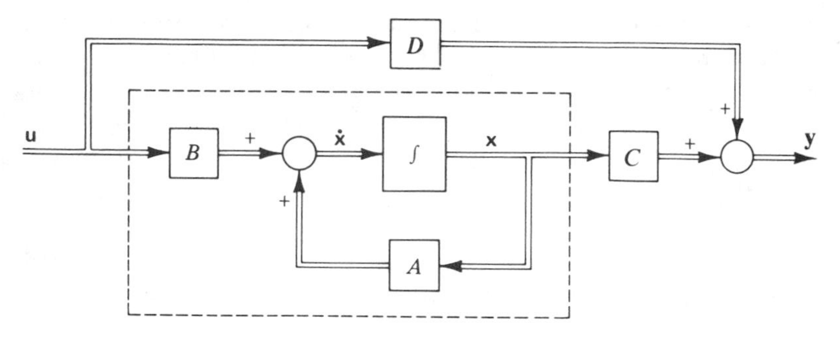

FIGURE 9.9
General state-space representation for a system.

solving the various state-space equations to obtain the $\dot{x}$ terms and then integrating each $\dot{x}$ term to obtain the corresponding state variable.

In Sect. 6.4, we showed how differential equations may be solved numerically. To summarize, Fig. 9.10 shows a plot of $\dot{x}$ versus time. The integral of $\dot{x}$ is

$$x = \int \dot{x}\, dt = x_0 + \int_{t_0}^{t_0 + \Delta t} \dot{x}\, dt$$

where x_0 is the value of x at the beginning of the interval (at time t_0). The area shown shaded is the integral of $\dot{x}$ from time t_0 to time $t_0 + \Delta t$. For a small time increment Δt, this area is closely approximated by the area of the trapezoid

$$0.5(\dot{x}_0 + \dot{x})\Delta t$$

where $\dot{x}_0$ is the value of $\dot{x}$ at the beginning of the interval and $\dot{x}$ is the value at the end. The equation for the value of x is

$$x = x_0 + 0.5(\dot{x}_0 + \dot{x})\ \Delta t$$

where x is the value at the end of the interval.

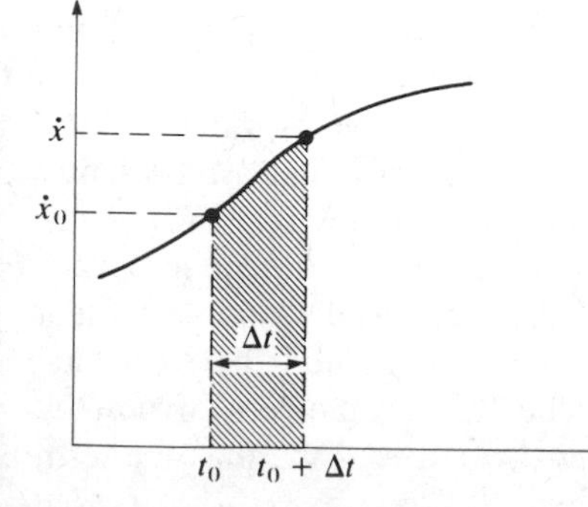

FIGURE 9.10
Plot of $\dot{x}$ versus time.

To illustrate this procedure, consider the differential equation given in Eq. (9.1). The state-space representation obtained by direct programming is given by Eq. (9.4):

$$\begin{aligned} \dot{x}_1 &= x_2 \\ \dot{x}_2 &= -2x_1 - 3x_2 + f(t) \\ y &= 3x_1 + x_2 \end{aligned} \tag{9.4}$$

The computer program for obtaining the response y for the case in which $f(t) = u(t)$ is a unit step function with all the initial conditions zero is

```
      X1 = 0.0
      X2 = 0.0
      DX2 = 0.0
      F = 1.0
      T = 0.0
      DT = 0.01
      DO 25 K=0, 400
      Y = 3.0*X1+X2
      IF (MOD(K,10).NE.0) GO TO 10
      WRITE (6,5) T,Y
 5    FORMAT (2F 10.2)
10    T = T+DT
      X10 = X1
      X20 = X2
      DX20 = DX2
      DX2 = F-2.0*X1-3.0*X2
      X2 = X20+0.5*(DX2+DX20)*DT
      X1 = X10+0.5*(X2+X20)*DT
25    CONTINUE
      END
```

In this program X1, X2, and DX2 represent x_1, x_2, and $\dot{x}_2$, respectively. The first three statements are the initial values. The next statement F = 1.0 is the forcing function $f(t) = u(t)$. The time increment for the integration interval is DT = 0.01 s. The DO loop solves the equations 400 times. Thus the duration of the solution is 400(0.01) = 4 s. The response is Y = 3.0*X1 + X2. Corresponding values of T and Y are now printed. The time for the next integration interval is T = T + DT. The value of X1 that is presently stored becomes the old value (beginning value) for the next interval, X10 = X1. Similarly, the values of X2 and DX2 that are presently stored become the old values (beginning values) for the next interval, X20 = X2 and DX20 = DX2. The new value of DX2 is obtained by the statement DX2 = F − 2.0*X1 − 3.0*X2. The new value of X2 is obtained by the statement X2 = X20 + 0.5*(DX2 + DX20)*DT. Note that the value X20 on the right side is the value of X2 at the beginning of the interval. The value of the derivative at the beginning of the interval is DX20, and the value at the end is DX2, thus the term

0.5*(DX2 + DX20)*DT is the trapezoidal integration of $\dot{x}_2$ during the interval. Similarly, the new value of X1 is obtained by the statement X1 = X10 + 0.5*(X2 + X20)*DT. Because $\dot{x}_1 = x_2$, the last term is the trapezoidal integration of $\dot{x}_1$ during the interval.

9.2 SIGNAL FLOW GRAPHS

A *signal flow graph* is a diagram which may be used to represent a control system. The signal flow graph representation for a system appears to be quite similar to the block-diagram representation. However, there are important differences. For more complex systems, the signal flow graph is easier to construct than the block diagram. System equations are more readily obtained from the signal flow graph than from the block diagram.

The left side of Fig. 9.11 shows the block-diagram representation, and the right side shows the equivalent signal flow graph. Each small circle in a signal flow

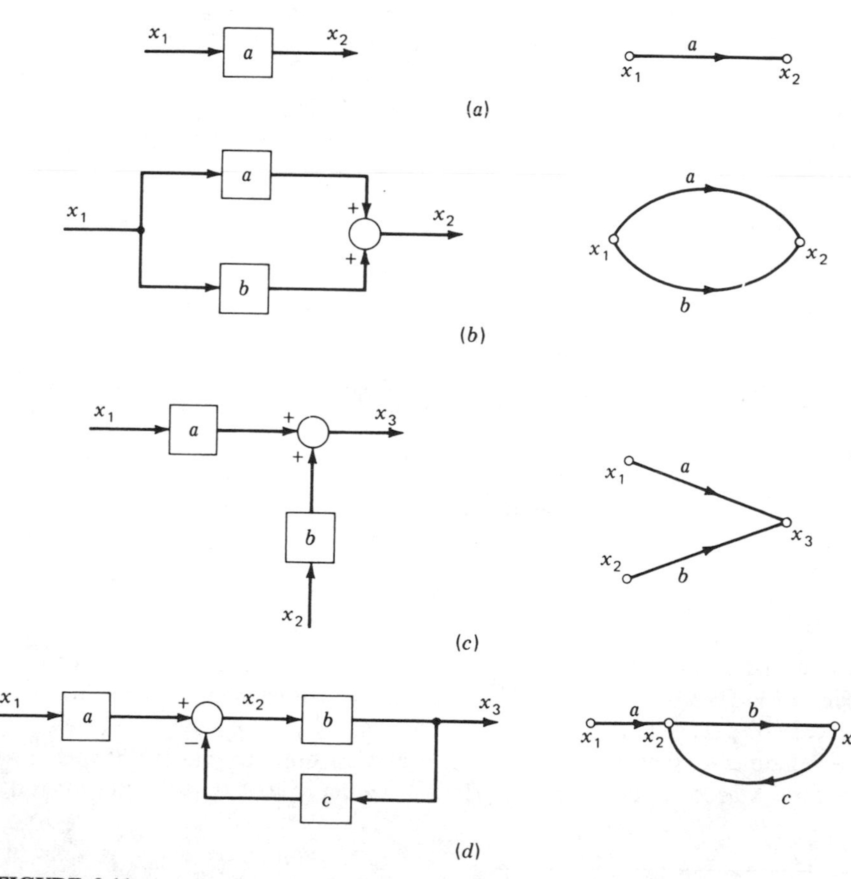

FIGURE 9.11
Block diagrams and corresponding signal flow graphs.

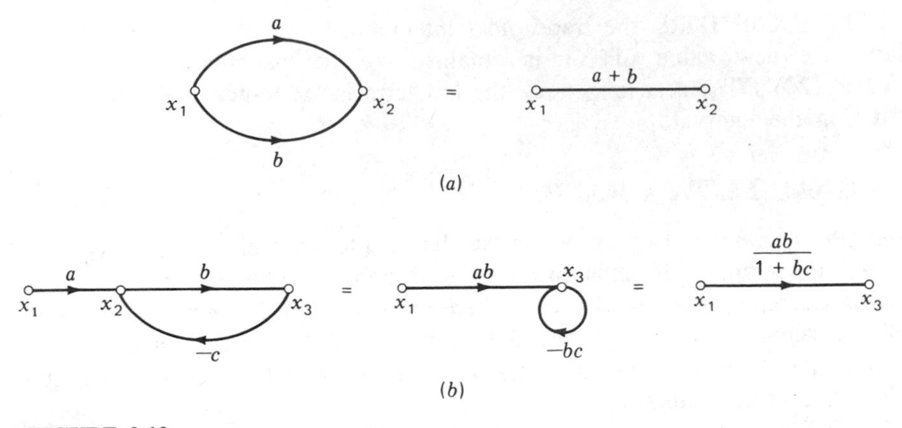

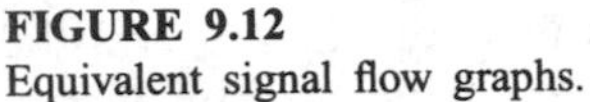

FIGURE 9.12
Equivalent signal flow graphs.

graph is called a *node*, which is a point that represents a variable or signal. The directed line segment joining two nodes is called a *branch*. The gain between two nodes is called the *transmittance*. In Fig. 9.11*a*, variable x_1 is multiplied by the transmittance a to yield output x_2 (that is, $x_2 = ax_1$). In Fig. 9.11*b*, output x_2 is the sum of ax_1 and bx_1 (that is, $x_2 = ax_1 + bx_1$). In Fig. 9.11*c*, the signal x_3 is the sum of ax_1 and bx_2. In Fig. 9.11*d*, the signal x_2 is $ax_1 - cx_3$, and the signal x_3 is bx_2.

As illustrated in Fig. 9.12*a*, the upper branch with a transmittance a and the lower branch with a transmittance b may be combined into one branch with a transmittance $a + b$ [note that $x_2 = ax_1 + bx_1 = (a + b)x_1$]. The equation for the left side of Fig. 9.11*d* is

$$x_3 = bx_2 = b(ax_1 - cx_3)$$

The equation for the equivalent signal flow graph shown in the middle of Fig. 9.12*b* is

$$x_3 = abx_1 - bcx_3$$

The basis for the signal flow representation shown on the right side of Fig. 9.12*b* is obtained by solving the preceding expression for x_3:

$$x_3 = \frac{ab}{1 + bc} x_1$$

The block-diagram representation for two basic control systems is shown on the left side of Fig. 9.13. The corresponding signal flow graph for each system is shown on the right side of Fig. 9.13.

The relationship between an input node and an output node is obtained by application of Mason's gain formula.* This formula yields directly the overall

---

* S. J. Mason, "Feedback Theory—Some Properties of Signal Flow Graphs," *Proc. IRE*, 41(9): 1144–1156, September 1953.

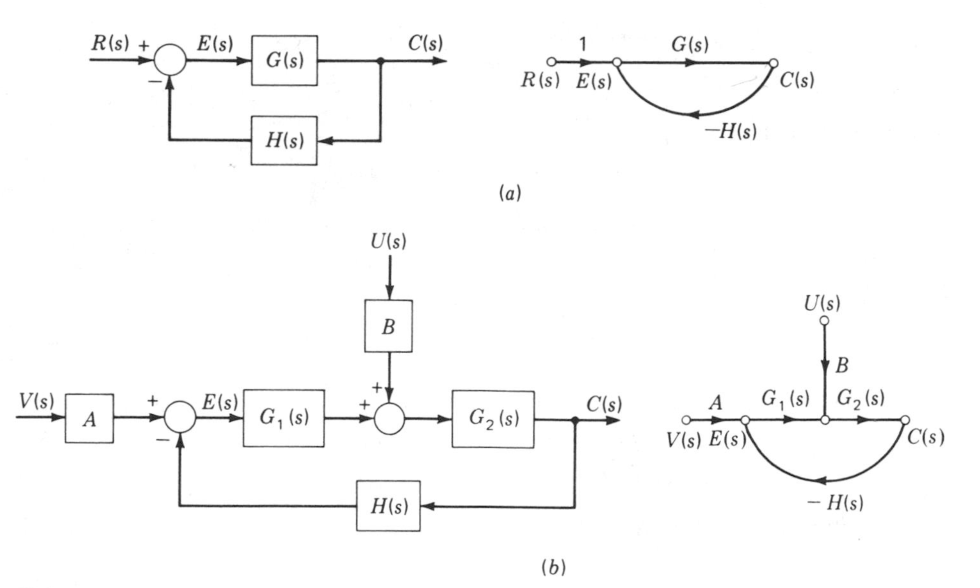

FIGURE 9.13
Basic systems.

transmittance from an input node to an output node

$$T = \frac{x_{\text{out}}}{x_{\text{in}}} = \sum_k \frac{T_k \Delta_k}{\Delta} \tag{9.24}$$

where T_k is the gain (transmittance) of the kth forward path from the input node x_{in} to the output node x_{out} and Δ is the determinant of the graph which is defined as follows:

$\Delta =$ 1 − (sum of all individual loop gains) + (sum of gain products of all combinations of two nontouching loops) − (sum of gain products of all combinations of three nontouching loops) + ⋯

$\Delta_k =$ determinant of graph in which all loops touching the kth forward path are removed (i.e., set equal to zero)

A forward path is any path which goes from the input node to the output node along which no node is passed through more than once. A loop is any path which originates and terminates at the same node along which no node is passed through more than once.

Touching loops are loops which have one or more nodes in common. Similarly, a loop which touches the kth forward path is one that has one or more nodes in common with the path.

The first step in applying Mason's gain formula is to identify the forward paths, from the input node to the output node, and all the loops in the system. For the signal flow graph shown in Fig. 9.13a, there is one forward path from the input node $R(s)$ to the output node $C(s)$, and there is only one loop. Thus,

$$T_1 = (1)G(s) = G(s) \qquad L_1 = G(s)[-H(s)] = -G(s)H(s)$$

The determinant of the graph is

$$\Delta = 1 - L_1 = 1 + G(s)H(s)$$

Because loop L_1 touches the forward path T_1, setting L_1 equal to zero in the equation for Δ gives $\Delta_1 = 1$. Application of Mason's gain formula yields the relationship

$$\frac{C(s)}{R(s)} = \frac{T_1\Delta_1}{\Delta} = \frac{G(s)}{1 + G(s)H(s)} \tag{9.25}$$

For the signal flow graph of Fig. 9.13*b*, there is one forward path from the input node $V(s)$ to the output node $C(s)$, and there is only one loop. Thus,

$$T_1 = AG_1(s)G_2(s) \qquad L_1 = -G_1(s)G_2(s)H(s)$$

The determinant of the graph is

$$\Delta = 1 - L_1 = 1 + G_1(s)G_2(s)H(s)$$

Because loop L_1 touches the forward path T_1, setting L_1 equal to zero in the equation for Δ gives $\Delta_1 = 1$. Application of Mason's gain formula gives

$$\frac{C(s)}{V(s)} = \frac{T_1\Delta_1}{\Delta} = \frac{AG_1(s)G_2(s)}{1 + G_1(s)G_2(s)H(s)} \tag{9.26}$$

By regarding $U(s)$ as the input node in Fig. 9.13*b* and $C(s)$ as the output node, it follows that

$$T_1 = BG_2(s) \qquad L_1 = -G_1(s)G_2(s)H(s)$$
$$\Delta = 1 - L_1 = 1 + G_1(s)G_2(s)H(s)$$

Because loop L_1 touches the forward path from $U(s)$ to $C(s)$, setting L_1 equal to zero in the expression for Δ yields $\Delta_1 = 1$. The overall relationship between $U(s)$ and $C(s)$ is

$$\frac{C(s)}{U(s)} = \frac{T_1\Delta_1}{\Delta} = \frac{BG_2(s)}{1 + G_1(s)G_2(s)H(s)} \tag{9.27}$$

An input node represents an independent source or signal entering the system. There are only two input nodes [$V(s)$ and $U(s)$] for the system in Fig. 9.13*b*. Nodes which are not input nodes may be regarded as output nodes. If $E(s)$ is regarded as an output node, note that the transmittance of the forward path from $V(s)$ to $E(s)$ is A. Because loop L_1 touches this path, it follows that

$$T_1 = A \qquad L_1 = -G_1(s)G_2(s)H(s)$$
$$\Delta = 1 - L_1 = 1 + G_1(s)G_2(s)H(s) \qquad \Delta_1 = 1$$

Application of Mason's gain formula yields

$$\frac{E(s)}{V(s)} = \frac{T_1\Delta_1}{\Delta} = \frac{A}{1 + G_1(s)G_2(s)H(s)} \tag{9.28}$$

The transmittance of the path from $U(s)$ to $E(s)$ is $-BG_2(s)H(s)$. Loop L_1 touches this path. Thus,

$$T_1 = -BG_2(s)H(s) \qquad\qquad L_1 = -G_1(s)G_2(s)H(s)$$
$$\Delta = 1 - L_1 = 1 + G_1(s)G_2(s)H(s) \qquad \Delta_1 = 1$$

Application of Mason's gain formula gives

$$\frac{E(s)}{U(s)} = \frac{T_1\Delta_1}{\Delta} = -\frac{BG_2(s)H(s)}{1 + G_1(s)G_2(s)H(s)} \tag{9.29}$$

The block diagram and corresponding signal flow graph for a more complex system are shown in Fig. 9.14. There are two forward paths from input $R(s)$ to output $C(s)$:

$$T_1 = G_1G_2G_3G_5 \qquad T_2 = G_4G_5$$

The gains for the three loops in the system are

$$L_1 = -G_1G_2G_3G_5H_1 \qquad L_2 = -G_2H_2 \qquad L_3 = -G_4G_5H_1$$

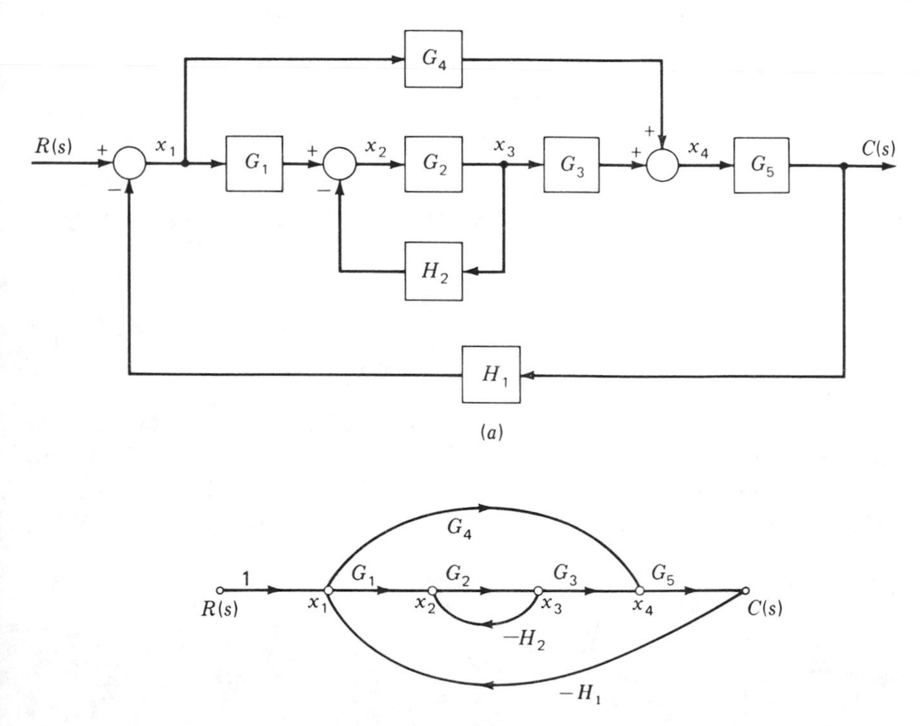

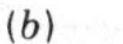
(b)

FIGURE 9.14
Block diagram and corresponding signal flow graph.

The numbering of the forward paths and loops is arbitrary. Because L_2 and L_3 are nontouching loops,

$$\begin{aligned}\Delta &= 1 - (L_1 + L_2 + L_3) + L_2L_3 \\ &= 1 + (G_1G_2G_3G_5H_1 + G_2H_2 + G_4G_5H_1) + G_2G_4G_5H_1H_2\end{aligned}$$

Because all three loops touch path T_1, setting $L_1 = L_2 = L_3 = 0$ in the preceding expression for Δ gives $\Delta_1 = 1$. Because loops L_1 and L_3 touch path T_2, setting $L_1 = L_3 = 0$ in the expression for Δ gives $\Delta_2 = 1 - L_2 = 1 + G_2H_2$. Substitution of these results into Mason's gain formula yields

$$\frac{C(s)}{R(s)} = \frac{T_1\Delta_1 + T_2\Delta_2}{\Delta} = \frac{G_1G_2G_3G_5 + (G_4G_5)(1 + G_2H_2)}{1 + (G_1G_2G_3 + G_4 + G_2G_4H_2)G_5H_1 + G_2H_2} \tag{9.30}$$

The signal flow graph corresponding to the block-diagram of Fig. 9.3 is shown in Fig. 9.15. The symbol $D^{-1} = 1/D$ indicates integration. There are two loops, $L_1 = -3D^{-1}$ and $L_2 = -2D^{-2}$. Because there are no nontouching loops,

$$\Delta = 1 - (L_1 + L_2) = 1 + 3D^{-1} + 2D^{-2}$$

There are two forward paths from input node $f(t)$ to output node $c(t)$. The transmittances of the paths are $T_1 = D^{-1}$ and $T_2 = 3D^{-2}$, respectively. Each forward path touches both loops, so that $\Delta_1 = \Delta_2 = 1$. Application of Mason's gain formula gives

$$y(t) = \frac{T_1\Delta_1 + T_2\Delta_2}{\Delta} f(t) = \frac{D^{-1} + 3D^{-2}}{1 + 3D^{-1} + 2D^{-2}} f(t) = \frac{D + 3}{D^2 + 3D + 2} f(t) \tag{9.1}$$

On block diagrams, the signal coming from a summing point and the system inputs and outputs are nodes. If the nodes are identified, Mason's gain formula may be applied directly to the block diagram.

Next, consider the simultaneous equations

$$x_1 = a_{11}x_1 + a_{12}x_2 + b_1u_1$$

$$x_2 = a_{21}x_1 + a_{22}x_2 + b_2u_2$$

The signal flow graph for the first equation is shown in Fig. 9.16*a*, and the graph for the second equation is shown in Fig. 9.16*b*. Combining these two graphs yields the complete signal flow graph for the system, shown in Fig. 9.16*c*. The inputs to the system are u_1 and u_2. The system has three loops. The gain for each loop is

$$L_1 = a_{11} \qquad L_2 = a_{12}a_{21} \qquad L_3 = a_{22}$$

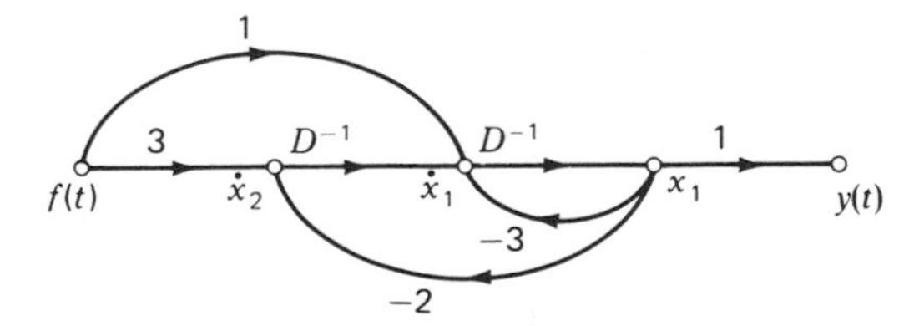

FIGURE 9.15
Signal flow graph.

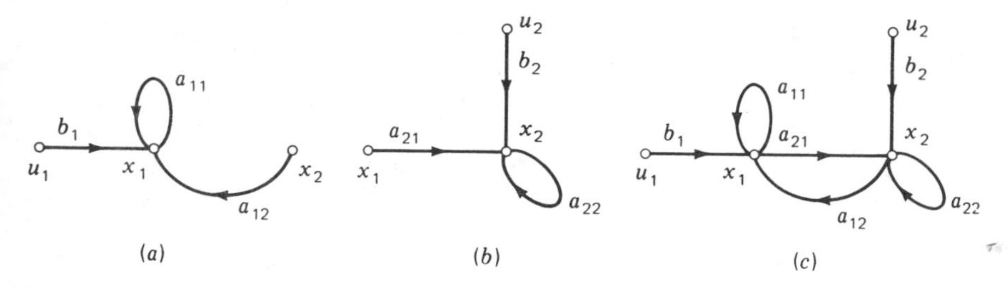

FIGURE 9.16
Signal flow graph for simultaneous equations.

To obtain the equation relating u_1 and x_2, first note that there is only one forward path from u_1 to x_2. The gain of this path is

$$T_1 = b_1 a_{21}$$

Loops L_1 and L_3 are nontouching. Thus, the determinant for the system is

$$\Delta = 1 - (L_1 + L_2 + L_3) + L_1 L_3 = 1 - (a_{11} + a_{12}a_{21} + a_{22}) + a_{11}a_{22}$$

Because all three loops touch the forward path, $\Delta_1 = 1$. Application of Mason's gain formula yields

$$\frac{x_2}{u_1} = \frac{T_1\Delta_1}{\Delta} = \frac{b_1 a_{21}}{1 - (a_{11} + a_{12}a_{21} + a_{22}) + a_{11}a_{22}}$$

To obtain the equation relating u_2 and x_2, first note that there is only one forward path from u_2 to x_2. The gain of this path is $T_1 = b_2$. Because loops L_2 and L_3 touch this path, setting $L_2 = L_3 = 0$ in the equation for Δ gives

$$\Delta_1 = 1 - L_1 = 1 - a_{11}$$

The overall relationship is now found to be

$$\frac{x_2}{u_2} = \frac{T_1\Delta_1}{\Delta} = \frac{b_2(1 - a_{11})}{1 - (a_{11} + a_{12}a_{21} + a_{22}) + a_{11}a_{22}}$$

The gain of the forward path from u_1 to x_1 is $T_1 = b_1$. Because loops L_1 and L_2 touch this path, setting $L_1 = L_2 = 0$ in the equation for Δ gives

$$\Delta_1 = 1 - L_3 = 1 - a_{22}$$

Application of Mason's gain formula shows that

$$\frac{x_1}{u_1} = \frac{T_1\Delta_1}{\Delta} = \frac{b_1(1 - a_{22})}{1 - (a_{11} + a_{12}a_{21} + a_{22}) + a_{11}a_{22}}$$

The gain of the forward path from u_2 to x_1 is $T_1 = a_{12}b_2$. Because all loops touch this forward path, $\Delta_1 = 1$. Thus it is found that

$$\frac{x_1}{u_2} = \frac{a_{12}b_2}{1 - (a_{11} + a_{12}a_{21} + a_{22}) + a_{11}a_{22}}$$

Signal flow graphs provide the basis for many powerful techniques employed in the design of control systems.

9.3 SOLUTION OF STATE-SPACE EQUATIONS

Consider the differential equation

$$\frac{dx}{dt} = ax \tag{9.31}$$

Integration yields

$$x(t) = x_0 + \int_0^t ax(\tau)\,d\tau \tag{9.32}$$

Replacing t by τ gives

$$x(\tau) = x_0 + \int_0^\tau ax(\tau)\,d\tau$$

Substituting this result into Eq. (9.32) shows that

$$\begin{aligned} x(t) &= x_0 + \int_0^t a\left[x_0 + \int_0^\tau ax(\tau)\,d\tau\right] d\tau \\ &= x_0 + x_0\int_0^t a\,d\tau + \int_0^t a\int_0^\tau ax(\tau)\,d\tau\,d\tau \end{aligned}$$

Continuing this process of replacing t by τ and substituting the result into Eq. (9.32) yields

$$x(t) = x_0\left(1 + \int_0^t a\,d\tau + \int_0^t a\int_0^\tau a\,d\tau\,d\tau + \int_0^t a\int_0^\tau a\int_0^\tau a\,d\tau\,d\tau\,d\tau\ + \cdots\right)$$

The results of the preceding integrations are

$$\begin{aligned} &\int_0^t a\,d\tau = at \\ &\int_0^t a\int_0^\tau a\,d\tau\,d\tau = \int_0^t a(a\tau)\,d\tau = \frac{a^2t^2}{2!} \\ &\int_0^t a\int_0^\tau a\int_0^\tau a\,d\tau\,d\tau\,d\tau = \int_0^t a\frac{a^2t^2}{2!}\,d\tau = \frac{a^3t^3}{3!} \end{aligned}$$

Thus, the solution of the differential equation given by Eq. (9.31) is

$$\begin{aligned} x(t) &= \left(1 + at + \frac{a^2t^2}{2!} + \frac{a^3t^3}{3!} + \cdots\right)x_0 \\ &= e^{at}x_0 \end{aligned} \tag{9.33}$$

The solution of a system of differential equations is obtained in a similar manner. Thus, consider the system

$$\dot{\mathbf{x}}(t) = A\mathbf{x}(t) \qquad \mathbf{x}(0) = \mathbf{x}_0 \tag{9.34}$$

The corresponding differential equations are

$$\begin{aligned}
\dot{x}_1 &= a_{11}x_1 + a_{12}x_2 + \cdots + a_{1n}x_n \\
\dot{x}_2 &= a_{21}x_1 + a_{22}x_2 + \cdots + a_{2n}x_n \\
&\vdots \\
\dot{x}_n &= a_{n1}x_1 + a_{n2}x_2 + \cdots + a_{nn}x_n
\end{aligned}$$

Integration of each equation gives

$$\begin{aligned}
x_1 - x_1(0) &= \int_0^t (a_{11}x_1 + a_{12}x_2 + \cdots + a_{1n}x_n)\, d\tau \\
x_2 - x_2(0) &= \int_0^t (a_{21}x_1 + a_{22}x_2 + \cdots + a_{2n}x_n)\, d\tau \\
&\vdots \\
x_n - x_n(0) &= \int_0^t (a_{n1}x_1 + a_{n2}x_2 + \cdots + a_{nn}x_n)\, d\tau
\end{aligned}$$

The corresponding matrix representation is

$$\mathbf{x}(t) = \mathbf{x}(0) + \int_0^t A\mathbf{x}(\tau)\, d\tau$$

Replacing t by τ and substituting this result back into the preceding give

$$\begin{aligned}
\mathbf{x}(t) &= \mathbf{x}(0) + \int_0^t A\left[\mathbf{x}(0) + \int_0^\tau A\mathbf{x}(\tau)\, d\tau\right] d\tau \\
&= \mathbf{x}(0) + \mathbf{x}(0)\int_0^t A\, d\tau + \int_0^t A \int_0^\tau A\mathbf{x}(\tau)\, d\tau\, d\tau
\end{aligned}$$

Continuing this process yields

$$\mathbf{x}(t) = \mathbf{x}(0)\left[I + \int_0^t A\, d\tau + \int_0^t A\int_0^\tau A\, d\tau\, d\tau + \int_0^t A\int_0^\tau A\int_0^\tau A\, d\tau\, d\tau\, d\tau + \cdots\right]$$

The result of each integration is

$$\begin{aligned}
&\int_0^t A\, d\tau = At \\
&\int_0^t A\int_0^\tau A\, d\tau = \int_0^t A(A\tau)\, d\tau = \frac{A^2t^2}{2!} \\
&\int_0^t A\int_0^\tau A\int_0^\tau A\, d\tau\, d\tau\, d\tau = \int_0^t A\frac{A^2t^2}{2!}\, d\tau = \frac{A^3t^3}{3!}
\end{aligned}$$

Thus the solution of the system of differential equations represented by Eq. (9.34) is

$$\begin{aligned}\mathbf{x}(t) &= \left(I + At + \frac{A^2t^2}{2!} + \frac{A^3t^3}{3!} + \cdots\right)\mathbf{x}(0) = e^{At}\mathbf{x}(0) \\ &= \mathbf{\Phi}(t)\mathbf{x}(0)\end{aligned} \tag{9.35}$$

where

$$\mathbf{\Phi}(t) = e^{At} = I + At + \frac{A^2t^2}{2!} + \frac{A^3t^3}{3!} + \cdots \tag{9.36}$$

The function of $\mathbf{\Phi}(t) = e^{At}$ is called the *state transition matrix*. The state transition matrix operates on the initial state of the system $\mathbf{x}(0)$ to yield the state of the system $\mathbf{x}(t)$ at any time t.

The exponential representation of the state transition matrix ($e^{At} = I + At + A^2t^2/2! + \cdots$) has properties similar to its scalar analog ($e^{at} = 1 + at + a^2t^2/2! + \cdots$). An important property is revealed by differentiation of the preceding expression:

$$\frac{d\mathbf{\Phi}(t)}{dt} = Ae^{At} = A\left(I + At + \frac{A^2t^2}{2!} + \cdots\right) \tag{9.37}$$

In using matrix methods, it is sometimes necessary to obtain the inverse of a matrix. The inverse A^{-1} of a matrix A is determined by

$$A^{-1} = \frac{\text{Adj } A}{|A|}$$

where Adj A is the adjoint of matrix A. The adjoint is the transpose of the cofactor matrix. Consider the matrix

$$A = \begin{bmatrix} 2 & 0 & 3 \\ 4 & 2 & 1 \\ 0 & 1 & 3 \end{bmatrix}$$

The adjoint is

$$\text{Adj } A = \begin{bmatrix} \begin{vmatrix} 2 & 1 \\ 1 & 3 \end{vmatrix} & -\begin{vmatrix} 4 & 1 \\ 0 & 3 \end{vmatrix} & \begin{vmatrix} 4 & 2 \\ 0 & 1 \end{vmatrix} \\ -\begin{vmatrix} 0 & 3 \\ 1 & 3 \end{vmatrix} & \begin{vmatrix} 2 & 3 \\ 0 & 3 \end{vmatrix} & -\begin{vmatrix} 2 & 0 \\ 0 & 1 \end{vmatrix} \\ \begin{vmatrix} 0 & 3 \\ 2 & 1 \end{vmatrix} & -\begin{vmatrix} 2 & 3 \\ 4 & 1 \end{vmatrix} & \begin{vmatrix} 2 & 0 \\ 4 & 2 \end{vmatrix} \end{bmatrix}^T = \begin{bmatrix} 5 & -12 & 4 \\ 3 & 6 & -2 \\ -6 & 10 & 4 \end{bmatrix}^T$$

Thus, the inverse is

$$A^{-1} = \frac{\begin{bmatrix} 5 & 3 & -6 \\ -12 & 6 & 10 \\ 4 & -2 & 4 \end{bmatrix}}{22}$$

When the initial state of the system is known at time t_0 rather than at time $t = 0$, carrying through the preceding analysis with t_0 as the lower limit of integration yields

$$\mathbf{x}(t) = \left(I + \int_{t_0}^{t} A\, d\tau + \int_{t_0}^{t} A \int_{t_0}^{\tau} A\, d\tau\, d\tau + \cdots\right)\mathbf{x}(t_0)$$

Evaluating the integrals shows that

$$\int_{t_0}^{t} A\, d\tau = A(t - t_0)$$

$$\int_{t_0}^{t} A \int_{t_0}^{\tau} A\, d\tau\, d\tau = \int_{t_0}^{t} A[A(\tau - t_0)]\, d\tau = A^2 \frac{(t - t_0)^2}{2!}$$

Thus,

$$\mathbf{x}(t) = \left[I + A(t - t_0) + A^2 \frac{(t - t_0)^2}{2!} + \cdots\right]\mathbf{x}(t_0) = e^{A(t - t_0)}\mathbf{x}(t_0)$$

$$= \mathbf{\Phi}(t - t_0)\mathbf{x}(t_0) \tag{9.38}$$

Properties of the state transition matrix are most readily obtained by using the exponential representation. This is illustrated as follows:

$$\begin{aligned}
&\mathbf{\Phi}(0) = e^{A0} = I \\
&\mathbf{\Phi}^{-1}(t) = (e^{At})^{-1} = e^{-At} = \Phi(-t) \\
&\mathbf{\Phi}^{k}(t) = (e^{At})^{k} = e^{A(kt)} = \Phi(kt) \\
&\mathbf{\Phi}(t_1 + t_2) = e^{A(t_1 + t_2)} = e^{At_1}e^{At_2} = \Phi(t_1)\Phi(t_2) = \Phi(t_2)\Phi(t_1) \\
&\mathbf{\Phi}(t_2 - t_1)\mathbf{\Phi}(t_1 - t_0) = e^{A(t_2 - t_1) + A(t_1 - t_0)} = e^{A(t_2 - t_0)}
\end{aligned} \tag{9.39}$$

Such relationships can save considerable time and computational effort.

9.4 METHODS OF COMPUTING $\mathbf{\Phi}(t)$

A major problem in solving differential equations by the state-space method is the determination of the state transition matrix. The infinite series representation for $\mathbf{\Phi}(t)$ involves lengthy computations and results in a very awkward form.

Laplace Transform Method

Consider the system of equations

$$\begin{aligned}
\dot{x}_1 &= a_{11}x_1 + a_{12}x_2 + \cdots + a_{1n}x_n \\
\dot{x}_2 &= a_{21}x_1 + a_{22}x_2 + \cdots + a_{2n}x_n \\
&\vdots \\
\dot{x}_n &= a_{n1}x_1 + a_{n2}x_2 + \cdots + a_{nn}x_n
\end{aligned}$$

The Laplace transform is

$$sX_1(s) - x_1(0) = a_{11}X_1(s) + a_{12}X_2(s) + \cdots + a_{1n}X_n(s)$$
$$sX_2(s) - x_2(0) = a_{21}X_1(s) + a_{22}X_2(s) + \cdots + a_{2n}X_n(s)$$
$$\vdots$$
$$sX_n(s) - x_n(0) = a_{n1}X_1(s) + a_{n2}X_2(s) + \cdots + a_{nn}X_n(s)$$

The matrix representation for each of the preceding systems of equations is

$$\dot{\mathbf{x}}(t) = A\mathbf{x}(t)$$

and

$$s\mathbf{X}(s) - \mathbf{x}(0) = A\mathbf{X}(s)$$

Note that the matrix differential equations transform in the same manner as ordinary differential equations. The last equation may be written in the form

$$[sI - A]\mathbf{X}(s) = \mathbf{x}(0)$$

Premultiplying by the inverse matrix $[sI - A]^{-1}$ yields

$$\mathbf{X}(s) = [sI - A]^{-1}\mathbf{x}(0)$$

Inverting yields for the desired solution

$$\mathbf{x}(t) = \mathscr{L}^{-1}[sI - A]^{-1}\mathbf{x}(0)$$

Comparison with Eq. (9.35) reveals that

$$\mathbf{\Phi}(t) = \mathscr{L}^{-1}[sI - A]^{-1} \tag{9.40}$$

Illustrative example 9.1. As shown in Fig. 9.17, Chrysler Corporation uses an optical gauging system to automatically verify the complete underbody dimensional specifications. This new laser gauging system uses sensors and cameras to ensure correct alignment of all sheet-metal body components for production of the Dodge Spirit and Plymouth Acclaim four-door family sedans. The operation of the closed-loop control of this highly computerized optical gauging system is described by

$$\ddot{y} + 5\dot{y} + 6y = f(t) \tag{9.41}$$

The solution when the right side is zero $[f(t) = 0]$ is called the *force-free response.* The next section shows that the methods used to obtain the force-free response can be readily extended to determine the response when $f(t)$ is not zero (i.e., the forced response). Use direct programming to determine the response of this system when $f(t) = 0$, and the initial conditions are $y_0 = 1$ and $\dot{y}_0 = 0$.

Solution. For direct programming, $y = x_1$ and $\dot{y} = \dot{x}_1 = x_2$. The resulting state-space representation for the differential equation is

$$\begin{aligned} \dot{x}_1 &= x_2 \\ \dot{x}_2 &= -6x_1 - 5x_2 \end{aligned} \tag{9.42}$$

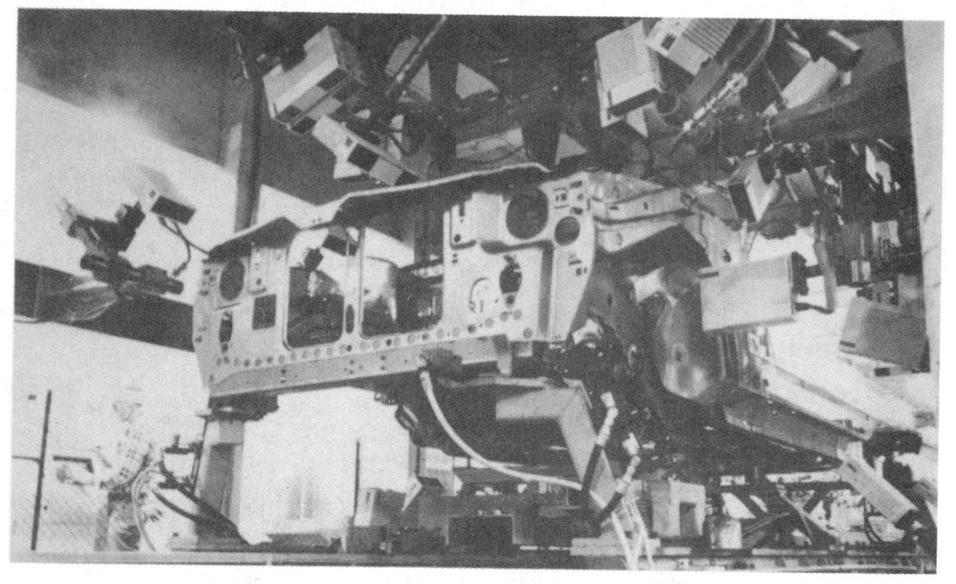

FIGURE 9.17
Optical laser gauging system. (Courtesy Chrysler Corporation.)

The Laplace transform is

$$sX_1(s) - x_1(0) = X_2(s)$$
$$sX_2(s) - x_2(0) = -6X_1(s) - 5X_2(s)$$

The matrix form is

$$\begin{bmatrix} s & 0 \\ 0 & s \end{bmatrix} \begin{bmatrix} X_1(s) \\ X_2(s) \end{bmatrix} = \begin{bmatrix} 0 & 1 \\ -6 & -5 \end{bmatrix} \begin{bmatrix} X_1(s) \\ X_2(s) \end{bmatrix} + \begin{bmatrix} x_1(0) \\ x_2(0) \end{bmatrix}$$
$$\left[\begin{bmatrix} s & 0 \\ 0 & s \end{bmatrix} - \begin{bmatrix} 0 & 1 \\ -6 & -5 \end{bmatrix}\right] \begin{bmatrix} X_1(s) \\ X_2(s) \end{bmatrix} = \begin{bmatrix} x_1(0) \\ x_2(0) \end{bmatrix}$$

Note that this has the general form

$$[sI - A]\mathbf{X}(s) = \mathbf{x}(0)$$

Premultiplying by $[sI - A]^{-1}$ gives

$$\mathbf{X}(s) = [sI - A]^{-1}\mathbf{x}(0)$$

The $[sI - A]$ matrix is

$$[sI - A] = \begin{bmatrix} s & -1 \\ 6 & s+5 \end{bmatrix}$$

The inverse is

$$\mathbf{\Phi}(s) = [sI - A]^{-1} = \frac{\begin{bmatrix} s+5 & 1 \\ -6 & s \end{bmatrix}}{s^2 + 5s + 6} \tag{9.43}$$

To invert, each element is written in its partial-fraction expansion form. For the first element

$$\frac{s+5}{(s+2)(s+3)} = \frac{3}{s+2} - \frac{2}{s+3}$$

Thus,

$$\mathbf{\Phi}(s) = [sI - A]^{-1} = \begin{bmatrix} \dfrac{3}{s+2} - \dfrac{2}{s+3} & \dfrac{1}{s+2} - \dfrac{1}{s+3} \\ \dfrac{-6}{s+2} + \dfrac{6}{s+3} & \dfrac{-2}{s+2} + \dfrac{3}{s+3} \end{bmatrix}$$

Inverting yields

$$\mathbf{\Phi}(t) = \mathscr{L}^{-1}[sI - A]^{-1} = \begin{bmatrix} 3e^{-2t} - 2e^{-3t} & e^{-2t} - e^{-3t} \\ -6e^{-2t} + 6e^{-3t} & -2e^{-2t} + 3e^{-3t} \end{bmatrix} \tag{9.44}$$

The solution $\mathbf{x}(t) = \mathbf{\Phi}(t)\mathbf{x}(0)$ is

$$\begin{bmatrix} x_1 \\ x_2 \end{bmatrix} = \begin{bmatrix} 3e^{-2t} - 2e^{-3t} & e^{-2t} - e^{-3t} \\ -6e^{-2t} + 6e^{-3t} & -2e^{-2t} + 3e^{-3t} \end{bmatrix} \begin{bmatrix} x_1(0) \\ x_2(0) \end{bmatrix} \tag{9.45}$$

Because $x_1 = y$ and $x_2 = \dot{x}_1 = \dot{y}$, the initial values are $x_1(0) = y_0 = 1$ and $x_2(0) = \dot{y}_0 = 0$. Substitution of these values into the preceding matrix expression (and noting that $y = x_1$ and $\dot{y} = x_2$) gives

$$\begin{aligned} y = x_1 &= 3e^{-2t} - 2e^{-3t} \\ \dot{y} = x_2 &= -6(e^{-2t} - e^{-3t}) \end{aligned} \tag{9.46}$$

Illustrative example 9.2. Repeat Illustrative example 9.1 except use series programming.

Solution. Comparison of Eq. (9.41) with the general form given by Eq. (9.9) shows that $a_1 = 5$, $a_2 = 6$, and $b_1 = b_2 = 0$. The resulting state-space representation is

$$\begin{bmatrix} \dot{x}_1 \\ \dot{x}_2 \end{bmatrix} = \begin{bmatrix} -5 & 1 \\ -6 & 0 \end{bmatrix} \begin{bmatrix} x_1 \\ x_2 \end{bmatrix} \tag{9.47}$$

where $x_1 = y$ and thus $\dot{x}_1 = \dot{y}$. Note that the first matrix relationship is $\dot{x}_1 = -5x_1 + x_2$. Hence for series programming $x_2 = \dot{x}_1 + 5x_1 = \dot{y} + 5y$ whereas for direct programming $x_2 = \dot{y}$. The $[sI - A]$ matrix is

$$[sI - A] = \begin{bmatrix} s+5 & -1 \\ 6 & s \end{bmatrix}$$

The inverse is

$$\mathbf{\Phi}(s) = [sI - A]^{-1} = \frac{\begin{bmatrix} s & 1 \\ -6 & s+5 \end{bmatrix}}{s^2 + 5s + 6}$$

$$= \begin{bmatrix} \dfrac{-2}{s+2} + \dfrac{3}{s+3} & \dfrac{1}{s+2} - \dfrac{1}{s+3} \\ \dfrac{-6}{s+2} + \dfrac{6}{s+3} & \dfrac{3}{s+2} - \dfrac{2}{s+3} \end{bmatrix} \tag{9.48}$$

Inverting to obtain the state transition matrix $\mathbf{\Phi}(t)$ yields for $\mathbf{x}(t) = \mathbf{\Phi}(t)\mathbf{x}(0)$

$$\begin{bmatrix} x_1 \\ x_2 \end{bmatrix} = \begin{bmatrix} -2e^{-2t} + 3e^{-3t} & e^{-2t} - e^{-3t} \\ -6e^{-2t} + 6e^{-3t} & 3e^{-2t} - 2e^{-3t} \end{bmatrix} \begin{bmatrix} x_1(0) \\ x_2(0) \end{bmatrix}$$

For $x_1 = y$ and $x_2 = \dot{y} + 5y$, then $x_1(0) = y_0 = 1$ and $x_2(0) = \dot{y}_0 + 5y_0 = 0 + 5 = 5$. Thus

$$\begin{aligned} x_1 &= (-2e^{-2t} + 3e^{-3t}) + 5(e^{-2t} - e^{-3t}) = 3e^{-2t} - 2e^{-3t} \\ x_2 &= (-6e^{-2t} + 6e^{-3t}) + 5(3e^{-2t} - 2e^{-3t}) = 9e^{-2t} - 4e^{-3t} \end{aligned} \tag{9.49}$$

For $y = x_1$ and $\dot{y} = x_2 - 5y$, then

$$y = 3e^{-2t} - 2e^{-3t}$$

and

$$\dot{y} = -6e^{-2t} + 6e^{-3t}$$

These are the same results for y and $\dot{y}$ as were obtained in Illustrative example 9.1.

Comparison of Illustrative examples 9.1 and 9.2 shows that the state transition matrix for a system is not unique, but rather depends on the particular state-space representation chosen.

Illustrative example 9.3. Figure 9.18 shows a picture of the Cintimatic 20VC (vertical changer) three-axis vertical CNC (computer numerical control) milling machine which is manufactured by the Cincinnati Milacron Company. The X, Y, and Z axes are each powered by a solid-state, maintenance-free dc servomotor. It is capable of three-axis contouring with repeatability to $\pm$ 150 millionths of an inch. This positional control system is described by Eq. (9.50). Use direct programming to determine the response of this system for the case in which $f(t) = 0$ and the initial conditions are $y_0 = 1$ and $\dot{y}_0 = 0$.

$$\ddot{y} + 10\dot{y} + 25y = f(t) \tag{9.50}$$

Solution. For direct programming $x_1 = y$ and $x_2 = \dot{x}_1 = \dot{y}$. The resulting state-space representation for this differential equation is

$$\begin{aligned} \dot{x}_1 &= x_2 \\ \dot{x}_2 &= -25x_1 - 10x_2 \end{aligned}$$

The corresponding matrix form $\dot{\mathbf{x}} = A\mathbf{x}$ is

$$\begin{bmatrix} \dot{x}_1 \\ \dot{x}_2 \end{bmatrix} = \begin{bmatrix} 0 & 1 \\ -25 & -10 \end{bmatrix} \begin{bmatrix} x_1 \\ x_2 \end{bmatrix}$$

The $[sI - A]$ matrix is

$$[sI - A] = \begin{bmatrix} s & -1 \\ 25 & s + 10 \end{bmatrix}$$

FIGURE 9.18
CNC milling machine. (Courtesy Cincinnati Milacron Company.)

Inverting gives

$$\mathbf{\Phi}(s) = [sI - A]^{-1} = \frac{\begin{bmatrix} s+10 & 1 \\ -25 & s \end{bmatrix}}{s^2 + 10s + 25} \tag{9.51}$$

Obtaining the partial-fraction expansion for each element and then inverting yield

$$\mathbf{\Phi}(t) = \begin{bmatrix} (1+5t)e^{-5t} & te^{-5t} \\ -25te^{-5t} & (1-5t)e^{-5t} \end{bmatrix} \tag{9.52}$$

The solution $\mathbf{x}(t) = \mathbf{\Phi}(t)\mathbf{x}(0)$ is

$$\begin{bmatrix} x_1 \\ x_2 \end{bmatrix} = \begin{bmatrix} (1+5t)e^{-5t} & te^{-5t} \\ -25te^{-5t} & (1-5t)e^{-5t} \end{bmatrix} \begin{bmatrix} x_1(0) \\ x_2(0) \end{bmatrix} \tag{9.53}$$

The initial values are $x_1(0) = y_0 = 1$ and $x_2(0) = \dot{y}_0 = 0$. Thus

$$\begin{aligned} y = x_1 &= (1+5t)e^{-5t} \\ \dot{y} = x_2 &= -25te^{-5t} \end{aligned} \tag{9.54}$$

The result for $\dot{y} = x_2 = \dot{x}_1$ is verified by differentiation of the $y = x_1$ equation.

Signal Flow Graph Method

The use of signal flow graphs saves considerable computational effort in determining the state transition matrix $\mathbf{\Phi}(t)$. The general solution for the force-free case is

$$\mathbf{x}(t) = \mathbf{\Phi}(t)\mathbf{x}(0)$$

The Laplace transform is

$$\mathbf{X}(s) = \mathbf{\Phi}(s)\mathbf{x}(0)$$

For a second-order system, this relationship is

$$\begin{bmatrix} X_1(s) \\ X_2(s) \end{bmatrix} = \begin{bmatrix} \phi_{11}(s) & \phi_{12}(s) \\ \phi_{21}(s) & \phi_{22}(s) \end{bmatrix} \begin{bmatrix} x_1(0) \\ x_2(0) \end{bmatrix}$$

Inverting yields

$$\begin{bmatrix} x_1(t) \\ x_2(t) \end{bmatrix} = \begin{bmatrix} \phi_{11}(t) & \phi_{12}(t) \\ \phi_{21}(t) & \phi_{22}(t) \end{bmatrix} \begin{bmatrix} x_1(0) \\ x_2(0) \end{bmatrix}$$

The matrix for an nth-order system is

$$\begin{bmatrix} X_1(s) \\ X_2(s) \\ \vdots \\ X_n(s) \end{bmatrix} = \begin{bmatrix} \phi_{11}(s) & \phi_{12}(s) & \cdots & \phi_{1n}(s) \\ \phi_{21}(s) & \phi_{22}(s) & \cdots & \phi_{2n}(s) \\ \cdots & \cdots & \cdots & \cdots \\ \phi_{n1}(s) & \phi_{n2}(s) & \cdots & \phi_{nn}(s) \end{bmatrix} \begin{bmatrix} x_1(0) \\ x_2(0) \\ \vdots \\ x_n(0) \end{bmatrix} \tag{9.55}$$

Inverting yields the desired result:

$$\begin{bmatrix} x_1(t) \\ x_2(t) \\ \vdots \\ x_n(t) \end{bmatrix} = \begin{bmatrix} \phi_{11}(t) & \phi_{12}(t) & \cdots & \phi_{1n}(t) \\ \phi_{21}(t) & \phi_{22}(t) & \cdots & \phi_{2n}(t) \\ \cdots & \cdots & \cdots & \cdots \\ \phi_{n1}(t) & \phi_{n2}(t) & \cdots & \phi_{nn}(t) \end{bmatrix} \begin{bmatrix} x_1(0) \\ x_2(0) \\ \vdots \\ x_n(0) \end{bmatrix} \tag{9.56}$$

Each element $\phi_{ij}(s)$ in Eq. (9.55) may be obtained directly from the signal flow graph. To construct the signal flow graph, first obtain the state-space form

$$\dot{\mathbf{x}}(t) = A\mathbf{x}(t)$$

Next take the Laplace transform

$$s\mathbf{X}(s) - \mathbf{x}(0) = A\mathbf{X}(s)$$

Transpose $\mathbf{x}(0)$ to the right side and then multiply through by s^{-1}. That is,

$$\mathbf{X}(s) = s^{-1}\mathbf{x}(0) + s^{-1}A\mathbf{X}(s)$$

The signal flow graph is now constructed in which the initial values of the state variables $\mathbf{x}(0)$ are the inputs and the Laplace transforms of the state variables $\mathbf{X}(s)$ are the outputs. Application of Mason's gain formula in which $x_j(0)$ is the input and

$X_i(s)$ is the output yields the relationship

$$\frac{X_i(s)}{x_j(0)} = \phi_{ij}(s)$$

Thus, each element of $\phi(s)$ in Eq. (9.55) may be obtained directly from this signal flow graph. Inverting each $\phi_{ij}(s)$ yields the corresponding element $\phi_{ij}(t)$ in Eq. (9.56).

Illustrative example 9.4. Use the signal flow graph method to determine the solution of Illustrative example 9.1.

Solution. The matrix relationships are

$$\dot{x}_1 = x_2$$
$$\dot{x}_2 = -6x_1 - 5x_2$$

To construct the signal flow graph, first take the Laplace transforms of these matrix relationships. Thus,

$$sX_1(s) - x_1(0) = X_2(s)$$
$$sX_2(s) - x_2(0) = -6X_1(s) - 5X_2(s)$$

Taking the initial-condition terms to the right side and dividing through by s give

$$X_1(s) = s^{-1}X_2(s) + s^{-1}x_1(0)$$
$$X_2(s) = s^{-1}[-6X_1(s) - 5X_2(s)] + s^{-1}x_2(0)$$

The signal flow graph for the first relationship is shown in Fig. 9.19*a*, and the graph for the second relationship is shown in Fig. 9.19*b*. These two graphs are combined to yield the signal flow for the system, shown in Fig. 9.19*c*. The graph has two loops

$$L_1 = -5s^{-1} \qquad L_2 = -6s^{-2}$$

Because there are no nontouching loops, the system determinant is

$$\Delta = 1 - (L_1 + L_2) = 1 + \frac{5}{s} + \frac{6}{s^2} = \frac{s^2 + 5s + 6}{s^2}$$

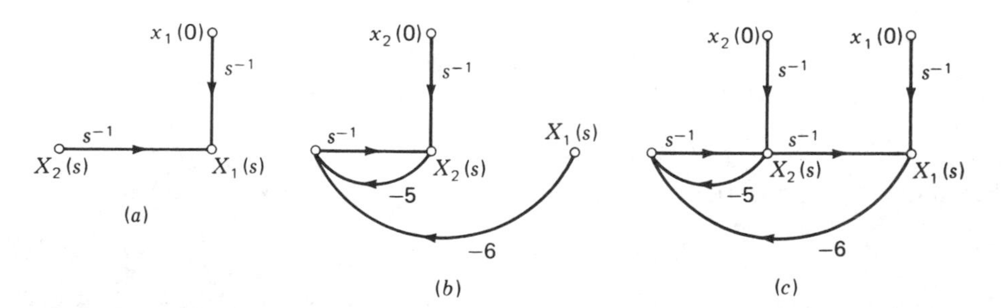

FIGURE 9.19
Signal flow graph.

The transfer function relating input $x_1(0)$ to output $X_1(s)$ is $\phi_{11}(s)$. The transmittance of the forward path from $x_1(0)$ to $X_1(s)$ is $T_1 = 1/s$. Because L_2 touches T_1, setting $L_2 = 0$ in the equation for Δ gives $\Delta_1 = 1 - L_1 = 1 + 5/s = (s + 5)/s$. The transfer function relating input $x_2(0)$ to output $X_1(s)$ is $\phi_{12}(s)$. The transmittance of the forward path from $x_2(0)$ to $X_1(s)$ is $T_2 = 1/s^2$. Because both L_1 and L_2 touch T_2, letting $L_1 = L_2 = 0$ in the equation for Δ gives $\Delta_2 = 1$. Thus, the equation for $X_1(s)$ is

$$X_1(s) = \phi_{11}(s)x_1(0) + \phi_{12}(s)x_2(0) = \frac{T_1\Delta_1}{\Delta} x_1(0) + \frac{T_2\Delta_2}{\Delta} x_2(0)$$
$$= \frac{s+5}{s^2+5s+6} x_1(0) + \frac{1}{s^2+5s+6} x_2(0)$$

The transfer function relating $x_1(0)$ to $X_2(s)$ is $\phi_{21}(s)$. The transmittance of this path is $T_3 = -6/s^2$. Because L_1 and L_2 touch this path, setting $L_1 = L_2 = 0$ in the equation for Δ gives $\Delta_3 = 1$. The transfer function relating $x_2(0)$ to $X_2(s)$ is $\phi_{22}(s)$. The transmittance of this path is $T_4 = 1/s$. Because L_1 and L_2 touch this path, $\Delta_4 = 1$. The resulting equation for $X_2(s)$ is

$$X_2(s) = \phi_{21}(s)x_1(0) + \phi_{22}(s)x_2(0) = \frac{T_3\Delta_3}{\Delta} x_1(0) + \frac{T_4\Delta_4}{\Delta} x_2(0)$$
$$= \frac{-6}{s^2+5s+6} x_1(0) + \frac{s}{s^2+5s+6} x_2(0)$$

The matrix form of the preceding relationships is

$$\begin{bmatrix} X_1(s) \\ X_2(s) \end{bmatrix} = \frac{\begin{bmatrix} s+5 & 1 \\ -6 & s \end{bmatrix}}{s^2+5s+6} \begin{bmatrix} x_1(0) \\ x_2(0) \end{bmatrix}$$

The first term on the right is the $\mathbf{\Phi}(s) = [sI - A]^{-1}$ matrix. This is the same result attained by the Laplace transform method [i.e., Eq. (9.43)]. Note that the signal flow graph method eliminates the need to obtain the matrix inverse.

Illustrative example 9.5. Use the signal flow graph method to determine the solution of Illustrative example 9.2.

Solution. The matrix relationships are

$$\dot{x}_1 = -5x_1 + x_2$$
$$\dot{x}_2 = -6x_1$$

The Laplace transforms of these matrix relationships are

$$sX_1(s) - x_1(0) = -5X_1(s) + X_2(s)$$
$$sX_2(s) - x_2(0) = -6X_1(s)$$

Taking the initial-condition terms to the right side and dividing through by s give

$$X_1(s) = s^{-1}[-5X_1(s) + X_2(s)] + s^{-1}x_1(0)$$
$$X_2(s) = s^{-1}[-6X_1(s)] + s^{-1}x_2(0)$$

The signal flow graph for the first relationship is shown in Fig. 9.20*a*, and the graph for the second relationship is shown in Fig. 9.20*b*. These two graphs are combined to

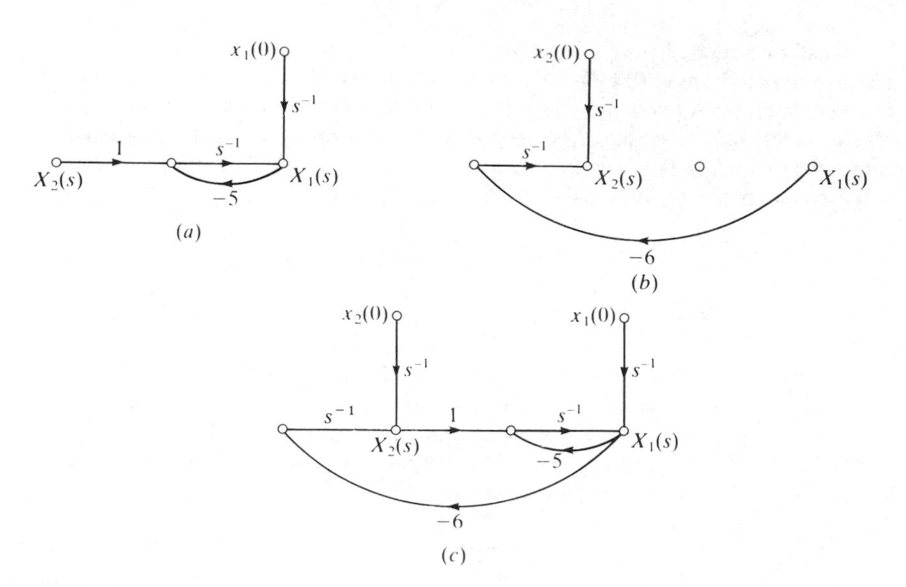

FIGURE 9.20
Signal flow graph.

yield the signal flow graph for the system shown in Fig. 9.20*c*. This graph has two loops

$$L_1 = -5s^{-1} \qquad L_2 = -6s^{-2}$$

The system determinant is

$$\Delta = 1 - (L_1 + L_2) = 1 + (5s^{-1} + 6s^{-2}) = \frac{s^2 + 5s + 6}{s^2}$$

Application of Mason's gain formula yields the following equation for $X_1(s)$ and $X_2(s)$:

$$\begin{bmatrix} X_1(s) \\ X_2(s) \end{bmatrix} = \frac{\begin{bmatrix} s & 1 \\ -6 & s+5 \end{bmatrix}}{s^2 + 5s + 6} \begin{bmatrix} x_1(0) \\ x_2(0) \end{bmatrix}$$

This is the same result for $\mathbf{X}(s) = \mathbf{\Phi}(s)\mathbf{x}(0)$ that was obtained in Illustrative example 9.2.

Illustrative example 9.6. Use the signal flow graph method to determine the state transition matrix for Illustrative example 9.3.

Solution. The matrix relationships are

$$\begin{aligned} \dot{x}_1 &= x_2 \\ \dot{x}_2 &= -25x_1 - 10x_2 \end{aligned}$$

The Laplace transforms of these relationships are

$$\begin{aligned} sX_1(s) - x_1(0) &= X_2(s) \\ sX_2(s) - x_2(0) &= -25X_1(s) - 10X_2(s) \end{aligned}$$

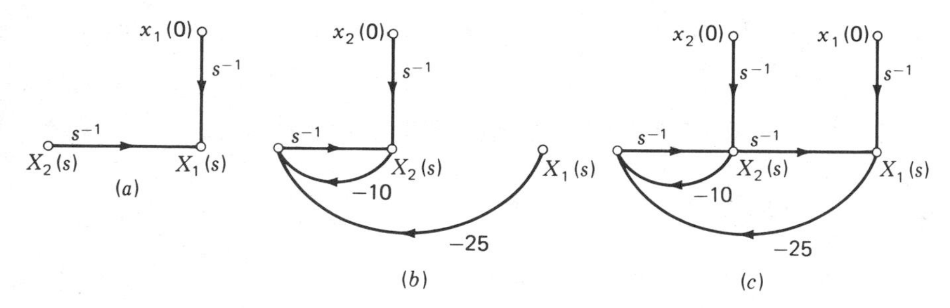

FIGURE 9.21
Signal flow graph.

Taking the initial conditions to the right side and dividing through by s give

$$X_1(s) = s^{-1}X_2(s) + s^{-1}x_1(0)$$
$$X_2(s) = s^{-1}[-25X_1(s) - 10X_2(s)] + s^{-1}x_2(0)$$

The signal flow graph for the first relationship is shown in Fig. 9.21*a*, and the graph for the second relationship is shown in Fig. 9.21*b*. Combining these two graphs yields the signal flow graph for the system, which is shown in Fig. 9.21*c*. This system has two loops

$$L_1 = -10s^{-1} \qquad L_2 = -25s^{-2}$$

The determinant for this system is

$$\Delta = 1 - (L_1 + L_2) = \frac{s^2 + 10s + 25}{s^2}$$

Application of Mason's gain formula yields the following matrix for $X_1(s)$ and $X_2(s)$:

$$\begin{bmatrix} X_1(s) \\ X_2(s) \end{bmatrix} = \frac{\begin{bmatrix} s+10 & 1 \\ -25 & s \end{bmatrix}}{(s+5)^2} \begin{bmatrix} x_1(0) \\ x_2(0) \end{bmatrix}$$

The first matrix on the right is the $\mathbf{\Phi}(s) = [sI - A]^{-1}$ matrix. This is the same result as that obtained by the Laplace transform method.

9.5 FORCED RESPONSE

The general representation for a system of differential equations with a forcing function is

$$\dot{\mathbf{x}}(t) = A\mathbf{x}(t) + \mathbf{b}f(t) \qquad \mathbf{x}(0) = \mathbf{x}_0 \tag{9.57}$$

To solve this system, first rewrite it in the form

$$\dot{\mathbf{x}}(t) - A\mathbf{x}(t) = \mathbf{b}f(t)$$

Multiplying through by e^{-At} shows that

$$e^{-At}[\dot{\mathbf{x}}(t) - A\mathbf{x}(t)] = \frac{d}{dt}[e^{-At}\mathbf{x}(t)] = e^{-At}\mathbf{b}f(t)$$

Integration between the limits of 0 and t of the two expressions on the right side of the preceding equation gives

$$e^{-At}\mathbf{x}(t) - \mathbf{x}(0) = \int_0^t e^{-A\tau}\mathbf{b}f(\tau)\,d\tau \tag{9.58}$$

Multiplying through by e^{At} yields the desired solution:

$$\mathbf{x}(t) = e^{At}\mathbf{x}(0) + \int_0^t e^{-A(t-\tau)}\mathbf{b}f(\tau)\,d\tau \tag{9.59}$$

Replacing e^{At} by $\mathbf{\Phi}(t)$ and $e^{A(t-\tau)}$ by $\mathbf{\Phi}(t-\tau) = \mathbf{\Phi}(t)\mathbf{\Phi}(-\tau)$ yields the form

$$\mathbf{x}(t) = \mathbf{\Phi}(t)\mathbf{x}(0) + \mathbf{\Phi}(t)\int_0^t \mathbf{\Phi}(-\tau)\mathbf{b}f(\tau)\,d\tau \tag{9.60}$$

The first term on the right side of Eq. (9.60) is recognized as the force-free response. The second term on the right is the response due to the forcing function $f(t)$. Although this second term may be evaluated by the integration process indicated, the Laplace transform method usually affords considerable simplification.

The Laplace transform of Eq. (9.57) is

$$s\mathbf{X}(s) - \mathbf{x}(0) = A\mathbf{X}(s) + \mathbf{b}F(s)$$

This may be written in the form

$$[sI - A]\mathbf{X}(s) = \mathbf{x}(0) + \mathbf{b}F(s)$$

Premultiplying by the inverse $[sI - A]^{-1}$ gives

$$\mathbf{X}(s) = [sI - A]^{-1}\mathbf{x}(0) + [sI - A]^{-1}\mathbf{b}F(s) \tag{9.61}$$

Inverting yields

$$\begin{aligned}\mathbf{x}(t) &= \mathscr{L}^{-1}[sI - A]^{-1}\mathbf{x}(0) + \mathscr{L}^{-1}[sI - A]^{-1}\mathbf{b}F(s)\\ &= \mathscr{L}^{-1}\mathbf{\Phi}(s)\mathbf{x}(0) + \mathscr{L}^{-1}\mathbf{\Phi}(s)\mathbf{b}F(s)\end{aligned} \tag{9.62}$$

Comparison with Eq. (9.60) shows that

$$\mathbf{\Phi}(t) = \mathscr{L}^{-1}[sI - A]^{-1} = \mathscr{L}^{-1}\mathbf{\Phi}(s) \tag{9.40}$$

and

$$\mathbf{\Phi}(t)\int_0^t \mathbf{\Phi}(-\tau)\mathbf{b}f(\tau)\,d\tau = \mathscr{L}^{-1}[sI - A]^{-1}\mathbf{b}F(s) = \mathscr{L}^{-1}\mathbf{\Phi}(s)\mathbf{b}F(s) \tag{9.63}$$

The first result was previously established by Eq. (9.40).

Illustrative example 9.7. The input-output relationship for a system is given by Eq. (9.64). Use direct programming to obtain the response of this system for the case in which the forcing function is a unit step function and all the initial conditions are zero.

$$\ddot{y} + 5\dot{y} + 6y = f'(t) + 6f(t) \tag{9.64}$$

Solution. The operator form for the differential equation is

$$y = \frac{D+6}{D^2+5D+6} f(t)$$

For direct programming x_1 is such that the numerator of the operator is unity.

$$x_1 = \frac{1}{D^2+5D+6} f(t) \tag{9.65}$$

Thus,

$$\ddot{x}_1 + 5\dot{x}_1 + 6x_1 = f(t)$$

For $\dot{x}_1 = x_2$, then $\ddot{x}_1 = \dot{x}_2$. The resulting state-space representation is

$$\begin{aligned}\dot{x}_1 &= x_2 \\ \dot{x}_2 &= -6x_1 - 5x_2 + f(t)\end{aligned}$$

and

$$y = 6x_1 + x_2$$

The corresponding matrix form is

$$\begin{bmatrix}\dot{x}_1 \\ \dot{x}_2\end{bmatrix} = \begin{bmatrix}0 & 1 \\ -6 & -5\end{bmatrix}\begin{bmatrix}x_1 \\ x_2\end{bmatrix} + \begin{bmatrix}0 \\ 1\end{bmatrix} f(t)$$

$$y = [6 \quad 1]\begin{bmatrix}x_1 \\ x_2\end{bmatrix}$$

Hence,

$$A = \begin{bmatrix}0 & 1 \\ -6 & -5\end{bmatrix} \qquad \mathbf{b} = \begin{bmatrix}0 \\ 1\end{bmatrix} \qquad \mathbf{c}^T = [6 \quad 1] \tag{9.66}$$

Because all the initial conditions are zero, then $y_0 = \dot{y}_0 = f(0) = 0$. To obtain the initial conditions in terms of x, first differentiate $y = 6x_1 + x_2$.

$$\begin{aligned}\dot{y} &= 6\dot{x}_1 + \dot{x}_2 = 6x_2 + [-6x_1 - 5x_2 + f(t)] \\ &= -6x_1 + x_2 + f(t)\end{aligned}$$

Solving for x_1 and x_2 in terms of y and $\dot{y}$ and then evaluating at time $t = 0$ give

$$x_1(0) = \frac{y_0 - \dot{y}_0 + f(0)}{12} \qquad x_2(0) = \frac{y_0 + \dot{y}_0 - f(0)}{2} \tag{9.67}$$

As would be expected when all the initial conditions are zero, the initial values of the state variables are also zero.

The $[sI - A]$ matrix is

$$[sI - A] = \begin{bmatrix}s & -1 \\ 6 & s+5\end{bmatrix}$$

The inverse is

$$\mathbf{\Phi}(s) = [sI - A]^{-1} = \frac{\begin{bmatrix}s+5 & 1 \\ -6 & s\end{bmatrix}}{s^2+5s+6}$$

Thus

$$\mathbf{\Phi}(s)\mathbf{b}F(s) = \frac{\begin{bmatrix} s+5 & 1 \\ -6 & s \end{bmatrix}}{s^2+5s+6}\begin{bmatrix} 0 \\ 1 \end{bmatrix}F(s) = \frac{\begin{bmatrix} 1 \\ s \end{bmatrix}}{s^2+5s+6}F(s)$$

Application of Eq. (9.62) gives

$$\begin{bmatrix} X_1(s) \\ X_2(s) \end{bmatrix} = \frac{\begin{bmatrix} s+5 & 1 \\ -6 & s \end{bmatrix}}{s^2+5s+6}\begin{bmatrix} x_1(0) \\ x_2(0) \end{bmatrix} + \frac{\begin{bmatrix} 1 \\ s \end{bmatrix}}{s^2+5s+6}F(s) \tag{9.68}$$

For $\mathbf{x}(0) = \mathbf{0}$ and $F(s) = 1/s$, then

$$\begin{bmatrix} x_1(t) \\ x_2(t) \end{bmatrix} = \mathscr{L}^{-1}\begin{bmatrix} \dfrac{1}{s(s+2)(s+3)} \\ \dfrac{1}{(s+2)(s+3)} \end{bmatrix} = \begin{bmatrix} \dfrac{1-3e^{-2t}+2e^{-3t}}{6} \\ e^{-2t}-e^{-3t} \end{bmatrix}$$

The desired solution is

$$y = 6x_1 + x_2 = (1-3e^{-2t}+2e^{-3t}) + (e^{-2t}-e^{-3t}) = 1 - 2e^{-2t} + e^{-3t} \tag{9.69}$$

Signal Flow Graph Method

To apply this method, first take the Laplace transform of the state-space equation $\dot{\mathbf{x}} = A\mathbf{x} + \mathbf{b}f(t)$. That is,

$$s\mathbf{X}(s) - \mathbf{x}(0) = A\mathbf{X}(s) + \mathbf{b}F(s)$$

The signal flow graph is then constructed by writing this in the form

$$\mathbf{X}(s) = s^{-1}\mathbf{x}(0) + s^{-1}A\mathbf{X}(s) + s^{-1}\mathbf{b}F(s)$$

The input nodes for the resulting graph are the initial conditions $\mathbf{x}(0)$ and the transform of the input $F(s)$. The output nodes are the transforms of the state variables $\mathbf{X}(s)$. Application of Mason's gain formula yields directly the following matrix:

$$\begin{bmatrix} X_1(s) \\ X_2(s) \\ \vdots \\ X_n(s) \end{bmatrix} = \begin{bmatrix} \phi_{11}(s) & \phi_{12}(s) & \cdots & \phi_{1n}(s) \\ \phi_{21}(s) & \phi_{22}(s) & \cdots & \phi_{2n}(s) \\ \cdots & \cdots & \cdots & \cdots \\ \phi_{n1}(s) & \phi_{n2}(s) & \cdots & \phi_{nn}(s) \end{bmatrix}\begin{bmatrix} x_1(0) \\ x_2(0) \\ \vdots \\ x_n(0) \end{bmatrix} + \begin{bmatrix} \psi_1(s) \\ \psi_2(s) \\ \vdots \\ \psi_n(s) \end{bmatrix}F(s) \tag{9.70}$$

Inverting yields the desired solution. Note that each element $\phi_{ij}(s)$ and $\psi_i(s)$ is obtained individually from the signal flow graph.

Illustrative example 9.8. Use the signal flow graph method to determine the solution of Illustrative example 9.7.

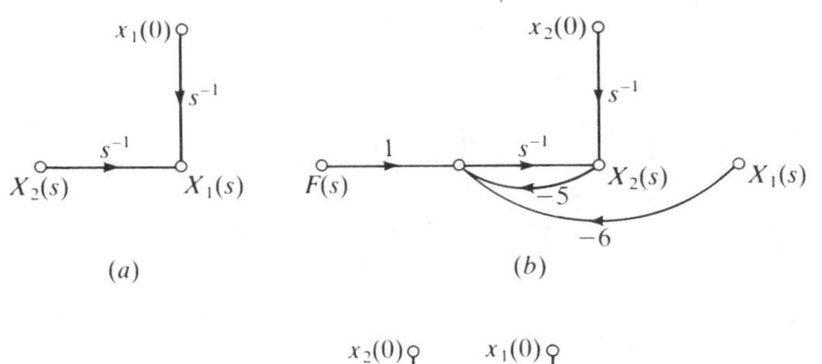

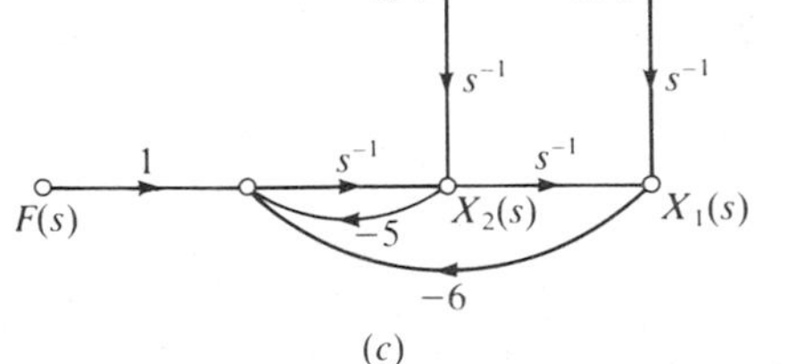

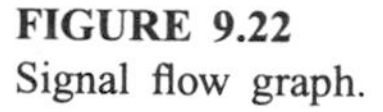

FIGURE 9.22
Signal flow graph.

Solution. The Laplace transform of the state-space equations is

$$sX_1(s) - x_1(0) = X_2(s)$$
$$sX_2(s) - x_2(0) = -6X_1(s) - 5X_2(s) + F(s)$$

Taking the initial conditions to the right side and dividing through by s give

$$X_1(s) = s^{-1}x_1(0) + s^{-1}X_2(s)$$
$$X_2(s) = s^{-1}x_2(0) + s^{-1}[-6X_1(s) - 5X_2(s) + F(s)]$$

The signal flow graph for the first equation is shown in Fig. 9.22*a*, and the graph for the second equation is shown in Fig. 9.22*b*. These graphs are combined to yield the resultant graph for the system, shown in Fig. 9.22*c*. Application of Mason's gain formula to this graph yields

$$\begin{bmatrix} X_1(s) \\ X_2(s) \end{bmatrix} = \begin{bmatrix} \dfrac{s+5}{s^2+5s+6} & \dfrac{1}{s^2+5s+6} \\ \dfrac{-6}{s^2+5s+6} & \dfrac{s}{s^2+5s+6} \end{bmatrix} \begin{bmatrix} x_1(0) \\ x_2(0) \end{bmatrix} + \begin{bmatrix} \dfrac{1}{s^2+5s+6} \\ \dfrac{s}{s^2+5s+6} \end{bmatrix} F(s)$$

The first matrix on the right side is the $[sI - A]^{-1}$ matrix, and the last matrix is the $[sI - A]^{-1}\mathbf{b}$ matrix. These are the same results as those obtained by the Laplace transform method in Eq. (9.68). A major feature of the signal flow graph method is that it eliminates the need to obtain the inverse matrix.

Illustrative example 9.9. Use the signal flow graph method to determine the solution of the differential equation

$$\ddot{c} + 6\dot{c} + 25c = 100f(t) \tag{9.71}$$

The forcing function $f(t)$ is a unit step function, and all initial conditions are zero.

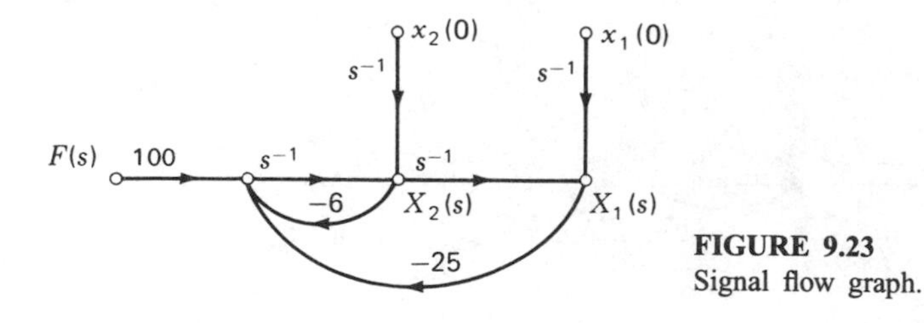

FIGURE 9.23
Signal flow graph.

Solution. If we let $c = x_1$ and $\dot{c} = \dot{x}_1 = x_2$, the state-space representation is

$$\dot{x}_1 = x_2$$
$$\dot{x}_2 = -25x_1 - 6x_2 + 100f(t)$$

Transforming gives

$$sX_1(s) - x_1(0) = X_2(s)$$
$$sX_2(s) - x_2(0) = -25X_1(s) - 6X_2(s) + 100F(s)$$

The form for constructing the signal flow graph is

$$X_1(s) = s^{-1}x_1(0) + s^{-1}X_2(s)$$
$$X_2(s) = s^{-1}x_2(0) + s^{-1}[-25X_1(s) - 6X_2(s) + 100F(s)]$$

The resulting signal flow graph is shown in Fig. 9.23.

Application of Mason's gain formula gives

$$\begin{bmatrix} X_1(s) \\ X_2(s) \end{bmatrix} = \frac{\begin{bmatrix} s+6 & 1 \\ -25 & s \end{bmatrix}}{s^2+6s+25} \begin{bmatrix} x_1(0) \\ x_2(0) \end{bmatrix} + \begin{bmatrix} 100 \\ 100s \end{bmatrix} \frac{F(s)}{s^2+6s+25} \tag{9.72}$$

For $x_1(0) = x_2(0) = 0$ and $F(s) = 1/s$, the inverse yields the desired result

$$\begin{aligned} c = x_1 &= 4 + 5e^{-3t}\sin(4t - 126.9°) \\ \dot{c} = x_2 &= 25e^{-3t}\sin 4t \end{aligned} \tag{9.73}$$

Note that no special considerations are needed in applying this method to complex conjugate roots.

9.6 TRANSFER FUNCTIONS

The state-space representation for a system to be controlled is shown in Fig. 9.24*a*. The system to be controlled is referred to as the *plant*. In state-space notation, the characteristics of the plant are given by Eqs. (9.20) and (9.21):

$$\dot{\mathbf{x}} = A\mathbf{x} + \mathbf{b}u \tag{9.20}$$

$$y = \mathbf{c}^T\mathbf{x} \tag{9.21}$$

The corresponding transfer function representation for the plant is shown in Fig. 9.24*b*. In Sec. 9.1, techniques for obtaining various state-space representations for a system were illustrated. We now show how the transfer function may be obtained

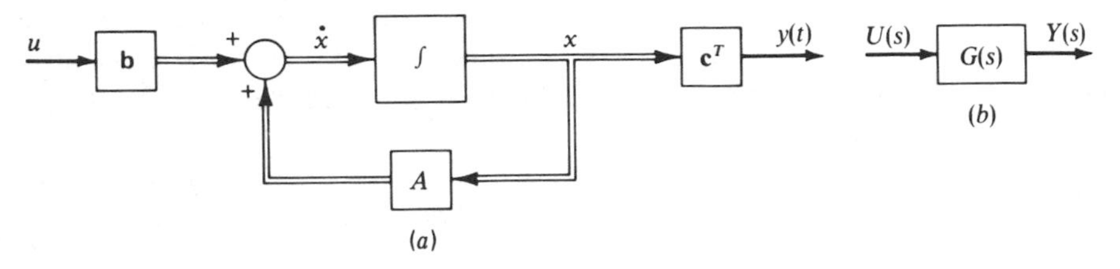

FIGURE 9.24
State-space representation and corresponding transfer function.

from the state-space representation. The Laplace transforms of the preceding matrix equations are

$$s\mathbf{X}(s) - \mathbf{x}(0) = A\mathbf{X}(s) + \mathbf{b}U(s)$$
$$Y(s) = \mathbf{c}^T\mathbf{X}(s)$$

In obtaining the transfer function, the initial conditions are ignored. Thus, the first relationship becomes

$$[sI - A]\mathbf{X}(s) = \mathbf{b}U(s)$$

Hence

$$\mathbf{X}(s) = [sI - A]^{-1}\mathbf{b}U(s) = \mathbf{\Phi}(s)\mathbf{b}U(s) \tag{9.74}$$

where $\mathbf{\Phi}(s) = [sI - A]^{-1}$ is the Laplace transform of the state transition matrix $\mathbf{\Phi}(t)$ for the plant. Substitution of this result into the second transformed equation gives

$$Y(s) = \mathbf{c}^T\mathbf{\Phi}(s)\mathbf{b}U(s) = G(s)U(s)$$

Hence

$$G(s) = \mathbf{c}^T\mathbf{\Phi}(s)\mathbf{b} \tag{9.75}$$

Because $\mathbf{\Phi}(s) = \text{Adj}(sI - A)/|sI - A|$, the preceding may be written in the form

$$G(s) = \frac{Y(s)}{U(s)} = \frac{\mathbf{c}^T[\text{Adj}(sI - A)]\mathbf{b}}{|sI - A|}$$

The numerator polynomial for $G(s)$ is $\mathbf{c}^T[\text{Adj}(sI - A)]\mathbf{b}$. The denominator, which is the characteristic function, is $|sI - A|$. The roots of the characteristic equation $|sI - A| = 0$ are the eigenvalues for the system.

For a given plant $G(s)$, there are an infinite number of state-space representations. The transformation $\mathbf{x} = T\mathbf{z}$ transforms Eqs. (9.20) and (9.21) into the form

$$\dot{\mathbf{z}} = T^{-1}AT\mathbf{z} + T^{-1}\mathbf{b}u$$
$$y = \mathbf{c}^T T\mathbf{z}$$

This new state-space representation for the plant may be written in the form

$$\dot{\mathbf{z}} = A^*\mathbf{z} + \mathbf{b}^*u$$
$$y = \mathbf{c}^{*T}\mathbf{z}$$

where $A^* = T^{-1}AT$, $\mathbf{b}^* = T^{-1}\mathbf{b}$, and $\mathbf{c}^{*T} = \mathbf{c}^T T$ or $\mathbf{c}^* = T^T\mathbf{c}$. Note that these equations have the same form as Eqs. (9.20) and (9.21) but are different matrix representations for the same plant. In this analysis, T is any matrix which possesses an inverse T^{-1}.

The characteristic function for the plant transfer function $G(s)$ is $|sI - A|$. Because the characteristic equation is a unique property of the system and does not depend on the particular matrix representation, it follows that

$$|sI - A| = |sI - A^*|$$

To prove the preceding relationship, note that

$$\begin{aligned} |sI - A^*| &= |sI - T^{-1}AT| = |sIT^{-1}T - T^{-1}AT| \\ &= |T^{-1}(sI - A)T| \end{aligned}$$

Because the determinant of a product is equal to the product of the determinants,

$$\begin{aligned} |sI - A^*| &= |T^{-1}|\,|sI - A|\,|T| = |T^{-1}|\,|T|\,|sI - A| \\ &= |T^{-1}T|\,|sI - A| = |sI - A| \end{aligned}$$

Similarly, the transfer function for the plant $G(s) = \mathbf{c}^T[sI - A]^{-1}\mathbf{b}$ is unaffected by the particular matrix representation for the plant. Thus,

$$\mathbf{c}^T[sI - A]^{-1}\mathbf{b} = \mathbf{c}^{*T}[sI - A^*]^{-1}\mathbf{b}^*$$

The preceding relationship is proved as follows:

$$\begin{aligned} \mathbf{c}^{*T}[sI - A^*]^{-1}\mathbf{b}^* &= \mathbf{c}^T T[sI - T^{-1}AT]^{-1}T^{-1}\mathbf{b} = \mathbf{c}^T[T(sI - T^{-1}AT)T^{-1}]^{-1}\mathbf{b} \\ &= \mathbf{c}^T[sTT^{-1} - A]^{-1}\mathbf{b} = \mathbf{c}^T[sI - A]^{-1}\mathbf{b} = G(s) \end{aligned}$$

From Eq. (9.66), for the case of Illustrative example 9.7,

$$\mathbf{c} = \begin{bmatrix} 6 \\ 1 \end{bmatrix} \qquad A = \begin{bmatrix} 0 & 1 \\ -6 & -5 \end{bmatrix} \qquad \mathbf{b} = \begin{bmatrix} 0 \\ 1 \end{bmatrix}$$

Thus

$$[sI - A] = \begin{bmatrix} s & -1 \\ 6 & s+5 \end{bmatrix} \qquad \text{Adj}\,[sI - A] = \begin{bmatrix} s+5 & 1 \\ -6 & s \end{bmatrix}$$

$$|sI - A| = \begin{vmatrix} s & -1 \\ 6 & s+5 \end{vmatrix} = s^2 + 5s + 6$$

Application of Eq. (9.75) yields

$$G(s) = \frac{[6 \quad 1]\begin{bmatrix} s+5 & 1 \\ -6 & s \end{bmatrix}\begin{bmatrix} 0 \\ 1 \end{bmatrix}}{s^2 + 5s + 6} = \frac{s+6}{s^2 + 5s + 6}$$

The differential equation for this system, Eq. (9.64), may be written in operator form:

$$c(t) = \frac{D+6}{D^2+5D+6} f(t)$$

Replacing the operator D by s verifies the result obtained for $G(s)$.

Closed-Loop Systems

A general block-diagram representation for a closed-loop system with linear state variable feedback is shown in Fig. 9.25*a*. The corresponding state-space representation is shown in Fig. 9.25*b*. For linear state variable feedback, the control input

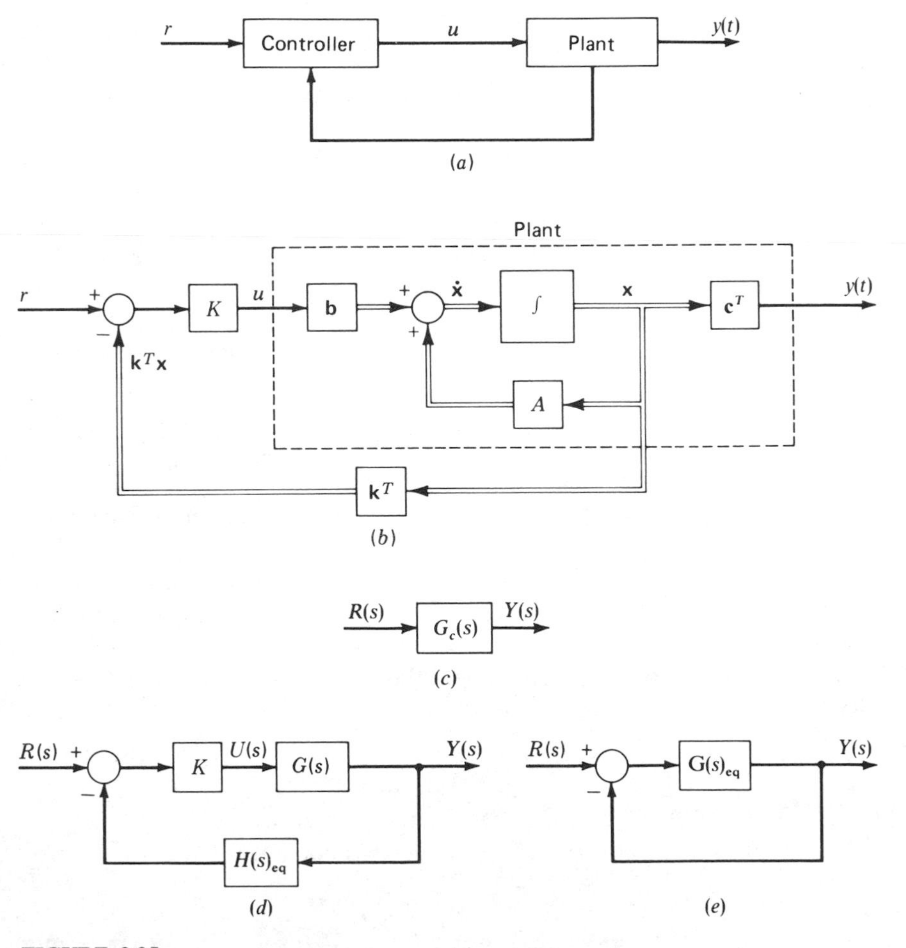

FIGURE 9.25
Closed-loop system with linear state variable feedback.

u is the difference between the reference input r and a weighted sum of the state variables. That is,

$$u = K[r - (k_1x_1 + k_2x_2 + \cdots + k_nx_n)]$$

In matrix notation, this relationship is

$$u = K(r - \mathbf{k}^T\mathbf{x})$$

To obtain the transfer function for the closed-loop system, first substitute u into Eq. (9.20). Thus,

$$\dot{\mathbf{x}} = (A - K\mathbf{b}\mathbf{k}^T)\mathbf{x} + K\mathbf{b}r$$
$$y = \mathbf{c}^T\mathbf{x}$$

Transforming each expression, then solving the first for $\mathbf{X}(s)$ and substituting this result into the second give

$$Y(s) = \mathbf{c}^T\mathbf{X}(s) = K\mathbf{c}^T[sI - (A - K\mathbf{b}\mathbf{k}^T)]^{-1}\mathbf{b}R(s)$$

The closed-loop transfer function $Y(s)/R(s)$ may be written in the form

$$\frac{Y(s)}{R(s)} = K\mathbf{c}^T[sI - A_c]^{-1}\mathbf{b} = K\mathbf{c}^T\mathbf{\Phi}_c(s)\mathbf{b} \tag{9.76}$$

where $A_c = [A - K\mathbf{b}\mathbf{k}^T]$ and $\mathbf{\Phi}_c(s) = [sI - A_c]^{-1}$ is the state transition matrix for the closed-loop system. Note that the state transition matrix for the closed-loop system is the same as that for the plant except that A is replaced by $A_c = [A - K\mathbf{b}\mathbf{k}^T]$. The transfer function representation $Y(s) = G_c(s)R(s)$ for the closed-loop system is shown in Fig. 9.25*c*. The closed-loop transfer function may be written in the form

$$G_c(s) = \frac{Y(s)}{R(s)} = \frac{K\mathbf{c}^T[\mathrm{Adj}(sI - A_c)]\mathbf{b}}{|sI - A_c|}$$

This shows that the denominator which is the characteristic equation for the closed-loop system is $|sI - A_c|$.

Figure 9.25*d* shows an equivalent block-diagram representation for the system of Fig. 9.25*b*. The feedforward elements are the same for both cases. The feedback signal which goes to the comparator in Fig. 9.25*b* is $\mathbf{k}^T\mathbf{x}$. The Laplace transform for this signal is $\mathbf{k}^T\mathbf{X}(s)$. For the equivalent system the signal going to the comparator is $H(s)_{\mathrm{eq}}Y(s)$. Thus

$$H(s)_{\mathrm{eq}}Y(s) = \mathbf{k}^T\mathbf{X}(s)$$

Solving for $H(s)_{\mathrm{eq}}$ shows that

$$H(s)_{\mathrm{eq}} = \frac{\mathbf{k}^T\mathbf{X}(s)}{Y(s)} = \frac{\mathbf{k}^T\mathbf{\Phi}(s)\mathbf{b}}{\mathbf{c}^T\mathbf{\Phi}(s)\mathbf{b}} \tag{9.77}$$

where for the plant $\mathbf{X}(s) = \mathbf{\Phi}(s)\mathbf{b}U(s)$ and $Y(s) = \mathbf{c}^T\mathbf{X}(s) = \mathbf{c}^T\mathbf{\Phi}(s)\mathbf{b}U(s)$. Figure 9.25*e* shows the unity-feedback system which is equivalent to Fig. 9.25*d*. The input-output relationships for Fig. 9.25*d* and *e* are

$$\frac{Y(s)}{R(s)} = \frac{KG(s)}{1 + KG(s)H(s)_{\text{eq}}} \qquad \frac{Y(s)}{R(s)} = \frac{G(s)_{\text{eq}}}{1 + G(s)_{\text{eq}}} \tag{9.78}$$

Equating these expressions and solving for $G(s)_{\text{eq}}$ yield

$$G(s)_{\text{eq}} = \frac{KG(s)}{1 + KG(s)H(s)_{\text{eq}} - KG(s)} = \frac{K\mathbf{c}^T\mathbf{\Phi}(s)\mathbf{b}}{1 + K(\mathbf{k} - \mathbf{c})^T\mathbf{\Phi}(s)\mathbf{b}} \tag{9.79}$$

where $G(s)H(s)_{\text{eq}} = [\mathbf{c}^T\mathbf{\Phi}(s)\mathbf{b}][\mathbf{k}^T\mathbf{\Phi}(s)\mathbf{b}]/[\mathbf{c}^T\mathbf{\Phi}(s)\mathbf{b}] = \mathbf{k}^T\mathbf{\Phi}(s)\mathbf{b}$. Thus, the transfer functions $G(s)$, $G(s)_{\text{eq}}$, and $H(s)_{\text{eq}}$ may be obtained from the state variable representation for a linear state-space feedback control system.

9.7 MULTIVARIABLE SYSTEMS

Transfer functions yield much insight into the synthesis process. In this section, we show how transfer functions are combined with state-space techniques to provide a powerful method for the synthesis of multivariable systems.* This method was developed by Dr. Joseph L. Peczkowski.

A multivariable system is one in which there is more than one input and more than one output. The system to be controlled is referred to as the *plant*. The state-space representation for the plant is given by Eqs. (9.22) and (9.23):

$$\dot{\mathbf{x}} = A\mathbf{x} + B\mathbf{u} \tag{9.22}$$

$$\mathbf{y} = C\mathbf{x} + D\mathbf{u} \tag{9.23}$$

The vector **u** represents the inputs to the plant, the vector **x** represents the state variables, and the vector **y** represents the outputs from the plant. For obtaining basic information such as the transfer function for a system, initial conditions are ignored. The Laplace transform for the preceding equations is

$$s\mathbf{X}(s) = A\mathbf{X}(s) + B\mathbf{U}(s)$$

$$\mathbf{Y}(s) = C\mathbf{X}(s) + D\mathbf{U}(s)$$

The corresponding block-diagram representation is shown in Fig. 9.26*a*. Note that $\mathbf{U}(s)$ is the input and $\mathbf{Y}(s)$ is the output. More than one signal travels along the

---

* J. L. Peczkowski, "Total Multivariable Synthesis with Transfer Functions," *Proceedings Bendix Control and Control Theory Symposium*, pp. 109–126, Bendix Executive Offices, Southfield, Michigan, April 1980.

J. L. Peczkowski and M. K. Sain, "Linear Multivariable Synthesis with Transfer Functions," in *Alternatives for Multivariable Control*, M. K. Sain, J. L. Peczkowski, and J. L. Melsa, eds., National Engineering Consortium, Chicago, 1978, pp. 71–87.

M. K. Sain and J. L. Peczkowski, "Nonlinear Control by Coordinated Feedback with Gas Turbine Applications," *Proceedings American Control Conference*, IEEE Service Center, Piscataway, New Jersey, June 1985.

J. L. Peczkowski and M. K. Sain, "Synthesis of System Responses: A Nonlinear Multivariable Control Design Approach," *Proceedings American Control Conference*, IEEE Catalog No. 85CH 2119-6, Library of Congress, No. 84-73269, June 1985.

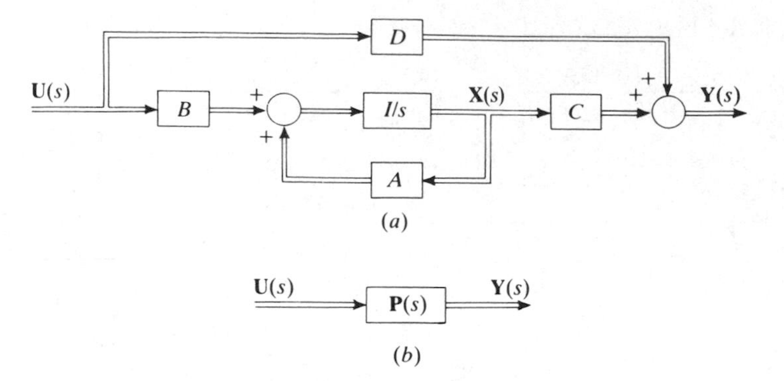

FIGURE 9.26
State-space representation for a plant.

double-width paths. The equivalent block-diagram representation in which $\mathbf{P}(s)$ is the transfer function relating the output $\mathbf{Y}(s)$ to the input $\mathbf{U}(s)$ is shown in Fig. 9.26*b*. Solving the first of the preceding equations for $\mathbf{X}(s)$ and then substituting this result into the expression for $\mathbf{Y}(s)$ give

$$\mathbf{X}(s) = [sI - A]^{-1} B\mathbf{U}(s)$$

and

$$\mathbf{Y}(s) = [C(sI - A)^{-1}B + D]\mathbf{U}(s) = \mathbf{P}(s)\mathbf{U}(s)$$

Thus

$$\mathbf{P}(s) = C(sI - A)^{-1}B + D \tag{9.80}$$

Because there may be many different state-space representations for a given plant, matrices A, B, C, and D are not unique. However, there is but one transfer function for a given plant. Thus, the transfer function matrix $\mathbf{P}(s)$ is unique.

Figure 9.27*a* shows a multivariable system. From this block diagram, we see that the individual equations are

$$\mathbf{E}(s) = \mathbf{R}(s) - \mathbf{Y}(s)$$
$$\mathbf{U}(s) = \mathbf{G}(s)\mathbf{E}(s)$$
$$\mathbf{Y}(s) = \mathbf{P}(s)\mathbf{U}(s)$$

For the case in which there are two system inputs R_1 and R_2, two plant inputs U_1 and U_2, and two system outputs Y_1 and Y_2, the preceding equations become

$$\begin{bmatrix} E_1(s) \\ E_2(s) \end{bmatrix} = \begin{bmatrix} R_1(s) \\ R_2(s) \end{bmatrix} - \begin{bmatrix} Y_1(s) \\ Y_2(s) \end{bmatrix}$$

$$\begin{bmatrix} U_1(s) \\ U_2(s) \end{bmatrix} = \begin{bmatrix} G_{11}(s) & G_{12}(s) \\ G_{21}(s) & G_{22}(s) \end{bmatrix} \begin{bmatrix} E_1(s) \\ E_2(s) \end{bmatrix}$$

$$\begin{bmatrix} Y_1(s) \\ Y_2(s) \end{bmatrix} = \begin{bmatrix} P_{11}(s) & P_{12}(s) \\ P_{21}(s) & P_{22}(s) \end{bmatrix} \begin{bmatrix} U_1(s) \\ U_2(s) \end{bmatrix}$$

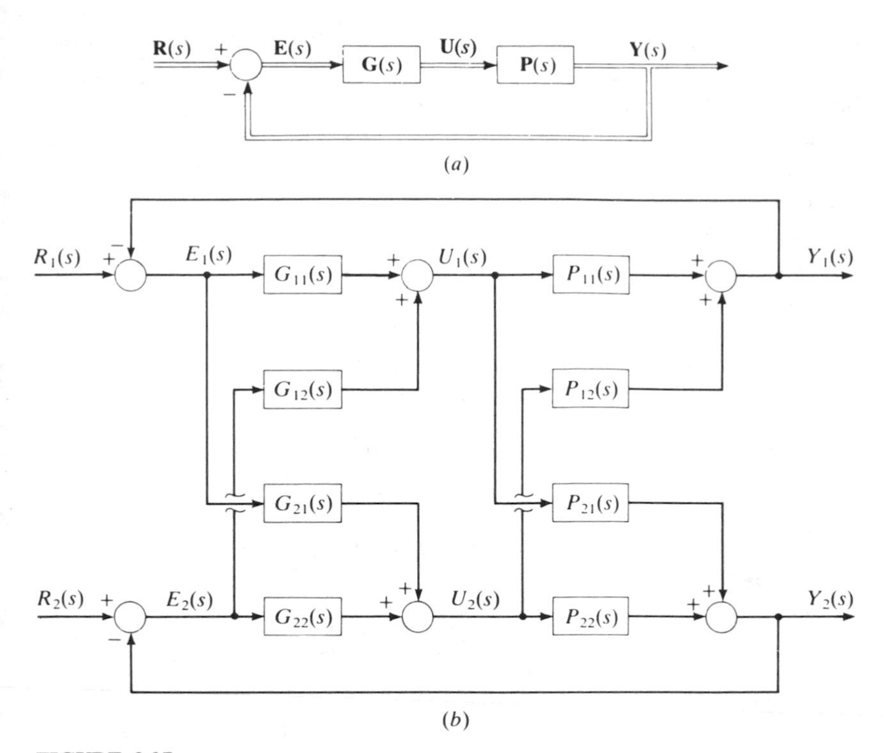

FIGURE 9.27
Multivariable system.

All the preceding relationships are contained in the block diagram shown in Fig. 9.27*b*. The equations for the $\mathbf{E}(s) = \mathbf{R}(s) - \mathbf{Y}(s)$ matrix are

$$E_1(s) = R_1(s) - Y_1(s)$$
$$E_2(s) = R_2(s) - Y_2(s)$$

The first summing point of the top loop yields $E_1(s)$, and the first summing point of the bottom loop yields $E_2(s)$. The equations for the $\mathbf{U}(s) = \mathbf{G}(s)\mathbf{E}(s)$ matrix are

$$U_1(s) = G_{11}(s)E_1(s) + G_{12}(s)E_2(s)$$
$$U_2(s) = G_{21}(s)E_1(s) + G_{22}(s)E_2(s)$$

The second summing point of the top loop yields $U_1(s)$, and the second summing point of the bottom loop yields $U_2(s)$. The equations for the $\mathbf{Y}(s) = \mathbf{P}(s)\mathbf{U}(s)$ matrix are

$$Y_1(s) = P_{11}(s)U_1(s) + P_{12}(s)U_2(s)$$
$$Y_2(s) = P_{21}(s)U_1(s) + P_{22}(s)U_2(s)$$

The last summing point of the top loop yields $Y_1(s)$, and the last summing point of the bottom loop yields $Y_2(s)$.

As an example of a multivariable system, consider the control of a jet engine. The system inputs would be the desired value of the engine speed and the desired value of the burner exit temperature. The outputs would be the actual value of the engine speed and the actual value of the burner exit temperature. The inputs to the plant (i.e., jet engine) would be the fuel flow to the engine and the nozzle jet area. The state variables ($x_1, x_2, \ldots$) would be various engine parameters such as engine speed, fuel flow, compressor discharge pressure, nozzle area, exit temperature, etc.

For a given system usually the plant characteristics are known. The problem then is to determine the characteristics of the controller to yield a desired relationship between the input and the output. The block-diagram representation for a closed-loop multivariable control system is shown in Fig. 9.28*a*. The equation for the feedforward portion of the system which relates $\mathbf{Y}(s)$ to $\mathbf{E}(s)$ is

$$\mathbf{Y}(s) = \mathbf{P}(s)\mathbf{U}(s) = \mathbf{P}(s)\mathbf{G}(s)\mathbf{E}(s) = \mathbf{P}(s)\mathbf{G}(s)[\mathbf{R}(s) - \mathbf{Y}(s)]$$

Solving for $\mathbf{Y}(s)$ shows that

$$[I + \mathbf{P}(s)\mathbf{G}(s)]\mathbf{Y}(s) = \mathbf{P}(s)\mathbf{G}(s)\mathbf{R}(s)$$

Thus,

$$\mathbf{Y}(s) = [I + \mathbf{P}(s)\mathbf{G}(s)]^{-1}\mathbf{P}(s)\mathbf{G}(s)\mathbf{R}(s) \tag{9.81}$$

An equivalent block-diagram representation for the control system of Fig. 9.28*a* is shown in Fig. 9.28*b*. The open-loop transfer function $\mathbf{T}(s)$ shown in Fig. 9.28*b* is called the *transmittance*. The transmittance $\mathbf{T}(s)$ relates the input to the output, i.e.,

$$\mathbf{Y}(s) = \mathbf{T}(s)\mathbf{R}(s) \tag{9.82}$$

Comparison of Eqs. (9.81) and (9.82) shows that

$$\mathbf{T}(s) = [I + \mathbf{P}(s)\mathbf{G}(s)]^{-1}\mathbf{P}(s)\mathbf{G}(s) \tag{9.83}$$

Equation (9.83) relates $\mathbf{G}(s)$, $\mathbf{T}(s)$, and $\mathbf{P}(s)$. To determine the equation for the controller characteristics $\mathbf{G}(s)$ in terms of the known plant characteristics $\mathbf{P}(s)$ and the desired transmittance $\mathbf{T}(s)$, first premultiply both sides of Eq. (9.83) by $[I + \mathbf{P}(s)\mathbf{G}(s)]$:

$$[I + \mathbf{P}(s)\mathbf{G}(s)]\mathbf{T}(s) = \mathbf{P}(s)\mathbf{G}(s)$$

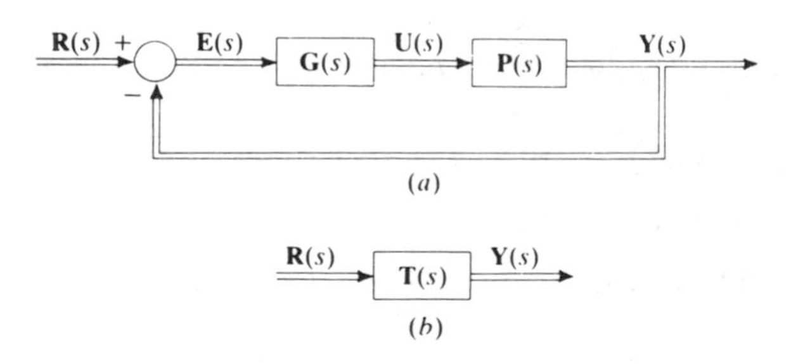

FIGURE 9.28
Closed-loop multivariable control system.

This may be rewritten as

$$\mathbf{T}(s) = \mathbf{P}(s)\mathbf{G}(s) - \mathbf{P}(s)\mathbf{G}(s)\mathbf{T}(s) = \mathbf{P}(s)\mathbf{G}(s)[I - \mathbf{T}(s)]$$

Premultiplying both sides by $\mathbf{P}^{-1}(s)$ and postmultiplying both sides by $[I - \mathbf{T}(s)]^{-1}$ yield $\mathbf{G}(s)$ in terms of $\mathbf{T}(s)$ and $\mathbf{P}(s)$:

$$\mathbf{G}(s) = \mathbf{P}^{-1}(s)\mathbf{T}(s)[I - \mathbf{T}(s)]^{-1} \tag{9.84}$$

This result enables the designer to determine the controller characteristics $\mathbf{G}(s)$ to yield a desired transmittance $\mathbf{T}(s)$ relating the output to the input for a given plant $\mathbf{P}(s)$.

The quantity $\mathbf{T}(s)[I - \mathbf{T}(s)]^{-1}$ in Eq. (9.84) is rather awkward and cumbersome. A more convenient form is now developed. Figure 9.29 shows an equivalent representation for the multivariable control system shown in Fig. 9.28*a*. Comparison of Fig. 9.29 and Fig. 9.28*a* shows that

$$\mathbf{Y}(s) = \mathbf{Q}(s)\mathbf{E}(s) = \mathbf{P}(s)\mathbf{G}(s)\mathbf{E}(s)$$

Thus

$$\mathbf{Q}(s) = \mathbf{P}(s)\mathbf{G}(s)$$

Premultiplying by $\mathbf{P}^{-1}(s)$ and then solving for $\mathbf{G}(s)$ give

$$\mathbf{G}(s) = \mathbf{P}^{-1}(s)\mathbf{Q}(s) \tag{9.85}$$

Comparison of Eqs. (9.84) and (9.85) shows that

$$\mathbf{Q}(s) = \mathbf{T}(s)[I - \mathbf{T}(s)]^{-1} \tag{9.86}$$

The matrix $\mathbf{Q}(s)$ is called the *performance matrix*. The performance matrix $\mathbf{Q}(s)$ is thus seen to depend on the response matrix $\mathbf{T}(s)$. When the number of inputs is equal to the number of outputs, $\mathbf{T}(s)$ is a square matrix. Furthermore, when the system is decoupled (one input controls one output, only), $\mathbf{T}(s)$ is a diagonal matrix.

$$\mathbf{T}(s) = \begin{bmatrix} t_{11} & 0 & \cdots & 0 \\ 0 & t_{22} & \cdots & 0 \\ \cdots & \cdots & \cdots & \cdots \\ 0 & 0 & \cdots & t_{nn} \end{bmatrix}$$

The matrix $\mathbf{Q}(s)$ has the same decoupling properties as $\mathbf{T}(s)$. That is, if $\mathbf{T}(s)$ is a diagonal, block diagonal, or triangular matrix, then $\mathbf{Q}(s)$ will also be a diagonal, block diagonal, or triangular matrix.

When $\mathbf{T}(s)$ and thus $\mathbf{Q}(s)$ are diagonal matrices, each element $q_{ii}(s)$ on the diagonal of $\mathbf{Q}(s)$ is related to the corresponding element $t_{ii}(s)$ on the diagonal of $\mathbf{T}(s)$

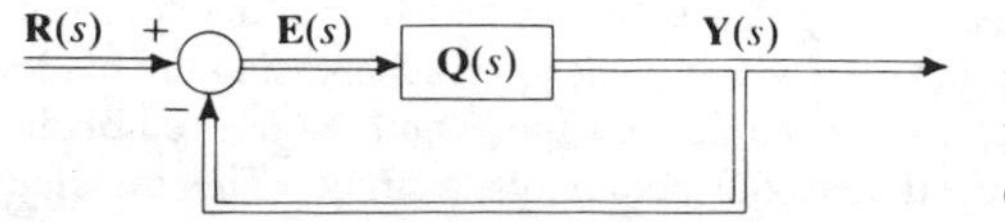

FIGURE 9.29
Equivalent representation for multivariable control system.

TABLE 9.1
Unity feedback, decoupled response

$q_{ii}(s)$	$t_{ii}(s)$	
$\dfrac{1}{\tau s}$	$\dfrac{1}{1+\tau s}$	
$\dfrac{1/(\tau_1+\tau_2)}{s\{1+[\tau_1\tau_2/(\tau_1+\tau_2)]s\}}$	$\dfrac{1}{(1+\tau_1 s)(1+\tau_2 s)}$	
$\dfrac{\omega_n/(2\zeta)}{s\{1+[1/(2\zeta\omega_n)]s\}}$	$\dfrac{\omega_n^2}{s^2+2\zeta\omega_n s+\omega_n^2}$	
$\dfrac{1/(\tau_1+\tau_2+\tau_3)}{s\{1+[\tau_1\tau_2/(\tau_1+\tau_2+\tau_3)]s\}(1+\tau_3 s)}$	$\dfrac{1}{(1+\tau_1 s)(1+\tau_2 s)(1+\tau_3 s)}$	$\tau_1>\tau_2>\tau_3$

by the relationship

$$q_{ii}(s) = \frac{t_{ii}(s)}{1 - t_{ii}(s)} \tag{9.87}$$

Table 9.1 shows corresponding values of elements of $\mathbf{Q}(s)$ and $\mathbf{T}(s)$.

Illustrative example 9.10. The transfer function for the plant of a single-input single-output system is

$$\mathbf{P}(s) = \frac{5}{1+0.5s}$$

Determine the controller characteristic $\mathbf{G}(s)$ such that the transmittance $T(s)$ is

$$\mathbf{T}(s) = \frac{1}{1+0.4s}$$

Solution. From Table 9.1 for $t_{ii}(s) = 1/(1+\tau s)$ then $q_{ii}(s) = 1/(\tau s)$, thus the performance matrix is

$$\mathbf{Q}(s) = \frac{1}{0.4s} = \frac{2.5}{s}$$

The desired controller characteristic is

$$\mathbf{G}(s) = \mathbf{P}^{-1}(s)\mathbf{Q}(s) = \frac{1+0.5s}{5}\frac{2.5}{s} = \frac{1+0.5s}{2s}$$

The resulting system is shown in Fig. 9.30. Notice that the numerator $(1 + 0.5s)$ of $G(s)$ cancels the denominator of $P(s)$. With this method, a considerable amount of cancellation is inherent. This tends to simplify the resulting system.

Illustrative example 9.11. A hydromatic propeller is one in which both the speed of the propeller and the blade angle are controlled. For a given speed, increasing the blade angle increases the power. The two inputs are the desired speed and the desired blade angle. The outputs are the actual speed and the actual blade angle. The transfer

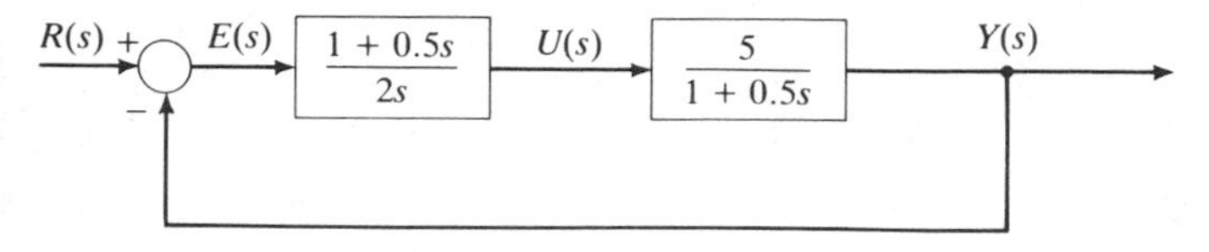

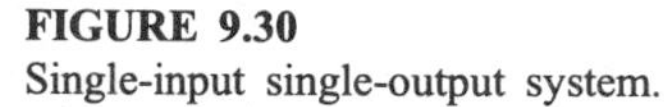

FIGURE 9.30
Single-input single-output system.

function for a hydromatic propeller is

$$\mathbf{P}(s)=\begin{bmatrix}\dfrac{5}{1+2s} & \dfrac{50}{1+2s}\\[2ex] \dfrac{1}{1+s} & \dfrac{-10}{1+s}\end{bmatrix}$$

The desired transmittance is

$$\mathbf{T}(s)=\begin{bmatrix}\dfrac{1}{1+0.5s} & 0\\[2ex] 0 & \dfrac{1}{1+0.2s}\end{bmatrix}$$

Determine the required controller $\mathbf{G}(s)$.

Solution. From Table 9.1, for $t_{ii}(s) = 1/(1+\tau s)$ then $q_{ii}(s) = 1/(\tau s)$. Thus, the performance matrix is

$$\mathbf{Q}(s)=\begin{bmatrix}\dfrac{2}{s} & 0\\[2ex] 0 & \dfrac{5}{s}\end{bmatrix}$$

The transfer function $\mathbf{G}(s)$ for the required controller is

$$\mathbf{G}(s)=\mathbf{P}^{-1}(s)\mathbf{Q}(s)=\begin{bmatrix}0.10(1+2s) & 0.50(1+s)\\ 0.01(1+2s) & -0.05(1+s)\end{bmatrix}\begin{bmatrix}\dfrac{2}{s} & 0\\[2ex] 0 & \dfrac{5}{s}\end{bmatrix}$$

$$=\begin{bmatrix}\dfrac{0.20(1+2s)}{s} & \dfrac{2.50(1+s)}{s}\\[2ex] \dfrac{0.02(1+2s)}{s} & \dfrac{-0.25(1+s)}{s}\end{bmatrix}$$

Sensitivity Analysis

In practice the transfer function of the plant will change in time due to aging, wearing of parts, etc. This will cause the transmittance $\mathbf{T}(s)$ to change also. For example, if $\mathbf{P}(s)$ for the system of Illustrative example 9.10 (see Fig. 9.30) were to

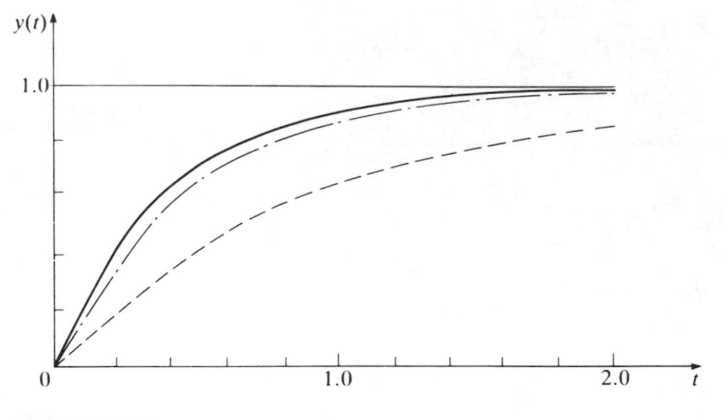

FIGURE 9.31
Step function response.

change from $5/(1 + 0.5s)$ to $2/(1 + 0.5s)$, then the transmittance would change from $1/(1 + 0.4s)$ to $1/(1 + s)$. The response of the original system to a unit step function is shown by the solid curve in Fig. 9.31. The response for the changed value of $\mathbf{P}(s)$ is shown by the dashed curve. The effect of variations in the plant characteristics $\mathbf{P}(s)$ upon the transmittance $\mathbf{T}(s)$ may be minimized by a sensitivity analysis.

Figure 9.32 shows the multiple-input multiple-output system of Fig. 9.28*a* in which the feedback element $\mathbf{H}(s)$ has been added. For a change $\Delta\mathbf{P}(s)$ in the transfer function of the plant, the new transfer function will be $\mathbf{P}(s) + \Delta\mathbf{P}(s)$. Thus the new system output will be

$$\begin{aligned}\mathbf{Y}(s) + \Delta\mathbf{Y}(s) &= [\mathbf{P}(s) + \Delta\mathbf{P}(s)]\mathbf{U}(s) = [\mathbf{P}(s) + \Delta\mathbf{P}(s)]\mathbf{G}(s)\mathbf{E}(s)\\ &= [\mathbf{P}(s) + \Delta\mathbf{P}(s)]\mathbf{G}(s)\{\mathbf{R}(s) - \mathbf{H}(s)[\mathbf{Y}(s) + \Delta\mathbf{Y}(s)]\}\end{aligned}$$

Algebraically rearranging shows that

$$\{I + [\mathbf{P}(s) + \Delta\mathbf{P}(s)]\mathbf{G}(s)\mathbf{H}(s)\}[\mathbf{Y}(s) + \Delta\mathbf{Y}(s)] = [\mathbf{P}(s) + \Delta\mathbf{P}(s)]\mathbf{G}(s)\mathbf{R}(s)$$

For $\Delta\mathbf{P}(s) = \mathbf{0}$, this becomes

$$[I + \mathbf{P}(s)\mathbf{G}(s)\mathbf{H}(s)]\mathbf{Y}(s) = \mathbf{P}(s)\mathbf{G}(s)\mathbf{R}(s)$$

The change $\Delta\mathbf{Y}(s)$ due to a change $\Delta\mathbf{P}(s)$ is

$$[I + \mathbf{P}(s)\mathbf{G}(s)\mathbf{H}(s)]\,\Delta\mathbf{Y}(s) = \Delta\mathbf{P}(s)\mathbf{G}(s)\mathbf{R}(s) \tag{9.88}$$

This result relates the change $\Delta\mathbf{Y}(s)$ due to a change $\Delta\mathbf{P}(s)$ in the transfer function of the plant.

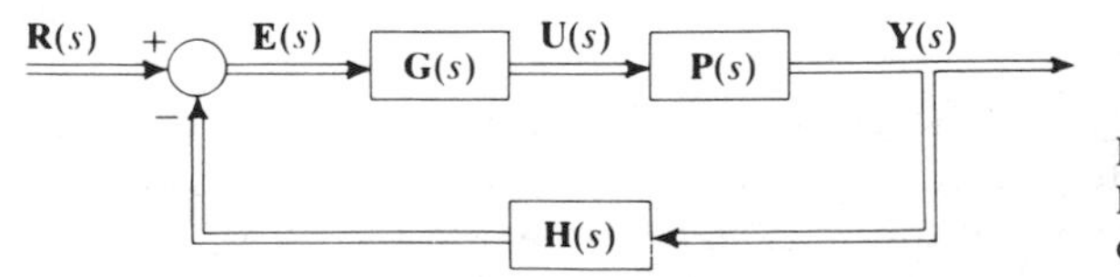

FIGURE 9.32
Multivariable system with feedback element $\mathbf{H}(s)$.

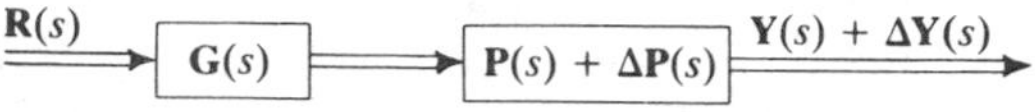

$\mathbf{R}(s) \Rightarrow \boxed{\mathbf{G}(s)} \Rightarrow \boxed{\mathbf{P}(s) + \Delta\mathbf{P}(s)} \Rightarrow \mathbf{Y}(s) + \Delta\mathbf{Y}(s)$

FIGURE 9.33
Multivariable system without feedback.

Figure 9.33 shows the system of Fig. 9.32 in which the feedback loop has been eliminated (i.e., the open-loop system). For this open-loop system the variation in the output due to a change in the plant is

$$[\mathbf{Y}(s) + \Delta\mathbf{Y}(s)]_{\mathrm{OL}} = [\mathbf{P}(s) + \Delta\mathbf{P}(s)]\mathbf{G}(s)\mathbf{R}(s)$$

The subscript "OL" represents the output if the system were an open-loop system. For $\Delta\mathbf{P}(s) = \mathbf{0}$, the output is

$$\mathbf{Y}(s)_{\mathrm{OL}} = \mathbf{P}(s)\mathbf{G}(s)\mathbf{R}(s)$$

The change in the open-loop output due to a change in the transfer function of the plant is

$$\Delta\mathbf{Y}(s)_{\mathrm{OL}} = \Delta\mathbf{P}(s)\ \mathbf{G}(s)\mathbf{R}(s) \tag{9.89}$$

The right sides of Eqs. (9.88) and (9.89) are identical. Thus

$$[I + \mathbf{P}(s)\mathbf{G}(s)\mathbf{H}(s)]\ \Delta\mathbf{Y}(s) = \Delta\mathbf{Y}(s)_{\mathrm{OL}}$$

Premultiplying by $[I + \mathbf{P}(s)\mathbf{G}(s)\mathbf{H}(s)]^{-1}$ and postmultiplying by $\Delta\mathbf{Y}(s)_{\mathrm{OL}}^{-1}$ show that

$$\Delta\mathbf{Y}(s)\ \Delta\mathbf{Y}(s)_{\mathrm{OL}}^{-1} = [I + \mathbf{P}(s)\mathbf{G}(s)\mathbf{H}(s)]^{-1}$$

For the case of a single-input single-output system, this becomes

$$\frac{\Delta\mathbf{Y}(s)}{\Delta\mathbf{Y}(s)_{\mathrm{OL}}} = [I + \mathbf{P}(s)\mathbf{G}(s)\mathbf{H}(s)]^{-1}$$

The left side is the ratio of the closed-loop response variation to the open-loop response variation. This ratio is the sensitivity $\mathbf{S}$. That is,

$$\mathbf{S} = [I + \mathbf{P}(s)\mathbf{G}(s)\mathbf{H}(s)]^{-1} \tag{9.90}$$

We now show how to select $\mathbf{H}(s)$ so that the system sensitivity is considerably less than it would be for a unity-feedback system. From Fig. 9.32, it follows that

$$\mathbf{Y}(s) = \mathbf{P}(s)\mathbf{G}(s)[\mathbf{R}(s) - \mathbf{H}(s)\mathbf{Y}(s)]$$

Hence

$$[I + \mathbf{P}(s)\mathbf{G}(s)\mathbf{H}(s)]\mathbf{Y}(s) = \mathbf{P}(s)\mathbf{G}(s)\mathbf{R}(s)$$

Premultiplying by $[I + \mathbf{P}(s)\mathbf{G}(s)\mathbf{H}(s)]^{-1} = \mathbf{S}$ shows that

$$\mathbf{Y}(s) = [I + \mathbf{P}(s)\mathbf{G}(s)\mathbf{H}(s)]^{-1}\mathbf{P}(s)\mathbf{G}(s)\mathbf{R}(s) = \mathbf{S}\mathbf{P}(s)\mathbf{G}(s)\mathbf{R}(s) \tag{9.91}$$

In Fig. 9.34*a* and *b* there are equivalent representations for the system of Fig. 9.32. From Fig. 9.34*b*, it follows that

$$\mathbf{Y}(s) = \mathbf{T}(s)\mathbf{R}(s) \tag{9.92}$$

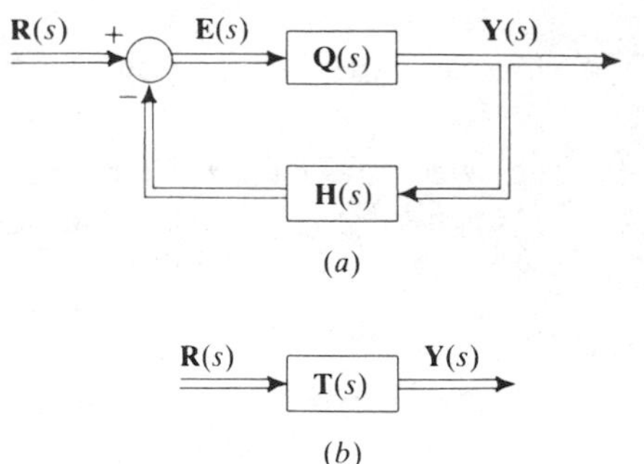

FIGURE 9.34
Equivalent representation for multivariable system with feedback element $\mathbf{H}(s)$.

Equating the right sides of Eqs. (9.91) and (9.92) shows that

$$\mathbf{SP}(s)\mathbf{G}(s) = \mathbf{T}(s) \tag{9.93}$$

Premultiplying by $\mathbf{S}^{-1}$ and then $\mathbf{P}^{-1}(s)$ gives

$$\mathbf{G}(s) = \mathbf{P}^{-1}(s)\mathbf{S}^{-1}\mathbf{T}(s) \tag{9.94}$$

Comparison of Fig. 9.34*a* and Fig. 9.32 shows that

$$\mathbf{Q}(s) = \mathbf{P}(s)\mathbf{G}(s) \tag{9.95}$$

Substitution of $\mathbf{G}(s)$ from Eq. (9.94) into Eq. (9.95) gives

$$\mathbf{Q}(s) = \mathbf{S}^{-1}\mathbf{T}(s) \tag{9.96}$$

Substitution of the right side of Eq. (9.96) into Eq. (9.94) shows that

$$\mathbf{G}(s) = \mathbf{P}^{-1}(s)\mathbf{Q}(s) \tag{9.97}$$

To obtain the equation for the required feedback elements $\mathbf{H}(s)$, first postmultiply Eq. (9.90) by $I + \mathbf{P}(s)\mathbf{G}(s)\mathbf{H}(s)$.

$$\mathbf{S}[I + \mathbf{P}(s)\mathbf{G}(s)\mathbf{H}(s)] = \mathbf{S} + \mathbf{SP}(s)\mathbf{G}(s)\mathbf{H}(s) = I$$

In accordance with Eq. (9.93), $\mathbf{SP}(s)\mathbf{G}(s)$ may be replaced by $\mathbf{T}(s)$. Thus,

$$\mathbf{S} = I - \mathbf{T}(s)\mathbf{H}(s) \tag{9.98}$$

Premultiplying by $\mathbf{T}^{-1}(s)$ and solving for $\mathbf{H}(s)$ give

$$\mathbf{H}(s) = \mathbf{T}^{-1}(s)[I - \mathbf{S}] \tag{9.99}$$

The usual design procedure is first to determine the sensitivity for a unity-feedback system. For $\mathbf{H}(s) = I$, Eq. (9.98) becomes

$$\mathbf{S} = I - \mathbf{T}(s) \tag{9.100}$$

By knowing the desired improvement in sensitivity, the new value of $\mathbf{S}$ may be determined. For a given plant, the plant characteristics $\mathbf{P}(s)$ are known. The transmittance $\mathbf{T}(s)$ is selected to achieve the desired performance characteristics from the input to the output. Thus, by knowing $\mathbf{S}$, $\mathbf{P}(s)$, and $\mathbf{T}(s)$, the required controller $\mathbf{G}(s)$ and feedback $\mathbf{H}(s)$ may be obtained from Eqs. (9.97) and (9.99).

Illustrative example 9.12. For the system of Illustrative example 9.10, determine $\mathbf{G}(s)$ and $\mathbf{H}(s)$ so as to decrease the sensitivity by a factor of 10.

Solution. For the unity-feedback system of Illustrative example 9.10, the sensitivity is

$$\mathbf{S} = I - \mathbf{T}(s) = 1 - \frac{1}{1+0.4s} = \frac{0.4s}{1+0.4s}$$

To decrease the sensitivity by a factor of 10, the new value of $\mathbf{S}$ is

$$\mathbf{S} = \frac{0.04s}{1+0.4s}$$

The required values for $\mathbf{Q}(s)$, $\mathbf{G}(s)$, and $\mathbf{H}(s)$ are

$$\mathbf{Q}(s) = \mathbf{S}^{-1}\mathbf{T}(s) = \frac{1+0.4s}{0.04s}\,\frac{1}{1+0.4s} = \frac{25}{s}$$

$$\mathbf{G}(s) = \mathbf{P}^{-1}(s)\mathbf{Q}(s) = \frac{1+0.5s}{5}\,\frac{25}{s} = \frac{5(1+0.5s)}{s}$$

$$\mathbf{H}(s) = \mathbf{T}^{-1}(s)[I - \mathbf{S}] = (1+0.4s)\left(1 - \frac{0.04s}{1+0.4s}\right) = 1 + 0.36s$$

If the plant transfer function changes from $5/(1 + 0.5s)$ to $2/(1 + 0.5s)$, the response of the resulting system to a unit step function is shown by the dot-dash curve of Fig. 9.31. The change in the response due to the change in the plant characteristics has been minimized considerably.

PROBLEMS

9.1. The dynamics of a helicopter in the hovering condition are described by the differential equation

$$y(t) = \frac{1}{(D+2)(D+3)} f(t)$$

Determine the computer diagram and state-space representation by
(*a*) Direct programming
(*b*) Parallel programming
(*c*) Series programming
(*d*) General programming

9.2. The motion of a driverless vehicle which follows a wire embedded in the floor is described by the differential equation

$$y(t) = \frac{2(D+5)}{(D+2)(D+3)} f(t)$$

Determine the computer diagram and state-space representation by
(*a*) Direct programming
(*b*) Parallel programming
(*c*) Series programming
(*d*) General programming

9.3. The speed of an automobile which is being controlled by the cruise control is described by the differential equation

$$y(t) = \frac{2(D+5)}{(D+2)(D+3)(D+4)} f(t)$$

Determine the computer diagram and state-space representation by

(*a*) Direct programming

(*b*) Parallel programming

(*c*) Series programming

(*d*) General programming

9.4. Determine the state-space representation for the feedback control system shown in

(*a*) Fig. P9.4*a*

(*b*) Fig. P9.4*b*

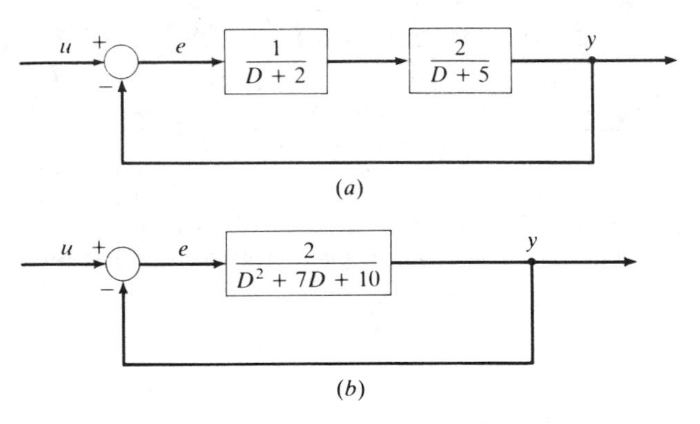

FIGURE P9.4

9.5. Determine the state-space representation for the feedback control system shown in

(*a*) Fig. P9.5*a*

(*b*) Fig. P9.5*b*

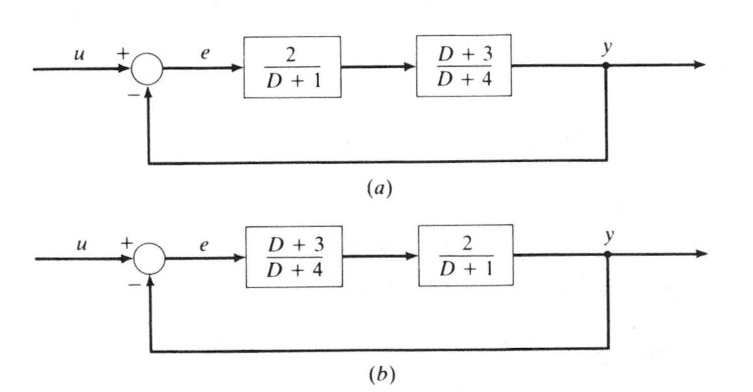

FIGURE P9.5

9.6. Determine the state-space representation for the feedback control system shown in
(*a*) Fig. P9.6*a*
(*b*) Fig. P9.6*b*

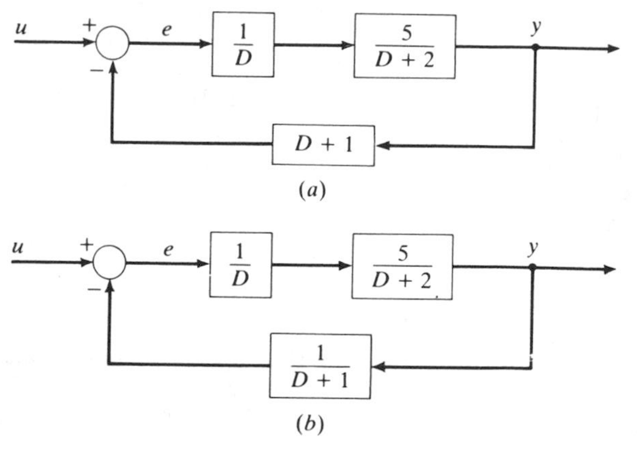

FIGURE P9.6

9.7. Determine the state-space representation for the feedback system shown in
(*a*) Fig. P9.7*a*
(*b*) Fig. P9.7*b*

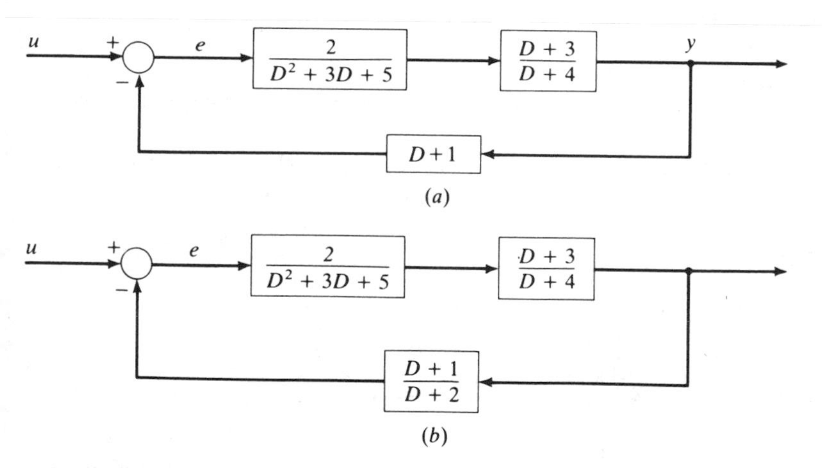

FIGURE P9.7

9.8. For the system of Prob. 9.1, construct the signal flow graph corresponding to the computer diagram obtained by
(*a*) Direct programming
(*b*) Parallel programming
(*c*) Series programming
(*d*) General programming
Apply Mason's gain formula to verify that the signal flow graph represents the differential equation.

9.9. For the system of Prob. 9.2, construct the signal flow graph corresponding to the computer diagram obtained by
(*a*) Direct programming
(*b*) Parallel programming
(*c*) Series programming
(*d*) General programming
Apply Mason's gain formula to verify that the signal flow graph represents the differential equation.

9.10. For the system of Prob. 9.3, construct the signal flow graph corresponding to the computer diagram obtained by
(*a*) Direct programming
(*b*) Parallel programming
(*c*) Series programming
(*d*) General programming
Apply Mason's gain formula to verify that the signal flow graph represents the differential equation.

9.11. Construct the signal flow graph for the system shown in
(*a*) Fig. P9.4*a*
(*b*) Fig. P9.4*b*
Apply Mason's gain formula to determine the differential equation relating output $y(t)$ to input $u(t)$.

9.12. Construct the signal flow graph for the system shown in
(*a*) Fig. P9.5*a*
(*b*) Fig. P9.5*b*
Apply Mason's gain formula to determine the differential equation relating output $y(t)$ to input $u(t)$.

9.13. Construct the signal flow graph corresponding to the block diagram shown in Fig. P9.13. Use Mason's gain formula to determine the equation relating
(*a*) $c(t)$ and $r(t)$
(*b*) $c(t)$ and $d(t)$

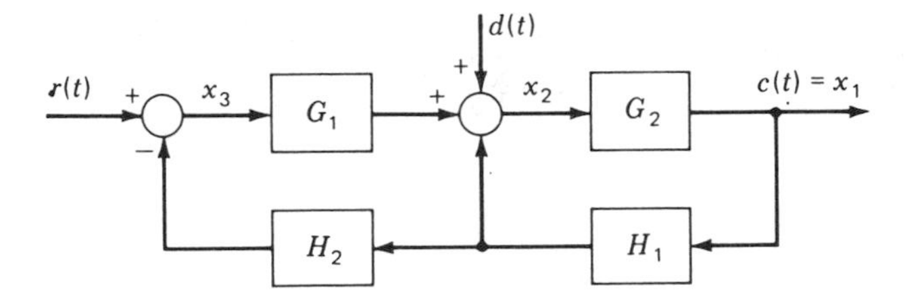

FIGURE P9.13

9.14. Construct the signal flow graph corresponding to the block diagram shown in Fig. P9.14. Use Mason's gain formula to determine the equation relating $c(t)$ and $r(t)$.

9.15. Determine the signal flow graph corresponding to the block diagram shown in Fig. P9.15. Use Mason's gain formula to determine the equation relating $c(t)$ and $r(t)$.

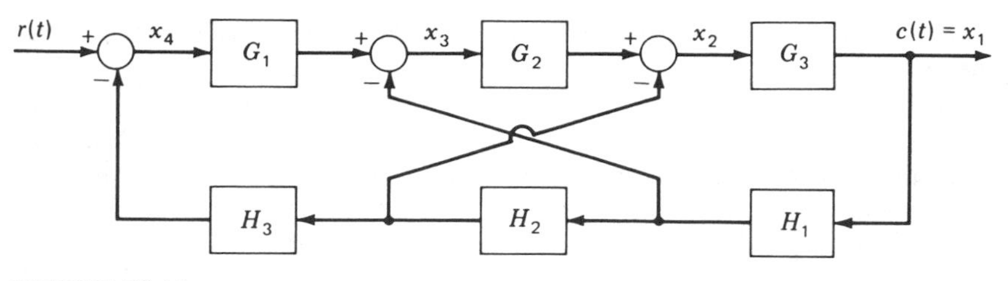

FIGURE P9.14

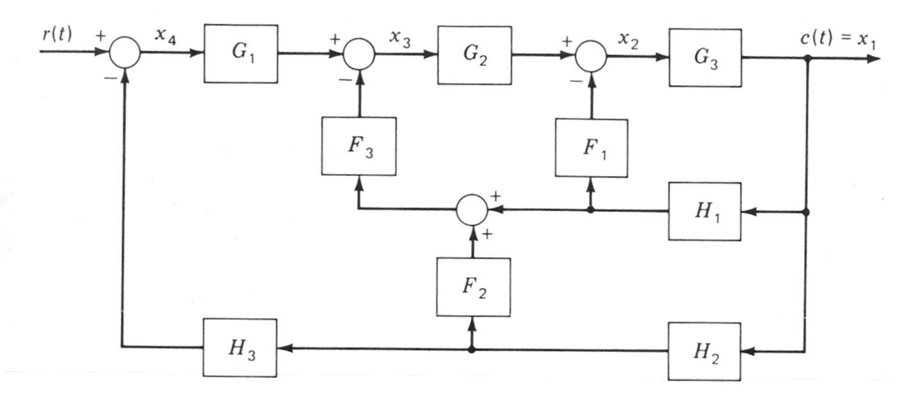

FIGURE P9.15

9.16. Determine the signal flow graph corresponding to the block diagram shown in
(*a*) Fig. 9.1*b*
(*b*) Fig. 9.2*b*
(*c*) Fig. 9.4
Use Mason's gain formula to determine the equation relating $c(t)$ and $r(t)$.

9.17. Obtain the signal flow graph for the following simultaneous equations:

$$x_1 = 4x_2$$
$$x_2 = -x_1 + f(t)$$

Determine the equation relating
(*a*) x_1 and $f(t)$
(*b*) x_2 and $f(t)$

9.18. Obtain the signal flow graph for the following simultaneous equations:

$$x_1 = 4x_2 + u_1$$
$$x_2 = -x_1 + u_2$$

Determine the equation relating
(*a*) x_1 to u_1 and u_2
(*b*) x_2 to u_1 and u_2

9.19. In Prob. 9.4, the state-space representation was obtained for the feedback control system shown in
(*a*) Fig. P9.4*a*
(*b*) Fig. P9.4*b*
Use this representation and the Laplace transform method to determine the response $y(t)$ when $u(t) = 0$ and the initial conditions are $y_0 = 2$ and $\dot{y}_0 = 0$.

9.20. The motion of a hydraulic actuator is described by the differential equation

$$\ddot{y} + 3\dot{y} + 2y = f(t)$$

Use direct programming and the Laplace transform method to determine the solution for the case in which $f(t) = 0$ and the initial conditions are $y_0 = 1$ and $\dot{y}_0 = 0$.

9.21. The motion of a numerically controlled machine tool is described by the differential equation

$$\ddot{y} + 7\dot{y} + 10y = f(t)$$

Use parallel programming and the Laplace transform method to determine the solution for the case in which $f(t) = 0$ and the initial conditions are $y_0 = 3$ and $\dot{y}_0 = 0$.

9.22. The motion of a robot arm which is controlled by an electric motor is described by

$$\ddot{y} + 6\dot{y} + 9y = f(t)$$

Use series programming and the Laplace transform method to determine the solution when $f(t) = 0$ and the initial conditions are $y_0 = 1$ and $\dot{y}_0 = 0$.

9.23. The motion of a ship which incorporates an automatic steering system is described by the differential equation

$$\ddot{y} + 4\dot{y} + 5y = f'(t) + 5f(t)$$

Use general programming and the Laplace transform method to determine the solution for the case in which $f(t) = 0$ and the initial conditions are $y_0 = 1$ and $\dot{y}_0 = 0$.

9.24. The motion of a numerically controlled lathe is described by the differential equation

$$(D+1)(D+2)(D+5) = f(t)$$

Use series programming and the Laplace transform method to determine the solution when $f(t) = 0$ and the initial conditions are $y_0 = 1$ and $\dot{y}_0 = \ddot{y}_0 = 0$.

9.25. Use the signal flow graph method to determine the solution of Prob. 9.20.

9.26. Use the signal flow graph method to determine the solution of Prob. 9.21.

9.27. Use the signal flow graph method to determine the solution of Prob. 9.22.

9.28. Use the signal flow graph method to determine the solution of Prob. 9.23.

9.29. Use the signal flow graph method to determine the solution of Prob. 9.24.

9.30. Repeat Prob. 9.20 except that $f(t) = 2$ and all the initial conditions are zero.

9.31. Repeat Prob. 9.21 except that $f(t) = 30$ and all the initial conditions are zero.

9.32. Repeat Prob. 9.22 except that $f(t) = 9$ and all the initial conditions are zero.

9.33. Repeat Prob. 9.23 except that $f(t) = 1$ and all the initial conditions are zero.

9.34. Repeat Prob. 9.24 except that $f(t) = 60$ and all the initial conditions are zero.

9.35. Consider the differential equation

$$(D+3)(D+4)y(t) = (D+6)f(t)$$

Use direct programming and the Laplace transform method to determine the solution when $f(t) = 2u(t)$ and the initial conditions are $y_0 = 1$, $\dot{y}_0 = 0$, and $f(0) = 0$.

9.36. The operation of a robot submarine is described by the differential equation

$$(D+2)(D+5)y(t) = (D+10)f(t)$$

Use direct programming and the Laplace transform method to determine the solution when $f(t) = u(t)$ and the initial conditions are $y_0 = 1$, $\dot{y}_0 = 0$, and $f(0) = 0$.

9.37. The operation of a numerical path-controlled lathe is described by the differential equation

$$(D^2 + 9D + 20)y(t) = (D+6)f(t)$$

Use series programming and the Laplace transform method to determine the response $y(t)$ when $f(t) = u(t)$ is a unit step function and the initial conditions are $y_0 = 1$, $\dot{y}_0 = 0$, and $f(0) = 0$.

9.38. The system for controlling the pitch of an airplane is described by the differential equation

$$(D+2)(D+5)y(t) = (D^2 + 3D + 10)f(t)$$

Use general programming and the Laplace transform method to determine the solution when $f(t) = 2$ and the initial conditions are $y_0 = 1$, $\dot{y}_0 = 0$, and $f(0) = f'(0) = 0$.

9.39. Use the signal flow graph method to determine the solution of Prob. 9.35.

9.40. Use the signal flow graph method to determine the solution of Prob. 9.36.

9.41. Use the signal flow graph method to determine the solution of Prob. 9.37.

9.42. For the system shown in Fig. 9.25*b*, $K = 2$ and $\mathbf{c}^T = \mathbf{k}^T = [1 \quad 0]$.

$$A = \begin{bmatrix} 0 & 1 \\ -2 & -3 \end{bmatrix} \qquad \mathbf{b} = \begin{bmatrix} 0 \\ 1 \end{bmatrix}$$

Determine the transfer function $G(s)$ for the plant, $H(s)_{eq}$, $G(s)_{eq}$, and the closed-loop transfer function $G_c(s)$.

9.43. For the system shown in Fig. 9.25*b*, $K = 2$ and $\mathbf{c}^T = \mathbf{k}^T = [-1 \quad 1]$.

$$A = \begin{bmatrix} -4 & 1 \\ 0 & -1 \end{bmatrix} \qquad \mathbf{b} = \begin{bmatrix} 0 \\ 1 \end{bmatrix}$$

Determine the transfer function $G(s)$ for the plant, $H(s)_{eq}$, $G(s)_{eq}$, and the closed-loop transfer function $G_c(s)$ for the system.

9.44. The state-space representation for the closed-loop control system shown in Fig. 9.25*b* is

$$A = \begin{bmatrix} 0 & 1 \\ 0 & -2 \end{bmatrix} \qquad \mathbf{b} = \begin{bmatrix} 0 \\ 1 \end{bmatrix} \qquad K = 1$$

$$\mathbf{c}^T = [5 \quad 0] \qquad \mathbf{k}^T = [5 \quad 5]$$

Determine the transfer function $G(s)$ for the plant, $H(s)_{eq}$, $G(s)_{eq}$, and the closed-loop transfer function $G_c(s)$.

9.45. The transfer function for a plant is

$$P(s) = \frac{5}{1+2s}$$

Determine the controller $G(s)$ such that the transmittance will be

$$T(s) = \frac{25}{s^2 + 5s + 25}$$

9.46. The transfer function for a plant is

$$\mathbf{P}(s) = \begin{bmatrix} \dfrac{4}{s+4} & -\dfrac{20}{s+4} \\ \dfrac{10}{s+2} & \dfrac{20}{s+2} \end{bmatrix}$$

The desired transmittance is

$$\mathbf{T}(s) = \begin{bmatrix} \dfrac{4}{s^2+2s+4} & 0 \\ 0 & \dfrac{1}{(1+0.2s)(1+0.5s)} \end{bmatrix}$$

Determine the required controller **G**(s).

9.47. For the system of Prob. 9.45, determine **G**(s) and **H**(s) so as to reduce the sensitivity by a factor of 5.

9.48. For the system of Prob. 9.46, determine **G**(s) and **H**(s) so as to reduce the sensitivity by a factor of 10.

CHAPTER 10

DIGITAL CONTROL SYSTEMS

Recently, numerous systems have been devised which utilize digital computers as control elements. Such systems are generally sampled-data systems. That is, the information fed into a digital computer is the value (sample) of the corresponding signal at some instant of time. The computed output remains unchanged until new information (another sample) is fed into the digital computer.

Another well-known application of the sampled-data process is the time-sharing of telemetered information from spacecraft. Instead of temperature at five different places in the vehicle and pressure at three different places being telemetered back on eight different channels, each signal uses the same channel for 1 s out of 8 s. A similar application occurs in the control of industrial processes in which system variables such as temperature, pressure, and flow are sampled periodically in order to make more effective use of the expensive control equipment, especially the digital computer. Another example is a guidance system that utilizes radar scanning in which a given sector, or region, is scanned once every revolution. Thus, in any given direction the signal is sampled at a rate equal to the scan rate of the radar. Sampling is inherent in some applications (e.g., radar scanning), it is necessary in other applications (e.g., to enter data into digital computers), and finally, it is desirable for still other applications (e.g., timesharing).

10.1 SAMPLED-DATA SYSTEMS

A schematic representation of a sampler switch is shown in Fig. 10.1*a*. The switch closes every T s to admit the input signal. A typical input signal is represented by the continuous function $f(t)$ shown in Fig. 10.1*b*. The shaded pulses represent the signal at the output of the sampler. When the duration is much shorter than the system time constants, the output of the sampler may be approximated by the train of impulses $f^*(t)$, as is illustrated in Fig. 10.1*c*. The term $f^*(t)$ is read "f star of t."

The area of each impulse is equal to the value of the input signal at the time $t = nT$ of the particular impulse. Thus, the area of the nth impulse which occurs at time $t = nT$ is $f(nT)$. The equation for the entire train of impulses is

$$f^*(t) = f(0)\delta(t) + f(T)\delta(t-T) + f(2T)\delta(t-2T) + \cdots$$

$$= \sum_{n=0}^{\infty} f(nT)\delta(t-nT) \tag{10.1}$$

where $\delta(t)$ is a unit impulse at $t = 0$ and $\delta(t - nT)$ is a unit impulse at $t = nT$.

The Laplace transform of the sampled signal is

$$F^*(s) = \mathscr{L}[f^*(t)] = f(0) + f(T)e^{-Ts} + f(2T)e^{-2Ts} + \cdots$$

$$= \sum_{n=0}^{\infty} f(nT)e^{-nTs} \tag{10.2}$$

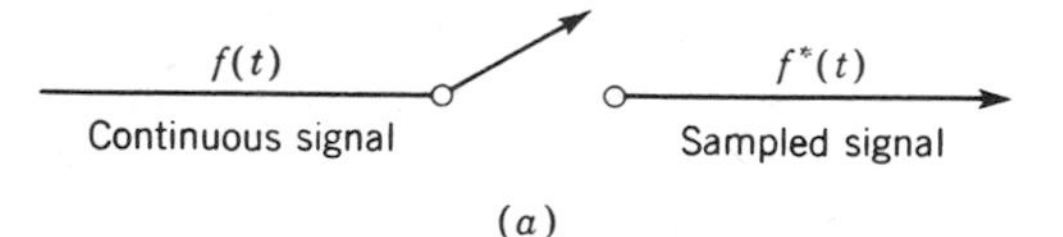

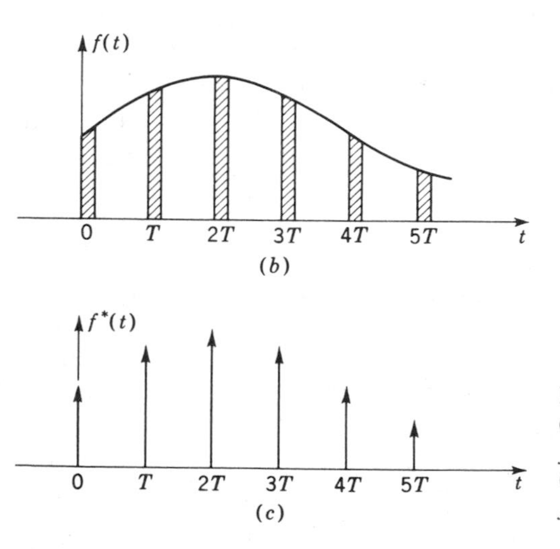

FIGURE 10.1
(*a*) Sampler switch, (*b*) continuous input $f(t)$ and output at switch (shaded pulses), (*c*) impulse approximation of switch output $f^*(t)$.

To illustrate the preceding concepts, consider the continuous input

$$f(t) = e^{-at}$$

The corresponding sampled signal $f^*(t)$ is

$$f^*(t) = \delta(t) + e^{-aT}\delta(t - T) + e^{-2aT}\delta(t - 2T) + \cdots = \sum_{n=0}^{\infty} e^{-anT}\delta(t - nT)$$

The Laplace transform of the continuous input $f(t)$ is

$$F(s) = \mathscr{L}[f(t)] = \frac{1}{s + a}$$

The Laplace transform of the sampled signal $f^*(t)$ is

$$\begin{aligned} F^*(s) &= \mathscr{L}[f^*(t)] = 1 + e^{-aT}e^{-sT} + e^{-2aT}e^{-2sT} + \cdots \\ &= 1 + e^{-(s+a)T} + e^{-2(s+a)T} + \cdots \\ &= \sum_{n=0}^{\infty} e^{-n(s+a)T} \end{aligned}$$

The Laplace transform of a sampled signal is an infinite series. In the following, we see that the z transform greatly simplifies the analysis of such systems.

10.2 THE z TRANSFORM

The simple substitution

$$z = e^{Ts}$$

converts the Laplace transform to the z transform. Making this substitution in Eq. (10.2) gives

$$\begin{aligned} Z[f^*(t)] = F(z) &= f(0) + \frac{f(T)}{z} + \frac{f(2T)}{z^2} + \cdots \\ &= \sum_{n=0}^{\infty} f(nT)z^{-n} \end{aligned} \tag{10.3}$$

where $F(z)$ designates the z transform of $f^*(t)$. Because only values of the signal at the sampling instants are considered, the z transform of $f(t)$ is the same as that of $f^*(t)$.

Illustrative example 10.1. Determine the z transform for a unit step function. For this function $f(nT) = 1$ for $n = 0, 1, 2, \ldots$; thus application of Eq. (10.3) gives

$$Z[u^*(t)] = 1 + \frac{1}{z} + \frac{1}{z^2} + \cdots = \frac{z}{z - 1} \tag{10.4}$$

This series is convergent for $|z| > 1$. In solving problems by z transforms, the term z acts as a dummy operator. There is no need to specify the values of z over which $F(z)$ is convergent. It suffices to know that such values exist.

Illustrative example 10.2. Determine the z transform of the exponential function e^{-at}. For this function $f(nT) = e^{-anT}$; thus

$$Z(e^{-at}) = 1 + \frac{e^{-aT}}{z} + \left(\frac{e^{-aT}}{z}\right)^2 + \cdots = \frac{z}{z - e^{-aT}} \tag{10.5}$$

Table 10.1 gives a partial listing of Laplace transforms and corresponding z transforms for commonly encountered functions.

When the Laplace transform of a function is known, the corresponding z transform may be obtained by the partial-fraction method.

Illustrative example 10.3. Determine the z transform for the function whose Laplace transform is

$$F(s) = \frac{1}{s(s+1)} = \frac{1}{s} - \frac{1}{s+1}$$

From Table 10.1, the z transform corresponding to $1/s$ is $z/(z - 1)$, and that corresponding to $1/(s + 1)$ is $z/(z - e^{-T})$. Thus,

$$F(z) = \frac{z}{z-1} - \frac{z}{z - e^{-T}} = \frac{z(1 - e^{-T})}{(z-1)(z - e^{-T})} \tag{10.6}$$

TABLE 10.1
z transforms

Time function	Laplace transform	z transform
$\delta(t)$	1	1
$u(t)$	$\frac{1}{s}$	$\frac{z}{z-1}$
t	$\frac{1}{s^2}$	$\frac{zT}{(z-1)^2}$
$\frac{t^2}{2}$	$\frac{1}{s^3}$	$\frac{z(z+1)T^2}{2(z-1)^3}$
e^{-at}	$\frac{1}{s+a}$	$\frac{z}{z-e^{-aT}}$
te^{-at}	$\frac{1}{(s+a)^2}$	$\frac{zTe^{-aT}}{(z-e^{-aT})^2}$
$a^{t/T}$	$\frac{1}{s-(1/T)(\ln a)}$	$\frac{z}{z-a}$ $(a > 0)$
$\sin \omega t$	$\frac{\omega}{s^2+\omega^2}$	$\frac{z \sin \omega T}{z^2 - 2z\cos \omega T + 1}$
$\cos \omega t$	$\frac{s}{s^2+\omega^2}$	$\frac{z^2 - z\cos \omega T}{z^2 - 2z\cos \omega T + 1}$

Illustrative example 10.4. Determine the z transform of cos ωt. It is known that the Laplace transform is $s/(s^2 + \omega^2)$. Performing a partial-fraction expansion gives

$$\mathcal{L}(\cos \omega t) = \frac{s}{s^2 + \omega^2} = \frac{\frac{1}{2}}{s + j\omega} + \frac{\frac{1}{2}}{s - j\omega}$$

The corresponding z transform is

$$Z(\cos \omega t) = \frac{1}{2}\left(\frac{z}{z - e^{-j\omega T}} + \frac{z}{z - e^{j\omega T}}\right) = \frac{z^2 - z(e^{j\omega T} + e^{-j\omega T})/2}{z^2 - z(e^{j\omega T} + e^{-j\omega T}) + 1}$$

$$= \frac{z^2 - z\cos \omega T}{z^2 - 2z\cos \omega T + 1} \qquad (10.7)$$

The Residue Method

This a powerful technique for obtaining z transforms. The z transform of $f^*(t)$ may be expressed in the form

$$F(z) = Z[f^*(t)] = \Sigma \text{ residues of } F(s)\frac{z}{z - e^{sT}} \text{ at poles of } F(s) \qquad (10.8)$$

When the denominator of $F(s)$ contains a linear factor of the form $s - r$ such that $F(s)$ has a first-order pole at $s = r$, the corresponding residue R is

$$R = \lim_{s \to r}(s - r)\left[F(s)\frac{z}{z - e^{sT}}\right] \qquad (10.9)$$

When $F(s)$ contains a repeated pole of order q, the residue is

$$R = \frac{1}{(q-1)!} \lim_{s \to r} \frac{d^{q-1}}{ds^{q-1}} \left[(s - r)^q F(s)\frac{z}{z - e^{sT}}\right] \qquad (10.10)$$

As illustrated by the following examples, determining residues is similar to evaluating the constants in a partial-fraction expansion.

Illustrative example 10.5. Determine the z transform of a unit step function. For $F(s) = 1/s$, there is but one pole at $s = 0$. The corresponding residue is

$$R = \lim_{s \to 0} s\left(\frac{1}{s}\frac{z}{z - e^{sT}}\right) = \frac{z}{z - 1}$$

This verifies the result of Eq. (10.4).

Illustrative example 10.6. Determine the z transform of e^{-at}. For this function, $F(s) = 1/(s + a)$, which has but one pole at $s = -a$. Thus,

$$R = \lim_{s \to -a}(s + a)\left(\frac{1}{s + a}\frac{z}{z - e^{sT}}\right) = \frac{z}{z - e^{-aT}}$$

This verifies the result of Eq. (10.5).

Illustrative example 10.7. Determine the z transform for the function whose Laplace transform is

$$F(s) = \frac{1}{s(s+1)}$$

The poles of $F(s)$ occur at $s = 0$ and $s = -1$. The residue due to the pole at $s = 0$ is

$$R_1 = \lim_{s \to 0} s \left[\frac{1}{s(s+1)} \frac{z}{z - e^{sT}} \right] = \frac{z}{z-1}$$

The residue due to the pole at $s = -1$ is

$$R_2 = \lim_{s \to -1} (s+1) \left[\frac{1}{s(s+1)} \frac{z}{z - e^{sT}} \right] = -\frac{z}{z - e^{-T}}$$

Adding these two residues verifies the result given by Eq. (10.6).

Illustrative example 10.8. Determine the z transform of $\cos \omega t$. The Laplace transform is

$$F(s) = \frac{s}{s^2 + \omega^2} = \frac{s}{(s - j\omega)(s + j\omega)}$$

The poles are at $s = j\omega$ and $s = -j\omega$. Thus,

$$R_1 = \left(\frac{s}{s + j\omega} \frac{z}{z - e^{sT}} \right)_{s = j\omega} = \frac{1}{2} \frac{z}{z - e^{j\omega T}}$$

$$R_2 = \left(\frac{s}{s - j\omega} \frac{z}{z - e^{sT}} \right)_{s = -j\omega} = \frac{1}{2} \frac{z}{z - e^{-j\omega T}}$$

Adding these two residues verifies the result given by Eq. (10.7).

Illustrative example 10.9. Determine the z transform corresponding to the function $f(t) = t$. The Laplace transform is

$$F(s) = \frac{1}{s^2}$$

This has a second-order pole at $s = 0$. Application of Eq. (10.10) gives

$$R = \frac{d}{ds} \left(\frac{z}{z - e^{sT}} \right)_{s=0} = \left[\frac{zTe^{sT}}{(z - e^{sT})^2} \right]_{s=0} = \frac{Tz}{(z-1)^2}$$

Theorems

The following basic theorems extend the usefulness of the z transform method. In addition, these theorems help us to better understand z transformations.

MULTIPLICATION BY e^{-at}. The z transform of $e^{-at} f(t)$ is

$$Z[e^{-at} f(t)] = F(ze^{aT}) \tag{10.11}$$

Thus, replacing z in $F(z)$ by ze^{aT} yields the transform for $e^{-at}f(t)$. Recall that the Laplace transform of $e^{-at}f(t)$ is equal to $F(s + a)$. That is, replacing s by $s + a$ gives the effect of multiplying by e^{-at}.

To prove this theorem, note from Eq. (10.3) that

$$Z[e^{-at}f(t)] = \sum_{n=0}^{\infty} f(nT)e^{-anT}z^{-n} = \sum_{n=0}^{\infty} f(nT)\,(ze^{aT})^{-n}$$

The right side is $F(z)$ with z replaced by ze^{at}.

MULTIPLICATION BY t. The z transform of $tf(t)$ is

$$Z[tf(t)] = -zT\,\frac{d}{dz}\,F(z) \tag{10.12}$$

To prove this theorem, note that

$$-zT\,\frac{d}{dz}\,F(z) = -zT\,\frac{d}{dz}\left[f(0) + \frac{f(T)}{z} + \frac{f(2T)}{z^2} + \frac{f(3T)}{z^3} + \cdots\right]$$

$$= T\,\frac{f(T)}{z} + 2T\,\frac{f(2T)}{z^2} + 3T\,\frac{f(3T)}{z^3} + \cdots$$

The right side is the z transform of $tf(t)$.

To illustrate this theorem, consider the function $f(t) = e^{-at}$, for which $F(z) = z/(z - e^{-at})$. Application of Eq. (10.12) shows that

$$Z[te^{-at}] = -zT\,\frac{d}{dz}\,\frac{z}{z - e^{-aT}} = \frac{zTe^{-aT}}{(z - e^{-aT})^2}$$

MULTIPLICATION BY $a^{t/T}$. The z transform of $a^{t/T}f(t)$ is

$$Z[a^{t/T}f(t)] = F\left(\frac{z}{a}\right) \tag{10.13}$$

To prove this theorem, replace z by z/a in the expression for $F(z)$. Thus

$$F\left(\frac{z}{a}\right) = f(0) + a\,\frac{f(T)}{z} + a^2\,\frac{f(2T)}{z^2} + \cdots$$

The right side is the z transform of $a^{t/T}f(t)$.

PARTIAL DIFFERENTIATION. This theorem states that

$$Z\left[\frac{\partial}{\partial a}\,[f(t,\, a)]\right] = \frac{\partial}{\partial a}\,[F(z,\, a)] \tag{10.14}$$

This theorem is useful for ascertaining additional z transforms. For example, it is known that the z transform of e^{at} is $z/(z - e^{aT})$. Thus

$$Z\left[\frac{\partial}{\partial a}\,e^{at}\right] = Z[te^{at}] = \frac{\partial}{\partial a}\,\frac{z}{z - e^{aT}} = \frac{Tze^{aT}}{(z - e^{aT})^2}$$

For the case in which $a = 0$

$$Z[t] = \frac{Tz}{(z-1)^2}$$

By proceeding in a similar manner, the z transforms for t^2, t^3, etc., are also obtained.

INITIAL-VALUE THEOREM. The area of the first impulse $f(0)$ of the sampled function $f^*(t)$ is

$$f(0) = \lim_{z \to \infty} F(z) \tag{10.15}$$

This theorem is verified directly by taking the limit as z approaches infinity in Eq. (10.3).

FINAL-VALUE THEOREM. The area of impulse $f(nT)$ as n becomes infinite is

$$f(\infty) = \lim_{z \to 1} \frac{z-1}{z} F(z) \tag{10.16}$$

To prove this theorem, consider the following sums S_n and S_{n-1}:

$$S_n = f(0) + \frac{f(T)}{z} + \cdots + \frac{f[(n-1)T]}{z^{n-1}} + \frac{f(nT)}{z^n}$$

$$S_{n-1} = f(0) + \frac{f(T)}{z} + \cdots + \frac{f[(n-1)T]}{z^{n-1}}$$

Dividing the second series by z and then subtracting the second from the first give

$$S_n - \frac{1}{z} S_{n-1} = \left(1 - \frac{1}{z}\right) f(0) + \cdots + \left(1 - \frac{1}{z}\right) \frac{f[(n-1)T]}{z^{n-1}} + \frac{f(nT)}{z^n}$$

Taking the limit as z approaches 1 gives

$$\lim_{z \to 1} \left(S_n - \frac{1}{z} S_{n-1}\right) = f(nT)$$

When n is very large, $S_{n-1} \approx S_n \approx F(z)$. Thus, the final-value theorem given by Eq. (10.16) is verified.

REAL TRANSLATION. The solid curve shown in Fig. 10.2a is the continuous function $f(t)$. The value at time $t = 0$ is $f(0)$, the value at $t = T$ is $f(T)$, etc. When the function is shifted to the right (delayed) by a time nT, as shown in Fig. 10.2b, the function $f(t - nT)$ results. The value of $f(t - nT)$ at time $t = nT$ is $f(0)$, the value at $t = (n+1)T$ is $f(T)$, etc. The z transform for $f(t)$ is

$$Z[f(t)] = f(0) + \frac{f(T)}{z} + \frac{f(2T)}{z^2} + \cdots$$

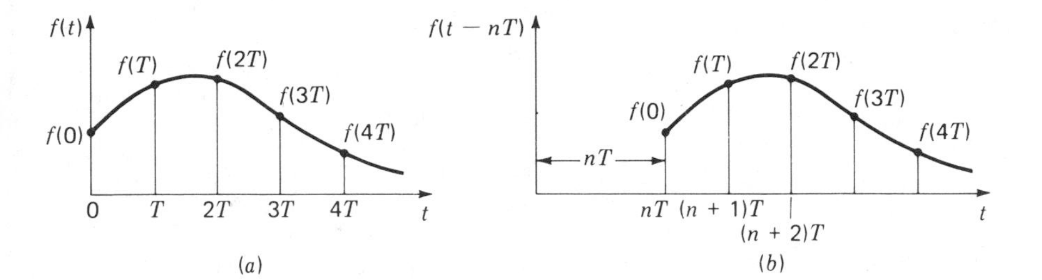

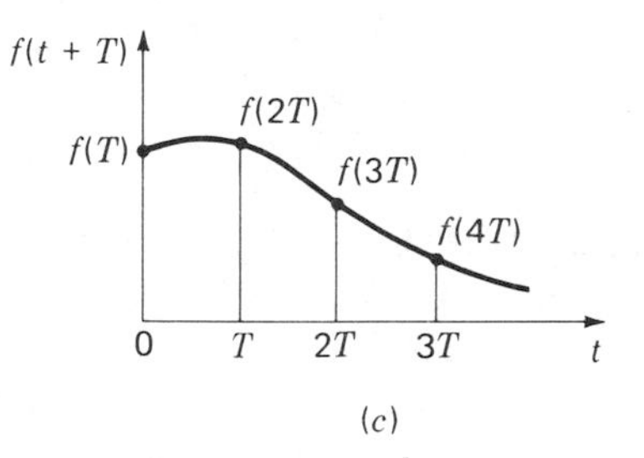

FIGURE 10.2
Translation of a continuous function $f(t)$.

The z transform for $f(t - nT)$ is

$$\begin{aligned} Z[f(t-nT)] &= \frac{f(0)}{z^n} + \frac{f(T)}{z^{n+1}} + \frac{f(2T)}{z^{n+2}} + \cdots \\ &= \frac{1}{z^n}\left[f(0) + \frac{f(T)}{z} + \frac{f(2T)}{z^2} + \cdots\right] \\ &= z^{-n}F(z) \end{aligned} \tag{10.17}$$

Thus, multiplying a z transform by z^{-n} has the effect of delaying the function by a time nT. For Laplace transforms, multiplication by $e^{-t_0 s}$ has the effect of delaying the function by a time t_0.

When the function of Fig. 10.2*a* is shifted to the left one period T, the function $f(t + T)$ shown in Fig. 10.2*c* results. The value of $f(t + T)$ at time $t = 0$ is $f(T)$, the value at $t = T$ is $f(2T)$, etc. The z transform for $f(t + T)$ is

$$Z[f(t+T)] = f(T) + \frac{f(2T)}{z} + \frac{f(3T)}{z^2} + \cdots$$

Multiplying through both sides by $1/z$ and adding $f(0)$ to both sides yield

$$z^{-1}Z[f(t+T)] + f(0) = f(0) + \frac{f(T)}{z} + \frac{f(2T)}{z^2} + \cdots = F(z)$$

Thus

$$Z[f(t+T)] = zF(z) - zf(0)$$

Similarly, it may be shown that

$$Z[f(t+2T)] = z^2F(z) - z^2f(0) - zf(T)$$

and in general

$$Z[f(t+nT)] = z^nF(z) - z^nf(0) - z^{n-1}f(T) - \cdots - zf[(n-1)T] \quad (10.18)$$

DISCRETE FUNCTIONS. The notation $f(k)$ may be used to represent $f^*(t)$. The plot of $f(k)$, shown dashed in Fig. 10.3*a*, is identical to that of $f^*(t)$ shown in Fig. 10.2*a*. The z transform for $f(k)$ is

$$Z[f(k)] = f(0) + \frac{f(1)}{z} + \frac{f(2)}{z^2} + \frac{f(3)}{z^3} + \cdots = F(z) \quad (10.19)$$

When $f(k)$ is delayed n sampling instants, the function $f(k-n)$ shown in Fig. 10.3*b* results. The value of $f(k-n)$ when $k = n$ is $f(0)$, the value when $k = n+1$ is $f(1)$, etc. The z transform of $f(k-n)$ is

$$\begin{aligned} Z[f(k-n)] &= \frac{f(0)}{z^n} + \frac{f(1)}{z^{n+1}} + \frac{f(2)}{z^{n+2}} + \cdots \\ &= \frac{1}{z^n}\left[f(0) + \frac{f(1)}{z} + \frac{f(2)}{z^2} + \cdots\right] \\ &= z^{-n}F(z) \end{aligned} \quad (10.20)$$

When the function $f(k)$ of Fig. 10.3*a* is shifted one sampling period to the left, the function $f(k+1)$ shown in Fig. 10.3*c* results. The value of $f(k+1)$ when $k = 0$ is

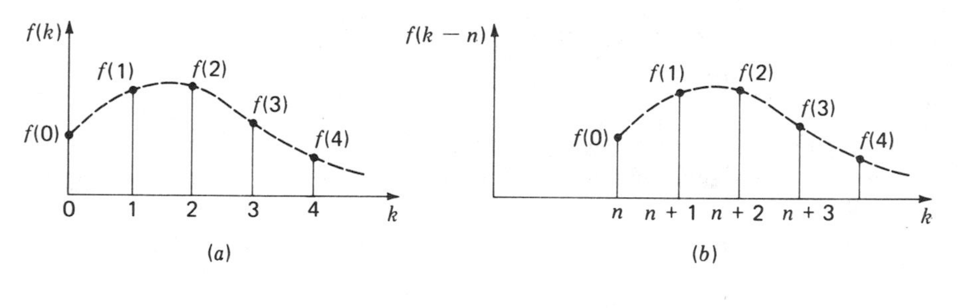

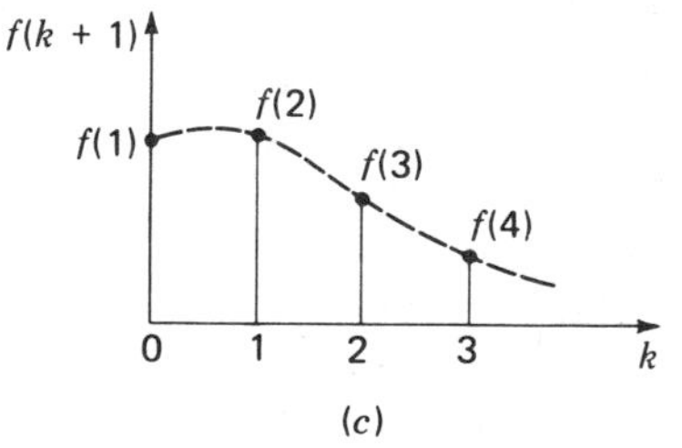

FIGURE 10.3
Translation of a discrete function $f(k)$.

$f(1)$, the value when $k = 1$ is $f(2)$, etc. The z transform of $f(k+1)$ is

$$Z[f(k+1)] = f(1) + \frac{f(2)}{z} + \frac{f(3)}{z^2} + \cdots$$

Multiplying through both sides by $1/z$ and adding $f(0)$ to both sides give

$$z^{-1}Z[f(k+1)] + f(0) = f(0) + \frac{f(1)}{z} + \frac{f(2)}{z^2} + \cdots = F(z)$$

Thus,

$$Z[f(k+1)] = zF(z) - zf(0)$$

Similarly, it follows that

$$Z[f(k+2)] = z^2F(z) - z^2f(0) - zf(1)$$

In general,

$$Z[f(k+n)] = z^nF(z) - z^nf(0) - z^{n-1}f(1) - \cdots - zf(n-1) \qquad (10.21)$$

As shown in the following, z transform theorems for $f(k)$ are quite similar to those for $f(t)$.

MULTIPLICATION BY a^k. The z transform of $a^kf(k)$ is

$$Z[a^kf(k)] = F\left(\frac{z}{a}\right) \qquad (10.22)$$

This theorem is readily proved by replacing z by z/a in Eq. (10.19). Thus

$$F\left(\frac{z}{a}\right) = f(0) + \frac{af(1)}{z} + \frac{a^2f(2)}{z^2} + \cdots$$

The right side is the z transform of $a^kf(k)$.

For a unit step function, $f(k) = 1$, and $F(z) = 1 + 1/z + 1/z^2 + \cdots = z/(z-1)$. Application of Eq. (10.22) shows that

$$Z[a^k] = F\left(\frac{z}{a}\right) = \frac{z/a}{z/a - 1} = \frac{z}{z-a} \qquad (10.23)$$

MULTIPLICATION BY k. The z transform of $kf(k)$ is

$$Z[kf(k)] = -z\frac{d}{dz}F(z) \qquad (10.24)$$

To verify this theorem, note that

$$-z\frac{d}{dz}F(z) = -z\frac{d}{dz}\left[f(0) + \frac{f(1)}{z} + \frac{f(2)}{z^2} + \frac{f(3)}{z^3} + \cdots\right]$$

$$= \frac{f(1)}{z} + \frac{2f(2)}{z^2} + \frac{3f(3)}{z^3} + \cdots$$

This result is the z transform of $kf(k)$.

TABLE 10.2
Properties of the z transform

$f(t)$ or $f(k)$	$Z[f(t)]$ or $Z[f(k)]$
$af(t)$ or $af(k)$	$aF(z)$
$f_1(t)+f_2(t)$ or $f_1(k)+f_2(k)$	$F_1(z)+F_2(z)$
$f(t-nT)$ or $f(k-n)$	$z^{-n}F(z)$
$f(t+T)$ or $f(k+1)$	$zF(z)-zf(0)$
$f(t+2T)$	$z^2F(z)-z^2f(0)-zf(T)$
$f(k+2)$	$z^2F(z)-z^2f(0)-zf(1)$
$f(t+nT)$	$z^nF(z)-z^nf(0)-z^{n-1}f(T)-\cdots-zf[(n-1)T]$
$f(k+n)$	$z^nF(z)-z^nf(0)-z^{n-1}f(1)-\cdots-zf(n-1)$
$tf(t)$	$-zT\frac{d}{dz}F(z)$
$kf(k)$	$-z\frac{d}{dz}F(z)$
$e^{-at}f(t)$	$F(ze^{aT})$
$e^{-ak}f(k)$	$F(ze^{a})$
$a^{t/T}f(t)$ or $a^kf(k)$	$F\left(\frac{z}{a}\right)$
$\frac{\partial}{\partial a}f(t,a)$	$\frac{\partial}{\partial a}F(z,a)$

To illustrate this theorem, consider the function $f(k)=a^k$ for which $F(z)=z/(z-a)$. Application of Eq. (10.24) gives

$$Z[ka^k]=-z\frac{d}{dz}\frac{z}{z-a}=\frac{az}{(z-a)^2} \tag{10.25}$$

A listing of z transform theorems and properties is given in Table 10.2.

10.3 INVERSE z TRANSFORMS

Inspection of Table 10.1 shows that z transforms tend to be more complicated than corresponding Laplace transforms. Fortunately, there are some relatively simple techniques for obtaining inverse z transforms.

Partial-Fraction Method

To illustrate this method, suppose we want to obtain the inverse z transform for the function

$$F(z)=\frac{(1-e^{-T})z}{(z-1)(z-e^{-T})} \tag{10.26}$$

Performing a partial-fraction expansion gives

$$F(z) = \left[\frac{1-e^{-T}}{(z-1)(z-e^{-T})}\right] z = \left(\frac{K_1}{z-1} + \frac{K_2}{z-e^{-T}}\right) z$$
$$= \frac{z}{z-1} - \frac{z}{z-e^{-T}}$$

From Table 10.1 the corresponding time function is

$$f(t) = 1 - e^{-t}$$

Thus, the inverse $f^*(t)$ is

$$f^*(t) = \sum_{n=0}^{\infty} (1 - e^{-nT})\delta(t - nT) \tag{10.27}$$

The z transform for a system may be written in the general form

$$C(z) = z\frac{A(z)}{B(z)} \tag{10.28}$$

For the case of distinct roots, $B(z)$ may be factored as

$$B(z) = (z - r_1)(z - r_2) \cdots (z - r_n)$$

The resulting partial-fraction expansion for $C(z)$ is

$$C(z) = z\left(\frac{K_1}{z-r_1} + \frac{K_2}{z-r_2} + \cdots + \frac{K_i}{z-r_i} + \cdots + \frac{K_n}{z-r_n}\right)$$

where

$$K_i = \lim_{s \to r_i} \left[(z - r_i)\frac{A(z)}{B(z)}\right]$$

Multiplying in the z factor gives

$$C(z) = \frac{zK_1}{z-r_1} + \frac{zK_2}{z-r_2} + \cdots + \frac{zK_i}{z-r_i} + \cdots + \frac{zK_n}{z-r_n} \tag{10.29}$$

Each of these terms has the same form. The inverse z transform of this form is obtained by inverting Eq. (10.23). That is,

$$Z^{-1}\left[\frac{z}{z-a}\right] = a^k$$

Thus, the inverse z transform of Eq. (10.29) is

$$c(k) = K_1 r_1^k + K_2 r_2^k + \cdots + K_i r_i^k + \cdots + K_n r_n^k \tag{10.30}$$

Illustrative example 10.10. The z transform for a digital control system is

$$C(z) = z\frac{z+2}{(z-1.0)(z+0.5)(z-0.2)}$$

Determine the inverse z transform.

Solution. The partial-fraction expansion is

$$C(z) = z\left(\frac{2.50}{z-1} + \frac{1.43}{z+0.5} - \frac{3.93}{z-0.2}\right)$$
$$= 2.50\,\frac{z}{z-1} + 1.43\,\frac{z}{z+0.5} - 3.93\,\frac{z}{z-0.2}$$

The inverse transform is

$$c(k) = 2.50 + 1.43(-0.5)^k - 3.93(0.2)^k$$

For a root r which is repeated twice, $B(z)$ has the form

$$B(z) = (z-r)^2\,(z-r_1)\,\cdots\,(z-r_{n-2})$$

The partial-fraction expansion for $C(z)$ is written as

$$C(z) = z\left[\frac{C_1 r}{(z-r)^2} + \frac{C_2}{z-r} + \frac{K_1}{z-r_1} + \cdots + \frac{K_{n-2}}{z-r_{n-2}}\right]$$
$$= \frac{zC_1 r}{(z-r)^2} + \frac{zC_2}{z-r} + \frac{zK_1}{z-r_1} + \cdots + \frac{zK_{n-2}}{z-r_{n-2}} \tag{10.31}$$

The inverse z transform of the first term is obtained by inverting Eq. (10.25).

$$Z^{-1}\left[\frac{za}{(z-a)^2}\right] = ka^k$$

The inverse z transform of Eq. (10.31) is

$$c(k) = (C_1 k + C_2)r^k + K_1 r_1^k + \cdots + K_{n-2} r_{n-2}^k \tag{10.32}$$

Note that the response for repeated roots has the same form as for continuous systems.

Illustrative example 10.11. The z transform for a digital control system is

$$C(z) = z\left[\frac{z-2}{(z-0.5)^2\,(z-1)}\right]$$

Determine the inverse z transform.

Solution. The partial-fraction expansion is

$$C(z) = z\left[\frac{3}{(z-0.5)^2} + \frac{4}{z-0.5} - \frac{4}{z-1.0}\right]$$
$$= \frac{6z(0.5)}{(z-0.5)^2} + \frac{4z}{z-0.5} - \frac{4z}{z-1.0}$$

Inverting yields

$$c(k) = (6k+4)\,(0.5)^k - 4$$

When $B(z)$ has complex conjugate roots $a \pm jb$, then $B(z)$ has the form

$$B(z) = (z^2 - 2az + a^2 + b^2)(z - r_1) \cdots (z - r_{n-2})$$

Thus,

$$C(z) = z \frac{A(z)}{[z - (a + jb)][z - (a - jb)](z - r_1) \cdots (z - r_{n-2})}$$

The partial-fraction expansion for $C(z)$ is written as

$$C(z) = z \left[\frac{K_c}{z - (a + jb)} + \frac{K_{-c}}{z - (a - jb)} + \frac{K_1}{z - r_1} + \cdots + \frac{K_{n-2}}{z - r_{n-2}} \right] \tag{10.33}$$

The partial-fraction constant K_c is

$$K_c = \lim_{z \to a + jb} [z - (a + jb)] \frac{A(z)}{[z - (a + jb)][z - (a - jb)](z - r_1) \cdots (z - r_{n-2})}$$

$$= \frac{1}{2jb} \left[\frac{A(z)}{(z - r_1) \cdots (z - r_{n-2})} \right]_{z = a + jb} = \frac{1}{2jb} K(a + jb) \tag{10.34}$$

where

$$K(a + jb) = \left[(z^2 - 2az + a^2 + b^2) \frac{A(z)}{B(z)} \right]_{z = a + jb}$$

The quantity in brackets is $A(z)/B(z)$ in which the quadratic term $(z^2 - 2az + a^2 + b^2)$ has been canceled from $B(z)$.

The constant K_{-c} is

$$K_{-c} = \lim_{z \to a - jb} [z - (a - jb)] \frac{A(z)}{[z - (a + jb)][z - (a - jb)](z - r_1) \cdots (z - r_{n-2})}$$

$$= -\frac{1}{2jb} \left[\frac{A(z)}{(z - r_1) \cdots (z - r_{n-2})} \right]_{z = a - jb} = -\frac{1}{2jb} K(a - jb) \tag{10.35}$$

where

$$K(a - jb) = \left[(z^2 - 2az + a^2 + b^2) \frac{A(z)}{B(z)} \right]_{z = a - jb}$$

The constants $K(a + jb)$ and $K(a - jb)$ are complex conjugates. As shown in Fig. 10.4, these constants may be expressed in the form

$$\begin{aligned} K(a + jb) &= |K(a + jb)| e^{j\alpha} \\ K(a - jb) &= |K(a + jb)| e^{-j\alpha} \end{aligned} \tag{10.36}$$

The length of both of these vectors is $|K(a + jb)| = |K(a - jb)|$. The angle of the $K(a + jb)$ vector is α, and the angle of the $K(a - jb)$ vector is $-\alpha$. The inverse z transform of Eq. (10.33) is

$$c(k) = K_c(a + jb)^k + K_{-c}(a - jb)^k + K_1 r_1^k + \cdots + K_{n-2} r_{n-2}^k \tag{10.37}$$

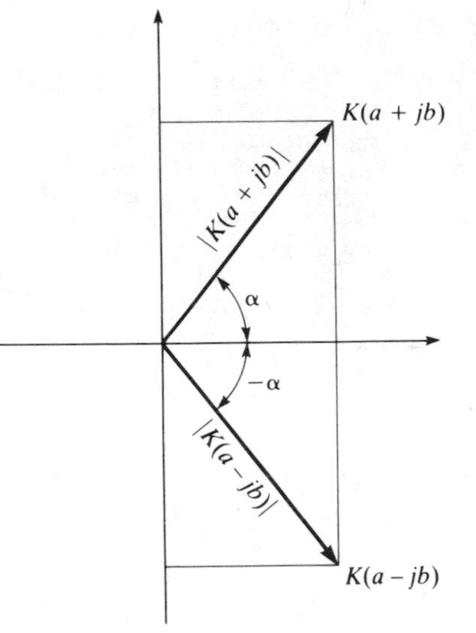

FIGURE 10.4
The vectors $K(a + jb)$ and $K(a - jb)$.

The vectors $a + jb$ and $a - jb$ are shown in Fig. 10.5. These vectors may be expressed in polar form as

$$a + jb = Re^{j\beta}$$

$$a - jb = Re^{-j\beta}$$

The length of both these vectors is $R = \sqrt{a^2 + b^2}$. The angle of the $a + jb$ vector is β, and the angle of the $a - jb$ vector is $-\beta$. Substitution of the polar representation

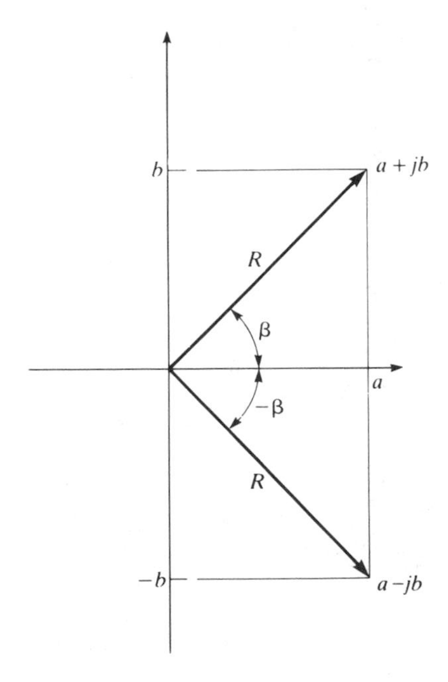

FIGURE 10.5
The vectors $a + jb$ and $a - jb$.

for $a + jb$ and $a - jb$ into Eq. (10.37) gives

$$c(k) = K_c R^k e^{jk\beta} + K_{-c} R^k e^{-jk\beta} + K_1 r_1^k + \cdots + K_{n-2} r_{n-2}^k$$

Substituting from Eqs. (10.34), (10.35), and (10.36) into the preceding yields

$$c(k) = R^k \frac{|K(a+jb)|}{b} \left(\frac{e^{j(k\beta+\alpha)} - e^{-j(k\beta+\alpha)}}{2j} \right) + K_1 r_1^k + \cdots + K_{n-2} r_{n-2}^k$$

$$= R^k \frac{|K(a+jb)|}{b} \sin(k\beta + \alpha) + K_1 r_1^k + \cdots + K_{n-2} r_{n-2}^k \qquad (10.38)$$

By noting that

$$|K(a+jb)| \sin(k\beta+\alpha) = |K(a+jb)| (\cos\alpha \sin k\beta + \sin\alpha \cos k\beta)$$
$$= A \sin k\beta + B \cos k\beta$$

where $A = |K(a+jb)| \cos\alpha$ is the horizontal component of the vector $K(a + jb)$ and $B = |K(a+jb)| \sin\alpha$ is the vertical component, then Eq. (10.38) may be written in the alternate form

$$c(k) = \frac{R^k}{b} (A \sin k\beta + B \cos k\beta) + K_1 r_1^k + \cdots + K_{n-2} r_{n-2}^k$$

Illustrative example 10.12. The z transform for a digital control system is

$$C(z) = z \frac{z - 0.7}{(z^2 + 0.6z + 0.25)(z + 0.5)}$$

Determine the inverse z transform.

Solution. For $-2a = 0.6$ and $a^2 + b^2 = 0.25$, then $a = -0.3$ and $b = \sqrt{0.25 - 0.09} = 0.4$. The polar form for $a + jb$ is $Re^{j\beta}$. That is, $a + jb = -0.3 + j0.4 = 0.5\underline{/126.9^\circ}$. Thus

$$R = 0.5 \qquad \text{and} \qquad \beta = 126.9^\circ$$

Evaluation of $K(a + jb)$ yields

$$K(a+jb) = \lim_{z \to -0.3+j0.4} (z^2 + 0.6z + 0.25) \left[\frac{z - 0.7}{(z^2 + 0.6z + 0.25)(z + 0.5)} \right]$$

$$= \left(\frac{z - 0.7}{z + 0.5} \right)_{z = -0.3 + j0.4} = \frac{-1.0 + j0.4}{0.2 + j0.4} = 2.41\underline{/94.8^\circ}$$

Thus,

$$|K(a+jb)| = 2.41 \qquad \text{and} \qquad \alpha = 94.8^\circ$$

The partial-fraction expansion for $C(z)$ is

$$C(z) = z \left[\frac{K_c}{z - (-0.3 + j0.4)} + \frac{K_{-c}}{z - (-0.3 - j0.4)} + \frac{K_1}{z + 0.5} \right]$$

Inverting yields the general form

$$c(k) = R^k \frac{|K(a+jb)|}{b} \sin(k\beta + \alpha) + K_1(-0.5)^k$$

The constant K_1 is

$$K_1 = \lim_{z \to -0.5} (z+0.5) \left[\frac{z-0.7}{(z^2+0.6z+0.25)(z+0.5)} \right] = \left(\frac{z-0.7}{z^2+0.6z+0.25} \right)_{z=-0.5}$$

$$= \frac{-1.2}{0.20} = -6.0$$

Substitution of the previously attained values into the general form yields for the solution

$$c(k) = (0.5)^k \frac{2.41}{0.4} \sin(126.9°k + 94.8°) - 6(-0.5)^k$$

Division Method

Dividing the numerator of $F(z)$ by the denominator yields a power series of the form

$$F(z) = C_0 + C_1 z^{-1} + C_2 z^{-2} + \cdots \tag{10.39}$$

The corresponding inverse is

$$f^*(t) = \sum_{n=0}^{\infty} C_n \delta(t - nT)$$

For example, the z transform for the function e^{-at} is $F(z) = z/(z - e^{-aT})$. Dividing the numerator of $F(z)$ by the denominator shows that

$$\begin{array}{r|l} & 1 + e^{-aT}z^{-1} + e^{-2aT}z^{-2} + \cdots \\ z - e^{-aT} & z \\ & \underline{z - e^{-aT}} \\ & e^{-aT} \\ & \underline{e^{-aT} - e^{-2aT}z^{-1}} \\ & e^{-2aT}z^{-1} \\ & \underline{e^{-2aT}z^{-1} + e^{-3aT}z^{-2}} \end{array}$$

Thus,

$$F(z) = \frac{z}{z - e^{-aT}} = 1 + e^{-aT}z^{-1} + e^{-2aT}z^{-2} + \cdots$$

The constant is the value of $f(t) = e^{-at}$ at time $t = 0$, the coefficient e^{-aT} of z^{-1} is the value of $f(t)$ at $t = T$, and the coefficient e^{-2aT} is the value of $f(t)$ at $t = 2T$. Inverting yields

$$f(nT) = \sum_{n=0}^{\infty} e^{-anT} \delta(t - nT)$$

Residue Method

To develop this method, first write the basic relationship for $F(z)$ in the form

$$F(z) = f(0) + \frac{f(T)}{z} + \cdots + \frac{f[(n-1)T]}{z^{n-1}} + \frac{f(nT)}{z^n} + \frac{f[(n+1)T]}{z^{n+1}} + \cdots$$

Multiplication by z^{n-1} gives

$$F(z)z^{n-1} = f(0)z^{n-1} + \cdots + f[(n-1)T] + \frac{f(nT)}{z} + \frac{f[(n+1)T]}{z^2} + \cdots$$

From complex variable theory it is known that the coefficient $f(nT)$ of the $1/z$ term of the preceding Laurent expansion is

$$\begin{aligned} f(nT) &= \frac{1}{2\pi j}\int_C F(z)z^{n-1}\,dz \\ &= \Sigma \text{ residues of } F(z)z^{n-1} \text{ at poles of } F(z)z^{n-1} \end{aligned} \tag{10.40}$$

The contour C is any closed path which encloses all poles of $F(z)z^{n-1}$.

Replacing the function $F(s)z/(z - e^{st})$ in Eqs. (10.9) and (10.10) by $F(z)z^{n-1}$ yields the following equations for determining the residues in Eq. (10.40). In particular, the residue due to a first-order pole at $z = r$ is

$$R = \lim_{z \to r} (z - r)\,[F(z)z^{n-1}] \tag{10.41}$$

Similarly, the residue due to a repeated pole of order q is

$$R = \frac{1}{(q-1)!} \lim_{z \to r} \frac{d^{q-1}}{dz^{q-1}} [(z-r)^q F(z)z^{n-1}] \tag{10.42}$$

Application of the residue method to determine the inverse of Eq. (10.26) gives

$$R_1 = \left[\frac{(1-e^{-T})z^n}{z - e^{-T}}\right]_{z=1} = 1$$

$$R_2 = \left[\frac{(1-e^{-T})z^n}{z - 1}\right]_{z=e^{-T}} = -e^{-nT}$$

Adding these residues gives $f(nT) = 1 - e^{-nT}$, which verifies the result given by Eq. (10.27).

As another example, determine the inverse z transform for the function

$$F(z) = \frac{Tz}{(z-1)^2}$$

This function has a second-order pole at $z = 1$; thus

$$R = \lim_{z \to 1} \frac{d}{dz} Tz^n = (nTz^{n-1})_{z=1} = nT$$

For $f(nT) = nT$, the corresponding time function is $f(t) = t$.

10.4 BLOCK-DIAGRAM ALGEBRA

In writing the transfer function for feedback control systems with sampling switches, one encounters some terms which are starred and others which are not. Thus, it is necessary to develop some mathematical techniques for handling such mixed terms. Figure 10.6*a* shows a sampling switch followed by a linear element whose transfer function is $G(s)$. The transformed equation for the output $Y(s)$ is

$$Y(s) = F^*(s)G(s) \tag{10.43}$$

For $0 < t < T$, the response $y(t)$ is that due to the first impulse at $t = 0$ of area $f(0)$. Thus, for this interval

$$y(t) = \mathscr{L}^{-1}[f(0)G(s)] = f(0)\mathscr{L}^{-1}[G(s)] = f(0)g(t) \tag{10.44}$$

where $g(t) = \mathscr{L}^{-1}[G(s)]$ is the response of the linear element to a unit impulse which occurs at $t = 0$.

For $T < t < 2T$, the response $y(t)$ is that due to the impulse at $t = 0$ plus that at $t = T$. For this interval, $F^*(s) = f(0) + f(T)e^{-Ts}$. Thus

$$Y(s) = [f(0) + f(T)e^{-Ts}]G(s)$$

Inverting gives

$$y(t) = f(0)g(t) + f(T)g(t - T) \tag{10.45}$$

where $g(t - T) = \mathscr{L}^{-1}[G(s)e^{-Ts}]$ is the response of the linear element to a unit impulse which occurs at time $t = T$. The response $y(t)$ for the interval $2T < t < 3T$ is

$$y(t) = f(0)g(t) + f(T)g(t - T) + f(2T)g(t - 2T)$$

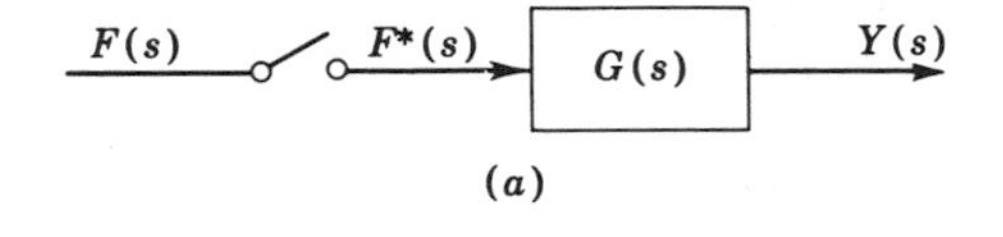

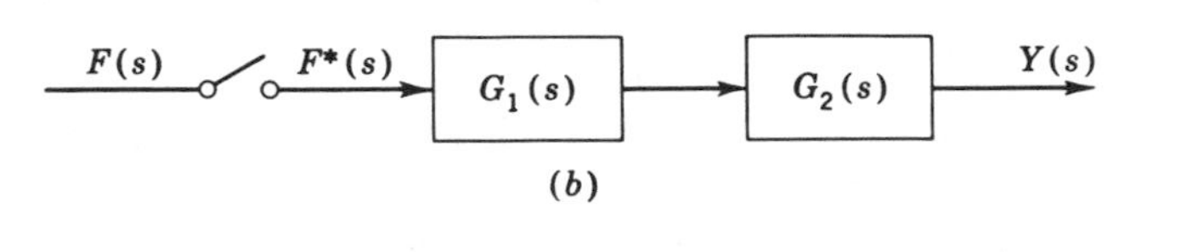

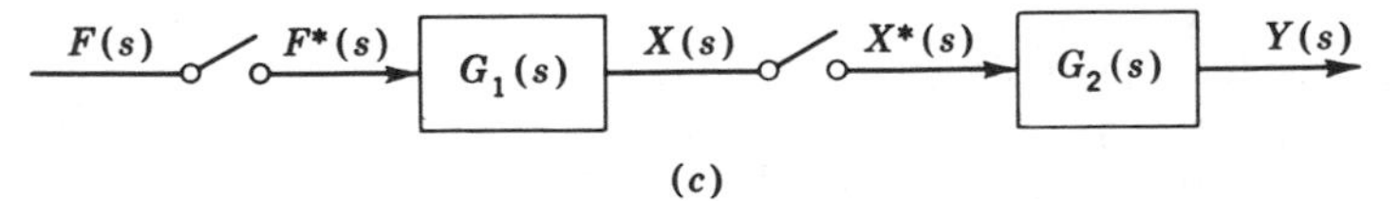

FIGURE 10.6
Sampler configurations.

In general, the response $y(t)$ is

$$y(t) = \sum_{n=0}^{\infty} f(nT)g(t - nT) \tag{10.46}$$

When n is such that $nT > t$, then $g(t - nT)$ is zero. That is, the impulse response is zero for negative time.

Taking the limit as t approaches 0 in Eq. (10.44), and the limit as t approaches T in Eq. (10.45), etc., yields the value of $y(nT)$ at the sampling instants.

$$\begin{aligned} y(0) &= f(0)g(0) \\ y(T) &= f(0)g(T) + f(T)g(0) \\ y(2T) &= f(0)g(2T) + f(T)g(T) + f(2T)g(0) \\ &\vdots \\ y(nT) &= f(0)g(nT) + f(T)g[(n-1)T] + \cdots \end{aligned} \tag{10.47}$$

Figure 10.7 shows a response function $y(t)$ which is discontinuous at the sampling instants. This occurs when the order of the denominator of $G(s)$ exceeds the order of the numerator by only 1. When the order of the denominator exceeds the order of the numerator by 2 or more, $y(t)$ is continuous. When $y(t)$ is continuous, Eq. (10.47) yields the values at the sampling instants. When $y(t)$ is discontinuous, Eq. (10.47) yields the values immediately after the sampling instants [that is, $y(0+)$, $y(T+)$, $y(2T+)$, ...]. Figure 10.1b shows that the values in Eq. (10.2) would also be the values immediately following the sampling instants if the function were discontinuous. Replacing f by y in Eq. (10.2) gives

$$Y^*(s) = y(0) + y(T)e^{-Ts} + y(2T)e^{-2Ts} + \cdots$$

Substitution of the values from Eq. (10.47) into the preceding expression gives

$$\begin{aligned} Y^*(s) &= f(0)[g(0) + g(T)e^{-Ts} + g(2T)e^{-2Ts} + \cdots] \\ &\quad + f(T)e^{-Ts}[g(0) + g(T)e^{-Ts} + g(2T)e^{-2Ts} + \cdots] \\ &\quad + f(2T)e^{-2Ts}[g(0) + g(T)e^{-Ts} + g(2T)e^{-2Ts} + \cdots] \\ &\quad + \cdots \\ &= [f(0) + f(T)e^{-Ts} + \cdots][g(0) + g(T)e^{-Ts} + \cdots] \end{aligned}$$

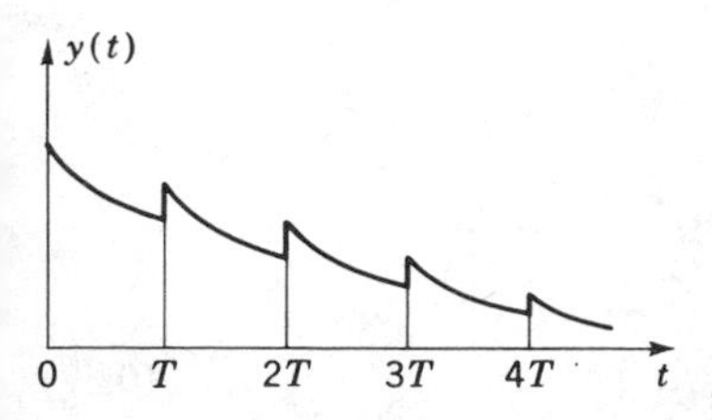

FIGURE 10.7
Response function that is discontinuous at the sampling instants.

Thus,

$$Y^*(s) = F^*(s)G^*(s) \tag{10.48}$$

The term $G^*(s)$ is called the *pulse transfer function* of the system.

Comparison of Eqs. (10.43) and (10.48) reveals a basic mathematical relationship for starring quantities. That is, starring both sides of Eq. (10.43) gives

$$[Y(s)]^* = Y^*(s)$$

and

$$[F^*(s)G(s)]^* = F^*(s)[G(s)]^* = F^*(s)G^*(s)$$

Letting $z = e^{sT}$ in Eq. (10.48) yields the z transform

$$Y(z) = F(z)G(z) \tag{10.49}$$

For the sampler configuration of Fig. 10.6*b*, the Laplace transform is

$$Y(s) = F^*(s)G_1(s)G_2(s)$$

Starring gives

$$Y^*(s) = F^*(s)[G_1(s)G_2(s)]^* = F^*(s)G_1G_2^*(s)$$

where

$$G_1G_2^*(s) = [G_1(s)G_2(s)]^*$$

The corresponding z transform is

$$Y(z) = F(z)G_1G_2(z) \tag{10.50}$$

Illustrative example 10.13. Determine the z transform for Fig. 10.6*b* when $G_1(s) = 1/s$ and $G_2(s) = 1/(s+1)$. The product $G_1(s)G_2(s)$ is

$$G_1(s)G_2(s) = \frac{1}{s(s+1)}$$

The z transform for this function is given by Eq. (10.6). That is,

$$G_1G_2(z) = \frac{z(1-e^{-T})}{(z-1)(z-e^{-T})} \tag{10.51}$$

For the sampler configuration shown in Fig. 10.6*c*, the Laplace relationships are

$$X(s) = F^*(s)G_1(s)$$
$$Y(s) = X^*(s)G_2(s)$$

Starring the first equation and then substituting this result for $X^*(s)$ into the second equation yield

$$Y(s) = F^*(s)G_1^*(s)G_2(s)$$

Starring gives

$$Y^*(s) = F^*(s)G_1^*(s)G_2^*(s)$$

The corresponding z transform is

$$Y(z) = F(z)G_1(z)G_2(z) \tag{10.52}$$

Illustrative example 10.14. Determine the z transform for Fig. 10.6c, when $G_1(s) = 1/s$ and $G_2(s) = 1/(s+1)$. From Eqs. (10.4) and (10.5), it follows that

$$G_1(z) = \frac{z}{z-1}$$

and

$$G_2(z) = \frac{z}{z-e^{-T}}$$

Thus,

$$G_1(z)G_2(z) = \frac{z^2}{(z-1)(z-e^{-T})} \tag{10.53}$$

From the preceding two examples note that

$$G_1G_2(z) \neq G_1(z)G_2(z) \tag{10.54}$$

The function $G_1G_2(z)$ is the z transform corresponding to the product $G_1(s)G_2(s)$, whereas the function $G_1(z)G_2(z)$ is the product of the z transform for $G_1(s)$ and the z transform for $G_2(s)$.

Two sampled-data feedback control systems are shown in Fig. 10.8. The general procedure for determining the transformed equation for a sampled-data system is as follows:

1. In addition to the actual system input $R(s)$, regard all switch outputs (starred quantities) as inputs.
2. In addition to the system output $C(s)$, regard all switch inputs as outputs.
3. Write equations for each output in terms of its inputs.
4. Star quantities as necessary in order to determine $C(z)$.

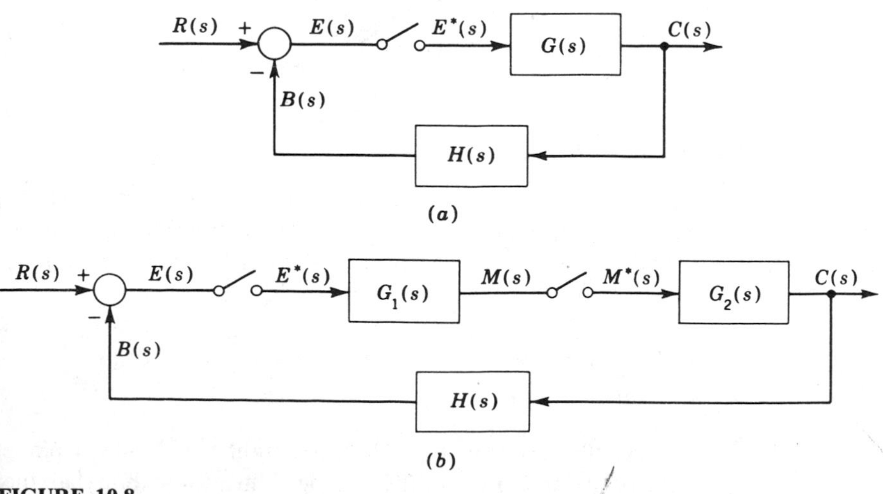

FIGURE 10.8
Sampled-data systems.

Application of this method to the system shown in Fig. 10.8*a* gives

$$C(s) = E^*(s)G(s)$$
$$E(s) = R(s) - E^*(s)G(s)H(s)$$

Starring gives

$$C^*(s) = E^*(s)G^*(s)$$
$$E^*(s) = R^*(s) - E^*(s)GH^*(s)$$

Solving the last equation for $E^*(s)$ and substituting into the first give

$$C^*(s) = \frac{G^*(s)}{1 + GH^*(s)} R^*(s)$$

The corresponding z transform is

$$C(z) = \frac{G(z)}{1 + GH(z)} R(z) \tag{10.55}$$

The equations relating the inputs and outputs of Fig. 10.8*b* are

$$C(s) = M^*(s)G_2(s)$$
$$M(s) = E^*(s)G_1(s)$$
$$E(s) = R(s) - M^*(s)G_2(s)H(s)$$

Starring all equations, then solving for $C^*(s)$ give

$$C^*(s) = \frac{G_1^*(s)G_2^*(s)}{1 + G_1^*(s)G_2H^*(s)} R^*(s)$$

The corresponding z transform is

$$C(z) = \frac{G_1(z)G_2(z)}{1 + G_1(z)G_2H(z)} R(z) \tag{10.56}$$

10.5 TRANSIENT RESPONSE

For continuous systems, it was found that a system is unstable if any root of the characteristic equation is in the right half of the s plane. This right half plane may be designated by $\sigma + j\omega$ in which $\sigma > 0$. The corresponding portion of the z plane is

$$z = e^{sT} = e^{\sigma T} e^{j\omega T}$$

The magnitude is

$$|z| = e^{\sigma T}$$

For $\sigma > 0$, then $|z| > 1$. As illustrated in Fig. 10.9, the right half of the s plane corresponds to the outside of the unit circle of the z plane. Thus, for stability, all the roots of the z-transformed characteristic equation must lie within the unit circle. The

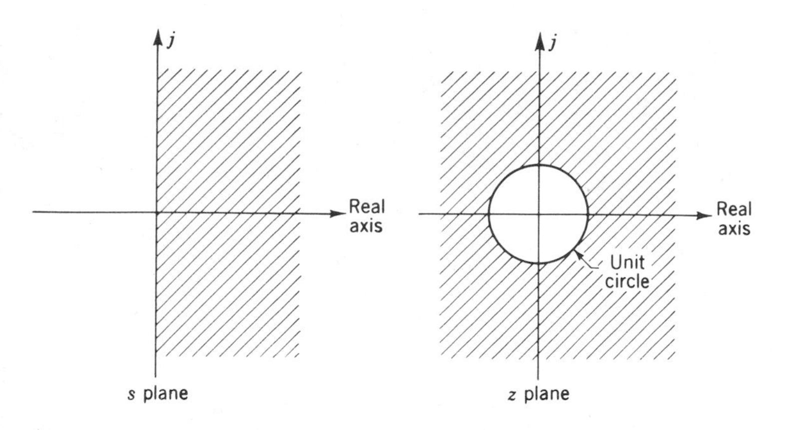

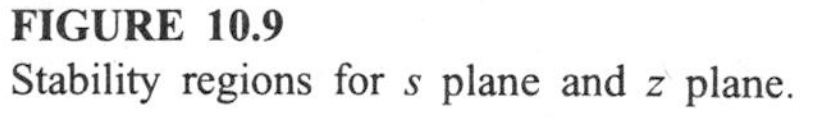
FIGURE 10.9
Stability regions for s plane and z plane.

general form of the response for distinct roots is given by Eq. (10.30). Note that if r_i is greater than 1, then r_i^k becomes infinite as k becomes infinite. Similarly, the response for the case of a repeated root is given by Eq. (10.32). Again, if r is greater than 1, the response becomes infinite as k becomes infinite. The response for the case of complex conjugate roots $a \pm jb$ is given by Eq. (10.38). The term $R = \sqrt{a^2 + b^2}$ is the distance from the origin to either root. When R is greater than 1, the roots lie outside the unit circle. For this case R^k becomes infinite as k becomes infinite. If any root of the characteristic equation lies outside the unit circle, the system is unstable.

Consider now the sampled-data system shown in Fig. 10.10. The z transform for the output $C(z)$ is

$$C(z) = \frac{G(z)}{1 + G(z)} R(z) = \frac{N_{G(z)}}{D_{G(z)} + N_{G(z)}} R(z) \tag{10.57}$$

where $G(z) = N_{G(z)}/D_{G(z)}$ in which $N_{G(z)}$ is the numerator of $G(z)$ and $D_{G(z)}$ is the denominator of $G(z)$. The z-transformed characteristic equation is

$$N_{G(z)} + D_{G(z)} = 0 \tag{10.58}$$

As was the case for continuous systems, the roots of the characteristic equation may be distinct, repeated, or complex conjugate. In Fig. 10.11, root r_1 is distinct, root r is repeated, and the roots $a \pm jb$ are complex conjugate.

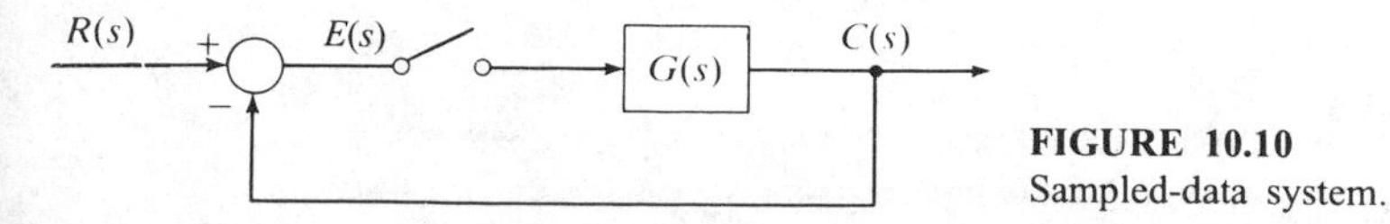

FIGURE 10.10
Sampled-data system.

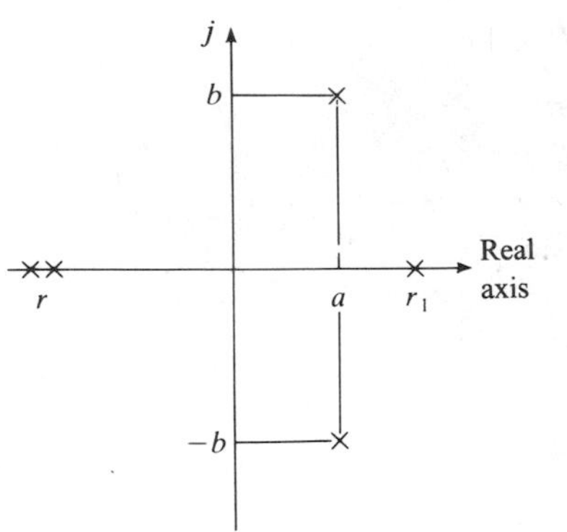

FIGURE 10.11
Distinct root r_1, repeated root r, and complex conjugate roots $a \pm jb$.

Figure 10.12 shows the digital control system for controlling the engine of the Boeing 767 and the McDonnell Douglas MD-11 commercial airplanes. The block-diagram representation is shown in Fig. 10.13. For this system

$$G(s) = \frac{K}{s(s+1)}$$

The partial-fraction expansion for $G(s)$ is

$$G(s) = K\left(\frac{1}{s} - \frac{1}{s+1}\right)$$

FIGURE 10.12
Digital engine control system. (Courtesy Allied-Signal Inc., Controls and Accessories.)

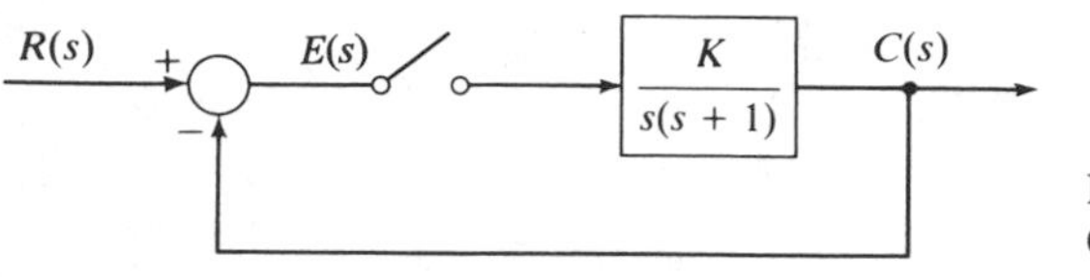

FIGURE 10.13
Computer-controlled system.

The corresponding z transform is

$$G(z) = K\left(\frac{z}{z-1} - \frac{z}{z-e^{-T}}\right) = K\,\frac{z(1-e^{-T})}{(z-1)(z-e^{-T})} \tag{10.59}$$

Thus, the characteristic equation for this system is

$$(z-1)(z-e^{-T}) + K(1-e^{-T})z = 0 \tag{10.60}$$

Routh's Criterion

To apply this criterion, it is necessary to transform the unit circle of the z plane to the vertical imaginary axis of the λ plane. This is accomplished by the transformation

$$\lambda = \frac{z+1}{z-1}$$

Solving for z gives

$$z = \frac{\lambda+1}{\lambda-1} \tag{10.61}$$

When the characteristic equation is expressed in terms of λ, then Routh's criterion may be applied in the same manner as for continuous systems.

Illustrative example 10.15. For a sampling period $T = 1$ s, determine the value of K such that the engine control system shown in Fig. 10.13 becomes unstable; that is, the roots of the characteristic equation lie on the unit circle of the z plane (i.e., the imaginary axis of the λ plane).

Solution. For $T = 1$, Eq. (10.60) becomes

$$(z-1)(z-0.368) + 0.632Kz = 0 \tag{10.62}$$

Using Eq. (10.61) to transform from the z plane to the λ plane gives

$$\frac{0.632K\lambda^2 + 1.264\lambda + (2.736 - 0.632K)}{(\lambda-1)^2} = 0$$

The Routh array is

$$\begin{array}{ccc} 0.632K & 2.736 - 0.632K & 0 \\ 1.264 & 0 & \\ 2.736 - 0.632K & & \end{array}$$

This system is unstable for $K < 0$ and for $K > 2.736/0.632 = 4.33$. The range of values of K such that this system is stable is

$$0 < K < 4.33$$

The characteristic equation for the continuous system is $s(s + 1) + K = 0$. The continuous system is stable for all positive values of K.

If in Illustrative example 10.15 the sampling rate is increased from 1 sample per second ($T = 1.0$) to 10 samples per second ($T = 0.1$), then the system will be unstable for $K > 40.1$. In general, making the sampling time shorter tends to make the system behave more as the corresponding continuous system does. Usually, stability is improved as the sampling rate is increased.

Root Locus

All the techniques for constructing root-locus plots for continuous systems (in the s plane) apply equally well to sampled-data systems (in the z plane). For example, for $T = 1$ the characteristic equation given by Eq. (10.60) becomes Eq. (10.62). This characteristic equation has an × at $z = 1$ and an × at $z = 0.368$. There is a ○ at $z = 0$. The complete root-locus plot may now be constructed as shown in Fig. 10.14. This root-locus plot crosses the unit circle at $z = -1$. The value of the gain ($0.632K$) at any point on a root-locus plot is the product of the distance from this point to each × divided by the distance from this point to each ○. Thus, the value of $0.632K$ at which the system becomes unstable is $0.632K = (1.368)(2)/(1)$, or

$$K = \frac{2.736}{0.632} = 4.33$$

This verifies the result ascertained by Routh's criterion.

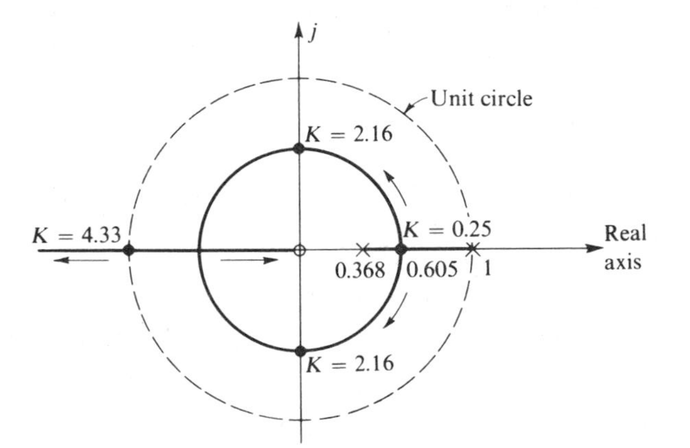

FIGURE 10.14
Root-locus plot for $(z - 1)(z - 0.368) + 0.632Kz = 0$.

Two horizontal lines of constant b are shown in the s plane of Fig. 10.15a. The corresponding paths in the z plane are radial straight lines.

$$z = e^{sT} = e^{(a \pm jb)T} = e^{aT} e^{\pm jbT}$$

The angle of inclination of these radial lines is $\theta = \pm bT$.

Two vertical lines of constant a are shown in the s plane of Fig. 10.15b. The corresponding paths in the z plane are circles of radius e^{aT}. For negative values of a,

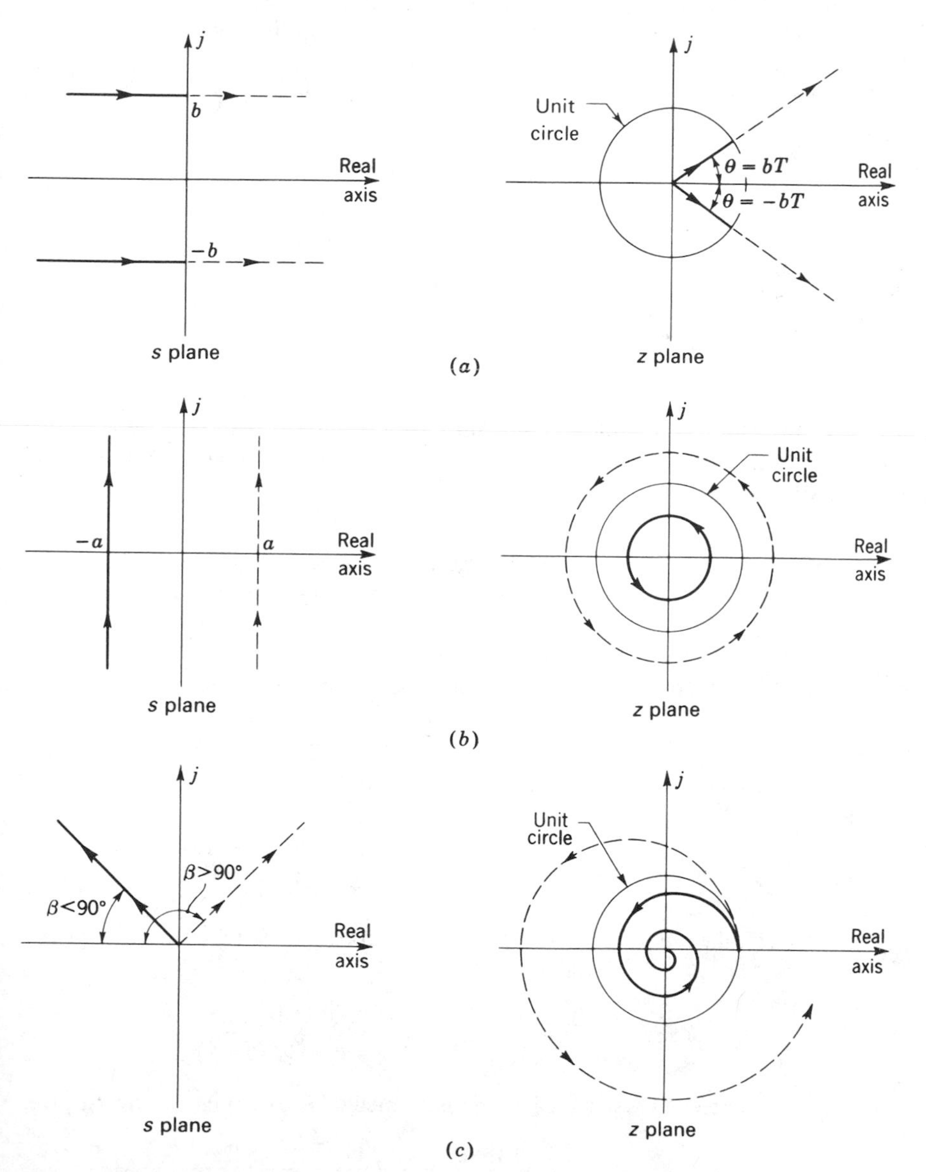

FIGURE 10.15
Corresponding paths in the s plane and z plane.

the circles are inside the unit circle of the z plane. For positive values of a, the circles lie outside the unit circle of the z plane.

Radial lines of constant damping ratio $\zeta = \cos\beta$ are shown in Fig. 10.15c. In polar coordinates, $s = a \pm jb = -\zeta\omega_n \pm j\omega_n\sqrt{1-\zeta^2}$. Thus

$$z = e^{-\zeta\omega_n T} \exp\left(\pm j\omega_n \sqrt{1-\zeta^2 T}\right)$$

The corresponding paths in the z plane are logarithmic spirals. For $\beta < 90°$ the spirals decay within the unit circle, and for $\beta > 90°$ the spirals grow outside the unit circle.

Consider now how a given point $z = re^{j\theta}$ in the z plane maps back into the s plane. For

$$z = re^{j\theta} = e^{sT} = e^{(a \pm jb)T}$$

$$\ln r + j\theta = aT \pm jbT \qquad -\pi < \theta < \pi$$

Equating real and imaginary parts shows that

$$\ln r = aT$$

$$\theta = \pm bT \qquad -\pi < \theta < \pi$$

This verifies the fact that a circle of constant radius r in the z plane is a vertical line of constant a in the s plane. Similarly, a ray at angle θ in the z plane is a horizontal line of constant b in the s plane.

Illustrative example 10.16. For the computer-controlled system shown in Fig. 10.13, determine the response to a unit step-function excitation for the case in which $T = 1$ and $K = 0.25$.

Solution. Substitution of these values into Eq. (10.59) gives

$$G(z) = \frac{0.158z}{(z-1)(z-0.368)}$$

Substitution into Eq. (10.57) gives

$$C(z) = \frac{G(z)}{1+G(z)} R(z) = \frac{0.158z}{(z-1)(z-0.368)+0.158z} R(z) \tag{10.63}$$

For $K = 0.25$, the root-locus plot of Fig. 10.14 shows that the characteristic equation has a repeated root at $z = 0.605$. Thus

$$C(z) = \frac{0.158z}{(z-0.605)^2} R(z) = \frac{0.158z^2}{(z-0.605)^2(z-1)}$$

where $R(z) = z/(z-1)$. Using the long-division method to determine the inverse gives

$$z^3 - 2.21z^2 + 1.58z - 0.368 \,\overline{)\, 0.158z^2} \quad \text{quotient: } 0.158z^{-1} + 0.349z^{-2} + 0.522z^{-3} + \cdots$$

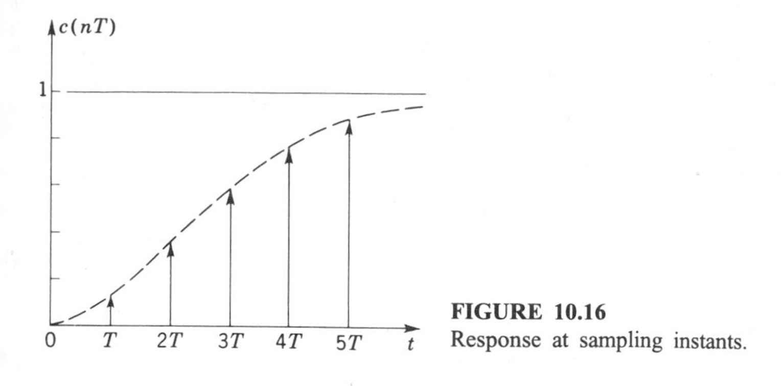

FIGURE 10.16
Response at sampling instants.

Because

$$C(z) = c(0) + c(T)z^{-1} + c(2T)z^{-2} + c(3T)z^{-3} + \cdots$$

then

$$\begin{aligned} c(0) &= 0 \\ c(T) &= 0.158 \\ c(2T) &= 0.349 \\ c(3T) &= 0.522 \end{aligned} \tag{10.64}$$

A plot of the response $c(nT)$ at the sampling instants is shown in Fig. 10.16.

The long-division method becomes quite cumbersome for computing $c(nT)$ for larger values of n. A more convenient procedure results from expressing the solution in the form of a difference equation.

Difference Equations

To determine the inverse z transform by this method, write the equation for $C(z)$, Eq. (10.63), in the form

$$C(z) = \frac{0.158z}{(z-1)(z-0.368) + 0.158z} R(z) = \frac{0.158z}{z^2 - 1.21z + 0.368} R(z)$$

Thus

$$C(z) - 1.21z^{-1}C(z) + 0.368z^{-2}C(z) = 0.158z^{-1}R(z)$$

Application of Eq. (10.20) to invert the preceding expression yields directly the difference equation

$$c(k) = 1.21c(k-1) - 0.368c(k-2) + 0.158r(k-1)$$

This difference equation gives the value $c(k)$ at the kth sampling instant in terms of values at preceding sampling instants. Application of this result to obtain the values at the sampling instants gives

$$c(0) = 0$$
$$c(1) = 0.158r(0) = 0.158$$
$$c(2) = 1.21c(1) + 0.158r(1) = 0.349$$
$$c(3) = 1.21c(2) - 0.368c(1) + 0.158r(2) = 0.522$$

Such recurrence relationships lend themselves very well to solution by a digital computer.

The response $c(k)$ at the sampling instants may also be obtained by performing a partial-fraction expansion and then inverting. Thus

$$C(z) = z\left[\frac{0.158z}{(z-1)(z-0.605)^2}\right] = z\left[\frac{K_1}{z-1} + \frac{C_1}{(z-0.605)^2} + \frac{C_2}{z-0.605}\right]$$

The partial-fraction expansion constants are $K_1 = 1.0$, $C_1 = -0.24 = -(0.39)(0.605)$, and $C_2 = -1.0$. Thus, $C(z)$ becomes

$$C(z) = \frac{z}{z-1} - \frac{z}{z-0.605} - 0.39\frac{0.605z}{(z-0.605)^2}$$

By noting that

$$Z^{-1}\left[\frac{z}{z-1}\right] = 1 \qquad Z^{-1}\left[\frac{z}{z-a}\right] = a^k \qquad Z^{-1}\left[\frac{az}{(z-a)^2}\right] = ka^k$$

the inverse is found to be

$$c(k) = 1.0 - (1.0 + 0.39k)(0.605^k)$$

With this method, the value of $c(k)$ at any sampling instant k may be calculated directly without the need to compute the value at all the preceding instants.

10.6 FILTERS

Sampled-data systems usually incorporate a filter, as illustrated in Fig. 10.17. A perfect filter would convert the sampled signal $f^*(t)$ back to the continuous input $f(t)$. That is, the output $y(t)$ of the filter would equal $f(t)$. If such a perfect filter were possible, then the sampled-data system would behave the same as the continuous system.

Zero-Order Hold

The most commonly used filter is that in which the value of the last sample is retained until the next sample is taken. This type of filter is called a *zero-order hold*,

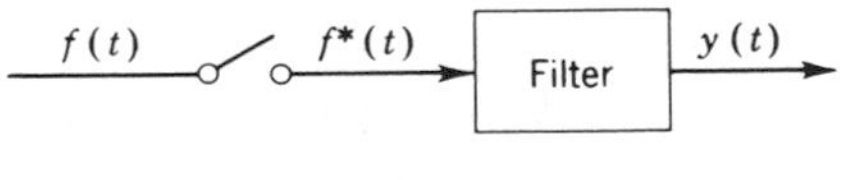

f(t)
f*(t)
Filter
y(t)

FIGURE 10.17
Schematic representation of a sampler and filter.

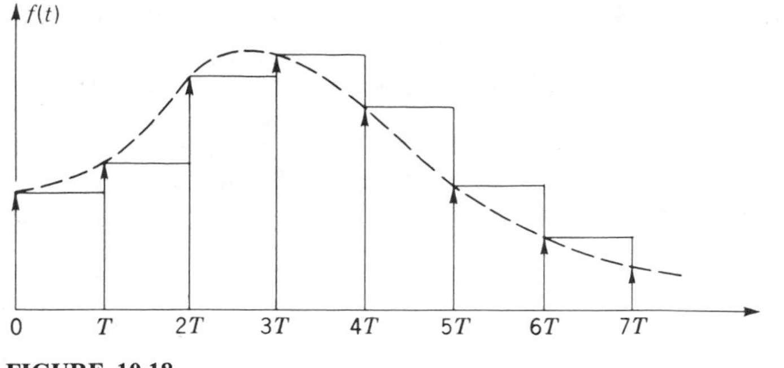

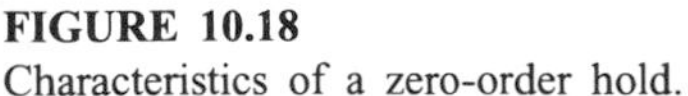
FIGURE 10.18
Characteristics of a zero-order hold.

or *boxcar generator*. The dashed curve in Fig. 10.18 represents the continuous function $f(t)$. The vertical arrows at the sampling instants are the impulses which represent the sampled signal $f^*(t)$. Because the zero-order hold retains the value of $f(t)$ at each sampling instant, $y(t)$ is the series of steps shown in Fig. 10.18. The equation for this series of steps (i.e., pulse functions) is

$$y(t) = f(0)[u(t) - u(t-T)] + f(T)[u(t-T) - u(t-2T)] + f(2T)[u(t-2T) - u(t-3T)] + \cdots$$

The Laplace transform is

$$Y(s) = f(0)\frac{1-e^{-Ts}}{s} + f(T)\frac{e^{-Ts} - e^{-2Ts}}{s} + f(2T)\frac{e^{-2Ts} - e^{-3Ts}}{s} + \cdots$$
$$= \frac{1-e^{-Ts}}{s}[f(0) + f(T)e^{-Ts} + f(2T)e^{-2Ts} + \cdots] = \frac{1-e^{-Ts}}{s}F^*(s)$$

This result shows that the Laplace transform for a zero-order hold is

$$\frac{1-e^{-Ts}}{s} \tag{10.65}$$

Suppose that a zero-order hold is added to the system of Fig. 10.13. The new block diagram for this system is shown in Fig. 10.19. The transfer function for the hold is included in the overall transfer function $G(s)$. Thus,

$$G(s) = (1 - e^{-Ts})\frac{K}{s^2(s+1)} \tag{10.66}$$

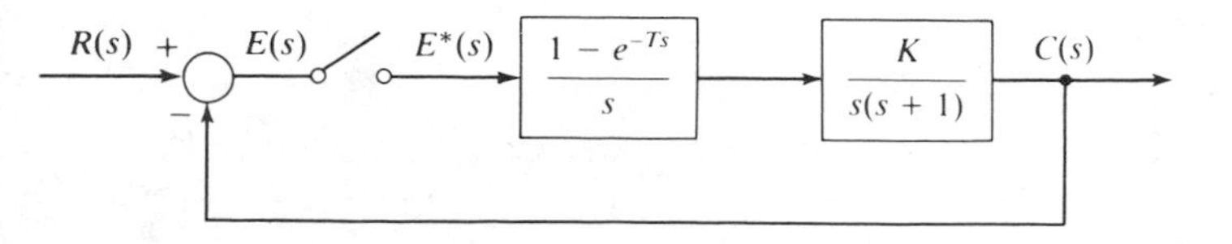

FIGURE 10.19
Sampled-data system with a zero-order hold.

To determine $G(z)$ when $G(s)$ contains a $1 - e^{-Ts}$ factor, first write Eq. (10.66) in the form

$$G(s) = G_1(s)G_2(s) \tag{10.67}$$

where $G_1(s) = 1 - e^{-Ts}$ and $G_2(s)$ is the remaining portion of $G(s)$. The function $G_1(s)$ is the Laplace transform of a unit impulse at the origin and a negative unit impulse at $t = T$. The corresponding time function $g_1(t)$ is shown in Fig. 10.20. Because this time function $g_1(t)$ exists only at the sampling instants, the sampled function $g_1^*(t)$ will be the same as $g_1(t)$. Thus,

$$G_1(s) = G_1^*(s)$$

Substitution of this result into Eq. (10.67) shows that

$$G(s) = G_1^*(s)G_2(s)$$

Starring gives

$$G^*(s) = G_1^*(s)G_2^*(s)$$

The corresponding z transform is

$$\begin{aligned} G(z) &= G_1(z)G_2(z) = (1 - z^{-1})G_2(z) \\ &= \frac{z-1}{z} G_2(z) \end{aligned} \tag{10.68}$$

For the case of Eq. (10.66),

$$G_2(s) = \frac{K}{s^2(s+1)} = K\left(\frac{1}{s^2} - \frac{1}{s} + \frac{1}{s+1}\right)$$

Thus,

$$G_2(z) = K\left[\frac{Tz}{(z-1)^2} - \frac{z}{z-1} + \frac{z}{z - e^{-T}}\right]$$

Substitution of this result into Eq. (10.68) gives

$$G(z) = K\left(\frac{T}{z-1} - 1 + \frac{z-1}{z-e^{-T}}\right) = K\left(\frac{T}{z-1} + \frac{e^{-T}-1}{z-e^{-T}}\right)$$

For $T = 1$, then $G(z)$ becomes

$$G(z) = \frac{0.368K(z + 0.717)}{(z-1)(z-0.368)} \tag{10.69}$$

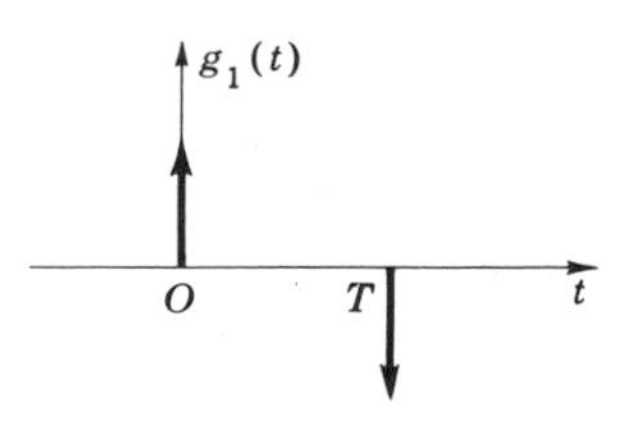

FIGURE 10.20
Time function $g_1(t) = \mathcal{L}^{-1}[G_1(s)] = \mathcal{L}^{-1}(1 + e^{-Ts})$.

The corresponding characteristic equation for this sampled-data system is

$$D_{G(z)} + N_{G(z)} = (z - 1)(z - 0.368) + 0.368K(z + 0.717) = 0 \qquad (10.70)$$

Replacing z by $(\lambda + 1)/(\lambda - 1)$ so that Routh's criterion may be applied gives

$$0.632K\lambda^2 + (1.264 - 0.528K)\lambda + (2.736 - 0.104K) = 0$$

The Routh array is

$$\begin{array}{cc} 0.632K & 2.736 - 0.104K \\ 1.264 - 0.528K & 0 \\ 2.736 - 0.104K & \end{array}$$

This system becomes unstable for $K < 0$ and for $K > 1.264/0.528 = 2.39$. Thus, the range of values of K for which the system is stable is

$$0 < K < 2.39$$

From the characteristic equation, Eq. (10.70), the root-locus plot may be constructed as shown in Fig. 10.21. The root-locus plot crosses the unit circle at $K = 2.39$. Repeated roots $z = 0.65$ occur at $K = 0.2$.

Now let us determine the response of this system to a unit step function for the case in which $K = 1$. From Eq. (10.69), it follows that

$$G(z) = \frac{0.368(z + 0.717)}{(z - 1)(z - 0.368)} = \frac{0.368(z + 0.717)}{z^2 - 1.368z + 0.368}$$

The z transform for the output is

$$C(z) = \frac{G(z)}{1 + G(z)} R(z)$$

$$= \frac{N_{G(z)}}{D_{G(z)} + N_{G(z)}} R(z) = \frac{0.368(z + 0.717)R(z)}{(z^2 - 1.368z + 0.368) + 0.368(z + 0.717)}$$

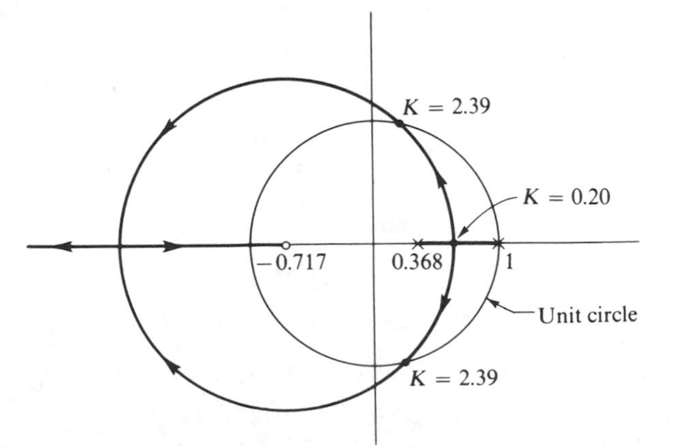

FIGURE 10.21
Root-locus plot for $(z - 1)(z - 0.368) + 0.368K(z + 0.717) = 0$.

Thus,

$$C(z) - 1.0z^{-1}C(z) + 0.632z^{-2}C(z) = 0.368z^{-1}R(z) + 0.264z^{-2}R(z)$$

Application of Eq. (10.20) yields the desired computational form

$$c(k) = c(k-1) - 0.632c(k-2) + 0.368r(k-1) + 0.264r(k-2)$$

Substitution of $c(k) = r(k) = 0$ for $k < 0$ and $r(k) = 1$ for $k \geq 0$ yields the following values for $c(k)$ at the sampling instants:

$$\begin{aligned} c(0) &= 0 \\ c(1) &= 0.368r(0) = 0.368 \\ c(2) &= c(1) - 0.632c(0) + 0.368r(1) + 0.264r(0) = 1.00 \\ c(3) &= c(2) - 0.632c(1) + 0.368r(2) + 0.264r(1) = 1.40 \end{aligned} \tag{10.71}$$

To determine the response $c(k)$ at the sampling instants by the partial-fraction expansion method, first write $C(z)$ in the form

$$C(z) = \frac{0.368(z + 0.717)R(z)}{z^2 - z + 0.632}$$

For $R(z) = z/(z - 1)$, then

$$\begin{aligned} C(z) &= z\,\frac{0.368(z+0.717)}{(z^2 - z + 0.632)\,(z-1)} \\ &= z\left[\frac{K_c}{z - (0.50 + j0.62)} + \frac{K_{-c}}{z - (0.50 - j0.62)} + \frac{K_1}{z-1}\right] \end{aligned}$$

Note that the quadratic term has complex conjugate roots $a \pm jb = 0.50 \pm j0.62$. Inverting yields the general form for the solution

$$c(k) = R^k\,\frac{|K(a+jb)|}{b}\,\sin(k\beta + \alpha) + K_1$$

where

$$K_1 = \left[\frac{0.368(z+0.717)}{z^2 - z + 0.632}\right]_{z=1} = 1.0$$

$$a + jb = 0.50 + j0.62 = 0.80\underline{/51.0^\circ} = R\underline{/\beta}$$

$$\begin{aligned} K(a+jb) = |K(a+jb)|\underline{/\alpha} &= \left[\frac{0.368(z+0.717)}{z-1}\right]_{z=0.50+j0.62} \\ &= \frac{0.368(1.217 + j0.62)}{-0.50 + j0.62} = \frac{0.368(1.366)\underline{/27.0^\circ}}{0.80\underline{/129.0^\circ}} \\ &= 0.63\underline{/-102.0^\circ} \end{aligned}$$

Thus

$$c(k) = 0.80^k \frac{0.63}{0.62} \sin(51.0k - 102.0) + 1$$
$$= 1 + 1.02(0.80)^k \sin(51.0k - 102.0)$$

Values of $c(k)$ are the same as those given by Eq. (10.71).

The time at the kth sampling instant is $t = kT$. Replacing $c(k)$ by $c(kT)$ yields values in terms of time.

First-Order Hold

The characteristics of a first-order hold are illustrated in Fig. 10.22*a*. For the region $nT < t < (n + 1)T$ the output is the straight line that is the extrapolation of the two preceding sampled values. That is,

$$f(t) = f(nT) + f'(nT)\,(t - nT) \qquad nT < t < (n + 1)T$$

where the slope $f'(nT)$ is

$$f'(nT) = \frac{f(nT) - f[(n-1)T]}{T}$$

The behavior of the first-order hold is illustrated in Fig. 10.22*b*. The dashed curve is the continuous function $f(t)$. The output of the first-order hold is the series of straight-line segments. For a first-order hold, the Laplace transform is

$$\left(\frac{1}{s} + \frac{1}{Ts^2}\right)(1 - e^{-Ts})$$

The entire class of higher-order hold filters is revealed by writing the power series approximation for $f(t)$ between the sampling instants $nT < t < (n + 1)T$. Thus

$$f(t) = f(nT) + f'(nT)\,(t - nT) + \frac{f''(nT)}{2!}(t - nT)^2 + \cdots$$

A zero-order hold approximates $f(t)$ by the first term, a first-order hold approximates $f(t)$ by the first two terms, a second-order hold approximates $f(t)$ by the

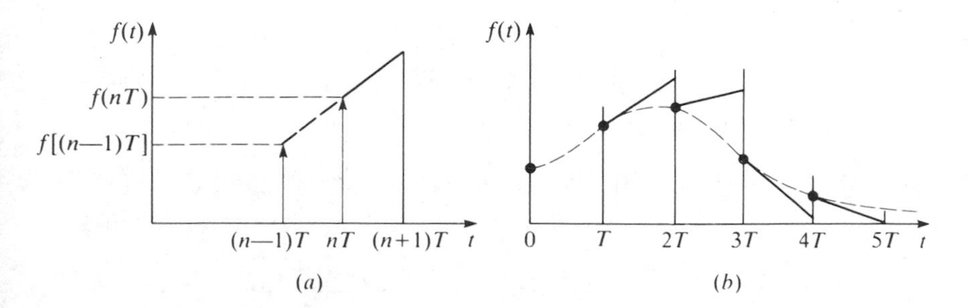

FIGURE 10.22
Characteristics and behavior of a first-order hold.

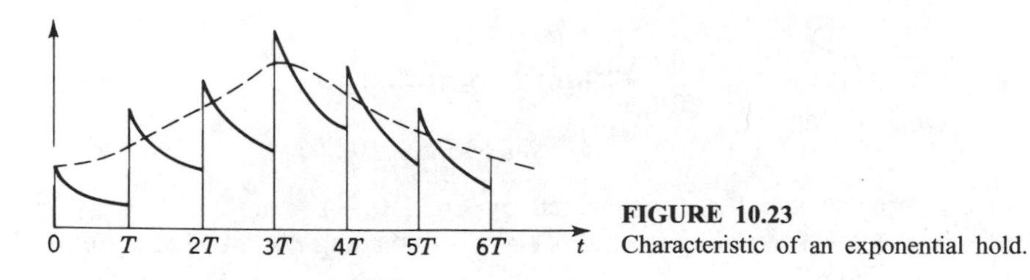

FIGURE 10.23
Characteristic of an exponential hold.

first three terms, etc. Because of practical considerations, first-order and higher-order holds are seldom used. By far, the zero-order hold is the most commonly used filter.

Exponential Hold

One of the simplest filters is the exponential hold. The transfer function for an exponential hold is

$$G(s) = \frac{1}{1 + \tau s}$$

Many components have this transfer function and thus may be used as exponential-hold filters. The output from an exponential hold is an exponential decay between sampling periods, as illustrated in Fig. 10.23. At each sampling instant, there is a step change or discontinuity between successive exponential decays. The amount of the step change at each sampling instant is equal to the value (area) of the impulse at the particular sampling instant.

10.7 DISCRETE DATA SYSTEMS

When a system deals strictly with discrete data, the system can be described by the following set of discrete state equations:

$$\mathbf{x}[(k+1)T] = A\mathbf{x}(kT) + \mathbf{b}f(kT) \tag{10.72}$$

Application of Eqs. (10.19) and (10.21) yields for the transform

$$z\mathbf{X}(z) - z\mathbf{x}(0) = A\mathbf{X}(z) + \mathbf{b}F(z)$$

or

$$[zI - A]\mathbf{X}(z) = z\mathbf{x}(0) + \mathbf{b}F(z)$$

Premultiplying both sides by $[zI - A]^{-1}$ gives

$$\mathbf{X}(z) = [zI - A]^{-1}z\mathbf{x}(0) + [zI - A]^{-1}\mathbf{b}F(z) = [zI - A]^{-1}[z\mathbf{x}(0) + \mathbf{b}F(z)] \tag{10.73}$$

Inverting yields

$$\begin{aligned}\mathbf{x}(kT) &= Z^{-1}\left([zI - A]^{-1}z\right)\mathbf{x}(0) + Z^{-1}\left\{[zI - A]^{-1}\mathbf{b}F(z)\right\} \\ &= Z^{-1}\left\{[zI - A]^{-1}[z\mathbf{x}(0) + \mathbf{b}F(z)]\right\}\end{aligned} \tag{10.74}$$

Discrete Data Representation for Continuous Systems

The state-space representation for a continuous system is

$$\dot{\mathbf{x}}(t) = A\mathbf{x}(t) + \mathbf{b}f(t)$$

The solution referred to time $t = 0$ is given by Eq. (9.59). That is,

$$\mathbf{x}(t) = e^{At}\mathbf{x}(0) + \int_0^t e^{-A(t-\tau)}\mathbf{b}f(\tau)\, d\tau \tag{9.59}$$

This is the state transition equation for a continuous system.

For the case in which the initial state is known at time $t = t_0$ rather than at time $t = 0$, the solution is obtained by first writing the state-space representation in the form

$$\dot{\mathbf{x}}(t) - A\mathbf{x}(t) = \mathbf{b}f(t)$$

Multiplying through by e^{-At} gives

$$e^{-At}[\dot{\mathbf{x}}(t) - A\mathbf{x}(t)] = \frac{d}{dt}\left[e^{-At}\mathbf{x}(t)\right] = e^{-At}\mathbf{b}f(t)$$

Integration between the limits t_0 and t of the two expressions on the right side of the preceding equation shows that

$$e^{-At}\mathbf{x}(t) - e^{-At_0}\mathbf{x}(t_0) = \int_{t_0}^t e^{-A\tau}\mathbf{b}f(\tau)\, d\tau$$

Multiplying through by e^{At} and solving for $\mathbf{x}(t)$ yield the desired solution referred to time t_0.

$$\mathbf{x}(t) = e^{A(t-t_0)}\mathbf{x}(t_0) + \int_{t_0}^t e^{A(t-\tau)}\mathbf{b}f(\tau)\, d\tau \tag{10.75}$$

For $t_0 = kT$ and $t = (k+1)T$, the preceding becomes

$$\mathbf{x}[(k+1)T] = e^{AT}\mathbf{x}(kT) + \int_{kT}^{(k+1)T} e^{A[(k+1)T-\tau]}\mathbf{b}f(\tau)\, d\tau$$

Assuming that $f(\tau)$ remains constant over the interval $kT < t < (k+1)T$, then $f(\tau)$ may be replaced by $f(kT)$. Letting $\lambda = (k+1)T - \tau$, in which case $d\tau = -d\lambda$, we see that

$$\mathbf{x}[(k+1)T] = e^{AT}\mathbf{x}(kT) + \int_0^T e^{A\lambda}\, d\lambda\, \mathbf{b}f(kT)$$

This result may be expressed in the form

$$\mathbf{x}[(k+1)T] = \mathbf{\Phi}(T)\mathbf{x}(kT) + G(T)f(kT) \tag{10.76}$$

where

$$\mathbf{\Phi}(T) = e^{AT} \qquad \text{and} \qquad G(T) = \left(\int_0^T e^{A\lambda}\, d\lambda\right)\mathbf{b}$$

This is the discrete data representation for the continuous system $\dot{\mathbf{x}} = A\mathbf{x} + \mathbf{b}f(t)$. Note that it has the same form as Eq. (10.72), in which $\mathbf{\Phi}(T)$ corresponds to A and $G(T)$ corresponds to $\mathbf{b}$.

Illustrative example 10.17. Obtain the discrete time state-space representation for the differential equation

$$\ddot{c} + 3\dot{c} + 2c = f(t)$$

Evaluate the resulting discrete time representation for a sampling period $T = 0.2$ s.

Solution. Letting $c = x_1$ and $\dot{c} = \dot{x}_1 = x_2$ yields the state-space representation

$$\begin{bmatrix} \dot{x}_1 \\ \dot{x}_2 \end{bmatrix} = \begin{bmatrix} 0 & 1 \\ -2 & -3 \end{bmatrix} \begin{bmatrix} x_1 \\ x_2 \end{bmatrix} + \begin{bmatrix} 0 \\ 1 \end{bmatrix} f(t)$$

The $\mathbf{\Phi}(t) = e^{At}$ matrix is

$$\mathbf{\Phi}(t) = \mathscr{L}^{-1}[sI - A]^{-1} = \mathscr{L}^{-1}\begin{bmatrix} s & -1 \\ 2 & s+3 \end{bmatrix}^{-1} = \mathscr{L}^{-1}\frac{\begin{bmatrix} s+3 & 1 \\ -2 & s \end{bmatrix}}{s^2 + 3s + 2}$$

$$= \begin{bmatrix} 2e^{-t} - e^{-2t} & e^{-t} - e^{-2t} \\ -2e^{-t} + 2e^{-2t} & -e^{-t} + 2e^{-2t} \end{bmatrix}$$

Replacing t by T yields the matrix $\mathbf{\Phi}(T)$. The $G(T)$ matrix is obtained as follows:

$$G(T) = \begin{bmatrix} \int_0^T (2e^{-\lambda} - e^{-2\lambda})\, d\lambda & \int_0^T (e^{-\lambda} - e^{-2\lambda})\, d\lambda \\ \int_0^T (-2e^{-\lambda} + 2e^{-2\lambda})\, d\lambda & \int_0^T (-e^{-\lambda} + 2e^{-2\lambda})\, d\lambda \end{bmatrix} \begin{bmatrix} 0 \\ 1 \end{bmatrix}$$

$$= \begin{bmatrix} \dfrac{1 - 2e^{-T} + e^{-2T}}{2} \\ e^{-T} - e^{-2T} \end{bmatrix}$$

The substitution of these results into Eq. (10.76) yields

$$\begin{bmatrix} x_1[(k+1)T] \\ x_2[(k+1)T] \end{bmatrix} = \begin{bmatrix} 2e^{-T} - e^{-2T} & e^{-T} - e^{-2T} \\ -2e^{-T} + 2e^{-2T} & -e^{-T} + 2e^{-2T} \end{bmatrix} \begin{bmatrix} x_1(kT) \\ x_2(kT) \end{bmatrix}$$
$$+ \begin{bmatrix} \dfrac{1 - 2e^{-T} - e^{-2T}}{2} \\ e^{-T} - e^{-2T} \end{bmatrix} f(kT)$$

Evaluation at $T = 0.2$ gives

$$\begin{bmatrix} x_1[(k+1)T] \\ x_2[(k+1)T] \end{bmatrix} = \begin{bmatrix} 0.968 & 0.149 \\ -0.298 & 0.521 \end{bmatrix} \begin{bmatrix} x_1(kT) \\ x_2(kT) \end{bmatrix} + \begin{bmatrix} 0.016 \\ 0.149 \end{bmatrix} f(kT)$$

Note that the form of this result is identical to Eq. (10.72), in which the first matrix on the right is the A matrix and the last matrix is the $\mathbf{b}$ matrix.

Response between Sampling Instants

An important feature of the state variable method is that it can be modified easily to determine the output between sampling instants. Thus, letting $t_0 = kT$ and $t = (k + \Delta)T$, where $0 < \Delta < 1$, in Eq. (10.75) gives

$$\mathbf{x}[(k+\Delta)T] = e^{A\Delta T}\mathbf{x}(kT) + \int_0^{\Delta T} e^{A\lambda}\, d\lambda\, \mathbf{b} f(kT)$$
$$= \mathbf{\Phi}(\Delta T)\mathbf{x}(kT) + G(\Delta T) f(kT) \tag{10.77}$$

where

$$\mathbf{\Phi}(\Delta T) = e^{A\Delta T} \qquad \text{and} \qquad G(\Delta T) = \left(\int_0^{\Delta T} e^{A\lambda}\, d\lambda\right)\mathbf{b}$$

The function $\mathbf{\Phi}(\Delta T)$ is obtained by replacing T by ΔT in the function $\mathbf{\Phi}(T)$. Similarly, $G(\Delta T)$ is obtained by replacing T by ΔT in the $G(T)$ function.

Illustrative example 10.18. For the system of Illustrative example 10.17, determine the discrete data representation for the response halfway between the sampling instants.

Solution. Replacing T by ΔT in the $\mathbf{\Phi}(T)$ and $G(T)$ matrices yields the $\mathbf{\Phi}(\Delta T)$ and $G(\Delta T)$ matrices. Application of Eq. (10.77) yields

$$\begin{bmatrix} x_1[(k+\Delta)T] \\ x_2[(k+\Delta)T] \end{bmatrix} = \begin{bmatrix} 2e^{-\Delta T} - e^{-2\Delta T} & e^{-\Delta T} - e^{-2\Delta T} \\ -2e^{-\Delta T} + 2e^{-2\Delta T} & -e^{-\Delta T} + 2e^{-2\Delta T} \end{bmatrix} \begin{bmatrix} x_1(kT) \\ x_2(kT) \end{bmatrix} + \begin{bmatrix} \dfrac{1 - 2e^{-\Delta T} + e^{-2\Delta T}}{2} \\ e^{-\Delta T} - e^{-2\Delta T} \end{bmatrix} f(kT)$$

For $\Delta = 0.5$ and $T = 0.2$, then $\Delta T = 0.1$. Thus

$$\begin{bmatrix} x_1[(k+0.5)T] \\ x_2[(k+0.5)T] \end{bmatrix} = \begin{bmatrix} 0.991 & 0.086 \\ -0.172 & 0.733 \end{bmatrix} \begin{bmatrix} x_1(kT) \\ x_2(kT) \end{bmatrix} + \begin{bmatrix} 0.009 \\ 0.086 \end{bmatrix} f(kT)$$

Note that to obtain the response between the sampling instants, it is necessary to use Eq. (10.76) to obtain the values $\mathbf{x}(kT)$ used in the preceding expression.

The values $f(kT)$ are the values of the input signal at the sampling instants. Letting Δ approach 1 in the term $\mathbf{x}[(k + \Delta)T]$ yields $\mathbf{x}[(k + 1)T]$. The values $\mathbf{x}[(k + 1)T]$ are the values for $\mathbf{x}$ just before the $(k + 1)$st sampling instant. Letting Δ approach zero yields the values of $\mathbf{x}$ just after the kth sampling instant. Inverting a z transform such as $\mathbf{X}(z)$ yields values of $\mathbf{x}$ just after each sampling instant.

10.8 SAMPLED-DATA CONTROL SYSTEMS

The pitch rate control of a high-speed aircraft is represented by the computer-controlled system shown in Fig. 10.24*a*. For this system the output of the sampler is $e^*(t)$. For the interval $kT \le t < (k + 1)T$, the output of the sampler is the unit

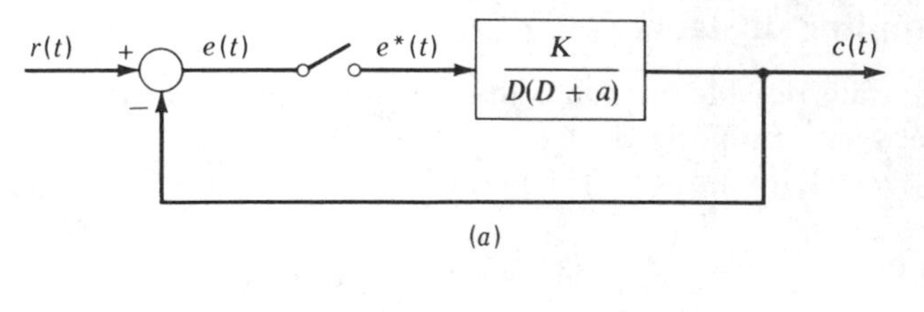

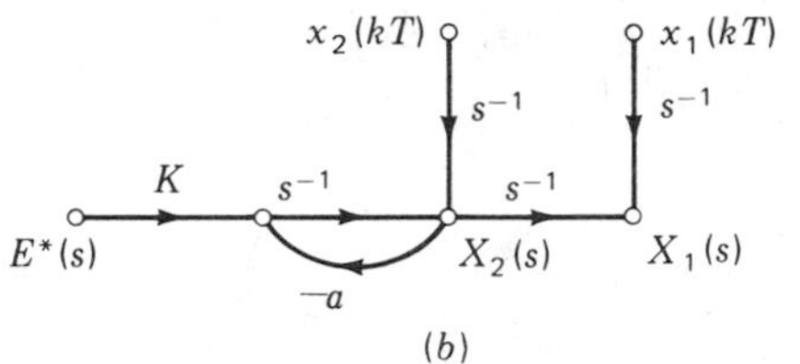

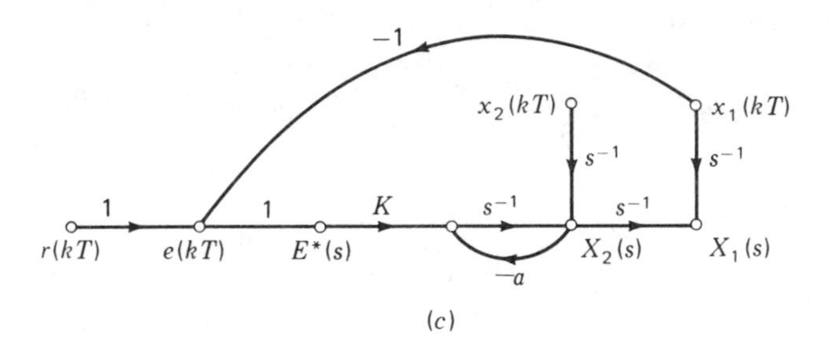

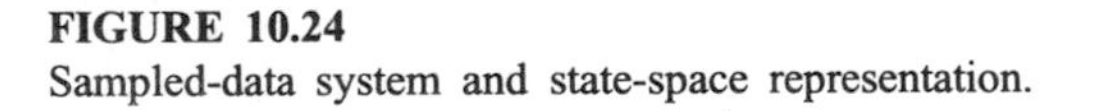

FIGURE 10.24
Sampled-data system and state-space representation.

impulse $e(kT)$. That is,

$$e^*(t) = e(kT)\delta(t - kT) \qquad kT \leq t < (k+1)T$$

The Laplace transform is

$$E^*(s) = e(kT)$$

The effect of the time delay $t_0 = kT$ will be accounted for later in the analysis by replacing t by $t - kT$ in the inverse Laplace transform. The differential equation for the continuous portion of the system is

$$c(t) = \frac{Ke^*(t)}{D(D+a)}$$

or

$$\ddot{c} + a\dot{c} = Ke^*(t)$$

By letting $c = x_1$ and $\dot{c} = \dot{x}_1 = x_2$, the state-space representation is

$$\begin{bmatrix} \dot{x}_1 \\ \dot{x}_2 \end{bmatrix} = \begin{bmatrix} 0 & 1 \\ 0 & -a \end{bmatrix} \begin{bmatrix} x_1 \\ x_2 \end{bmatrix} + \begin{bmatrix} 0 \\ K \end{bmatrix} e^*(t)$$

The Laplace transform for each of these relationships is

$$sX_1(s) - x_1(kT) = X_2(s)$$
$$sX_2(s) - x_2(kT) = -aX_2(s) + KE^*(s)$$

The signal flow graph for the preceding relationships is shown in Fig. 10.24*b*. The error signal at time kT is

$$e(kT) = r(kT) - c(kT) = r(kT) - x_1(kT)$$

The signal flow graph for the entire system may now be completed as shown in Fig. 10.24*c*.

Application of Mason's gain formula gives

$$\begin{bmatrix} X_1(s) \\ X_2(s) \end{bmatrix} = \begin{bmatrix} \dfrac{1}{s} - \dfrac{K}{s(s+a)} & \dfrac{1}{s(s+a)} \\ \dfrac{-K}{s+a} & \dfrac{1}{s+a} \end{bmatrix} \begin{bmatrix} x_1(kT) \\ x_2(kT) \end{bmatrix} + \begin{bmatrix} \dfrac{K}{s(s+a)} \\ \dfrac{K}{s+a} \end{bmatrix} r(kT)$$

Because of the time delay $t_0 = kT$, after the inverse Laplace transform is obtained, t is replaced by $t - kT$. Thus,

$$\begin{bmatrix} x_1(t) \\ x_2(t) \end{bmatrix} = \begin{bmatrix} 1 - \dfrac{K}{a}(1 - e^{-at^*}) & \dfrac{1}{a}(1 - e^{-at^*}) \\ -Ke^{-at^*} & e^{-at^*} \end{bmatrix} \begin{bmatrix} x_1(kT) \\ x_2(kT) \end{bmatrix} + \begin{bmatrix} \dfrac{K}{a}(1 - e^{-at^*}) \\ Ke^{-at^*} \end{bmatrix} r(kT)$$

where $t^* = t - kT$. The value of the output at the sampling instants is obtained by letting $t = (k + 1)T$, in which case $t^* = t - kT = (k + 1)T - kT = T$. For $a = 1$, the preceding becomes

$$\begin{bmatrix} x_1[(k+1)T] \\ x_2[(k+1)T] \end{bmatrix} = \begin{bmatrix} 1 - K(1 - e^{-T}) & 1 - e^{-T} \\ -Ke^{-T} & e^{-T} \end{bmatrix} \begin{bmatrix} x_1(kT) \\ x_2(kT) \end{bmatrix} + \begin{bmatrix} K(1 - e^{-T}) \\ Ke^{-T} \end{bmatrix} r(kT)$$

For $K = 0.25$ and for a sampling period $T = 1$, these equations become

$$\begin{bmatrix} x_1[(k+1)T] \\ x_2[(k+1)T] \end{bmatrix} = \begin{bmatrix} 0.842 & 0.632 \\ -0.092 & 0.368 \end{bmatrix} \begin{bmatrix} x_1(kT) \\ x_2(kT) \end{bmatrix} + \begin{bmatrix} 0.158 \\ 0.092 \end{bmatrix} r(kT)$$

For $x_1(0) = c(0) = 0$, $x_2(0) = \dot{x}_1(0) = \dot{c}(0) = 0$, $r(kT) = 0$ for $k < 0$, and $r(kT) = 1$ for $k \geq 0$, this problem is the same as Illustrative example 10.16. The preceding equation yields the values $x_1(0) = 0$, $x_1(T) = 0.158$, $x_1(2T) = 0.349$, and $x_1(3T) = 0.522$. These values correspond to the values $c(0)$, $c(T)$, $c(2T)$, and $c(3T)$, respectively, of Eq. (10.64). To obtain the response between sampling instants, let $t = (k + \Delta)T$, in which case $t^* = t - kT = \Delta T$. Thus

$$\begin{bmatrix} x_1[(k+\Delta)T] \\ x_2[(k+\Delta)T] \end{bmatrix} = \begin{bmatrix} 1 - K(1 - e^{-\Delta T}) & 1 - e^{-\Delta T} \\ -Ke^{-\Delta T} & e^{-\Delta T} \end{bmatrix} \begin{bmatrix} x_1(kT) \\ x_2(kT) \end{bmatrix} + \begin{bmatrix} K(1 - e^{-\Delta T}) \\ Ke^{-\Delta T} \end{bmatrix} r(kT)$$

The response at the midpoint is obtained by letting $\Delta = 0.5$. Thus, for $K = 0.25$ and $T = 1$

$$\begin{bmatrix} x_1[(k+0.5)T] \\ x_2[(k+0.5)T] \end{bmatrix} = \begin{bmatrix} 0.902 & 0.393 \\ -0.152 & 0.606 \end{bmatrix} \begin{bmatrix} x_1(kT) \\ x_2(kT) \end{bmatrix} + \begin{bmatrix} 0.098 \\ 0.152 \end{bmatrix} r(kT)$$

Because $c = x_1$, values midway between the sampling instants are $c(0.5T) = 0.098$, $c(1.5T) = 0.277$, $c(2.5T) = 0.458$, etc.

Zero-Order Hold

The system shown in Fig. 10.25*a* is the same as that of Fig. 10.24*a* except that a zero-order hold has been added after the sampler. The output $m(t)$ of the zero-order hold is constant between sampling instants. That is,

$$m(t) = e(kT) \qquad kT \le t < (k+1)T$$

The Laplace transform is

$$M(s) = \frac{e(kT)}{s} \tag{10.78}$$

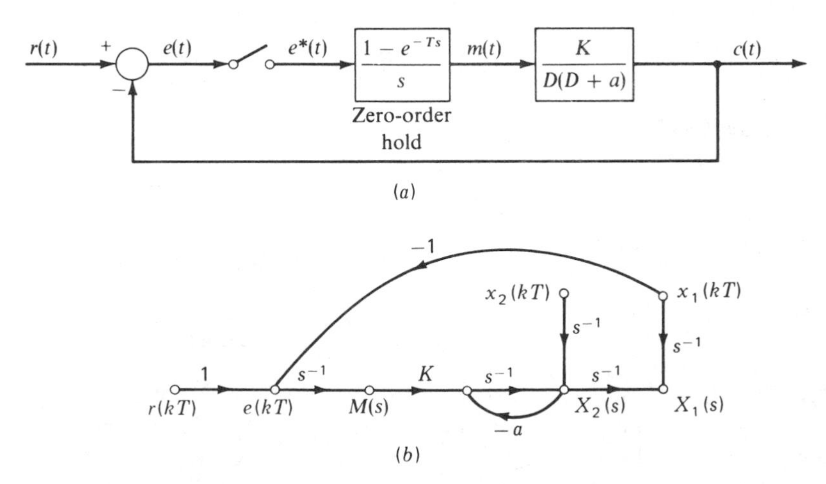

FIGURE 10.25
Sampled-data system with zero-order hold and state-space representation.

The effect of the time delay $t_0 = kT$ will be accounted for later by replacing t by $t - kT$. The error signal at time kT is

$$e(kT) = r(kT) - c(kT) = r(kT) - x_1(kT) \tag{10.79}$$

The signal flow graph for the plant is the same as that shown in Fig. 10.24*b* except that $E^*(s)$ is replaced by $M(s)$. Incorporation of the signal flow representation for Eqs. (10.78) and (10.79) with that for the plant yields the system graph shown in Fig. 10.25*b*. Application of Mason's gain formula to this signal flow graph yields

$$\begin{bmatrix} X_1(s) \\ X_2(s) \end{bmatrix} = \begin{bmatrix} \dfrac{1}{s} - \dfrac{K}{s^2(s+a)} & \dfrac{1}{s(s+a)} \\ \dfrac{-K}{s(s+a)} & \dfrac{1}{s+a} \end{bmatrix} \begin{bmatrix} x_1(kT) \\ x_2(kT) \end{bmatrix} + \begin{bmatrix} \dfrac{K}{s^2(s+a)} \\ \dfrac{K}{s+a} \end{bmatrix} r(kT)$$

Because of the time delay $t_0 = kT$, after the inverse Laplace transform has been obtained, t is replaced by $t - kT$. For $a = 1$, the resulting inverse is

$$\begin{bmatrix} x_1(t) \\ x_2(t) \end{bmatrix} = \begin{bmatrix} 1 - K(t^* - 1 + e^{-t^*}) & 1 - e^{-t^*} \\ -K(1 - e^{-t^*}) & e^{-t^*} \end{bmatrix} \begin{bmatrix} x_1(kT) \\ x_2(kT) \end{bmatrix} + \begin{bmatrix} K(t^* - 1 + e^{-t^*}) \\ K(1 - e^{-t^*}) \end{bmatrix} r(kT)$$

where $t^* = t - kT$. The value of the output at the sampling instants is obtained by letting $t = (k+1)T$ in which case $t^* = (k+1)T - kT = T$. Replacing t by $(k+1)T$ and t^* by T in the preceding gives

$$\begin{bmatrix} x_1[(k+1)T] \\ x_2[(k+1)T] \end{bmatrix} = \begin{bmatrix} 1 - K(T - 1 + e^{-T}) & 1 - e^{-T} \\ -K(1 - e^{-T}) & e^{-T} \end{bmatrix} \begin{bmatrix} x_1(kT) \\ x_2(kT) \end{bmatrix} + \begin{bmatrix} K(T - 1 + e^{-T}) \\ K(1 - e^{-T}) \end{bmatrix} r(kT)$$

For $K = 1$ and $T = 1$, the response at the sampling instants is

$$\begin{bmatrix} x_1[(k+1)T] \\ x_2[(k+1)T] \end{bmatrix} = \begin{bmatrix} 0.632 & 0.632 \\ -0.632 & 0.368 \end{bmatrix} \begin{bmatrix} x_1(kT) \\ x_2(kT) \end{bmatrix} + \begin{bmatrix} 0.368 \\ 0.632 \end{bmatrix} r(kT)$$

For $x_1(0) = c(0) = 0$, $x_2(0) = \dot{x}_1(0) = \dot{c}(0) = 0$, $r(kT) = 0$ for $k < 0$, and $r(kT) = 1$ for $k \geq 0$, this problem is the same as that worked in Sec. 10.6 whose solution is given by Eqs. (10.71). The preceding matrix yields the values $x_1(0) = 0$, $x_1(T) = 0.368$, $x_1(2T) = 1.0$, and $x_1(3T) = 1.4$. These values correspond to the values $c(0) = 0$, $c(T) = 0.368$, $c(2T) = 1.0$, and $c(3T) = 1.4$, respectively, of Eq. (10.71).

The response between sampling instants is obtained by replacing t by $(k + \Delta)T$, so that $t^* = \Delta T$. The result for $K = 1$, $\Delta = 0.5$, and $T = 1$ is

$$\begin{bmatrix} x_1[(k+\Delta)T] \\ x_2[(k+\Delta)T] \end{bmatrix} = \begin{bmatrix} 0.893 & 0.393 \\ -0.393 & 0.607 \end{bmatrix} \begin{bmatrix} x_1(kT) \\ x_2(kT) \end{bmatrix} + \begin{bmatrix} 0.107 \\ 0.393 \end{bmatrix} r(kT)$$

Because $c = x_1$, then $c(0.5T) = x_1(0.5T) = 0.107$, $c(1.5T) = x_1(1.5) = 0.684$, and $c(2.5T) = x_1(2.5T) = 1.248$. Note that in obtaining the response between sampling instants, it is necessary to know the values $x_1(kT)$ and $x_2(kT)$ at the beginning of each sampling instant. In this method, the values $x_1(kT)$, $x_2(kT)$, etc., are the values just before the sampling instants, whereas the values obtained by taking the inverse z transform are the values just after the sampling instants. If no discontinuity occurs at the sampling instants, these values are the same. In summary, letting $t^* = T$ yields the values for obtaining the response at the sampling instants $\mathbf{x}[(k+1)T]$, and letting $t^* = \Delta T$ yields the values for obtaining the response between the sampling instants $\mathbf{x}[(k+\Delta)T]$.

Sampled-data systems are readily programmed for solution on a digital computer. The equation relating $c(t)$ to the sampled signal $e^*(t)$ for the system shown in Fig. 10.24a is

$$c(t) = \frac{K}{D(D+a)} e^*(t)$$

For direct programming, x_1 is the differential equation that results when the numerator is replaced by 1. That is,

$$x_1 = \frac{1}{D(D+a)} e^*(t)$$

For $a = 1$, the resulting state-space representation is

$$\begin{aligned} \dot{x}_1 &= x_2 \\ \dot{x}_2 &= -x_2 + e^*(t) \\ c &= Kx_1 \end{aligned}$$

The area of the pulse representing $e^*(t)$ is equal to the value of $e(t)$ at the sampling instant. By designating Δt as the integration interval (width of pulse), then

$$e^*(t)\,\Delta t = e(t)$$

or

$$e^*(t) = \frac{e(t)}{\Delta t}$$

This is the computer program for obtaining the system response $c(t)$ for the case in which $r(t) = u(t)$ is a unit step function, $K = 0.25$, and all initial conditions are zero.

```
REAL K
X1 = 0.0
X2 = 0.0
DX2 = 0.0
DT = 0.01
T = 0.0
K = 0.25
```

```
      R = 1.0
      DO 25 I = 1, 5
      DO 20 J = 0, 99
      C = K*X1
      E = R-C
      ESTAR = 0.0
      IF (J.NE.0) GO TO 5
      ESTAR = E/DT
    5 IF (MOD(J,10).NE.0) GO TO 15
      WRITE (6,10) T, E, C
   10 FORMAT (3F10.2)
   15 T = T+DT
      X10=X1
      X20 = X2
      DX20 = DX2
      DX2 = ESTAR-X2
      X2 = 0.5*DT*(DX2+DX20)+X20
      X1 = 0.5*DT*(X2+X20)+X10
   20 CONTINUE
   25 CONTINUE
      END
```

In this program X1, X2, DX2, R, C, E, and ESTAR represent x_1, x_2, $\dot{x}_2$, $r(t)$, $c(t)$, $e(t)$, and $e^*(t)$, respectively.

The integration interval is DT = 0.01, and the starting value of time is T = 0.0. The statements before the first DO loop initialize the system. For the first DO loop when I = 1, the solution is being executed during the first sampling period. For I = 2, the solution is being executed during the second sampling period, etc. The second DO loop (for J = 0 to 99) executes the solution 100 times during each sampling period. For DT = 0.01, the sampling period is 100(0.01) = 1 s. Next, period J = 0. When J = 0, ESTAR = E/DT. For the remainder of the sampling interval J is not equal to 0, so ESTAR = 0. Corresponding values of T, E, and C are now printed. The T = T + DT statement sets the value of time for the end of the next integration interval. The X10 = X1 statement sets the present value of X1 as the starting value for the new integration interval. Similarly, the X20 = X2 and the DX20 = DX2 statements set the present values of X2 and DX2 as the starting values for the new integration interval. The value of DX2 at the end of this integration interval is obtained by the statement DX2 = ESTAR−X2. The new value of X2 is obtained by the statement X2 = 0.5*DT*(DX2 + DX20) + X20. The value of X20 on the right side is the value at the beginning of the integration interval. The value of the derivative at the beginning of the interval is DX20, and the value at the end is DX2, thus the term 0.5*DT*(DX2 + DX20) + X20 is the trapezoidal integration of $\dot{x}_2$ during the interval. Similarly, the new value of X1 is obtained by the statement X1 = 0.5*DT*(X2 + X20) + X10. Because $\dot{x}_1 = x_2$, the right side is the trapezoidal integration of $\dot{x}_1$ during the interval.

The system shown in Fig. 10.25*a* is the same as that shown in Fig. 10.24*a* except that a zero-order hold has been added. It is a relatively easy matter to change the computer program for the case of a zero-order hold. The input to the plant is now $m(t)$ rather than $e^*(t)$. The three consecutive statements ESTAR = 0.0, IF (J.NE.0) GO TO 5, and ESTAR = E/DT are now replaced by two new statements

```
IF(J.NE.0) GO TO 5
M = E
```

where $m(t)$ is designated as M. At the start of the sampling interval J = 0. Thus M takes on the value E = $e(t)$ at the beginning of the sampling period. It then retains this value until the beginning of the next sampling period when J again equals 0. The statement DX2 = ESTAR−X2 is replaced by DX2 = M−X2; the statement REAL K is replaced by REAL K, M; and K is 1.0 rather than 0.25.

10.9 COMPUTER-CONTROLLED SYSTEMS

A schematic diagram of a system controlled by a digital computer is shown in Fig. 10.26. At each sampling instant, the digital controller samples the error signal $e(t)$. The controller operates on this sampled value $e^*(t)$ and previous sampled values to obtain an output $m^*(t)$. This value of $m^*(t)$ is then retained by the zero-order hold until a new value is computed at the next sampling instant.

That is, during the kth sampling interval $m(t)$ retains the value $m^*(t) = m(kT)$ at the beginning of the interval.

$$m(t) = m(kT) \qquad kT \leq t < (k+1)T$$

By designating the Laplace transform of the zero-order hold and plant as $G(s)$, it follows from Fig. 10.26 that

$$C(s) = G(s)M^*(s)$$

Starring yields

$$C^*(s) = G^*(s)M^*(s)$$

Hence

$$C(z) = G(z)M(z) \tag{10.80}$$

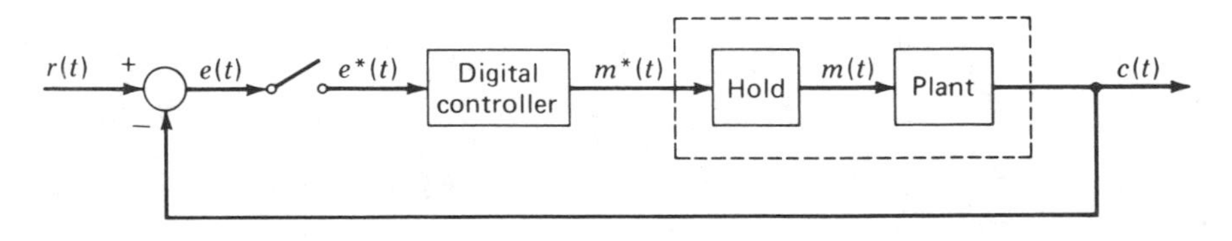

FIGURE 10.26
Digital control system.

The z transform for the digital controller is

$$M(z) = D(z)E(z) \tag{10.81}$$

The Laplace transform for the comparator equation $[e(t) = r(t) - c(t)]$ is

$$E(s) = R(s) - C(s)$$

The corresponding z transform is

$$E(z) = R(z) - C(z) \tag{10.82}$$

The z transform for the output $C(z)$ is obtained as follows:

$$C(z) = G(z)M(z) = G(z)D(z)E(z)$$
$$= G(z)D(z)[R(z) - C(z)]$$

Thus

$$C(z) = \frac{G(z)D(z)}{1 + G(z)D(z)} R(z)$$

Solving for the z transform of the controller gives

$$D(z) = \frac{C(z)}{G(z)[R(z) - C(z)]} = \frac{1}{G(z)} \frac{C(z)/R(z)}{1 - C(z)/R(z)} \tag{10.83}$$

This result shows that the controller characteristics $D(z)$ may be obtained by knowing the plant and hold characteristics $G(z)$ and the desired response $C(z)$ to a given input $R(z)$.

Figure 10.27 shows the CO_2 robotic laser weld system at the General Motors Cadillac Luxury Car Division's Grand Blanc plant. Precise path accuracy and speed of the robot are maintained by a digital controller. The controller also controls the operation of the laser and monitors all operating parameters such as gas flow and temperatures. In addition, it provides a safety interlock to ensure the safety of the operator. This system provides a very high quality weld which eliminates the need for any additional metal finishing.

Illustrative example 10.19. The transfer function for a plant is $(s + 2)/[s(s + 1)]$. Determine the characteristics of a digital controller such that the response of the system to a unit step function will be $c(t) = 5(1 - e^{-2t})$. The sampling period is $T = 1.0$ s.

Solution. For the plant and zero-order hold

$$G(s) = G_1(s)G_2(s) = (1 - e^{-Ts}) \frac{s + 2}{s^2(s + 1)}$$

where

$$G_1(s) = 1 - e^{-Ts}$$

and

$$G_2(s) = \frac{s + 2}{s^2(s + 1)} = \frac{2}{s^2} - \frac{1}{s} + \frac{1}{s + 1}$$

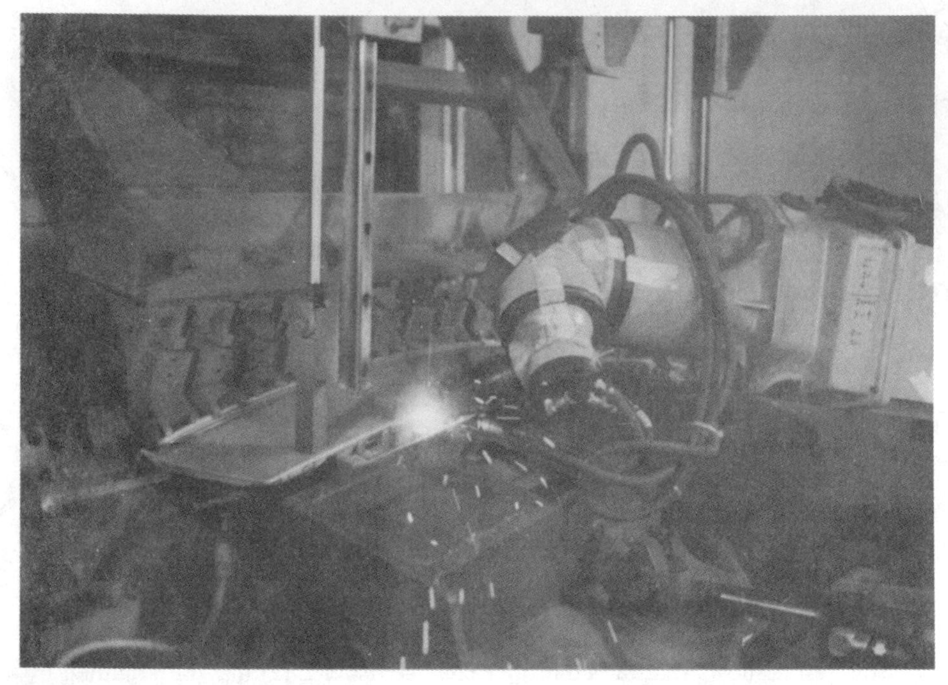

FIGURE 10.27
Robotic laser weld system. (Courtesy Cadillac Division, General Motors Corp.)

Application of Eq. (10.68) yields

$$G(z) = \frac{z-1}{z} G_2(z) = \frac{z-1}{z}\left[\frac{2z}{(z-1)^2} - \frac{z}{z-1} + \frac{z}{z-0.368}\right]$$
$$= \frac{1.368z - 0.104}{z^2 - 1.368z + 0.368}$$

For

$$C(s) = 5\left(\frac{1}{s} - \frac{1}{s+2}\right)$$

then

$$C(z) = 5\left(\frac{z}{z-1} - \frac{z}{z-0.135}\right) = \frac{4.32z}{(z-1)(z-0.135)}$$

Thus

$$\frac{C(z)}{R(z)} = \frac{4.32}{z-0.135} \qquad \text{and} \qquad 1 - \frac{C(z)}{R(z)} = \frac{z-4.45}{z-0.135}$$

Substitution of the preceding results into Eq. (10.83) gives

$$D(z) = \frac{M(z)}{E(z)} = \frac{z^2 - 1.368z + 0.368}{1.368z - 0.104}\,\frac{4.32}{z-4.45} = \frac{4.32z^2 - 5.91z + 1.59}{1.368z^2 - 6.202z + 0.46}$$

Dividing numerator and denominator by $1.368z^2$ and cross-multiplying give

$$M(z) - 4.53z^{-1}M(z) + 0.34z^{-2}M(z) = 3.16E(z) - 4.32z^{-1}E(z) + 1.16z^{-2}E(z)$$

Inverting yields the desired controller characteristics

$$m(k) = 4.53m(k-1) - 0.34m(k-2) + 3.16e(k) - 4.32e(k-1) + 1.16e(k-2)$$

Discrete Equivalent of Analog Controllers

One of the most versatile and widely used controllers for continuous systems is the one which provides proportional plus integral action. The equation of operation for this controller is

$$m(t) = \left(K_1 + K_2 \frac{1}{D}\right) e(t) \tag{10.84}$$

The contribution of the various actions is adjusted by varying constants K_1 and K_2. The discrete equivalent of this controller is commonly employed as the digital controller for process control systems. Using rectangular integration to approximate the integral yields for the discrete form

$$m(k) = K_1 e(k) + K_2 T \sum_{n=1}^{k} e(n) + m(0)$$

Generally it is more accurate and more convenient to work with the change in m in going from one sampling period to the next. Thus replacing k by $k-1$ in the preceding and then subtracting yield

$$\Delta m(k) = m(k) - m(k-1) = K_1[e(k) - e(k-1)] + K_2 T e(k)$$

Note that the preceding equation eliminates the need to know the initial value $m(0)$. The z transform of this equation is

$$M(z) - z^{-1}M(z) = K_1[E(z) - z^{-1}E(z)] + K_2 T E(z)$$

or

$$D(z) = \frac{M(z)}{E(z)} = K_1 + K_2 T \frac{z}{z-1}$$

The Laplace transform for the continuous controller is

$$D(s) = \frac{M(s)}{E(s)} = K_1 + K_2 T \frac{1}{s}$$

Note that the $z/(z-1)$ term for the discrete controller corresponds to integration in the equivalent continuous controller (i.e., the $1/s$ term).

Optimum Response

Now we show how the characteristics of a controller may be selected to yield optimum response. Any one of numerous optimum-performance criteria may be

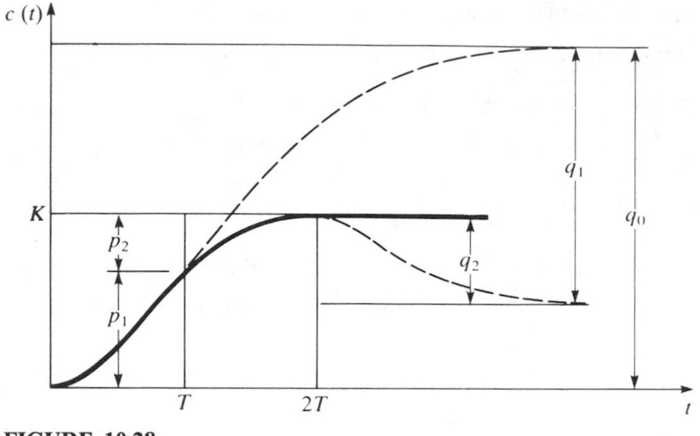

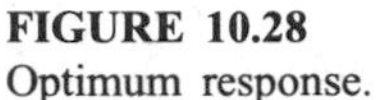
FIGURE 10.28
Optimum response.

used. In the following, the method developed by Kalman* for obtaining the controller characteristics to yield a desired transient behavior is described. Figure 10.26 shows a schematic diagram of the overall system. For illustrative purposes, assume that the transfer function of the plant is $ab/[(s + a)(s + b)]$. The transient behavior is considered optimum when the system responds to a step function in minimum time with no overshoot. In addition, there should be no steady-state error.

In Fig. 10.28, the optimum response $c(t)$ is indicated by a heavy line. The unit step input is $r(t) = 1$. During the first interval $0 \leq t < T$, the input to the plant $m(t) = q_0$ is large so as to accelerate the system rapidly toward its final value. The first dotted extension in Fig. 10.28 shows the subsequent response if the plant input is not changed after the first sampling period. Thus, at time $t = T$, it is necessary to begin to decelerate the response so as to achieve the desired final value $c(\infty)$. During the second interval $T \leq t < 2T$, the plant input is changed by an amount q_1, so that $m(t) = q_0 + q_1$. As drawn in Fig. 10.28, the term q_1 is a negative constant. The response $c(t)$ reaches the final value at the end of the second interval $t = 2T$. The second dotted extension shows the path of the response if the plant input is not changed after the second interval. Thus for the third and subsequent intervals, the plant input is $q_0 + q_1 + q_2 = m(t)$.

A desirable feature of this performance criterion is that the error is reduced exactly to zero after two periods. The error then remains zero not only at the sampling points but also in between. This criterion thus eliminates the possibility of ripple (i.e., undesirable oscillations between sampling instants).

Because $m(t)$ is the output of a zero-order hold, it follows that the z transform of the sampled input $m^*(t)$ is

$$M(z) = q_0 + (q_0 + q_1)z^{-1} + (q_0 + q_1 + q_2)\,(z^{-2} + z^{-3} + \cdots)$$

* R. E. Kalman (discussion of): "Sampled-Data Processing Techniques for Feedback Control Systems," A. R. Bergon and J. R. Ragazzini, *Trans. AIEE*, 73(2): 245–246 (1954).

Dividing $M(z)$ by the z transform of the unit step input $[R(z) = 1 + z^{-1} + z^{-2} + \cdots]$ gives

$$\frac{M(z)}{R(z)} = q_0 + q_1 z^{-1} + q_2 z^{-2} \tag{10.85}$$

Similarly, from Fig. 10.28 the z transform of the response at the sampling instants is

$$C(z) = p_1 z^{-1} + (p_1 + p_2)(z^{-2} + z^{-3} + z^{-4} + \cdots)$$

Dividing by $R(z)$ shows that

$$\frac{C(z)}{R(z)} = p_1 z^{-1} + p_2 z^{-2} \tag{10.86}$$

The pulse transfer function for the hold and plant is

$$G(z) = \frac{C(z)/R(z)}{M(z)/R(z)} = \frac{p_1 z^{-1} + p_2 z^{-2}}{q_0 + q_1 z^{-1} + q_2 z^{-2}} \tag{10.87}$$

The pulse transfer function for the digital controller is

$$D(z) = \frac{M(z)}{E(z)} = \frac{M(z)}{R(z) - C(z)} = \frac{M(z)/R(z)}{1 - C(z)/R(z)} = \frac{q_0 + q_1 z^{-1} + q_2 z^{-2}}{1 - p_1 z^{-1} - p_2 z^{-2}} \tag{10.88}$$

Inverting yields

$$m(k) = p_1 m(k-1) + p_2 m(k-2) + q_0 e(k) + q_1 e(k-1) + q_2 e(k-2)$$

This is the desired relationship for the digital controller which results in an optimum response for the system. As indicated in the preceding analysis, the controller for a digital control system is programmed to yield a desired behavior. Because of the flexibility of programming, a digital controller is more versatile than a controller for a continuous system, which is limited to some action described by a differential operator.

Application of the final-value theorem shows that the final steady-state value of $c(t)$ is

$$\begin{aligned} c(\infty) &= \lim_{z \to 1} \frac{z-1}{z} C(z) = \frac{z-1}{z} \left[p_1 z^{-1} + (p_1 + p_2)(z^{-2} + z^{-3} + \cdots)\right]_{z=1} \\ &= \frac{z-1}{z} \left[(p_1 z^{-1} + p_2 z^{-2})(1 + z^{-1} + z^{-2} + \cdots)\right]_{z=1} \\ &= (p_1 z^{-1} + p_2 z^{-2})_{z=1} = p_1 + p_2 \end{aligned}$$

Similarly, the final steady-state value of $m(t)$ is

$$\begin{aligned} m(\infty) &= \lim_{z \to 1} \frac{z-1}{z} M(z) \\ &= \frac{z-1}{z} \left[q_0 + (q_0 + q_1) z^{-1} + (q_0 + q_1 + q_2)(z^{-2} + z^{-3} + z^{-4} + \cdots)\right]_{z=1} \\ &= \frac{z-1}{z} \left[(q_0 + q_1 z^{-1} + q_2 z^{-2})(1 + z^{-1} + z^{-2} + \cdots)\right]_{z=1} \\ &= (q_0 + q_1 z^{-1} + q_2 z^{-2})_{z=1} = q_0 + q_1 + q_2 \end{aligned}$$

The ratio of the steady-state value of the system output $c(\infty)$ to the system input $r(\infty)$ is the closed-loop steady-state gain K for the system

$$K = \frac{c(\infty)}{r(\infty)} = p_1 + p_2 \tag{10.89}$$

The ratio of the steady-state value of the plant output $c(\infty)$ to the plant input is the steady-state gain K_p for the plant

$$K_p = \frac{c(\infty)}{m(\infty)} = \frac{p_1 + p_2}{q_0 + q_1 + q_2} \tag{10.90}$$

For a second-order plant, $G(z)$ will have the form

$$G(z) = \frac{a_1 z^{-1} + a_2 z^{-2}}{b_0 + b_1 z^{-1} + b_2 z^{-2}}$$

In order that the system have the desired steady-state gain

$$K = p_1 + p_2 = k(a_1 + a_2)$$

or

$$k = \frac{K}{a_1 + a_2} \tag{10.91}$$

Multiplication of the numerator and denominator of $G(z)$ by k yields directly the form given by Eq. (10.87) in which the numerator coefficients are the p's and the denominator coefficients are the q's.

Illustrative example 10.20. Repeat Illustrative example 10.19, except determine the difference equation for the digital controller to yield optimum response with a closed-loop steady-state gain of 5.

Solution. In Illustrative example 10.19, it was found that the z transform for the zero-order hold and plant is

$$G(z) = \frac{1.368z - 0.104}{z^2 - 1.368z + 0.368}$$

The sum of the numerator coefficients is $a_1 + a_2 = 1.368 - 0.104 = 1.264$. To obtain the desired closed-loop steady-state gain, it is necessary that $K = p_1 + p_2 = 5.0$. The scale factor k is

$$k = \frac{K}{a_1 + a_2} = \frac{5}{1.264} = 3.96$$

Thus, multiplying both the numerator and denominator of $G(z)$ by 3.96 yields

$$G(z) = \frac{5.42z - 0.42}{3.96z^2 - 5.42z + 1.46} = \frac{5.42z^{-1} - 0.42z^{-2}}{3.96 - 5.42z^{-1} + 1.46z^{-2}}$$

Comparison with Eq. (10.87) shows that $p_1 = 5.42$, $p_2 = -0.42$, $q_0 = 3.96$, $q_1 = -5.42$, and $q_2 = 1.46$. Note that $p_1 + p_2 = 5.42 - 0.42 = 5$. The resulting equation for the digital controller is

$$m(k) = 5.42m(k-1) - 0.42m(k-2) + 3.96e(k) - 5.42e(k-1) + 1.46e(k-2)$$

To obtain the computer program for this system, first write the equation relating $c(t)$ to $m(t)$.

$$c(t) = \frac{D+2}{D(D+1)} m(t)$$

For direct programming

$$x_1 = \frac{1}{D(D+1)} m(t)$$

The resulting state-space representation is

$$\dot{x}_1 = x_2$$
$$\dot{x}_2 = -x_2 + m(t)$$
$$c = 2x_1 + x_2$$

The computer program is

```
   REAL MK,MK1,MK2
   X1 = 0.0
   X2 = 0.0
   DX2 = 0.0
   EK = 0.0
   EK1 = 0.0
   MK = 0.0
   MK1 = 0.0
   DT = 0.01
   T = 0.0
   R = 1.0
   DO 25 I = 1,5
   DO 20 J = 0,99
   C = 2.0*X1+X2
   E = R-C
   IF (J.NE.0) GO TO 5
   EK2 = EK1
   EK1 = EK
   EK = E
   MK2 = MK1
   MK1 = MK
   MK = 5.42*MK1-0.42*MK2+3.96*EK-5.42*EK1+1.46*EK2
 5 IF (MOD(J,10).NE.0) GO TO 15
   WRITE (6,10) T, E, EK, MK, C
10 FORMAT (5F10.2)
15 T = T+DT
   X10 = X1
   X20 = X2
```

```
      DX20 = DX2
      DX2 = MK-X2
      X2 = 0.5*DT*(DX2+DX20)+X20
      X1 = 0.5*DT*(X2+X20)+X10
   20 CONTINUE
   25 CONTINUE
      END
```

In this program EK, EK1, EK2, MK, MK1, and MK2 represent $e(k)$, $e(k-1)$, $e(k-2)$, $m(k)$, $m(k-1)$, and $m(k-2)$, respectively. The statements before the first DO loop initialize the system. For the first DO loop, when I = 1, the solution is being executed during the first sampling period. For I = 2, the solution is being executed during the second sampling period, etc. The second DO loop executes the solution 100 times during each sampling period. Values of $c(t)$ and $e(t)$ are determined for each integration interval DT = 0.01. Because J = 0 at the beginning of the sampling period, the IF statement sets $e(k)$ equal to the value of $e(t)$ at the beginning of the sampling period. For the remainder of the sampling period J is not equal to 0 so that $e(k)$ retains this value. The value $e(k-1)$ is set equal to the value of $e(k)$ for the preceding sampling period, and $e(k-2)$ is set equal to the preceding value of $e(k-1)$. Similarly, the value of $m(k-1)$ is set equal to the preceding value of $m(k)$, and the value of $m(k-2)$ is set equal to the preceding value of $m(k-1)$. The present value of $m(k)$ is calculated from the equation for the digital controller. The remainder of the program uses trapezoidal integration to solve the state-space equations. Any integration method such as the fourth-order Runge-Kutta method can be used.

For the case of a third-order plant, $M(z)$ is

$$M(z) = q_0 + (q_0 + q_1)z^{-1} + (q_0 + q_1 + q_2)z^{-2} + (q_0 + q_1 + q_2 + q_3)(z^{-3} + z^{-4} + z^{-5} + \cdots)$$

Division by $R(z)$ yields

$$\frac{M(z)}{R(z)} = q_0 + q_1 z^{-1} + q_2 z^{-2} + q_3 z^{-3}$$

The z transform for the response at the sampling instants is

$$C(z) = p_1 z^{-1} + (p_1 + p_2)z^{-2} + (p_1 + p_2 + p_3)(z^{-3} + z^{-4} + z^{-5} + \cdots)$$

Dividing by $R(z)$ shows that

$$\frac{C(z)}{R(z)} = p_1 z^{-1} + p_2 z^{-2} + p_3 z^{-3}$$

The pulse transfer function of the hold and plant is

$$G(z) = \frac{C(z)/R(z)}{M(z)/R(z)} = \frac{p_1 z^{-1} + p_2 z^{-2} + p_3 z^{-3}}{q_0 + q_1 z^{-1} + q_2 z^{-2} + q_3 z^{-3}} \tag{10.92}$$

The closed-loop steady-state gain for the system is

$$K = \frac{c(\infty)}{r(\infty)} = p_1 + p_2 + p_3$$

The steady-state gain for the plant is

$$K_p = \frac{c(\infty)}{m(\infty)} = \frac{p_1 + p_2 + p_3}{q_0 + q_1 + q_2 + q_3}$$

For a third-order plant, $G(z)$ has the form

$$G(z) = \frac{a_1 z^{-1} + a_2 z^{-2} + a_3 z^{-3}}{b_0 + b_1 z^{-1} + b_2 z^{-2} + b_3 z^{-3}}$$

To obtain the desired steady-state gain, then

$$K = p_1 + p_2 + p_3 = k(a_1 + a_2 + a_3) \tag{10.93}$$

or

$$k = \frac{K}{a_1 + a_2 + a_3}$$

Multiplication of the numerator and denominator of $G(z)$ by k yields directly the form in which the numerator coefficients are the p's and the denominator coefficients are the q's. That is,

$$G(z) = \frac{C(z)/R(z)}{M(z)/R(z)} = \frac{p_1 z^{-1} + p_2 z^{-2} + p_3 z^{-3}}{q_0 + q_1 z^{-1} + q_2 z^{-2} + q_3 z^{-3}}$$

Hence

$$D(z) = \frac{M(z)}{E(z)} = \frac{M(z)/R(z)}{1 - C(z)/R(z)} = \frac{q_0 + q_1 z^{-1} + q_2 z^{-2} + q_3 z^{-3}}{1 - p_1 z^{-1} - p_2 z^{-2} - p_3 z^{-3}}$$

For a plant of order n, then $G(z)$ has the form

$$G(z) = \frac{a_1 z^{-1} + a_2 z^{-2} + \cdots + a_n z^{-n}}{b_0 + b_1 z^{-1} + b_2 z^{-2} + \cdots + b_n z^{-n}}$$

The steady-state gain is

$$K = p_1 + p_2 + \cdots + p_n = k(a_1 + a_2 + \cdots + a_n) \tag{10.94}$$

Hence,

$$k = \frac{K}{a_1 + a_2 + \cdots + a_n}$$

Multiplying the numerator and denominator of $G(z)$ by k yields the desired form in which the numerator coefficients are the p's and the denominator coefficients are the q's. Because

$$D(z) = \frac{M(z)}{E(z)} = \frac{q_0 + q_1 z^{-1} + \cdots + q_n z^{-n}}{1 - p_1 z^{-1} - p_2 z^{-2} - \cdots - p_n z^{-n}} \tag{10.95}$$

then

$$m(k) = p_1 m(k-1) + p_2 m(k-2) + \cdots + p_n m(k-n) + q_0 e(k) + q_1 e(k-1) + \cdots + q_n e(k-n)$$

This is the required relationship for the digital controller to yield optimum response.

PROBLEMS

10.1. For each of the following functions write

$$f^*(t) = \sum_{n=0}^{\infty} f(nT)\delta(t-nT)$$

and then take the Laplace transform to obtain $F^*(s) = \mathscr{L}[f^*(t)]$. Finally, determine the z transform $F(z)$.

(*a*) $f(t) = e^{at}$ (*b*) $f(t) = a^{t/T}$

10.2. For each of the following functions, determine the z transform $F(z)$.

$$(a)\quad f(t) = \cosh \omega t = \frac{e^{\omega t} + e^{-\omega t}}{2}$$

$$(b)\quad f(t) = \sin \omega t = \frac{e^{j\omega t} - e^{-j\omega t}}{2j}$$

10.3. Use Table 10.1 to obtain the z transform corresponding to each of the following Laplace transforms.

$$(a)\quad F(s) = \frac{a}{s(s+a)} \qquad (b)\quad F(s) = \frac{\omega}{s^2 - \omega^2}$$

10.4. Determine the Laplace transform of each of the following functions, and then use Table 10.1 to obtain the corresponding z transform.

(*a*) $f(t) = \sin \omega t$ (*b*) $f(t) = \cos \omega t$

10.5. The z transform for a unit step function is

$$Z[u(t)] = \frac{z}{z-1}$$

Use appropriate transform theorems to determine the z transform for each of the following functions.

(*a*) $f(t) = t$ (*b*) $f(t) = t^2$ (*c*) $f(t) = e^{at}t^2$

10.6. Given that the z transform for a unit step function is

$$Z[u(t)] = \frac{z}{z-1}$$

Use appropriate transform theorems to determine the z transform for each of the following functions.

(*a*) $f(t) = e^{at}$ (*b*) $f(t) = te^{at}$ (*c*) $f(t) = t^2e^{at}$

10.7. Use the partial-fraction method to determine the inverse z transform of each of the following.

$$(a)\quad C(z) = z\frac{z+1}{(z-0.6)(z+0.2)} \qquad (b)\quad C(z) = z\frac{2.9z-1.7}{(z-1)(z+0.5)(z-0.2)}$$

10.8. Use the partial-fraction method to determine the inverse z transform of each of the following.

$$(a)\quad C(z) = z\frac{9z-4}{(z-2)^2(z+5)} \qquad (b)\quad C(z) = z\frac{0.5z+0.02}{(z+0.1)^2(z+0.4)}$$

10.9. Use the partial-fraction method to determine the inverse z transform of each of the following.

$$(a)\quad C(z) = z\frac{z+0.5}{(z^2-z+0.5)(z-0.5)}$$

$$(b)\quad C(z) = z\frac{z+0.4}{(z^2+0.8z+0.25)(z+0.7)}$$

10.10. Use the residue method to determine the z transform corresponding to each of these Laplace transforms:

$$(a)\quad F(s) = \frac{s+3}{(s+1)(s+2)} \qquad (b)\quad F(s) = \frac{1}{s^2(s+1)}$$

10.11. Use the residue method to determine the z transform of each of the following functions:

(a) $f(t) = e^{at}$ (b) $f(t) = t$ (c) $f(t) = te^{at}$

Verify the resulting answers by using the method of residues to invert.

10.12. Use the method of residues to invert each of the following.

$$(a)\quad F(z) = \frac{z(e^T - e^{-T})}{(z-e^T)(z-e^{-T})} \qquad (b)\quad F(z) = \frac{zT^2(z+1)}{(z-1)^3}$$

10.13. Use block-diagram algebra to determine $C(z)$ for each of the systems shown in Fig. P10.13.

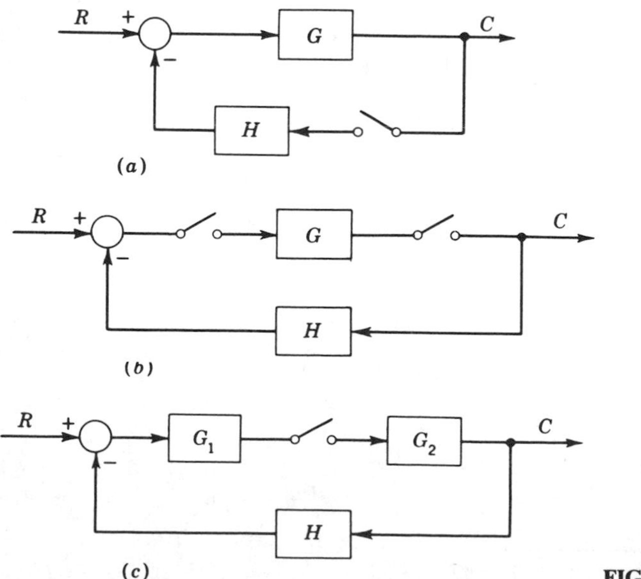

FIGURE P10.13

10.14. Figure P10.14 shows a sampled-data system whose sampling rate is $T = 1$ s.
(*a*) Apply Routh's criterion to determine whether the system becomes unstable.
(*b*) Construct the root-locus plot for the system.

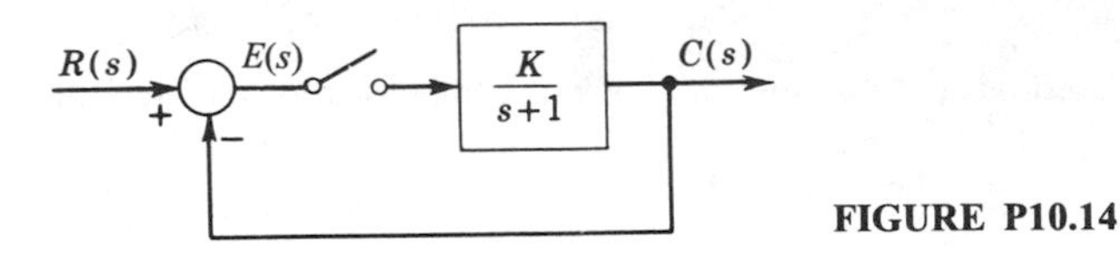

FIGURE P10.14

10.15. Determine the unit step-function response for the system of Prob. 10.14 for the case in which $K = 1$.

10.16. For the system shown in Fig. 10.13, it is desired to use a sampling rate $T = 0.2$ s rather than $T = 1$ s. Determine the value of K at which the system becomes unstable by using Routh's criterion and by constructing the root-locus plot.

10.17. Figure 10.14 shows the root-locus plot for the system of Fig. 10.13 when $T = 1$ s. Determine the unit step-function response when $K = 2.16$.

10.18. Figure P10.18 shows a sampled-data system whose sampling rate is $T = 1$ s.
(*a*) Apply Routh's criterion to determine the value of K at which the system becomes unstable.
(*b*) Construct the root-locus plot for this system. From this plot determine the value of K at which the system becomes unstable.

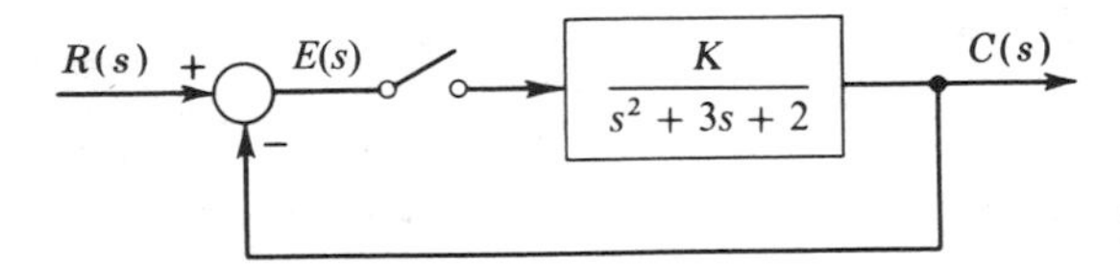

FIGURE P10.18

10.19. Repeat Prob. 10.18, except $T = 2$ s.

10.20. For the system shown in Fig. P10.20, the sampling period is $T = 1$ s. Determine the z transform for $C(z)$. Construct the root-locus plot, and then determine the value of the gain K at which the system becomes unstable.

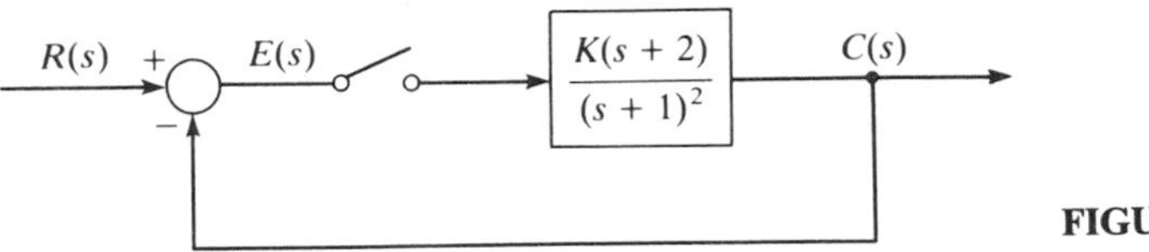

FIGURE P10.20

10.21. For the system shown in Fig. P10.21, the sampling period is $T = 1$ s. Determine the z transform for $C(z)$. Construct the root-locus plot, and then determine the value of the gain K at which the system becomes unstable.

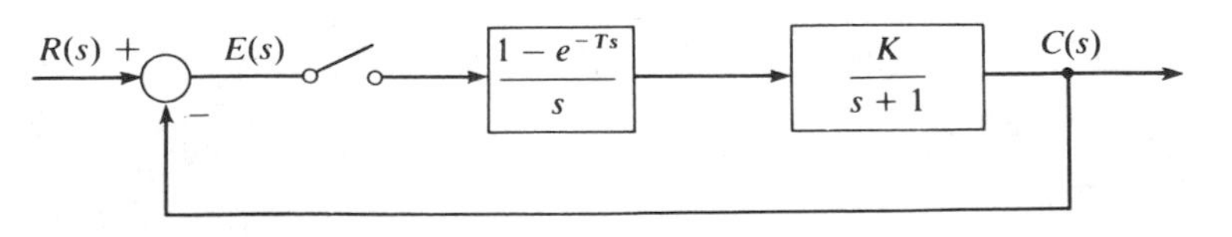

FIGURE P10.21

10.22. For the system shown in Fig. P10.22, the sampling period is $T = 1$ s. Determine the z transform for $C(z)$. Construct the root-locus plot, and then determine the value of the gain K at which the system becomes unstable.

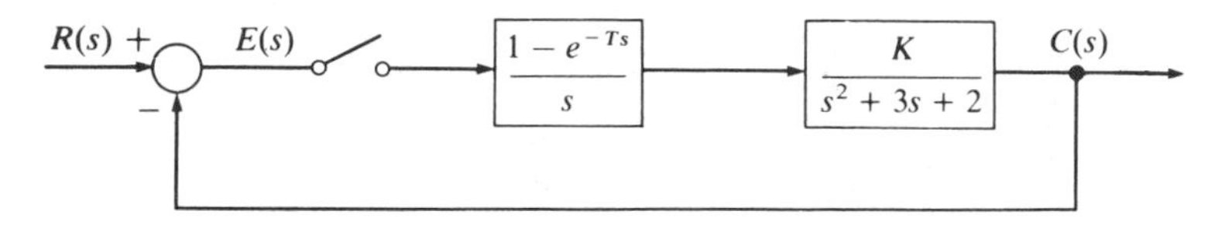

FIGURE P10.22

10.23. For the system shown in Fig. P10.23, the sampling period is $T = 1$ s. Determine the z transform for $C(z)$. Construct the root-locus plot, and then determine the value of the gain K at which the system becomes unstable.

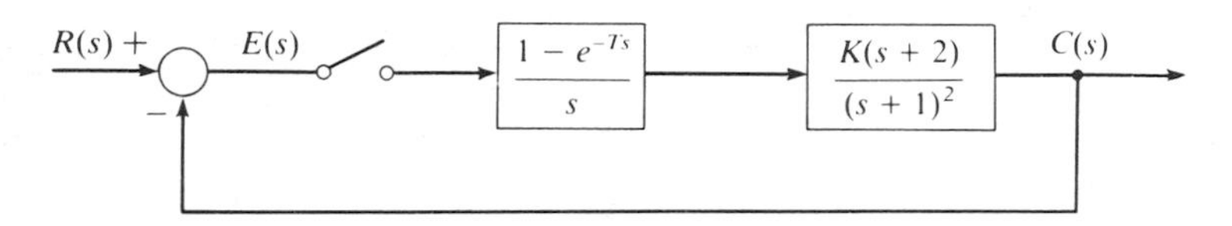

FIGURE P10.23

10.24. The system shown in Fig. P10.24 has an exponential hold $1/(1 + \tau s)$. For a sampling period $T = 1$ s, determine the value of the gain K at which the system becomes unstable when $\tau = 0.2$.

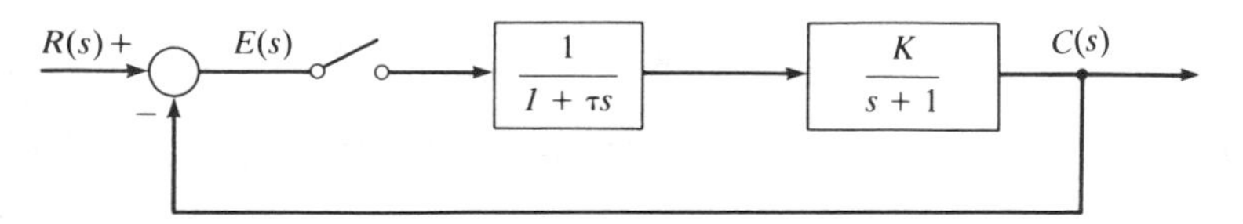

FIGURE P10.24

10.25. The system shown in Fig. P10.24 has an exponential hold $1/(1 + \tau s)$. For a sampling period $T = 1$ s, determine the value of the gain K at which the system becomes unstable when $\tau = 1.0$. Determine the unit step-function response when $K = 4.0$ and when $K = 5.08$.

10.26. Determine the discrete time state-space representation for the response at the sampling instants and for the response midway between the sampling instants for $T = 0.2$ s for each of the following continuous systems.

(*a*) $\dot{c} + c = f(t)$ (*b*) $\ddot{c} + 2\dot{c} + c = f(t)$

Evaluate the response at $t = 0.1, 0.2, 0.3$, and 0.4 s when $f(t)$ is a unit step function and all the initial conditions are zero.

10.27. For the system shown in Fig. P10.27, the input $f(t)$ is a unit step function and $T = 1$ s. Use the z transform method to determine the response at each of the first two sampling instants. Use the state-space method to determine the response just before and just after each of the first two sampling instants.

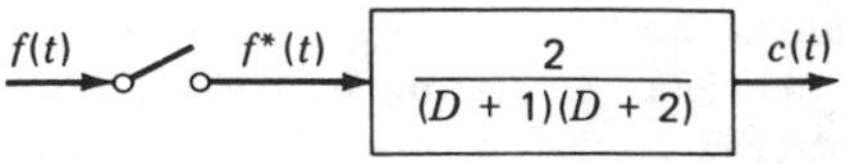

FIGURE P10.27

10.28. Repeat Prob. 10.27 except use the system shown in Fig. P10.28.

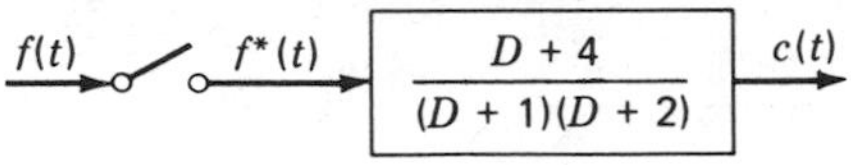

FIGURE P10.28

10.29. Repeat Prob. 10.27 except use the system shown in Fig. P10.29.

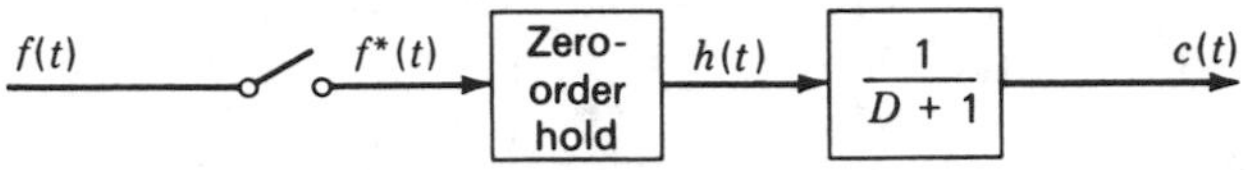

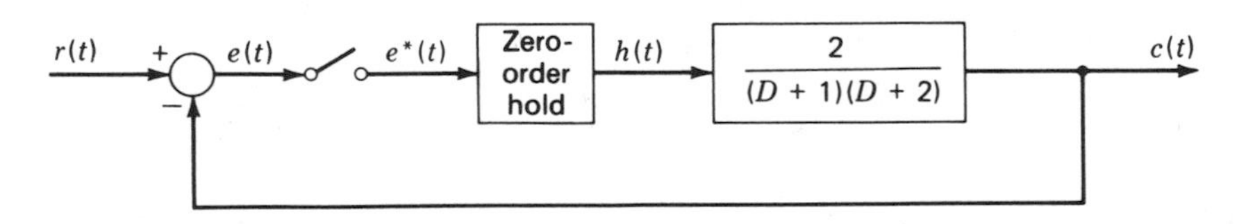

FIGURE P10.29

10.30. For the system shown in Fig. P10.30, the input $r(t)$ is a unit step function and $T = 1$ s. Use the z transform method to determine the response at each of the first two sampling instants. Use the state-space method to determine the response just before and just after each of the first two sampling instants.

FIGURE P10.30

10.31. For the system of Fig. 10.26, the transfer function for the plant is $4/(s+1)(s+2)$ and the sampling period is $T = 1$ s. Determine the characteristics of the digital controller so that the response of the system to a unit step function will be $c(t) = 5(1 - e^{-2t})$.

10.32. For the system of Fig. 10.26, the transfer function for the plant is $4(s+1)/[s(s+2)]$ and the sampling period is $T = 1$ s. Determine the characteristics of the digital controller so that the response of the system to a unit step function will be $c(t) = 10(1 - e^{-t})$.

10.33. For the system of Fig. 10.26, the transfer function for the plant is $(s+2)/(s+1)^2$, and the sampling period is $T = 1$ s. Determine the characteristics of the digital controller so that the response of the system to a unit step function will be $c(t) = 4(1 - e^{-t})$.

10.34. For the system of Fig. 10.26, the transfer function for the plant is $4/[(s+1)(s+2)]$, and the sampling period is $T = 1$ s. The closed-loop steady-state gain for the system is to be 5. Determine the characteristics of the digital controller to yield the optimum response.

10.35. For the system of Fig. 10.26, the transfer function for the plant is $4(s + 1)/[s(s + 2)]$, and the sampling period is $T = 1$ s. The closed-loop steady-state gain for the system is to be 10. Determine the characteristics of the digital controller to yield the optimum response.

10.36. For the system of Fig. 10.26, the transfer function for the plant is $(s + 2)/(s + 1)^2$, and the sampling period is $T = 1$ s. The closed-loop steady-state gain for the system is to be 4. Determine the characteristics of the digital controller to yield the optimum response.

CHAPTER 11

FREQUENCY-RESPONSE METHODS

Frequency-response methods provide a convenient means for investigating the dynamic behavior of control systems. By frequency response is meant the response of a system to a sinusoidal input $f = f_0 \sin \omega t$. A characteristic of linear systems is that after the effect of the initial transients has "died out," the response also becomes a sinusoid with the same angular velocity ω as the input. As illustrated in Fig. 11.1, the response $y = y_0 \sin(\omega t + \phi)$ is displaced some phase angle ϕ from the input, and the amplitude y_0 is different from that of the input f_0. Both the phase angle ϕ and the amplitude ratio y_0/f_0 are functions of the angular velocity ω of the input signal. Graphs of ϕ versus ω and of the amplitude ratio y_0/f_0 versus ω form the basis for frequency-response methods.

11.1 FREQUENCY RESPONSE

Because frequency-response methods are based on a knowledge of ϕ versus ω and y_0/f_0 versus ω, we now show how these quantities may be determined directly by substitution of $j\omega$ for D in the operational form of the differential equation for the system. The general operational form of a differential equation is

$$y(t) = \frac{(a_m D^m + a_{m-1} D^{m-1} + \cdots + a_1 D + a_0) f(t)}{D^n + b_{n-1} D^{n-1} + \cdots + b_1 D + b_0} = \frac{L_m(D) f(t)}{L_n(D)} \tag{11.1}$$

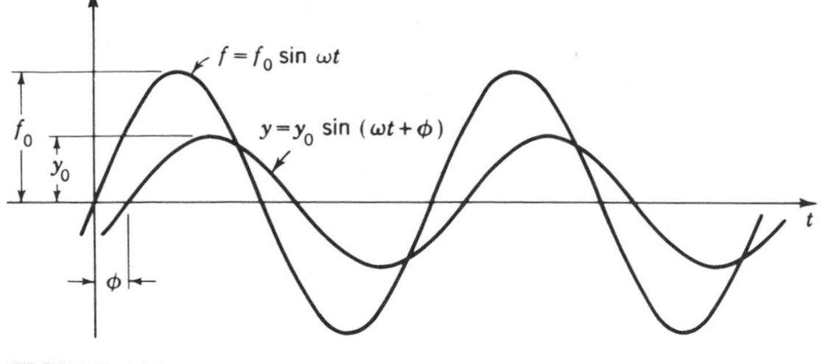

FIGURE 11.1
Sinusoidal response.

The transform of the preceding is

$$Y(s) = \frac{L_m(s)F(s) + I(s)}{L_n(s)} = \frac{L_m(s)N_{F(s)}}{L_n(s)D_{F(s)}} + \frac{I(s)}{L_n(s)} \tag{11.2}$$

The transform for the input is

$$F(s) = \mathscr{L}(f_0 \sin \omega t) = \frac{\omega f_0}{s^2 + \omega^2} = \frac{N_{F(s)}}{D_{F(s)}} \tag{11.3}$$

Expanding Eq. (11.2) in a partial-fraction expansion and noting that $D_{F(s)} = s^2 + \omega^2 = (s - j\omega)(s + j\omega)$ yield

$$Y(s) = \frac{K_1}{s - r_1} + \cdots + \frac{K_n}{s - r_n} + \frac{K_c}{s - j\omega} + \frac{K_{-c}}{s + j\omega} + \frac{I_1}{s - r_1} + \cdots + \frac{I_n}{s - r_n} \tag{11.4}$$

where $r_1, r_2, \ldots, r_n$ are the zeros of $L_n(s)$; and $K_1, K_2, \ldots, K_n$, K_c, and K_{-c} are the constants which arise from the partial-fraction expansion of $L_m(s)N_{F(s)}/L_n(s)D_{F(s)}$; and $I_1, I_2, \ldots, I_n$ are the constants which arise from the partial-fraction expansion of $I(s)/L_n(s)$. Inverting Eq. (11.4) gives

$$y(t) = (K_1 + I_1)^{r_1 t} + \cdots + (K_n + I_n)e^{r_n t} + \frac{1}{b}|K(a + jb)|e^{at} \sin(bt + \phi) \tag{11.5}$$

For a stable system, $r_1, r_2, \ldots, r_n$ must have negative real parts, so that after sufficient time the effect of these terms becomes negligible. Thus, for stable systems, the steady-state (hence the subscript "ss") sinusoidal response $y(t)_{ss}$ is determined by the last term of Eq. (11.5). For the quadratic $s^2 + \omega^2$, it follows that $a = 0$ and $b = \omega$. Thus

$$y(t)_{ss} = \frac{1}{\omega}|K(j\omega)| \sin(\omega t + \phi) \tag{11.6}$$

The terms $|K(j\omega)|$ and $\phi = \measuredangle K(j\omega)$ are evaluated as follows:

$$K(j\omega) = \lim_{s \to j\omega} \left[(s^2 + \omega^2) \frac{L_m(s) N_{F(s)}}{L_n(s) D_{F(s)}} \right] \tag{11.7}$$

$$= \lim_{s \to j\omega} \frac{(s^2 + \omega^2) L_m(s) \omega f_0}{L_n(s)\,(s^2 + \omega^2)}$$

$$= \frac{L_m(j\omega)\omega f_0}{L_n(j\omega)} \tag{11.8}$$

The terms $L_m(j\omega)$ and $L_n(j\omega)$ are obtained by substituting $j\omega$ for D in $L_m(D)$ and $L_n(D)$. From Eq. (11.8) it follows that

$$|K(j\omega)| = \left| \frac{L_m(j\omega)}{L_n(j\omega)} \right| \omega f_0 \tag{11.9}$$

and

$$\phi = \measuredangle \frac{L_m(j\omega)}{L_n(j\omega)} \omega f_0 = \measuredangle \frac{L_m(j\omega)}{L_n(j\omega)} \tag{11.10}$$

Substitution of Eq. (11.9) into Eq. (11.6) yields

$$y(t)_{ss} = \left| \frac{L_m(j\omega)}{L_n(j\omega)} \right| f_0 \sin(\omega t + \phi) = y_0 \sin(\omega t + \phi) \tag{11.11}$$

From Eq. (11.11), the amplitude y_0 of the steady-state response $y(t)_{ss}$ is seen to be

$$y_0 = \left| \frac{L_m(j\omega)}{L_n(j\omega)} \right| f_0 \tag{11.12}$$

The amplitude ratio is

$$\frac{y_0}{f_0} = \left| \frac{L_m(j\omega)}{L_n(j\omega)} \right| \tag{11.13}$$

Illustrative example 11.1. A system is described by the differential equation

$$\frac{dy}{dt} + y = f(t) \qquad \text{or} \qquad y = \frac{1}{D+1} f(t)$$

Determine the response $y(t)$ to the sinusoidal input $f(t) = f_0 \sin \omega t$.

Solution. The transform of the differential equation is

$$sY(s) - y(0) + Y(s) = F(s)$$

or

$$Y(s) = \frac{F(s)}{s+1} + \frac{y(0)}{s+1} = \frac{\omega f_0}{(s+1)\,(s^2 + \omega^2)} + \frac{y(0)}{s+1}$$

The partial-fraction expansion is

$$Y(s) = \frac{K_1}{s+1} + \frac{K_c}{s-j\omega} + \frac{K_{-c}}{s+j\omega} + \frac{I_1}{s+1}$$

where $I_1 = y(0)$. Inverting yields

$$y(t) = (K_1 + I_1)e^{-t} + \frac{1}{\omega}\,|K(j\omega)|\sin(\omega t + \phi)$$

where

$$K_1 = \lim_{s \to -1} \frac{(s+1)\omega f_0}{(s+1)(s^2+\omega^2)} = \frac{\omega f_0}{\omega^2+1}$$

$$K(j\omega) = \lim_{s \to j\omega} \frac{(s^2+\omega^2)\omega f_0}{(s+1)(s^2+\omega^2)} = \frac{\omega f_0}{1+j\omega}$$

$$|K(j\omega)| = \frac{\omega f_0}{\sqrt{1+\omega^2}}$$

$$\phi = \measuredangle K(j\omega) = -\tan^{-1}\omega$$

The resulting transient response is

$$y(t) = \left[\frac{\omega f_0}{\omega^2+1} + y(0)\right] e^{-t} + \frac{f_0}{\sqrt{1+\omega^2}}\sin(\omega t + \phi)$$

The first term is the initial transient which "dies out" because of the decaying exponential e^{-t}. The second term is the steady-state frequency response $y(t)_{ss} = y_0 \sin(\omega t + \phi)$, where

$$y_0 = \frac{f_0}{\sqrt{1+\omega^2}} \qquad \text{and} \qquad \phi = -\tan^{-1}\omega$$

The amplitude ratio and phase shift may be verified directly by application of Eqs. (11.10) and (11.13). From the differential equation it follows that the differential operator is $1/(D+1)$. Replacing D by $j\omega$ shows that

$$\frac{y_0}{f_0} = \left|\frac{L_m(j\omega)}{L_n(j\omega)}\right| = \left|\frac{1}{j\omega+1}\right| = \frac{1}{\sqrt{1+\omega^2}}$$

$$\phi = \measuredangle \frac{L_m(j\omega)}{L_n(j\omega)} = \measuredangle \frac{1}{j\omega+1} = -\tan^{-1}\omega$$

We now show how to obtain plots of the amplitude ratio y_0/f_0 versus ω and plots of the phase angle ϕ versus ω. Consider the first-order linear differential equation

$$y(t) = \frac{1}{1+\tau D} f(t) \tag{11.14}$$

The substitution of $j\omega$ for D in the differential operator gives

$$\frac{L_m(j\omega)}{L_n(j\omega)} = \frac{1}{1+j\tau\omega} \tag{11.15}$$

The ratio of the amplitude of the output to that of the input is

$$\frac{y_0}{f_0} = \left|\frac{L_m(j\omega)}{L_n(j\omega)}\right| = \frac{1}{\sqrt{1+(\tau\omega)^2}} \tag{11.16}$$

The phase angle ϕ is

$$\phi = \measuredangle \frac{1}{1+j\tau\omega} \tag{11.17}$$

The angle of the numerator is $\measuredangle(1 + j0) = \tan^{-1} 0 = 0$. Thus, subtracting the angle of the denominator from the angle of the numerator gives

$$\phi = 0 - \measuredangle(1+j\tau\omega) = -\tan^{-1}\tau\omega \tag{11.18}$$

Figure 11.2 shows a graph of the amplitude ratio y_0/f_0 versus ω and a graph of the phase angle ϕ as a function of the angular velocity ω of the forcing function. When ω is small, the amplitude of the output is almost equal to that of the input and the phase angle ϕ is quite small. The output cannot keep up with the input at higher frequencies, and it begins to lag behind the input. For $\omega > 1/\tau$ this effect becomes very pronounced.

From Fig. 11.2, notice that, for a given ω, the amplitude ratio has a given value, and also there is a certain phase angle ϕ between the input and the output. This amplitude ratio $|L_m(j\omega)/L_n(j\omega)|$ and the phase angle $\measuredangle L_m(j\omega)/L_n(j\omega)$ determine a vector $L_m(j\omega)/L_n(j\omega)$. The path of the tip of this vector (vector loci) values of ω from 0 to ∞ is shown in Fig. 11.3. This is called a *polar plot*. The polar plot in Fig. 11.3 conveys the same information as the two separate curves shown in

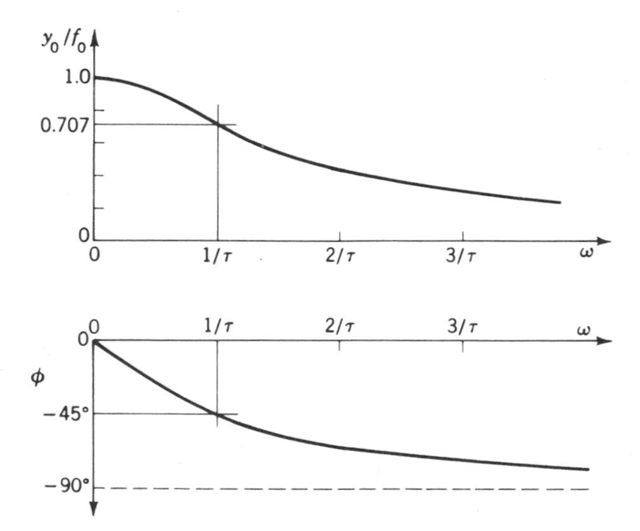

FIGURE 11.2
Response curves for the function $1/(1 + j\tau\omega)$.

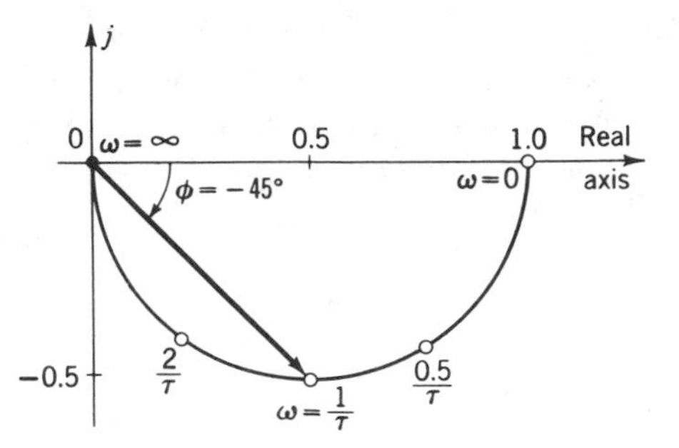

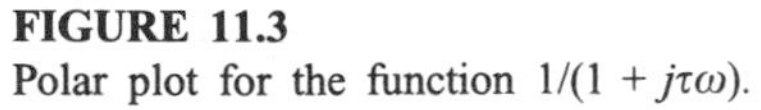
FIGURE 11.3
Polar plot for the function $1/(1 + j\tau\omega)$.

Fig. 11.2. The polar plot for this first-order system is a semicircle, as shown in Fig. 11.3. At $\omega = 0$, the length of the vector is 1, and the phase angle is zero. At $\omega = 1/\tau$ the phase angle $\phi = -45°$, and the length of the vector is 0.707.

Consider the general first-order system

$$y(t) = \frac{K}{1 + \tau D} f(t)$$

Replacing D by $j\omega$ in the differential operator shows that

$$\frac{L_m(j\omega)}{L_n(j\omega)} = \frac{K}{1 + j\tau\omega}$$

To prove that the polar plot for this first-order system is a circle, multiply the numerator and the denominator by the complex conjugate of the denominator:

$$\frac{K}{1 + j\tau\omega}\,\frac{1 - j\tau\omega}{1 - j\tau\omega} = \frac{K}{1 + \tau^2\omega^2} - \frac{jK\tau\omega}{1 + \tau^2\omega^2} = x + jy$$

Here $x = K/(1 + \tau^2\omega^2)$ is the real part, and $y = -K\tau\omega/(1 + \tau^2\omega^2)$ is the imaginary part. Now, note that

$$\left(x - \frac{K}{2}\right)^2 + y^2 = \left(\frac{K}{1 + \tau^2\omega^2} - \frac{K}{2}\right)^2 + \left(\frac{K\tau\omega}{1 + \tau^2\omega^2}\right)^2 = \left(\frac{K}{2}\right)^2$$

This is recognized as the equation for a circle with center at $(K/2, 0)$ and radius $R = K/2$.

The operation of the feedforward part of a control system is given by

$$c(t) = G(D)e(t) \tag{11.19}$$

When the actuating signal $e(t)$ is a sinusoid $[e(t) = e_0 \sin \omega t]$, then the controlled variable $c(t)$ is also a sinusoid $[c(t) = c_0 \sin(\omega t + \phi)]$. The amplitude ratio c_0/e_0 and the phase angle ϕ are

$$\begin{aligned} \frac{c_0}{e_0} &= |G(j\omega)| \\ \phi &= \measuredangle G(j\omega) \end{aligned} \tag{11.20}$$

Thus $G(j\omega)$ completely describes the frequency response of the feedforward elements. Similarly, the response of the feedback elements is determined by $H(j\omega)$.

The transfer function is the same as the differential operator except that D is replaced by s. Thus either the substitution of $j\omega$ for D in the differential operator or the substitution of $j\omega$ for s in the transfer function gives the vector equation for evaluating the frequency response.

11.2 LOGARITHMIC REPRESENTATION

Frequency-response methods are based on the response $G(j\omega)$ of the feedforward elements and $H(j\omega)$ of the feedback elements. The transfer functions for these quantities are $G(s)$ and $H(s)$, respectively. These quantities are usually obtained in factored form and are composed of multiples or ratios of one or more of the following terms: s, $1 + \tau s$, $(s^2 + 2\zeta\omega_n s + \omega_n^2)/\omega_n^2$. The substitution of $j\omega$ for s means that $G(j\omega)$ or $H(j\omega)$ will be composed of terms such as $j\omega$, $1 + j\tau\omega$, or $(\omega_n^2 - \omega^2 + j2\zeta\omega_n\omega)/\omega_n^2$. To obtain the resulting frequency response, the multiplication of such terms is simplified by the use of logarithms. For example, let the value of $G(s)$ be given by the equation

$$G(s) = \frac{K}{s(1 + \tau s)} \tag{11.21}$$

Substituting $j\omega$ for s yields

$$G(j\omega) = \frac{K}{j\omega(1 + j\tau\omega)} \tag{11.22}$$

The magnitude is

$$|G(j\omega)| = \frac{K}{|j\omega||1 + j\tau\omega|} = \frac{K}{\omega\sqrt{1 + \tau^2\omega^2}} \tag{11.23}$$

The logarithm of the preceding expression is

$$\log\left|\frac{G(j\omega)}{K}\right| = -\log\omega - \frac{1}{2}\log(1 + \tau^2\omega^2) \tag{11.24}$$

The angle ϕ is

$$\begin{aligned}\phi = \measuredangle G(j\omega) &= -\measuredangle j\omega - \measuredangle(1 + j\tau\omega)\\ &= -90^\circ - \tan^{-1}\tau\omega\end{aligned} \tag{11.25}$$

The contribution due to the $1/(j\omega)$ term is shown in Fig. 11.4. The equation for the amplitude is $\log|1/(j\omega)| = -\log\omega$. When ω is 1, the value of $-\log\omega$ is 0; and when ω changes by multiples of 10, the value of $-\log\omega$ changes in increments of 1. The slope of this straight-line logarithmic response shown in Fig. 11.4 is -1 log unit per decade. A *decade* is the horizontal distance on the frequency scale from any value of ω to 10ω. Thus, the distance from $\omega = 1$ to $\omega = 10$ or from $\omega = 3$ to $\omega = 30$, etc., is a decade.

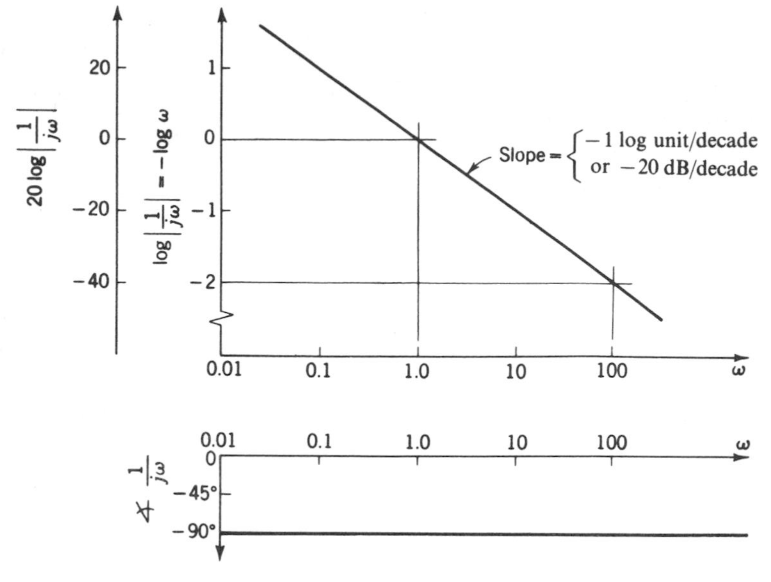

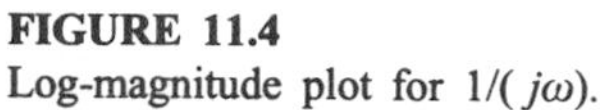

FIGURE 11.4
Log-magnitude plot for $1/(j\omega)$.

The vertical log-magnitude scale is sometimes expressed in decibels. To convert the vertical log-magnitude scale of Fig. 11.4 to decibel units, it is necessary only to multiply by a factor of 20. In decibels, the slope is -20 dB per decade rather than -1 log unit per decade. In this text, the log-magnitude scale is expressed directly in logarithmic units rather than in decibels.

The contribution due to $\log |1/(1 + j\tau\omega)|$ is shown in Fig. 11.5. For small values of ω such than $\omega \ll 1/\tau$,

$$-\tfrac{1}{2}\log(1 + \tau^2\omega^2) \approx -\tfrac{1}{2}\log 1 = 0 \tag{11.26}$$

This is the equation for the low-frequency asymptote to the exact curve, as is shown in Fig. 11.5. For $\omega \gg 1/\tau$,

$$-\tfrac{1}{2}\log(1 + \tau^2\omega^2) \approx -\tfrac{1}{2}\log \tau^2\omega^2 = -\log \tau\omega \tag{11.27}$$

This is the high-frequency asymptote. When $\omega = 1/\tau$, the value of $-\log\, \tau\omega$ is zero; and for $\omega = 10/\tau$, the value is -1; etc. The slope of this high-frequency asymptote is thus seen to be -1 log unit per decade, or -20 dB per decade. The break frequency ($\omega = 1/\tau$) is located directly under the intersection of the two asymptotes. For most preliminary design work, the asymptotes are sufficiently close to the exact curve that the extra effort involved in using the exact curve is generally not warranted. The maximum error occurs at the "break frequency" and is $0 - \tfrac{1}{2}\log\,(1 + 1) = -0.303/2 = -0.1515$. Notice that Fig. 11.5 is applicable for any value of τ.

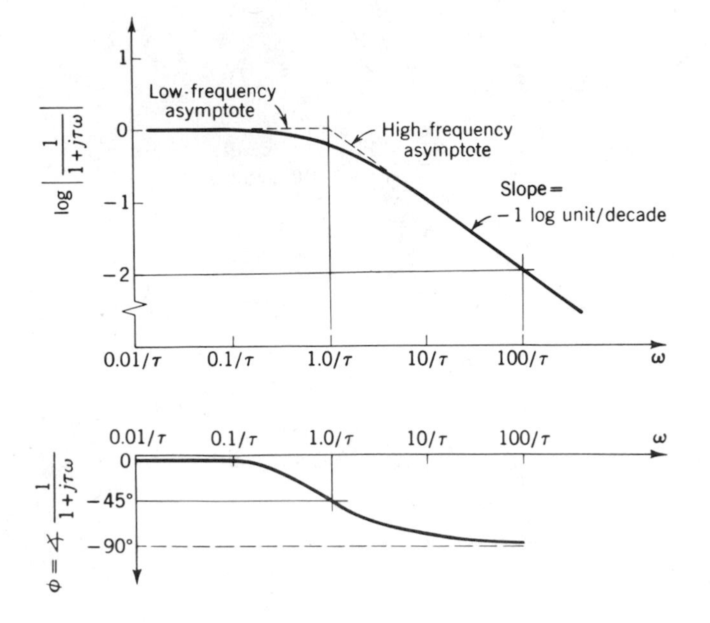

FIGURE 11.5
Log-magnitude plot for $1/(1 + j\tau\omega)$.

The phase-angle curve is obtained by solving the equation

$$\phi = -\tan^{-1} \tau\omega \tag{11.28}$$

Figure 11.6 shows graphically how Figs. 11.4 and 11.5 may be added to solve Eqs. (11.24) and (11.25). For numerical purposes, it is assumed that the value of τ is 0.1 s. At any frequency ω, the resulting log is the sum of the log of each term. For example, when $\omega = 1.0$, $\log|1/(j\omega)| = 0$ and $\log|1/(1 + 0.1j\omega)| = 0$. The sum which is $\log|G(j\omega)/K| = 0$. At $\omega = 100$, $\log|1/(j\omega)| = -2$ and $\log|1/(1 + 0.1j\omega)| = -1$. The sum which is $\log|G(j\omega)/K| = -3$. Also note that the resulting slope is the sum of the slope of each term. For $\omega < 1/\tau = 10$, the slope of the $\log|1/(j\omega)|$ term is -1, and the slope of the $\log|1/(1 + 0.1j\omega)|$ term is 0. The sum is the slope of the $\log|G(j\omega)/K|$ term which is -1. For $\omega > 1/\tau = 10$, the slope of the $\log|1/j\omega|$ term is -1, and the slope of the $\log|1/(1 + 0.1j\omega)|$ term is -1. The sum is the slope of the $\log|G(j\omega)/K|$ term, which is -2. Similarly, at any frequency ω, the resulting angle is the sum of the angles of each term. Such graphs of the log magnitude and angle versus log frequency are called *log-magnitude plots*, or *Bode plots*. H. W. Bode* made many contributions to the development of frequency-response techniques.

* H. W. Bode, *Network Analysis and Feedback Amplifier Design*, D. Van Nostrand Company, Inc., Princeton, N.J., 1945.

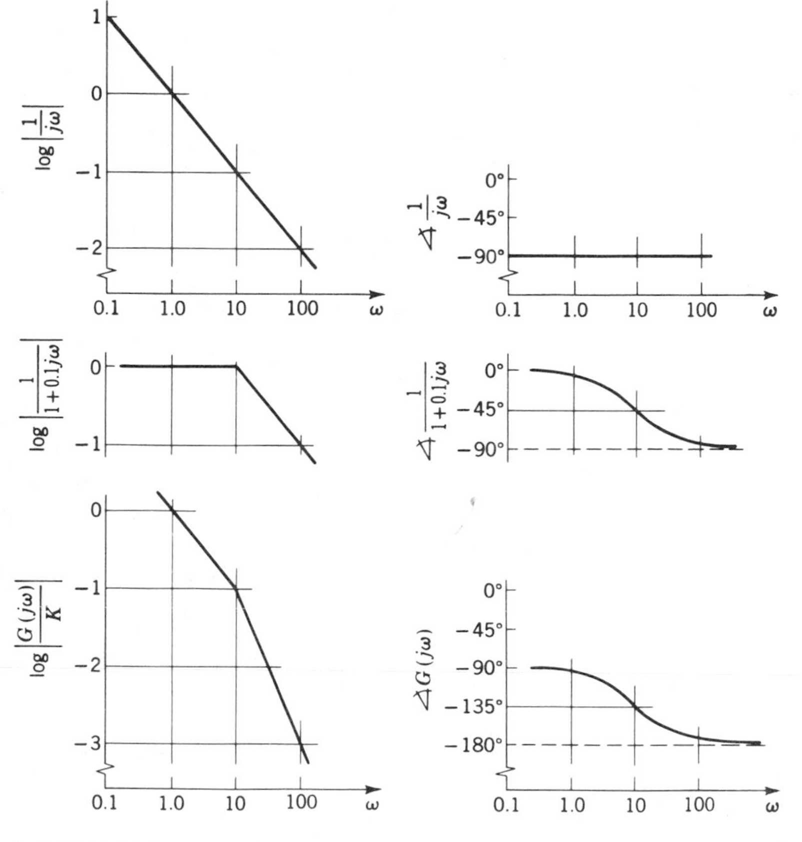

FIGURE 11.6
Log-magnitude plot for $1/[j\omega(1 + 0.1j\omega)]$.

A feature of the logarithmic method is that if the term appears in the numerator rather than the denominator, it is necessary merely to change the sign of the amplitude and phase-angle scales in tabulating the result. For example, if Eq. (11.22) were of the form

$$G(j\omega) = \frac{K(1 + j\tau\omega)}{j\omega}$$

then

$$\log\frac{|G(j\omega)|}{K} = -\log\omega + \frac{1}{2}\log(1 + \tau^2\omega^2) \tag{11.29}$$

and

$$\measuredangle G(j\omega) = -90° + \tan^{-1}\tau\omega \tag{11.30}$$

By comparing Eqs. (11.29) and (11.30) with Eqs. (11.24) and (11.25), we see that only the sign of the term which went from the denominator to the

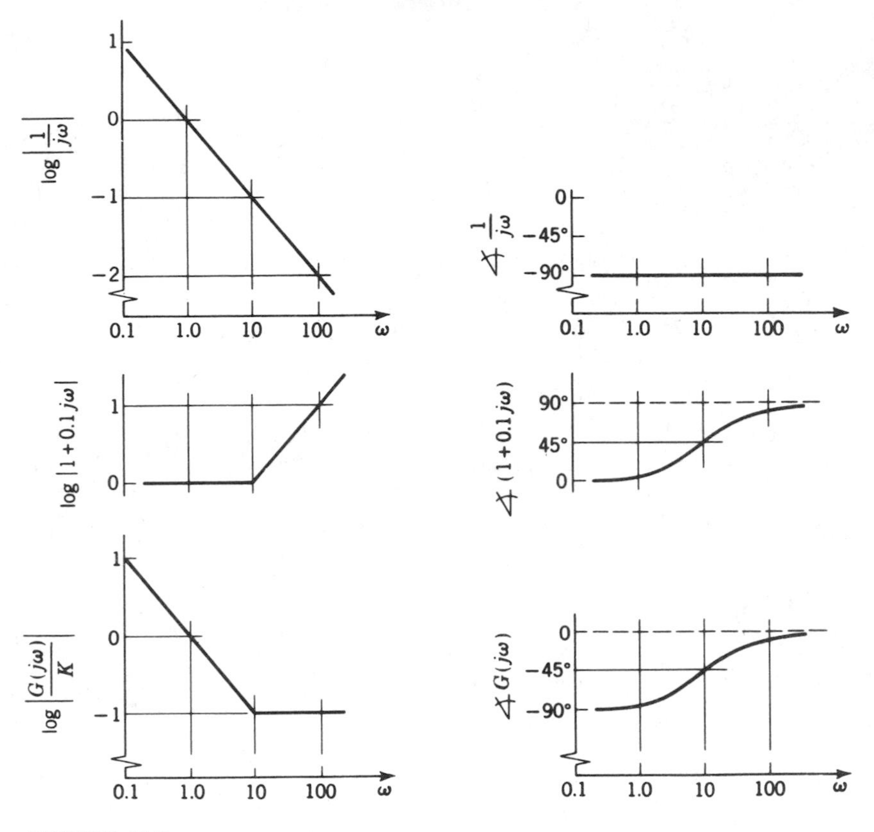

FIGURE 11.7
Log-magnitude plot for $G(j\omega)/K = (1 + 0.1j\omega)/(j\omega)$.

numerator has changed. Figure 11.7 shows the log-magnitude plot for $G(j\omega)/K = (1 + j\tau\omega)/(j\omega)$, in which the value of τ is assumed to be 0.1 s. Comparison of the log-magnitude plot for $1 + j\tau\omega$ of Fig. 11.7 with that for $1/(1 + j\tau\omega)$ of Fig. 11.6 shows that the sign of the logarithm of the amplitude ratio and the sign of the phase angle have been changed.

The third type of term which occurs is of the form $(\omega_n^2 - \omega^2 + j2\zeta\omega_n\omega)/\omega_n^2$. By using a generalized graph, this term may be treated in a manner analogous to that described for evaluating terms of the form $1 + j\tau\omega$. Consider the function

$$G(j\omega) = \frac{K_1}{(1 + j\tau\omega)(\omega_n^2 - \omega^2 + j2\zeta\omega_n\omega)}$$

This may be rewritten in the form

$$G(j\omega) = \frac{K}{(1 + j\tau\omega)\left[(\omega_n^2 - \omega^2) + j2\zeta\omega_n\omega\right]/\omega_n^2} \tag{11.31}$$

where $K = K_1/\omega_n^2$. The logarithm of the magnitude of the quadratic term is

$$\log\left|\frac{1}{1-(\omega/\omega_n)^2+j2\zeta(\omega/\omega_n)}\right| = \log\frac{1}{\sqrt{[1-(\omega/\omega_n)^2]^2+[2\zeta(\omega/\omega_n)]^2}}$$

$$= -\frac{1}{2}\log\left\{\left[1-\left(\frac{\omega}{\omega_n}\right)^2\right]^2+\left(2\zeta\frac{\omega}{\omega_n}\right)^2\right\} \tag{11.32}$$

For small values of ω/ω_n such that $\omega/\omega_n \ll 1$, Eq. (11.32) becomes

$$-\tfrac{1}{2}\log 1 = 0$$

This is the equation for the low-frequency asymptote, which is a horizontal straight line like that obtained for the $1 + j\tau\omega$ term.

The equation for the high-frequency asymptote is obtained by noting that, for $\omega/\omega_n \gg 1$, it follows that

$$\left[1-\left(\frac{\omega}{\omega_n}\right)^2\right]^2 \approx \left(\frac{\omega}{\omega_n}\right)^4 \gg \left(2\zeta\frac{\omega}{\omega_n}\right)^2$$

Thus, for $\omega/\omega_n \gg 1$, Eq. (11.32) becomes

$$-\frac{1}{2}\log\left(\frac{\omega}{\omega_n}\right)^4 = -2\log\frac{\omega}{\omega_n} \tag{11.33}$$

The slope of the high-frequency asymptote is -2 log units per decade, or -40 dB per decade, and this asymptote intersects the low-frequency asymptote at $\omega/\omega_n = 1$.

The value of the phase angle is

$$\phi = -\tan^{-1}\frac{2\zeta(\omega/\omega_n)}{1-(\omega/\omega_n)^2} \tag{11.34}$$

The nondimensional curves for the logarithm of the amplitude and the phase angle, as given by Eqs. (11.32) and (11.34), respectively, are shown in Fig. 11.8. The curves for the reciprocal of this function are obtained by merely changing the sign of the amplitude and phase-angle scales.

The log-magnitude plot of Fig. 11.8 shows that the amplitude ratio attains its maximum value in the vicinity of $\omega/\omega_n = 1$. The frequency ω at which this maximum occurs is called the *resonant* or *peak frequency* ω_p. The equation for $G(j\omega)$ is

$$G(j\omega) = \frac{1}{1-(\omega/\omega_n)^2+j2\zeta(\omega/\omega_n)}$$

The magnitude of $G(j\omega)$ is

$$|G(j\omega)| = \frac{1}{\sqrt{[1-(\omega/\omega_n)^2]^2+[2\zeta(\omega/\omega_n)]^2}} \tag{11.35}$$

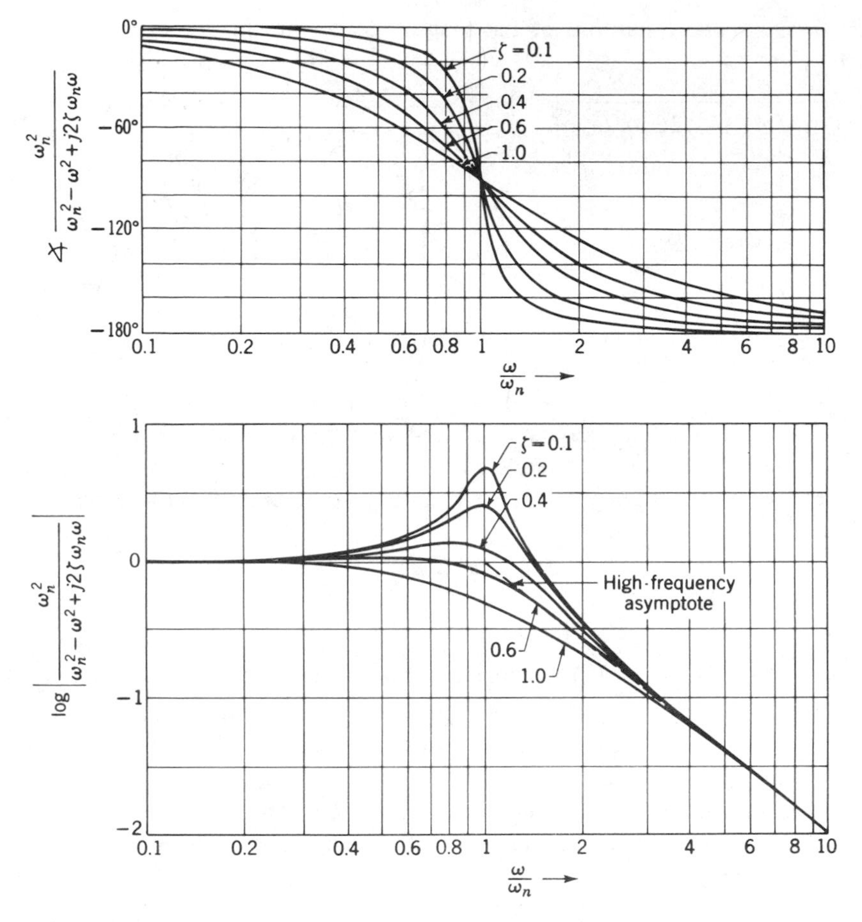

FIGURE 11.8
Log-magnitude plot for $1/[1 - (\omega/\omega_n)^2 + j2\zeta\omega/\omega_n]$.

The maximum value of $|G(j\omega)|$ occurs when the denominator is a minimum. The quantity under the radical is a function of the frequency ratio ω/ω_n:

$$f(\omega/\omega_n) = \left[1 - \left(\frac{\omega}{\omega_n}\right)^2\right]^2 + \left(2\zeta\frac{\omega}{\omega_n}\right)^2$$

The frequency $\omega = \omega_p$ at which this function attains its minimum value is where

$$\frac{df(\omega/\omega_n)}{d(\omega/\omega_n)} = 2\left[1 - \left(\frac{\omega}{\omega_n}\right)^2\right]\left(-2\frac{\omega}{\omega_n}\right) + 2\left(2\zeta\frac{\omega}{\omega_n}\right)(2\zeta) = 0$$

or

$$\left(\frac{\omega}{\omega_n}\right)^2 = 1 - 2\zeta^2$$

Hence, the peak frequency $\omega = \omega_p$ is

$$\omega_p = \omega_n\sqrt{1-2\zeta^2} \qquad 0 \le \zeta \le 0.707 \tag{11.36}$$

The corresponding amplitude ratio at this frequency is the maximum or peak value M_p. This peak value is obtained by substituting $\omega = \omega_p$ into Eq. (11.35). That is,

$$M_p = |G(j\omega_p)| = \frac{1}{2\zeta\sqrt{1-\zeta^2}} \qquad 0 \le \zeta \le 0.707 \tag{11.37}$$

In summary, for any function which is composed of multiples of terms such as $j\omega$, $1 + j\tau\omega$, and $(\omega_n^2 - \omega^2 + j2\zeta\omega_n\omega)/\omega_n^2$, the value of $\log|G(j\omega)|$ and $\measuredangle G(j\omega)$ at any given angular velocity ω of the driving sine wave is the sum of the contribution due to each term, which is obtained from Figs. 11.4, 11.5, or 11.8. Thus, having $|G(j\omega)|$ and $\measuredangle G(j\omega)$, one may construct a Bode plot or a polar plot of $G(j\omega)$, as was demonstrated by Fig. 11.3.

To write a computer program for obtaining $|G(j\omega)|$ and $\measuredangle G(j\omega)$, first obtain the magnitude and angle of each term in $G(j\omega)$. The product of the magnitude of each term is $|G(j\omega)|$, and the sum of the angle of each term is $\measuredangle G(j\omega)$. To illustrate this procedure, consider the transfer function

$$G(s) = \frac{10}{s(1+0.25s)\,(s^2+10s+64)/64}$$

Replacing s by $j\omega$ gives

$$G(j\omega) = \frac{10}{j\omega(1+0.25j\omega)\,(64-\omega^2+j10\omega)/64}$$

This may be written in the form

$$G(j\omega) = 10G_1(j\omega)G_2(j\omega)G_3(j\omega)$$

where

$$G_1(j\omega) = \frac{1}{j\omega}$$

$$G_2(j\omega) = \frac{1}{1+0.25j\omega}$$

$$G_3(j\omega) = \frac{64}{64-\omega^2+j10\omega}$$

The magnitudes and angles are

$$|G_1(j\omega)| = \frac{1}{\omega} \qquad \measuredangle G_1(j\omega) = -90.0^\circ = \frac{-\pi}{2}$$

$$|G_2(j\omega)| = \frac{1}{\sqrt{1+(0.25\omega)^2}} \qquad \measuredangle G_2(j\omega) = -\tan^{-1} 0.25\omega$$

$$|G_3(j\omega)| = \frac{64}{\sqrt{(64-\omega^2)^2+(10\omega)^2}} \qquad \measuredangle G_3(j\omega) = -\tan^{-1}\frac{10\omega}{64-\omega^2}$$

The computer program is as follows:

```
      W = 1.0
      DO 25 K = 1,6
      G1 = 1.0/W
      AG1 = -3.1416/2.0
      G2 = 1.0/SQRT(1.0+(0.25*W)**2)
      AG2 = -ATAN2(0.25*W,1.0)
      G3 = 64.0/SQRT((64.0 - W**2)**2+(10.0*W)**2)
      AG3 = -ATAN2(10.0*W,64.0 - W**2)
      G = 10.0*G1*G2*G3
      AG = AG1+AG2+AG3
      X = G*COS(AG)
      Y = G*SIN(AG)
      AGD = 180.0*AG/3.1416
      WRITE(6,10) W,G,AGD,X,Y
   10 FORMAT(5F10.2)
      W = 2.0*W
   25 CONTINUE
      END
```

In this program, W = ω, G1 = $|G_1(j\omega)|$, AG1 = $\measuredangle G_1(j\omega)$, G2 = $|G_2(j\omega)|$, AG2 = $\measuredangle G_2(j\omega)$, G3 = $|G_3(j\omega)|$, AG3 = $\measuredangle G_3(j\omega)$, G = $|G(j\omega)|$, and AG = $\measuredangle G(j\omega)$. The statement ATAN2(Y,X) returns the angle whose tangent is y/x. This statement puts the angle in the correct quadrant, as determined by the sign of y and the sign of x. The real component of $G(j\omega)$ is X = G*COS(AG), and the imaginary component is Y = G*SIN(AG). The computer works with angles in radian measure. The statement AGD = 180.0*AG/3.1416 changes the angle of G from radians to degrees. The first time through the DO loop, W = ω = 1. Thereafter, W = 2W so that the corresponding values of ω, $|G(j\omega)|$, $\measuredangle G(j\omega)$, X, and Y are printed for values of $\omega = 1, 2, 4, 8, 16$, and 32. Many computers have plot routines that enable them to plot corresponding values of X and Y, which is the polar plot. The Bode plot is obtained by plotting $\log|G(j\omega)|$ versus ω and plotting $\measuredangle G(j\omega)$ versus ω.

Experimental Determination of Frequency Response

A feature of frequency-response methods is that the responses $G(j\omega)$ and $H(j\omega)$ may be determined experimentally. For example, at a given frequency ω, the value of $G(j\omega)$ is obtained by exciting the feedforward elements with a sinusoidal input of angular velocity ω and then measuring the ratio of the amplitude of the output to that of the input and measuring the phase angle ϕ. By repeating this process for a wide range of values of ω, the frequency response is obtained. The response $H(j\omega)$ is similarly obtained by sinusoidally exciting the feedback elements.

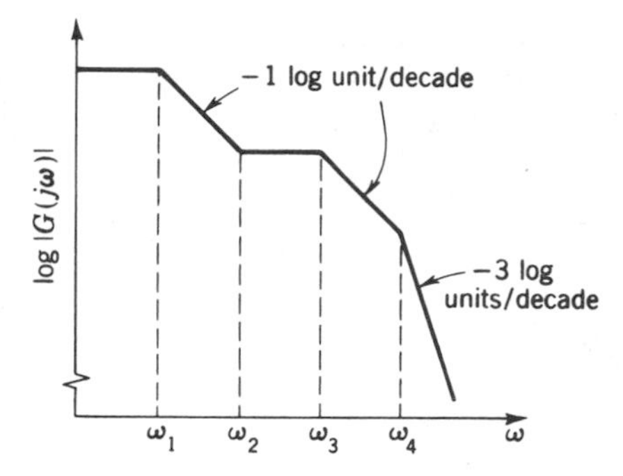

FIGURE 11.9
Experimentally determined log-magnitude plot.

The asymptotes of an experimentally determined log-magnitude plot are shown in Fig. 11.9. Because of the change of slope at ω_1, there is a term $1/(1+j\tau_1\omega)$, where $\tau_1 = 1/\omega_1$ in the frequency-response equation. At the angular velocity ω_2 there is a net increase of +1 log unit per decade, and thus the term $1+j\tau_2\omega$, where $\tau_2 = 1/\omega_2$, appears in the numerator of the response expression. The change in slope at ω_3 is indicative of the term $1/(1+j\tau_3\omega)$, where $\tau_3 = 1/\omega_3$. Because the slope changes by -2 log units per decade at ω_4, there is a quadratic term in the denominator. The value of the break frequency ω_4 is equal to the natural frequency ω_n for the quadratic. That is, from Fig. 11.9, it follows that at the break point for a quadratic, $\omega/\omega_n = \omega_4/\omega_n = 1$, or $\omega_n = \omega_4$. To determine the damping ratio ζ, it is necessary to compare the exact response curve for the component which causes this quadratic term in $G(j\omega)$ to the general response curves of Fig. 11.8. By knowing M_p, the damping ratio ζ may be determined from Eq. (11.37).

From the preceding, it follows that the frequency response for $G(j\omega)$ is

$$G(j\omega) = \frac{K(1+j\tau_2\omega)}{(1+j\tau_1\omega)(1+j\tau_3\omega)\left[\left(\omega_4^2-\omega^2+j2\zeta\omega_4\omega\right)/\omega_4^2\right]} \tag{11.38}$$

In the next section, we show how the value of K can be determined directly from the low-frequency portion of the log-magnitude plot.

After the equation for the frequency response has been experimentally determined, it is a simple matter to substitute s for $j\omega$ to obtain the transfer function. The substitution of s for $j\omega$ in Eq. (11.38) gives

$$G(s) = \frac{K(1+\tau_2 s)}{(1+\tau_1 s)(1+\tau_3 s)\left[\left(s^2+2\zeta\omega_4 s+\omega_4^2\right)/\omega_4^2\right]} \tag{11.39}$$

The substitution of D for s gives the differential equation of operation.

The magnitude of $1+j\tau\omega$ is the same as that of $1-j\tau\omega$. However, as the phase angle for $1+j\tau\omega$ goes from 0° to 90°, the phase angle for $1-j\tau\omega$ goes from 0° to $-90°$. Most systems are minimum-phase systems; i.e., all factors are in the form $j\omega$, $1+j\tau\omega$, or $(\omega_n^2-\omega^2+j2\zeta\omega_n\omega)/\omega_n^2$. For minimum-phase systems, as ω becomes infinite, the phase angle is $\phi = -90°(n-m)$, where n is the order of the denominator and m is that of the numerator. Non-minimum-phase systems may be detected from the phase-angle plot, because as ω becomes infinite,

$\phi \neq -90°(n - \text{m})$. For either minimum- or non-minimum-phase systems, the slope of the log-magnitude diagram at high frequencies is $-(n - \text{m})$ log units per decade. All the frequency-response techniques to be discussed in this text are equally valid for minimum- or non-minimum-phase systems.

11.3 EVALUATING THE GAIN K

In general, a transfer function may be expressed in the form

$$G(s) = \frac{K_n\{(1 + \tau_a s) \cdots [(s^2 + 2\zeta_a\omega_{n_a}s + \omega_{n_a}^2)/\omega_{n_a}^2] \cdots\}}{s^n\{(1 + \tau_1 s) \cdots [(s^2 + 2\zeta_1\omega_{n_1}s + \omega_{n_1}^2)/\omega_{n_1}^2] \cdots\}} \tag{11.40}$$

where K_n is the overall gain of the transfer function $G(s)$ and n is the power to which the s term in the denominator is raised. Usually, the value of n is 0, 1, or 2. The first time constant in the numerator is τ_a, the second τ_b, etc. The natural frequency for the first quadratic term in the numerator is ω_{n_a}, and the damping ratio is ζ_a. Similarly, τ_1 is the first time constant that appears in the denominator, τ_2 is the second, etc. The natural frequency for the first quadratic term in the denominator is ω_{n_1}, and its damping ratio is ζ_1.

Substitution of $j\omega$ for s in Eq. (11.40) gives

$$G(j\omega) = \frac{K_n\{(1 + j\tau_a\omega) \cdots [(\omega_{n_a}^2 - \omega^2 + j2\zeta_a\omega_{n_a}\omega)/\omega_{n_a}^2] \cdots\}}{(j\omega)^n\{(1 + j\tau_1\omega) \cdots [(\omega_{n_1}^2 - \omega^2 + j2\zeta_1\omega_{n_1}\omega)/\omega_{n_1}^2] \cdots\}} \tag{11.41}$$

For small values of ω, all the terms inside the braces of the preceding expression approach 1, so that

$$G(j\omega) = \frac{K_n}{(j\omega)^n} \qquad \omega \approx 0 \tag{11.42}$$

As indicated from the preceding equation, the gain K_n can be determined from the low-frequency portion of the log-magnitude plot. The techniques for evaluating K_n for $n = 0$, 1, or 2 are described in the following.

$n = 0$. When n is zero, there is no $j\omega$ term in $G(j\omega)$. Thus, Eq. (11.42) becomes

$$G(j\omega) = K_0 \qquad \omega \approx 0 \tag{11.43}$$

A typical log-magnitude plot for this case is shown in Fig. 11.10*a*. For small values of ω the low-frequency asymptote has a constant value $G(j\omega) = K_0$.

$n = 1$. For small values of ω and for n equal to 1, it follows from Eq. (11.42) that

$$G(j\omega) = \frac{K_1}{j\omega} \qquad \omega \approx 0 \tag{11.44}$$

The logarithm of the magnitude of $G(j\omega)$ is

$$\log|G(j\omega)| = \log K_1 - \log\omega \qquad \omega \approx 0 \tag{11.45}$$

From Eq. (11.45) it follows that as ω changes by a factor of 10, $\log|G(j\omega)|$ changes by -1. Thus for $\omega \approx 0$, the slope of the curve of $\log|G(j\omega)|$ versus ω is

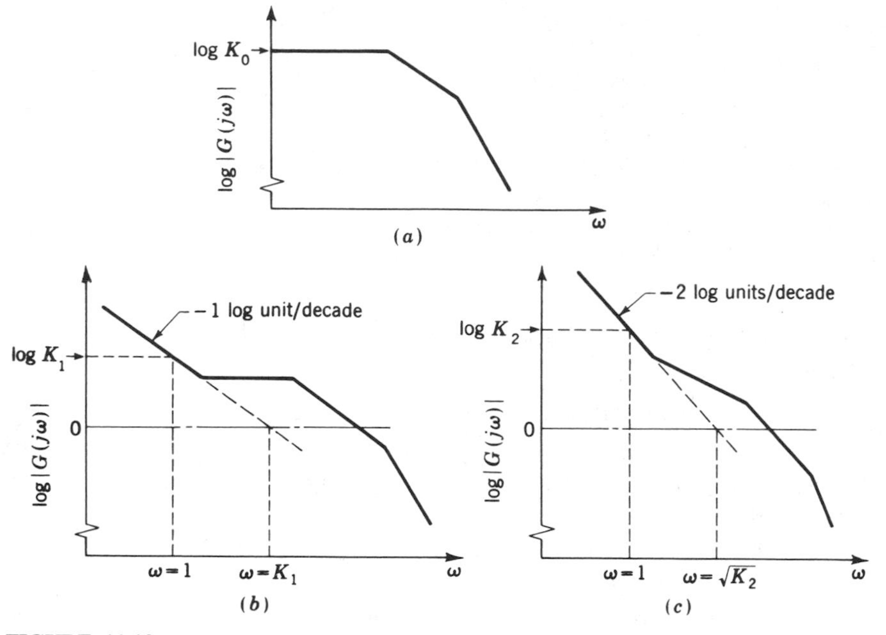

FIGURE 11.10
Log-magnitude plots: (*a*) $n = 0$, (*b*) $n = 1$, (*c*) $n = 2$.

-1 log unit per decade. A typical log-magnitude plot for the case in which $n = 1$ is shown in Fig. 11.10*b*. Note that the low-frequency slope of -1 log unit per decade or its extension intersects the horizontal line through $\log|G(j\omega)| = 0$ at the point where $\omega = K_1$. This fact follows directly from Eq. (11.45) by noting that when $\log|G(j\omega)| = 0$, then $\log\omega = \log K_1$ or simply $\omega = K_1$. In addition, for ω equal to 1, Eq. (11.45) becomes $\log|G(j\omega)| = \log K_1$. Thus, as shown in Fig. 11.10*b*, a vertical line through $\omega = 1$ intersects the low-frequency asymptote or its extension at the value $\log K_1$.

$n = 2$. The low-frequency equation for this case is

$$G(j\omega) = \frac{K_2}{(j\omega)^2} = -\frac{K_2}{\omega^2} \qquad \omega \approx 0 \tag{11.46}$$

The logarithm of the magnitude of $G(j\omega)$ is

$$\log|G(j\omega)| = \log K_2 - 2\log\omega \qquad \omega \approx 0 \tag{11.47}$$

A typical log-magnitude diagram for $n = 2$ is illustrated in Fig. 11.10*c*. The slope at low frequencies is -2 log units per decade. From Eq. (11.47), it follows that when $\log|G(j\omega)| = 0$, then $\log\omega = \frac{1}{2}\log K_2$, or $\omega = \sqrt{K_2}$. Thus, the low-frequency asymptote or its extension intersects the horizontal line $\log|G(j\omega)| = 0$ at the frequency $\omega = \sqrt{K_2}$. When $\omega = 1$, then Eq. (11.47) becomes $\log|G(j\omega)| = \log K_2$. Thus a vertical line through $\omega = 1$ intersects the low-frequency asymptote or its extension at $\log K_2$.

Replacing $j\omega$ by s in Eq. (11.42) and solving for K_n show that

$$K_n = \lim_{s \to 0} s^n G(s)$$

Comparison of the result for $n = 0$ with Eq. (5.81) shows that $K_0 = K_p$ is the positional error constant. Similarly, comparison of the result for $n = 1$ with Eq. (5.82) shows that $K_1 = K_v$ is the velocity error constant. Finally, comparison of the result for $n = 2$ with Eq. (5.83) shows that $K_2 = K_a$ is the acceleration error constant.

11.4 EQUIVALENT UNITY-FEEDBACK SYSTEMS

Much simplification is afforded in the application of frequency-response methods to systems having unity feedback. A control system having feedback elements $H(D)$ can usually be represented by an equivalent unity-feedback system, as illustrated in Fig. 11.11. For the case in which $H(D)$ is a constant, the equivalent unity-feedback system is readily obtained by moving the constant $H(D)$ to the input side of the main loop. The systems represented by Figs. 3.37 and 4.13 have a constant term $H(D) = C_4$ in the feedback path. Moving C_4 to the input side of the main loop yields the unity-feedback systems shown in Fig. 11.12*a* and *b*, respectively.

To obtain the equivalent unity-feedback system when $H(D)$ is not a constant, first write $H(D)$ in the form $H(D) = C[1 + H_1(D)]$. The constant C may now be taken out of the feedback path, and the remaining term $1 + H_1(D)$ may be represented as shown in Fig. 11.13 by two separate paths. The design of such systems in which there is an inner, or minor, feedback path is often facilitated by the use of inverse polar plots, as discussed in Sec. 12.7.

In obtaining the equivalent unity-feedback system, only constant terms are to be taken outside the main loop. The fact that $r(t)$ is equal to some constant times the command signal does not affect the basic dynamic behavior of the system.

System Type

When a system is represented in its equivalent unity-feedback form, the value of n in $G(s)$ as indicated by Eq. (11.40) has a predominant effect upon the behavior of the system. When $n = 0$, the system is designated as a *type 0 system*. A type 1 system is one for which $n = 1$, a type 2 system is one for which $n = 2$, etc.

A type 0 system results when there is no integration, as in a proportional control. For an integral control in which there is one integrator in the feedforward elements, n is equal to 1. A type 2 system has two integrations in the feedforward

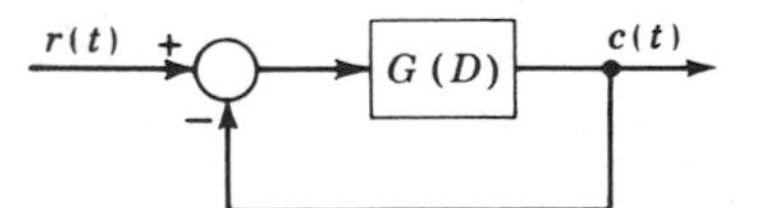

FIGURE 11.11
Unity-feedback system.

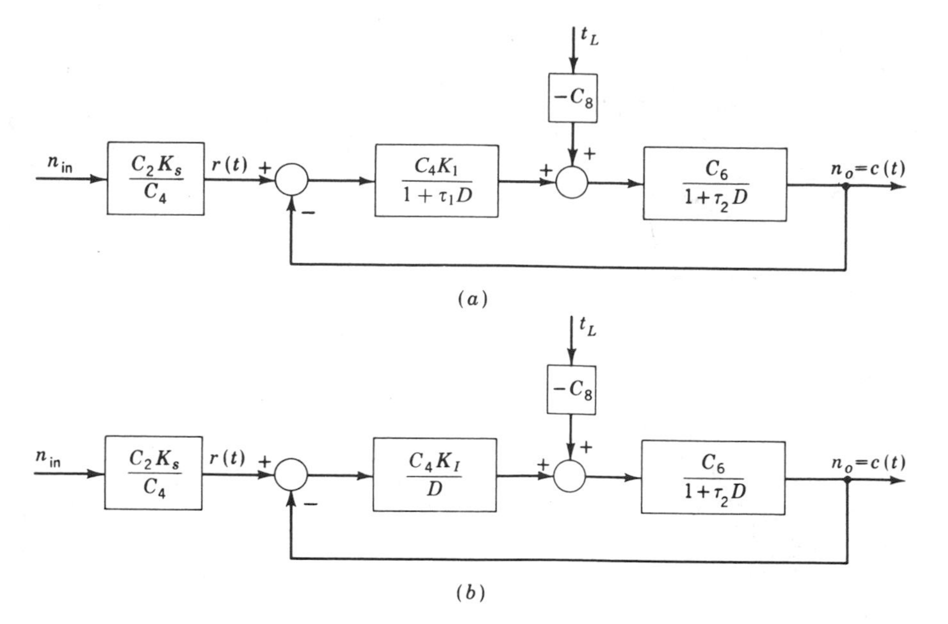

FIGURE 11.12
Equivalent unity-feedback systems when $H(D)$ is a constant.

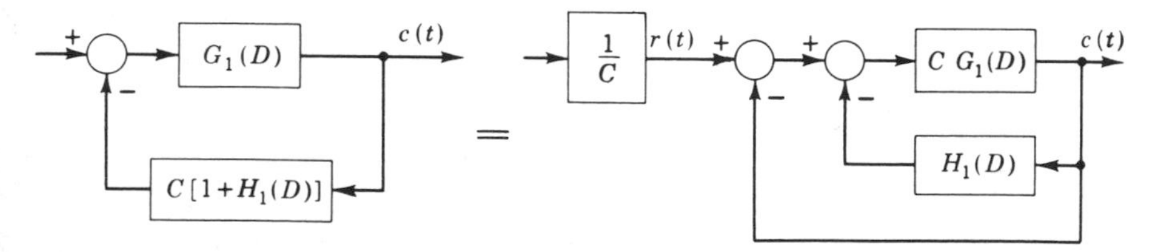

FIGURE 11.13
Equivalent unity-feedback system when $H(D)$ is not a constant.

elements, etc. The system in Fig. 11.12*a* is a type 0 system, and the system in Fig. 11.12*b* is a type 1 system.

11.5 POLAR PLOTS

Vector loci or polar plots are better suited for the solution of certain control problems than are log-magnitude plots, and vice versa. As explained later, other methods of representing frequency-response information are the log-modulus, or Nichols, plot and the inverse polar plot. A control engineer must be familiar with all these means of plotting frequency-response data in order to be able to select the method which is best suited to a particular problem.

The polar plots for a number of commonly encountered functions are shown in Fig. 11.14. Polar plots may often be roughly sketched by knowing the location at

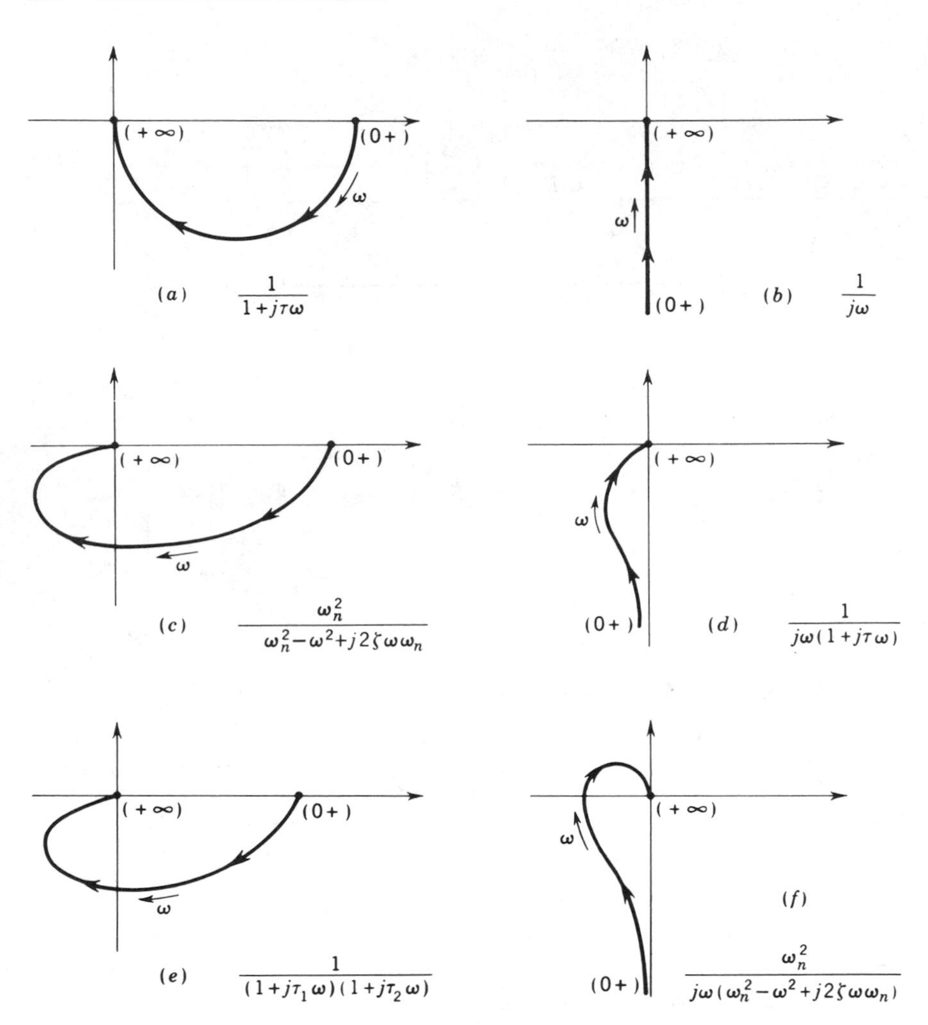

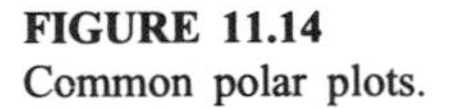
FIGURE 11.14
Common polar plots.

low frequencies ($\omega \to 0$) and at high frequencies ($\omega \to \infty$). For example, in Fig. 11.14*a*

$$\left.\frac{1}{1+j\tau\omega}\right|_{\omega=0+} = 1$$
$$\left.\frac{1}{1+j\tau\omega}\right|_{\omega=\infty} \approx \left.\frac{1}{j\tau\omega}\right|_{\omega=+\infty} = \left.\frac{e^{-j90^\circ}}{\tau\omega}\right|_{\omega=+\infty} = (0+)e^{-j90^\circ} \tag{11.48}$$

Thus, the locus of Fig. 11.14*a* begins at the +1 point on the positive real axis ($\phi = 0^\circ$), and as ω approaches ∞, the locus approaches 0 along the -90° axis. The low- and high-frequency values for Fig. 11.14*b* are

$$\left.\frac{1}{j\omega}\right|_{\omega=0+} = \left.\frac{1}{\omega}e^{-j90^\circ}\right|_{\omega=0+} = (+\infty)e^{-j90^\circ}$$
$$\left.\frac{1}{j\omega}\right|_{\omega=+\infty} = \left.\frac{1}{\omega}e^{-j90^\circ}\right|_{\omega=+\infty} = (0+)e^{-j90^\circ} \tag{11.49}$$

For Fig. 11.14*c*

$$\left.\frac{1}{1-(\omega/\omega_n)^2 + j2\zeta(\omega/\omega_n)}\right|_{\omega=0+} = 1$$
$$\left.\frac{1}{1-(\omega/\omega_n)^2 + j2\zeta(\omega/\omega_n)}\right|_{\omega=+\infty} \approx \left.\frac{1}{-(\omega/\omega_n)^2}\right|_{\omega=+\infty} \tag{11.50}$$
$$= -\left.\left(\frac{\omega_n}{\omega}\right)^2\right|_{\omega=+\infty} = (0+)e^{-j180^\circ}$$

The limiting value for a function composed of multiples of the preceding terms is equal to the product of the contribution of each term. For example, in Fig. 11.14*d*

$$\left.\frac{1}{j\omega(1+j\tau\omega)}\right|_{\omega=0+} = \left.\frac{1}{j\omega}\right|_{\omega=0+} \left.\frac{1}{1+j\tau\omega}\right|_{\omega=0+} = (+\infty)e^{-j90^\circ}$$
$$\left.\frac{1}{j\omega(1+j\tau\omega)}\right|_{\omega=+\infty} = \left.\frac{1}{j\omega}\right|_{\omega=+\infty} \left.\frac{1}{1+j\tau\omega}\right|_{\omega=+\infty} = (0+)e^{-j90^\circ}(0+)e^{-j90^\circ}$$
$$= (0+)e^{-j180^\circ} \tag{11.51}$$

Application of this technique will verify the results shown in Fig. 11.14*e* and *f*.

The low- and high-frequency locations are summarized in Table 11.1. The reciprocal of any term in Table 11.1 is obtained by changing the sign of the phase angle and taking the reciprocal of the magnitude. At low frequencies, there is a phase shift with the $1/(j\omega)$ term only. Thus, for a type 0 system, the polar plot originates on the positive real axis, where ϕ is zero. Typical polar plots for type 0, 1,

TABLE 11.1

ω	$\frac{1}{j\omega}$	$\frac{1}{1+j\tau\omega}$	$\frac{\omega_n^2}{\omega_n^2-\omega^2+j2\zeta\omega\omega_n}$
$0+$	$\infty\angle -90^\circ$	$1\angle 0^\circ$	$1\angle 0^\circ$
$+\infty$	$0\angle -90^\circ$	$0\angle 90^\circ$	$0\angle -180^\circ$

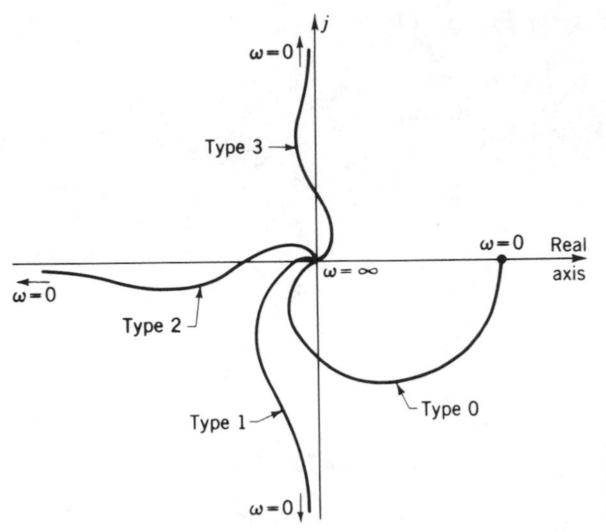

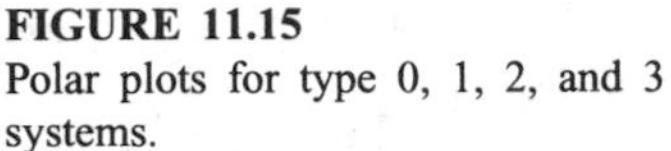

FIGURE 11.15
Polar plots for type 0, 1, 2, and 3 systems.

2, and 3 systems are shown in Fig. 11.15. A type 1 system begins at infinity on the negative 90° axis, a type 2 system begins at infinity on the negative 180° axis, etc. For most systems, the polar plot terminates at the origin when $\omega = \infty$.

11.6 *M* AND α CIRCLES

Frequency-response methods make extensive use of the open-loop frequency response $G(j\omega)$. The open-loop response is the response that would be obtained if the feedback path were disconnected at the comparator (i.e., opened).

For unity-feedback systems, the closed-loop frequency response $C(j\omega)/R(j\omega)$ is related to the open-loop response $G(j\omega)$ by the equation

$$\frac{C(j\omega)}{R(j\omega)} = \frac{G(j\omega)}{1 + G(j\omega)} \tag{11.52}$$

Figure 11.16 shows a typical $G(j\omega)$ plot. The vector from the origin to a point on the curve is $G(j\omega)$, and the vector from the point $-1 + j0$ to the same point on this

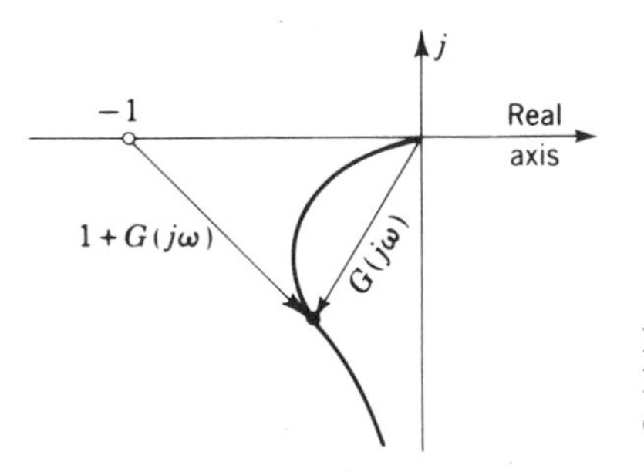

FIGURE 11.16
Determination of closed-loop frequency response from open-loop response.

curve is $1 + G(j\omega)$. The ratio of these two vectors is the closed-loop frequency response for the value of ω at that point. This shows that every point on the $G(j\omega)$ plane corresponds to a certain value of $C(j\omega)/R(j\omega)$. The magnitude of the ratio of the amplitude of the output sinusoid to the input is designated by the symbol $M = |C(j\omega)/R(j\omega)|$.

In Fig. 11.17, we see that the locus of lines of constant M are circles on the $G(j\omega)$ plane. The proof of this follows: Consider any point $G(j\omega) = x + jy$ in the $G(j\omega)$ plane of Fig. 11.17. The closed-loop frequency response is

$$\frac{C(j\omega)}{R(j\omega)} = \frac{x+jy}{1+x+jy} \tag{11.53}$$

The magnitude of the preceding equation is

$$M = \left|\frac{C(j\omega)}{R(j\omega)}\right| = \left(\frac{x^2+y^2}{1+2x+x^2+y^2}\right)^{1/2} \tag{11.54}$$

Squaring and cross-multiplying give

$$x^2(M^2-1) + 2xM^2 + y^2(M^2-1) = -M^2$$

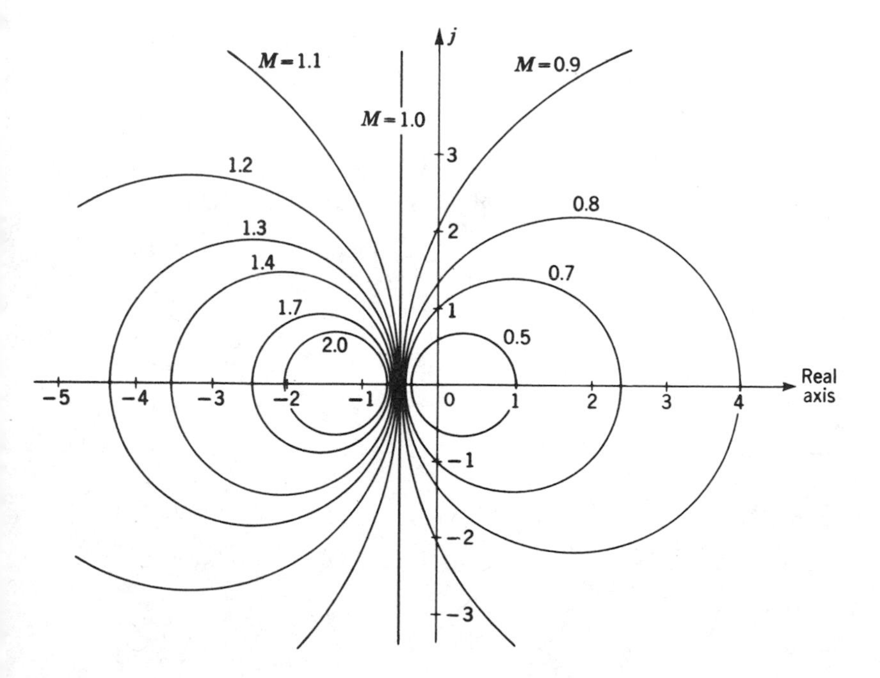

FIGURE 11.17
Constant-M circles.

Dividing by $M^2 - 1$ and completing the square by adding $M^4/(M^2 - 1)^2$ to both sides yield

$$x^2 + \frac{2xM^2}{M^2 - 1} + \frac{M^4}{(M^2 - 1)^2} + y^2 = \frac{M^4}{(M^2 - 1)^2} - \frac{M^2}{M^2 - 1}$$

Thus,

$$\left(x + \frac{M^2}{M^2 - 1}\right)^2 + y^2 = \frac{M^2}{(M^2 - 1)^2} \tag{11.55}$$

Equation (11.55) is the equation of a circle, as shown in Fig. 11.18, with center at

$$x = \frac{-M^2}{M^2 - 1} \qquad y = 0 \tag{11.56}$$

and radius

$$r = \left|\frac{M}{M^2 - 1}\right| \tag{11.57}$$

The closed-loop frequency response may be expressed in the form

$$\frac{C(j\omega)}{R(j\omega)} = Me^{j\alpha} \tag{11.58}$$

where $M = |C(j\omega)/R(j\omega)|$ and $\alpha = \measuredangle C(j\omega)/R(j\omega)$.

The loci of lines of constant phase angle α for the closed-loop response are also circles. The circles of constant α are shown in Fig. 11.19. From Eq. (11.53),

$$\alpha = \measuredangle \frac{x + jy}{1 + x + jy}$$

Multiplying the numerator and denominator by the complex conjugate of the denominator gives

$$\alpha = \measuredangle \frac{x^2 + x + y^2 + jy}{(1 + x)^2 + y^2}$$

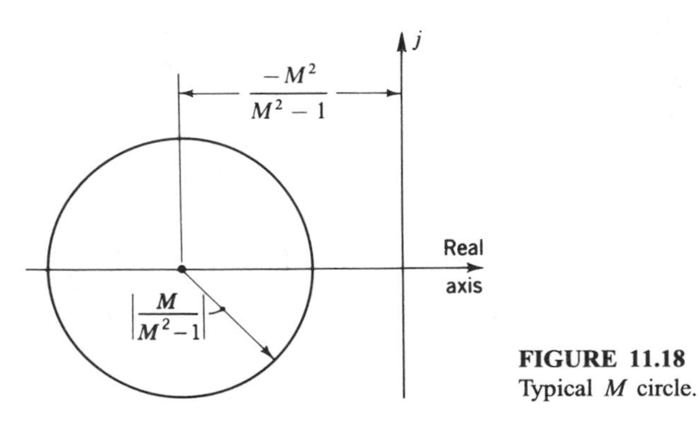

FIGURE 11.18
Typical M circle.

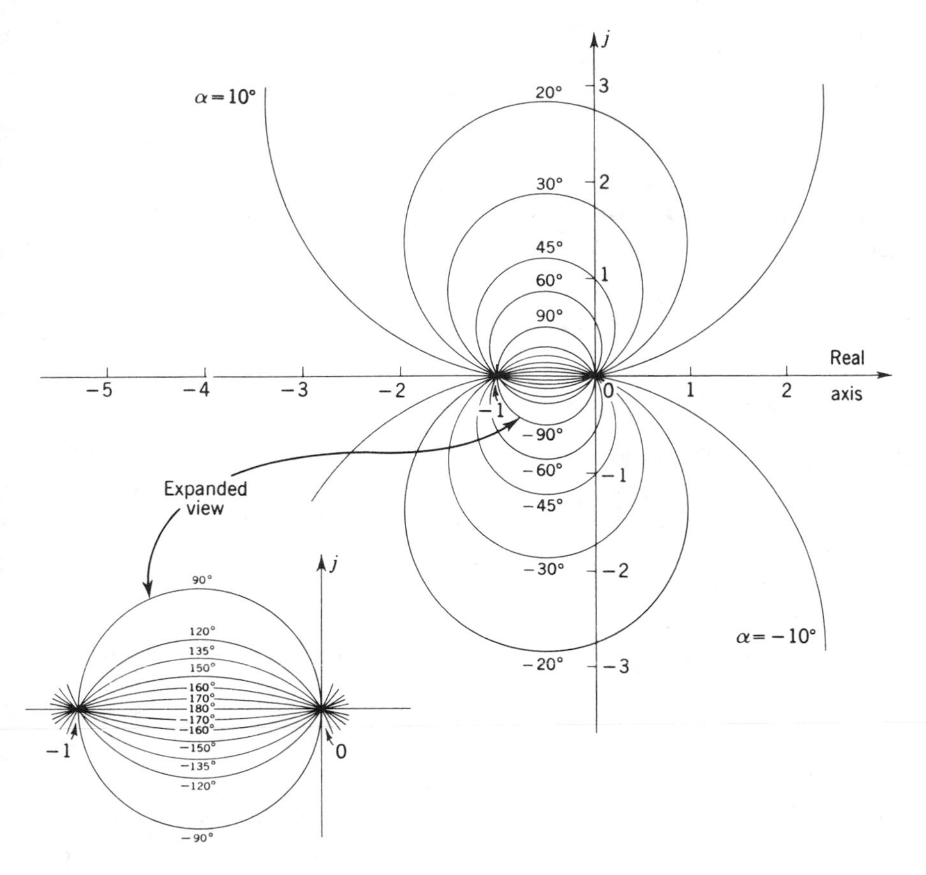

FIGURE 11.19
Constant-α circles.

Letting $N = \tan\alpha$, we get

$$N = \tan\alpha = \frac{y}{x^2 + x + y^2}$$

This may be written in the form

$$x^2 + x + y^2 - \frac{y}{N} = 0$$

Completing the square gives

$$\left(x + \frac{1}{2}\right)^2 + \left(y - \frac{1}{2N}\right)^2 = \frac{N^2 + 1}{4N^2}$$

The centers of these circles are located at

$$x = -\frac{1}{2} \qquad \text{and} \qquad y = \frac{1}{2N} \tag{11.59}$$

The radius of each circle is

$$r = \frac{1}{2N}\sqrt{N^2 + 1} \tag{11.60}$$

11.7 CORRELATION BETWEEN TRANSIENT AND FREQUENCY RESPONSE

The transient response of a system can be ascertained from the frequency response. The correlation for a type 1 system is developed first.

Type 1 Systems

Figure 11.20 shows the block-diagram representation for a second-order type 1 system. As mentioned in Sec. 11.4, a type 1 system has a lone s term in the denominator of $G(s)$. The closed-loop transfer function $C(s)/R(s)$ is

$$\frac{C(s)}{R(s)} = \frac{K_1/\tau}{s^2 + (1/\tau)s + K_1/\tau} = \frac{\omega_n^2}{s^2 + 2\zeta\omega_n s + \omega_n^2} \tag{11.61}$$

The closed-loop response is described by a second-order differential equation in which $\omega_n^2 = K_1/\tau$ and $2\zeta\omega_n = 1/\tau$, or $\zeta = 1/(2\sqrt{K_1\tau})$. Thus ζ and ω_n completely describe the transient behavior. To correlate the transient behavior with the frequency response, note that

$$M = \left|\frac{C(j\omega)}{R(j\omega)}\right| = \frac{\omega_n^2}{\left|(\omega_n^2 - \omega^2) + j(2\zeta\omega\omega_n)\right|} = \frac{1}{\sqrt{[1 - (\omega/\omega_n)^2]^2 + (2\zeta\omega/\omega_n)^2}} \tag{11.62}$$

The value of ω at which M is a maximum is obtained by differentiating Eq. (11.62) with respect to ω and then setting the resulting expression equal to zero. Thus,

$$\frac{\omega_p}{\omega_n} = \sqrt{1 - 2\zeta^2} \qquad 0 \le \zeta \le 0.707 \tag{11.63}$$

where ω_p is the value of ω at which M attains its peak or maximum value. Substituting ω_p/ω_n for ω/ω_n in Eq. (11.62) gives the peak or maximum value of M, which is designated M_p.

$$M_p = \frac{1}{2\zeta\sqrt{1 - \zeta^2}} \qquad 0 \le \zeta \le 0.707 \tag{11.64}$$

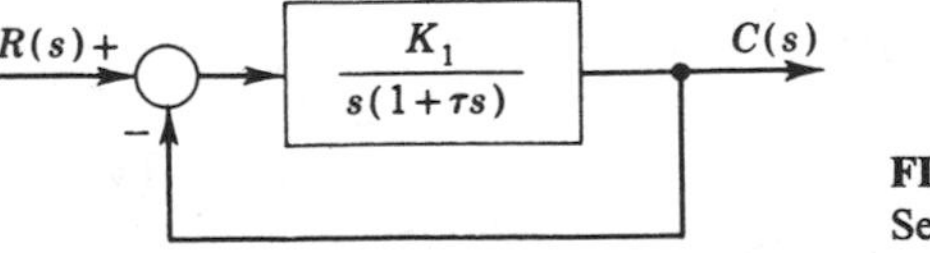

FIGURE 11.20
Second-order type 1 system.

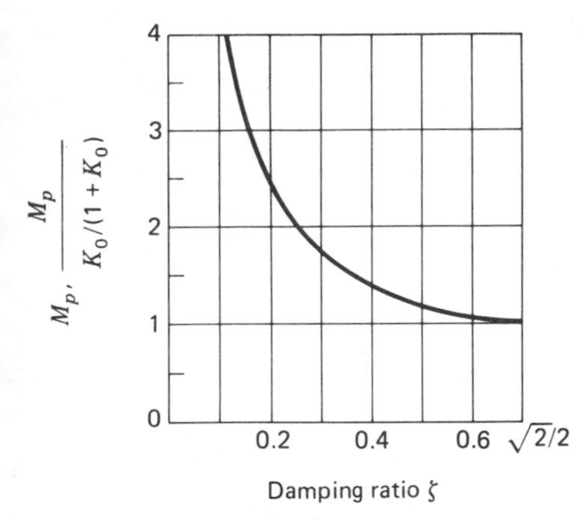

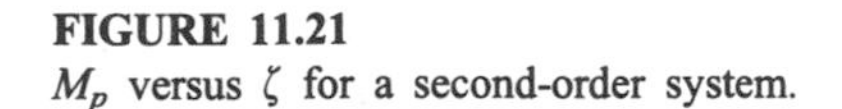

FIGURE 11.21
M_p versus ζ for a second-order system.

The preceding result has significance only for $0 \leq \zeta \leq 0.707$, in which case $M_p \geq 1$. Figure 11.21 shows a plot of M_p versus ζ.

Often we want to know the damping ratio ζ that corresponds to a certain value of M_p. Solving Eq. (11.64) for the damping ratio ζ shows that

$$\zeta = \left[\frac{1}{2}\left(1 - \sqrt{1 - \frac{1}{M_p^2}}\right)\right]^{1/2} \qquad M_p \geq 1 \tag{11.65}$$

As previously discussed, the zeros of the characteristic function which are located nearest the imaginary axis have a predominant effect upon the transient behavior of higher-order systems. Thus, the transient behavior of a higher-order type 1 system for which $M_p \geq 1$ (the value of M_p may be obtained from a polar plot for the system) may be approximated by a second-order system whose damping ratio ζ, as obtained from Eq. (11.65), corresponds to the value of M_p for the system.

Illustrative example 11.2. Figure 11.22 shows the polar plot for a type 1 system for which $M_p = 1.6$ and $\omega_p = 3$. Determine the values of ζ and ω_n to be used for approximating the transient behavior.

Solution. From Eq. (11.65), it follows that for $M_p = 1.6$ the corresponding value of ζ is 0.33. The value of ω_n may now be ascertained from Eq. (11.63).

$$\omega_n = \frac{\omega_p}{\sqrt{1 - 2\zeta^2}} = \frac{3}{\sqrt{1 - 2(0.33)^2}} = 3.4$$

Notice from Fig. 6.9 that for $\zeta > 0.707$ ($M_p < 1$) there is no overshoot of the response to a step change in the input. This is similar to the type of response that is obtained from a first-order system. Thus when M_p is less than 1, the transient response is approximated by an equivalent first-order system.

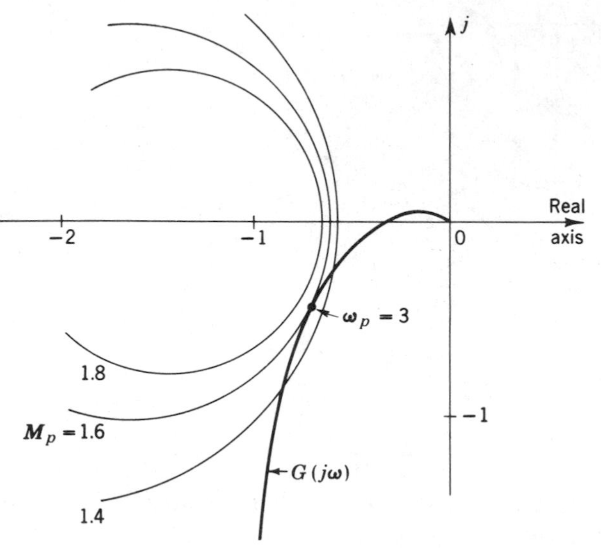

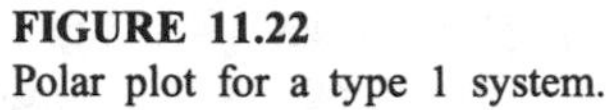
FIGURE 11.22
Polar plot for a type 1 system.

Figure 11.23 shows a first-order type 1 system. The correlation between the frequency and the transient response for this system is obtained by first writing the equation for the closed-loop response

$$\frac{C(s)}{R(s)} = \frac{K_1}{K_1 + s} = \frac{1}{1 + (1/K_1)s} = \frac{1}{1 + \tau_c s}$$

where $\tau_c = 1/K_1$ is the closed-loop time constant. Figure 11.24 shows a plot of the open-loop frequency response

$$G(j\omega) = \frac{K_1}{j\omega}$$

When the magnitude of $G(j\omega)$ is 1, then

$$|G(j\omega)| = \frac{K_1}{|j\omega_c|} = \frac{K_1}{\omega_c} = 1$$

where ω_c designates the value of ω at which $|G(j\omega)| = 1$. That is, ω_c is the value of ω at which the $G(j\omega)$ plot crosses the unit circle, as illustrated in Fig. 11.24. Because $K_1 = \omega_c$ and $\tau_c = 1/K_1$, then

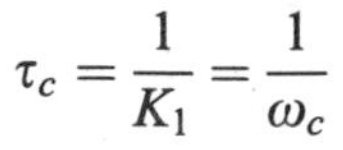

$$\tau_c = \frac{1}{K_1} = \frac{1}{\omega_c}$$

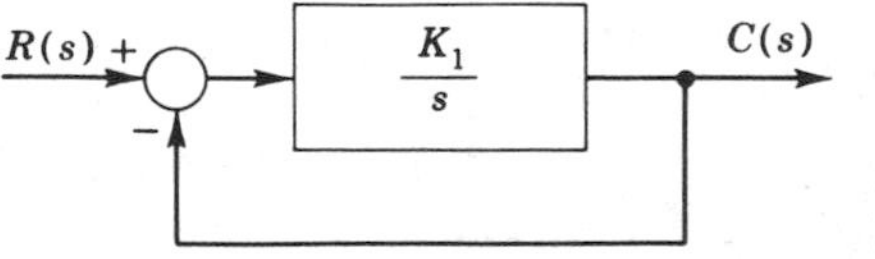

FIGURE 11.23
First-order type 1 system.

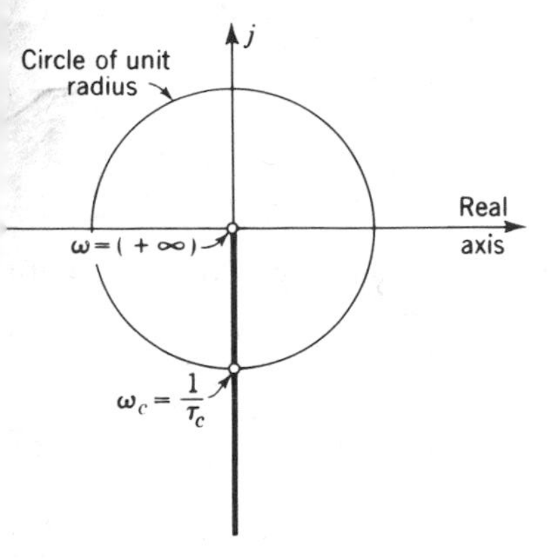

FIGURE 11.24
Polar plot for $G(j\omega) = K_1 j\omega$.

Thus, for type 1 systems in which $M_p < 1$, an indication of the equivalent time constant τ_c is obtained by taking the reciprocal of the angular frequency ω_c at which the $G(j\omega)$ plot crosses the unit circle.

The preceding results correlating transient and frequency response for type 1 systems are summarized in Table 11.2.

For $M_p < 1$, the response is approximated by a first-order system in which τ_c describes the transient behavior. For $M_p \geq 1$, the response is approximated by a second-order system in which ζ and ω_n describe the transient behavior.

Although the preceding correlation criteria are developed for type 1 systems, these criteria yield good approximations for type 2 systems, type 3 systems, etc. In the next section, we show that the preceding correlation criteria should be modified somewhat for the case of type 0 systems.

Type 0 Systems

Figure 11.25 shows the block-diagram representation for a second-order type 0 system. The closed-loop transfer function $C(s)/R(s)$ is

$$\frac{C(s)}{R(s)} = \frac{K_0/b}{s^2 + (a/b)s + (1 + K_0)/b} = \frac{[K_0/(1 + K_0)]\omega_n^2}{s^2 + 2\zeta\omega_n s + \omega_n^2}$$

TABLE 11.2
Correlation criteria for type 1 systems

$M_p < 1$	$M_p > 1$
$\tau_c = \dfrac{1}{\omega_c}$	$M_p = \dfrac{1}{2\zeta\sqrt{1-\zeta^2}}$
	$\dfrac{\omega_p}{\omega_n} = \sqrt{1-2\zeta^2}$

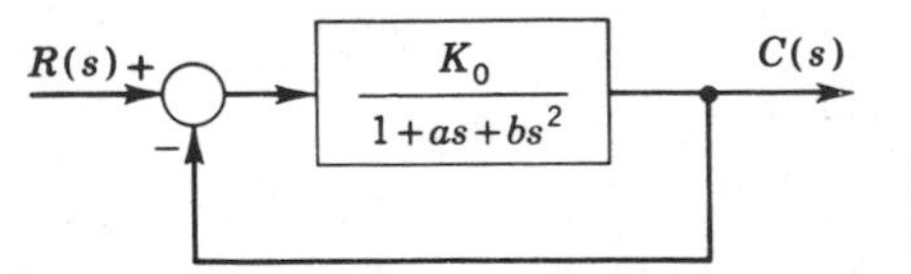

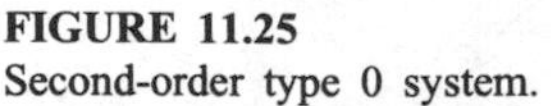

FIGURE 11.25
Second-order type 0 system.

where $\omega_n^2 = (1 + K_0)/b$ and $2\zeta\omega_n = a/b$. To correlate the frequency response with the transient response, note that

$$M = \left|\frac{C(j\omega)}{R(j\omega)}\right| = \frac{K_0/(1+K_0)}{|1-(\omega/\omega_n)^2 + 2j\zeta(\omega/\omega_n)|} = \frac{K_0/(1+K_0)}{\sqrt{[1-(\omega/\omega_n)^2]^2 + (2\zeta\omega/\omega_n)^2}} \tag{11.66}$$

Differentiating to determine the value of ω at which M is a maximum gives the same result as for the corresponding type 1 system, i.e.,

$$\omega_p = \omega_n\sqrt{1-2\zeta^2} \qquad 0 \le \zeta \le 0.707 \tag{11.63}$$

Substituting ω_p for ω in Eq. (11.66) gives

$$M_p = \frac{K_0/(1+K_0)}{2\zeta\sqrt{1-\zeta^2}} \qquad 0 \le \zeta \le 0.707$$

or

$$\frac{M_p}{K_0/(1+K_0)} = \frac{1}{2\zeta\sqrt{1-\zeta^2}} \qquad \frac{M_p}{K_0/(1+K_0)} \ge 1 \tag{11.67}$$

Thus, for type 0 systems, the ordinate $M_p/[K_0/(1 + K_0)]$ of Fig. 11.21 is to be used. Note that for large values of K_0, the preceding criterion becomes the same as that for a type 1 system.

Solving Eq. (11.67) for the damping ratio ζ that corresponds to a certain value of M_p and K_0 shows that

$$\zeta = \left[\frac{1}{2}\left(1 - \sqrt{1 - \left(\frac{K}{M_p}\right)^2}\right)\right]^{1/2} \qquad \frac{M_p}{K} \ge 1 \tag{11.68}$$

where $K = K_0/(1 + K_0)$.

Illustrative example 11.3. Figure 11.26 shows the polar plot for a type 0 system in which $M_p = 1.6$, $\omega_p = 3$, and $K_0 = 4$. Determine the values of ζ and ω_n to be used in approximating the transient behavior of this system.

Solution. From Eq. (11.68), it follows that for $K/M_p = [K_0/(1 + K_0)]/M_p = 0.8/1.6 = 0.5$, the corresponding value of ζ is 0.26. The value of ω_n is now ascertained from Eq. (11.63). Thus,

$$\omega_n = \frac{\omega_p}{\sqrt{1-2\zeta^2}} = \frac{3}{\sqrt{1-2(0.26)^2}} = 3.22$$

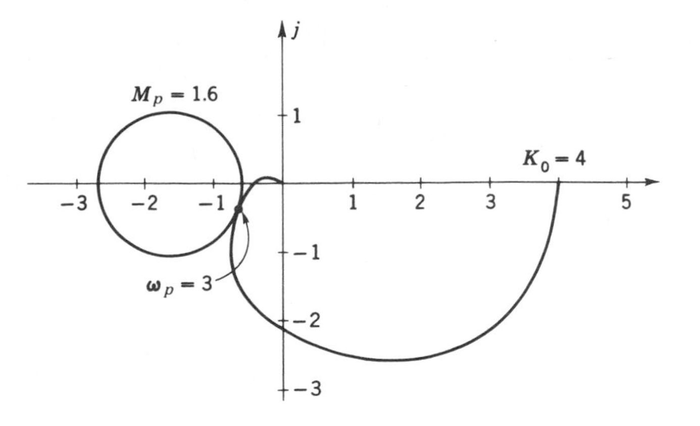

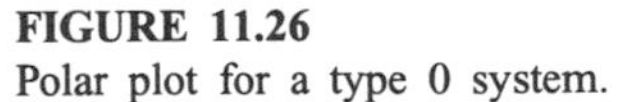

FIGURE 11.26
Polar plot for a type 0 system.

For a type 0 system in which $M_p/[K_0/(1 + K_0)] < 1$, the transient behavior is best described by a first-order system. Figure 11.27 shows a first-order type 0 system. The closed-loop transfer function is

$$\frac{C(s)}{R(s)} = \frac{K_0}{1 + K_0 + \tau s} = \frac{K_0/(1 + K_0)}{1 + \tau_c s}$$

where $\tau_c = \tau/(1 + K_0)$ is the time constant for the closed-loop response. Figure 11.28 shows a plot of the open-loop frequency response

$$G(j\omega) = \frac{K_0}{1 + j\tau\omega}$$

This plot crosses a circle of radius r at $\omega = \omega_c$. That is,

$$|G(j\omega)| = \frac{K_0}{\sqrt{1 + (\tau\omega_c)^2}} = r \tag{11.69}$$

The value of r such that ω_c is the reciprocal of the closed-loop time constant is obtained by substituting $\omega_c = 1/\tau_c = (1 + K_0)/\tau$ into Eq. (11.69). This gives

$$K_0^2 = r^2[1 + (1 + K_0)^2]$$

or

$$r = \frac{K_0}{\sqrt{1 + (1 + K_0)^2}}$$

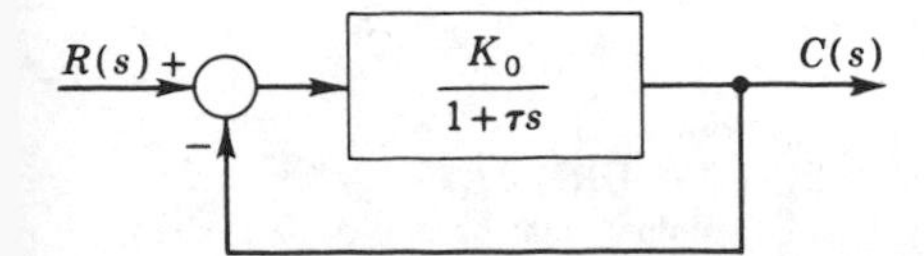

FIGURE 11.27
First-order type 0 system.

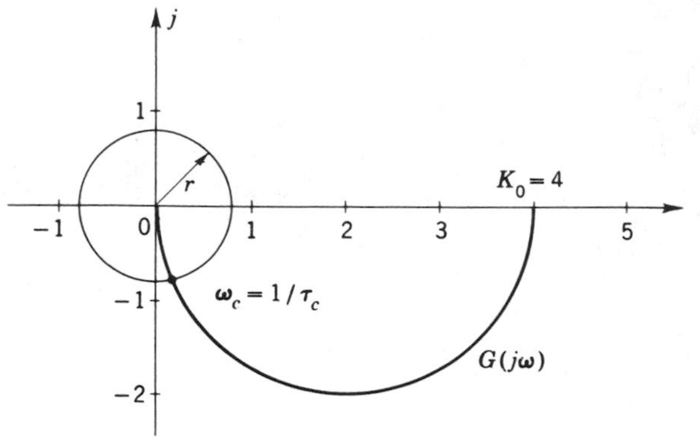

FIGURE 11.28
Polar plot for $G(j\omega) = K_0/(1 + j\omega)$.

For large values of K_0, the radius r approaches the unit circle and the preceding criterion becomes the same as that for a type 1 system.

Illustrative example 11.4. Figure 11.29 shows the polar plot of a type 0 system. Determine the value of τ_c to be used in approximating the transient behavior of this system.

Solution. From the polar plot, we see that $K_0 = 4$. The value of r is

$$r = \frac{4}{\sqrt{1 + 5^2}} = \frac{4}{\sqrt{26}} = 0.785$$

The polar plot shows that $\omega_c = 2.5$. Thus

$$\tau_c = \frac{1}{\omega_c} = 0.4$$

The preceding correlation criteria for type 0 systems are summarized in Table 11.3.

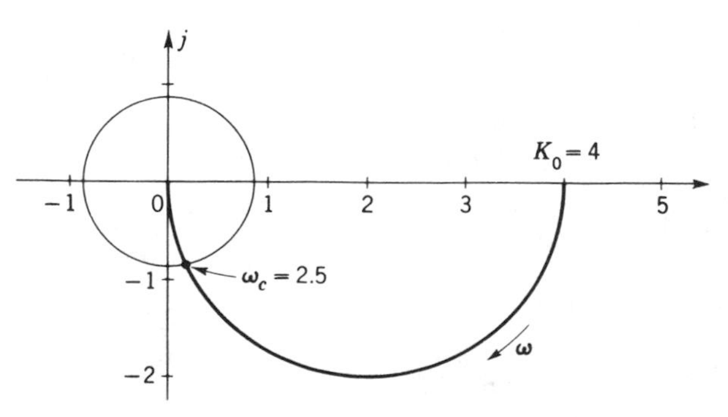

FIGURE 11.29
Polar plot for a type 0 system.

TABLE 11.3
Correlation criteria for type 0 systems

$\dfrac{M_p}{K_0/(1+K_0)} < 1$	$\dfrac{M_p}{K_0/(1+K_0)} \geq 1$
$\tau_c = \dfrac{1}{\omega_c}$	$\dfrac{M_p}{K_0/(1+K_0)} = \dfrac{1}{2\zeta\sqrt{1-\zeta^2}}$
$r = \dfrac{K_0}{\sqrt{1+(1+K_0)^2}}$	$\dfrac{\omega_p}{\omega_n} = \sqrt{1-2\zeta^2}$

Bandwidth

The closed-loop transfer function for the system of Fig. 11.27 is

$$\frac{C(s)}{R(s)} = \frac{K_0}{(1+K_0)+\tau s} = \frac{K_0/(1+K_0)}{1+[\tau/(1+K_0)]s} = \frac{K}{1+\tau_c s}$$

where $K = K_0/(1 + K_0)$ is the steady-state gain and $\tau_c = \tau/(1 + K_0)$ is the closed-loop time constant. Replacing s by $j\omega$ yields for the closed-loop frequency response

$$\frac{C(j\omega)}{R(j\omega)} = \frac{K}{1+j\tau_c\omega}$$

The amplitude ratio M of the closed-loop frequency response is

$$M(j\omega) = \left|\frac{C(j\omega)}{R(j\omega)}\right| = \frac{K}{\sqrt{1+(\tau_c\omega)^2}}$$

A plot of the amplitude ratio is shown in Fig. 11.30*a*. For small values of ω the amplitude ratio is

$$M(j0) = \left|\frac{C(j\omega)}{R(j\omega)}\right|_{\omega=0} = K$$

The frequency $\omega = \omega_b$ at which the amplitude ratio decreases to $\sqrt{2}/2 = 0.707$ times its value for low frequencies $[M(j\omega) = 0.707M(j0)]$ is

$$M(j\omega) = \frac{K}{\sqrt{1+(\tau_c\omega_b)^2}} = 0.707K$$

Squaring both sides and solving for ω_b give

$$\omega_b = \frac{1}{\tau_c}$$

The frequency ω_b is called the *bandwidth*. For $\tau_c = 1$, the bandwidth is $\omega_b =$ 1 rad/s. For $\tau_c = 0.1$, the bandwidth is $\omega_b = 1/0.1 = 10$ rad/s. The bandwidth is a

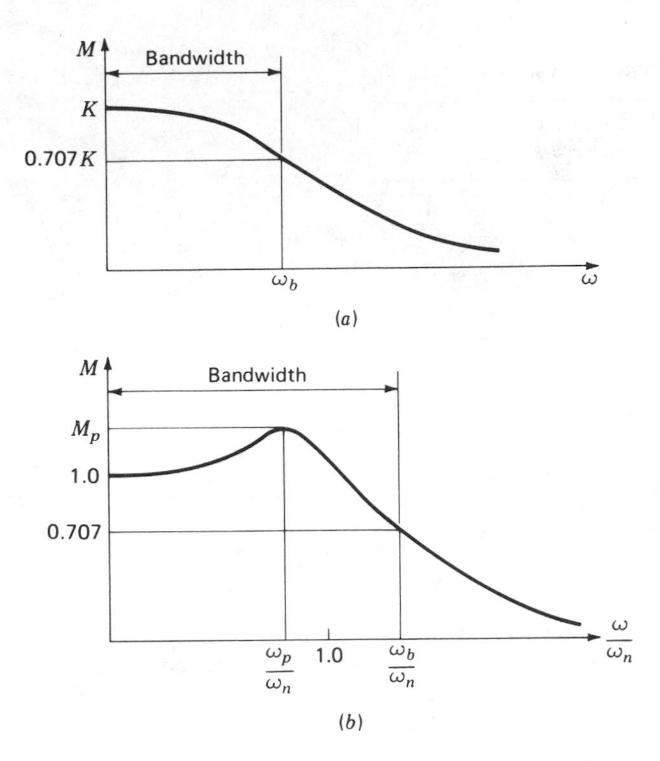

FIGURE 11.30
Closed-loop frequency response for (*a*) a type 0 system and (*b*) a type 1 system.

measure of the frequency range $(0 < \omega < \omega_b)$ over which the closed-loop amplitude ratio is greater than $\sqrt{2}/2$ times its value for low frequencies. The response follows the input reasonably well for frequencies within the bandwidth. Note from Fig. 11.30*a* that the amplitude ratio decreases rapidly for frequencies greater than ω_b.

The bandwidth provides a measure of the speed of response of a system. In general, the larger the bandwidth, the faster is the speed of response. For the example under consideration, the larger bandwidth $\omega_b = 10$ rad/s corresponds to the faster-responding system $\tau_c = 0.1$.

For the system of Fig. 11.20, the closed-loop transfer function is given by Eq. (11.61) and the amplitude ratio M of the closed-loop frequency response is given by Eq. (11.62). That is,

$$\frac{C(s)}{R(s)} = \frac{K_1/\tau}{s^2 + (1/\tau)s + K_1/\tau} = \frac{\omega_n^2}{s^2 + 2\zeta\omega_n s + \omega_n^2} \tag{11.61}$$

$$M(j\omega) = \left|\frac{C(j\omega)}{R(j\omega)}\right| = \frac{1}{\sqrt{[1 - (\omega/\omega_n)^2]^2 + (2\zeta\omega/\omega_n)^2}} \tag{11.62}$$

where $\omega_n^2 = K_1/\tau$ and $\zeta = 1/\left(2\sqrt{K_1\tau}\right)$. A plot of this closed-loop amplitude ratio is shown in Fig. 11.30*b*. The peak value M_p is determined by Eq. (11.64), and the frequency at which the peak value occurs is determined by Eq. (11.63). For numerical purposes, if $\zeta = 0.4$, then $M_p = 1.36$ and $\omega_p/\omega_n = 0.82$.

For small values of ω, the amplitude ratio is

$$M(j0) = M(j\omega)|_{\omega=0} = \left|\frac{C(j\omega)}{R(j\omega)}\right|_{\omega=0} = 1$$

The frequency $\omega = \omega_b$ at which the amplitude ratio decreases to 0.707 times its value for low frequencies is

$$M(j\omega) = \frac{1}{\sqrt{[1-(\omega_b/\omega_n)^2]^2 + (2\zeta\omega_b/\omega_n)^2}} = 0.707M(j0) = 0.707$$

Squaring and solving for $(\omega_b/\omega_n)^2$ give

$$\left(\frac{\omega_b}{\omega_n}\right)^2 = (1-2\zeta^2) + \sqrt{2-4\zeta^2(1-\zeta^2)}$$

For $\zeta = 0, 0.4$, and 0.707, the corresponding values of ω_b/ω_n are 1.55, 1.36, and 1.0.

11.8 DETERMINING THE GAIN K TO YIELD A DESIRED M_p

Figure 11.31 shows a typical polar plot of $G(j\omega)$. If the gain K of the original system is doubled, the value of $G(j\omega)$ is doubled at every point. As shown in Fig. 11.31, it is not necessary to change the shape of the polar plot; it suffices merely to change the scale by multiplying the old scale by the factor 2. Values of this new scale are shown in parentheses. We now show how the gain K can be adjusted so that the polar plot $G(j\omega)$ will be tangent to any desired $M_p > 1$ circle. This, in effect, is determining the gain K so that the system will have a desired M_p.

An M circle is shown in Fig. 11.32. From Eqs. (11.56) and (11.57), it follows that the center is located on the real axis at $-M^2/(M^2-1)$ and the radius is

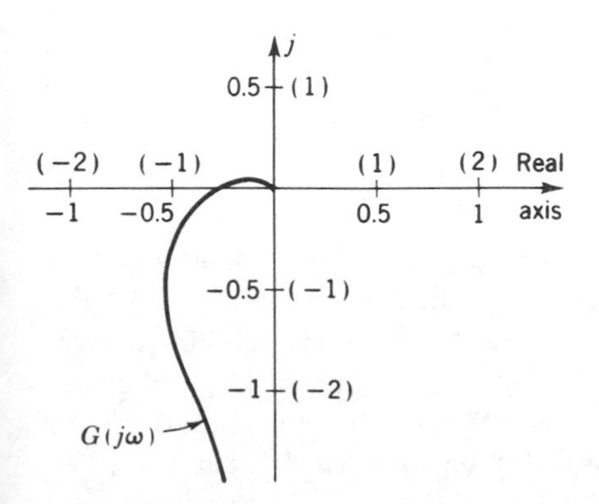

FIGURE 11.31
Typical polar plot.

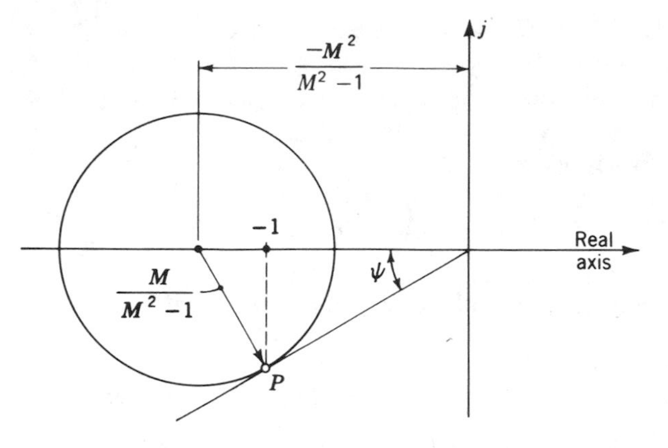

FIGURE 11.32
Tangent to an M circle.

$M/(M^2 - 1)$. The line drawn from the origin, tangent to the M circle at the point P, has an included angle of ψ. The value of $\sin\psi$ is

$$\sin\psi = \frac{1}{M} \tag{11.70}$$

A characteristic feature of the point of tangency P is that a line drawn from point P perpendicular to the negative real axis intersects this axis at the -1 point. This characteristic may be proved from the geometry of Fig. 11.32.

The procedure for determining the gain K so that $G(j\omega)$ will have a desired value of M_p is as follows:

1. Draw the polar plot for $G(j\omega)/K$.
2. Draw the tangent line to the desired M_p circle [Eq. (11.70)].
3. Draw the circle with center on the negative real axis that is tangent to both the $G(j\omega)/K$ plot and the tangent line, as shown in Fig. 11.33.
4. Erect the perpendicular to the negative real axis from point P, the point of tangency of this circle and the tangent line. This perpendicular intersects the negative real axis at a value $-A$.
5. So that the circle drawn in step 3 will correspond to the desired M_p circle, this point should be -1 rather than $-A$. The desired gain is that value of K which changes the scale so that this does become the -1 point; thus $K(-A) = -1$ or $K = 1/A$.

As illustrated in Fig. 11.33, the perpendicular drawn from point P to the negative real axis intersects the negative real axis at a value of -0.05. However, this value should be -1. Multiplication of the scale by a factor of 20 (that is, $-0.05 \times 20 = -1$), as shown in Fig. 11.33 by the numbers in parentheses, converts this point to the -1 point. Thus, the original function should have a gain of 20 in order that the circle drawn will be the desired M_p circle.

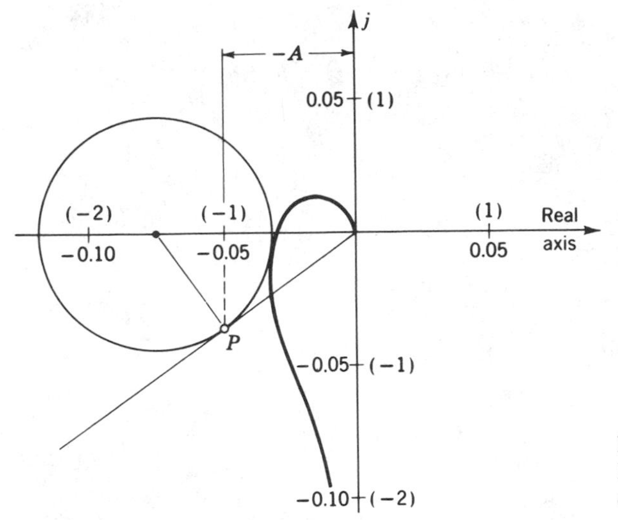

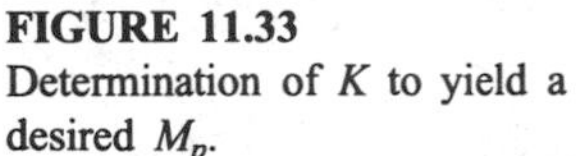

FIGURE 11.33
Determination of K to yield a desired M_p.

Illustrative example 11.5. Figure 11.34 shows an astronaut using a nitrogen-propelled, hand-controlled device called the *Manning maneuvering unit* (MMU). This device allows for much greater mobility than that afforded to previous space walkers who had to use restrictive tethers. Determine the gain K such that this unity-feedback system for which $G(s) = K/[s(1 + 0.1s)]$ will have a peak value $M_p = 1.4$. What values of ζ and ω_n should be used in approximating the transient behavior of this system?

Solution. First construct the polar plot for

$$\frac{G(j\omega)}{K} = \frac{1}{j\omega(1 + 0.1j\omega)}$$

as shown in Fig. 11.35. The value of ψ is obtained from Eq. (11.70).

$$\psi = \sin^{-1}\frac{1}{M_p} = \sin^{-1}\frac{1}{1.4} = 45.6° \tag{11.71}$$

By trial and error, the circle which is tangent to both the $G(j\omega)/K$ plot and the line drawn at the angle $\psi = 45.6°$ is determined. The perpendicular drawn from point P intersects the negative real axis at the point $-A = -0.06$, or $A = 0.06$. Thus, the gain $K = 1/0.06 = 16.7$.

From Eq. (11.65), for $M_p = 1.4$ the corresponding value of ζ is 0.387. From Fig. 11.35, $\omega_p = 10.8$; hence from Eq. (11.63) the value of ω_n is

$$\omega_n = \frac{10.8}{\sqrt{1 - 2(0.387)^2}} = 12.9$$

Because the transfer function for this second-order system is known, the results of the preceding correlation between the frequency response and the

FIGURE 11.34
McCandless Flies First "Solo" in Space (National Aeronautics and Space Administration).

transient response may be checked analytically. The open-loop transfer function is

$$G(s) = \frac{K}{s(1 + 0.1s)} = \frac{10K}{s(s + 10)}$$

The corresponding closed-loop transfer function for this unity-feedback system is

$$\frac{C(s)}{R(s)} = \frac{G(s)}{1 + G(s)} = \frac{10K}{s^2 + 10s + 10K}$$

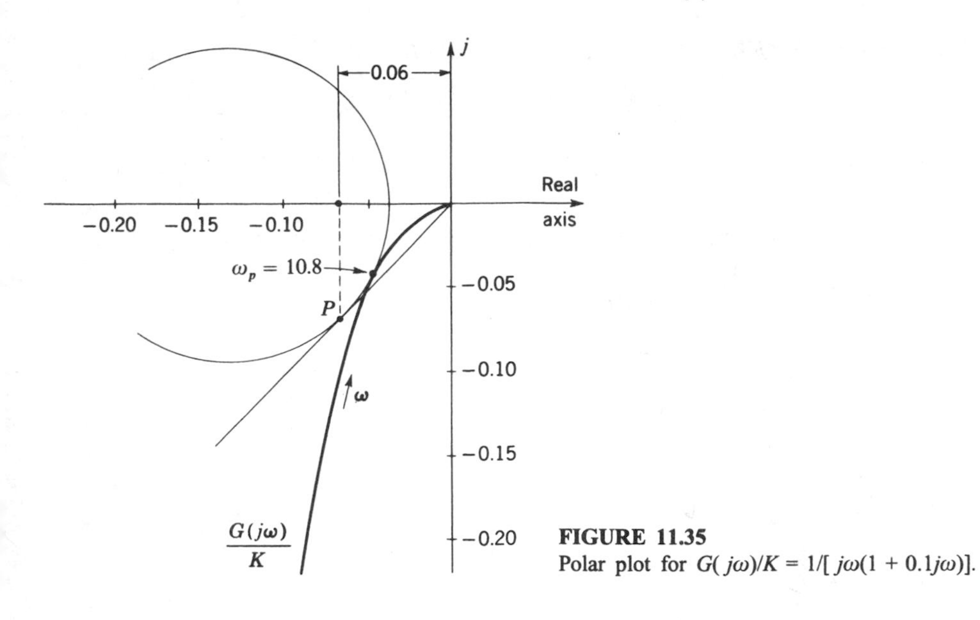

FIGURE 11.35
Polar plot for $G(j\omega)/K = 1/[j\omega(1 + 0.1j\omega)]$.

From the characteristic function $s^2 + 10s + 10K$, it follows that $\omega_n^2 = 10K$ and $2\zeta\omega_n = 10$. Thus,

$$\omega_n = \sqrt{10K} \qquad \text{and} \qquad \zeta = \frac{10}{2\omega_n} = \frac{1}{2}\sqrt{\frac{10}{K}}$$

Replacing s by $j\omega$ in the expression for the closed-loop transfer function gives

$$\frac{C(j\omega)}{R(j\omega)} = \frac{10K}{10K - \omega^2 + j10\omega}$$

The amplitude ratio M is

$$M = \left|\frac{C(j\omega)}{R(j\omega)}\right| = \frac{10K}{\sqrt{(10K - \omega^2)^2 + (10\omega)^2}}$$

Differentiating the denominator with respect to ω and setting it equal to zero yield the frequency ω_p at which M is a maximum. Thus

$$2(10K - \omega^2)(-2\omega) + 2(10\omega)(10) = 0$$

or

$$\omega = \omega_p = \sqrt{10K - 50}$$

Substitution of this value of $\omega = \omega_p$ into the expression for M yields for the peak value of the amplitude ratio M

$$M_p = \frac{10K}{\sqrt{50^2 + 100(10K - 50)}} = \frac{10K}{\sqrt{1000K - 2500}}$$

The value of K such that $M_p = 1.4$ is $K = 100/6 = 16.7$. Thus

$$\omega_n = \sqrt{10K} = \sqrt{167} = 12.9$$

$$\zeta = \frac{1}{2}\sqrt{\frac{10}{K}} = \frac{\sqrt{0.6}}{2} = 0.387$$

Figure 11.36*a* shows the log-magnitude plot of the open-loop frequency response

$$G(j\omega) = \frac{16.7}{j\omega(1 + 0.1j\omega)}$$

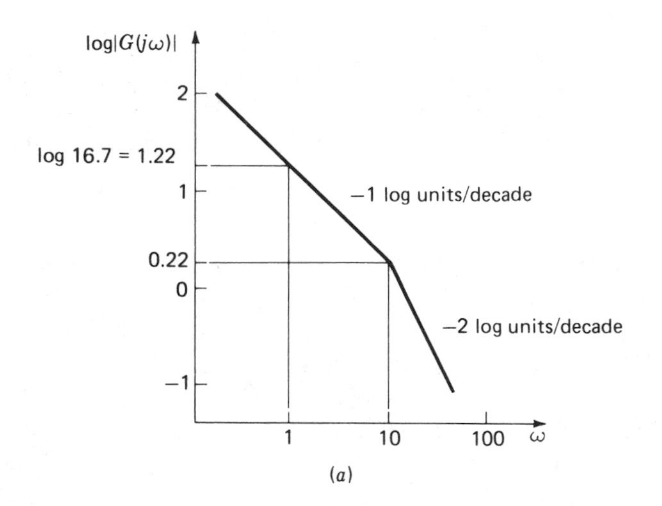

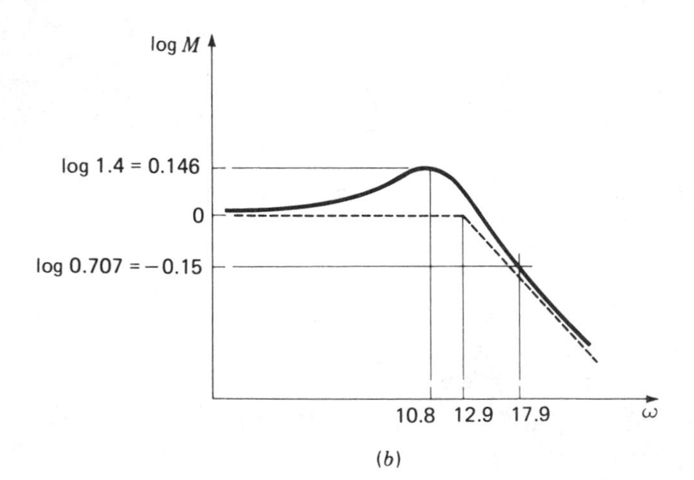

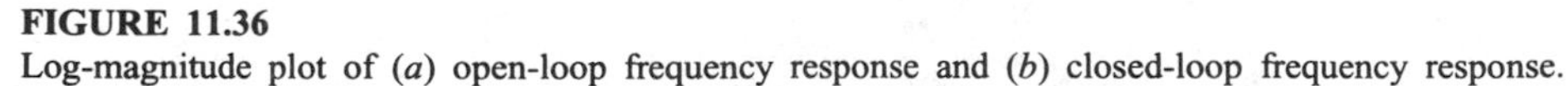

FIGURE 11.36
Log-magnitude plot of (*a*) open-loop frequency response and (*b*) closed-loop frequency response.

The low-frequency asymptote is

$$|G(j\omega)| \approx \frac{16.7}{\omega} \qquad \omega \approx 0$$

The point on this asymptote at $\omega = 1$ is located by the coordinate $\log|G(j\omega)| = \log 16.7 = 1.22$. The slope changes from -1 to -2 at the break frequency $1/\tau = 1/0.1 = 10$ rad/s. The velocity error constant is $K_v = 16.7$. Figure 11.36*b* shows the amplitude ratio M for the closed-loop frequency response

$$M = \left|\frac{C(j\omega)}{R(j\omega)}\right| = \frac{16.7}{|j\omega(1+0.1j\omega)+16.7|}$$

$$= \frac{16.7}{\sqrt{(16.7-0.1\omega^2)^2+\omega^2}}$$

For small values of ω, the amplitude ratio M approaches unity ($\log 1 = 0$). The peak value $M_p = 1.4$ ($\log 1.4 = 0.146$) occurs at the frequency $\omega_p = 10.8$. The low-frequency asymptote (slope = 0) and the high frequency asymptote (slope = -2) intersect at the natural frequency $\omega_n = 12.9$. The bandwidth is the value of ω such that M is 0.707 times its low-frequency value of 1.0. Thus, letting $M = 0.707$ in the preceding equation and solving for ω yield $\omega = \omega_b = 17.9$ rad/s.

Illustrative example 11.6. Figure 5.15 shows NASA's shuttle-borne observatory Astro, which will explore the invisible ultraviolet and x-ray universe. The system for controlling the ultraviolet imaging telescope (UIT) is the unity-feedback system shown in Fig. 11.37*a* for which the transfer function $G(s)$ is

$$G(s) = \frac{100}{(1+0.1s)(s^2+8s+25)} = \frac{4}{(1+0.1s)(s^2+8s+25)/25}$$

The corresponding frequency response is

$$G(j\omega) = \frac{4}{(1+0.1j\omega)[1-\omega^2/25+j(8\omega/25)]}$$

(*a*) Determine the factor K_c by which the gain of the system should be changed so that the resulting system will have an M_p of 1.4.

(*b*) What values of ζ and ω_n should be used to approximate the transient behavior of this resulting system?

Solution. (*a*) For an M_p of 1.4, the tangent line is constructed at an angle of $\psi = 45.6°$. As shown in Fig. 11.37*b*, the perpendicular from point P to the real axis intersects the real axis at the point -1.85. In order that this be the -1 point, the original scale must be changed by a factor of $1/1.85 = 0.54$. This is the factor K_c by which the gain of the system should be changed. Thus,

$$\text{New gain} = (K_c)\,(\text{original gain}) = (0.54)\,(4) = 2.16$$

(*b*) For this type 0 system,

$$\frac{M_p}{K_0/(1+K_0)} = \frac{1.4}{2.16/3.16} = 2.05$$

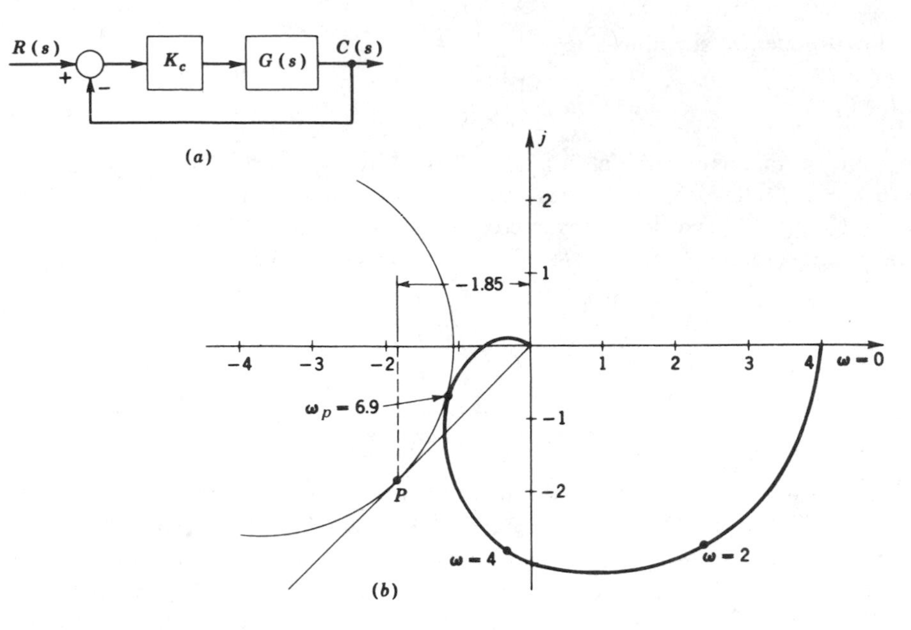

FIGURE 11.37
Determination of the change of gain K.

From Eq. (11.68), it follows that for $[K_0/(1 + K_0)]/M_p = 1/2.05 = 0.488$ the corresponding value of ζ is 0.25. The polar plot shows that $\omega_p = 6.9$. Hence from Eq. (11.63)

$$\omega_n = \frac{6.9}{\sqrt{1 - 2(0.25)^2}} = 7.37$$

This last example shows that if the $G(j\omega)$ plot is employed rather than the $G(j\omega)/K$ plot, then $-1/A$ is equal to the factor K_c by which the gain should be changed in order to obtain a desired M_p.

Log-Modulus Plots

In addition to log-magnitude and polar plots, another method of representing frequency-response information is via log-modulus, or Nichols, plots.* The log-modulus curve is a plot of $\log|G(j\omega)|$ versus $\measuredangle G(j\omega)$ for various values of angular velocity ω. The log-modulus curve for $G(j\omega)/K = 1/[j\omega(1 + 0.1j\omega)]$ is constructed in Fig. 11.38. From the equation for $G(j\omega)/K$ we see that for a given angular velocity ω, the corresponding phase angle and amplitude ratio can be

* H. M. James, N. B. Nichols, and R. S. Phillips, *Theory of Servomechanisms*, McGraw-Hill, New York, 1947.

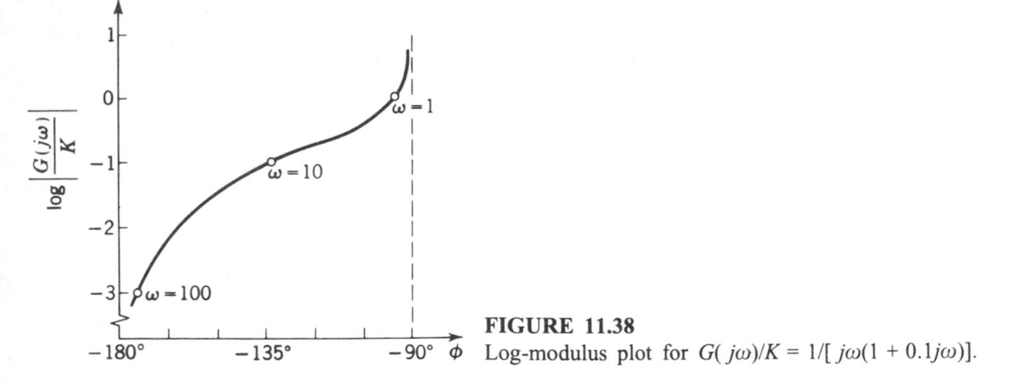

FIGURE 11.38
Log-modulus plot for $G(j\omega)/K = 1/[j\omega(1 + 0.1j\omega)]$.

obtained. These values determine one point on the log-modulus plot. By repeating this process for other values of ω, the log-modulus plot of $G(j\omega)/K$ shown in Fig. 11.38 is obtained. Every point on the log-modulus plot corresponds to a certain value of $G(j\omega)$. Lines of constant M and constant α which are circles on the polar plot (Figs. 11.17 and 11.19) become contours when drawn on the log-modulus plot. These M and α contours are shown in Fig. 11.39. We now show how the log-modulus techniques may be used to determine the gain K so that a system will have a desired value of M_p.

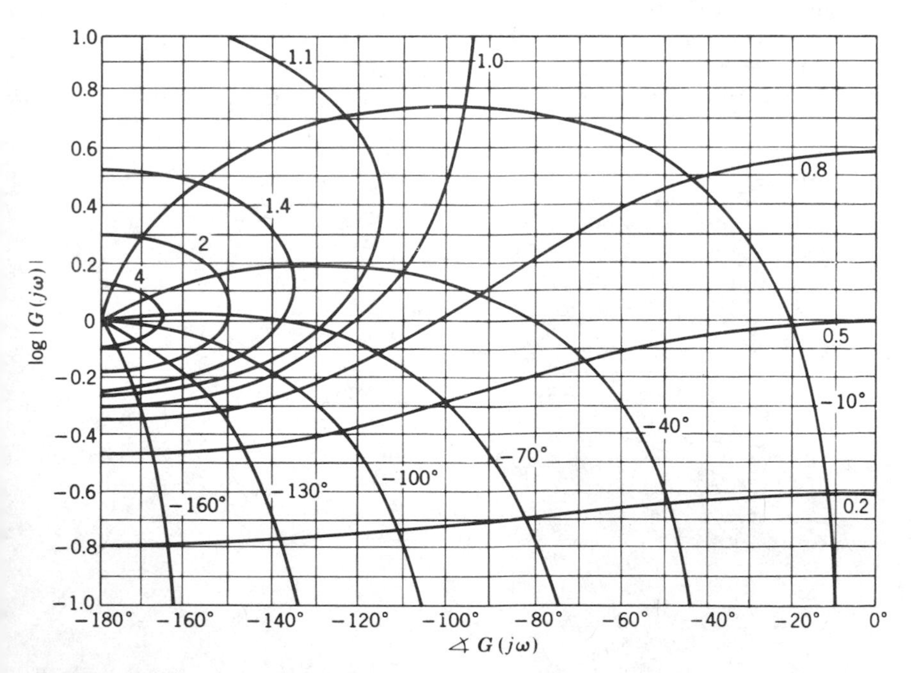

FIGURE 11.39
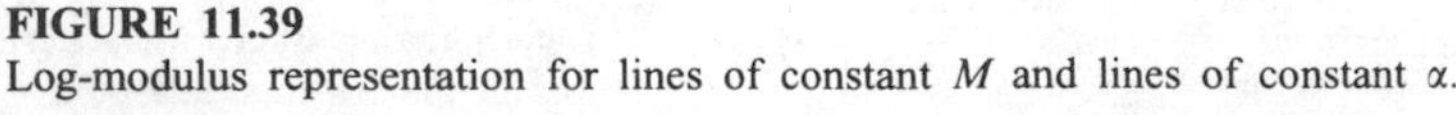
Log-modulus representation for lines of constant M and lines of constant α.

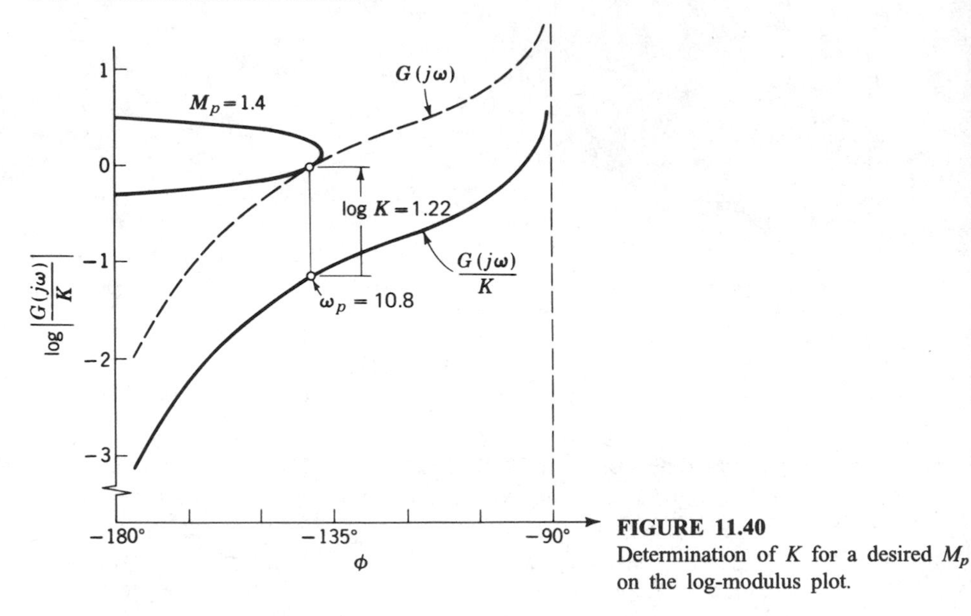

FIGURE 11.40
Determination of K for a desired M_p on the log-modulus plot.

Illustrative example 11.7. Repeat Illustrative example 11.5, except use log-modulus techniques.

Solution. Figure 11.40 shows the log-modulus plot of $G(j\omega)/K$ corresponding to the polar plot of Fig. 11.35. Changing the gain K does not affect the phase angle, but merely moves the log-modulus curve vertically up for $K > 1$ and down for $K < 1$. In Fig. 11.40 note that the original function represented by the solid line must be moved up 1.22 log units so that it will be tangent to the desired M_p contour. Because

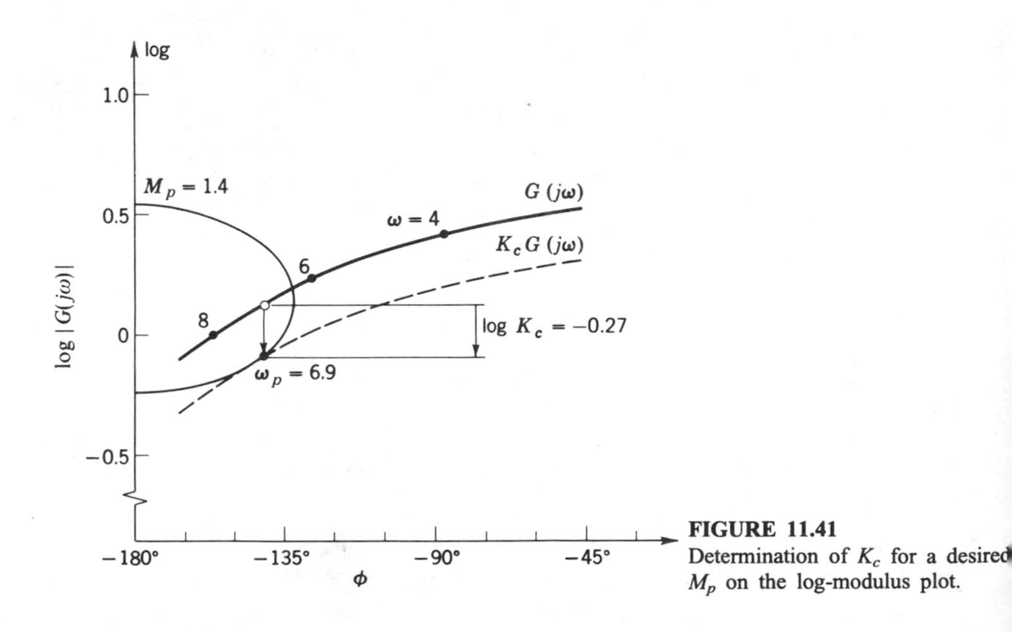

FIGURE 11.41
Determination of K_c for a desired M_p on the log-modulus plot.

$\log K = 1.22$, it follows that the required gain is $K = 16.7$. The values of ζ and ω_n are found as before.

Illustrative example 11.8. Repeat Illustrative example 11.6, except use log-modulus techniques.

Solution. Figure 11.41 shows the log-modulus plot of $G(j\omega)$ corresponding to the polar plot of Fig. 11.37*b*. Because $\log K_c = -0.27$, then $K_c = 0.54$.

PROBLEMS

11.1. Each of the mechanical systems shown in Fig. P11.1 is excited sinusoidally by a force $f = f_0 \sin \omega t$. For each system, determine:
(*a*) The equation for the amplitude ratio y_0/f_0
(*b*) The equation for the phase shift ϕ

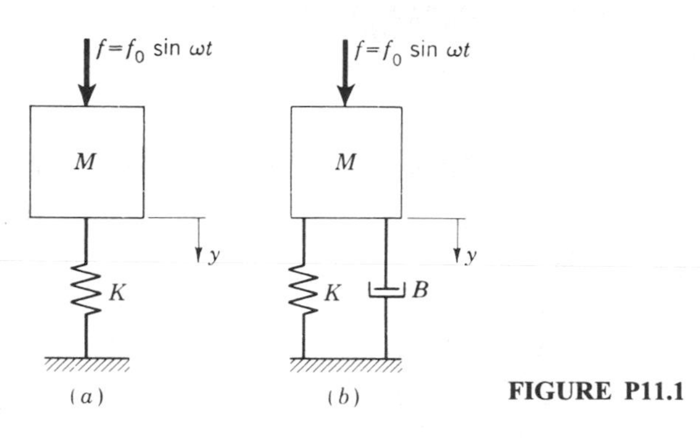

FIGURE P11.1

11.2. Each of the mechanical systems shown in Fig. P11.2 is excited sinusoidally by a motion of the support $x = x_0 \sin \omega t$. For each system, determine:
(*a*) The equation for the amplitude ratio y_0/x_0
(*b*) The equation for the phase shift ϕ

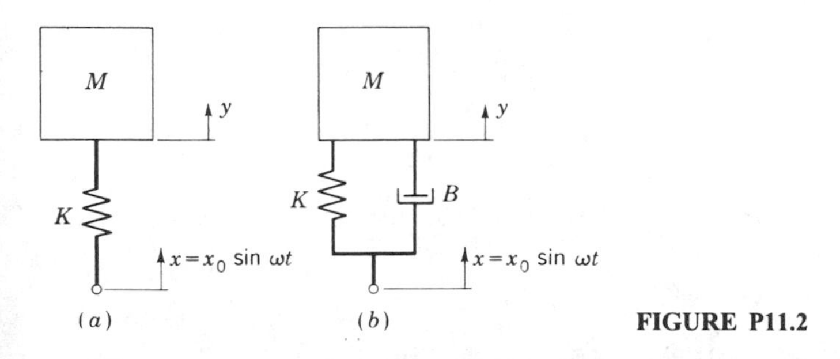

FIGURE P11.2

11.3. Figure P11.3 shows a seismic instrument. The motion of the pen relative to the base is $z = y - x$. Show that the differential equation relating z to the motion of the base is

$$(MD^2 + BD + K)z = -MD^2x$$

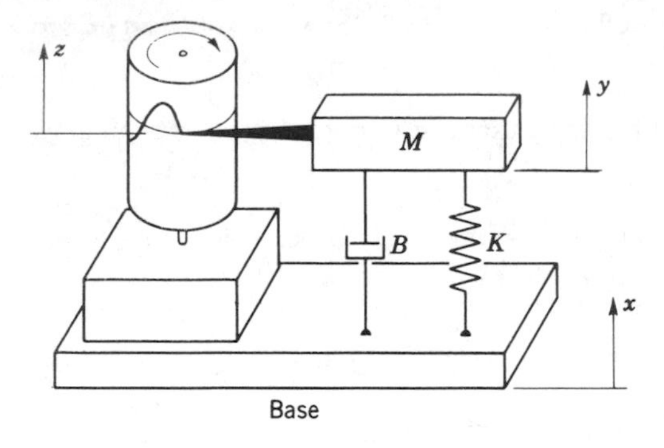

FIGURE P11.3

(*a*) Determine the equation for the amplitude ratio z_0/x_0 for the case in which the motion of the base is $x = x_0 \sin \omega t$.

(*b*) Show that for very large values of ω/ω_n, $z_0/x_0 = -1$. Such an instrument which may be used to measure displacements is called a *vibrometer*.

(*c*) Show that for small values of ω/ω_n,

$$\frac{z_0}{a_0} = \frac{1}{\omega_n^2}$$

where $\ddot{x} = -x_0\omega^2 \sin \omega t = a_0 \sin \omega t$. Such an instrument which measures acceleration is called an *accelerometer*.

11.4. For each of the functions given below, evaluate $|G(j\omega)|$ and $\phi = \measuredangle G(j\omega)$ for $\omega = 4(1/0.25) = 16$, $\omega = 2(1/0.25) = 8$, $\omega = 1/0.25 = 4$, $\omega = (\frac{1}{2})(1/0.25) = 2$, and $\omega = (\frac{1}{4})(1/0.25) = 1$. [Note that values of ω in the vicinity of the break frequency (1/0.25) yield the most significant information.] Construct the exact log-magnitude plot for each function, and sketch in the asymptotes.

$$(a)\quad G(j\omega) = \frac{10}{1 + 0.25j\omega} \qquad (b)\quad G(j\omega) = \frac{10}{j\omega(1 + 0.25j\omega)}$$

$$(c)\quad G(j\omega) = \frac{10}{(1 + 0.25j\omega)^2}$$

11.5. For the quadratic given below, evaluate $|G(j\omega)|$ and $\phi = \measuredangle G(j\omega)$ for values of $\omega/\omega_n = 4$, $\omega/\omega_n = 2$, $\omega/\omega_n = 1$, $\omega/\omega_n = \frac{1}{2}$, and $\omega/\omega_n = \frac{1}{4}$ and for $\zeta = 0.4$. Repeat for the case in which $\zeta = 0.1$. Construct the exact log-magnitude plot, and sketch in the asymptotes.

$$G(j\omega) = \frac{10}{1 - (\omega/\omega_n)^2 + j2\zeta(\omega/\omega_n)}$$

11.6. Construct the polar plot for each of the $G(j\omega)$ functions given in Probs. 11.4 and 11.5.

11.7. For each of the following functions:

$$G(s) = \frac{10}{(1 + s)(1 + 0.1s)}$$

$$G(s) = \frac{10(1 + s)}{s(s^2 + 8s + 100)/100}$$

(*a*) Draw the asymptotes for the log magnitude plot.

(*b*) Sketch the polar plot.

11.8. Sketch the asymptotes of the log-magnitude plot and the polar plot for each of the following transfer functions.

$$(a)\quad G(s) = \frac{10(1+s)}{(1+10s)(1+0.1s)}$$

$$(b)\quad G(s) = \frac{10(1+s)}{s(1+0.1s)^2}$$

11.9. For each of the transfer functions given in Prob. 11.7, write a computer program to determine $|G(j\omega)|$, $\measuredangle G(j\omega)$, $\log \omega$, $\log |G(j\omega)|$, X, and Y when $\omega = 0.25$, 0.50, 1.0, 2.0, 4.0, 8.0, 16.0, and 32.0.

11.10. For each of the transfer functions given in Prob. 11.8, write a computer program to determine $|G(j\omega)|$, $\measuredangle G(j\omega)$, $\log \omega$, $\log |G(j\omega)|$, X, and Y when $\omega = 0.125$, 0.25, 0.5, 1.0, 2.0, 4.0, 8.0, 16.0, and 32.0.

11.11. The asymptotes of the log-magnitude plot for two $G(j\omega)$ functions are shown in Fig. P11.11. Determine the equation for $G(j\omega)$, and evaluate the gain K for each case.

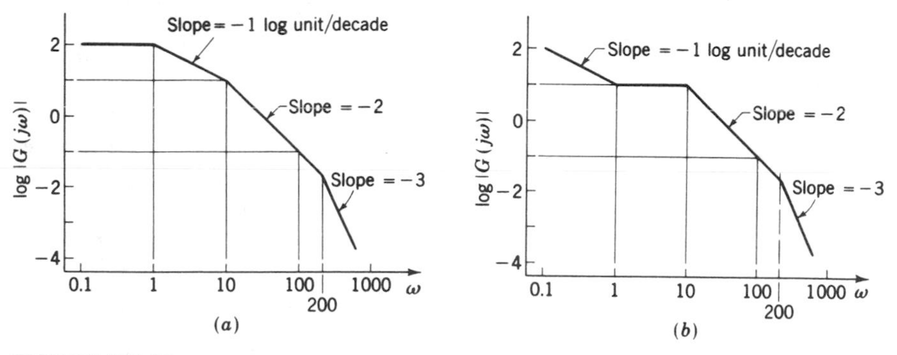

FIGURE P11.11

11.12. The asymptotes of the log-magnitude plot for two $G(j\omega)$ functions are shown in Fig. P11.12. Determine the equation for $G(j\omega)$, and evaluate the gain K for each case.

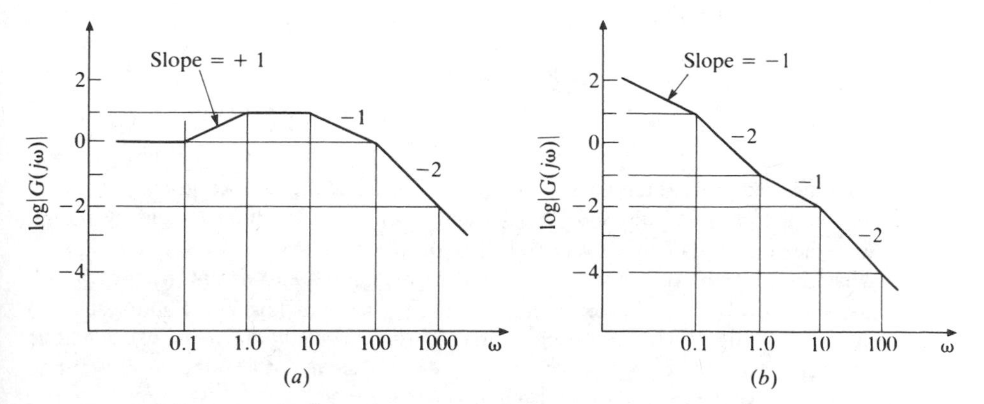

FIGURE P11.12

11.13. For the system shown in Fig. P11.13, the frequency-response curves for $G_1(j\omega)$ and $G_2(j\omega)$ were determined experimentally. Construct the log-magnitude plot for $G(j\omega) = G_1(j\omega)G_2(j\omega)$. Determine the equation for $G(j\omega)$, and evaluate the gain K.

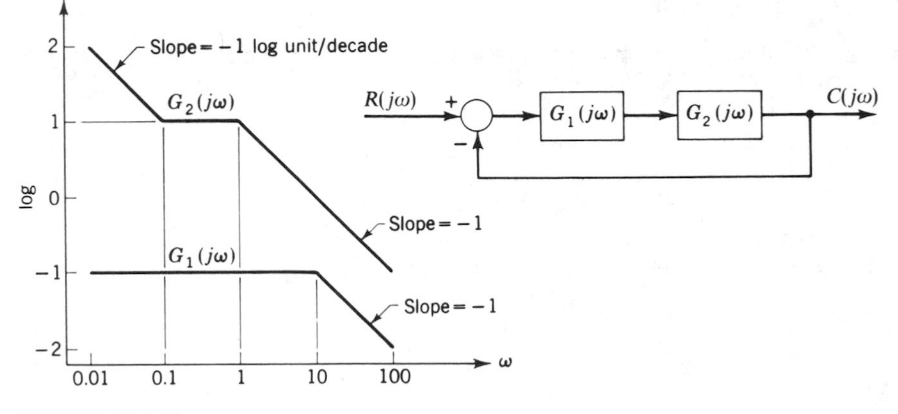

FIGURE P11.13

11.14. Convert each of the systems shown in Fig. P11.14 to an equivalent unity-feedback system.

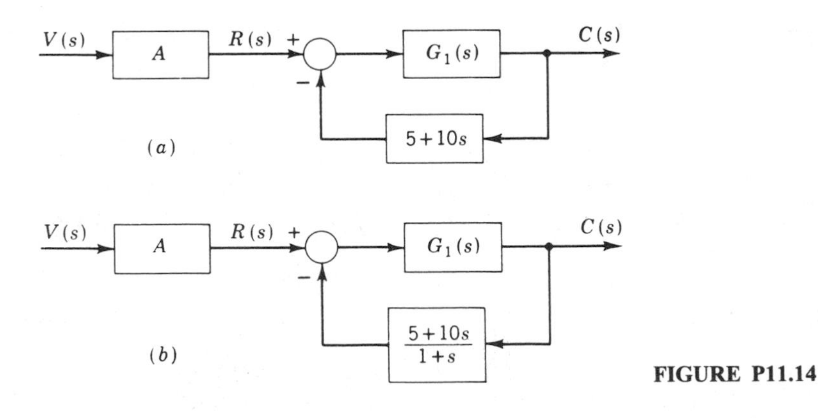

FIGURE P11.14

11.15. From the polar plot of a type 1 system, it is found that $M_p = 1.4$ and $\omega_p = 5$. What are the approximate values of the damping ratio ζ and natural frequency ω_n?

11.16. From the polar plot of a type 0 system, it is found that $K_0 = 4$, $M_p = 1.4$, and $\omega_p = 5$. What are the approximate values of the damping ratio ζ and natural frequency ω_n?

11.17. The polar plots of $G(j\omega)$ for two unity-feedback systems (system A and system B) are shown in Fig. P11.17. For each system, determine whether the response would be better approximated by a first- or a second-order system and also the corresponding values of τ or ζ and ω_n to be used.

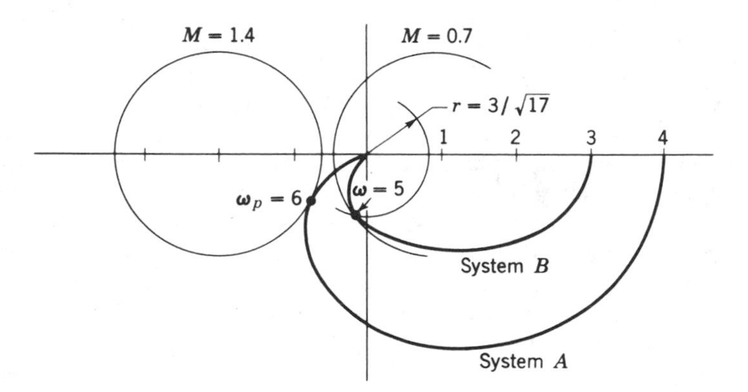

FIGURE P11.17

11.18. For the system shown in Fig. P11.18, write the characteristic equation and then determine the value of K such that $\omega_n = 1$. What is the corresponding value of the damping ratio ζ? Sketch the asymptotes of the log-magnitude plot for both the open-loop and the closed-loop frequency response. Obtain the equation for M and differentiate to obtain the value $\omega = \omega_p$ at which $M = M_p$ has its peak value. Determine the peak value M_p. Will Eqs. (11.63) and (11.65) yield exact or approximate values for ω_n and ζ for this case?

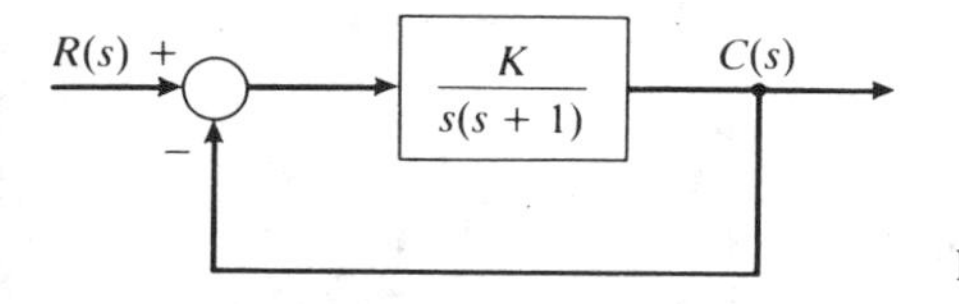

FIGURE P11.18

11.19. For the system shown in Fig. P11.19, write the characteristic equation and then determine the value of K such that $\omega_n = 3$. What is the corresponding value of the damping ratio ζ? Sketch the asymptotes of the log-magnitude plot for both the open-loop and the closed-loop frequency response. Determine ω_p and M_p. Will Eqs. (11.63) and (11.68) yield exact or approximate values for ω_n and ζ for this case?

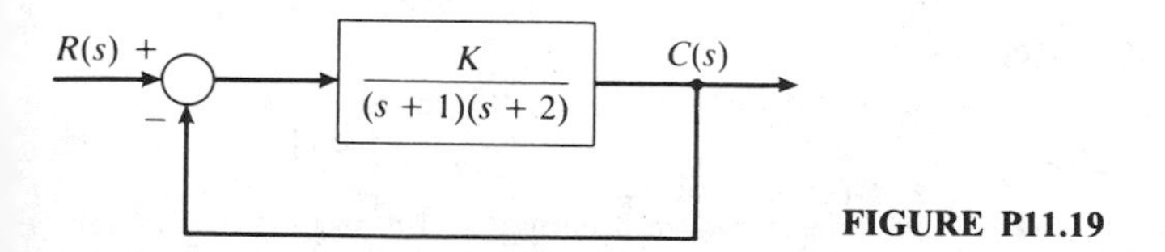

FIGURE P11.19

11.20. (*a*) It is desired to have an M_p of 1.4 for a unity-feedback type 0 control system in which $K_0 = 5$. From the polar plot of $G(j\omega)$, it is found that $-A = -\frac{1}{3}$ and $\omega_p = 4$. Determine the factor by which the gain should be changed to yield the desired value of M_p. What are the approximate damping ratio and the natural frequency for the resulting system?

(*b*) Repeat part *a* for a type 1 system.

11.21. For the system shown in Fig. P11.21,

$$G(s) = \frac{3}{(s^2 + 2\zeta\omega_n s + \omega_n^2)/\omega_n^2} = \frac{3}{(s^2 + 8s + 4)/4}$$

(*a*) Determine the closed-loop transfer function $C(s)/R(s)$. What are the corresponding values of ζ and ω_n for this closed-loop transfer function?

(*b*) Construct the polar plot for $G(j\omega)$, and then determine the factor K_c by which the gain should be changed in order to yield an M_p of $2/\sqrt{3}$ [$\psi = \sin^{-1}(\sqrt{3}/2) = 60°$]. By using the correlation criterion between frequency response and transient response, determine the approximate values of ζ and ω_n for the resulting closed-loop system.

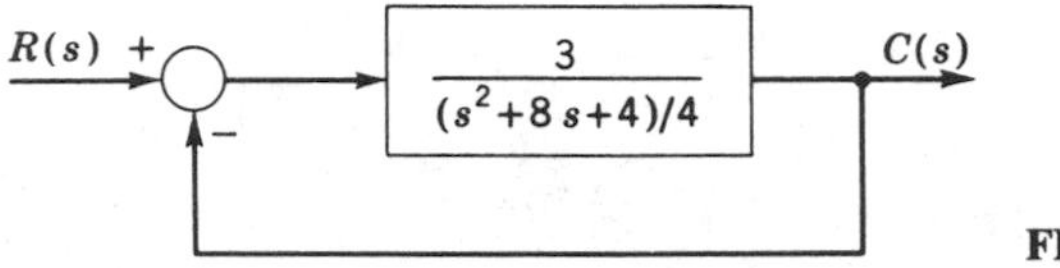

FIGURE P11.21

11.22. By the use of polar plots, determine the value of K to yield an M_p of 1.4 for each of the following unity-feedback systems.

(*a*) $G(s) = \dfrac{K}{s(1 + 0.25s)}$ (*b*) $G(s) = \dfrac{K}{s(0.25s^2 + 0.40s + 1)}$

(*c*) $G(s) = \dfrac{K}{(1 + 0.25s)(0.25s^2 + 0.40s + 1)}$

11.23. Repeat Prob. 11.22, except use log-modulus plots rather than polar plots.

11.24. For each system of Prob. 11.22, let $K = 4$, and then determine the factor K_c by which the gain should be changed in order to obtain an M_p of 1.4.

11.25. Repeat Prob. 11.24, except use log-modulus plots rather than polar plots.

11.26. The open-loop transfer function for a unity-feedback system is

$$G(s) = \frac{K}{s(s + 4)}$$

From the polar plot determine the gain K such that $M_p = 1.365$. What are the corresponding values of the damping ratio ζ, natural frequency ω_n, and bandwidth ω_b?

11.27. Repeat Prob. 11.26, except $M_p = 1.155$.

11.28. The open-loop transfer function for a unity-feedback system is

$$G(s) = \frac{K}{(s + 4)^2}$$

From the polar plot determine the gain K such that $M_p = 1.365$. What are the corresponding values of the damping ratio ζ, natural frequency ω_n, and bandwidth ω_b?

11.29. Repeat Prob. 11.28, except $M_p = 1.155$.

CHAPTER 12

SYSTEM COMPENSATION

Additional insight into the correlation between the shape of a polar plot and the dynamic behavior of a system is gained via the Nyquist stability criterion.* For many design problems, it is necessary not only to change the gain K, as discussed in the preceding chapter, but also to reshape the polar plot. In this chapter, the significance of the Nyquist stability criterion is discussed, and then it is shown how system performance may be improved by reshaping the polar plot.

12.1 NYQUIST STABILITY CRITERION

Consider the function

$$F(s) = s - r$$

The vector from the origin to point r and the vector from the origin to point s are indicated in Fig. 12.1a. The term $s - r$ is represented by the vector from r to s. Point s travels around the closed path indicated as the path of values of s. That is, s goes along the path to point s_1, then to s_2, and then back to its original position. Point r is a fixed point located inside the path of values of s. The dotted vector from r to s_1 indicates the location of the $s - r$ vector when s is located at point s_1. Similarly, the dotted vector from r to s_2 represents $s - r$ when s is at point s_2. Figure 12.1a shows that when r is located inside the path of values of s, the vector $s - r$

* H. Nyquist, "Regeneration Theory," *Bell System Tech. J.*, 11:126–147 (1932).

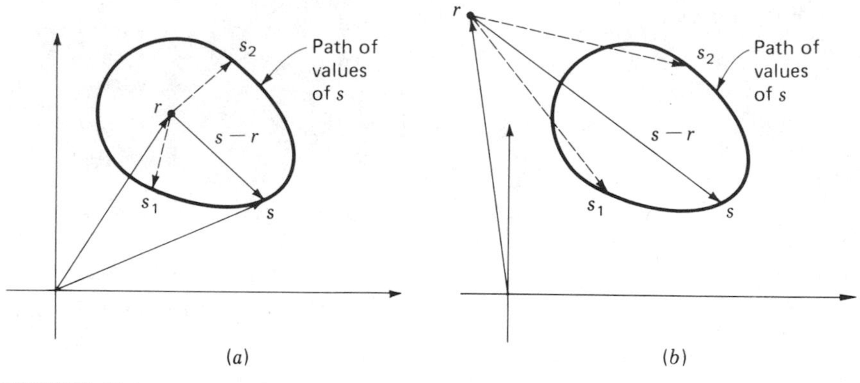

FIGURE 12.1
Rotation of $s - r$ vector when point r is (a) inside path of values of s, (b) outside path of values of s.

rotates through one complete revolution as s traverses the path from s to s_1 to s_2 and then back to s. Thus, the angle of the function $F(s) = s - r$ increases by 2π rad (1 r) as s traverses the path of values of s.

For the reciprocal $1/(s - r)$, the angle is the negative of the angle for the term $s - r$. Thus, the angle of the reciprocal decreases by 2π rad (1 r) as s traverses the path.

The case in which r is located outside the path is shown in Fig. 12.1b. The vector $s - r$ goes from point r to s. The angle of $s - r$ changes slightly as s travels from its initial position to point s_1. In traveling from s_1 to s_2, the angle of the function returns to its initial value and then changes slightly in the opposite direction. In returning to its starting position the angle of vector $s - r$ returns to its initial value. Thus, as s traverses the path of values, the vector $s - r$ eventually returns to its initial position, so that there is no net rotation of the vector.

Consider the function

$$F(s) = \frac{(s - z_1)(s - z_2) \cdots (s - z_m)}{(s - p_1)(s - p_2) \cdots (s - p_n)}$$

where $z_1, z_2, \ldots, z_m$ are the zeros of the function and $p_1, p_2, \ldots, p_n$ are the poles of the function. For each zero of $F(s)$ located within the path of the values of s, the angle of the function $F(s)$ increases by 2π rad (1 r) as s traverses the path. For each pole of $F(s)$ located within the path of values, the angle decreases by 2π rad as s traverses the path. Poles and zeros outside the path do not affect the angle.

The following general equation may be formulated:

$$N = Z - P \tag{12.1}$$

where P = number of poles of $F(s)$ located inside path of values for s
Z = number of zeros of $F(s)$ located inside path of values for s
N = net number of encirclements of origin of $F(s)$ plane

When the net number of encirclements N is in the same sense as motion around the path of the s contour, an excess of zeros Z is indicated. The opposite sense signifies

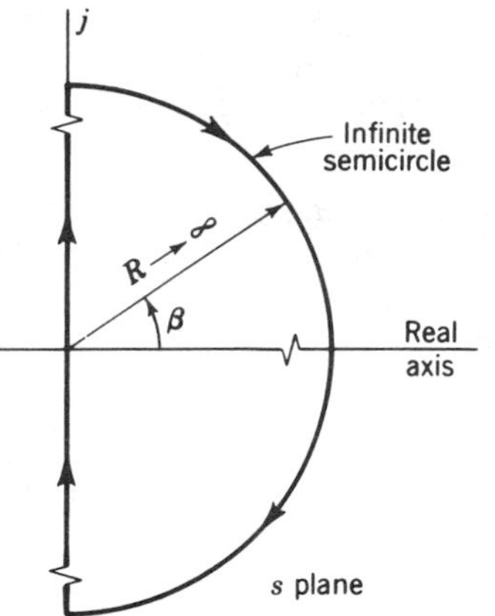

FIGURE 12.2
Path of values of s that encloses the entire right half plane.

an excess of poles P. From Eq. (12.1), it follows that when N is in the same sense, N is a positive number. Similarly, when N is in the opposite sense, N is a negative number.* Generally, the path of values for s is traversed in a clockwise direction. Thus, for a net number of clockwise encirclements, N is positive and there is an excess of zeros. For a net number of counterclockwise encirclements, N is negative and there is an excess of poles.

For control work, the path of values for s is usually taken as shown in Fig. 12.2. This contour is seen to proceed from the origin up the imaginary axis to infinity; then an infinite semicircle ($R \to \infty$) sweeps around to the bottom of the imaginary axis, whence it returns to the origin. This contour, in effect, encloses the entire right half plane.

As discussed in Chap. 6, a system is basically unstable if any zeros of the characteristic function are located in the right half plane. Note that

$$1 + G(s)H(s) = 1 + \frac{N_{G(s)}N_{H(s)}}{D_{G(s)}D_{H(s)}} = \frac{D_{G(s)}D_{H(s)} + N_{G(s)}N_{H(s)}}{D_{G(s)}D_{H(s)}} \tag{12.2}$$

It is apparent that the zeros of $1 + G(s)H(s)$ are the zeros of the characteristic function $D_{G(s)}D_{H(s)} + N_{G(s)}N_{H(s)}$. Similarly, the poles of $1 + G(s)H(s)$ are the zeros of $D_{G(s)}D_{H(s)}$. By letting s assume the values indicated along the contour of Fig. 12.2, and letting $F(s) = 1 + G(s)H(s)$, Eq. (12.1) becomes

$$Z = N + P \tag{12.3}$$

where Z = number of zeros of characteristic function [i.e., zeros of $1 + G(s)H(s)$] in the right half plane

P = number of zeros of $D_{G(s)}D_{H(s)}$ [i.e., poles of $1 + G(s)H(s)$] in the right half plane

N = net number of encirclements of the origin of the $1 + G(s)H(s)$ map.

---

* Often Eq. (12.1) is written in the form $N = P - Z$. When this form is used, N is negative for a net number of encirclements in the same sense and is positive for the opposite sense.

Because $G(s)H(s)$ is usually obtained in factored form, it is more convenient to construct the map for $G(s)H(s)$ than for $1 + G(s)H(s)$. The effect of adding +1 to each point of the $G(s)H(s)$ map to obtain the $1 + G(s)H(s)$ map is accomplished simply by adding +1 to the scale of the real axis, as shown by the numbers in parentheses in Fig. 12.3*b*. We see that the -1 point of the $G(s)H(s)$ map corresponds to the origin of the $1 + G(s)H(s)$ map. Thus, N is equal to the net number of encirclements of the -1 point of the $G(s)H(s)$ plot. The application of the Nyquist stability criterion to type 0, type 1, and type 2 systems is now illustrated.

Type 0 System

Consider the following type 0 system:

$$G(s)H(s) = \frac{52}{(s+2)(s^2+2s+5)} = \frac{52/10}{(0.5s+1)(0.2s^2+0.4s+1)} \tag{12.4}$$

Figure 12.3*a* shows the path of values of s that encloses the entire right half plane. Along the positive imaginary axis (indicated by the heavy line), s equals $j\omega$. Thus, for this region Eq. (12.4) becomes

$$G(j\omega)H(j\omega) = \frac{5.2}{(1+0.5j\omega)(1-0.2\omega^2+0.4j\omega)}$$

This is the frequency response for the system. It is indicated by the heavy line in Fig. 12.3*b*. For values of s along the infinite semicircle of Fig. 12.3*a*, Eq. (12.4)

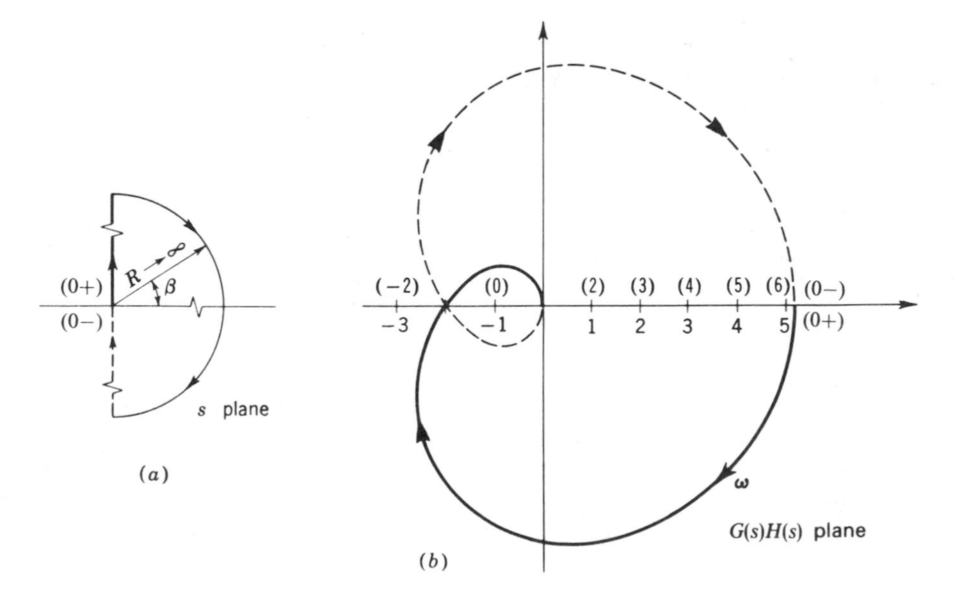

FIGURE 12.3
(*a*) Path of values of s, (*b*) map of $G(s)H(s)$ for a type 0 system.

becomes

$$G(s)H(s) = \lim_{R\to\infty} \frac{52}{(Re^{j\beta}+2)\,(R^2e^{2j\beta}+2Re^{j\beta}+5)} \approx \lim_{R\to\infty} \frac{52}{R^3e^{3j\beta}} = 0$$

Thus, the $G(s)H(s)$ plot runs into the origin for values of s on the infinite semicircle.

For values of s on the negative imaginary axis (indicated by the dotted line), $s = -j\omega$. Because $G(-j\omega)H(-j\omega)$ is the complex conjugate of $G(j\omega)H(j\omega)$, the plot for $G(-j\omega)H(-j\omega)$ is obtained by reflecting the frequency response $G(j\omega)H(j\omega)$ about the real (horizontal) axis. The resulting $G(-j\omega)H(-j\omega)$ plot is shown dotted in Fig. 12.3*b*.

The poles of $1 + G(s)H(s)$ are the poles of $G(s)H(s)$. Because there are no poles of Eq. (12.4) in the right half plane (i.e., within the path of values of s), then $P = 0$. There are two clockwise encirclements of the -1 point of Fig. 12.3*b*; hence, $N = 2$. Note that the -1 point of the $G(s)H(s)$ plot is the origin of the $1 + G(s)H(s)$ plot. Application of the Nyquist stability criterion to determine Z gives

$$Z = N + P = 2$$

Because there are two zeros of the characteristic function located in the right half plane, this system is basically unstable.

In general, for any type 0 system, the frequency response $G(j\omega)H(j\omega)$ begins on the real axis for $\omega = 0$ and terminates at the origin for $\omega = \infty$. The $G(s)H(s)$ plot remains at the origin as the path of values of s traverses the infinite semicircle. For values of s along the negative imaginary axis (that is, $s = -j\omega$), then $G(-j\omega)H(-j\omega)$ is the complex conjugate of the frequency response $G(j\omega)H(j\omega)$.

At the intersection of a polar plot and the real axis, the imaginary part of $G(j\omega)H(j\omega)$ is zero. The frequency at which a polar plot crosses the real axis is obtained by setting the imaginary part of $G(j\omega)H(j\omega)$ equal to zero. Substitution of this value of ω into the expression for $G(j\omega)H(j\omega)$ yields the location at which the polar plot crosses the real axis. Similarly, the location at which the polar plot crosses the imaginary axis is obtained by setting the real part of $G(j\omega)H(j\omega)$ equal to zero.

To illustrate this procedure, consider the system whose polar plot is shown in Fig. 12.3. The frequency response $G(j\omega)H(j\omega)$ may be written as

$$\begin{aligned} G(j\omega)H(j\omega) &= \frac{5.2}{(1+0.5j\omega)(1-0.2\omega^2+0.4j\omega)} \\ &= \frac{5.2}{(1-0.4\omega^2)+j\omega(0.9-0.1\omega^2)} \end{aligned}$$

Multiplying the numerator and denominator by the complex conjugate of the denominator gives

$$G(j\omega)H(j\omega) = 5.2\frac{(1-0.4\omega^2)-j\omega(0.9-0.1\omega^2)}{(1-0.4\omega^2)+\omega^2(0.9-0.1\omega^2)^2}$$

The imaginary part is zero when $0.9 - 0.1\omega^2 = 0$, or $\omega = 3$. Substitution of this value of ω into the expression for $G(j\omega)H(j\omega)$ yields the value at which

$G(j\omega)H(j\omega)$ crosses the real axis:

$$G(j3)H(j3) = \frac{5.2}{1 - 0.4(3)^2 + j(0)} = \frac{5.2}{1 - 3.6} = -2.0$$

The real part is zero when $1 - 0.4\omega^2 = 0$, or $\omega = \sqrt{2.5} = 1.58$. This is the value of ω at which the polar plot crosses the imaginary axis:

$$G(j1.58)H(j1.58) = \frac{5.2}{0 + j(1.58)(0.9 - 0.25)} = -\frac{j5.2}{1.027} = -j5.06$$

Type 1 System

For a type 1 system, there is an s term in the denominator of $G(s)H(s)$. For example, consider the system

$$G(s)H(s) = \frac{40}{s(s+1)(s+4)} = \frac{10}{s(1+s)(1+0.25s)} \tag{12.5}$$

In Fig. 12.4a note that the assumed path of values of s excludes the origin. The reason for this will be made apparent in the following development.

The portion of the plot for $s = j\omega$ as ω takes on values from 0+ to $+\infty$ is indicated by the heavy line. This is the frequency response for the system

$$G(j\omega)H(j\omega) = \frac{10}{j\omega(1 + j\omega)(1 + 0.25j\omega)}$$

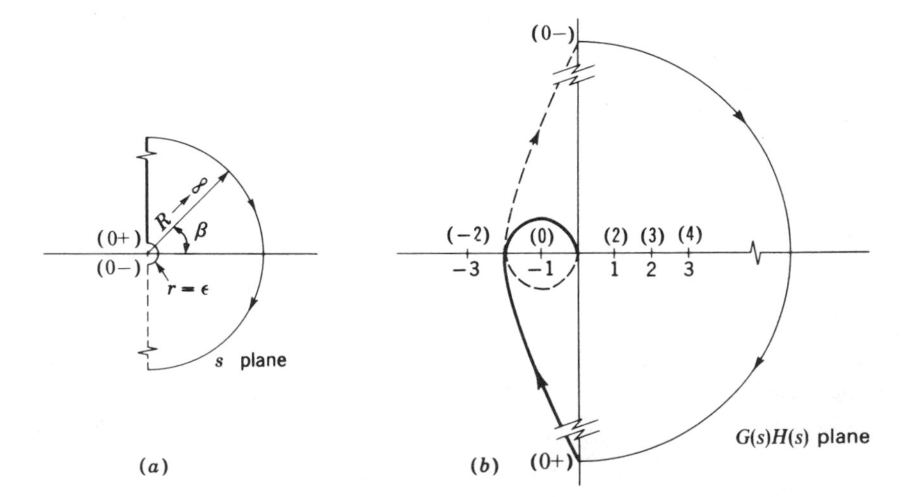

FIGURE 12.4
(a) Path of values of s that excludes pole at origin, (b) map of $G(s)H(s)$ for a type 1 system.

For values of s on the infinite semicircle, the $G(s)H(s)$ plot runs into the origin, i.e.,

$$G(s)H(s) = \lim_{R\to\infty} \frac{40}{Re^{j\beta}(Re^{j\beta}+1)(Re^{j\beta}+4)} \approx \lim_{R\to\infty} \frac{40}{R^3e^{3j\beta}} = 0$$

The dotted portion of the $G(s)H(s)$ plot corresponds to $s = -j\omega$, where $G(-j\omega)H(-j\omega)$ is the complex conjugate of $G(j\omega)H(j\omega)$. To complete the Nyquist plot, it is necessary to connect the ends corresponding to $\omega = 0-$ and $\omega = 0+$. If the path of values of s were to run through the origin, then for $s = 0$ the function $G(s)H(s)$ in Eq. (12.5) would be infinite and there would be no indication of how to join the $\omega = 0-$ and $\omega = 0+$ ends of the $G(s)H(s)$ plot in Fig. 12.4*b*. Hence a small semicircle of radius ε is constructed about the origin, as illustrated in Fig. 12.4*a*. The equation for $G(s)H(s)$ is obtained by substitution of $s = \varepsilon e^{j\beta}$, where β varies from $-90°$ to $0°$ to $+90°$ as s traverses the small semicircle. Hence

$$G(s)H(s) = \lim_{\varepsilon\to 0} \frac{40}{\varepsilon e^{j\beta}(\varepsilon e^{j\beta}+1)(\varepsilon e^{j\beta}+4)} \approx \frac{40}{\varepsilon e^{j\beta}(1)(4)} \approx \frac{10}{\varepsilon}\, e^{-j\beta} \approx \infty(e^{-j\beta}) \tag{12.6}$$

Thus $G(s)H(s)$ is an infinite semicircle. As s traverses the small semicircle from $0-$ to $0+$, then β goes from $-90°$ to $0°$ to $+90°$. The corresponding infinite semicircle goes from $+90°$ to $0°$ to $-90°$ (i.e., the angle of the infinite semicircle is $-\beta$). The complete Nyquist plot is shown in Fig. 12.4*b*.

Because the origin has been excluded from the path of values of s, from Eq. (12.5) it follows that there are no poles of $G(s)H(s)$ within the path of values of s, and thus $P = 0$. As indicated in Fig. 12.4*b*, there are two clockwise encirclements of the -1 point, so that $N = 2$. Application of the Nyquist stability criterion gives

$$Z = N + P = 2$$

Thus, there are two roots of the characteristic equation in the right half plane. Consequently the system is unstable.

The general procedure for constructing the complete Nyquist plot for a type 1 system is summarized as follows. A type 1 system always has a pole at the origin, so that it is necessary to exclude the origin from the path of values of s, as indicated in Fig. 12.4*a*. The frequency response $G(j\omega)H(j\omega)$ corresponds to values of $s = j\omega$ as the path of values of s traverses the positive imaginary axis from $0+$ to $+\infty$. For values of s on the infinite semicircle, the $G(s)H(s)$ plot runs into the origin. As s traverses the negative imaginary axis, $G(s)H(s) = G(-j\omega)H(-j\omega)$ is the complex conjugate of the frequency response.

This complex conjugate of the frequency response is obtained by "flipping" the frequency-response plot about the horizontal axis. To complete the Nyquist plot, the ends at $\omega = 0-$ and $\omega = 0+$ are joined by assuming values of s on the small semicircle of radius ε. Equation (12.6) holds for any type 1 system, i.e.,

$$G(s)H(s) = \infty(e^{-j\beta})$$

This is an infinite semicircle.

In the application of the Nyquist stability criterion it would make no difference if the origin were included in the path of values of s, as shown in Fig. 12.5*a*, or if it were excluded, as shown in Fig. 12.4*a*. For the case in which the origin is included, β varies from $-90°$ to $-180°$ to $-270°$ as s traverses the small semicircle. Thus from Eq. (12.6), the infinite semicircle goes from 90° to 180° to 270°. The resultant $G(s)H(s)$ plot is shown in Fig. 12.5*b*. For this case, $N = +1$. Because $s = 0$ is a pole of $G(s)H(s)$ which is now included in the path of values of s, $P = 1$. Hence

$$Z = N + P = 1 + 1 = 2$$

Thus, the same result ($Z = 2$) is obtained whether the pole at the origin is included in or excluded from the path of values of s. When a pole of $G(s)H(s)$ occurs at the origin ($s = 0$), it is customary to exclude the origin from the path of values of s, as indicated in Fig. 12.4*a*.

A direct method for determining the net encirclements N of the -1 point is to draw a radial line from the -1 point out through the plot as illustrated in Fig. 12.5*b*. Note the direction of the arrow on the loci at each point where the radial line crosses the loci 3 times. At two crossings the direction of the arrow is such as to rotate the radial line in a clockwise direction. At the third crossing, the direction is such as to rotate the radial line in a counterclockwise direction. The value of N is the number of crossings at which the arrow tends to rotate the radial line clockwise minus the number of locus crossings at which the arrow tends to rotate the radial line

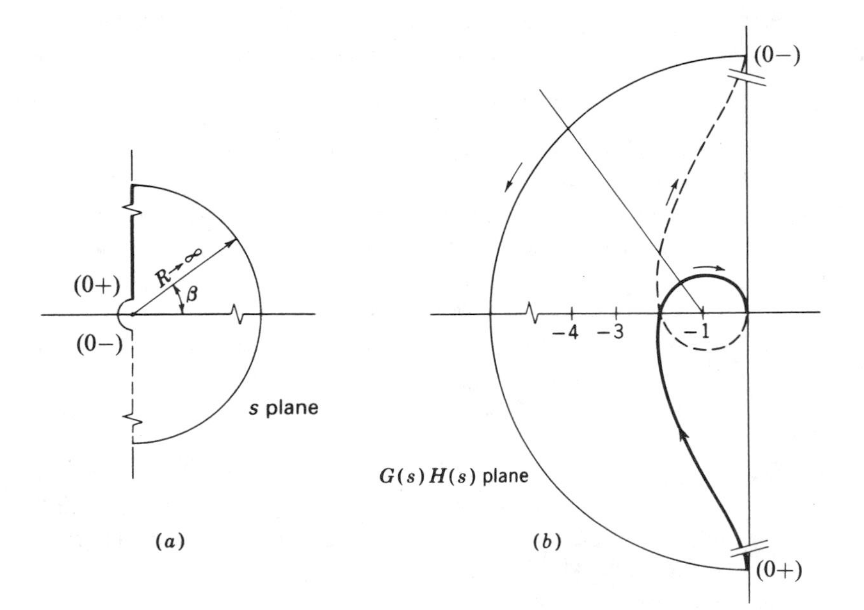

FIGURE 12.5
(*a*) Path of values of s that includes pole at origin, (*b*) map of $G(s)H(s)$ for a type 1 system.

counterclockwise. Thus, for this case $N = 2 - 1 = 1$. The radial line may be drawn at any angle from the -1 point.

Type 2 System

A type 2 system is characterized by the fact that there is a double pole at the origin. For example, consider the system

$$G(s)H(s) = \frac{4(1+s)}{s^2(1+0.1s)} \tag{12.7}$$

As for a type 1 system, the path of values of s is taken to exclude the origin, as shown in Fig. 12.6*a*. The frequency-response plot for this system is indicated by the heavy line in Fig. 12.6*b*. The $G(s)H(s)$ plot runs into the origin as s traverses the infinite semicircle. The $G(-j\omega)H(-j\omega)$ plot is merely the complex conjugate of the frequency response.

For values of s on the small semicircle which excludes the origin,

$$G(s)H(s) = \lim_{\varepsilon \to 0} \frac{4(1+\varepsilon e^{j\beta})}{\varepsilon^2 e^{2j\beta}(1+0.1\varepsilon e^{j\beta})} \approx \frac{4}{\varepsilon^2} e^{-j2\beta} \approx \infty(e^{-j2\beta}) \tag{12.8}$$

Equation (12.8) is valid for any type 2 system. As s traverses the small semicircle from $0-$ to $0+$, then β goes from $-90°$ to $0°$ to $+90°$. The corresponding infinite semicircle changes from $+180°$ to $0°$ to $-180°$ (i.e., the angle of the infinite semicircle is -2β).

For the $G(s)H(s)$ plot described by Eq. (12.7), there are no poles within the path of values of s, so that P is zero. Because there are no encirclements of the -1 point of Fig. 12.6*b*, N is also zero. Thus, $Z = N + P = 0$, so that the system is basically stable.

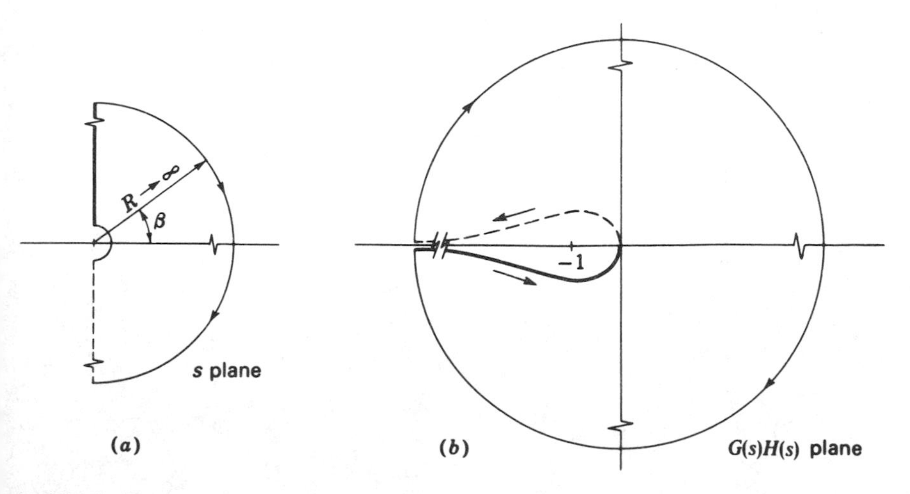

FIGURE 12.6
(*a*) Path of values of s, (*b*) map of $G(s)H(s)$ for a type 2 system.

When the polar plot for $G(s)H(s)$ goes through the -1 point, the system is on the borderline between being stable and unstable. For this case the characteristic equation has roots on the imaginary axis. For example, from Fig. 12.3*b*, note that the polar plot crosses the negative real axis at a value of -2. If the gain were halved, the polar plot would go through the -1 point. For this case, the characteristic equation becomes

$$(s+2)(s^2+2s+5)+26 = s^3+4s^2+9s+36 = (s+4)(s^2+9) \qquad (12.9)$$

Two roots, $s = \pm j3$, lie on the imaginary axis. Similarly, from Fig. 12.4*b*, note that the polar plot crosses the negative real axis at -2 also. Halving the gain yields for the characteristic equation

$$s(s+1)(s+4)+20 = s^3+5s^2+4s+20 = (s+5)(s^2+4) \qquad (12.10)$$

Two roots, $s = \pm j2$, lie on the imaginary axis.

For most systems, the open loop is stable. For this case, all the zeros of $D_G D_H$ lie in the left half plane ($P = 0$). Thus, the Nyquist stability criterion reduces to

$$Z = N$$

For a stable system ($Z = 0$), there are no net encirclements N of the -1 point.

In applying frequency-response methods, one always works with the open-loop response. Note that the closed loop may be stable even though the open loop is unstable, and vice versa.

12.2 GAIN MARGIN AND PHASE MARGIN

From the preceding discussion, the -1 point of the $G(s)H(s)$ map was seen to have great significance with regard to the stability of a system. Figure 12.7 shows a typical $G(j\omega)H(j\omega)$ plot in the vicinity of the -1 point. If the gain were multiplied

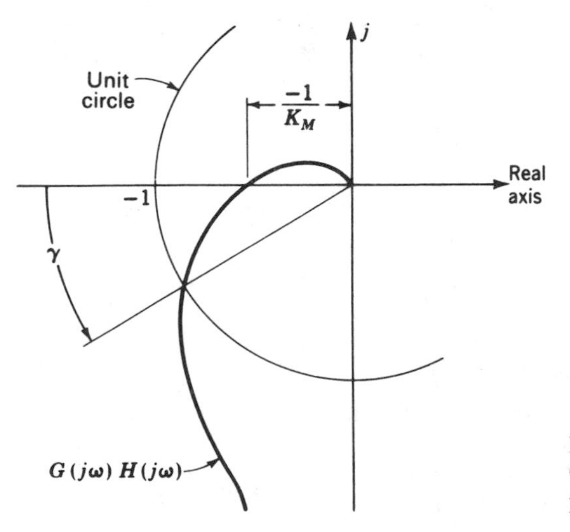

FIGURE 12.7
Gain margin and phase margin on the polar plot.

by an amount K_M, called the *gain margin*, then the $G(j\omega)H(j\omega)$ plot would go through the -1 point. Thus, the gain margin is an indication of how much the gain can be increased before the curve goes through the critical point.

The angle γ in Fig. 12.7 is the angle measured from the negative real axis to where the polar plot crosses a circle of unit radius. If angle γ is zero, the polar plot goes through the -1 point. Angle γ, called the *phase margin*, is thus another indication or measure of the closeness of the polar plot to the critical point. Note that a positive phase margin indicates a stable system, as does a gain margin greater than 1.

Log-Modulus Plots

Figure 12.8 shows the log-modulus plot corresponding to Fig. 12.7. At the -1 point, $|G(j\omega)H(j\omega)| = 1$ and $\phi = \measuredangle G(j\omega)H(j\omega) = -180°$. Thus, on the log-modulus plot, the -1 point is located at the intersection of the ordinate $\log|G(j\omega)H(j\omega)| = \log 1 = 0$ and the abscissa $\phi = -180°$. From Fig. 12.8, the vertical distance that the $G(j\omega)H(j\omega)$ plot may be raised before it goes through the -1 point is $\log K_M$.

The value of the phase margin may be obtained from a log-modulus plot as follows: The horizontal line in Fig. 12.8 of $\log|G(j\omega)H(j\omega)| = \log 1 = 0$ corresponds to the unit circle of Fig. 12.7. Angle γ between the point where this horizontal line intersects the $G(j\omega)H(j\omega)$ plot and the value $\phi = -180°$ is the phase margin.

Illustrative example 12.1. The log-modulus plot of $G(j\omega)$ for a unity-feedback system is shown in Fig. 12.9. What are the values of the gain margin and the phase margin for this system? By what factor K_c should the gain of the system be changed so that M_p will be 1.4? What are the new values for the gain margin and the phase margin?

Solution. From the $G(j\omega)$ plot for the original system, it follows that $\log K_M = 0.2$, or $K_M = 1.59$, and $\gamma = 19°$. The factor K_c is obtained by moving the $G(j\omega)$ locus straight down until the new locus $K_cG(j\omega)$ is tangent to the desired $M_p = 1.4$ contour.

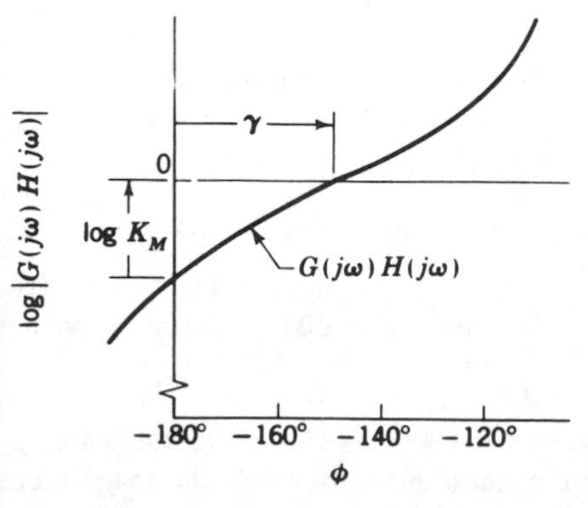

FIGURE 12.8
Gain margin and phase margin on the log-modulus plot.

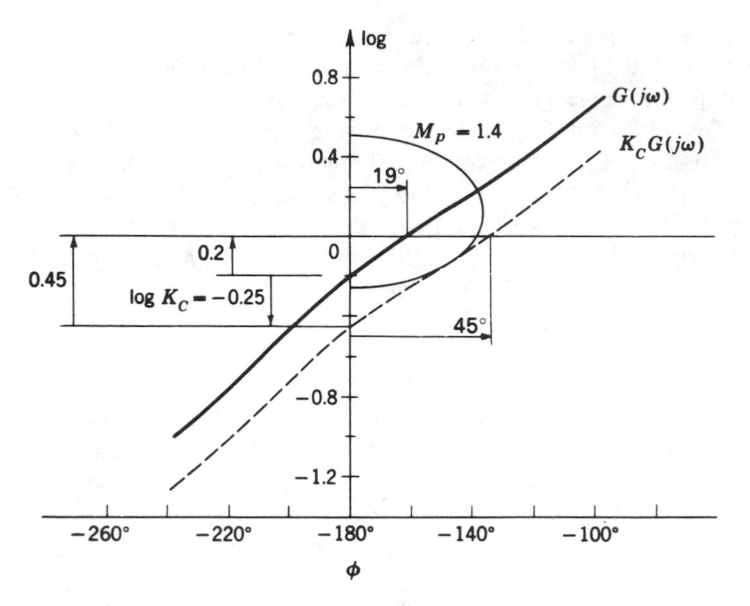

FIGURE 12.9
Log-modulus plot for a system.

From Fig. 12.9, it follows that

$$\log K_c = -0.25 \qquad \text{or} \qquad K_c = 0.563$$

The new gain margin is log K_M = +0.45 or K_M = 2.81, and the new phase margin is $\gamma = 45°$.

Bode Plots

Figure 12.10 shows a Bode plot. By entering the phase angle plot at $\phi = -180°$, note the corresponding value of $\log |G(j\omega)H(j\omega)| = -0.3$. Hence, the gain of the system may be increased by +0.3 log unit before the $G(j\omega)H(j\omega)$ curve goes through the -1 point. Thus, the value of log K_M may be represented graphically as in Fig. 12.10.

The unit circle of the polar plot corresponds to the horizontal line through the log-magnitude plot at $\log |G(j\omega)H(j\omega)| = \log 1 = 0$. The phase margin γ is the difference between the corresponding angle and the angle $\phi = -180°$, as illustrated in the angle plot of Fig. 12.10.

Illustrative example 12.2. Figure 12.11 shows the Bode plot for a control system. By what factor K_c should the gain of the system be changed so that the resulting phase margin is 45°?

Solution. Entering the angle plot at $\phi = -180° + \gamma = -135°$, the corresponding value of $\log |G(j\omega)H(j\omega)|$ is 0.3. To have a phase margin of 45°, the magnitude

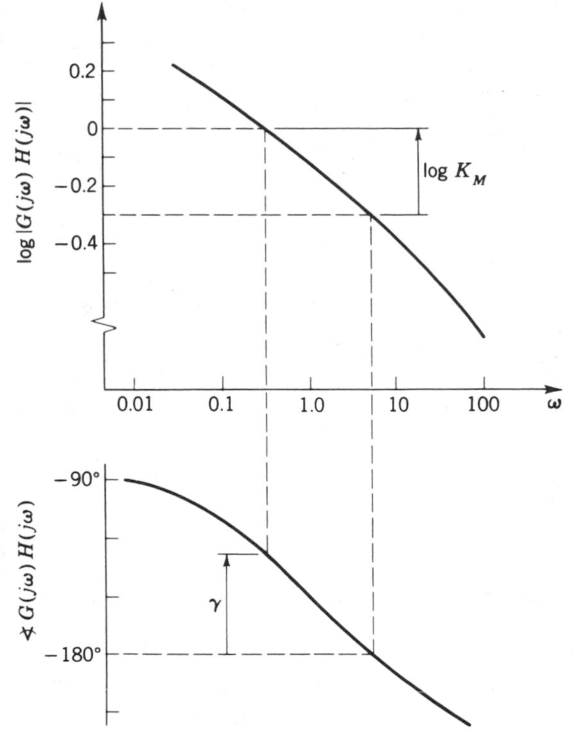

FIGURE 12.10
Gain margin and phase margin on the Bode plot.

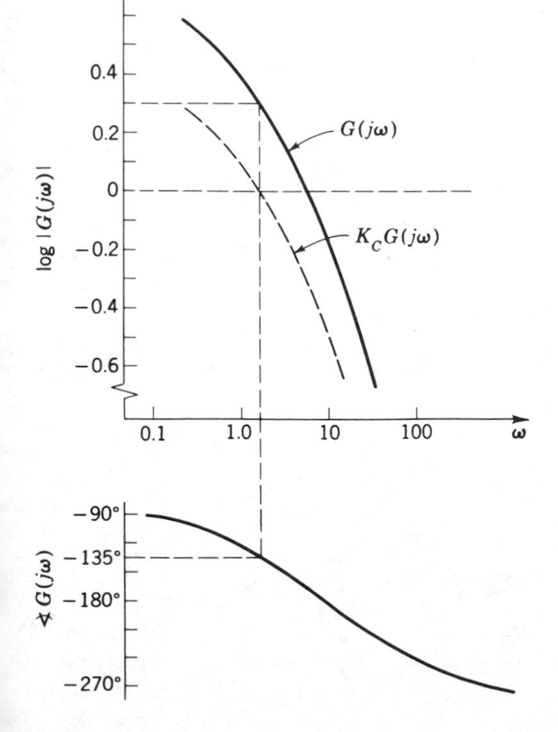

FIGURE 12.11
Bode plot for a system.

plot must be lowered by 0.3 log unit. Thus,

$$\log K_c = -0.3$$

or

$$K_c = 0.5$$

In Sec. 11.7 the transient-response parameters ζ and ω_n were related to the frequency-response parameters M_p and ω_p. This correlation is limited to use with polar or log-modulus plots. That is, M circles may be drawn on polar plots and M contours on log-modulus plots, but it is impossible to construct M lines on Bode plots because the magnitude and phase angle are on separate plots. When one is designing with Bode plots, the transient and frequency responses are correlated by relating ζ and ω_n to the phase margin.

Correlation between ζ, ω_n, and Phase Margin

We now show that the transient-response parameters ζ and ω_n may also be related to the phase margin γ. Although the following analysis is for a second-order type 1 system, the results yield good approximations for higher-order systems and for systems of other type numbers, such as a type 0 or a type 2 system.

The closed-loop transfer function for the system of Fig. 11.20 is

$$\frac{C(s)}{R(s)} = \frac{K/\tau}{s^2 + (1/\tau)s + K/\tau} = \frac{\omega_n^2}{s^2 + 2\zeta\omega_n s + \omega_n^2} \tag{12.11}$$

where

$$\omega_n^2 = \frac{K}{\tau} \qquad \text{and} \qquad 2\zeta\omega_n = \frac{1}{\tau}$$

The open-loop frequency response is

$$G(j\omega) = \frac{K}{j\omega(1 + j\tau\omega)}$$

The frequency ω_c at which the $G(j\omega)$ plot crosses the unit circle is called the *crossover frequency.* When $|G(j\omega)| = 1$, then

$$|G(j\omega)| = \frac{K}{\omega_c\sqrt{1 + \tau^2\omega_c^2}} = 1$$

Squaring gives

$$\tau^2\omega_c^4 + \omega_c^2 - K^2 = 0$$

or

$$\left(\frac{\tau}{K}\right)^2 \omega_c^4 + \frac{1}{K^2}\,\omega_c^2 - 1 = 0$$

Because $\tau/K = 1/\omega_n^2$ and $K = \omega_n^2 \tau = \omega_n^2/(2\zeta\omega_n) = \omega_n/(2\zeta)$, then

$$\left(\frac{\omega_c}{\omega_n}\right)^4 + 4\zeta^2 \left(\frac{\omega_c}{\omega_n}\right)^2 - 1 = 0$$

Application of the quadratic equation yields

$$\left(\frac{\omega_c}{\omega_n}\right)^2 = \sqrt{4\zeta^4 + 1} - 2\zeta^2 \tag{12.12}$$

The preceding equation relates ζ and the ratio ω_c/ω_n. A plot of this relationship is shown in Fig. 12.12*a*.

The phase angle is

$$\gamma = 180° + \measuredangle G(j\omega_c)$$

where $\measuredangle G(j\omega_c)$ is the angle of $G(j\omega)$ at which the polar plot crosses the unit circle. That is,

$$\measuredangle G(j\omega_c) = 0° - 90° - \tan^{-1} \tau\omega_c$$

Hence

$$\gamma = 90° - \tan^{-1} \tau\omega_c = \tan^{-1} \frac{1}{\tau\omega_c} = \tan^{-1} \frac{2\zeta}{\omega_c/\omega_n}$$

$$= \tan^{-1} \frac{2\zeta}{\sqrt{-2\zeta^2 + \sqrt{4\zeta^4 + 1}}} \tag{12.13}$$

Solving Eq. (12.13) for ζ in terms of γ gives

$$\zeta = \frac{(\tan\gamma)\sqrt{\cos\gamma}}{2} \tag{12.14}$$

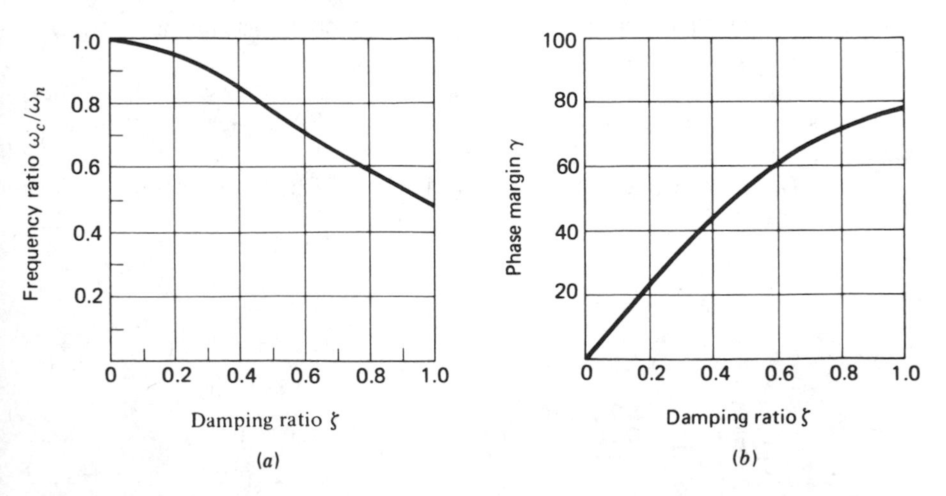

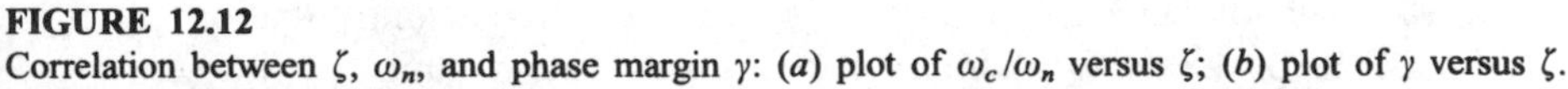

FIGURE 12.12
Correlation between ζ, ω_n, and phase margin γ: (*a*) plot of ω_c/ω_n versus ζ; (*b*) plot of γ versus ζ.

Figure 12.12*b* shows a plot of γ versus ζ. If γ is known, then the value of ζ may be determined from this plot. The corresponding value of ω_c/ω_n may be ascertained from Fig. 12.12*a*.

Illustrative example 12.3. Determine the values of ζ and ω_n for the system shown by the dotted line in Fig. 12.11.

Solution. For this system, the phase margin γ is 45°. From Eq. (12.14) or from Fig. 12.12*b*, the corresponding value of ζ is 0.42. Next, from Eq. (12.12) or from Fig. 12.12*a*, the ratio ω_c/ω_n is found to be 0.84. The frequency ω_c is the value of ω at which $\log|G(j\omega)H(j\omega)| = \log 1 = 0$. From Fig. 12.11, it is found that $\omega_c = 1.7$, and thus $\omega_n = 2.03$.

12.3 LEAD COMPENSATION

In Sec. 11.8, we showed how the gain K is selected in order to obtain a desired value of M_p. A change in the gain K in effect changes the scale factor of the polar plot but does not change the basic shape of the plot. In the design of control systems, it is often necessary to change the shape of the polar plot in order to achieve the desired dynamic performance. A common means of doing this is to insert elements in series with the feedforward portion of the control. This method of compensating the performance of the control system is called *series compensation.*

In general, the frequency-response characteristics of a component which is used to provide series compensation are such that the output of the component either lags or leads the input. In some cases, it is advantageous to use a component in which the output lags the input for a certain range of frequencies and then the output leads the input for other frequencies. This is known as *lag-lead series compensation.* A component which is used to provide series compensation is sometimes referred to as a *series equalizer.*

In this section, the design of series lead compensators is discussed. The next two sections consider series lag compensators and series lag-lead compensators, respectively.

Figure 12.13 shows both an electrical and a mechanical component which have the general phase-lead characteristic given by

$$\frac{E_o(s)}{E_{\text{in}}(s)} = \frac{Y(s)}{X(s)} = \frac{\tau_2}{\tau_1}\frac{1+\tau_1 s}{1+\tau_2 s} \qquad \tau_1 > \tau_2$$

The frequency response for the preceding transfer function is

$$\frac{E_o(j\omega)}{E_{\text{in}}(j\omega)} = \frac{Y(j\omega)}{X(j\omega)} = \frac{\tau_2}{\tau_1}\frac{1+j\tau_1\omega}{1+j\tau_2\omega} \qquad \tau_1 > \tau_2 \tag{12.15}$$

Because the steady-state gain ($\omega \approx 0$) is τ_2/τ_1, additional amplification equal to τ_1/τ_2 must be provided to maintain the original system gain.

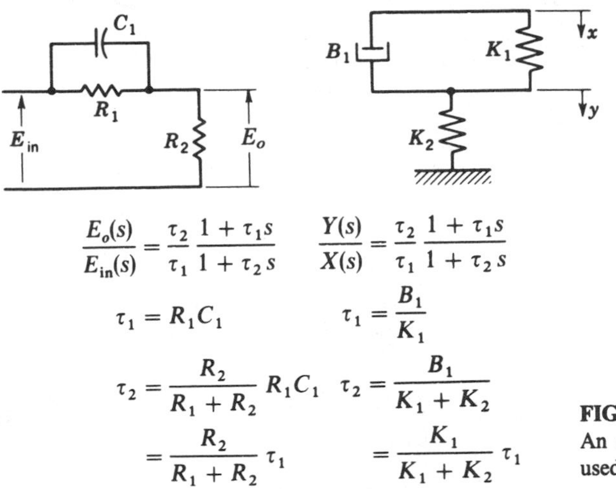

FIGURE 12.13
An electrical and a mechanical circuit used to obtain phase lead.

The construction of the log-magnitude plot for $(1 + j\tau_1\omega)/(1 + j\tau_2\omega)$ in which $\tau_2 = \tau_1/10$ is illustrated in Fig. 12.14. For the numerator, the break frequency occurs at $\omega = 1/\tau_1$, and for the denominator the break frequency occurs at $\omega = 1/\tau_2 = 10/\tau_1$. Adding the plot for the numerator to that for the denominator yields the resultant plot for phase lead.

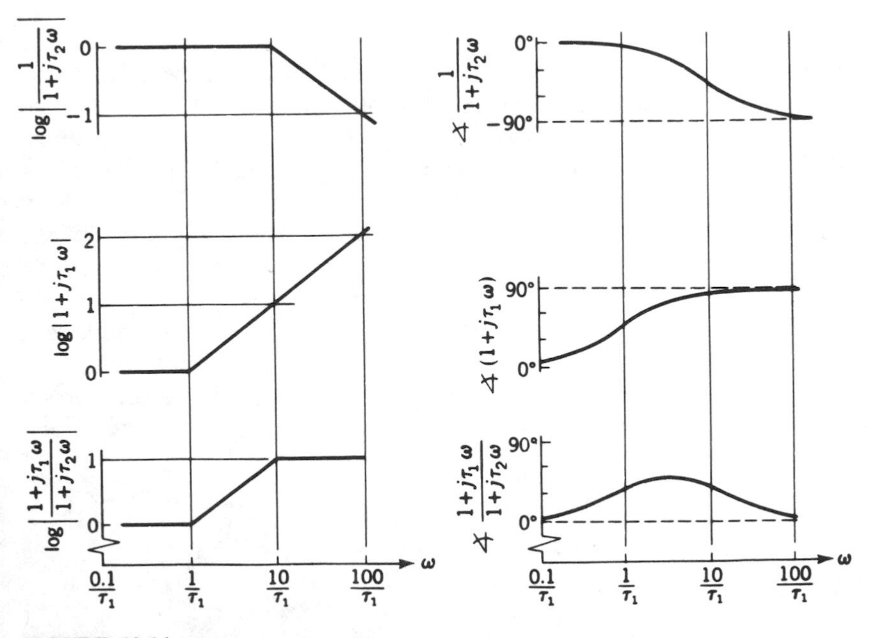

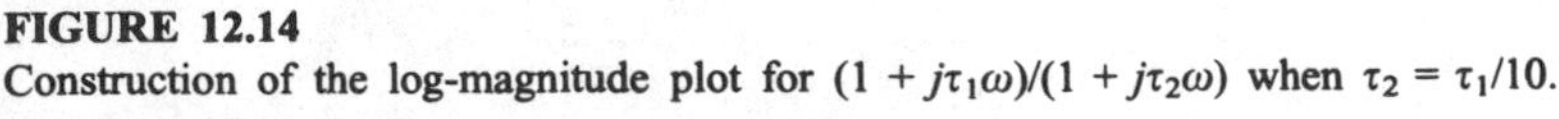

FIGURE 12.14
Construction of the log-magnitude plot for $(1 + j\tau_1\omega)/(1 + j\tau_2\omega)$ when $\tau_2 = \tau_1/10$.

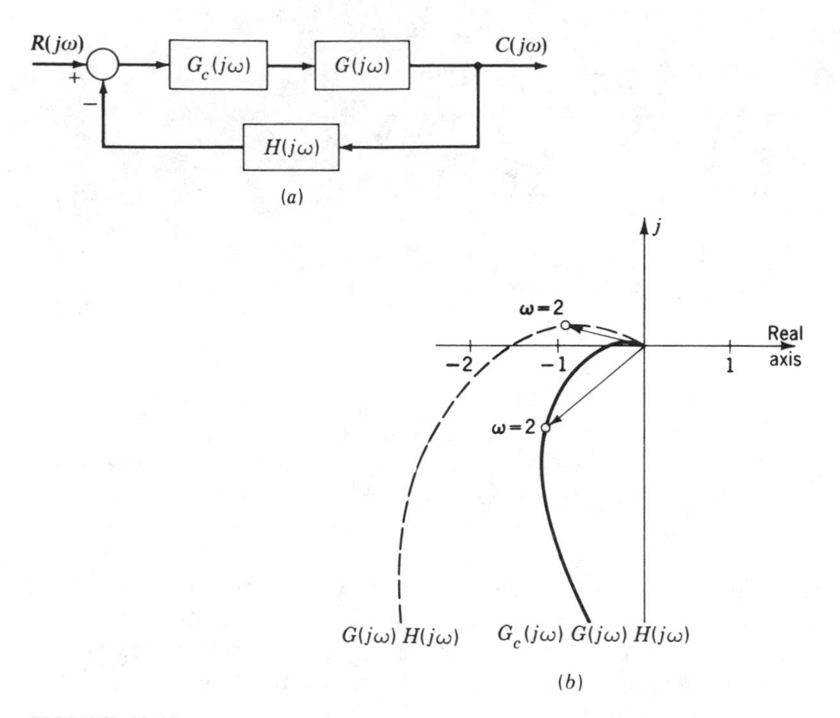

FIGURE 12.15
Use of phase lead to reshape a polar plot.

The use of a phase-lead component placed in series with the feedforward portion of a control system to improve stability is now illustrated. In Fig. 12.15 the dashed curve is the frequency response $G(j\omega)H(j\omega)$ of the uncompensated control system. This control is one that would inherently be unstable. The addition of the lead compensator $G_c(j\omega)$ to reshape the high-frequency portion of the polar plot is shown by the solid-line curve of Fig. 12.15. Note that lead compensation rotates a typical vector such as that for $\omega = 2$ in a counterclockwise direction away from the -1 point. Also note that the length of this vector is increased. Because of the counterclockwise rotation of a typical vector, the effect of lead compensation is to increase ω_p. This tends to increase the speed of response of the system.

To select a lead compensator, it is necessary to be familiar with some of the general properties which are now derived. The phase shift ϕ due to a lead compensator is

$$\phi = \measuredangle \frac{1 + j\tau_1\omega}{1 + j\tau_2\omega} = \tan^{-1}\tau_1\omega - \tan^{-1}\tau_2\omega \tag{12.16}$$

Differentiating with respect to ω gives

$$\frac{d\phi}{d\omega} = \frac{\tau_1}{1 + (\tau_1\omega)^2} - \frac{\tau_2}{1 + (\tau_2\omega)^2}$$

Setting this derivative equal to zero and solving for ω yield the frequency $\omega = \omega_m$ at which the phase shift is a maximum. That is,

$$\tau_1[1 + (\tau_2\omega_m)^2] - \tau_2[1 + (\tau_1\omega_m)^2] = 0$$
$$(\tau_1 - \tau_2)(1 - \tau_1\tau_2\omega_m^2) = 0$$
$$\omega_m^2 = \frac{1}{\tau_1\tau_2}$$

Figure 12.16*a* shows the log-magnitude plot for a lead compensator. Note that the maximum phase shift ϕ_m occurs at frequency $\omega_m = 1/\sqrt{\tau_1\tau_2}$. This frequency at which the maximum phase shift occurs is located at the midpoint between the break frequencies $\omega_1 = 1/\tau_1$ and $\omega_2 = 1/\tau_2$. Because frequency is plotted to a logarithmic scale, the midpoint is

$$\log \omega_m = \tfrac{1}{2}(\log \omega_1 + \log \omega_2) = \tfrac{1}{2}\log \omega_1\omega_2 = \log \frac{1}{\sqrt{\tau_1\tau_2}}$$

Hence

$$\omega_m = 1/\sqrt{\tau_1\tau_2} \tag{12.17}$$

The value of the maximum phase shift ϕ_m is obtained by substituting $\omega = \omega_m = 1/\sqrt{\tau_1\tau_2}$ into Eq. (12.16):

$$\phi_m = \measuredangle \frac{1 + j\sqrt{\tau_1/\tau_2}}{1 + j\sqrt{\tau_2/\tau_1}} \frac{1 - j\sqrt{\tau_2/\tau_1}}{1 - j\sqrt{\tau_2/\tau_1}} = \measuredangle \frac{1 + 1 + j\left(\sqrt{\tau_1/\tau_2} - \sqrt{\tau_2/\tau_1}\right)}{1 + \tau_2/\tau_1}$$
$$= \tan^{-1} \frac{\sqrt{\tau_1/\tau_2} - \sqrt{\tau_2/\tau_1}}{2} = \tan^{-1} \frac{(\tau_1/\tau_2) - 1}{2\sqrt{\tau_1/\tau_2}} \tag{12.18}$$

This result shows that ϕ_m is a function of the ratio τ_1/τ_2. Figure 12.16*b* shows a plot of ϕ_m versus log (τ_1/τ_2). The double scale for the horizontal axis displays both log (τ_1/τ_2) and τ_1/τ_2.

In accordance with Eq. (12.18), ϕ_m may be regarded as the angle of a right triangle whose opposite side is $\tau_1/\tau_2 - 1$ and whose adjacent side is $2\sqrt{\tau_1/\tau_2}$. From this triangle, it follows that

$$\sin \phi_m = \frac{\tau_1/\tau_2 - 1}{\tau_1/\tau_2 + 1}$$

Solving for the ratio τ_1/τ_2 shows that

$$\frac{\tau_1}{\tau_2} = \frac{1 + \sin \phi_m}{1 - \sin \phi_m} \tag{12.19}$$

To select a lead compensator, it is necessary to specify the values of both τ_1 and τ_2. Because of the two unknowns τ_1 and τ_2, the selection of a lead compensator to achieve a desired design specification is basically a trial-and-error process.

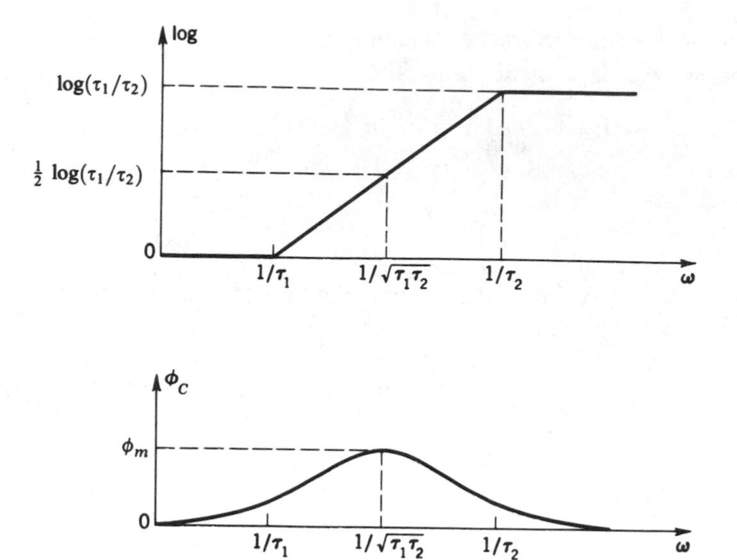

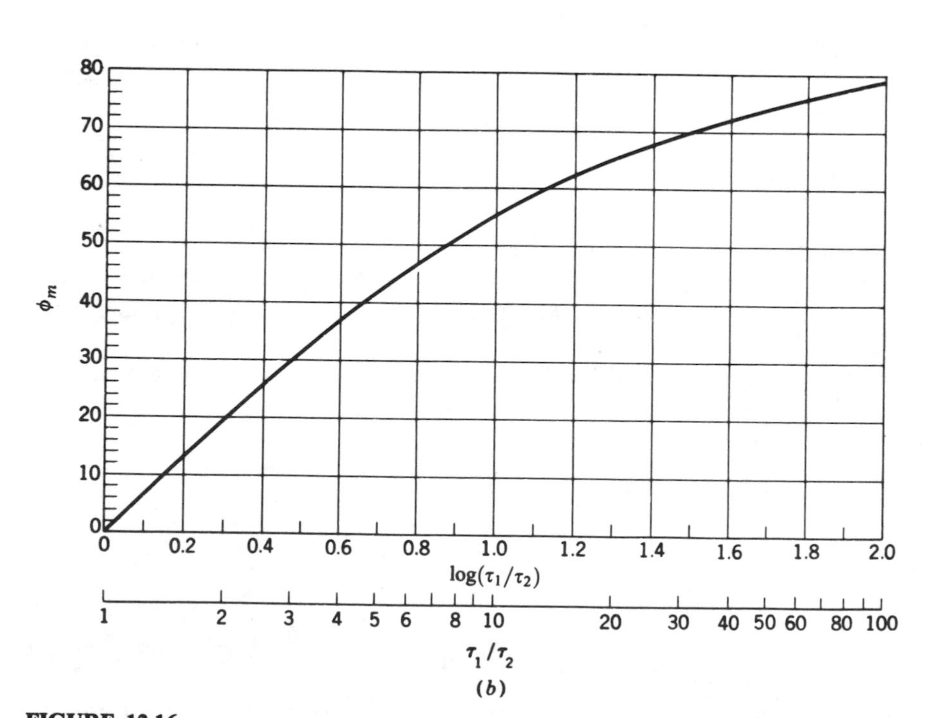

FIGURE 12.16
Lead compensator characteristics: (*a*) log-magnitude diagram, (*b*) maximum phase shift ϕ_m versus log (τ_1/τ_2).

However, a systematic procedure which rapidly converges is described in the following steps:

1. Determine the phase margin for the uncompensated system
2. Select a value for ϕ_m which is the difference between the desired phase margin and the value obtained in step 1, plus a small additional amount (say 5°).
3. Determine the ratio τ_1/τ_2 from Fig. 12.16*b*.
4. Determine the frequency where the log magnitude for the uncompensated system is $-0.5 \log (\tau_1/\tau_2)$. Use this frequency for ω_m.
5. Because the phase-lead compensator provides a gain of $0.5 \log (\tau_1/\tau_2)$ at $\omega_m = 1/\sqrt{\tau_1\tau_2}$, this will be the frequency where the compensated system crosses the unit circle. Determine the resulting phase margin for the compensated system.

If the phase margin is too small, increase the value τ_1/τ_2 and if it is too large decrease the ratio τ_1/τ_2 and then repeat the steps. The following example illustrates the application of this procedure.

Illustrative example 12.4. The X-15 aircraft, which is a rocket-powered, hypersonic vehicle that can travel in excess of mach 6 and is capable of going into space (leaving the sensible atmosphere), is shown in Fig. 12.17. The log-modulus plot of $G(j\omega)$ for

FIGURE 12.17
The X-15 aircraft. (Courtesy U.S. Air Force.)

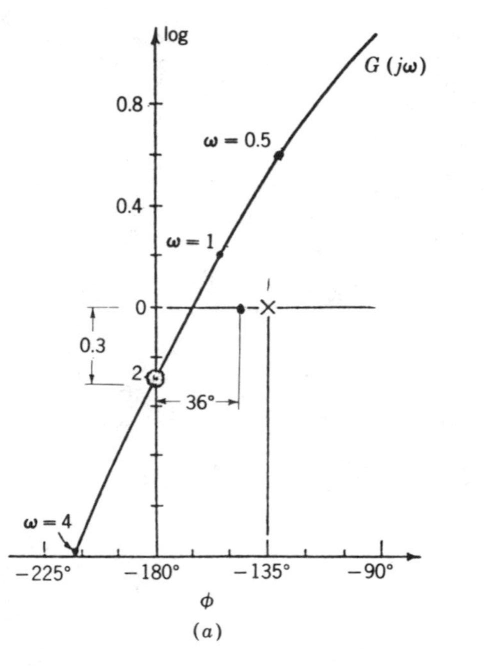

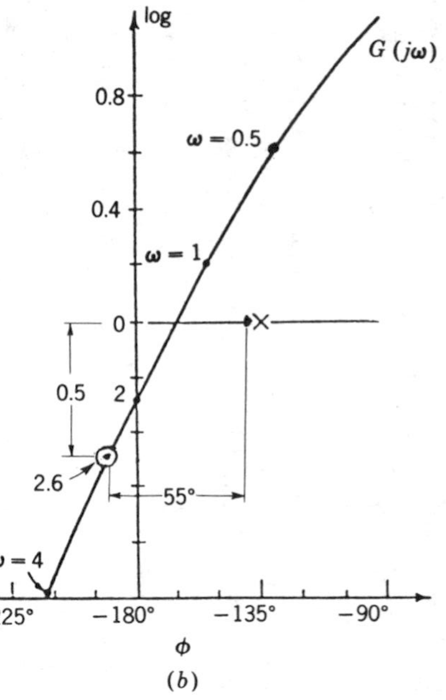

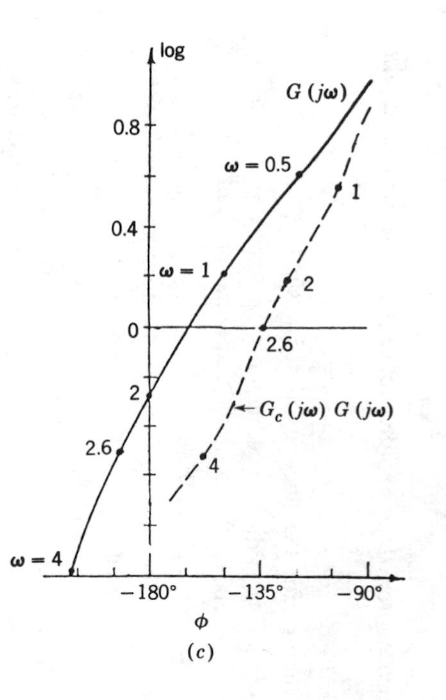

FIGURE 12.18
Use of series lead compensation to obtain a desired phase margin.

the unity-feedback attitude control system for this vehicle is shown in Fig. 12.18*a*. Determine the values of τ_1 and τ_2 for a series lead compensator such that the compensated system will have a phase margin of 45° ± 3°. What are the approximate values of ζ and ω_n for the resulting compensated system?

Solution. The phase margin for the uncompensated system $G(j\omega)$ shown in Fig. 12.18*a* is 14°. The difference between the desired phase margin 45° and that for the uncompensated system is 31°. The first estimate for ϕ_m is $31° + 5° = 36°$. Fig. 12.16*b* shows that for $\phi_m = 36°$, then $\tau_1/\tau_2 = 4$ and $\log(\tau_1/\tau_2) = 0.6$. At frequency $\omega_m = 1/\sqrt{\tau_1\tau_2}$, the lead compensator increases the gain by a factor $\frac{1}{2}\log(\tau_1/\tau_2) = 0.3$, and the phase angle is increased by $\phi_m = 36°$. The × in Fig. 12.18*a* is located at the point through which the log-modulus plot should pass to yield a phase margin of 45°. Thus, by measuring down −0.3 log unit from the × as shown in Fig. 12.18*a*, when the lead compensator is added, the point indicated by the circle which corresponds to ω_m will be raised vertically 0.3 log unit to the zero horizontal axis and shifted right 36°. The new location is indicated by the solid dot. To have a phase margin of 45°, the solid dot should be located at the ×. Thus, it is necessary to assume a larger value of τ_1/τ_2. The result for $\tau_1/\tau_2 = 10$ is illustrated in Fig. 12.18*b*. Note that this yields a phase margin of 43°, which is satisfactory. Because of the rapid convergence of this process, usually two or three trials suffice for obtaining the desired ratio τ_1/τ_2. The desired values of τ_1 and τ_2 are now computed from the relationships $\tau_1/\tau_2 = 10$ and $\omega_c = \omega_m = 1/\sqrt{\tau_1\tau_2} = 2.6$. Hence,

$$\tau_1 = 1.2 \qquad \text{and} \qquad \tau_2 = 0.12$$

The complete log-modulus plot for the resulting $G_c(j\omega)G(j\omega)$ system may now be constructed as illustrated in Fig. 12.18*c*.

For a phase margin of 43°, Eq. (12.14) or Fig. 12.12*b* shows that the approximate value of ζ is 0.40. From Eq. (12.12) or from Fig. 12.12*a*, it is found that $\omega_c/\omega_n = 0.85$. Thus, the approximate value of the natural frequency ω_n is 3.1.

By proceeding in a similar manner, this procedure may be used to obtain a desired gain margin or a desired value of M_p.

12.4 LAG COMPENSATION

The output lags the input for any component which has a transfer function of the form

$$\frac{E_o(s)}{E_{\text{in}}(s)} = \frac{Y(s)}{X(s)} = \frac{1+\tau_2 s}{1+\tau_1 s} \qquad \tau_1 > \tau_2 \tag{12.20}$$

The frequency response for the preceding transfer function is

$$\frac{E_o(j\omega)}{E_{\text{in}}(j\omega)} = \frac{Y(j\omega)}{X(j\omega)} = \frac{1+j\tau_2\omega}{1+j\tau_1\omega} \qquad \tau_1 > \tau_2 \tag{12.21}$$

Figure 12.19 shows both an electrical and a mechanical component in which the output lags the input, as described by Eq. (12.21). The construction of the log-magnitude plot of Eq. (12.21) for the case in which $\tau_2 = \tau_1/10$ is illustrated in Fig. 12.20. For the term $1/(1 + j\tau_1\omega)$, the break frequency occurs at $\omega = 1/\tau_1$, and for the numerator $1 + j\tau_2\omega$ the break frequency occurs at $\omega = 1/\tau_2 = 10/\tau_1$. The addition of

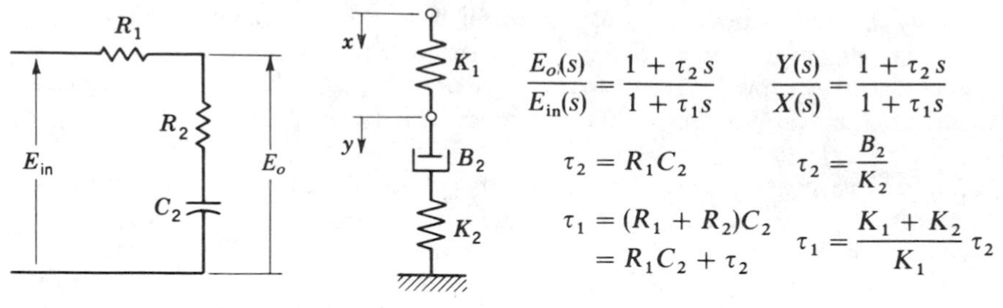

FIGURE 12.19
An electrical and a mechanical circuit used to provide lag compensation.

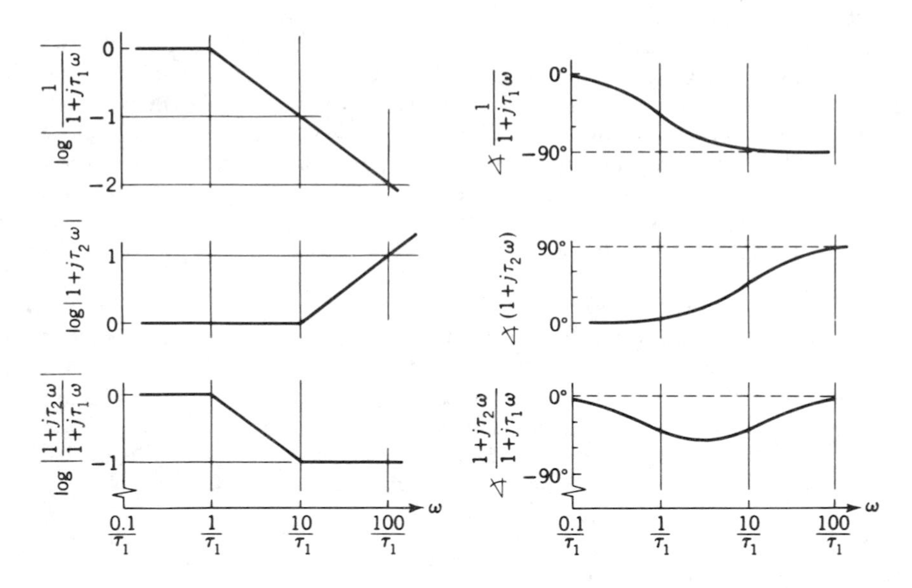

FIGURE 12.20
Construction of the log-magnitude plot for $(1 + j\tau_2\omega)/(1 + j\tau_1\omega)$ when $\tau_2 = \tau_1/10$.

the log-magnitude plot for the numerator to that for the denominator yields the resulting plot for $(1 + j\tau_1\omega)/(1 + j\tau_1\omega)$.

The use of a phase-lag component placed in series with the feedforward portion of a control system to improve stability is now illustrated. In Fig. 12.21 the dashed curve is the frequency response $G(j\omega)H(j\omega)$ of the uncompensated control system. This control is one that would inherently be unstable. The addition of series compensation $G_c(j\omega)$ reshapes the polar plot, as shown by the solid-line curve of Fig. 12.21. The resultant system has good dynamic response. Note that the effect of lag compensation is to shorten a typical vector such as that for $\omega = 0.5$ and to rotate it in a clockwise direction. The shortening is due to the *attenuation*. By attenuation

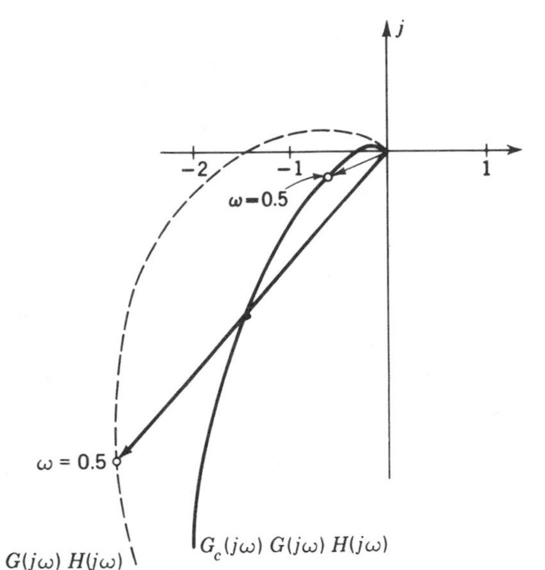

FIGURE 12.21
Use of phase lag to reshape a polar plot.

is meant multiplication by a factor less than 1. The attenuation caused by the use of lag compensation can be seen from Fig. 12.22*a*. The greater the spread in time constants τ_1 and τ_2, the more pronounced is the attenuation which occurs at higher frequencies. Series lag compensation has little effect on the low-frequency portion of the curve. By reshaping the polar plot, it has been possible to achieve good dynamic performance without changing the value of the gain K. Although it would have been possible to make this system stable by decreasing only the gain K, errors caused by friction, hysteresis, backlash, etc., tend to predominate as the gain is decreased; thus, in general, the higher the value of K, the more accurate the control system.

Figure 12.22*a* shows the log-magnitude plot for a lag compensator. Comparison of Figs. 12.16*a* and 12.22*a* shows that the phase-shift characteristics of a lag compensator are the same as those for a lead compensator except that the sign of the phase angle is negative. Similarly, the log-magnitude characteristics for a lag compensator are the same as those for a lead compensator except that the sign is reversed.

The negative phase shift associated with lag compensation is usually undesirable. The effectiveness of lag compensation is attributed to the attenuation which occurs at higher frequencies. In Fig. 12.22*a*, note that when $\omega = 10/\tau_2$, then the negative phase shift is very small. The solid-line curve in Fig. 12.22*b* is a plot of the value of the negative phase shift at $\omega = 10/\tau_2$. A significant difference between selecting a lag compensator and selecting a lead compensator is that for the lag compensator the region of the small phase shift ($\omega = 10/\tau_2$) is located in the vicinity of the point of interest, whereas for a lead compensator the region of maximum phase shift ($\omega = \omega_m$) is located in the vicinity of the point of interest.

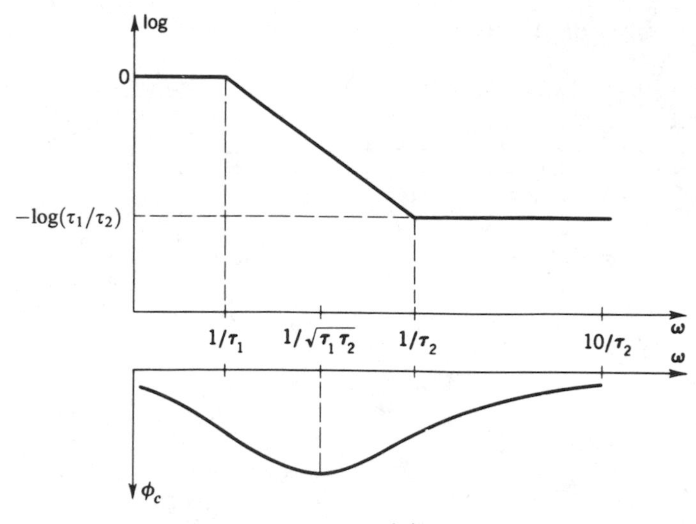

(a)

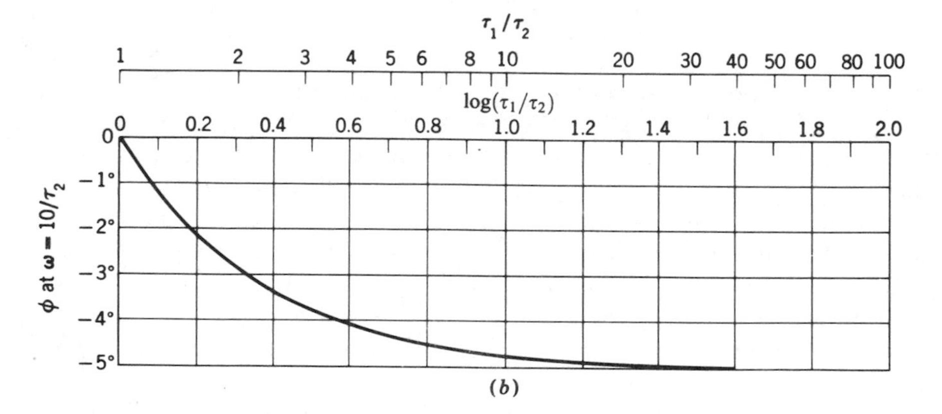

(b)

FIGURE 12.22
Lag compensator characteristics: (*a*) log-magnitude plot, (*b*) phase shift ϕ at $\omega = 10/\tau_2$.

A procedure for determining the lag compensator for obtaining a desired dynamic performance is described in the following steps:

1. Add 5° to the desired phase margin and then subtract 180° from this result.
2. For the uncompensated system, determine the value of the $\log |G(j\omega)H(j\omega)|$ at the angle determined in step 1.
3. Set this value of $\log |G(j\omega)H(j\omega)|$ equal to $\log \tau_1/\tau_2$. When the lag compensator is added to the uncompensated system, this will be the point where the resultant system crosses the unit circle (i.e., $\log |G(j\omega)H(j\omega)| - \log \tau_1/\tau_2 = 0$). This will be the $10/\tau_2$ frequency.

This procedure is illustrated by the following example.

Illustrative example 12.5. Repeat Illustrative example 12.4, except use a lag rather than a lead compensator. The log-modulus plot of $G(j\omega)$ is shown in Fig. 12.23*a*.

Solution. For a phase margin of 45°, the resultant $G_c(j\omega)G(j\omega)$ plot must pass through the point indicated by the $\times$ in Fig. 12.23*a*. From Fig. 12.22*b*, note that for values of $\tau_1/\tau_2 > 4$, the phase shift introduced by a lag compensator at $\omega = 10/\tau_2$ is about $-5°$. The angle in step 1 is $(45° + 5°) - 180° = -130°$ which is 5° to the right of the $\times$. Thus, a vertical line is erected 5° to the right of the $\times$, as shown in Fig. 12.23*a*. The intersection of this vertical line with the $G(j\omega)$ plot is indicated by a circle. In order that the compensator move this point to the $\times$, $-\log(\tau_1/\tau_2) = -0.6$, and thus $\tau_1/\tau_2 = 4.0$. Figure 12.23*a* shows that the frequency $\omega = 10/\tau_2$ is 0.5. The particular values of τ_1 and τ_2 are now computed from the relationships $\tau_1/\tau_2 = 4$ and $10/\tau_2 = 0.5$. Thus,

$$\tau_1 = 80 \qquad \text{and} \qquad \tau_2 = 20$$

With these values of τ_1 and τ_2, the log-modulus plot for the resultant compensated system $G_c(j\omega)G(j\omega)$ may now be constructed, as shown in Fig. 12.23*b*.

For a phase margin of 45°, it follows from Eq. (12.14) or from Fig. 12.12*b* that the approximate value of ζ is 0.42. Similarly, from Eq. (12.12) or from Fig. 12.12*a* the ratio ω_c/ω_n is found to be 0.84. Because the crossover frequency ω_c is 0.5, the approximate natural frequency ω_n is found to be 0.6.

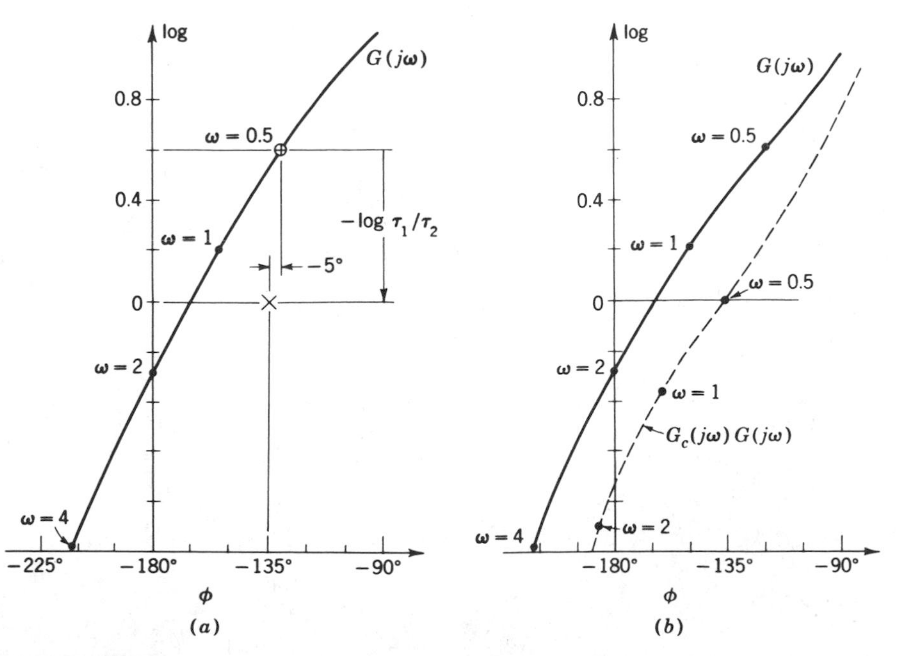

FIGURE 12.23
Use of series lag compensation to obtain a desired phase margin.

Note that the values of τ_1 and τ_2 for this lag compensator are larger than the values of τ_1 and τ_2 for the corresponding lead compensator. Thus, the system with lag compensation is slower than the corresponding system with lead compensation. This fact is also substantiated by noting that the approximate natural frequency for the system with lag compensation is considerably smaller ($\omega_n = 0.6$) than that for the system with lead compensation ($\omega_n = 3.1$).

By proceeding in a similar manner, it is possible to select lag compensators to yield a desired M_p or a desired gain margin. When a compensator is chosen to yield a desired phase margin or a desired gain margin, one may use either Bode plots or log-modulus plots. Because it is not possible to construct M contours on Bode plots, it is necessary to use log-modulus plots for obtaining a compensator to yield a desired value of M_p.

12.5 LAG-LEAD COMPENSATION

A lag-lead compensator is a series combination of a lag and a lead network. The general transfer function for a lag-lead compensator is

$$\frac{E_o(s)}{E_{\text{in}}(s)} = \frac{Y(s)}{X(s)} = \frac{1 + c\tau_2 s}{1 + c\tau_1 s}\frac{1 + \tau_1 s}{1 + \tau_2 s} \qquad \tau_1 > \tau_2 \tag{12.22}$$

where $c > 1$ is a constant. The substitution of $j\omega$ for s gives

$$\frac{E_o(j\omega)}{E_{\text{in}}(j\omega)} = \frac{Y(j\omega)}{X(j\omega)} = \frac{1 + jc\tau_2\omega}{1 + jc\tau_1\omega}\frac{1 + j\tau_1\omega}{1 + j\tau_2\omega} \qquad \tau_1 > \tau_2 \tag{12.23}$$

Rather than using a lag and a lead compensator in series, it is possible to use a single compensator, as shown in Fig. 12.24.

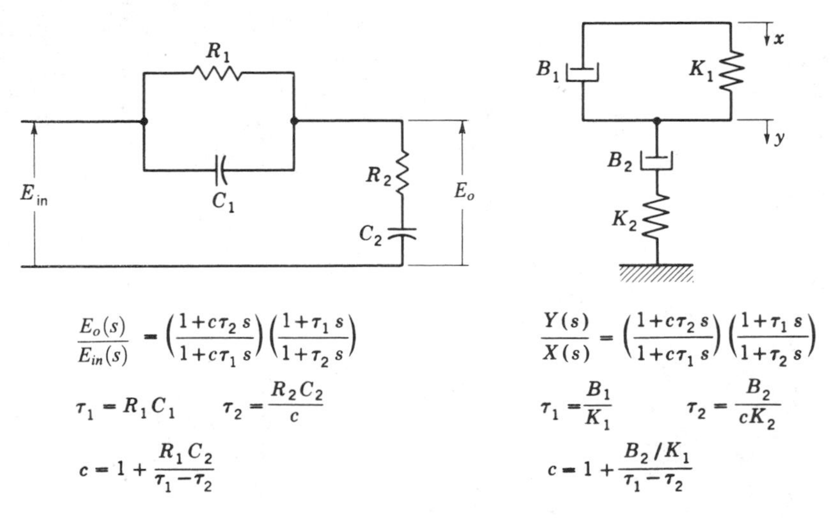

FIGURE 12.24
An electrical and a mechanical circuit used to provide lag-lead compensation.

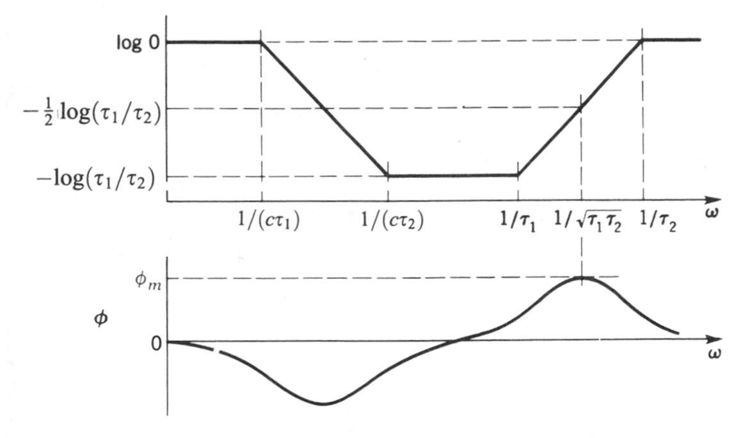

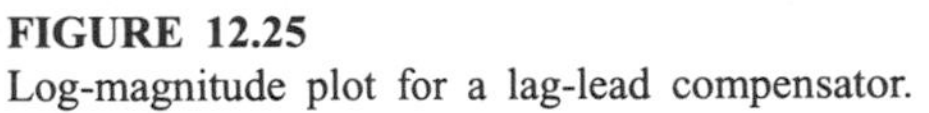

FIGURE 12.25
Log-magnitude plot for a lag-lead compensator.

The log-magnitude plot for a typical lag-lead compensator is shown in Fig. 12.25. Because $c\tau_1 > c\tau_2 > \tau_1 > \tau_2$, the first break frequency occurs at $1/(c\tau_1)$. This break frequency belongs to a denominator term, so that the magnitude plot has a slope of -1 log unit per decade between $1/(c\tau_1)$ and $1/(c\tau_2)$. The second break frequency $1/(c\tau_2)$ is associated with a numerator term, and thus the magnitude plot again becomes horizontal. The third break frequency $1/\tau_1$ also occurs in the numerator. This results in a slope of $+1$ log unit per decade in the region from $1/\tau_1$ to $1/\tau_2$. Finally the break frequency $1/\tau_2$, which occurs in the denominator, causes the magnitude curve to become horizontal again. Note that the maximum phase shift ϕ_m occurs at $\omega = 1/\sqrt{\tau_1\tau_2}$, and the corresponding attenuation is $-\frac{1}{2}\log(\tau_1/\tau_2)$. This is the same as for a lead compensator only (see Fig. 12.16*a*) except that the sign of the log of the magnitude is negative. This feature makes the lag-lead compensator considerably more effective than the lead compensator only.

Illustrative example 12.6. Repeat Illustrative example 12.4, except use a lag-lead compensator rather than a lead compensator.

Solution. Figure 12.26*a* shows the log-modulus plot for the uncompensated system $G(j\omega)$. As a first trial, it is assumed that $\tau_1/\tau_2 = 2$; thus $\phi_m = 20°$ and $-\frac{1}{2}\log(\tau_1/\tau_2) = -0.15$ log unit. The circle in Fig. 12.26*a* will be shifted downward -0.15 log unit, and it will be moved to the right 20°. The resulting location is indicated by the $\times$. The corresponding phase margin is $\gamma = 45°$.

The particular values of τ_1 and τ_2 for this lag-lead compensator are determined from the relationships $\tau_1/\tau_2 = 2$ and $\omega_m = 1/\sqrt{\tau_1\tau_2} = 1.1$. Thus, $\tau_1 = 1.30$ and $\tau_2 = 0.65$.

A factor of 5 provides a reasonable separation between the second and third break frequencies (see Fig. 12.25), that is, $1/\tau_1 = 5[1/(c\tau_2)]$. Solving this relationship for c gives

$$c = 5\,\frac{\tau_1}{\tau_2} = 10$$

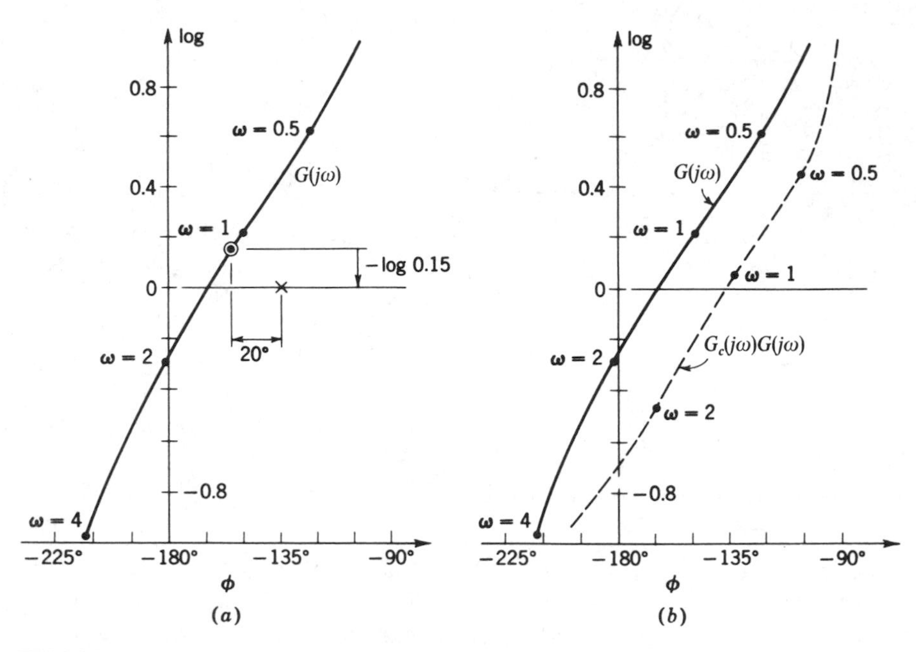

FIGURE 12.26
Use of series lag-lead compensation to obtain a desired phase margin.

With the compensator specified, the log-modulus plot for the resulting compensated system may now be constructed, as shown in Fig. 12.26*b*. As indicated by Fig. 12.25, the lag-lead compensation affects neither the low- nor the high-frequency regions but rather the midfrequency region.

12.6 INTERNAL FEEDBACK

Another method commonly used to alter frequency-response characteristics involves providing a separate internal-feedback path about certain components. In employing log-magnitude plots to investigate the effect of internal feedback, the use of a few approximations affords much simplification. This approximate analysis in effect puts the designer in the right ballpark. In the latter design stages, it may be desirable to make an exact analysis. The approximations which are used to evaluate the effect of placing a feedback element $H_1(s)$ around a component $G_1(s)$, as shown in Fig. 12.27, are that when $|G_1(j\omega)H_1(j\omega)| \ll 1$,

$$G(j\omega) = \frac{G_1(j\omega)}{1 + G_1(j\omega)H_1(j\omega)} \approx G_1(j\omega) \tag{12.24}$$

For the case when $|G_1(j\omega)H_1(j\omega)| \gg 1$,

$$G(j\omega) = \frac{G_1(j\omega)}{1 + G_1(j\omega)H_1(j\omega)} \approx \frac{G_1(j\omega)}{G_1(j\omega)H_1(j\omega)} = \frac{1}{H_1(j\omega)} \tag{12.25}$$

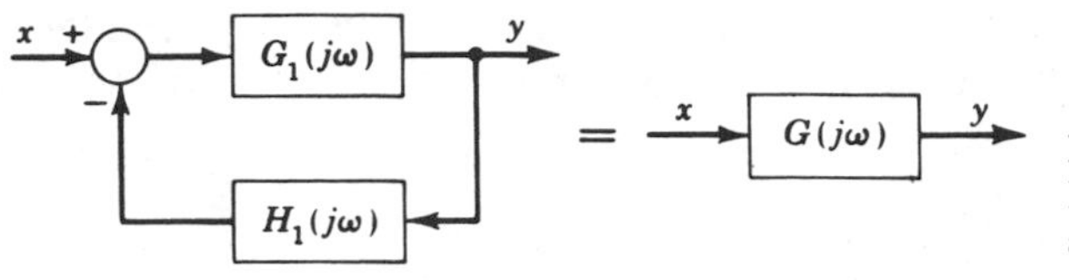

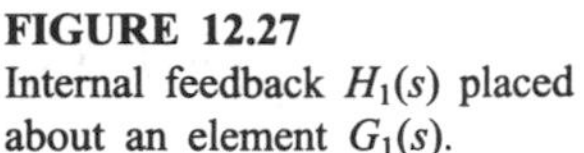
FIGURE 12.27
Internal feedback $H_1(s)$ placed about an element $G_1(s)$.

Note that when $|G_1(j\omega)H_1(j\omega)| \ll 1$, then $|G_1(j\omega)| \ll |1/H_1(j\omega)|$. Thus Eq. (12.24) indicates that $G(j\omega)$ is to be approximated by $G_1(j\omega)$ when $|G_1(j\omega)| \ll |1/H_1(j\omega)|$. Similarly, when $|G_1(j\omega)H_1(j\omega)| \gg 1$, then $|1/H_1(j\omega)| \ll |G_1(j\omega)|$. Thus, Eq. (12.25) indicates that $G(j\omega)$ is to be approximated by $1/H_1(j\omega)$ when $|1/H_1(j\omega)| \ll |G_1(j\omega)|$. In summary, $G(j\omega)$ is always approximated by the smaller of $G_1(j\omega)$ or $1/H_1(j\omega)$.

When $|G_1(j\omega)H_1(j\omega)| = 1$, then $|G_1(j\omega)| = |1/H_1(j\omega)|$. Thus, the intersection of the log-magnitude plot for $|G_1(j\omega)|$ and that for $|1/H_1(j\omega)|$ determines the point at which $|G_1(j\omega)H_1(j\omega)| = 1$. Usually at high frequencies $|G_1(j\omega)H_1(j\omega)| \ll 1$, in which case $|G_1(j\omega)| \ll |1/H_1(j\omega)|$. Thus, the high-frequency response is approximated by $G_1(j\omega)$, and the low-frequency response is approximated by $1/H_1(j\omega)$.

In summary, this approximation converts an internal-feedback path to an approximate open-loop element for which a standard analysis can be made. With the use of inverse polar plots, as is described next, no approximations are employed.

12.7 INVERSE POLAR PLOTS

A plot of the function $G^{-1}(j\omega) = 1/G(j\omega)$ is called an *inverse polar plot.* Figure 12.28 shows a typical inverse polar plot for the function $G^{-1}(j\omega)$. At any frequency ω, the vector from the origin to a point on the plot defines the vector $G^{-1}(j\omega)$ for that frequency. The length of the vector is $|G^{-1}(j\omega)| = |1/G(j\omega)|$, and the angle is

$$\measuredangle G^{-1}(j\omega) = \measuredangle \frac{1}{G(j\omega)} = -\measuredangle G(j\omega)$$

A plot of M circles and α lines for inverse polar plots is accomplished by first taking the reciprocal of Eq. (11.52), i.e.,

$$\frac{R(j\omega)}{C(j\omega)} = \frac{1 + G(j\omega)}{G(j\omega)} = G^{-1}(j\omega) + 1 \tag{12.26}$$

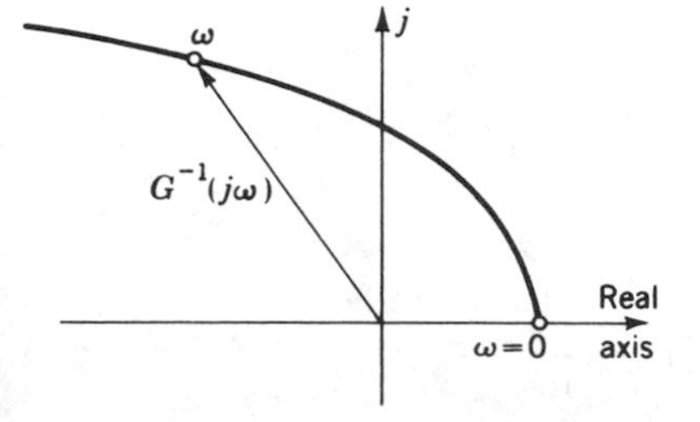

FIGURE 12.28
Typical inverse polar plot $G^{-1}(j\omega)$.

A typical vector $G^{-1}(j\omega)$ as shown in Fig. 12.27 may be written in the general form $G^{-1}(j\omega) = x + jy$. Substitution of this general representation for $G^{-1}(j\omega)$ into Eq. (12.26) gives

$$\frac{R(j\omega)}{C(j\omega)} = x + 1 + jy \qquad (12.27)$$

Because $M = |C(j\omega)/R(j\omega)|$, from the magnitude of Eq. (12.27) it follows that

$$\left|\frac{R(j\omega)}{C(j\omega)}\right| = \frac{1}{M} = \sqrt{(x+1)^2 + y^2}$$

Squaring this result gives

$$(x+1)^2 + y^2 = \frac{1}{M^2} \qquad (12.28)$$

Thus, on the inverse plane, lines of constant M are circles of radius $1/M$. The center of the concentric M circles is at the point $x = -1$ and $y = 0$, that is, the -1 point. A plot of these M circles on the inverse plane is shown in Fig. 12.29. Because the reciprocal of -1 is still -1, this point has the same significance for an inverse polar plot as for a direct polar plot. Polar plots are referred to as direct polar plots when it is necessary to distinguish them from inverse polar plots.

The lines of constant $\alpha = \measuredangle [C(j\omega)/R(j\omega)] = -\measuredangle [R(j\omega)/C(j\omega)]$ are determined from Eq. (12.27) as follows:

$$\alpha = -\measuredangle \frac{R(j\omega)}{C(j\omega)} = -\tan^{-1}\frac{y}{x+1} \qquad (12.29)$$

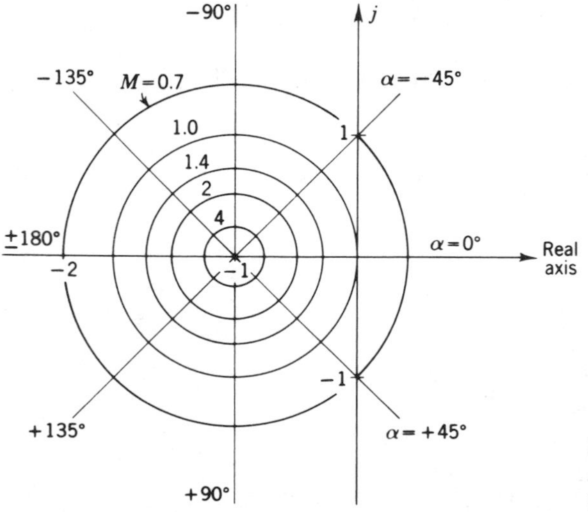

FIGURE 12.29
M circles and α rays on the inverse plane.

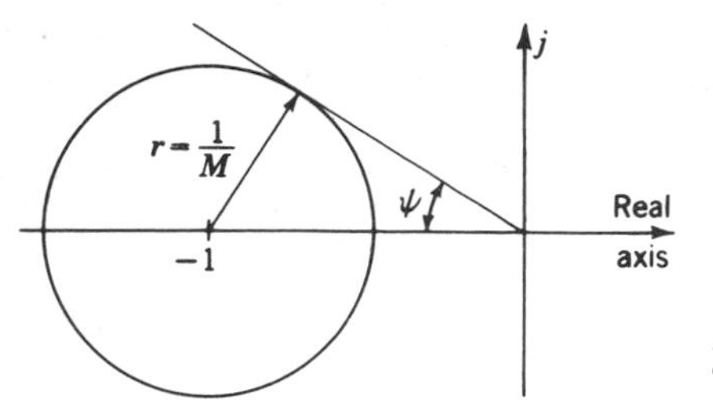

FIGURE 12.30
Tangent line to an M circle.

When plotted on the inverse plane, as shown in Fig. 12.29, lines of constant α are radial straight lines which pass through the $-1 + j0$ point.

As illustrated in Fig. 12.30, the angle ψ of a radial line drawn from the origin tangent to any M circle is

$$\sin \psi = \frac{1}{M} \tag{12.30}$$

The use of the inverse polar plot for determining the gain K to yield a desired value of M_p is similar to that for the direct polar plot. Consider the same function $G(j\omega) = K/[j\omega(1 + 0.1j\omega)]$ discussed in Sec. 11.8. The plot of the inverse function

$$\frac{K}{G(j\omega)} = KG^{-1}(j\omega) = j\omega(1 + 0.1j\omega) \tag{12.31}$$

is shown in Fig. 12.31. For a desired $M_p = 1.4$, angle ψ of the tangent line is

$$\psi = \sin^{-1} \frac{1}{M_p} = \sin^{-1} \frac{1}{1.4} = 45.6° \tag{12.32}$$

Next construct by trial and error the circle which is tangent to both the $KG^{-1}(j\omega)$ plot and the tangent line. In order for this circle to be the desired M_p circle, its center must be at the -1 point. From Fig. 12.31, we see that the center is at $-A = -16.7$.

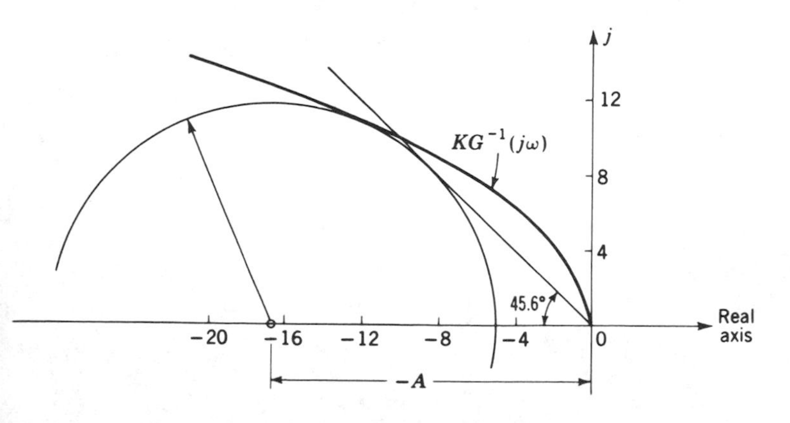

FIGURE 12.31
Inverse polar plot $KG^{-1}(j\omega) = j\omega(1 + 0.1j\omega)$.

To convert this to the -1 point, the scale factor must be multiplied by $1/A = 1/16.7 = 0.06$. The resulting function $G^{-1}(j\omega)$, which is tangent to the desired M circle, is

$$G^{-1}(j\omega) = \frac{1}{A} KG^{-1}(j\omega) = 0.06j\omega(1 + 0.1j\omega)$$

or

$$G(j\omega) = \frac{16.7}{j\omega(1 + 0.1j\omega)}$$

Thus, the value of A is the required gain K.

The general procedure for obtaining K by use of the inverse polar plot is as follows:

1. Plot the inverse function $KG^{-1}(j\omega)$.
2. Construct the tangent line in accordance with Eq. (12.30).
3. By trial and error, determine the circle which is tangent to both the $KG^{-1}(j\omega)$ plot and the tangent line.
4. The desired gain is $K = A$.

When the function $G^{-1}(j\omega)$ is plotted rather than $KG^{-1}(j\omega)$, then A is equal to the factor K_c by which the gain should be changed to yield the desired M_p.

Illustrative example 12.7. Consider the function

$$G(j\omega) = \frac{5}{j\omega(1 + 0.1j\omega)}$$

The inverse polar plot $G^{-1}(j\omega)$ is shown in Fig. 12.32. Determine the factor K_c by which the gain should be changed so that the resulting plot is tangent to the $M_p = 1.4$ circle.

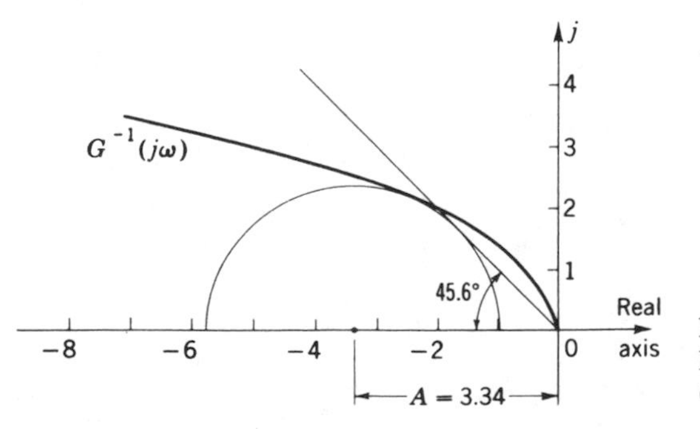

FIGURE 12.32
Inverse polar plot of $G^{-1}(j\omega) = j\omega(1 + 0.1j\omega)/5$.

Solution. In order that the point $-A$ in Fig. 12.32 be the -1 point, the scale factor must be multiplied by $1/A$. The resulting function is

$$\frac{1}{A}\,G^{-1}(j\omega) = \frac{1}{AG(j\omega)} = \frac{1}{K_cG(j\omega)}$$

or

$$K_cG(j\omega) = \frac{(3.34)\,(5)}{j\omega(1+0.1j\omega)} = \frac{16.7}{j\omega(1+0.1j\omega)}$$

The preceding analysis shows that $K_c = A$.

The major advantage of using the inverse plane is realized for systems with internal feedback. The reciprocal of Eq. (12.24) is

$$G^{-1}(j\omega) = \frac{1+G_1(j\omega)H_1(j\omega)}{G_1(j\omega)} = G_1^{-1}(j\omega) + H_1(j\omega) \tag{12.33}$$

As illustrated in Fig. 12.33, the vectors $G_1^{-1}(j\omega)$ and $H_1(j\omega)$ may be added as vector quantities to yield $G^{-1}(j\omega)$.

Illustrative example 12.8. For the system shown in Fig. 12.34*a*, the value of K_1 is 10 and K_2 is 2.5. From the inverse polar plot $1/[K_2G_1(j\omega)]$ of Fig. 12.34*b*, the value of M_p is found to be 5. It is desired to obtain an M_p of 1.4 by the use of an internal feedback-loop as shown in Fig. 12.35*a*. What feedback element $H_1(j\omega)$ should be used?

Solution. The equation for the inverse polar plot with internal feedback is

$$\frac{1}{G(j\omega)} = \frac{1}{K_2}\,\frac{1+G_1(j\omega)H_1(j\omega)}{G_1(j\omega)} = \frac{1}{K_2G_1(j\omega)} + \frac{H_1(j\omega)}{K_2} \tag{12.34}$$

The quantity $H_1(j\omega)/K_2$ must be such that when it is added to $1/[K_2G_1(j\omega)]$, the resulting plot will be tangent to the $M = 1.4$ circle. To obtain this, it is necessary only to raise the original plot vertically. We see from Fig. 12.33 that the horizontal component of $H_1(j\omega)$ is its real part $\mathscr{R}H_1(j\omega)$; and the vertical component is its imaginary part $\mathscr{I}H_1(j\omega)$. It is thus necessary only that $H_1(j\omega)$ be purely imaginary,

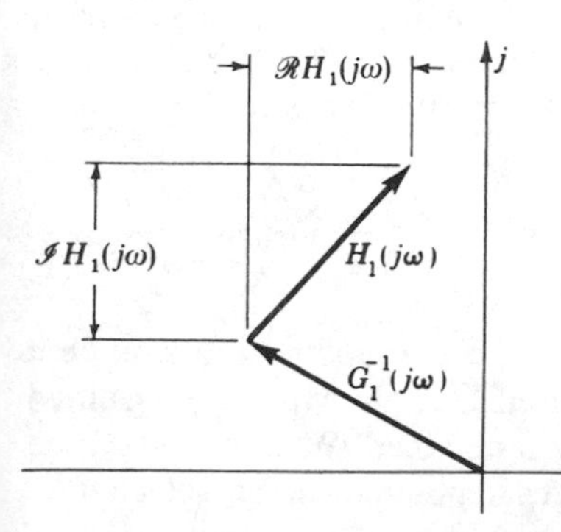

FIGURE 12.33
Vector addition of $G_1^{-1}(j\omega)$ and $H_1(j\omega)$ to yield $G^{-1}(j\omega)$.

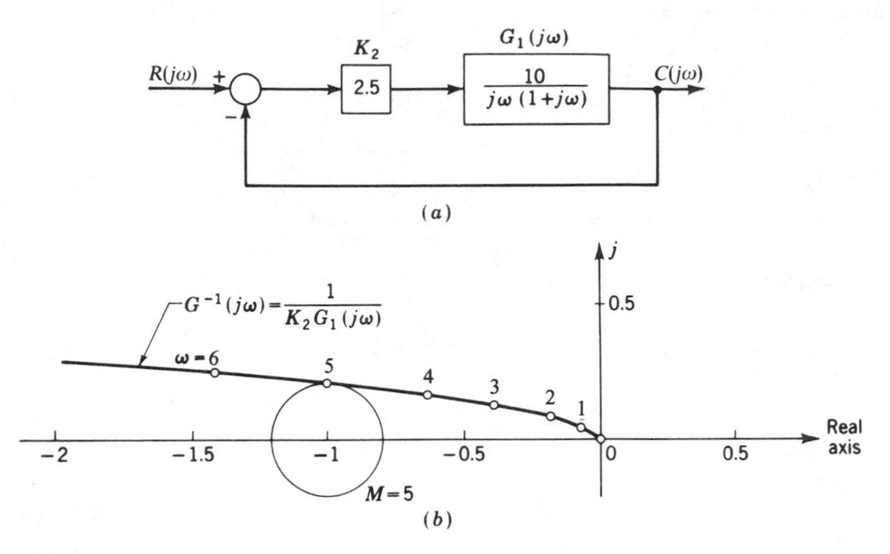

FIGURE 12.34
System without internal feedback.

i.e., of the form $H_1(s) = \beta s$ or $H_1(j\omega) = j\beta\omega$. In Fig. 12.35*b*, note that, at $\omega = 5$, the addition of $H_1(j5)/K_2 = 0.5j$ to the $1/[K_2G_1(j5)]$ vector causes the resulting curve to pass through the top of the $M = 1.4$ circle. For this case, the value of β is $j\beta 5/2.5 = 0.5j$, or $\beta = 0.25$. The resulting curve may now be constructed as indicated by the dashed line in Fig. 12.35*b*. Because this curve is not tangent to the $M = 1.4$ circle, another trial value must be taken. From the dashed loci of Fig. 12.35*b*, it now appears that the point of tangency is more likely to occur in the neighborhood of $\omega = 4.5$. The addition of $H_1(j4.5)/K_2 = 0.47j$ to $1/[K_2G_1(j4.5)]$ causes the new resulting curve to be tangent to the desired M circle. In this case $j\beta 4.5/2.5 = 0.47j$, or $\beta = 0.26$. Thus, the desired result is $H_1(j\omega) = 0.26j\omega$.

When $H_1(j\omega)$ is a constant, $H_1(j\omega)$ is entirely real. As may be seen from Fig. 12.33, the effect of a constant $H_1(j\omega)$ is to shift the inverse plot horizontally to the right. Suppose in the preceding problem that we desired to increase the speed of response by having ω_p equal 6 rather than 4.5. After a few trial values for $H_1(j6)/K_2$ are assumed, it is found that $H_1(j6)/K_2 = 0.6 + 0.4j$ makes the resultant plot tangent to the $M = 1.4$ circle at $\omega = 6$, as illustrated in Fig. 12.36. Because $H_1(j6)/K_2 = 0.6 + 0.4j = 0.6 + j\beta 6$, then $\beta = 0.4/6 = 0.067$, whence the required $H_1(j\omega)$ is $K_2(0.6 + j\beta\omega) = 1.5 + j0.167\omega$. Thus,

$$H_1(s) = 1.5 + 0.167s \tag{12.35}$$

The general procedure followed in this illustrative example was to assume a value of $H_1(j\omega)$ which makes a point lie on the desired M circle. From this assumed value, the general equation for the resulting plot was obtained. When the assumed point is not the point of tangency, then another trial point must be selected.

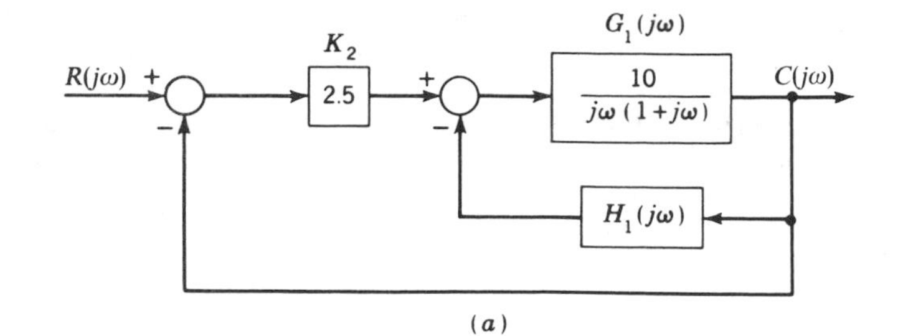

(a)

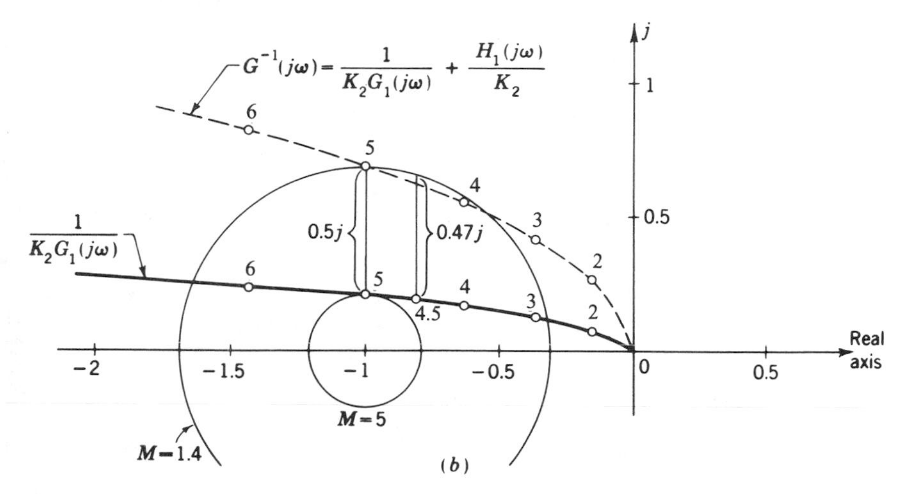

(b)

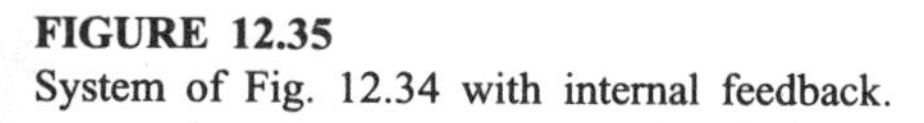

FIGURE 12.35
System of Fig. 12.34 with internal feedback.

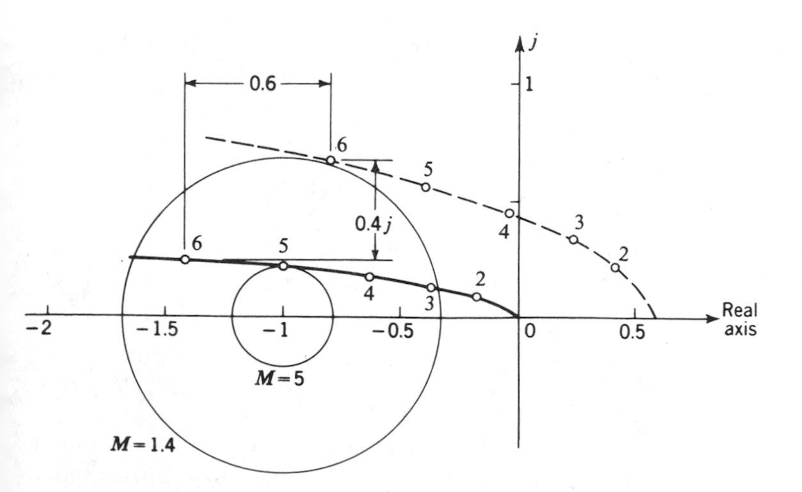

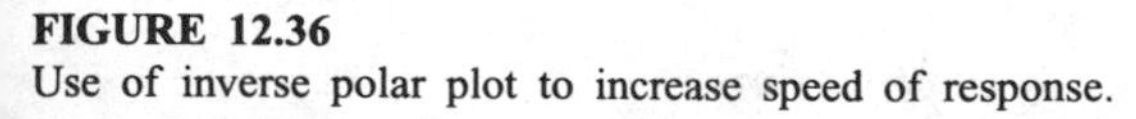

FIGURE 12.36
Use of inverse polar plot to increase speed of response.

12.8 STABILITY CRITERIA IN THE INVERSE PLANE

The inverse function $1/[G(s)H(s)]$ may be expressed in terms of the numerator and denominator terms as follows:

$$\frac{1}{G(s)H(s)} = \frac{D_G D_H}{N_G N_H}$$

Figure 12.37 shows the complete inverse polar plot for the function

$$G(s)H(s) = \frac{10(s+2)}{s(s+1)(s+4)}$$

or

$$\frac{1}{G(s)H(s)} = \frac{s(s+1)(s+4)}{10(s+2)}$$

As s proceeds up the vertical (imaginary) axis of Fig. 12.37a, then the corresponding inverse plot in Fig. 12.37b is indicated by the heavy line.

For values of s on the infinite semicircle $s = Re^{j\beta}$, then

$$\frac{1}{G(s)H(s)} \approx \frac{R^3 e^{3j\beta}}{Re^{j\beta}} = R^2 e^{2j\beta}$$

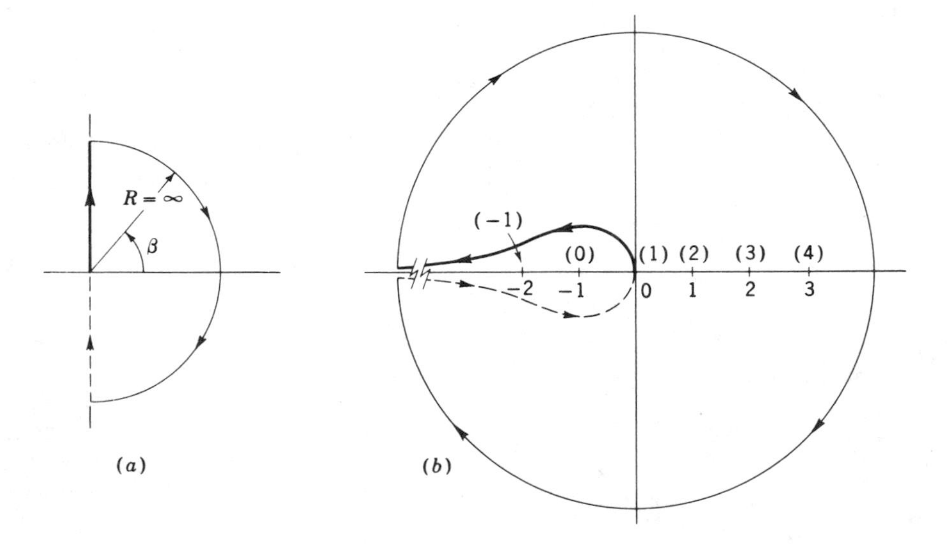

FIGURE 12.37
Inverse polar plot for $1/[G(s)H(s)] = s(s+1)(s+4)/[10(s+2)]$.

In general, if n is the highest order of s in the denominator of $G(s)H(s)$ and m is the highest order of s in the numerator of $G(s)H(s)$, then for values of s on the infinite semicircle

$$\frac{1}{G(s)H(s)} = R^{n-m}e^{j(n-m)\beta}$$

Because β changes by an angle of $-\pi$, the corresponding inverse plot transcribes an infinite semicircle which goes through the angle $-(n - m)\pi$. The negative imaginary axis $s = -j\omega$ is the complex conjugate of the positive imaginary axis $s = j\omega$; thus the portion shown dotted is the mirror image of the heavy-line portion.

The effect of adding +1 to the $1/[G(s)H(s)]$ plot is obtained by the new scale shown in parentheses above the real axis in Fig. 12.37*b*. Thus, the -1 point of the $1/[G(s)H(s)]$ plot is the origin of the $1 + 1/[G(s)H(s)]$ plot. With respect to the numerator and denominator terms, this new plot is

$$1 + \frac{1}{G(s)H(s)} = 1 + \frac{D_G D_H}{N_G N_H} = \frac{N_G N_H + D_G D_H}{N_G N_H}$$

Application of the Nyquist stability criterion to the inverse plot shows that

$$Z = N + P$$

where Z = number of zeros of characteristic function $N_G N_H + D_G D_H$ in the right half plane

P = number of zeros of $N_G N_H$ in the right half plane

N = net number of encirclements of the origin of the $1 + 1/[G(s)H(s)]$ plot (that is, -1 point of the $1/[G(s)H(s)]$ plot)

Note that N and Z have the same meaning for inverse plots as for direct plots. However, P is the number of zeros of $N_G N_H$ located within the path of values of s for inverse polar plots, whereas P is the number of zeros of $D_G D_H$ located within the path of values of s for direct polar plots. For the system of Fig. 12.37, $N = 0$ and $P = 0$. Application of the Nyquist stability criterion shows that $Z = N + P = 0$. Hence, the system is stable.

Illustrative example 12.9. Determine the inverse polar plot and the value of Z for the type 0 system whose direct polar plot is shown in Fig. 12.3.

Solution. For a type 0 system, the inverse polar plot begins on the positive real axis at the value $1/K$. The heavy-line portion of Fig. 12.38 is the portion of the inverse plot for $s = j\omega$. The dotted line is the portion for $s = -j\omega$. From Eq. (12.4), note that $n - m = 3$. Thus, the inverse plot transcribes an infinite semicircle which goes through the angle $-(n - m)\pi = -3\pi$. For the system of Fig. 12.38, $N = 2$ and $P = 0$. Application of the Nyquist stability criterion gives $Z = N + P = 2$. Thus, the system is basically unstable, as was previously determined by using the direct polar plot.

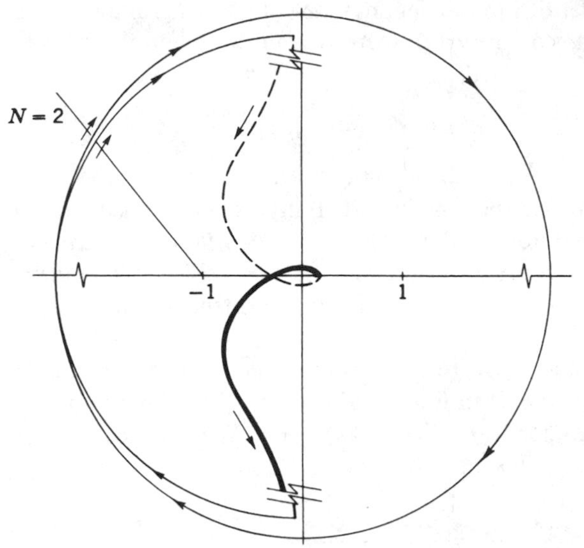

FIGURE 12.38
Inverse polar plot for $1/[G(s)H(s)] = (s + 2)(s^2 + 2s + 5)/52$.

On direct polar plots, instability is indicated when the frequency response crosses the negative real axis outside the -1 point (e.g., see Figs. 12.3 and 12.4). On inverse polar plots, instability is indicated when the frequency response crosses the negative real axis inside the -1 point (e.g., see Fig. 12.38).

PROBLEMS

12.1. For each of the $G(s)H(s)$ plots shown in Fig. P12.1, the path of values for s is the same as that shown in Fig. 12.2. For each plot, determine the number of roots of the characteristic equation which are located in the right half plane ($P = 0$ in all cases).

(*a*) For each stable system, determine the factor by which the gain should be changed so that the system will just become unstable.

(*b*) For each unstable system, determine the factor by which the gain should be changed so that the system will just become stable.

12.2. For each of the following unity-feedback systems, sketch the complete $G(s)$ plot and determine the number of roots of the characteristic equation that lie in the right half plane.

$$(a)\quad G(s) = \frac{10}{(1 + 0.25s)\,(0.25s^2 + 0.40s + 1)}$$

$$(b)\quad G(s) = \frac{10}{s(1 + 0.25s)}$$

$$(c)\quad G(s) = \frac{10}{s^2(1 + 0.25s)}$$

$$(d)\quad G(s) = \frac{10(1 + s)}{s^2(1 + 0.25s)}$$

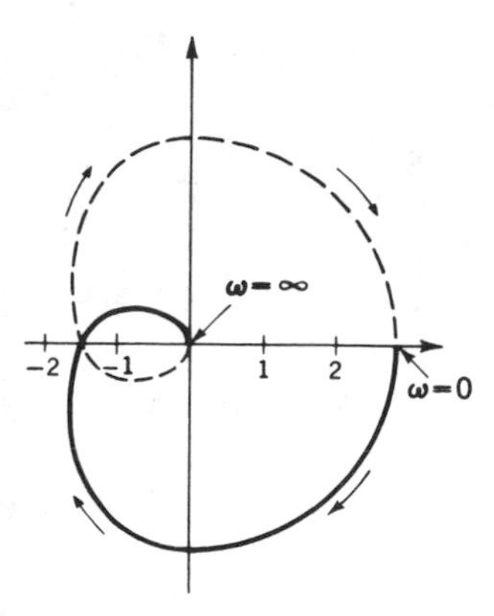

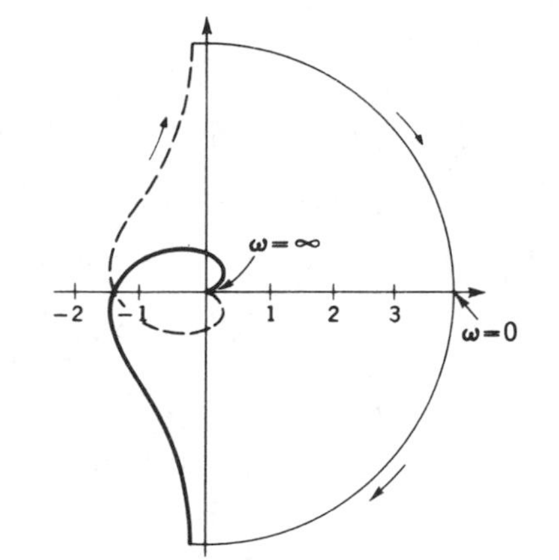

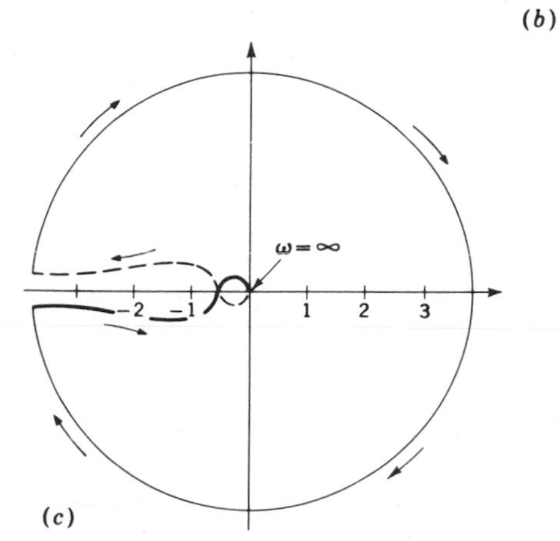

FIGURE P12.1

12.3. Sketch the log-magnitude plot for the system given in Prob. 12.2*a*. For this system, determine:

(*a*) The gain margin and the phase margin

(*b*) The factor by which the gain should be changed to yield a gain margin of 5

(*c*) The factor by which the gain should be changed to yield a phase margin of 40°

12.4. The open-loop transfer function for a unity-feedback system is

$$G(s) = \frac{2}{s(1+s)\,(1+2s)}$$

The Bode plot for this system is shown in Fig. P12.4.

(*a*) Determine the factor K_c by which the gain should be changed so that the resulting system will have a phase margin of 45°.

(*b*) Determine the factor K_c by which the gain should be changed so that the resulting system will have a gain margin of 5.

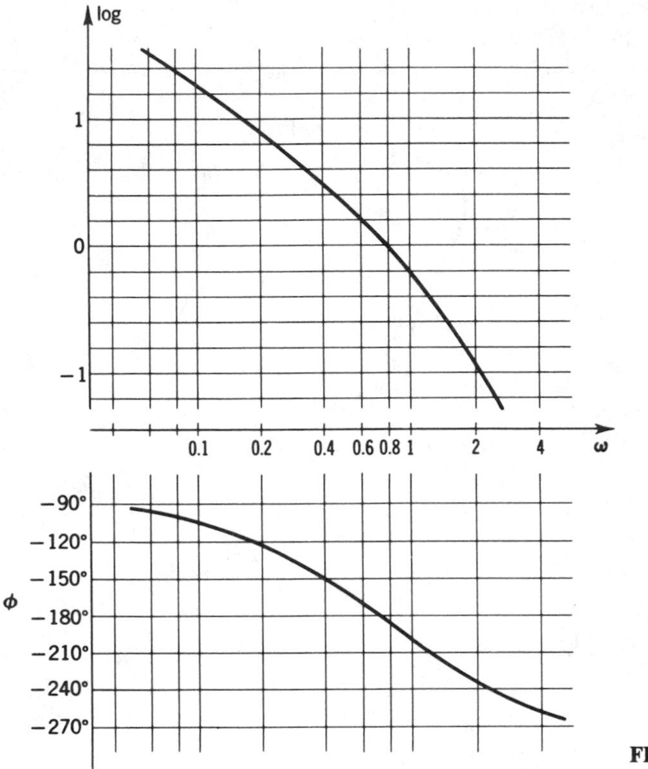

FIGURE P12.4

12.5. The Nichols plot of $G(j\omega)$ for a unity-feedback system is shown in Fig. P12.5.
(*a*) Determine the gain margin and the phase margin for this system.
(*b*) By what factor K_c should the gain K of the system be changed so that the system will have a phase margin of 45°?

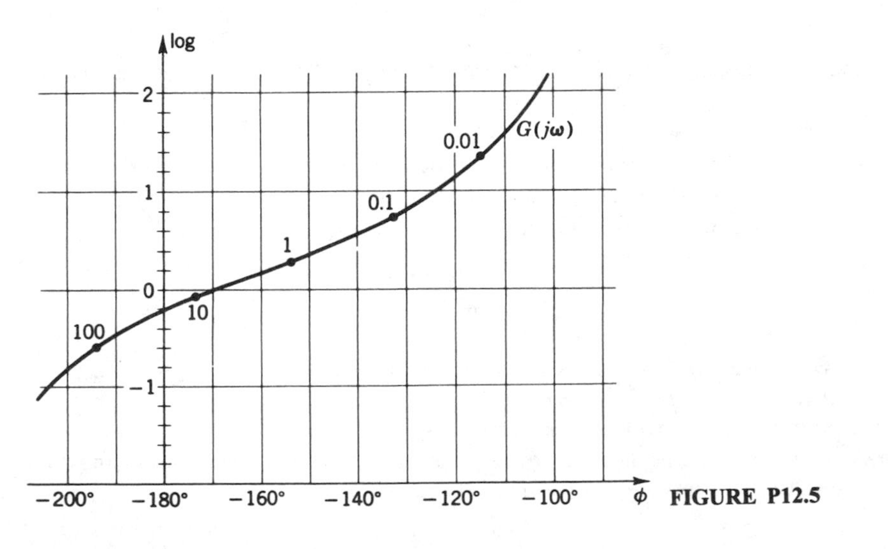

FIGURE P12.5

12.6. A plot of $G(j\omega)$ for a unity-feedback system is shown in Fig. P12.6. What are the values of the gain margin and the phase margin for this system? By what factor K_c should the gain of the system be changed so that M_p will be 1.4? What are the new values of the gain margin and the phase margin?

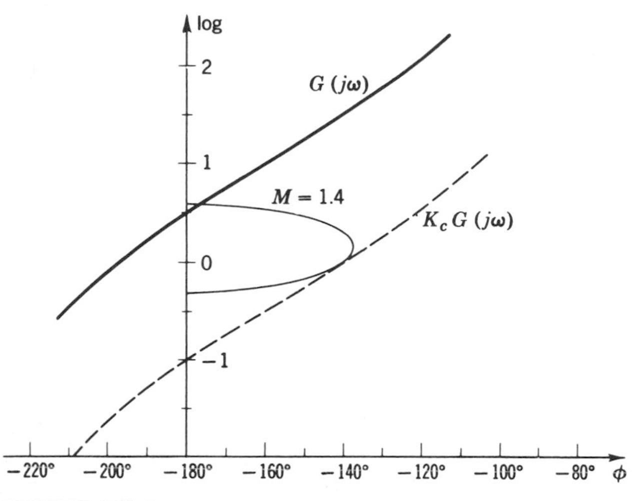

FIGURE P12.6

12.7. From the polar plot of a type 0 system ($K_0 = 4$), notice that $G(j5)H(j5) = 1.0\underline{/-145^\circ}$. What values of ζ and ω_n should be used to approximate the transient response of this system?

12.8. From the polar plot of a type 0 system ($K_0 = 4$), it is found that $M_p = 1.4$ and $\omega_p = 3$. What are the approximate values of the damping ratio ζ, natural frequency ω_n, phase margin γ, and crossover frequency ω_c?

12.9. The open-loop transfer function for a control system is

$$G(s)H(s) = \frac{10}{(1 + 0.5s)\,(s^2 + 6s + 16)/16}$$

Corresponding values of ω, $|G(j\omega)H(j\omega)|$, and $\measuredangle G(j\omega)H(j\omega)$ are as follows:

ω	$\lvert G(j\omega)H(j\omega)\rvert$	$\measuredangle G(j\omega)H(j\omega)$
0	10.00	0.0
2.0	6.67	−9.0
3.0	4.60	−125.1
3.5	3.72	−140.1
4.0	2.98	−153.4
4.5	2.38	−165.0
5.0	1.90	−174.9
∞	0.00	−270.0

Determine the factor K_c by which the gain should be changed to yield a phase margin of 45°. What are the approximate values of ζ and ω_n for the resulting system?

12.10. The open-loop transfer function for a control system is

$$G(s)H(s) = \frac{10}{s(1 + 0.25s)(1 + 0.5s)}$$

Corresponding values of ω, $|G(j\omega)H(j\omega)|$, and $\measuredangle G(j\omega)H(j\omega)$ are as follows:

ω	$\lvert G(j\omega)H(j\omega)\rvert$	$\measuredangle G(j\omega)H(j\omega)$
0.6	15.79	−115.2
1.0	8.68	−130.6
1.4	5.52	−144.3
1.8	3.77	−156.2
2.2	2.68	−166.5
2.6	1.97	−175.5
3.0	1.48	−183.2
∞	0.00	−270.0

Determine the value of K to yield a phase margin of 45°. What are the approximate values of ζ and ω_n for the resulting system?

12.11. The open-loop transfer function for a control system is

$$G(s)H(s) = \frac{K(s + 3)}{s^2 - 2s + 10}$$

Determine the values of $G(j\omega)H(j\omega)/K$ for $\omega = 0, 1, 2, 3, 4, 5$, and 6. Construct the entire Nyquist plot and then determine:

(*a*) The range of values of K such that the system is stable
(*b*) The value of K to yield a phase margin of 45°

12.12. The open-loop transfer function for a control system is

$$G(s)H(s) = \frac{K(s + 2)}{(s - 1)(s + 1)}$$

Determine the values of $G(j\omega)H(j\omega)/K$ for $\omega = 0, 1, 2, 3$, and 4. Construct the entire Nyquist plot, and then determine:

(*a*) The range of values of K such that the system is stable
(*b*) The value of K to yield a phase margin of 45°

12.13. The open-loop transfer function for a control system is

$$G(s)H(s) = \frac{K(s + 2)}{s(s - 1)}$$

Determine the values of $G(j\omega)H(j\omega)/K$ for $\omega = 0, 1, 2, 3, 4$, and 5. Sketch the complete polar plot and then determine:

(*a*) The range of values of K such that the system is stable
(*b*) The value of K to yield a phase margin of 45°

12.14. The log-modulus plot for the pitch rate control system for the space shuttle is shown in Fig. P12.14. Determine the values of τ_1 and τ_2 for a series lead compensator such that the resulting system will have a phase margin of 35° ± 3°.

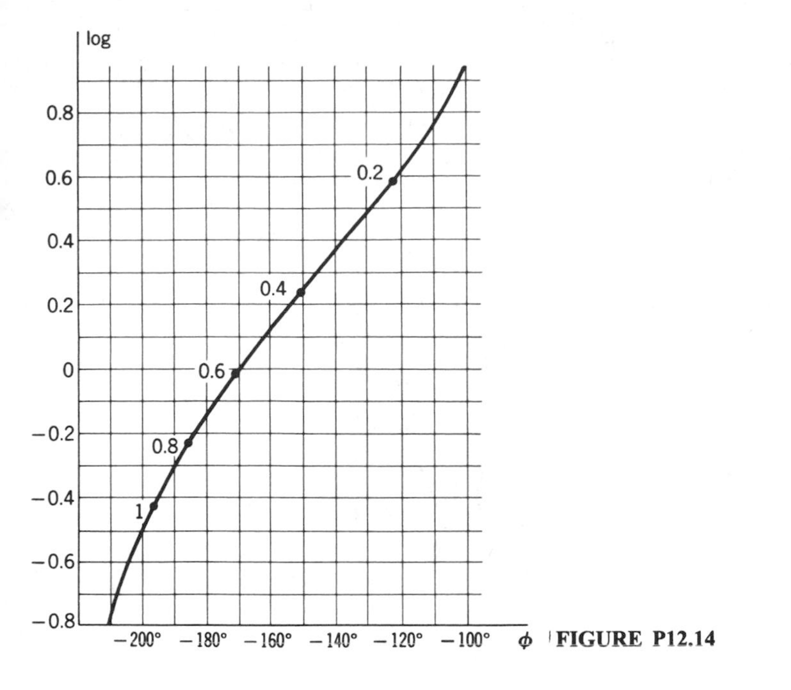

FIGURE P12.14

12.15. Repeat Prob. 12.14 except use a lag compensator.

12.16. Repeat Prob. 12.14 except use a lag-lead compensator.

12.17. The log-magnitude plot shown in Fig. P12.17 is for the unity-feedback system

$$G(s) = \frac{4}{(s+1)\,(s^2+8s+25)/25}$$

(*a*) Find the values of τ_1 and τ_2 for a series lag-lead compensator such that the resulting system will have a phase margin of 45° ± 3°.

(*b*) Use Fig. P12.17 to construct the log-modulus plot for this system. Using this plot, verify the results of part *a*.

12.18. The log-magnitude plot for the attitude control system for the Lunar Exploratory Module (LEM) is shown in Fig. P12.18. For this unity-feedback system

$$G(s) = \frac{2.5}{s(1+0.25s)\,(1+s)}$$

Determine the values of τ_1 and τ_2 for a series lead compensator $(1 + \tau_1 s)/(1 + \tau_2 s)$ such that the resulting system will have a phase margin of 45° ± 5°.

12.19. Repeat Prob. 12.18 except use a lag compensator.

12.20. Repeat Prob. 12.18 except use a lag-lead compensator.

12.21. Construct the approximate log-magnitude plots for each of the two systems shown in Fig. P12.21. For each system, write the equation for the open-loop transfer function corresponding to the asymptotes. To obtain the exact transfer function for each system, use block-diagram algebra to eliminate the minor feedback loops. Compare the exact and approximate transfer functions.

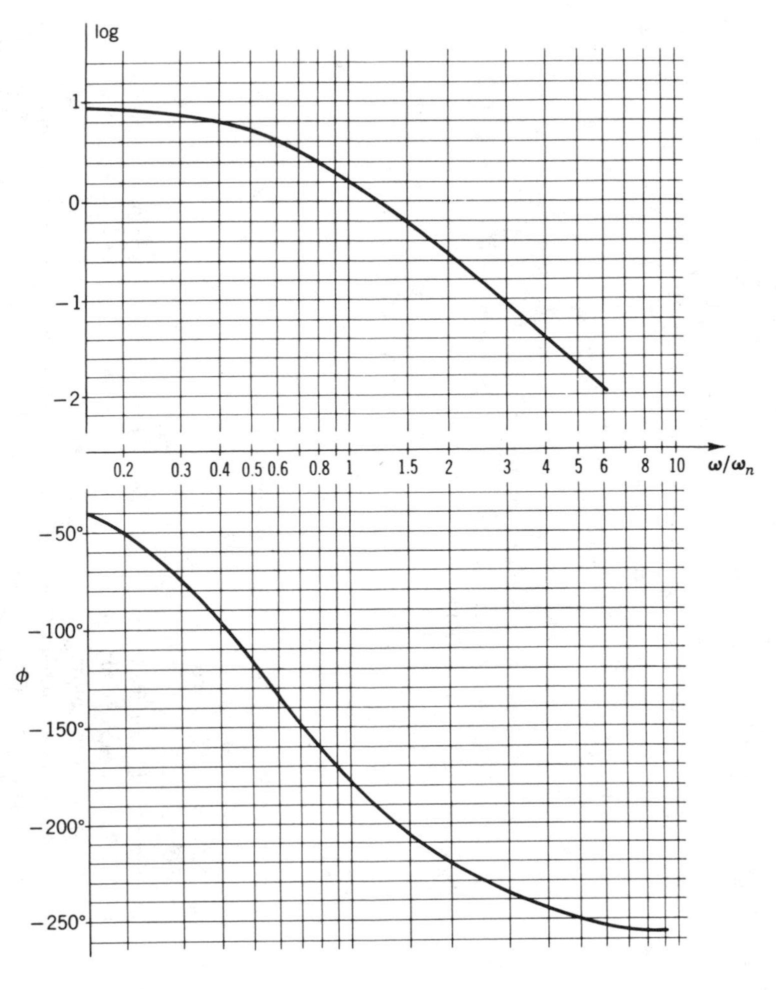

FIGURE P12.17

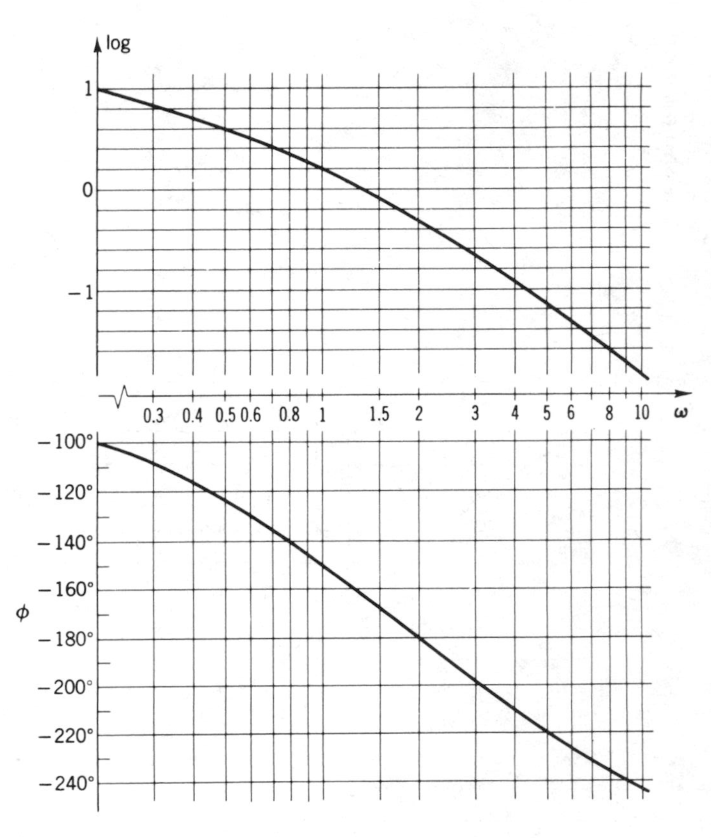

FIGURE P12.18

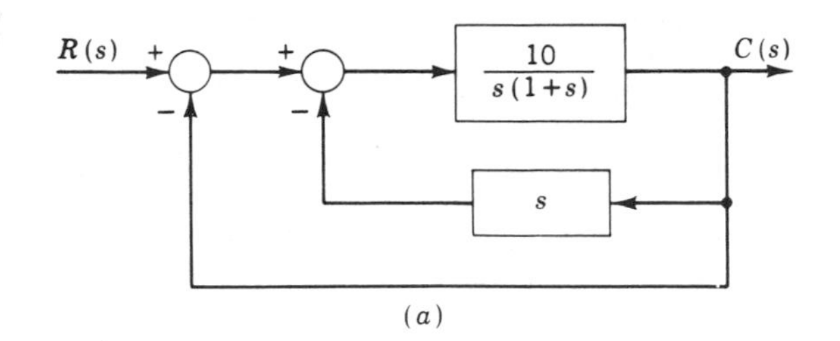

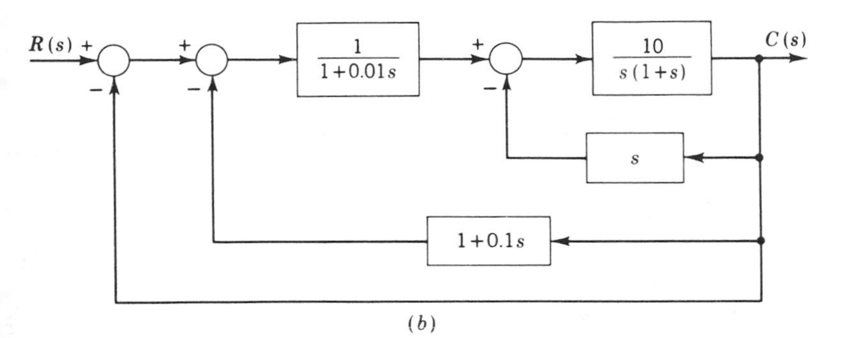

FIGURE P12.21

12.22. Repeat Prob. 11.22, but use inverse polar plots to determine the value of the gain K to yield an M_p of 1.4.

12.23. The inverse polar plot for the unity-feedback system

$$G_1(s) = \frac{2.5}{s(1 + 0.25s)\,(1 + s)} = \frac{10}{s(s + 1)\,(s + 4)}$$

is shown in Fig. P12.23. For an internal-feedback compensation of the form $H_1(s) = \alpha + \beta s$, determine the values of α and β such that the resulting system will have an M_p of 1.4 and ω_p occurs at $\omega = 2$.

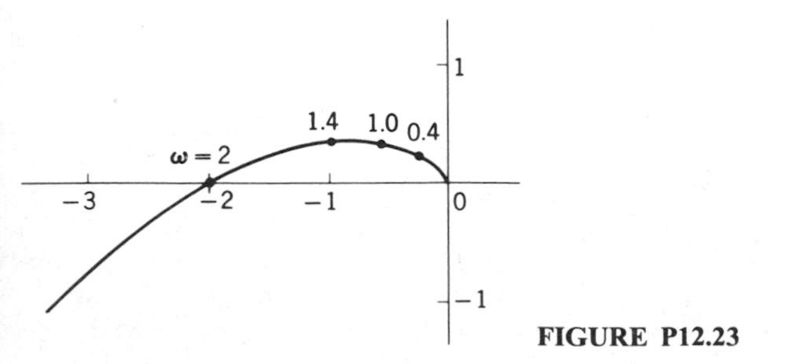

FIGURE P12.23

12.24. Use inverse polar plots to determine the value of β such that the system of Fig. P12.24 will have an M_p of 1.4.

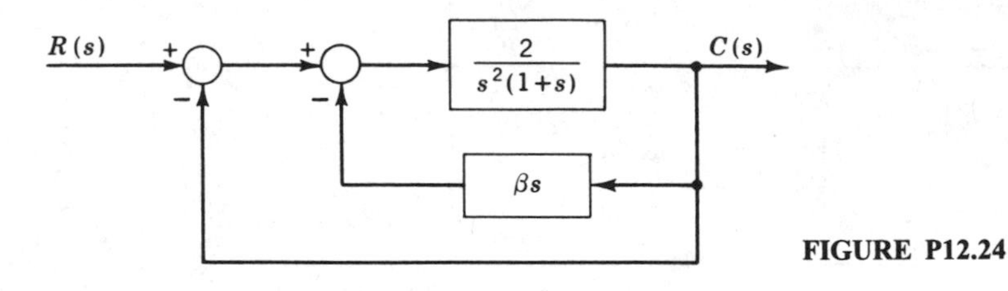

FIGURE P12.24

12.25. For the system shown in Fig. P12.25, determine the value of K_1 to yield an M_p of 1.4.

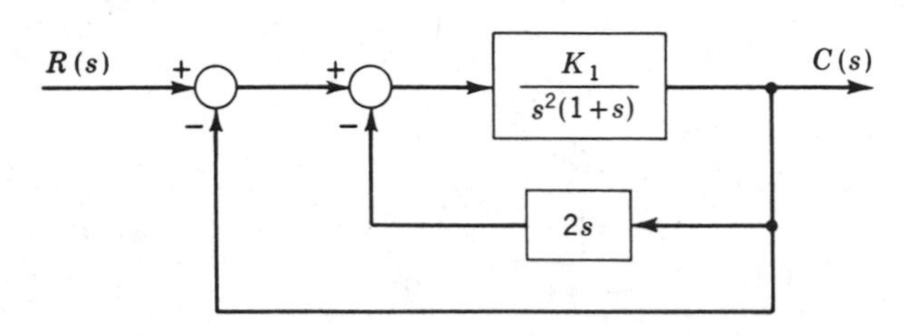

FIGURE P12.25

12.26. Repeat Prob. 12.2, except use the inverse plane.
12.27. Repeat Prob. 12.3, except use the inverse plane.

APPENDIX A

EQUILIBRIUM FLOW

Many pneumatic components use two orifices in series to obtain a controlled pressure in the chamber between the orifices. If the component is available, the chamber pressure may be experimentally determined for various operating conditions. However, in the initial design stages, before any parts have been manufactured, it is desirable to be able to predict the value of the chamber pressure. This may be accomplished by use of the nondimensional family of curves shown in Fig. A.1*a*. As illustrated by the insert above Fig. A.1*a*, pressures P_1, P_2, and P_3 represent the inlet pressure, chamber pressure, and discharge pressure, respectively. The symbol A_1 is the area of the first orifice times its coefficient of discharge, and A_2 is the area of the second orifice times its coefficient of discharge.

Usually, the overall pressure ratio P_1/P_3 is known, and the area ratio A_2/A_1 is known, so that the ratio P_2/P_1 can be found from Fig. A.1*a*. The value of the chamber pressure P_2 is then computed as the product of the ratio P_2/P_1 and the value of the inlet pressure P_1. In using Fig. A.1*a*, it is necessary to use absolute pressure. Because these are nondimensional curves, any consistent set of units may be used.

The derivation of this nondimensional family of curves is accomplished as follows: Assuming that the fluid is a perfect gas and that the kinetic energy at the inlet is negligible compared with other terms in the energy equation, the mass rate of

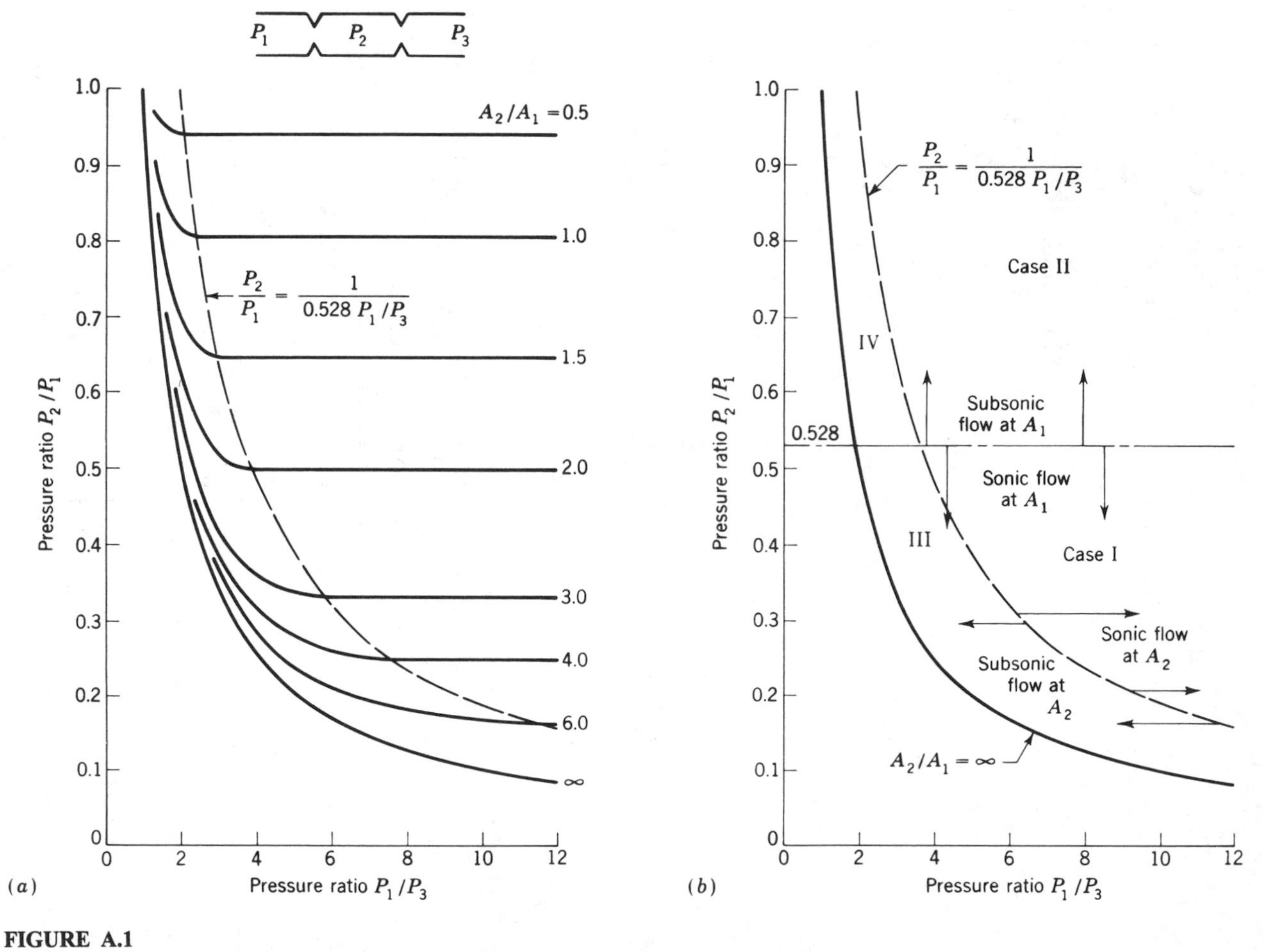

FIGURE A.1
Nondimensional curve for determining equilibrium conditions for flow through two orifices in series.

flow through the first orifice is

$$M_1 = \frac{A_1 P_1}{\sqrt{T_1}} \left\{ 2g_c \frac{k}{k-1} \frac{1}{R} \left[\left(\frac{P_{1_t}}{P_1} \right)^{2/k} - \left(\frac{P_{1_t}}{P_1} \right)^{(k+1)/k} \right] \right\}^{1/2} \tag{A.1}$$

where T_1 = stagnation temperature at inlet
g_c = gravitational conversion factor
k = ratio of specific heat at constant pressure to that at constant volume
P_{1_t} = throat pressure at first orifice
R = gas constant

By replacing the subscript 1 in Eq. (A.1) by 2 and the subscript 1_t by 2_t, the equation for the mass rate of flow through the second orifice is obtained. For equilibrium to exist, the mass rate of flow in must equal that out, so that

$$\frac{A_1 P_1}{\sqrt{T_1}} B_1^{1/2} = \frac{A_2 P_2}{\sqrt{T_2}} B_2^{1/2} \tag{A.2}$$

where

$$B_1 = \left(\frac{P_{1_t}}{P_1} \right)^{2/k} - \left(\frac{P_{1_t}}{P_1} \right)^{(k+1)/k}$$

$$B_2 = \left(\frac{P_{2_t}}{P_2} \right)^{2/k} - \left(\frac{P_{2_t}}{P_2} \right)^{(k+1)/k}$$

Because there is little time for heat transfer to take place, the flow may be considered to be adiabatic, so that $T_1 = T_2$. Thus Eq. (A.2) becomes

$$A_1 P_1 B_1^{1/2} = A_2 P_2 B_2^{1/2} \tag{A.3}$$

In the following analysis, it is assumed that the fluid is air, for which $k = 1.4$, and the critical pressure ratio is $P_2/P_1 = P_3/P_2 = 0.528$. By using the appropriate value of k and the critical ratio, this analysis is applicable for any gas.

When sonic flow exists at the first orifice, $B_1^{1/2} = 0.259$, and similarly for sonic flow at the second orifice $B_2^{1/2} = 0.259$. Thus for sonic flow at both orifices Eq. (A.3) reduces to

$$\frac{P_2}{P_1} = \frac{1}{A_2/A_1} \tag{A.4}$$

When sonic flow exists at the first orifice,

$$\frac{P_2}{P_1} \leq 0.528$$

Above the line $P_2/P_1 = 0.528$ shown in Fig. A.1*b*, subsonic flow exists at A_1, and below this line sonic flow exists at A_1.

The equation for the line of separation between subsonic and sonic flows at A_2 is obtained by noting that the critical ratio is $P_3/P_2 = 0.528$; thus

$$\frac{P_2}{P_1} = \frac{1}{(P_3/P_2)(P_1/P_3)} = \frac{1}{0.528 P_1/P_3} \tag{A.5}$$

The curve defined by the preceding equation is shown in Fig. A.1*b*. To the right of this curve sonic flow exists at A_2, and to the left of this curve subsonic flow exists at A_2. The regions in which each of the four possible combinations of sonic or subsonic flow exists at the first and second orifices are shown in Fig. A.1*b*:

Case I: sonic flow at both orifices
Case II: subsonic flow at the first orifice and sonic flow at the second
Case III: sonic flow at the first orifice and subsonic flow at the second
Case IV: subsonic flow at both orifices

For case I, it follows from Eq. (A.4) that the lines of constant values of A_2/A_1 are horizontal straight lines, as shown in Fig. A.1*a*. For case II, $P_2 = P_{1_t}$ and $B_2^{1/2} = 0.259$, so that Eq. (A.3) reduces to

$$\frac{P_2}{P_1} \frac{1}{B_1^{1/2}} = \frac{1}{0.259 A_2/A_1} \tag{A.6}$$

For a given area ratio A_2/A_1, there is but one value of P_2/P_1 which makes the left side of Eq. (A.6) equal to the right side. Thus, for case II, lines of constant A_2/A_1 are also horizontal.

By applying these techniques to case III and case IV, the complete family of curves shown in Fig. A.1*a* is obtained. This method of analysis may be extended to determine equilibrium flow conditions for three or more orifices in series.

Equation (A.1) is an awkward form for computing partial derivatives. However, for the case of sonic flow, this reduces to

$$M_1 = \frac{0.53}{\sqrt{T_1}} A_1 P_1 \tag{A.7}$$

where the preceding equation has units of pounds-mass (lb_m), pounds-force (lb_f), feet, degrees Rankine, and seconds. For the usual design case in which the stagnation temperature of the inlet air is 60°F or 520°R, then $M_1 = 0.0074 A_1 P_1$.

For the case of subsonic flow $P_{1_t} = P_2$, Eq. (A.1) may be approximated by noting that a plot of the function B_1 versus $(P_1 - P_2)P_2/P_1^2$ is very nearly a straight line. The slope of this line is such that

$$B_1 \approx \frac{0.261(P_1 - P_2)P_2}{P_1^2} \tag{A.8}$$

Substitution of the preceding approximation into Eq. (A.1) gives

$$\begin{aligned} M_1 &\approx \frac{2.06}{\sqrt{T_1}} A_1 \sqrt{0.261(P_1 - P_2)P_2} \\ &= \frac{1.05}{\sqrt{T_1}} A_1 \sqrt{(P_1 - P_2)P_2} \end{aligned} \tag{A.9}$$

where $\sqrt{(2g_c/R)[k/(k-1)]}) = 2.06$. For a stagnation temperature of 520°R, Eq. (A.9) is

$$M_1 = 0.046A_1\sqrt{(P_1 - P_2)P_2} \tag{A.10}$$

The length units cancel in Eqs. (A.7) to (A.10); hence any length unit may be used.

Because the curves shown in Fig. A.1 are nondimensional, they are valid for SI units as well as British gravitational units. When SI units (kg, m, N, s, K) are used, the g_c term does not appear in Eq. (A.1). Equations (A.2) to (A.6) are valid for SI units as well as British gravitational units. For SI units the constant in Eq. (A.8) remains the same. The constant 2.06 in Eq. (A.9) becomes 0.156, and the constant 1.05 becomes 0.08. Finally, the constant 0.046 in Eq. (A.10) becomes 0.0047 for SI units.

APPENDIX B

FOURIER SERIES, FOURIER INTEGRAL, AND THE LAPLACE TRANSFORM

A greater understanding of the Laplace transform $F(s)$ of a time function $f(t)$ may be obtained by examining the similarities which exist between Laplace transforms and the more familiar Fourier series and Fourier integral.

B.1 FOURIER SERIES

A periodic function as shown in Fig. B.1 may be represented by the series

$$f(t) = K + \sum_{n=1}^{\infty} (A_n \cos n\omega_0 t + B_n \sin n\omega_0 t) \tag{B.1}$$

where $\omega_0 = 2\pi/T$, in which T is the period. The constant K is evaluated as follows: Integration of each term in Eq. (B.1) over a complete period causes each term in the

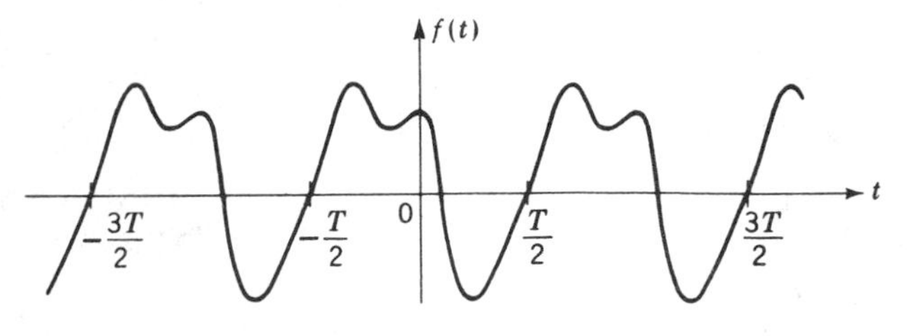

FIGURE B.1
Periodic function.

summation on the right side to vanish. Thus

$$\int_{-T/2}^{T/2} f(t)\, dt = K \int_{-T/2}^{T/2} dt + 0 = KT$$

or

$$K = \frac{1}{T} \int_{-T/2}^{T/2} f(t)\, dt \tag{B.2}$$

The value of K is equal to the average value of the function over a period.

To evaluate A_n, each term of Eq. (B.1) is multiplied by cos $m\omega_0 t$ and then integrated over a period. Because

$$\int_{-T/2}^{T/2} \cos n\omega_0 t \cos m\omega_0 t\, dt = \begin{cases} 0 & \text{for } m \neq n \\ \dfrac{T}{2} & \text{for } m = n \end{cases} \tag{B.3}$$

and

$$\int_{-T/2}^{T/2} \sin n\omega_0 t \cos m\omega_0 t\, dt = 0 \tag{B.4}$$

it follows that

$$A_n = \frac{2}{T} \int_{-T/2}^{T/2} f(t) \cos n\omega_0 t\, dt \tag{B.5}$$

Similarly, multiplication of each term of Eq. (B.1) by sin $m\omega_0 t$ and integration over the period yield the following result for B_n:

$$B_n = \frac{2}{T} \int_{-T/2}^{T/2} f(t) \sin n\omega_0 t\, dt \tag{B.6}$$

Equation (B.1) may be telescoped into a more convenient form by using Eqs. (5.22) and (5.23) to express the cosine and sine in exponential form.

$$A_n \cos n\omega_0 t = \frac{A_n}{2} \left(e^{jn\omega_0 t} + e^{-jn\omega_0 t}\right)$$

$$B_n \sin n\omega_0 t = -j\, \frac{B_n}{2} \left(e^{jn\omega_0 t} - e^{-jn\omega_0 t}\right)$$

Thus

$$f(t) = K + \tfrac{1}{2} \sum_{n=1}^{\infty} (A_n - jB_n)e^{jn\omega_0 t} + (A_n + jB_n)e^{-jn\omega_0 t} \tag{B.7}$$

By also writing Eqs. (B.5) and (B.6) in exponential form, we see that

$$\begin{aligned} A_n - jB_n &= \frac{2}{T} \int_{-T/2}^{T/2} f(t)\,(\cos n\omega_0 t - j \sin n\omega_0 t)\, dt \\ &= \frac{2}{T} \int_{-T/2}^{T/2} f(t) e^{-jn\omega_0 t}\, dt \end{aligned} \tag{B.8}$$

and

$$\begin{aligned} A_n + jB_n &= \frac{2}{T} \int_{-T/2}^{T/2} f(t)\,(\cos n\omega_0 t + j \sin n\omega_0 t)\, dt \\ &= \frac{2}{T} \int_{-T/2}^{T/2} f(t) e^{jn\omega_0 t}\, dt \end{aligned} \tag{B.9}$$

Substitution of the preceding results into Eq. (B.7) gives

$$\begin{aligned} f(t) = K &+ \frac{1}{T} \sum_{n=1}^{\infty} e^{jn\omega_0 t} \int_{-T/2}^{T/2} f(t) e^{-jn\omega_0 t}\, dt \\ &+ \frac{1}{T} \sum_{n=1}^{\infty} e^{-jn\omega_0 t} \int_{-T/2}^{T/2} f(t) e^{jn\omega_0 t}\, dt \end{aligned} \tag{B.10}$$

By noting that the last summation is unaltered by changing the sign of n, the Fourier series becomes

$$\begin{aligned} f(t) = K &+ \sum_{n=1}^{\infty} \frac{e^{jn\omega_0 t}}{T} \int_{-T/2}^{T/2} f(t) e^{-jn\omega_0 t}\, dt \\ &+ \sum_{n=-1}^{-\infty} \frac{e^{jn\omega_0 t}}{T} \int_{-T/2}^{T/2} f(t) e^{-jn\omega_0 t}\, dt \end{aligned} \tag{B.11}$$

Because the value of the summation for $n = 0$ is K,

$$f(t) = \sum_{n=-\infty}^{\infty} \frac{e^{jn\omega_0 t}}{T} \int_{-T/2}^{T/2} f(t) e^{-jn\omega_0 t}\, dt \tag{B.12}$$

Equation (B.12) is frequently written in the form

$$f(t) = \sum_{n=-\infty}^{\infty} C_n e^{jn\omega_0 t} \tag{B.13}$$

where

$$C_n = \frac{1}{T}\int_{-T/2}^{T/2} f(t)e^{-jn\omega_0 t}dt$$

B.2 FOURIER INTEGRAL

As the period T becomes infinite, the Fourier series expression given by Eq. (B.12) is

$$f(t) = \lim_{T\to\infty}\left[\sum_{n=-\infty}^{\infty}\frac{e^{jn\omega_0 t}}{T}\int_{-T/2}^{T/2} f(t)e^{-jn\omega_0 t}\,dt\right] \tag{B.14}$$

For large values of T it is more appropriate to use the following notation:

$$\lim_{T\to\infty}\omega_0 = \lim_{T\to\infty}\frac{2\pi}{T} = \Delta\omega$$

and

$$\lim_{T\to\infty} n\omega_0 = n\,\Delta\omega = \omega$$

Thus, Eq. (B.14) becomes

$$f(t) = \lim_{\substack{\Delta\omega\to 0\\ T\to\infty}}\left[\frac{1}{2\pi}\sum_{n=-\infty}^{\infty} e^{j\omega t}\,\Delta\omega\int_{-T/2}^{T/2} f(t)e^{-j\omega t}\,dt\right] \tag{B.15}$$

The limit of Eq. (B.15) is the Fourier integral

$$f(t) = \frac{1}{2\pi}\int_{-\infty}^{\infty} e^{j\omega t}\left[\int_{-\infty}^{\infty} f(t)e^{-j\omega t}\,dt\right]d\omega \tag{B.16}$$

The Fourier integral is frequently expressed by the Fourier transform pair

$$f(t) = \frac{1}{2\pi}\int_{-\infty}^{\infty} F(j\omega)e^{j\omega t}\,d\omega \tag{B.17}$$

$$F(j\omega) = \int_{-\infty}^{\infty} f(t)e^{-j\omega t}\,dt \tag{B.18}$$

Equation (B.18) is referred to as the *direct Fourier transform*, and Eq. (B.17) is the *inverse Fourier transform*.

For most physical problems, we want to know the solution for $t > 0$. Thus, if the initial conditions are known, then the lower limit of integration in Eq. (B.18) may be taken as zero.

To illustrate the use of the Fourier transform, consider the function

$$f(t) = e^{at} \qquad t \geq 0 \tag{B.19}$$

Application of the direct Fourier transform gives

$$F(j\omega) = \int_0^\infty e^{at}e^{-j\omega t}\,dt = \left.\frac{e^{at}e^{-j\omega t}}{a - j\omega}\right|_0^\infty \tag{B.20}$$

If the exponent a is less than zero, the preceding becomes

$$F(j\omega) = 0 - \frac{1}{a - j\omega} = \frac{1}{j\omega - a} \qquad a < 0 \tag{B.21}$$

However, if the exponent a is positive, e^{at} becomes infinite when evaluated at $t = \infty$, and thus $F(j\omega)$ diverges.

B.3 LAPLACE TRANSFORM

To extend the usefulness of the Fourier transform so that it is applicable to divergent functions, a converging factor $e^{-\sigma t}$ is introduced. Thus, the general transform equation is

$$\begin{aligned} F(\sigma + j\omega) &= \int_0^\infty f(t)e^{-\sigma t}e^{-j\omega t}\,dt \\ &= \int_0^\infty f(t)e^{-(\sigma + j\omega)t}\,dt \end{aligned} \tag{B.22}$$

The transform for the time function given by Eq. (B.19) is

$$\begin{aligned} F(\sigma + j\omega) &= \left.\frac{e^{(a - \sigma - j\omega)t}}{a - \sigma - j\omega}\right|_0^\infty \\ &= \frac{1}{\sigma + j\omega - a} \qquad \sigma > a \end{aligned} \tag{B.23}$$

The preceding equation converges when σ is greater than a. To ensure convergence of the Fourier transform, it was necessary that $\int_0^\infty |f(t)|\,dt < \infty$. However, the transform indicated by Eq. (B.22) converges when $\int_0^\infty |f(t)|e^{-\sigma t}\,dt < \infty$ for some finite σ.

The substitution of $s = \sigma + j\omega$ and $F(s) = F(\sigma + j\omega)$ into Eq. (B.22) yields the Laplace transform equation

$$F(s) = \int_0^\infty f(t)e^{-st}\,dt \tag{B.24}$$

It is only necessary that some finite value of σ exists such that $\int_0^\infty |f(t)|e^{-\sigma t}\,dt < \infty$ in order to verify the existence of the transform indicated by Eq. (B.24). For most functions $f(t)$ encountered in engineering work, the transform is convergent. Also note that to solve differential equations by Laplace transforms, it is not necessary to determine the value or values of σ over which $F(s)$ is convergent. It suffices to know that such a value or values of σ exist.

In effect, Eq. (B.24) is the result of substituting s for $j\omega$ and $F(s)$ for $F(j\omega)$ in Eq. (B.18). The use of these same substitutions in Eq. (B.17) yields the inverse

Laplace transform. That is,

$$f(t) = \frac{1}{2\pi j} \int_{\sigma - j\infty}^{\sigma + j\infty} F(s) e^{st}\, ds \tag{B.25}$$

The new limits of integration are obtained by noting that when $\omega = \pm\infty$, then $s = \sigma \pm j\omega = \sigma \pm j\infty$. Equation (B.25) is a line integral for which the path of integration is a vertical line which is displaced a distance σ from the imaginary axis, as shown in Fig. B.2. For convergence, it is necessary that σ be such that all the values of s which make $F(s)$ infinite [i.e., poles of $F(s)$] lie to the left of the vertical line shown in Fig. B.2.

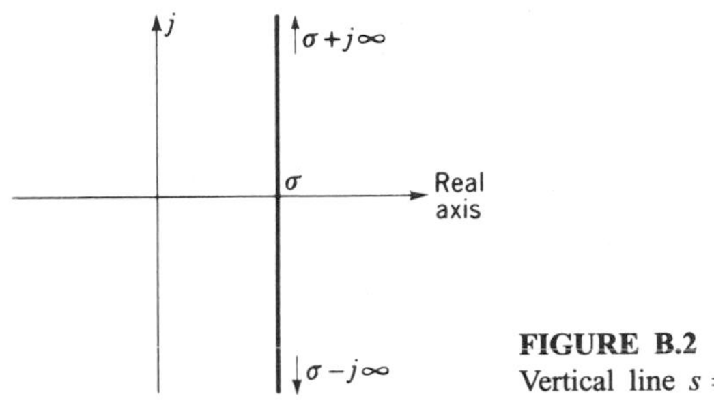

FIGURE B.2
Vertical line $s = \sigma + j\omega$.

INDEX

ac motors, 91–92
Acceleration error constant, 204–208, 520
Accelerometer, 548
Actuating signal, 4–7
Actuator:
  hydraulic, 70–75
  pneumatic, 79–84
Allied-Signal, Inc., Controls and Accessories, 95
α contours:
  on direct polar plot, 524–528
  on inverse polar plot, 584–585
  on log-modulus plot, 544–546
Amplifiers:
  hydraulic, 70–75
  jet pipe, 77–79
  operational, 339–345, 357–359
  pneumatic, 79–84
Analog computers, 337–363
  computer diagrams for, 342–363
  direct programming in, 342–353, 377–379
  general procedure for, 349–353
  initial conditions in, 343–348
  integration in, 340–342, 345
  multiplication by a constant in, 339–340
  operational amplifier in, 339–345, 357–359
  potentiometer in, 357–358
  simulation by, 358–363
  summation in, 340–342
  time scale in, 353–357
Analog-type controllers, 9
Analogies, 31–37
  direct, 31–35
  inverse, 31–37
Angle of arrival, 293–307
Angle condition, 285-307
Angle of departure, 292–307
Astro observatory, 189–190, 543
Asymptotes:
  intersection point, 288–290
  of log-magnitude plots, 508–520, 564–566
  of root-locus plots, 285–307
Attenuation, 576–577
Auxiliary equation, 260–261

Balanced valve, 70
Baldor, 85
Bandwidth, 535–537, 543
Bell Helicopter Textron, Inc., 302–303
Bergen, A.R., 490
Block diagrams, 5–7
  algebra of, 92–95
  of general system, 102–104
  for sampled-data systems, 458–462
Bode, H.W., 510
Bode plots, 510, 564–566
Boeing Company, 7–8, 95, 464
Breakaway point, 290–310

Break frequency, 509, 516–518
Break-in point, 291–310
British gravitational system, 37, 60–64

Capacitance, electrical, 22–25
Cascaded networks (*see* Series compensation)
Characteristic equation, 163
  (*see also* Characteristic function)
Characteristic function, 163, 216–232
  of closed-loop control system, 252–255
  effect of external disturbance on, 250–255
  predicting transient response from, 250–255
  relation to system stability, 250–255, 262–263
  of sampled-data systems, 462–469
  zeros of (*see* Zeros of characteristic function)
Chrysler Corporation, 248–250, 402–403
Charge, 23
Cincinnati Milacron Co., 405–406
Closed-loop system, 4–7
  characteristic function of, 252–255
  frequency response of, 525–535
  standard terminology of, 102–104
  state-space representation of, 384–388, 419–421
CNC milling machine, 405–406
Comdyna, Inc., 352–353
Command signal, 125–130
Comparator, 3–7
Compensated isochronous control, 143–144
  (*see also* Proportional plus integral controller)
Compensation (*see* Internal feedback; Series compensation)
Computer controlled systems, 486–496
Computer programs for:
  digital control systems, 484–496
  frequency response, 515–516
  linear control systems, 233–242
  non-linear control systems, 242–243
  root-locus plots, 318–319
  sampled-data systems, 479–486
  state-space representation, 388–391
Control elements, 5–7
Controlled variable, 3–7, 125–130
Controller, 125–147, 255–257
  integral, 125, 133–134, 139–141, 183–184
  proportional, 134–139, 181–183
    plus derivative, 146
    plus integral, 141–145
Convolution integral, 200–203
Correlation of transient and frequency response, 528–535
Critical damping, 228–233
Cross-over frequency, 566–568, 575, 579

Damped natural frequency, 226–228
Damper:
  rotational, 21–22
  translational, 16–18
Damping ratio, 227–233
Dashpot (*see* Damper)
dc motors, 85–89
Decade, 508
Decibel, 509
Decrement, logarithmic, 232–233
Degrees of freedom, 19–21
Derivative controller, 146
  (*see also* Controller)
Derivative time, 146
Difference equations, 469–470
Differential equations, 162–167
Differential equations, solution methods:
  analog computer 342–353
  classical methods, 162–167
  digital computer, 233–243
  Laplace transform, 167–208
  state-space methods, 398–416
Digital control systems, 439–496
  computer-controlled systems, 486–496
  computer program for, 484–496
  discrete data systems, 476–479
  sampled-data systems, 479–486
Direct analog, 31–35
Direct polar plots (*see* Polar plots)
Direct programming, 342–353, 377–379
Discrete data systems, 476–479
Discrete functions, 448–450
Disturbance, external, 5, 102–104, 125–134, 252–255
Droop controller, 137
  (*see also* Proportional controller)
Dual of mechanical circuit, 50

Eigenvalues, 417
  (*see also* Zeros of characteristic function)
Electric motors (*see* Motors, electric)
Electrical components, 22–23
Encoder, 9
Equalization (*see* Series compensation)
Equilibrium, 131–134
  (*see also* Steady-state analysis)
Equilibrium flow, 601–605
Equivalent unity feedback systems, 520–521
Error constants, 204–208
  acceleration, 204–208
  positional, 204–208
  velocity, 204–208
Error signal, 3-7

Euler's equations, 171
Evans, W.R., 279
Experimental determination:
  of frequency response, 516–518
  of gain constant, 516–518
  of time constant, 516–518
  of transfer function, 516–518
Exponential hold, 476
Exponential response, 77
Exponentially decaying function, 171
External disturbance, 5, 102–104, 125–134, 252–255

Feedback control system, 3–7
  general block diagram for, 92–95, 103–104
Feedback elements, 4–7
Feedforward elements, 6–7
Filters, 470–476
  exponential hold, 476
  first-order hold, 475–476
  zero-order hold, 470–475
Final-value theorem:
  for the Laplace transform, 179–180
  for the z transform, 446
First-order hold, 475–476
First-order systems:
  frequency response of, 504–511
  transient response of, 75–77, 165–166, 187, 192–195
Flapper valve, 79–83
Floating controller, 141
  (*see also* Integral controller)
Fluid analogs, 40–43
Flyball governor, 96–100
Force-current analog, 32
Force-free response, 402–411
Force-voltage analog, 31
Forced response, 411–419
  by Laplace transform method, 411–414
  by signal-flow graph method, 414–419
Forcing function, 163
Fourier integral, 609–610
Fourier series, 606–609
Fourier transform, 609
Frequency of oscillation, 226–233
  natural:
    damped, 226–228
    undamped, 227–233
Frequency response, 502–547
  closed-loop, 525–535
  computer program for, 515–516
  correlation with transient response, 528–535
  determining gain constant from, 518–520, 537–544
Frequency response (*Cont.*)
  experimental determination of, 516–518
  of first-order systems, 504–511
  obtaining desired M_p from, 537–544
  open-loop, 524–528
  representation, means of:
    inverse polar plots, 583–589
    log-magnitude plots, 508–520
    log-modulus plots, 544–547
    polar plots, 506–508, 521–524
  of second-order systems, 512–515, 528–533

Gain adjustment, determination of:
  direct polar plots, 537–544
  inverse polar plots, 585–589
  log-magnitude plots, 508–520
  log-modulus plots, 544–547
Gain constant, 508–515, 537–544
  experimental determination of, 516–518
Gain margin, 562–566
General programming, 382–384
General Motors, 487
Governor, flyball, 96–100
Grounded-chair representation, 29–30

Hydraulic amplifier, 70–75
Hydraulic integrator, 71
Hydraulic servomotor, 73–75
Hydraulic systems, 70–75

Impulse function, 170
Impedance:
  electrical, 24–26
  mechanical, 27-29
Impulse function, 170
Impulse response, 202, 254–255
Inductance, electrical, 22–24
Inertial guidance, 8
Initial conditions, 185–188
Initial-value theorem:
  for the Laplace transform, 180
  for the z transform, 446
Integral controller, 125, 133–134, 139–141, 183–184
  (*see also* Controller)
Integrators:
  analog computer, 340–342, 345
  hydraulic, 71
Integral time, 145
Integrodifferential equations, 191–192
Internalrse analog, 31–37
Inverse Laplace transforms, 167–169, 185–187, 218–227

Inverse polar plots, 583–589
  α contours on, 584–585
  M contours on, 583–585
  Nyquist stability criterion in, 590–592
Inverse *z* transforms, 450–457
  by division method, 456
  by partial-fraction method, 450–456
  by residue method, 457
Isochronous controller, 141
  (*see also* Integral controller)

James, H.M., 544
Jet pipe amplifier, 77–79

Kalman, R.E., 490

Lag compensation, 575–580
Lag-lead compensation, 580–582
Laplace transform properties, 173–179
  change of scale, 178
  differentiation, 174–177
  integration, 178–179
  multiplication:
    by e^{at}, 177–178
    by t, 177
  real translation, 173–174
    table of, 173
Laplace transform theorems:
  final-value, 179–180
  initial-value, 180
  linearity, 179
Laplace transforms, 167–208
  for arbitrary functions, 200–203
  for common functions, 168–173
  for convolution integral, 200–203
  definition of, 168
  derivation of, 610–611
  general procedures for, 192–198
  inverse, 167–169, 185–187, 218–227
  of piecewise continuous functions, 198–200
  of sampled-data systems, 440–443
  of state-transition matrix, 401–406
  tables of, 169, 173
  vertical shifting theorem for, 194, 196–198
Laser weld system, 487–488
Lead compensation, 568–575
L'Hopital's rule, 170
Linear control systems, 55–66, 162
  computer program for, 233–242
Linear operator, 13
Linearization, 55–70
  of nonlinear functions, 55–66
  of operating curves, 66–70
Loci equations, 310–312
Log-magnitude plots, 508–520, 564–566
Log-modulus plots, 544–547, 563–564
Logarithmic decrement, 232–233
Lunar exploratory module, 597

M contours:
  on direct polar plot, 524–528
  gain adjustment by, 537–544
  on inverse polar plot, 583–585
  on log-modulus plot, 544–547
Magnitude condition, 284–307
Manipulated variable, 125–130
Manning maneuvering unit, 539–540
Mason, S.J., 392
Mason's gain formula, 392–398
Mass, 17–18
Mass-spring-damper system:
  parallel, 27–28
  series, 18–19, 26
McDonnell Douglas, 464
Mechanical components:
  rotational, 21–22
  translational, 16–19
Melsa, J.L., 421
Microprocessor, 9
Minimum-phase systems, 517
Minor feedback loop, 93, 101, 582–583
  (*see also* Internal feedback)
Mobile laser ranging system, 296
Motors, electric, 85–89
  ac, two phase, 91–92
  dc:
    armature-controlled, 87–89
    field-controlled, 85–87
M_p adjustment:
  on direct polar plots, 529–530, 537–544
  on inverse polar plots, 585–589
  on log-modulus plots, 544–547
Multivariable systems, 421–431

National Aeronautics and Space Administration, 189–190, 540
National Institute of Standards and Technology, 9–10
Natural frequency, 227–233
Naval Command, Control and Ocean Surveillance Center, 11
Newton's method, 307–310
Nichols, N.B., 544
Nichols plots, 544–547
  (*see also* Log-modulus plots)
Non-linear control systems:
  computer program for, 242–243

Nonminimum-phase systems, 517–518
Nyquist, H., 553
Nyquist stability criterion, 553–562
  in inverse plane, 590–592

Open-loop system, 4–5, 133–134
  frequency response of, 524–528
  steady-state characteristics of, 133–134
  transfer function of, 283–284
Operational amplifier, 339–345, 357–359
Operational notation, 13–15
Optimum response, 489–496
Orifices in series, 601–605
Oscillation, frequency of, 226–233
Overshoot, percent of, 245–248

Parallel combinations, 24–28, 31
  of electrical elements, 24–25
  laws of, 24–28, 31
  of mechanical elements, 27–28
Parallel programming, 379–380
Partial-fraction expansion:
  for distinct roots, 164–166, 218–219
  for repeated roots, 219–221
Peak frequency, 513
Peak time, 244–245
Peczkowski, Joseph L., 421
Pendulum, 64–65
Performance matrix, 425–427, 430
Perturbation method, 106
Phase margin, 562
  correlation with transient response, 566–568
Phillips, R.S., 544
PID controllers, 146, 255–257
Piecewise continuous function, 198–200
Pneumatic systems, 79–84
  equilibrium flow in, 601–605
Polar plots:
  direct, 506–508, 521–524
  inverse, 583–589
Poles, 283–284
Positional error constant, 204–208, 520
Positional servomechanism remote control, 90–91
Potentiometer, 357–358
Power-amplifying device, 3–5
Process control systems, 489
Programming, 377–384
  direct, 377–379
  general, 382–384
  parallel, 379–380
  series, 380–382

Proportional controller, 134–139, 255–257
  plus integral, 141–145, 255–257
  (*see also* Controller)
    plus derivative, 146
Proportional plus integral controller, 141–145, 255–257
Pulse function, 170
Pulse transfer function, 460–462, 491–493
Puma robot, 255–256
Puritan-Bennett, 79–81

Quadripole, 261

Ragazzini, J.R., 490
Ramp function, 198
Raven, F.H., 168, 186, 218
Real translation:
  for the Laplace transform, 173–174
  for the z transform, 446–448
Reference input, 3–7, 125–130
Reference operating point, 18–19, 125–130
Regulator, 6
Remainder theorem, 307–310
Reset controller, 141
  (*see also* Integral controller)
Residue method, 443–444, 457
Resistance, electrical, 22–23
Resonant frequency, 513
Response function, 163
Response matrix, 425
Ripple, 490
Rise time, 243–250
Robot welder, 248–250
Roots of characteristic equation (*see* Zeros of characteristic function)
Root-locus method, 279–322
  computer solution, 318–319
  construction techniques:
    angle condition, 285–307
    angle of arrival, 293–307
    angle of departure, 292–307
    asymptotes, 285–307
    breakaway point, 290–310
    break-in point, 291–310
    general procedure, 295–307
    location on real axis, 285–307
    magnitude condition, 284–307
    for negative gain, 313–318
  determination of stability by, 279–285
  equations for, 310–312
  roots of characteristic equation by, 279–285
  for sampled-data systems, 463–469
  sensitivity analysis by, 319
  transient response from, 279–285
  variation of parameters by, 312–322

Routh, E.J., 257
Routh's stability criterion:
  for continuous systems, 257–262
  for sampled-data systems, 465–466
Runge-Kutta, fourth-order method, 236–243

Sain, M.K., 421
Sampled-data systems, 440–450, 479–486
  characteristic function of, 462–469
  computer program for, 479–486
  filters for, 470–476
  Laplace transform of, 440–443
  transient response in, 462–469
    by difference equations, 469–470
    by root-locus, 466–469
    by Routh's criterion, 465–466
  z transform of, 441–450
Sandia National Laboratories, 246–247
Scale factors, 34–37
Second-order systems:
  frequency response of, 512–515, 528–533
  transient response of, 227–233, 243–250, 254
Seismic instrument, 547–548
Sensitivity analysis, 319–322, 427–431
Series combinations:
  of electrical elements, 24
  laws of, 24, 31
  of mechanical elements, 26
Series compensation:
  phase lag, 575–580
  phase lag-lead, 580–582
  phase lead, 568–575
Series equalizer, 568
Series programming, 380–382
Servomechanism, 6
Settling time, 243–250
SI units, 38, 61–64, 605
Signal-flow graphs, 391–398, 407–411, 414–419
Simulation, 358–363
Sinusoidal function, 171–173
Sinusoidal response (*see* Frequency response)
Solar thermal power generator, 246–247
South Bend Controls, Inc., 79–80
South Bend Lathe, 9–10
Speed control system, 95–103
Spirule, 294–295
Spring:
  mechanical, 16
  torsional, 21
Stability:
  determination methods:
    characteristic function, 250–255, 262–263
    impulse response, 254–255
    Nyquist criterion, 553–562
Stability (*Cont.*)
    root-locus, 279–285
    Routh's criterion, 257–262
  in inverse plane, 590–592
  table correlating location of zeros and response functions, 263
Stable platform, 8
State-space methods, 377–416
  computer solution of, 388–391
  determination of transfer function from, 416–419
  for multivariable systems, 421–431
  programming of, 377–388
  solution of differential equations by, 398–416
  system representation by, 377–388
State-space representation, 238
State-transition matrix, 398–431
  by Laplace transforms, 401–406
  methods for obtaining, 398-401
  by signal-flow graph, 407–411
State variables, 377–384
Static loop sensitivity, 279
Stäubli Unimation, 255–256
Steady-state analysis, 123–147
  of derivative controller, 146
  of integral controller, 125, 133–134, 139
  of proportional controller, 134–139
    plus integral controller, 141
Steady-state constants, 123–131
Steady-state gain, 77
Step function, 76–77, 168–169
Step-function response:
  of first-order system, 76–77
  of second-order system, 230–232
Superposition, 185, 252–254
Synthetic division, 309–310
System type, 520–521, 528–535, 556–562
Systems:
  closed-loop, 4–7, 524–535
  equivalent unity feedback, 520–521
  open-loop, 4–5, 133–134, 524–535
  types of, 520–521, 528–535, 556–562

Tachometer, dc, 90
Thermal analogs, 38–40
Time constant, 77, 516–518
Time scale, 35–37, 353–357
Transfer function, 217, 416–419
  experimental determination of, 516–518
  of open-loop control systems, 283–284
  from state-space representation, 416–419
Transforms (*see* Laplace transforms; *z* transforms)

Transient response, 123, 162–208, 216–265,
from characteristic function, 250–255
correlation with frequency response, 528–535
correlation with phase margin, 566–568
from differential equations, 162–167
of first-order systems, 75–77, 165–166, 187, 192–195
from frequency response, 528–535
general form of, 250–257
from Laplace transforms, 167–208
from root-locus, 279–285
of sampled-data systems, 462–469
of second-order systems, 227–233, 243–250, 254
specifications, 243–250
Transient response specifications, 243–250
Translation:
for the Laplace transform, 173–174
for the z transform, 446–449
Translational mechanical components, 16–19
Transmittance, 319–322, 424–427
Trapezoidal integration, 233–236
Tuning controllers, 146
Type, system, 520–521, 528–535, 556–562

Ultraviolet imaging telescope, 543
Undamped natural frequency, 228
Unit impulse, 170
Unit step function, 76–77, 168–169
Unit vector, 172–173
United States Air Force, 182
U.S. Navy, Ocean Surveillance Center, 11
Unity-feedback system, 5, 520–521

Variation of parameters, 312–322
Vector loci, 506–507, 521–524
Vector loci plots (*see* Polar plots)
Velocity error constant, 204–208, 520
Vertical shifting theorem, 194, 196–198
Vibrometer, 548
Viscous damper, 16–17

Walking beam linkage, 73–75
Ward-Leonard system, 89

X-15 aircraft, 573

YF · 22 airplane, 181–184

z transforms, 441–457
inverse, 450–457
by residue method, 443–444, 457
table of, 442
properties, 450
theorems, 444–450
Zero-order hold, 470–475, 482–489
Zeros of characteristic function, 218–226
complex conjugate, 221–226
distinct, 218–219
repeated, 219–221
(*see also* Characteristic function)